SOLUTIONS MANUAL AND STUDY GUIDE

ORGANIC CHEMISTRY

ROBERT J. OUELLETTE

J. DAVID RAWN

PREPARED BY
ROBERT J. OUELLETTE & MARY H. BAILEY
The Ohio State University

PRENTICE HALL Upper Saddle River, NJ 07458

Acquisitions Editor: *John Challice*
Production Editor: *James Buckley*
Cover Designer: *Paul Gourhan*
Production Supervisor: *Joan Eurell*
Art Director: *Heather Scott*

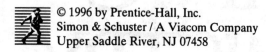
Printed in the United States of America

10 9 8 7 6 5 4 3 2 1

ISBN 0-13-238684-4

Prentice-Hall International (UK) Limited, *London*
Prentice-Hall of Australia Pty. Limited, *Sydney*
Prentice-Hall Canada, Inc., *Toronto*
Prentice-Hall Hispanoamericana, S.A., *Mexico*
Prentice-Hall of India Private Limited, *New Delhi*
Prentice-Hall of Japan, Inc., *Tokyo*
Simon & Schuster Asia Pte. Ltd., *Singapore*
Editora Prentice-Hall do Brasil, Ltda., *Rio de Janeiro*

Table of Contents

Preface   v

Introduction   How to Succeed in Organic Chemistry by Really Trying   vii
                   How to Use Curved Arrows   xi

Chapter 1   Atoms, Molecules and Bonding
                   Keys to the Chapter   1
                   Solutions to Exercises   9

Chapter 2   Structure and Properties of Organic Molecules
                   Keys to the Chapter   15
                   Solutions to Exercises   21

Chapter 3   Chemical Energetics
                   Keys to the Chapter   27
                   Solutions to Exercises   33

Chapter 4   Alkanes and Cycloalkanes: Structure and Properties
                   Keys to the Chapter   41
                   Solutions to Exercises   49

Chapter 5   Reactions of Alkanes and Cycloalkanes
                   Keys to the Chapter   61
                   Summary of Reactions   64
                   Solutions to Exercises   65

Chapter 6   Alkenes: Structure and Properties
                   Keys to the Chapter   71
                   Summary of Reactions   76
                   Solutions to Exercises   77

Chapter 7   Addition Reactions of Alkenes
                   Keys to the Chapter   83
                   Summary of Reactions   87
                   Solutions to Exercises   89

Chapter 8   Haloalkanes and Alcohols
                   Keys to the Chapter   101

	Summary of Reactions	108
	Solutions to Exercises	110

Chapter 9   Stereochemistry
| | Keys to the Chapter | 123 |
| | Solutions to Exercises | 128 |

Chapter 10   Nucleophilic Substitution and Elimination Reactions
| | Keys to the Chapter | 135 |
| | Solutions to Exercises | 138 |

Chapter 11   Alkynes
	Keys to the Chapter	145
	Summary of Reactions	148
	Solutions to Exercises	150

Chapter 12   Dienes and Allylic Compounds
	Keys to the Chapter	157
	Summary of Reactions	160
	Solutions to Exercises	161

Chapter 13   Arenes and Aromaticity
	Keys to the Chapter	173
	Summary of Reactions	176
	Solutions to Exercises	178

Chapter 14   Electrophilic Aromatic Substitution
	Keys to the Chapter	187
	Summary of Reactions	190
	Solutions to Exercises	193

Chapter 15   Spectroscopy and Structure Determination
| | Keys to the Chapter | 205 |
| | Solutions to Exercises | 211 |

Chapter 16   Alcohols: Reactions and Synthesis
	Keys to the Chapter	221
	Summary of Reactions	226
	Solutions to Exercises	229

Chapter 17   Ethers and Epoxides
	Keys to the Chapter	245
	Summary of Reactions	248
	Solutions to Exercises	251

Chapter 18   Aldehydes and Ketones

        Keys to the Chapter   263
        Summary of Reactions   267
        Solutions to Exercises   270

Chapter 19   Aldehydes and Ketones: Nucleophilic Addition Reactions

        Keys to the Chapter   279
        Summary of Reactions   283
        Solutions to Exercises   286

Chapter 20   Carbohydrates

        Keys to the Chapter   299
        Summary of Reactions   303
        Solutions to Exercises   305

Chapter 21   Carboxylic Acids

        Keys to the Chapter   319
        Summary of Reactions   323
        Solutions to Exercises   327

Chapter 22   Carboxylic Acid Derivatives

        Keys to the Chapter   341
        Summary of Reactions   345
        Solutions to Exercises   349

Chapter 23   Enols and Enolates: Condensation Reactions

        Keys to the Chapter   363
        Summary of Reactions   369
        Solutions to Exercises   373

Chapter 24   Lipids

        Keys to the Chapter   397
        Solutions to Exercises   400

Chapter 25   Amines and Amides

        Keys to the Chapter   403
        Summary of Reactions   408
        Solutions to Exercises   411

Chapter 26   Amino Acids and Proteins

        Keys to the Chapter   425
        Solutions to Exercises   430

Chapter 27   Aryl Halides, Phenols, and Anilines
                            Keys to the Chapter   439
                            Summary of Reactions   442
                            Solutions to Exercises   445

Chapter 28   Pericyclic Reactions
                            Keys to the Chapter   457
                            Solutions to Exercises   461

Chapter 29   Synthetic Polymers
                            Keys to the Chapter   469
                            Solutions to Exercises   473

Multiple Choice Examinations

                            Examination 1 (Chapters 1 through 3)   481
                            Examination 2 (Chapters 4 through 6)   491
                            Examination 3 (Chapters 7 through 12)   497
                            Examination 4 (Chapter 13 through 15)   507
                            Examination 5 (Chapter 16 through 20)   517
                            Examination 6 (Chapters 21 through 24)   527
                            Examination 7 (Chapters 25 through 29)   533

Preface

Welcome to the study of organic chemistry. We have written this book as a companion to the textbook *Organic Chemistry* by Robert J. Ouellette and J. David Rawn. Before you dig in, we'd like to offer an overview of this manual and give you some advice on how to get the most out of it.

The first part of the General Introduction provides information that is important for a good start in organic chemistry. Your instructor will likely agree with the formula for success we outline. It is really important that you take seriously your efforts to learn organic chemistry. We and your instructor want you to enjoy organic chemistry and to perform to your full potential.

The second part of the General Introduction explains the use of the curved arrow convention used to write reaction mechanisms. You should refer to this material when you first encounter the use of curved arrows in Chapter 2. As more reaction mechanisms are introduced in later chapters, you may find it useful from time to time to refer back to the general discussion of the use of arrows in tracking the location of electrons.

This remainder of the book is divided into chapters, each of which consists of two principal parts—a study guide and a solutions manual. At the end of the book there are also some sample multiple choice tests to give you additional practice.

The study guide portion of each chapter provides a synopsis of the corresponding chapter in *Organic Chemistry*. It provides a summary of the material covered in the chapter in the text, arranged by section. Within each section there are comments not only on the significance of that section but also on how it relates to material previously covered. The goal is to reinforce your understanding of the basic principles that interrelate the various parts of organic chemistry. In short, the study guide prompts you to tie things together as you go along.

The solutions manual portion of each chapter of this book gives answers to all Exercises at the end of the chapters in *Organic Chemistry* . Some of the answers simply give a name or the requested structure. In those cases, you need only check your answer if you made a mistake. For other questions there is a discussion of the thought process and relevant facts that you should have considered to arrive at an answer. The discussion indicates how this material fits in with other parts of the text. If you have trouble with a question, you should go back to reexamine this previously introduced material. Finally, some questions require interpolation or extrapolation to "explain" material that you have not learned but that can be predicted based on principles that you should know. In short, these questions are often designed to teach you how to "think" in organic chemistry terms. On chemistry examinations you may simply have to write down a name or a structure. However, most organic examinations require you to explain certain facts. You have to learn how to assemble your resources and select the relevant principles to answer the question.

Finally, a little advice. Really try to answer a problem before you "sneak a peek" at the solution. Try to determine what principles are involved and then reread the appropriate sections in the text or the study guide. If you look at the answer too soon, the temptation is to say "Yeah, I knew that." As a result, you have missed the opportunity to develop your problem-solving skills. Furthermore, the question is no longer useful as a learning tool. Don't make the same mistake twice. The Exercises in the text are paired in terms of the concepts or skills required to solve problems. You shouldn't have to look at the answer for the even-numbered one after having done so for the odd-numbered one.

Seven sample multiple-choice examinations are available at the end of this book. To make most efficient use of these examinations, you should wait until the indicated chapters have been covered. However, don't start doing the examination until you are fully prepared. As in the case of answering the exercises at the end of the text chapters, really try to do the examination before you look at the answer.

If you use both the study guide to review the facts and the solutions manual to check your answers, you'll be better prepared to test your skills on an exam. We hope you'll find this book a useful learning tool.

Robert J. Ouellette
Mary H. Bailey
Department of Chemistry
The Ohio State University
Columbus, OH 43210

General Introduction

How to Succeed In Organic Chemistry by Really Trying

You come to this course with a background in general chemistry, and you might expect to perform at a similar level in organic chemistry. Unfortunately, that is not always the case. General chemistry covers a range of subjects, not all of which are interrelated. For example, you can do gas law calculations without any understanding of periodic trends of the elements. Whole chapters, such as the one on electrochemistry, are unrelated to other chapters, such as the one on chemical kinetics. As a result, your grade reflects how well you did overall on a variety of subjects, but you may have done very poorly in one or more areas. In fact, if you stop and think about the "big picture" of chemistry as presented in general chemistry, you may conclude that there was none. You developed skills at doing problems in many areas. Which of these will be important in organic chemistry?

First of all, you will encounter fewer quantitative questions in your organic chemistry course. Second, you won't even touch on some subject areas, such as electrochemistry. So, what skills are important? Well, the answer is that structure and bonding, which was presented early in the first term of your general chemistry course, is exceedingly important, as are acid-base concepts. However, you'll find that it is the Lewis concept of acids and bases that is more important in organic chemistry. The Lewis concept was probably covered on one page in your general chemistry text, but we'll go into much more detail about it here. There are a few other such links of knowledge that you are going to have pull together and integrate to prepare for your study of organic chemistry.

You are probably aware of the reputation of organic chemistry as a very difficult course that requires a lot of memorization and that the course serves to "weed out" pre-professional students. The part about being difficult is true, but it is made more difficult by students who don't take to heart the advice given in the following discussion of how to succeed in organic chemistry. Thus, students eliminate themselves because they don't prepare early and study continuously during the term.

You are about to learn that there is a vertical integration of knowledge required to understand organic chemistry. Unlike general chemistry, you do have to have a command of the subject matter in preceding chapters in order to really understand the current chapter that you are studying. There are basic principles that underlie each apparently new subject presented. Only by constantly reviewing these principles and applying them to each new subject area will you succeed in the study of organic chemistry. Tie things together continually. You may succeed in memorizing facts initially, but eventually you will find studying organic chemistry increasingly difficult. Ask questions such as:

Where have I seen this compound or reaction before?
What features in this reaction are common to other reactions that I have seen before?
What principles relate this reaction to certain other reactions that I have seen before?

Each year, instructors in organic chemistry courses have to advise struggling students who are disappointed with their results in the first examination. Invariably students express dismay at their performance and offer several reasons why they could not possibly have done so poorly. These observations are summarized by some of the following statements.

1. I understand the material when you explain it, but I couldn't show you what I knew on the examination.
2. I studied for this exam this weekend and spent 15 hours reading and doing problems.
3. I outlined the chapters and prepared flash cards.

The tempting response to the first statement is that chemistry is not a spectator sport. You can't learn to be proficient at anything by watching others do it. You have to practice things on your own before game day.

The second statement illustrates a trap into which students fall as they bounce from one cram session to another in their courses. There may be subjects where a concentrated period of study may get you through the exam with a moderate grade. However, organic chemistry is not one of them. Fifteen one-hour study periods over 15 days would give better results! You have to "learn" some facts, concepts, and applications as you go along so that you can use that information in the interpretation of subsequently encountered material.

The physical act of filling a textbook with various colored marking pens is not related to learning. In fact, as the fraction of outlined material increases, the process becomes less useful. Just "reading" the outlined material over and over again without relating it to supporting arguments results in "memorization" and doesn't provide the understanding required to sort out concepts and patterns related to the "facts". Flash cards, while useful, have to be used realistically. Just memorizing the "answer" on the reverse side of the card containing the question gives only that one-to-one correspondence.

So, what do you have to do to succeed in organic chemistry? Unlike business, where the book is titled "How to Succeed in Business Without Really Trying", the equivalent book for organic chemistry should be titled "How to Succeed in Organic Chemistry by Really Trying". Unfortunately, many students don't recognize that truism until later in the course. Now is the time to consider what you need to do to meet your goals in the course. There is no single approach that works. However, there are several approaches that, if followed every day, will greatly improve your chances of success.

This Study Guide and Solutions Manual to accompany *Organic Chemistry* provides a synopsis of each chapter and insights into the relationship between current material and past material that you should have learned. In addition, there are pieces of advice on how to organize and approach the different subjects. The Study Guide section consist of two parts—Keys to the Chapter and Summary of Reactions. The Solutions Manual section contains structures of compounds as well as detailed discussions of what you should consider in arriving at many of the answers. However, don't look at this material until you have honestly attempted to solve the exercises on your own. Really try to do a problem before you "sneak a peek" at the answer in the solutions manual. Try to determine what principles are involved and then reread the appropriate sections in the text or the study guide.

Attend all lectures in this course. Too often students underrate the value of regular class attendance and skip lectures or ask other individuals to "take notes." Such reckless disregard for your responsibilities endangers your academic health (grade point average). Organic Chemistry is a very complex subject, and the textbook contains a great deal of material that cannot simply be memorized. You need guidance on the relative importance of various subjects as well as the experience of the lecturer in interrelating apparently disparate information from widely separated sections and chapters.

Lectures provide you with a different way to learn. Learning by reading is quite different than learning by listening. You need the benefits of both methods. Moreover, the lecturer's intonation and repetition of facts should provide you with "clues" as to what will be on the exam. You shouldn't have to ask what will be on the exam. An actively involved student who attends lectures should know!

You will take many pages of notes in this class. In addition, with the permission of the instructor, you may record the lecture. However, the class should be more than a period of physically (and accurately) copying material presented by the lecturer into your notebook. You are going to have to think about what is said as well. In order to do so, you have to know something about the subject of the lecture before entering the classroom. Look at the course syllabus or ask the instructor what might be covered in the next lecture. Then read over that material before the next class. This allows you to determine what you understand and what you don't. Then you'll be prepared to listen for a few "words of wisdom" from the lecturer to help clear up the difficult points and put that item in perspective with other material. Incidentally, it is obvious to the lecturer who has not taken this advice. The symptoms are furious note-taking and a troubled look during the lecture.

This textbook should be read not only in advance of lecture but after lecture as well. The subject matter is not light reading. In fact, you will have to have considerable discipline to study the text. You will have to constantly refer to prior material, because, unlike general chemistry, the subject matter of organic chemistry is strongly integrated. If you start missing concepts early in the course, further study will become more difficult as the term progresses.

Stay on top of things. Don't cram for an exam in another course and let chemistry slide for a while. Keep up with your reading. Study your notes in advance and after lecture. Work on problems in the chapter as well as the exercises at the end of the chapter. Organic chemistry is a highly structured discipline, and each level must be well understood in order to learn the next level. The least efficient way of studying is selected by some of each generation of students in organic chemistry. They study 10-15-20 hours on a weekend before the exam. We know of no case where this technique has benefited the student. It is difficult to concentrate on any subject for more than a couple of hours—and for organic chemistry, the time period may be even less. After some point you are only reading, and the rate of learning decreases. Prolonged study times may provide you with a rationalization for why you aren't responsible for your poor performance. But wouldn't you really be happier if you actually succeeded by taking the following advice? Study for 1-2 hours each day. Yes, we really mean it.

Work as many problems as possible as part of your studying. You have to be active in testing what you have learned by doing problems. Do some problems each day that correspond to the most recent lecture material. Don't give up too soon and look up the answer. You'll only get as much out of a problem as you put into it. Looking at the correct solution should be done only as a check for your answer. Finally, be realistic. If you couldn't do the problem, you probably won't be able to do related problems that will appear on exams. If you don't understand vital underlying concepts, seeing a specific answer doesn't provide you with the insight needed to work other related problems.

Get help early and often. The lecturer or graduate teaching assistants (GTA) can help you solve problems that you have earnestly attempted to work. Explain what you know and where you are stuck. Then the lecturer can provide the missing link or outline the principle that you may not understand. Asking someone to just solve problems is a waste of time if you haven't seriously tried to work on it. Furthermore, they can't possibly make up for difficulties as a result of missed classes or inadequate study time. You shouldn't expect the instructor to give you a private lecture.

Attend recitations or review sessions if they are offered. You can always derive some benefit from these sessions if you go prepared to be an active participant and to ask questions. Don't expect to derive benefit from questions asked by others. Don't expect these sessions to be a substitute for lecture not attended or for inadequate study time.

Now it is all up to you. You can succeed if you really try. Start developing solid study habits now. Try to really "understand" the material when you first encounter it so that you can create a solid foundation before going forward to new material. That foundation will enable you to master the new material and decrease the amount of "memory" required. If you do as we suggest, organic chemistry won't be as difficult as it is reputed to be.

How to Use Curved Arrows

Chemists use a curved arrow convention to keep track of electrons as they move to form new bonds or break old bonds. The tail of the arrow is placed at the point where the electrons that are involved in the reaction are located, and the head of the arrow points to the place where they will be located at the end of that step of the reaction. You should pay close attention to how these curved arrows are used, because they are used through the text. In addition, it is likely that you will have to use the arrows yourself in examinations. You should be comfortable enough to use these arrows to write mechanisms for compounds and reactions with which you may be unfamiliar.

It is also necessary to understand where the charge is located in the intermediates pictured in mechanisms and to show it. You can do this in several ways. First of all, independent of whether arrows are used or not, charge must be conserved. In other words, the sum of the charges of the reactants must equal the sum of the charges of the products. Second, you can determine the charge at any site by accounting for the direction of electron flow. In other words, what atom "owns" the electrons initially and what atom owns them at the end of the process? Finally, based on your developing experience in viewing structures, you should eventually be able to write a charge based on the number of bonds and electron pairs about the atom. For example, if an oxygen atom has three covalent bonds and one electron pair, it must have a positive charge. It may help to think of common simple analogous species. In the case of the described oxygen atom, that is the situation in the hydronium ion, H_3O^+. Similarly, if a nitrogen atom has four covalent bonds, it must have a positive charge, as is the case for the ammonium ion, NH_4^+.

Examples Using Single Arrows

A bond may form between a species that acts as an electron pair donor (Lewis base) and a species that acts as an electron pair acceptor. The tail of the curved arrow is placed at the electron pair of the Lewis base, and the head of the arrow is drawn to the Lewis acid. Consider the following two cases.

Both reactions proceed by the same mechanism and the flow of electrons is from a Lewis base to a carbocation. In each case a covalent bond is formed using the electron pair of the Lewis base. Originally the Lewis base had control of both electrons, but in the covalent bond one of the electrons "belongs" to the carbon atom. As a result that carbon atom is no longer charged. The charge on the Lewis base portion of the product becomes more positive because it has "lost" one electron as the result of bond formation. In the case of chlorine that change is from -1 to 0. In the case of oxygen the change is from 0 to +1.

Bond cleavage most often occurs heterolytically, and the shared electron pair is moved to one of the two atoms of the original bond. As a result the charge of one of the original two atoms becomes more positive and the other becomes more negative. The tail of the curved arrow is placed at the bonding pair, and the head of the arrow is pointed toward the atom to which the electrons move—usually the more electronegative atom. This process is illustrated by the following two reactions.

$$CH_3-\overset{\overset{\displaystyle H}{|}}{\underset{\underset{\displaystyle CH_3}{|}}{C}}-\ddot{B}r: \longrightarrow CH_3-\overset{\overset{\displaystyle H}{|}}{\underset{\underset{\displaystyle CH_3}{|}}{C}}{}^+ \;+\; :\ddot{B}r:^-$$

$$CH_3-\overset{\overset{\displaystyle H}{|}}{\underset{\underset{\displaystyle CH_3}{|}}{C}}-\overset{+}{\ddot{O}}-H \longrightarrow CH_3-\overset{\overset{\displaystyle H}{|}}{\underset{\underset{\displaystyle CH_3}{|}}{C}}{}^+ \;+\; :\underset{\underset{\displaystyle H}{|}}{\ddot{O}}-H$$

These two heterolytic cleavage reactions are mechanistically identical and are the reverse of the type of heterogenic reactions shown on the previous page. Note that the carbon atom become positively charged because it "loses" the electron that it contributed to the original covalent bond. The charge of the second atom becomes more negative. In the case of bromine the charge changes from 0 to -1, and in the case of oxygen the charge changes from +1 to 0.

The arrow as shown in both reactions seems to point to the lone pair of the electronegative atom. However, it is the atom that receives the electrons! There is often a limited amount of space, and the atoms and electron pairs surrounding the atom clutter the environment. For that reason, it is acceptable to allow the arrow head to point in the general direction of the atom receiving the electron pair, as shown in the following two reactions..

$$CH_3-\overset{\overset{\displaystyle H}{|}}{\underset{\underset{\displaystyle CH_3}{|}}{C}}-\overset{+}{\ddot{O}}-H \longrightarrow CH_3-\overset{\overset{\displaystyle H}{|}}{\underset{\underset{\displaystyle CH_3}{|}}{C}}{}^+ \;+\; :\underset{\underset{\displaystyle H}{|}}{\ddot{O}}-H$$

$$CH_3-\overset{\overset{\displaystyle H}{|}}{\underset{\underset{\displaystyle CH_3}{|}}{C}}-\ddot{B}r: \longrightarrow CH_3-\overset{\overset{\displaystyle H}{|}}{\underset{\underset{\displaystyle CH_3}{|}}{C}}{}^+ \;+\; :\ddot{B}r:^-$$

A bonding electron pair may move from one atom to another within a single species. In such cases, the atom or group of atoms moves with the bonding pair to the new site. This process occurs in rearrangement of carbocations. Again we have to keep track of charges. If a group and its bonding pair move, then the original site must become more positive because it loses the electron that is contributed to the covalent bond. The site to which the group moves then becomes more negative.

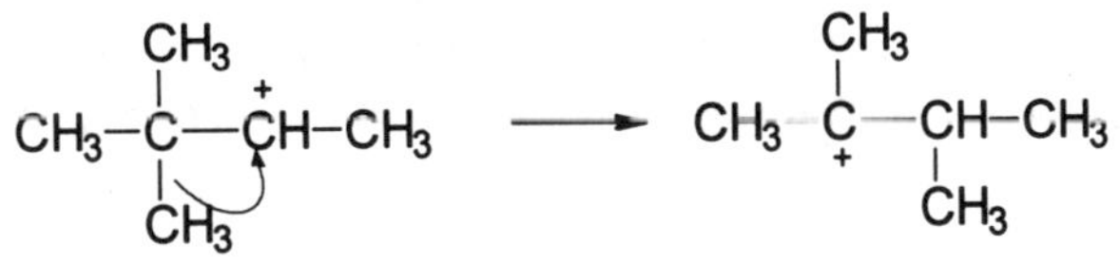

The second carbon from the left is electrically neutral initially and become positive in the rearranged product. The third carbon from the left is positively charged initially and becomes neutral in the product.

Examples Using Two Arrows

Examples of reactions using two arrows are many and include a number of processes where an atom or group of atoms is transferred intermolecularly. A bond must be broken, requiring one arrow, and a second bond must be formed, requiring a second arrow. The simplest examples are nucleophilic displacement reactions.

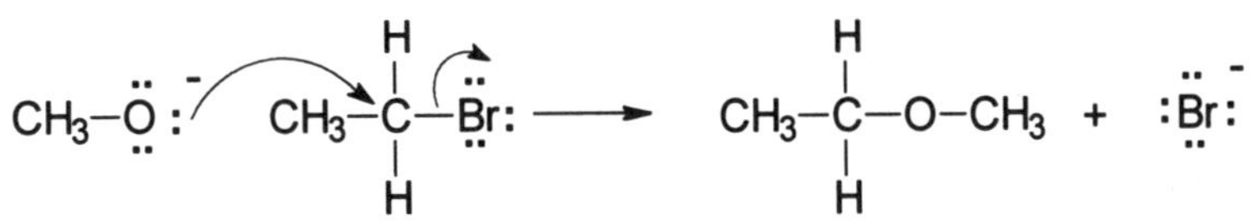

The lone pair electrons of the oxygen atom become a bonding pair, and as a result the oxygen no longer owns both electrons. The oxygen atom becomes more positive as its charge changes from -1 to 0. The bonding pair electrons of the C—Br bond move to form the bromide ion, which now has control of both electrons. The bromine atom becomes more negative as its charge changes from 0 to -1. Note that the charge of the carbon atom is unchanged because it still has four covalent bonds.

The electron pair in a π bond may be moved to form a covalent bond to an atom or group of atoms. This movement is shown by one arrow in which the tail is located at the π bond and the arrow head is directed to the atom or group of atoms that eventually bonds to one of the atoms of the original π bond. If the atom or group of atoms is bonded to another atom then that bond must be broken and a second arrow is required. Consider the addition of a proton from the hydronium ion to ethylene.

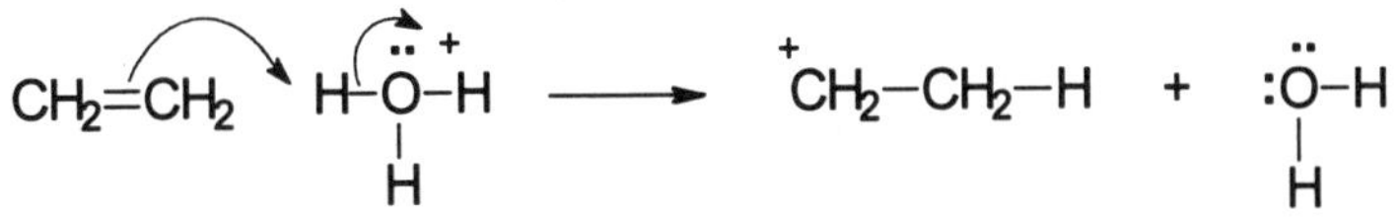

The carbon atom of the original double bond not forming the bond to the hydrogen atom has lost control of the electron that it contributed to the π bond, so it develops a positive charge. The other carbon atom remains electrically neutral because it has four covalent bonds in both the reactant and product. The bonding pair electrons of the H—O bond both belong to the oxygen atom in the product, so the charge of that atom changes from +1 to 0.

Multistep mechanisms are common, as for example the steps that follow the proton transfer reaction depicted above. The carbocation acts as a Lewis acid and combines with water, which acts as a Lewis base. Only one arrow is required in a process that is similar to one shown in the first part of the previous section.

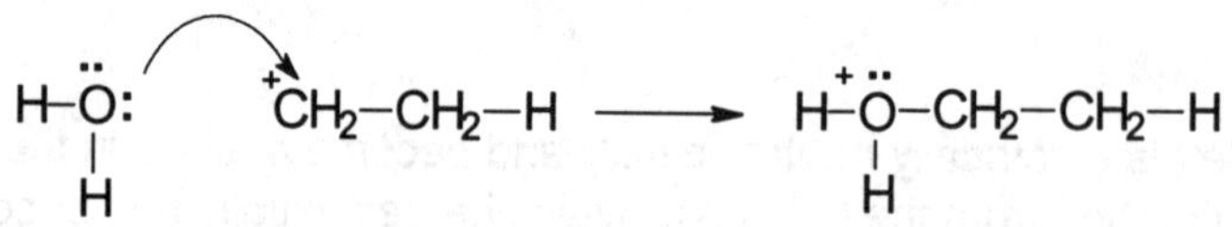

$$H-\overset{\cdot\cdot}{\underset{\underset{H}{|}}{O}}: \qquad \overset{+}{C}H_2-CH_2-H \longrightarrow H-\overset{+}{\underset{\underset{H}{|}}{\overset{\cdot\cdot}{O}}}-CH_2-CH_2-H$$

Originally water had control of both electrons, but in the covalent bond one of the electrons "belongs" to the carbon atom. As a result, that carbon atom is no longer positively charged. The charge on the Lewis base portion of the product becomes more positive, because it has "lost" one electron as the result of bond formation The charge of the oxygen atom changes from 0 to +1.

A third step is required to give the final neutral addition product. A proton is transferred from the oxygen atom of the conjugate acid of ethanol to the oxygen atom of water. Two arrows are required because one bond is made and one bond is broken.

$$H-\overset{\cdot\cdot}{\underset{\underset{H}{|}}{O}}: \qquad H-\overset{\cdot\cdot}{\underset{\underset{H}{|}}{O}}{}^{+}\!CH_2-CH_2-H \longrightarrow H-\overset{\cdot\cdot}{\underset{\underset{H}{|}}{O}}-H \; + \; \overset{\cdot\cdot}{\underset{\underset{H}{|}}{O}}-CH_2-CH_2-H$$

The carbonyl group is a second functional group having a π bond and therefore an electron pair that may be moved. When the carbonyl carbon atom is attacked by a nucleophile to form a covalent bond to the carbon atom it is necessary to move the electrons of the π bond to the oxygen atom. This movement is shown by one arrow in which the tail is located at the π bond and the arrow head is directed to the oxygen atom. Of course we also have to use an arrow showing the movement of the electron pair of the nucleophile toward the carbonyl carbon atom. Consider the addition of cyanide ion to ethanal (acetaldehyde).

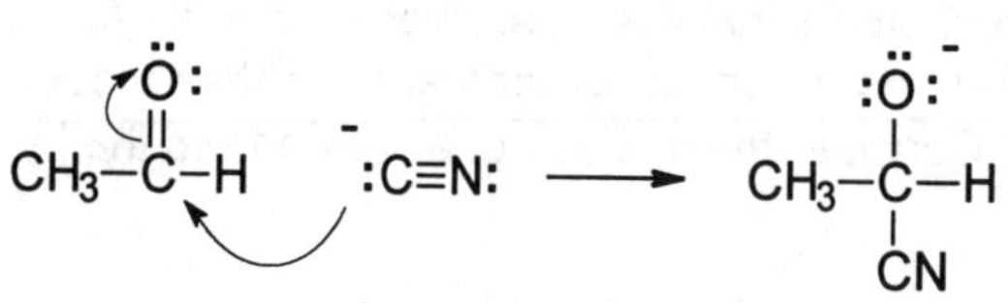

$$CH_3-\overset{\overset{\displaystyle \cdot\cdot}{O}:}{\underset{\displaystyle |}{\overset{\displaystyle ||}{C}}}-H \qquad :C\equiv N: {}^{-} \longrightarrow CH_3-\overset{\overset{\displaystyle :\overset{\cdot\cdot}{O}:{}^{-}}{|}}{\underset{\underset{\displaystyle CN}{|}}{C}}-H$$

The charge of the carbon atom of the nucleophile (CN⁻) changes from -1 to 0 as a result of the conversion of a nonbonding pair of electrons belonging only to one carbon atom to a bonding pair shared by two carbon atoms. Note that, in order to form that bond, a bond from the carbonyl carbon atom to the carbonyl oxygen atom must be broken. The charge of the oxygen atom changes from 0 to -1 as the result of conversion of a bonding pair of electrons, of which only one electron was contributed by oxygen, to a nonbonding pair of electrons which belong solely to oxygen.

Again, this process is just one step of a multistep mechanism. In the subsequent step a proton is transferred from an acid to the negatively charged oxygen atom.

$$\text{CH}_3-\overset{\overset{\displaystyle :\ddot{\text{O}}:^-}{\|}}{\underset{\underset{\displaystyle \text{CN}}{|}}{\text{C}}}-\text{H} \quad \text{H}-\text{C}\equiv\text{N}: \longrightarrow \text{CH}_3-\overset{\overset{\displaystyle :\ddot{\text{O}}-\text{H}}{\|}}{\underset{\underset{\displaystyle \text{CN}}{|}}{\text{C}}}-\text{H} \ + \ ^-:\text{C}\equiv\text{N}:$$

A different sequence of events may occur in which a net addition to a carbonyl group occurs. The hydration of ethanal provides one such example. First the carbonyl oxygen atom is protonated in a process that is shown using two arrows.

$$\text{CH}_3-\overset{\overset{\displaystyle \ddot{\text{O}}:}{\|}}{\text{C}}-\text{H} \quad \text{H}-\overset{+}{\underset{\underset{\displaystyle \text{H}}{|}}{\text{O}}}-\text{H} \longrightarrow \text{CH}_3-\overset{\overset{\displaystyle \overset{..\,+}{\text{O}}-\text{H}}{\|}}{\text{C}}-\text{H} \ + \ :\overset{..}{\underset{\underset{\displaystyle \text{H}}{|}}{\text{O}}}-\text{H}$$

The transfer of the proton results in an increase in positive charge of the carbonyl oxygen atom and a decrease in the charge of the oxygen atom of the hydronium ion as it is converted to water. Next, a nucleophile such as water can attack the carbonyl carbon atom.

$$\text{CH}_3-\overset{\overset{\displaystyle \overset{..\,+}{\text{O}}-\text{H}}{\|}}{\text{C}}-\text{H} \quad :\overset{..}{\underset{\underset{\displaystyle \text{H}}{|}}{\text{O}}}-\text{H} \longrightarrow \text{CH}_3-\overset{\overset{\displaystyle :\ddot{\text{O}}-\text{H}}{|}}{\underset{\underset{\displaystyle :\overset{+}{\underset{\underset{\displaystyle \text{H}}{|}}{\text{O}}}-\text{H}}{|}}{\text{C}}}-\text{H}$$

The charge of the oxygen atom of the nucleophile (H_2O) changes from 0 to -1 as a result of the conversion of a nonbonding pair of electrons belonging only to oxygen to a bonding pair shared with a carbon atom. Note that, in order to form that bond, a bond from the carbonyl carbon atom to the carbonyl oxygen atom must be broken. The charge of the oxygen atom changes from +1 to 0 as the result of conversion of a bonding pair of electrons, of which only one electron was contributed by oxygen, to a nonbonding pair of electrons which belong solely to oxygen.

In the final step to complete the addition process, a proton must be transferred from the conjugate acid of hydrate to water.

$$\text{H}-\overset{..}{\underset{\underset{\displaystyle \text{H}}{|}}{\text{O}}}: \quad \text{CH}_3-\overset{\overset{\displaystyle :\ddot{\text{O}}-\text{H}}{|}}{\underset{\underset{\displaystyle :\overset{+}{\underset{\underset{\displaystyle \text{H}}{|}}{\text{O}}}-\text{H}}{|}}{\text{C}}}-\text{H} \longrightarrow \text{CH}_3-\overset{\overset{\displaystyle :\ddot{\text{O}}-\text{H}}{|}}{\underset{\underset{\displaystyle :\ddot{\text{O}}-\text{H}}{|}}{\text{C}}}-\text{H} \ + \ \text{H}-\overset{..\,+}{\underset{\underset{\displaystyle \text{H}}{|}}{\text{O}}}-\text{H}$$

Acid derivatives also undergo initial addition reactions under either basic or acidic conditions. Under basic conditions, a nucleophile attacks the carbonyl carbon atom in a step similar to that shown for the reaction of cyanide ion with ethanal. Under acidic conditions, protonation of the carbonyl oxygen atom occurs as shown in the hydration of ethanal. The difference in the mechanism results from the subsequent reaction of the tetrahedral intermediate, which ejects a leaving group. First consider the attack of hydroxide ion on the carbonyl group of methyl ethanoate. The charge on the oxygen atom of the hydroxide changes from -1 to 0, while that of the oxygen atom of the carbonyl group changes from 0 to -1.

$$CH_3-\overset{\overset{\displaystyle :\ddot{O}:}{\|}}{C}-\ddot{O}-CH_3 \quad {}^{-}:\ddot{O}-H \quad \longrightarrow \quad CH_3-\overset{\overset{\displaystyle :\ddot{O}:^{-}}{|}}{\underset{\underset{\displaystyle :\ddot{O}-H}{|}}{C}}-\ddot{O}-CH_3$$

This anion now can eject a methoxide ion. A nonbonding electron pair of oxygen is moved to form the π bond again. This process is shown with one arrow. A second arrow is used to show the loss of the methoxide ion.

$$CH_3-\overset{\overset{\displaystyle :\ddot{O}:^{-}}{|}}{\underset{\underset{\displaystyle :\ddot{O}-H}{|}}{C}}-\ddot{O}-CH_3 \quad \longrightarrow \quad CH_3-\overset{\overset{\displaystyle \ddot{O}:}{\|}}{C}-\ddot{O}-H \quad + \quad {}^{-}:\ddot{O}-CH_3$$

Examples using three arrows

Examples of reactions using three arrows include elimination reactions. Two atoms or groups of atoms are removed from adjacent carbon atoms and a carbon-carbon single bond is converted into a double bond. A bond must be broken when one atom or group of atoms is removed. One arrow is used to form a bond to the removed group and a second arrow is used to convert the single bond into a double bond. A third arrow is used to show the cleavage of the bond of the second groups eliminated.

$$(CH_3)_3C-\ddot{O}:^{-} \quad H-CH_2-CH_2-\ddot{B}r: \quad \longrightarrow \quad CH_2{=}CH_2 \quad + \quad (CH_3)_3C-\ddot{O}-H \quad + \quad :\ddot{B}r:^{-}$$

The lone pair electrons of the oxygen atom become a bonding pair, and as a result the oxygen no longer owns both electrons. The oxygen atom becomes more positive as its charge changes from -1 to 0. The bonding pair electrons of the C—Br bond move to form the bromide ion, which now has control of both electrons. The bromine atom becomes more negative as its charge changes from 0 to -1. Note that the charge of the carbon atoms are unchanged because each still has four covalent bonds.

1

Atoms, Molecules, and Bonding

This chapter reviews the concepts of atomic structure and bonding that you learned in your first chemistry course, commonly referred to as General Chemistry. However, don't assume that you really know the material and as a result casually skip through the chapter. This course builds on a vertical assimilation of and use of knowledge to create a complex edifice. You will build a dangerous structure if your foundation is of sand rather than concrete. Unfortunately, generations of students have found out this truth too late to prevent the ultimate disaster of having to drop the course or fail it. The material is really not difficult and you should know it well.

Keys to the Chapter

1.1 Inorganic and Organic Compounds

This introduction merely informs you that organic compounds have different properties than inorganic compounds, which you will shortly learn is a consequence of the differences in chemical bonds. So, you are informed that you have to know something about atoms before you can understand bonding of atoms. In fact, this is just the beginning of a continuing story. You have to know everything that preceded the new information in order to understand that information.

1.2 Atomic Structure

Before you can understand organic chemistry, you have to have a firm grasp of certain aspects of atomic structure and the relationship between electron configuration and the properties of atoms. This section should be read carefully. Don't pass over it because you have "been there - done that" in General Chemistry. You should know, without reference to the periodic table, the **electron configuration** of the common elements in organic molecules and where they are located with respect to each other in the periodic table.

1.3 Atomic Properties

Two periodic trends are important to understanding the physical and chemical properties of organic compounds. They are electronegativity and atomic radius.

You should remember that **electronegativity**, which controls the type of bonds that atoms form, increases from left to right in a period and from bottom to top in a Group. There are few atoms that you need to consider, so you should remember that the order ofelectronegativities is C < N < O for the three most common elements in organic molecules, excluding hydrogen. Note that the values differ by 0.5 between neighboring elements in this part of the second period. Note also that there is a more pronounced difference between second and third period elements. Thus, fluorine and chlorine differ by 1.0, as do oxygen and sulfur. You will also use the relationship I < Br < Cl a lot in the course. Sulfur is encountered less frequently in organic compounds.

The differences in **atomic radii** for carbon, nitrogen, and oxygen are small but not unimportant. The differences in the sizes of the halogens are more significant and play a role in their chemistry.

1.4 Energy and Bond Formation

Make sure that you understand the sign conventions for energy changes and the related descriptive terms **exothermic** and **endothermic**. Enthalpy changes are used continually in the text. Recall **Hess's Law** from General Chemistry, and remember that you can add chemical equations and their related state functions such as the **enthalpy change**, symbolized by $\Delta H°$, to obtain a desired equation and hence to calculate its enthalpy change. However, don't make the mistake of assuming that the actual chemical reactions must go by the series of steps selected to make the calculation. The pathway of the chemical reaction is usually different than the selected steps used in Hess's Law.

1.5 Types of Bonds

There are two main classes of bonds. **Ionic bonds** predominated in General Chemistry, but covalent bonds are much more important in organic chemistry. You should be familiar with the idea of balance of charge in ionic compounds. When positive and negative ions combine to form an ionic compound, the charges of the cations and anions must be balanced to give a neutral compound. For ionic compounds, the cation is named first and then the anion. Thus ammonium sulfide contains NH_4^+ and S^{2-}. Two ammonium ions are required to balance the charge of one sulfide ion, so the formula of ammonium sulfide is $(NH_4)_2S$. Parentheses enclose a polyatomic ion when a formula unit contains two or more of that ion, and the subscript is placed outside the parentheses.

Carbon does not ordinarily form anions and cations, and if they do you will find that they are very reactive. The more common way that carbon bonds to other elements is by sharing of pairs of electrons in **covalent bonds**. You should be able to recognize and write simple **Lewis structures** of the limited number of atoms found in organic compounds. The stability of these structures is attributed to the **octet rule** that states that second row elements tend to form associations of atoms with eight electrons (both shared and unshared) in the valence shell of all atoms of the molecule. A series of steps used to write Lewis structures is given in Section 1.6 of the text.

One or more pairs of electrons can be shared between carbon atoms. Single, double, and triple bonds exist by sharing one, two, and three pairs of electrons, respectively. In applying the octet rule, the bonding electrons are double counted. That is, each atom "has" the bonding electrons, so they count toward the total of eight for each atom.

With the exception of bonds to carbon and to hydrogen, the bonds of carbon to other elements are **polar covalent**. The degree of polarity depends on the difference in the electronegativity values of the bonded atoms. The direction of the bond moment is indicated by an arrow with a cross at the end opposite the arrow head. The symbols δ^+ and δ^- are used to indicate the partially positive and partially negative atoms of the bonded atoms.

Coordinate covalent bonds are not common in organic chemistry, and you will be alerted to this type of bond when it occurs. These bonds form when both bonding electrons are derived from a single atom.

1.6 Strategy for Writing Lewis Structures

You may be able to write Lewis structure intuitively and may be tempted to skip over all the steps listed in this section. However, it is extremely important that you know first how many electrons are involved in a molecule based on its constituent atoms as well as where they are located. In other words you have to be a good bookkeeper. One of the downfalls for students in the continuing study of organic chemistry is due to the failure to account for electrons in molecules and how they are redistributed in chemical reactions between molecules.

Consider vinyl chloride, C_2H_3Cl, which is used to produce polymers for commercial products such as PVC pipes. Its Lewis structure is written based on the fact that carbon, hydrogen, and chlorine atoms contain 4, 1, and 7 valence electrons, respectively. Thus the molecule contains $2(4) + 3(1) + 7 = 18$ valence electrons. The basic skeleton of the molecule is:

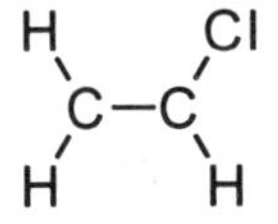

This leaves $18 - 10 = 8$ valence electrons to complete the structure. Each carbon atom still needs 2 more electrons to complete its octet, and the chlorine atom needs 6. Thus 10 electrons are needed, but only 8 are available. A deficiency of 2 electrons is solved by forming a double bond between the two carbon atoms.

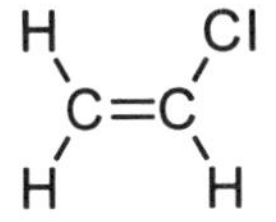

The remaining six electrons are assigned as nonbonded pairs to complete the octet of the chlorine atom.

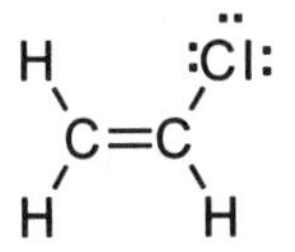

1.7 Formal Charge

Some guidelines for determining formal charges are given in this section in the text. Following the rules requires again that you have a firm grasp of the number of electrons associated with the common atoms found in organic molecules. Once you have done a few calculations using the formal equations, you should be able to determine formal charges "in your head". Moreover, you should eventually recognize that there is a formal charge on an atom just based on a visual inspection of the number of bonding and nonbonding electrons about an atom. For example, if nitrogen has three bonds—regardless of the combination of single, double, or triple bonds—and a pair of electrons, then it has no formal charge. If there are four bonds to nitrogen—regardless of the combination of single, double, or triple bonds—the nitrogen atom has a formal +1 charge. Similarly, if oxygen has two bonds—regardless of the combination of single or double bonds—and two pairs of electrons, then it has no formal charge. If there are three bonds to oxygen—regardless of the combination of single or double bonds—the oxygen atom has a formal +1 charge. Thus you should recognize that the following example contains oxygen with a +1 formal charge and that the species has a net +1 charge.

$$
\begin{array}{c}
\text{H} \\
\diagdown \\
\text{C} = \ddot{\text{O}} : \\
\diagup \quad \diagdown \\
\text{H} \qquad \text{H}
\end{array}
$$

You will have to learn to quickly identify the formal charge in reactive species and be able to keep track of the changes that result from the movement of electrons within the species and between two or more species—a subject discussed in Chapter 2.

1.8 Molecular Geometry

The "trends" in bond lengths as a function of constituent atoms and the number of bonds are listed in this section. You should "know" this material. An understanding of the material will be developed at numerous places later in the text.

The conventions showing bonds as "in the plane", "above the plane", and "below the plane" of the printed page are described. The ability to view three dimensions in two-dimensional representations varies considerably between individuals. In this course you will have to learn how to view molecular structures from a variety of angles. You should consider purchasing a molecular model kit to assist you in learning this skill.

1.9 Resonance Theory

For most compounds, one Lewis structure is sufficient to describe the distribution of electrons and the types of bonds in each compound. However, for some species a single Lewis structure is not definitive. **Resonance theory** is a bookkeeping device to describe the delocalization of electrons, giving structures that cannot be adequately described by a single Lewis structure. Some of the electrons are not restricted to regions between only two nuclei. Such bonding is described using two or more resonance contributors that differ only in the location of the electrons. The positions of the nuclei are unchanged. The actual structure of a molecule that is pictured by resonance structures has characteristics of all the resonance contributors. However, those resonance contributors are merely attempts to represent reality and in fact do not exist. In most cases, a single resonance contributor will be used to describe the process of the "movement" of electrons in chemical reactions. However, remember that other resonance structures are always implied even though they are not explicitly written.

Curved arrows are used to show the movement of electrons to transform one resonance contributor into another. The electrons move from the position indicated by the tail of the arrow toward the position shown by the head. This process is a skill that you will have to master.

The degree to which various resonance forms contribute to the actual structure in terms of the properties of the bonds and the location of charge is not the same for all resonance forms. The overriding first rule is that the Lewis octet must be considered as a first priority. After that, the location of charge on atoms of appropriate electronegativity can be considered.

1.10 Valence-Shell Electron-Pair Repulsion Theory

Like charges repel each other, so the electron pairs surrounding a central atom in a molecule should repel each other and move as far apart as possible. We use the theory of **valence-shell electron-pair repulsion** to predict the shapes of molecules. This concept is very much like what you learned in General Chemistry, but you now have to be concerned with only three types of molecular geometries—tetrahedral, trigonal planar, and linear.

Using **VSEPR** theory requires that regions of electron density be considered regardless of how many electrons are contained in the region. Thus, a single bonded pair or two pairs of electrons in a double bond are considered as "equal". In general then,

1. Two regions containing electrons around a central atom are 180° apart, producing a linear arrangement.

2. Three regions containing electrons around a central atom are 120° apart, producing a trigonal planar arrangement.

3. Four regions containing electrons around a central atom are 109.5° apart, producing a tetrahedral arrangement.

The electron pairs around a central atom may be bonding electrons or nonbonding electrons, and both kinds of valence-shell electron pairs must be considered in determining the shape of a molecule. When all of the electron pairs are arranged to minimize repulsion, we look at the molecule to see how the atoms are arranged in relation to each other. The geometric arrangement of the atoms determines the bond angles.

Consider the general structure of an isocyanate group in a structure, where R represents the "rest" of the carbon structure.

$$R-\overset{..}{N}=C=\overset{..}{\overset{..}{O}}:$$

The nitrogen atom has three regions containing electrons around it. They are a single bond, a double bond, and a nonbonded pair of electrons. So, these features will have a trigonal planar arrangement, and the R-N=C bond angle is 120°. The isocyanate carbon atom has two groups of electrons around it—two double bonds—so they will have a linear arrangement. The N=C=O bond angle is 180°.

1.11 Dipole Moments

The polarity of a molecule is given by its dipole moment. You will not have to be concerned with the value of the dipole moment, but you will often have to compare two or more molecules to rank their relative polarities. This skill requires that you understand two concepts—the polarity of individual bonds and the arrangement of those bonds in the molecule. Bond moments appropriately arranged can cancel one another and there is no net resultant dipole moment. In other molecules, bond moments may reinforce each other or partially cancel, causing a net resultant dipole moment.

1.12 Molecular Orbital Theory

Atomic orbitals, as you learned in General Chemistry, are convenient ways to represent the distribution (commonly thought of as location) of electrons in atoms. The combination of atomic orbitals to give molecular orbitals, which represent the distribution of electrons over two or more atoms, was also presented in General Chemistry, probably for molecules that consisted of only two atoms. Thus, this section is largely review material. However, in later chapters, this material will be revisited and the concept of molecular orbitals extended to molecules containing many atoms. The important concepts are summarized as follows:

1.   The number of molecular orbitals must equal the number of atomic orbitals used to "create" them.

2.   Molecular orbitals, as well as atomic orbitals, are represented by wave functions whose value may be positive or negative and is a function of geometry.

3.   There are two types of molecular orbitals, called sigma (σ) and pi (π).

4.   Molecular orbitals may be bonding or antibonding.

1.13 The Hydrogen Molecule

This section states that there are two ways that s orbitals can combine to give molecular orbitals. Bonding molecular orbitals are of lower energy than the constituent atomic orbitals. Antibonding molecular orbitals are of higher energy. In general, as in the case of the hydrogen molecule, the available electrons in the original atomic orbitals are sufficient in number to fill only the bonding molecular orbitals. Formation of a bond is exothermic because the bonding molecular orbital is of lower energy than the separate atomic orbitals.

1.14 Bonding in Carbon Compounds

The strongest bonds between carbon atoms and other atoms are σ bonds that result from overlap of atomic orbitals along the internuclear axis. Sideways overlap of p orbitals leads to a less stable π bond.

Atomic orbitals are combined (mixed) to give **hybridized** atomic orbitals. These orbitals are pictured to account for the geometry of molecules and properties of molecules. Although the concept was introduced early in your General Chemistry course, it was largely forgotten in subsequent material of this very diverse course. Such is not the case in organic chemistry. Orbital hybridization is at the heart of the structure of organic compounds, and you will have to understand the concept and be able to use it to understand the reactivity of organic compounds.

1.15 sp³ Hybridization of Carbon in Methane

Yes, this is another subject that you learned in General Chemistry, and it was probably covered on one page or so as it is here. However, understanding the hybridization about a carbon atom with four single bonds is now important, and you will need to know what is meant by the term sp³ and understand its consequences. In general, the mental picture used to "construct" hybridized orbitals consists of two parts. First, it "costs" energy to form the hybridized orbitals—that is, they are of higher energy than atomic orbitals. Second, more energy is released by formation of bonds using hybridized orbitals than atomic orbitals. Third, the net result is the formation of stronger bonds by hybridized orbitals.

The term **%s character**, used to describe the contribution of the atomic orbitals to a hybridized orbital, is one that you will use time and time again in explaining the electronic properties of bonds. It is one of the myriad concepts that is easy to overlook now and not understand later when it is important. Don't let this happen to you!

1.16 sp³ Hybridization of Carbon in Ethane

Ethane and other organic compounds containing four single bonds to carbon atoms consist of sigma bonds arranged at tetrahedral angles to one another. The important point to note in this section is that groups of atoms can rotate about a sigma bond and not disrupt the bond. The resulting **conformations** are just different temporary arrangements of atoms that still maintain their bonding arrangement amongst the constituent atoms. This phenomenon is considered in greater detail in Chapter 4.

1.17 sp² Hybridization of Carbon in Ethylene

The **sp² hybrid orbitals** of carbon occur in compounds such as ethylene that contain a double bond. The overlap of these orbitals with one another or with other orbitals such as an s orbital of hydrogen gives a sigma bond. The sp² hybrid orbitals are at 120° to one another. They have 33% s character because they are "formed" from one s orbital and two p orbitals. More importantly, each time there are three sp² hybrid orbitals about a carbon atom, there is also one remaining p orbital that forms a π bond with a neighboring atom, as in the case of another carbon atom in ethylene. The bond energy of the π bond in ethylene is less than that of the σ bond. Thus, the bond dissociation energy of a π bond is less than twice that of two σ bonds.

1.18 sp Hybridization of Carbon in Acetylene

The **sp hybrid orbitals** of carbon occur in compounds such as acetylene that contain a triple bond. The overlap of these orbitals with one another or with other orbitals such as an s orbital of hydrogen gives a sigma bond. The sp hybrid orbitals are at 180° to one another. They have 50% s character because they are "formed" from one s orbital and one p orbital. Each time there are two sp hybrid orbitals about a carbon atom, there are also two remaining p orbitals that form two π bonds with a neighboring atom, as in the case of another carbon atom in acetylene.

1.19 Effect of Hybridization on Bond Length

With increasing %s character, the electrons within a hybrid orbital are held closer to the nucleus of the atom. As a consequence, the bond lengths are shorter and the strength of the bond increases as shown for the C—H bonds in ethane, ethylene, and acetylene. **This is a concept that will permeate the study of organic chemistry.** Don't overlook this effect on bond strengths in later chapters. The effect of hybridization is seen in the bond dissociation energies of the single bond of any atom bonded to carbon.

1.20 Hybridization of Nitrogen

Hybridization is not a phenomenon restricted to carbon. It applies to other atoms as well. The only difference is in the number of electrons that are distributed in the orbitals. Nitrogen, a Group VA element, has five valence electrons.

In sp^3 hybridized nitrogen, the five electrons are distributed according to Hund's rule to give two electrons in one orbital and one each in the remaining three orbitals. The nonbonded electron pair is pictured as directed to the corner of a tetrahedron, as are the bonded pairs. However, the shape of such molecules is pyramidal, like ammonia, because it is the position of the atoms, not the electron pairs, that identifies the molecular shape.

In sp^2 hybridized nitrogen atoms, the five electrons are distributed to give two electrons in one orbital and one each in the remaining two orbitals. The fifth electron is in a p orbital, which we will see can combine to form π bonds. Note that the geometry of the bonds that can form using sp^2 hybridized orbitals will be at 120°, as will the angle between the bonds and the nonbonding electron pair.

In sp hybridized nitrogen atoms, the five electrons are distributed to give two electrons in one orbital and one in the other orbital. The remaining two electrons are in the two p orbitals, which we will see can combine to form π bonds. Note that the bond that can form using the sp hybridized orbital will be at 180° to the nonbonding electron pair.

1.21 Hybridization of Oxygen

The difference between the hybridization of oxygen compared to nitrogen and carbon is in the number of electrons that are distributed in the orbitals. Oxygen, a Group VIA element, has six valence electrons.

In sp^3 hybridized oxygen atoms, the six electrons are distributed according to Hund's rule to give two electrons in each of two orbitals and one each in the remaining two orbitals. The nonbonded electron pairs are pictured as directed to the corners of a tetrahedron, as are the bonded pairs. However, the shape of molecules like water is angular.

In sp^2 hybridized oxygen atoms, the six electrons are distributed to give two electrons in two orbitals and one in the remaining orbital. The sixth electron is in a p orbital, which we will see can combine to form π bonds. Note that the geometry of the bonds that can form using sp^2 hybridized orbitals will be at 120°, as will the angle between the bonds and the two nonbonding electron pairs.

Solutions to Exercises

1.1   **(a)** 5   **(b)** 7   **(c)** 4   **(d)** 6   **(e)** 7   **(f)** 7   **(g)** 6   **(h)** 5

1.2   **(a)** Cl; Br   **(b)** O; S   **(c)** N; C   **(d)** O; N

(e) Br; I   **(f)** F; C   **(g)** O; C   **(h)** O; I

1.3 $Ca(H_2PO_4)_2$

1.4 $Mg_3(AsO_3)_2$

1.5   **(a)** 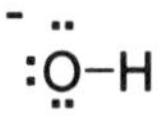 $:\overset{\cdot\cdot}{\underset{\cdot\cdot}{O}}-H$ ⁻   **(b)** ⁻$:C\equiv N:$ **(c)** $H-\overset{\cdot\cdot}{O}{}^{+}-H$ with H below   **(d)** 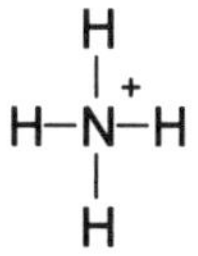 $H-\overset{H}{\underset{H}{N}}{}^{+}-H$ **(e)** 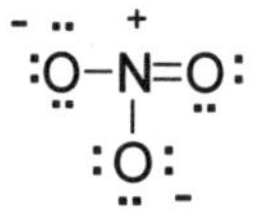 $:\overset{\cdot\cdot}{O}-N=\overset{\cdot\cdot}{O}:$ with $:\overset{\cdot\cdot}{O}:{}^{-}$ below N

1.6   **(a)** 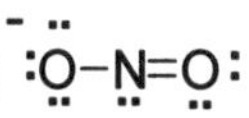 ⁻$:\overset{\cdot\cdot}{O}-N=\overset{\cdot\cdot}{O}:$ **(b)** 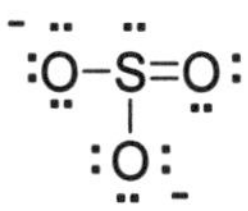 ⁻$:\overset{\cdot\cdot}{O}-\overset{\overset{\cdot\cdot}{O}:}{\underset{:O:{}^{-}}{S}}=\overset{\cdot\cdot}{O}:$ **(c)** 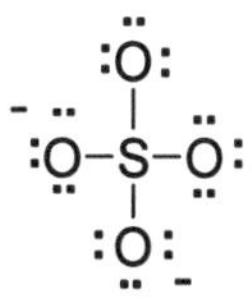 ⁻$:\overset{\cdot\cdot}{O}-\overset{:O:}{\underset{:O:{}^{-}}{S}}-\overset{\cdot\cdot}{O}:$⁻   **(d)** 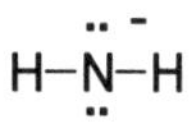 $H-\overset{\cdot\cdot}{N}-H$ ⁻   **(e)** 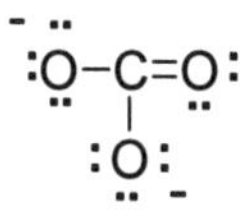 ⁻$:\overset{\cdot\cdot}{O}-\overset{:O:{}^{-}}{C}=\overset{\cdot\cdot}{O}:$

1.7   **(a)** 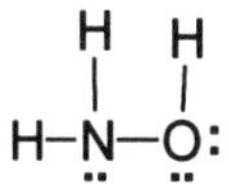 $H-\overset{H\ \ H}{\underset{\cdot\cdot}{N}-\overset{\cdot\cdot}{O}}:$ **(b)** 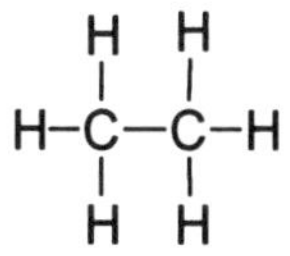 $H-\overset{H\ H}{\underset{H\ H}{C-C}}-H$ **(c)** $H-\overset{H\ H}{\underset{H}{C-\overset{\cdot\cdot}{O}}}:$

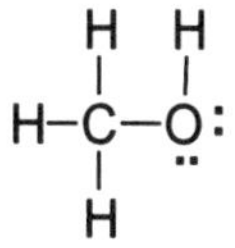

(d) 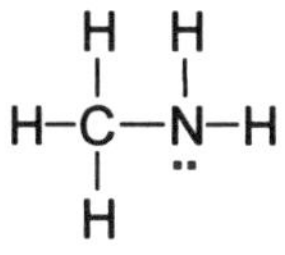 $H-\overset{H\ \ H}{\underset{H}{C-\overset{\cdot\cdot}{N}}}-H$ **(e)** 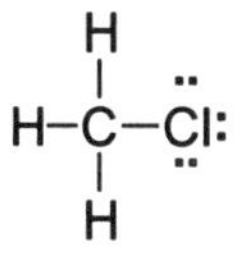 $H-\overset{H}{\underset{H}{C}}-\overset{\cdot\cdot}{\underset{\cdot\cdot}{Cl}}:$ **(f)** 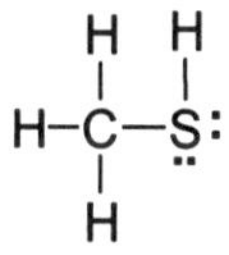 $H-\overset{H\ H}{\underset{H}{C}}-\overset{\cdot\cdot}{S}:$

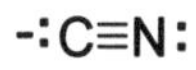

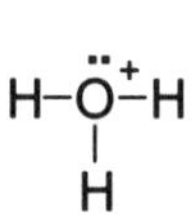

1.8   **(a)** 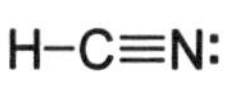 $H-C\equiv N:$ **(b)** $H-\overset{\cdot\cdot}{N}=\overset{\cdot\cdot}{N}-H$ **(c)** $H-\overset{H\ H}{C}=N:$

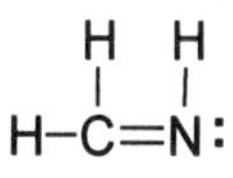

(d) 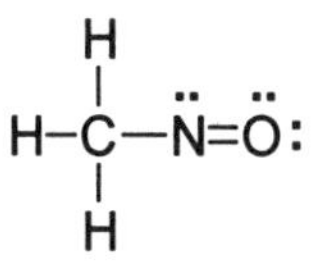 $H-\overset{H}{\underset{H}{C}}-\overset{\cdot\cdot}{N}=\overset{\cdot\cdot}{O}:$ **(e)** 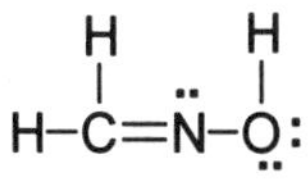 $H-\overset{H\ \ H}{C}=\overset{\cdot\cdot}{N}-\overset{\cdot\cdot}{O}:$ **(f)** 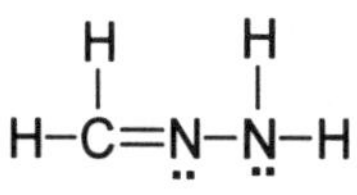 $H-\overset{H\ \ H}{C}=N-\overset{H}{\underset{\cdot\cdot}{N}}-H$

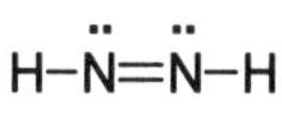

1.9

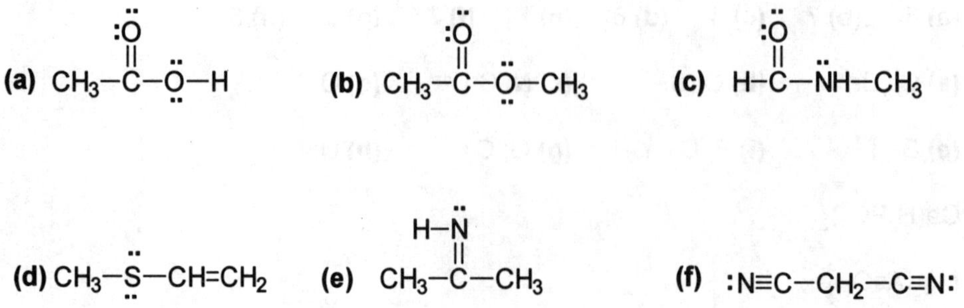

(a) $CH_3-\overset{\ddot{\text{O}}}{\overset{\|}{C}}-\ddot{\text{O}}-H$

(b) $CH_3-\overset{\ddot{\text{O}}}{\overset{\|}{C}}-\ddot{\text{O}}-CH_3$

(c) $H-\overset{\ddot{\text{O}}}{\overset{\|}{C}}-\overset{}{\ddot{N}}H-CH_3$

(d) $CH_3-\ddot{\underset{\cdot\cdot}{S}}-CH=CH_2$

(e) $CH_3-\overset{H-\ddot{N}}{\overset{\|}{C}}-CH_3$

(f) $:N{\equiv}C-CH_2-C{\equiv}N:$

1.10

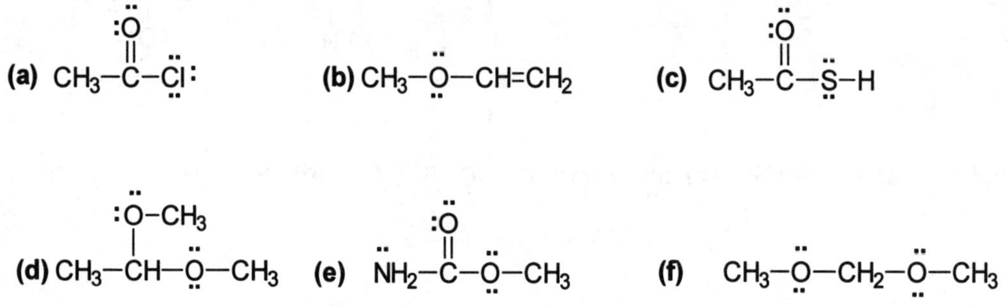

(a) $CH_3-\overset{\ddot{\text{O}}}{\overset{\|}{C}}-\ddot{\underset{\cdot\cdot}{Cl}}:$

(b) $CH_3-\ddot{\underset{\cdot\cdot}{O}}-CH=CH_2$

(c) $CH_3-\overset{\ddot{\text{O}}}{\overset{\|}{C}}-\ddot{S}-H$

(d) $CH_3-\overset{:\ddot{O}-CH_3}{\underset{|}{C}H}-\ddot{\underset{\cdot\cdot}{O}}-CH_3$

(e) $NH_2-\overset{\ddot{\text{O}}}{\overset{\|}{C}}-\ddot{\underset{\cdot\cdot}{O}}-CH_3$

(f) $CH_3-\ddot{\underset{\cdot\cdot}{O}}-CH_2-\ddot{\underset{\cdot\cdot}{O}}-CH_3$

1.11

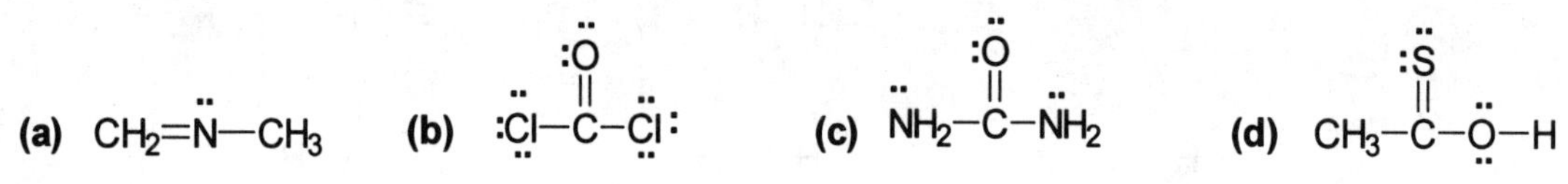

(a) $CH_2{=}\overset{}{\ddot{N}}-CH_3$

(b) $:\ddot{\underset{\cdot\cdot}{Cl}}-\overset{\ddot{\text{O}}}{\overset{\|}{C}}-\ddot{\underset{\cdot\cdot}{Cl}}:$

(c) $NH_2-\overset{\ddot{\text{O}}}{\overset{\|}{C}}-NH_2$

(d) $CH_3-\overset{:\ddot{S}}{\overset{\|}{C}}-\ddot{\underset{\cdot\cdot}{O}}-H$

1.12

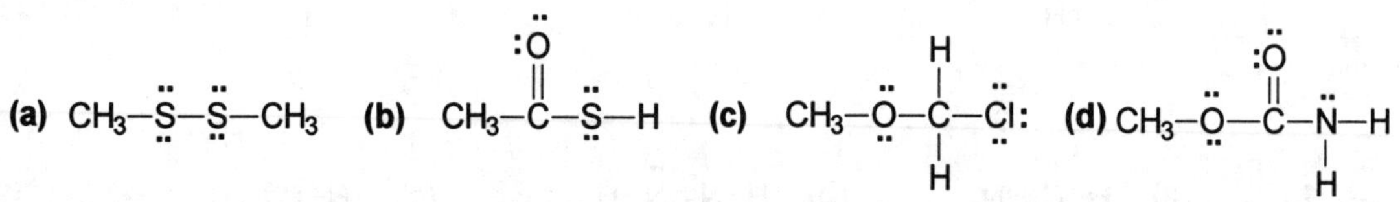

(a) $CH_3-\ddot{\underset{\cdot\cdot}{S}}-\ddot{\underset{\cdot\cdot}{S}}-CH_3$

(b) $CH_3-\overset{\ddot{\text{O}}}{\overset{\|}{C}}-\ddot{\underset{\cdot\cdot}{S}}-H$

(c) $CH_3-\ddot{\underset{\cdot\cdot}{O}}-\overset{H}{\underset{H}{\overset{|}{\underset{|}{C}}}}-\ddot{\underset{\cdot\cdot}{Cl}}:$

(d) $CH_3-\ddot{\underset{\cdot\cdot}{O}}-\overset{\ddot{\text{O}}}{\overset{\|}{C}}-\overset{}{\underset{H}{\overset{|}{N}}}-H$

1.13

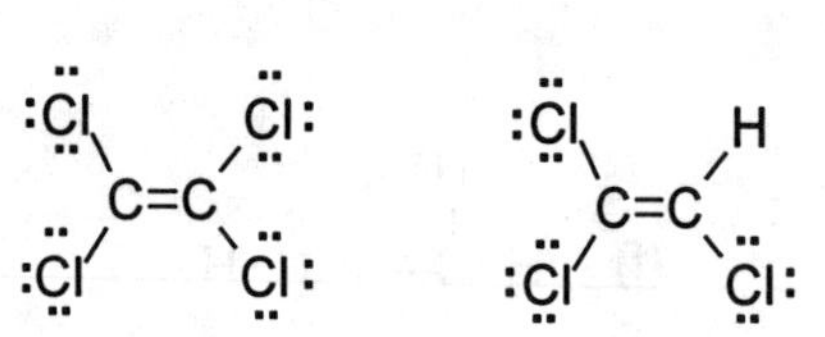

$\underset{:\ddot{\underset{\cdot\cdot}{Cl}}}{\overset{:\ddot{Cl}}{}}\overset{}{C}=C\underset{\ddot{\underset{\cdot\cdot}{Cl}}:}{\overset{\ddot{Cl}:}{}}$

$\underset{:\ddot{\underset{\cdot\cdot}{Cl}}}{\overset{:\ddot{Cl}}{}}\overset{}{C}=C\underset{\ddot{\underset{\cdot\cdot}{Cl}}:}{\overset{H}{}}$

1.14

$\underset{H}{\overset{H}{}}C=C\underset{C{\equiv}N:}{\overset{H}{}}$

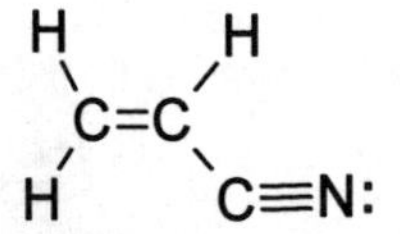

1.15   (a) none of the atoms has a formal charge
(b) nitrogen is +1; carbon is -1
(c) nitrogen is +1; oxygen is -1
(d) nitrogen atoms from left to right have 0, +1, and -1 formal charges

1.16   (a) oxygen is +1; boron is -1
(b) nitrogen is +1; aluminum is -1
(c) nitrogen is +1; singly bonded oxygen atom is -1
(d) phosphorus is +1; oxygen atom on the right is -1

1.17   (a) carbon is -1; oxygen is +1; total charge is 0
(b) nitrogen is zero; oxygen is +1; total charge is +1
(c) carbon is -1; nitrogen is 0; total charge is -1
(d) both carbon atoms are -1; total charge is -2

1.18   (a) central nitrogen atom is +1; the other nitrogen atoms are each -1; the total
charge is -1
(b) nitrogen atom is +1; both oxygen atoms are 0; the total charge is +1

1.19   (a) +1   (b) +1   (c) +1   (d) -1   (e) +1   (f) -1

1.20   (a) -1   (b) 0   (c) -1   (d) 0   (e) 0   (f) -1

1.21   The nitrogen atom is +1; the total charge is +1.

1.22   The phosphorus atom is +1.

1.23
$$:\overset{..}{\underset{..}{S}}-C\equiv N: \quad \longleftrightarrow \quad :\overset{..}{S}=C=\overset{..}{N}:^{-}$$

1.24   The pairs in both (a) and (b) represent contributing resonance forms.

(a)
$$:\overset{..}{\underset{..}{N}}-N\equiv N: \quad \longleftrightarrow \quad :\overset{..}{N}=N=\overset{..}{N}:$$

(b)
$$H-C\equiv N-\overset{..}{\underset{..}{O}}: \quad \longleftrightarrow \quad H-\overset{..}{C}=N=\overset{..}{\underset{..}{O}}$$

1.25
$$CH_3-\overset{\overset{\displaystyle :\overset{..}{O}:^{-}}{|}}{C}=\overset{+}{N}H_2$$

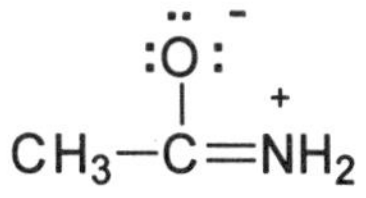

1.26
$$CH_3-\overset{\overset{\displaystyle :\overset{..}{O}:^{-}}{|}}{C}=\overset{..}{\underset{..}{O}}:$$

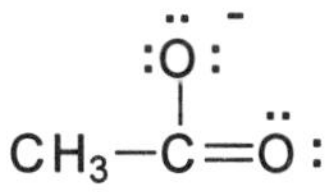

1.27   The alternate resonance form is structurally equivalent to the given resonance form and both contribute equally.

$$\overset{+}{C}H_2-CH=CH_2$$

1.28   The alternate resonance form has a negative charge on the carbon atom rather than the nitrogen atom. Because nitrogen is more electronegative then carbon, the original resonance form contributes to a larger extent.

$$\begin{array}{c} H \\ \diagdown \\ \diagup \\ H \end{array} \overset{+}{C}=\overset{..}{N}=\overset{-}{N}: \quad\longleftrightarrow\quad \begin{array}{c} H \\ \diagdown \\ \diagup \\ H \end{array} \overset{-}{C}-\overset{..}{\underset{}{N}}\overset{+}{\equiv}N:$$

1.29   **(a)** 180°   **(b)** 109°   **(c)** 109°   **(d)** 180°   **(e)** 180°

1.30   **(a)** 109°   **(b)** 109°   **(c)** 109°   **(d)** 109°

1.31   **(a)** 120°   **(b)** 120°   **(c)** 120°

1.32   **(a)** 109°   **(b)** 120°   **(c)** 120°

1.33   The carbon-fluorine bond is much shorter than the carbon-chlorine bond.

1.34   HS > HN > HO; the difference in electronegativity of the atoms in the O-H bond is larger than that of the atoms in the N-H bond. There is a substantially smaller difference in electronegativity of the atoms in the S-H bond and the bond has a small polarity.

1.35   The C=O and C=S bond moments are not equal, so they don't cancel each other. The net dipole moment is toward the more electronegative oxygen atom.

1.36   Acetone has the larger dipole moment, because the bond moments of the two C-Cl bonds are in opposition to the C=O bond moment in phosgene.

1.37   The net resultant of the bond moments of the carbon-bromine bonds in the cis compound is toward the side of the molecule containing the two carbon-bromine bonds. The bond moments of the carbon-bromine bonds in the trans compound are in opposition and therefore cancel one another.

1.38   The two bond moments must be in opposition to one another to give a resultant that is less the large bond moment value. The bond moment of the carbon-chlorine bond is toward the chlorine atom. Thus the bond moment of the carbon-nitrogen bond must be toward nitrogen.

1.39   **(a)** from left to right - sp^3, sp^2 **(b)** from left to right- sp^3, sp^2, sp^2
        (c) from left to right - sp^3, sp^2 **(d)** from left to right- sp^3, sp^2, sp^3

1.40   **(a)** from left to right - sp^2, sp^3 **(b)** from left to right- sp^3, sp^2, sp^2
        (c) from left to right - sp^3, sp^2, sp^3 **(d)** from left to right- sp, sp^3, sp

1.41   **(a)** sp^2
        (b) sp^3
        (c) double bonded oxygen is sp^2; single bonded is sp^3
        (d) double bonded oxygen is sp^2; single bonded is sp^3

1.42   **(a)** sp^3 **(b)** sp^3 **(c)** sp^2 **(d)** sp

1.43   The carbocation is sp^2 hybridized and the bond angles are 120°. The carbanion is sp^3 hybridized and the bond angles are 109°.

1.44   The three pairs of electrons - one nonbonded and two bonded - are in a common plane at 120° to one another.

1.45   The carbon atom is sp hybridized. The oxygen atoms are sp^2 hybridized.

$$:\ddot{O}=C=\ddot{O}:$$

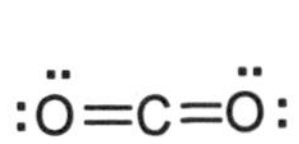

1.46   The nitrogen atom is sp hybridized. The oxygen atoms are sp^2 hybridized.

$$:\ddot{O}=\overset{+}{N}=\ddot{O}:$$

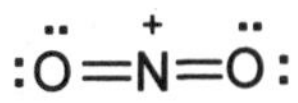

1.47   The carbon atom is sp^2 hybridized.

$$:\ddot{C}l-\overset{\overset{\ddot{O}}{\|}}{C}-\ddot{C}l:$$

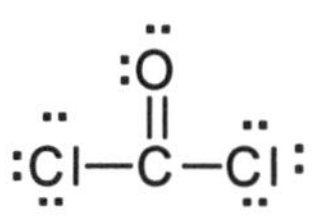

1.48   The carbon atom is sp^2 hybridized. The double bonded oxygen atom is sp^2; single bonded oxygen atom is sp^3.

1.49   Bond lengths between common sets of atoms tend to be the same and do not depend markedly on the other atoms of the structure.

1.50   The C=N bond is shorter than the C=C bond, because the atomic radius of nitrogen is smaller than the atomic radius of carbon.

1.51 $:NH_2-\ddot{O}-H$ $:\ddot{O}=\overset{+}{N}=\ddot{O}:$

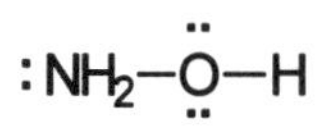

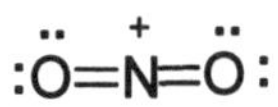

The nitronium ion has N=O bonds and bond lengths decrease with an increased number of bonding pairs of electrons.

1.52   Each of the three contributing resonance forms in CF_3^+ has a C=F bond which contributes to the overall shortening of the carbon-fluorine bonds in the resonance hybrid structure.

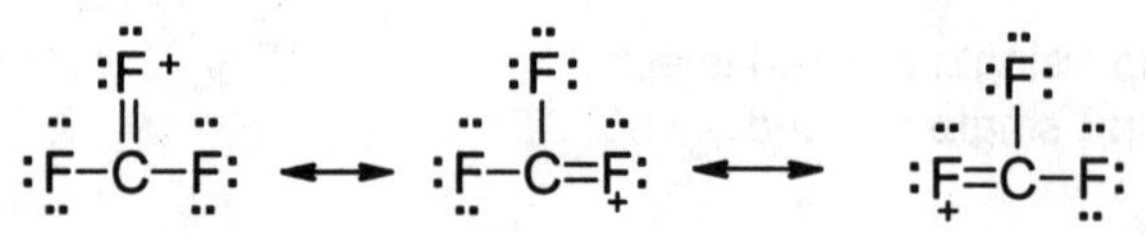

1.53   The bonds are sp^3-sp^3 and sp^3-sp^2 hybridized, respectively. An sp^2 hybridized atom holds the bonding pair of electrons closer to the nucleus and leads to a shortening of the bond.

1.54   The bond to the CH_3 group should also be 142 pm. The bond of oxygen to the CH group should be shorter than 142, because an sp^2 hybridized atom holds the bonding pair of electrons closer to the nucleus and leads to a shortening of the bond.

1.55   **(a)** 109°   **(b)** 120°

1.56   120°

1.57   The hybridization of both the carbon atoms in ethylene and the nitrogen atoms in diimide is sp^2. The H-N-N bond angle is 120°.

1.58   The H-C-H bond angle is 120°. The C-C-C bond angle is 180°. The hybridization of both of the terminal atoms is sp^2; the hybridization of the central carbon atom is sp.

1.59   109°

1.60   180°

2

Structure and Properties of Organic Molecules

In the last chapter, we reviewed the fundamentals of bonding and how hybrid orbitals are used to depict the bonding about carbon atoms in simple molecules. Now we expand on those concepts and really start to examine somewhat more complex structures and how molecules with certain types of bond react. If you casually drifted through the previous chapter based on your understanding of bonding from your General Chemistry course, then you may have already developed the unfortunate habit of overlooking ideas that you really have to learn well. There is a lot of material in this chapter that you really should regard as "new", because either it wasn't discussed in General Chemistry or it was discussed with a very different focus than in this Chapter. The central organization of organic chemistry—the structure of functional groups—is presented first, and then the physical and chemical properties of such groups are discussed.

Keys to the Chapter

2.1 Functional Groups

Functional groups are structural features of organic compounds other than carbon-carbon single bonds and carbon-hydrogen single bonds. Multiple bonds between carbon atoms as well as bonds from carbon to atoms such as oxygen, nitrogen, sulfur, and the halogens are functional groups. You should learn the features of the functional groups in this Section. In particular, pay attention to not only the composition of the functional groups, but the structure and bonding as well.

As we proceed with our study of organic chemistry these functional groups will be shown to behave chemically in ways that are predictable based on the number and type of bonds to carbon in each functional group. The chemistry of organic molecules depends on the functional groups that they contain. In the study of language, you can't understand the nuances of a sentence without knowing the meaning of the words. Likewise, the chemistry presented throughout this text won't make much sense unless you can use functional groups as easily as you can use simple words when you talk.

The only functional groups that do not contain atoms other than carbon and hydrogen contain carbon-carbon multiple bonds, as in ethylene and acetylene. Benzene, which also contains multiple bonds, is a representative of a group of "special" compounds called aromatic hydrocarbons.

2.2 Functional Groups Containing Oxygen

There are several types of functional groups containing oxygen in addition to carbon and hydrogen. Compounds with a carbon—oxygen and an oxygen—hydrogen bond are **alcohols**. Compounds with two carbon—oxygen bonds are **ethers**.

The oxygen atom forms double bonds to carbon in a **carbonyl group** in several functional groups. If the remaining two single bonds are to other carbon atoms, the compound is a **ketone**. If there is one single bond to a carbon atom and one to a hydrogen atom, the compound is an **aldehyde**.

Compounds with a single bond from an oxygen atom to a carbonyl group are found in **carboxylic acids** and **esters**. In carboxylic acids, the second bond to that oxygen atom is to a hydrogen atom; in esters it is to another carbon atom. Note that a carboxylic acid is not an aldehyde or a ketone, nor is it an alcohol. Both the carbonyl group and the hydroxyl group together are considered as a single functional group when they share a common carbon atom.

2.3 Functional Groups Containing Nitrogen

Nitrogen-containing functional groups will not be covered is as much detail as oxygen-containing functional groups. However, note that nitrogen can form functional groups containing single bonds in **amines**, double bonds in **imines**, and triple bonds in **nitriles**. A nitrogen atom bonded to a carbonyl group is an **amide**. The amide nitrogen atom may be bonded to any combination of hydrogen atoms or carbon atoms, and the functional group is still an amide.

2.4 Functional Groups Containing Sulfur or Halogens

Sulfur occurs in functional groups that parallel those of alcohols and ethers. The compounds are **thiols** and **thioethers**, respectively. Halogens can be bonded to sp^3 hybridized carbon atoms or to the sp^2-hybridized carbon atom of a carbonyl group.

2.5 Structural Formulas

Molecular formulas identify the total number of atoms of each element in a molecule. They tell nothing about the structure of the molecule. **Structural formulas** show how the atoms in the molecule are arranged and which atoms are bonded to each other. A complete structural formula shows every bond. A condensed formula abbreviates the structure by omitting some or all of the bonds and indicating the number of atoms bonded to each carbon atom with subscripts.

There are a number of conventions that are used to represent structures in varying degrees of detail and in shorthand form. Make sure that you understand the foundation laid in this section so that you can extend the method of writing structural formulas to more complex molecules. In general, you certainly must make sure that each atom has its appropriate number of bonds and that those bonds make chemical sense.

Condensed structural formulas leave out some bonds, and the bonded atoms are written close to each other. It is understood what is meant by the manner in which the structure is written. The method is part of the "rules of the game". Make sure that you understand the rules or you will certainly not play the game well. In general, atoms bonded to a carbon atom are usually written right after the carbon atom.

2.6 Bond-Line Structures

Have you every wondered how a secretary who can take shorthand can keep up with a person and record every detail? The same question could be raised for the court stenographer who keeps up with all the speeches in a trial. In this section you are introduced to a "shorthand" skill that is helpful in recording the details of chemical structure.   Learn the conventions given in this section.   Remember to mentally place a carbon atom at every intersection of two or more lines and at the end of every line. Also remember that there are four bonds to every carbon atom. The bonds from one carbon atom to other carbon atoms and to atoms of other elements are easy to identify; the bonds to hydrogen atoms are not visible in the bond-line structure and must be carefully accounted for. You will really come to appreciate this highly stylized method of representing structures when you start to write carbon after carbon after carbon in even medium-sized compounds. Bond-line structures are a better and faster way to record structural formulas than writing both atoms and bonds.

Don't get confused by the "hieroglyphics" of the complex structures shown in the problems in this section. Relax and learn to focus on the parts of the molecule that are the functional group. That is what you will have to do throughout the text. Remember that the chemistry occurs at the functional groups. Consider for example the structure of Diphepanol, which is used as a cough suppressant. Can you identify the functional groups? Can you write its molecular formula?

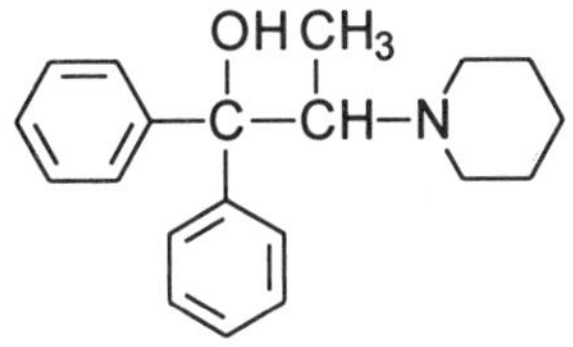

The oxygen atom contained in the molecule exists as the hydroxyl group, so the functional group is an alcohol. The nitrogen atom is bonded only to carbon atoms, so it is an amine. The molecular formula is $C_{20}H_{25}NO$.

Let's try another one. What are the oxygen-containing functional groups in the herbicide 2,4-D given by the following structure?

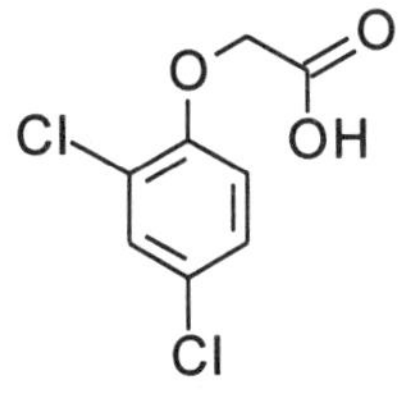

One of the oxygen atoms is present as a carbonyl group and a second as a hydroxyl group. They are both bonded to the same carbon atom, a characteristic of carboxylic acids. The third oxygen atom is bonded to two carbon atoms, a characteristic of an ether.

2.7 Isomers

This section gives you the first glimpse at why knowing the composition of a compound is not enough to establish its structure. For all but the simplest of molecules, a group of atoms can usually be bonded in several ways to give different structures called **constitutional isomers**. Recognizing this fact is simple, but actually distinguishing between structures that are isomers and those that are merely different representations of the same molecule is another matter.

There are many ways to write the structural formula of an organic compound. Two structural formulas with the same molecular formula may look so different that they appear at first glance to represent isomers. Carefully check the bonding sequence in each formula. If the bonding sequence is identical, the structural formulas represent two views of the same compound. If the bonding sequence is different, the two structural formulas represent isomers. If you list the sequence of the connectivity of the atoms you should be able to avoid the mistake of calling two different representations of the same molecule isomers. It is also necessary to avoid writing duplicate structures of the same compound when your lecturer asks you on a test to write all the isomers of a particular molecular formula. You will probably lose points for duplicate structures. Connectivity is a bit like the foot bone connected to the ankle bone connected to the leg bone, etc.

2.8 Structure and Properties

The three types of intermolecular attractive forces are London forces, dipole-dipole forces and hydrogen bonding. An understanding of these attractive forces allows you to make predictions about a number of the physical properties of organic compounds. In general, intermolecular forces affect vapor pressure, boiling point, and other physical properties. The stronger the forces of attraction, the harder it is to separate the molecules from each other - thus, the vapor pressure is decreased and the normal boiling point is increased.

London forces are weak electrostatic attractions existing between atoms and molecules due to instantaneous dipoles that result from distortion of the electron clouds. The strength of London forces is related to the polarizability of the electrons about atoms and the bonds between atoms. In general electrons that are farther from the atomic nucleus are more easily polarized. Thus, sulfur is more polarizable than is oxygen.

Dipole-dipole forces are electrostatic attractions between polar molecules and are stronger, in general, than London forces. These forces arise from the unequal sharing of electrons between atoms of different electronegativity. The attraction is between the partially positive end of one molecule and the partially negative end of another molecule. The polarity depends on the shape of the molecule.

Hydrogen bonding is a much stronger type of intermolecular attraction than London or dipole-dipole forces. **Hydrogen bonding** occurs in compounds where hydrogen is covalently bonded to a highly electronegative element (F, O, or N) and is attracted to another highly electronegative atom that has an unshared pair of electrons. Note that whereas intermolecular hydrogen bonding results in an increase in boiling point, intramolecular hydrogen bonding does not. Intramolecular hydrogen bonding is common when the number of atoms in a ring of atoms is five or six.

2.9 Solubility

The solution process is dependent on the intermolecular forces present in the solute and the solvent, and those that can result between solvent and solute molecules. The generalization that "like dissolves like" was fine in General Chemistry, but is not quite as useful in organic chemistry, because there is a larger variety of structures in organic molecules. So, you really have to know your functional groups and the types of bonds present in those groups. Otherwise you won't know whether two compounds are "like" or not.

The **dielectric constant** of a solvent is a measure of the ability of the solvent to dissolve polar substances. Table 2.1 in the text gives the most common solvents encountered in this text. Now all you have to do is to recognize functional groups of possible solutes.

2.10 Chemical Properties

There are millions of compounds and billions and billions of possible reactions. However, you can cope with the complexities of organic chemistry if you have been really listening to your lecturer, reading the text, and working problems. The subject is highly integrated, and the knowledge is vertically arrayed. In other words, there are a few principles that you have to learn, and each of them depends on what was introduced prior to a particular subject. In this section you learn that the study of chemical reactions is best done by recognizing reaction types. As you encounter each new reaction, classify it and relate it to the other reactions that you have encountered in earlier chapters,

2.11 Acid-Base Reactions

At this point you may think that you haven't really left your General Chemistry course. However, look again. You learned about the transfer of protons from one site to another in reactions studied in General Chemistry. The same process occurs in organic molecules, but now the focus is on the electron pair.

A **curved arrow** convention is introduced which considers the movement of electrons from the tail of the arrow to a point indicated by the arrowhead. In fact, you really were exposed to this approach when you learned about **Lewis acids** and **Lewis bases**, which are electron pair acceptors and electron pair donors, respectively. Unfortunately the discussion was probably limited to one page in your General Chemistry text. This concept of acid-base reactions is what organic chemistry is all about. In fact, you will eventually come to recognize that most organic reactions can be described as "have pair-will share". In organic reactions, one species with a nonbonded (or bonded) pair of electrons "donates" an electron pair to an electron-deficient species by forming a covalent bond between the two species. So, don't skip over this deceptively simple section.

2.12 Oxidation-Reduction Reactions

Generally organic oxidation-reduction reactions are not evaluated by assigning oxidation numbers. Instead, we look at the change in the number of bonds to hydrogen and/or oxygen. Oxidation is accompanied by an increase in the number of oxygen atoms and/or a decrease in the number of hydrogen atoms. Conversely, reduction is accompanied by an increase in the number of hydrogen atoms and/or a decrease in the number of oxygen atoms. The symbols [O] and [H] represent an unspecified oxidizing agent and reducing agent, respectively.

Consider the following reaction that occurs in the metabolism of THC, a depressant drug. What type of reaction has occurred?

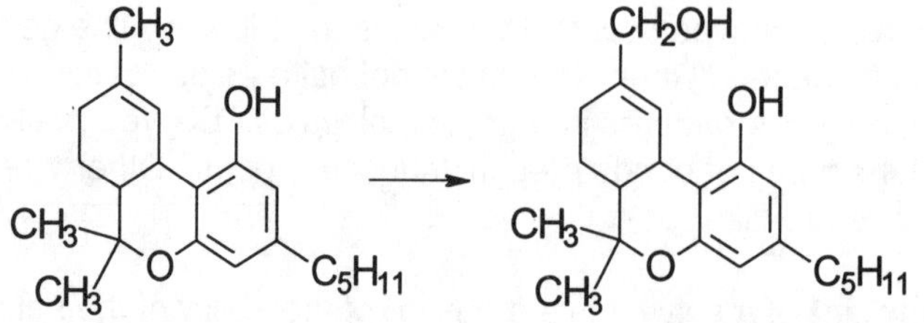

You should see that the CH_3 group at the top of the structure is changed to a CH_2OH group. Because the molecule has gained an oxygen atom, oxidation has occurred.

If a compound gains both hydrogen and oxygen in a chemical reaction, it is necessary to determine whether the two "balance" each other. It takes an increase of two hydrogen atoms to balance one oxygen atom. Thus, if the elements of water are added to a molecule, there is no oxidation or reduction. However, if two oxygen atoms and two hydrogen atoms are added as a result of a reaction, then there is a net oxidation. This is the case for the following reaction.

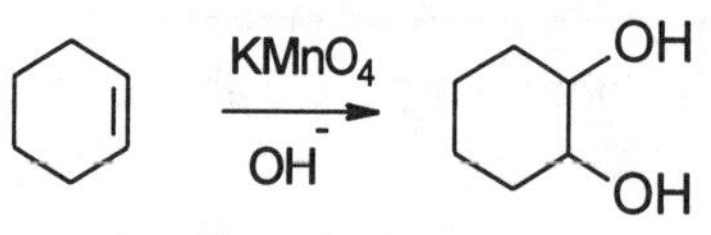

Do the Exercises on pages 92-94 in the text. If you can readily spot what has changed in the reactions of Exercises 2.39 and 2.40, then you are well on your way to developing insights into organic reactions. If not, then seek assistance!

2.13 Other Classes of Organic Reactions

Some of the types of organic reactions are a little easier to understand than others. You will encounter many examples of **addition**, **elimination,** and **substitution** reactions in the early chapters of the text. Fortunately, these are among the easiest to recognize. **Hydrolysis** and **condensation** reactions aren't encountered until the midpoint of the text. So, concentrate on the first three types of reactions. Rearrangement reactions are not encountered as frequently, but that doesn't mean that they are unimportant. They are among the most difficult to visualize because even the carbon skeleton may rearrange.

Make sure that you do the Problems and Exercises associated with this section. You must learn to focus on what is different between the reactants and products. Then you must associate this change with the terminology used to describe that change.

Solutions to Exercises

2.1   **(a)** amide   **(b)** ketone and double bond   **(c)** amide and benzene ring

2.2   **(a)** three ethers and an ester   **(b)** three ethers, ketone and ester
         (c) two amides, carboxylic acid

2.3   **(a)** C_5H_{12} **(b)** C_4H_{10} **(c)** C_4H_8
       (d) C_4H_6 **(e)** C_5H_{10} **(f)** C_5H_8

2.4   **(a)** C_9H_{20} **(b)** C_7H_{16} **(c)** C_4H_6
       (d) C_5H_8 **(e)** C_6H_{12} **(f)** C_5H_{10}

2.5   **(a)** $C_3H_6Cl_2$ **(b)** $C_3H_6Cl_2$ **(c)** $C_2H_4Br_2$
       (d) $C_3H_5Br_3$ **(e)** $C_3H_5F_3$ **(f)** $C_3H_5F_3$

2.6   **(a)** C_3H_8O **(b)** $C_4H_{10}O$ **(c)** C_2H_6S
       (d) C_3H_8S **(e)** C_3H_9N **(f)** C_3H_9N

2.7   **(a)** $Br{-}CH_2{-}CH_2{-}Br$ **(b)** $CH_3{-}CH_2{-}CH_2{-}CH_2{-}CH_3$
       (c) $CH_3{-}CH_2{-}CH_2{-}SH$ **(d)** $CH_3{-}CH_2{-}CH_2{-}NH_2$

2.8   **(a)** $CH_3{-}CH_2{-}CH_2{-}NH{-}CH_3$ **(b)** $CH_3{-}CH_2{-}CH_2{-}O{-}CH_3$
       (c) $CH_3{-}CH_2{-}CH_2{-}CH_2{-}CCl_3$ **(d)** $CH_3{-}CH_2{-}NH{-}CH_2{-}CH_3$

2.9   **(a)** $BrCH_2CH_2Br$ **(b)** $CH_3CH_2CH_2CH_2CH_3$
       (c) $CH_3CH_2CH_2SH$ **(d)** $CH_3CH_2CH_2NH_2$

2.10   **(a)** $CH_3CH_2CH_2NHCH_3$ **(b)** $CH_3CH_2CH_2OCH_3$
       (c) $CH_3CH_2CH_2CH_2CCl_3$ **(d)** $CH_3CH_2NHCH_2CH_3$

2.11   **(a)**

$$
\begin{array}{c}
\ \ \ \overset{\textstyle H}{|}\ \ \overset{\textstyle H}{|}\ \ \overset{\textstyle H}{|}\ \ \overset{\textstyle H}{|} \\
H{-}C{-}C{-}C{-}C{-}H \\
\ \ \ \underset{\textstyle H}{|}\ \ \underset{\textstyle H}{|}\ \ \underset{\textstyle H}{|}\ \ \underset{\textstyle H}{|}
\end{array}
$$

(b)

$$
\begin{array}{c}
\ \ \ \overset{\textstyle H}{|}\ \ \overset{\textstyle H}{|}\ \ \overset{\textstyle H}{|} \\
H{-}C{-}C{-}C{-}Cl \\
\ \ \ \underset{\textstyle H}{|}\ \ \underset{\textstyle H}{|}\ \ \underset{\textstyle H}{|}
\end{array}
$$

(c)

$$
\begin{array}{c}
\ \ \ \overset{\textstyle H}{|}\ \ \overset{\textstyle Cl}{|}\ \ \overset{\textstyle H}{|}\ \ \overset{\textstyle H}{|} \\
H{-}C{-}C{-}C{-}C{-}H \\
\ \ \ \underset{\textstyle H}{|}\ \ \underset{\textstyle H}{|}\ \ \underset{\textstyle H}{|}\ \ \underset{\textstyle H}{|}
\end{array}
$$

(d)

$$
\begin{array}{c}
\ \ \ \overset{\textstyle H}{|}\ \ \overset{\textstyle H}{|}\ \ \overset{\textstyle Br}{|}\ \ \overset{\textstyle H}{|} \\
H{-}C{-}C{-}C{-}C{-}H \\
\ \ \ \underset{\textstyle H}{|}\ \ \underset{\textstyle H}{|}\ \ \underset{\textstyle H}{|}\ \ \underset{\textstyle H}{|}
\end{array}
$$

(e)

$$
\begin{array}{c}
\ \ \ \overset{\textstyle H}{|}\ \ \overset{\textstyle H}{|}\ \ \overset{\textstyle Br}{|} \\
H{-}C{-}C{-}C{-}Br \\
\ \ \ \underset{\textstyle H}{|}\ \ \underset{\textstyle H}{|}\ \ \underset{\textstyle H}{|}
\end{array}
$$

(f)

$$
\begin{array}{c}
\ \ \ \overset{\textstyle H}{|}\ \ \overset{\textstyle Br}{|}\ \ \overset{\textstyle H}{|}\ \ \overset{\textstyle H}{|}\ \ \overset{\textstyle H}{|} \\
H{-}C{-}C{-}C{-}C{-}C{-}H \\
\ \ \ \underset{\textstyle H}{|}\ \ \underset{\textstyle Br}{|}\ \ \underset{\textstyle H}{|}\ \ \underset{\textstyle H}{|}\ \ \underset{\textstyle H}{|}
\end{array}
$$

2.12   **(a)**

$$
\begin{array}{c}
\ \ \ \overset{\textstyle H}{|}\ \ \overset{\textstyle H}{|}\ \ \overset{\textstyle H}{|} \\
H{-}C{-}C{-}C{-}H \\
\ \ \ \underset{\textstyle H}{|}\ \ \underset{\textstyle H}{|}\ \ \underset{\textstyle H}{|}
\end{array}
$$

(b)

$$
\begin{array}{c}
\ \ \ \overset{\textstyle H}{|}\ \ \overset{\textstyle H}{|}\ \ \overset{\textstyle Cl}{|} \\
H{-}C{-}C{-}C{-}Cl \\
\ \ \ \underset{\textstyle H}{|}\ \ \underset{\textstyle H}{|}\ \ \underset{\textstyle H}{|}
\end{array}
$$

(c)

$$
\begin{array}{c}
\ \ \ \overset{\textstyle H}{|}\ \ \overset{\textstyle H}{|}\ \ \overset{\textstyle H}{|}\ \ \overset{\textstyle H}{|} \\
H{-}C{-}C{-}C{-}C{-}S \\
\ \ \ \underset{\textstyle H}{|}\ \ \underset{\textstyle H}{|}\ \ \underset{\textstyle H}{|}\ \ \underset{\textstyle H}{|}
\end{array}
$$

(d) 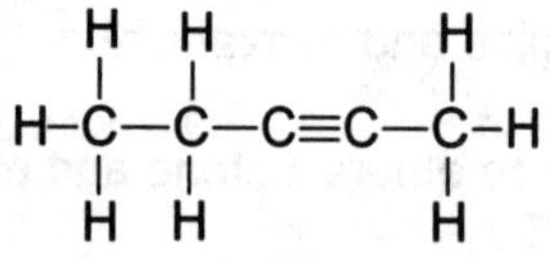
```
    H  H       H
    |  |       |
H — C — C — C ≡ C — C — H
    |  |       |
    H  H       H
```
(e) 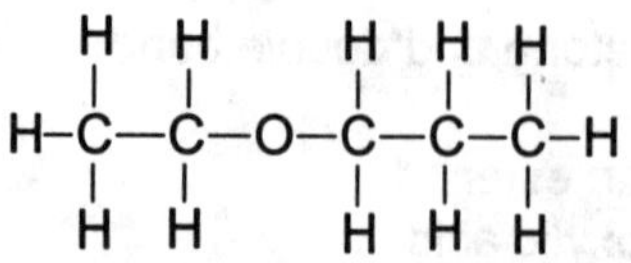
```
    H  H     H  H  H
    |  |     |  |  |
H — C — C — O — C — C — C — H
    |  |     |  |  |
    H  H     H  H  H
```
(f) 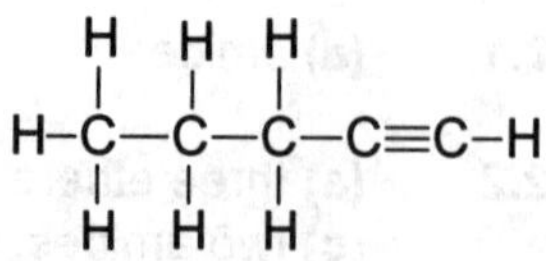
```
    H  H  H
    |  |  |
H — C — C — C — C ≡ C — H
    |  |  |
    H  H  H
```

2.13   **(a)** $C_{10}H_{18}O$ **(b)** $C_7H_{16}O$ **(c)** $C_8H_{17}Br$ **(d)** $C_{11}H_{18}$

2.14   **(a)** $C_{10}H_{12}O$ **(b)** $C_9H_{11}N$ **(c)** $C_{10}H_{16}$ **(d)** $C_4H_6O_2$

2.15   **(a)** $C_6H_{12}S$ **(b)** $C_{12}H_{20}O$ **(c)** $C_{15}H_{22}O$

2.16   **(a)** $C_9H_6O_2$ **(b)** $C_{10}H_{18}O$ **(c)** $C_{19}H_{28}O_2$

2.17   **(a)** different representations   **(b)** different representations   **(c)** isomers

2.18   **(a)** isomers   **(b)** isomers   **(c)** isomers

2.19   **(a)** 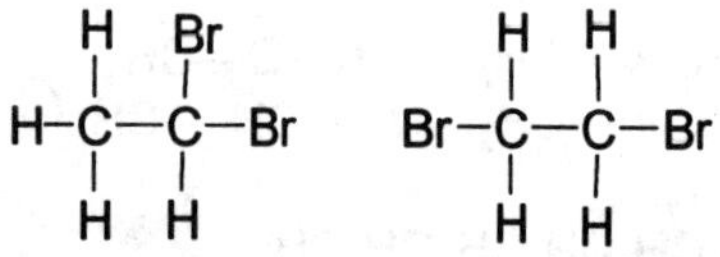
```
    H  Br
    |  |
H — C — C — Br        Br — C — C — Br
    |  |
    H  H
```

(b)
```
    H  H                    H        H
    |  |                    |        |
H — C — C — O          H — C — O — C — H
    |  |  |                |        |
    H  H  H                H        H
```

(c)
```
    H  Br
    |  |
H — C — C — Cl        Br — C — C — Cl
    |  |
    H  H
```

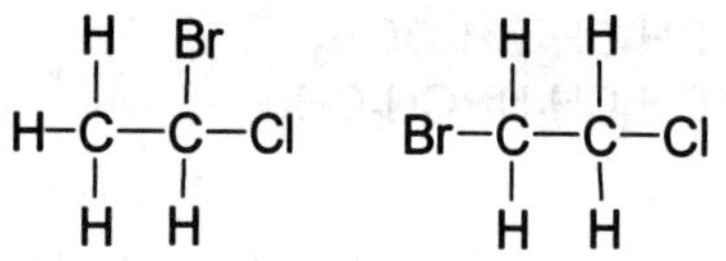

(d) 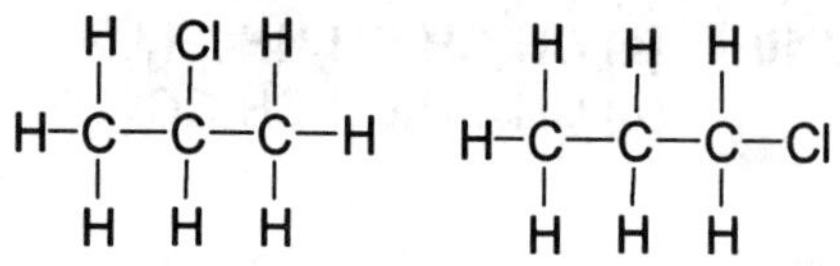
```
    H  Cl  H              H  H  H
    |  |   |              |  |  |
H — C — C — C — H    H — C — C — C — Cl
    |  |   |              |  |  |
    H  H   H              H  H  H
```

(e) 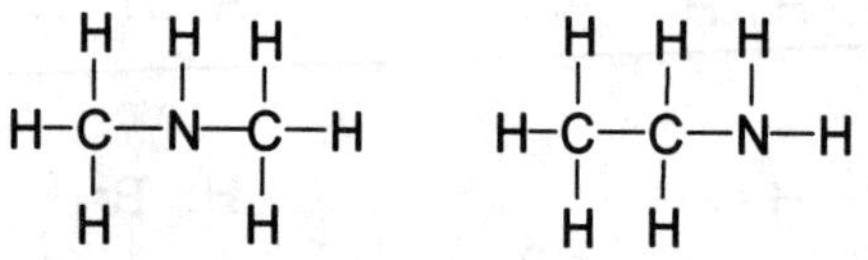
```
    H  H  H              H  H  H
    |  |  |              |  |  |
H — C — N — C — H    H — C — C — N — H
    |     |              |  |
    H     H              H  H
```

(f) 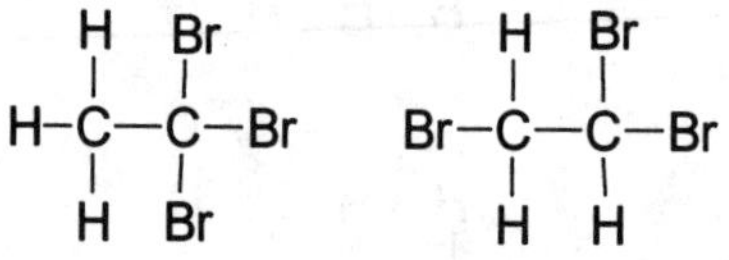
```
    H  Br
    |  |
H — C — C — Br        Br — C — C — Br
    |  |
    H  Br
```

2.20   **(a)** 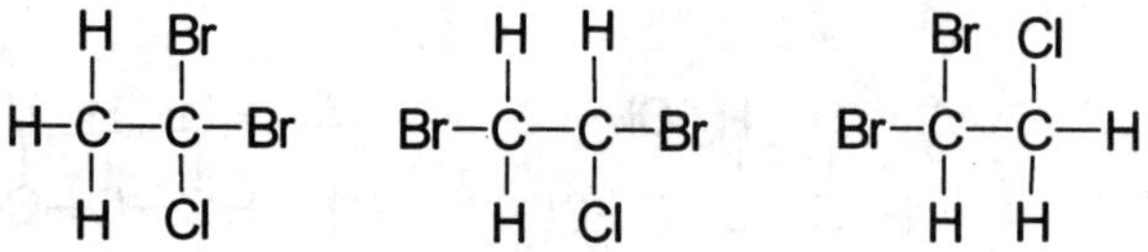
```
    H  Br                H  H              Br Cl
    |  |                 |  |              |  |
H — C — C — Br    Br — C — C — Br    Br — C — C — H
    |  |                 |  |              |  |
    H  Cl                H  Cl             H  H
```

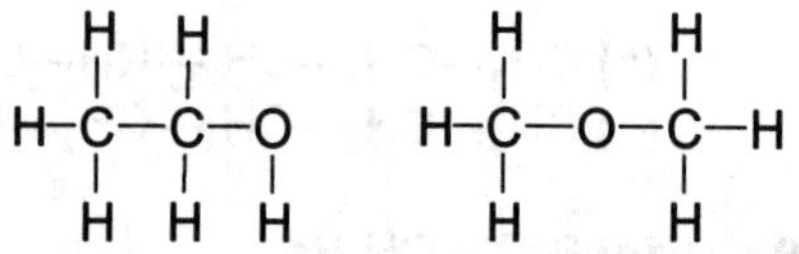

(b)

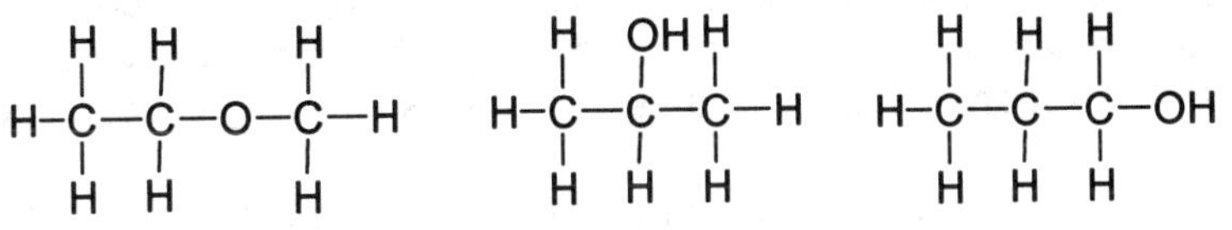

(c)

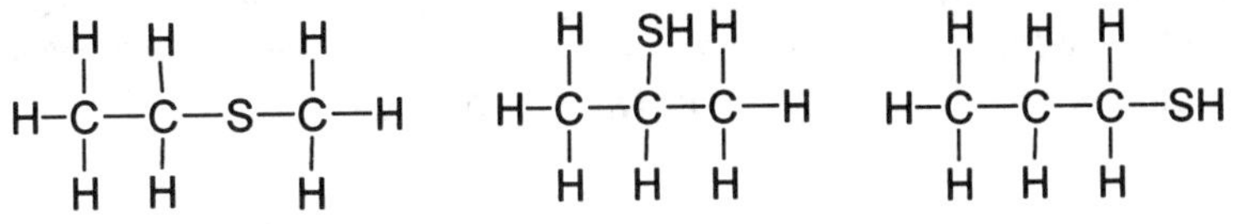

2.21 **(a)** The second structure is more compact and has a smaller surface area, resulting in weaker London forces.

(b) The second structure is more compact and has a smaller surface area, resulting in weaker London forces.

(c) The second structure lacks an N-H bond and cannot form hydrogen bonds.

(d) The second structure lacks an O-H bond and cannot form hydrogen bonds.

2.22 **(a)** Compound with S-H bonds cannot form hydrogen bonds. Both molecules are similarly shaped.

(b) Both molecules are similarly shaped and contain five atoms in a chain.

(c) Both molecules are similarly shaped with five atoms in a chain and a single atom appended to each chain.

(d) Both molecules are similarly shaped and each has an N-H bond to form hydrogen bonds.

2.23 The sulfur compound (III) has strong London forces due to the polarizability of the sulfur. The ether (II) has two polar C-O bonds compared to only one polar C-F bond in compound (I).

2.24 All three compounds have similar molecular weights and are similarly shaped. The dialcohol (III) forms stronger hydrogen bonds than the diamine (II). The polar difluoro compound (I) cannot form hydrogen bonds.

2.25 The bond moments of compound II are opposed to one another and there is no net dipole moment. The other two compounds have bond moments which are additive, so compounds (I) and (II) are polar.

2.26 The bond moments of the two O-C bonds at the "top" of structure (I) cancel the bond moments of the two O-C bonds at the "bottom" of the structure. Compound (II) is polar.

2.27 Propylene glycol has two hydroxyl groups per molecule, so it can form more hydrogen bonds with water as compared to 1-butanol.

2.28   Butanoic acid has an O-H group that can serve as a hydrogen bond donor to water, as well as two oxygen atoms that can serve as hydrogen bond acceptors. Ethyl ethanoate has only the two oxygen atoms, so it can serve only as a hydrogen bond acceptor.

2.29   The boiling point is indicative of an intramolecular hydrogen bond. The hydrogen of the hydroxyl group can bond to the oxygen atom of the carbonyl group.

2.30   The first compound has a distinctly lower boiling point due to the intramolecular hydrogen bond between the hydroxyl hydrogen atom and the chlorine atom. Only intermolecular hydrogen bonding is possible for the other two compounds.

2.31   (a), (b), (c), (d), (e), (f)

2.32   (a), (b), (c), (d), (e), (f)

2.33   (a), (b), (c), (d)

2.34 (a) 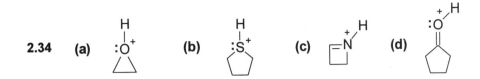 (b) (c) (d)

2.35 (a) CH_3-CH_2-Cl is the Lewis base; $AlCl_3$ is the Lewis acid.
(b) CH_3-CH_2-SH is the Lewis acid; CH_3-O^- is the Lewis base
(c) CH_3-CH_2-OH is the Lewis acid; NH_2^- is the Lewis base
(d) $(CH_3)_2N^-$ is the Lewis base; CH_3-OH is the Lewis acid

2.36 (a) $(CH_3)_2O$ is the Lewis base; HI is the Lewis acid.
(b) CH_3-CH_2^+ is the Lewis acid; H_2O is the Lewis base
(c) CH_3-CH=CH_2 is the Lewis base; HBr is the Lewis acid
(d) CH_3-C≡CH is the Lewis acid; CH_3-NH^- is the Lewis base

2.37 (a) reduction, because the molecule gains hydrogen atoms
(b) oxidation, because the two molecules lose hydrogen atoms
(c) oxidation, because the molecule gains an oxygen atom
(d) reduction, because the molecule loses an oxygen atom

2.38 (a) The molecule gains two hydrogen atoms and one oxygen atom.
(b) The molecule gains two hydrogen atoms and one oxygen atom.
(c) An electronegative nitrogen atom is replaced by an electronegative oxygen atom.
(d) The central carbon atom has two bonds to oxygen in both molecules.

2.39 (a) Oxidation occurs as a result of the gain of one oxygen atom.
(b) Oxidation occurs as a result of change of the CH_3 on the left side into a CO_2H group.
(c) Reduction as oxygen in the NO_2 group is replaced by hydrogen in the NH_2 group.

2.40 (a) Reduction occurs as a result of loss of oxygen at the sulfur atom.
(b) Oxidation occurs as a result of change of the CH_3 on the left side into a CO_2H group.
(c) Reduction as a result of gain of hydrogen atoms at the sulfur atoms.

2.41 (a) Addition of CH_3OH; the hydrogen adds to the oxygen atom and the CH_3O group to the carbon atom. The required reagent is CH_3OH.
(b) Substitution reaction in which bromide ion is replaced by a CN^- ion.
(c) Elimination reaction as a result of loss of water.
(d) Isomerization reaction in which carbon bonds are changed as well as the location of hydrogen atoms.

2.42 (a) Condensation of two molecules with the formation of water.
(b) Hydrolysis reaction forming two molecules. The OH part of water is located at the CH_3 group; the H part is located at the sulfur group.
(c) Substitution of OH by OCH_3. The reagent is CH_3OH.
(d) Addition reaction of H_2O.

2.43 Elimination of two atoms of chlorine to form $ZnCl_2$.

2.44   Elimination of two atoms of hydrogen to form H_2.

2.45   A reduction occurs as the result of addition of hydrogen atoms to the two nitrogen atoms.

2.46   An oxidation occurs as a result of change of the CH_3 on the left side into a CO_2H group.

2.47   The first step can be classified as an oxidation or a substitution reaction. The second step is an elimination reaction that forms HCl.

2.48   Elimination of water occurs in the first step. The second step is a reduction that occurs by addition of two hydrogen atoms.

2.49   Substitution of hydrogen by chlorine in the first step. Addition of HO and Cl occurs at the double bond in the second step. The third step is an elimination reaction yielding HCl even though the involved atoms are not adjacent to each other.

2.50   Addition of Cl_2 occurs in the first step; elimination of HCl occurs in the second step.

2.51   **(a)** Reduction occurs as a result of loss of two atoms of hydrogen.

   (b) Addition of water occurs with H at one carbon atom and OH at an adjacent carbon atom.

   (c) Oxidation occurs as the result of loss of hydrogen atoms.

2.52   **(a)** Oxidation occurs as the result of loss of hydrogen atoms at the top carbon atom of the structure.

   (b) Rearrangement of a phosphorus-containing group from one oxygen atom to another.

   (c) Elimination reaction yielding water as a second product.

3

Chemical Energetics

There are some areas of General Chemistry that you may not have enjoyed. For most students it is the math associated with chapters on kinetics and thermodynamics. Now, you face the same subjects again. However, there are a few bright spots. First, you are seeing it for the second time and it will probably make more sense. Second, the approach is not as detailed. It is not important that we have quantitative information about all reactions in organic chemistry. The approach is to examine the features of chemical structures and the changes that they undergo and then make some educated guesses about which reactions are likely and which ones are not.

Keys to the Chapter

3.1 Equilibria and Thermodynamics

Much of the discussion of organic chemical reactions centers on the "driving force". That term really refers to two related **thermodynamic** quantities—the magnitude of the equilibrium constant and the change in free energy for the reaction. Note that no matter how great the driving force, a reaction may occur at a very slow rate. Kinetics and thermodynamics are entirely different concepts. **Kinetics** is a study of the speed of a reaction and the determination of its **mechanism**.

3.2 Chemical Equilibrium

The definition of the equilibrium constant and its magnitude are no different from what you learned in your General Chemistry course. In this text you will learn which reactions have large equilibrium constants and which do not. In the case of reactions with small equilibrium constants, the reaction conditions are usually adjusted to shift the position of equilibrium by taking advantage of **LeChâtelier's principle**. That principle is much more important than you might have assumed based on its limited use in General Chemistry. You will use it many times in organic chemistry. In general, equilibrium systems respond in a contrary fashion. Add something and the system "tries" to get rid of it. Take something away and the system "tries" to replace it.

3.3 Equilibria in Acid-Base Reactions

The properties of acids are characterized by K_a and pK_a values. Stronger acids have large K_a values and small pK_a values. You won't have to learn all the numbers, but as you go along you will be expected to understand the relative order of acidity of various classes of compounds, and the lecturer may expect you to learn a few common values. For example, alcohols and carboxylic acids have pK_a values in the 16 and 5 range, respectively.

The properties of bases are characterized by K_b and pK_b values. Stronger bases have large K_b values and small pK_b values.

Acid-base equilibrium reactions proceed to favor the weaker of the two possible acids (or the weaker of the two bases). Once you know which side of the equation for the reaction is favored, you can turn to estimating the equilibrium constant. The equilibrium constant for the overall reaction is given by a quotient of two equilibrium constants. That means that you simply need to determine the number of powers of 10 by which the equilibrium constants differ. If that difference is 5 powers of ten, for example, then the equilibrium constant is either 10^{-5} or 10^5 depending on your analysis of whether the reaction is favorable or unfavorable.

3.4 Structure and Acidity

This section gives an overview of several structural features that affect the K_a of an acid. There are four factors—periodic trends, resonance effects, inductive effects, and hybridization effects. There won't be any others, but you will use these factors many times in studying the various types of functional groups and their reactions. The trick is to know when to use which factor to explain some facts that your lecturer cites on an examination. Of the four factors, resonance effects and inductive effects are widely encountered. **Resonance effects** are recognized by the presence of one or more multiple bonds alternating with single bonds in a molecule. **Inductive effects** are recognized by the presence of an electronegative atom in the vicinity of the reaction site. Pay close attention each time these effects are discussed, both to improve your understanding and to increase your ability to predict the results of these effects on a given reaction.

Recall that the concept of % s character was first introduced in Chapter 1, and you were told that this concept would permeate your study of organic chemistry. As you learn about hybridization effects in this section, you'll get a first look at the dramatic effect the % s character has on the acidity of hydrocarbons with different hybridization.

3.5 Equilibrium and Thermodynamics

The relationship between the **Gibbs free energy**, **enthalpy**, and **entropy** are given in this section. In general, you will not do extensive calculations using these quantities, but it is important to understand qualitatively how they contribute to the spontaneity of a reaction. In other words, is the driving force of a reaction the result of an enthalpy or an entropy contribution? You may find it useful to learn a few quantities. For example, an equilibrium constant of 10 at 25°C corresponds to a $\Delta G°$ of approximately -5.5 kJ mole^{-1}. Thus, a reaction with a $\Delta G°$ of approximately -16 kJ mole^{-1}, which is 3(-5.5 kJ mole^{-1}), would have an equilibrium constant of 10^3 at 25°C.

3.6 Enthalpy Changes in Chemical Reactions

The enthalpy change for a reaction must be considered to gauge its contribution to the driving force of a reaction. That change reflects the difference in the bond energies of the bonds in the reactant compared to the bonds in the product. If the bonds in the product are stronger than those in the reactant, the reaction is exothermic.

3.7 Bond Dissociation Energies

The general term bond energy is often misunderstood. More specifically we use the term **bond dissociation energy** to describe the energy required (an endothermic process) to break a bond and form two atomic or molecular fragments, each with one electron of the original shared pair. Thus, a very stable bond has a large bond dissociation energy—more energy must be added to cleave the bond. A high bond dissociation energy means that the bond (and molecule) is of low energy and stable. So, don't misinterpret what we mean when we say that a bond has high bond energy. It doesn't mean that the bond has a lot of energy and is about to fly apart because it is unstable.

Bond energies depend on the number of bonds. Even though π bonds are weaker than σ bonds, a double bond, which consists of a σ and a π bond, is stronger than a single bond because there are two bonds.

Hybridization effects on bond energies should be put in perspective with hybridization effects on acidity. The two processes are different. In the case of bond dissociation, one of the two bonding electrons that are held close to a carbon atom with a large % s character must be "retrieved" for the cleavage to occur. Remember that each fragment must depart with an electron. In the case of acid-base reactions, the electron pair that is held close to the atom with a large % s character makes it easier to "leave". So, even though the reactions occur with different facility, the phenomena are governed by the same effect. Compounds with hydrogen bonded to atoms with large % s character have high bond dissociation energies (are more stable) but have higher acidities (are more prone to removal of a proton by a base).

3.8 Estimating $\Delta H°$ from Bond Energies

Which organic reactions occur and which do not can be "explained" by estimating the difference in enthalpy content of the reactants and products based on the number and type of bonds. Remember that breaking bonds "costs" energy and making bonds "gives up" energy. So, focus on the bonds that change in reactions and ignore the rest.

3.9 Entropy Changes in Chemical Reactions

The entropy change for a reaction depends on two features—the $S°$ of the reactants and the products and the number of moles of products relative to reactants. Differences in the $S°$ of the reactants and the products often contribute little to the $\Delta G°$ as compared to $\Delta H°$. In fact, the spontaneity of a reaction is often determined by calculating or estimating $\Delta H°_{rxn}$ and assuming that ΔS_{rxn} is approximately zero. However, that assumption is invalid if the number of moles of products is not equal to the number of moles of reactants. The general rule of thumb is that a difference of one mole of product relative to reactant corresponds to a difference of approximately 125 J mole^{-1} K^{-1}. Whether the quantity is positive or negative depends on whether there are more or fewer moles of product relative to reactant.

3.10 Contribution of $\Delta H°_{rxn}$ and $\Delta S°_{rxn}$ to $\Delta G°_{rxn}$

The importance of $\Delta H°_{rxn}$ and $\Delta S°_{rxn}$ depends on the temperature of the reaction. At low temperatures $\Delta S°_{rxn}$ is less important. Of course, the ultimate effect of $\Delta S°_{rxn}$ on $\Delta G°_{rxn}$ depends on the value of $\Delta H°_{rxn}$. Note that at 25°C the rule of thumb +125 J mole^{-1} K^{-1} for a difference of one more mole of product relative to reactant corresponds to -37.5 kJ mole^{-1} in the -T$\Delta S°$ term. You should be able to do a "rough" calculation of the effect of $\Delta S°_{rxn}$ relative to $\Delta H°_{rxn}$.

3.11 Kinetics of Reactions

As illustrated by the reactions listed in this section, the speed of a reaction is unrelated to thermodynamic quantities such as the listed $\Delta H°_{rxn}$ values. An understanding of why some reactions are fast and some are slow depends on kinetic studies and the formulation of a reaction mechanism. Again, General Chemistry courses provide very little indication of how important reaction mechanisms are. The concept was used in the chapter on chemical kinetics and then probably was never mentioned again. Such is not the case in organic chemistry. The text is replete with reaction mechanisms.

3.12 Reaction Mechanisms

A mechanism is the series of steps that occur as a reactant is converted to a product. Seeing the reaction mechanism helps us understand how the transformation takes place. The complexity of mechanisms covers a wide range from simple one-step mechanisms to complex multistep mechanisms involving a series of intermediates.

Classifying the type of bond cleavage in a particular mechanism requires us to look carefully at the reactant and product, to identify which bond breaks and how it breaks. Specifically, we need to track the electrons in the bond as it breaks. Bond cleavage is **homolytic** ("same-breaking") if the two resulting fragments each retain one electron from the bond; fragments containing an unpaired electron are called **radicals** and are highly reactive. The decomposition of tetraethyl lead, a compound formerly used as a gasoline additive, produces radicals as a result of homolytic cleavage of a weak Pb—C bond.

$$CH_3CH_2-\underset{\underset{CH_2CH_3}{|}}{\overset{\overset{CH_2CH_3}{|}}{Pb}}-CH_2CH_3 \xrightarrow{\text{heat}} CH_3CH_2-\underset{\underset{CH_2CH_3}{|}}{\overset{\overset{CH_2CH_3}{|}}{Pb}}\cdot \quad + \quad \cdot CH_2CH_3$$

Bond cleavage is **heterolytic** ("different-breaking") if both electrons in the bond stay with one fragment; the fragment retaining the electron pair acquires a -1 charge and the other fragment a +1 charge. When a carbon atom is bonded to a highly electronegative element such as Cl, heterolytic cleavage gives the electron to the highly electronegative element and leaves the carbon atom or chain with a positive charge. Thus the carbon fragment is called a **carbocation**. A carbocation is formed from certain compounds containing carbon-bromine bonds, as shown by the following equation.

$$CH_3-\underset{\underset{CH_3}{|}}{\overset{\overset{CH_3}{|}}{C}}-\ddot{\underset{\cdot\cdot}{Br}}: \xrightarrow{H_2O} CH_3-\underset{\underset{CH_3}{|}}{\overset{\overset{CH_3}{|}}{C}}{}^+ \quad + \quad :\ddot{\underset{\cdot\cdot}{Br}}:{}^-$$

When a carbon atom is bonded to an electropositive element such as H, heterolytic cleavage releases the electropositive species as a cation and leaves the electron with the carbon fragment, which becomes a **carbanion**.

3.13 Structure and Stability of Carbon Intermediates

The stability of a carbon intermediate depends on the number of electrons about the carbon atom and the identity of the attached groups. In this section, the discussion is limited to the effect of carbon groups attached to the **carbocation**, **radical**, or **carbanion** center. The carbon groups are regarded as electron donating in an inductive sense. Thus, both carbocations and radicals, which are electron deficient, are stabilized by larger numbers of alkyl groups. Carbanions already have a sufficient number of electrons, and the supply of additional electron density by attached carbon groups is counterproductive. Thus, the order of stability of carbanions is opposite to that of carbocations. Carbocations are the more commonly encountered carbon intermediates in the text, so you should remember that tertiary intermediates are more stable than secondary intermediates, which are in turn more stable than primary intermediates.

3.14 Bond Formation from Reactive Intermediates

Bonds may form by a **homogenic** process but this is less common than bond formation by a **heterogenic** process. Remember the phrase—have pair, will share. In organic reactions the species with the pair is a **nucleophile**, and it is sought by a species lacking sufficient electrons, known as an **electrophile**. A curved arrow notation is used to show the "movement" of electrons from the nucleophile toward the electrophile. This notation indicates the original location of the electron pair at the start of the arrow and the destination of the electrons at the head of the arrow.

3.15 Representative Mechanisms

Reaction mechanisms cannot be ascertained by looking at a balanced chemical equation. However, there are a limited number of representative mechanisms that you will encounter, and you will soon learn what to expect from certain combinations of reactants. The nucleophilic substitution mechanism is common, as illustrated by the following equation.

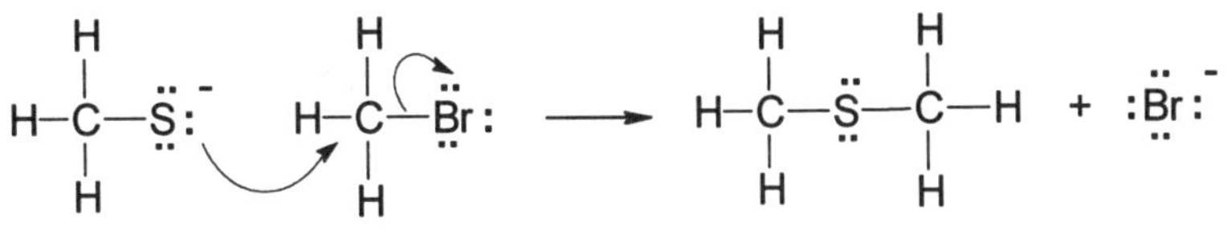

It is characterized by attack of a nucleophile with its pair of electrons forming a bond to carbon, while a leaving group departs with a pair of electrons that was formerly in a bond. Once you understand these facts, then you can predict the reaction mechanism of a host of reactions for various nucleophiles and leaving groups. For example, a reaction given by the following equation should be expected to occur by the same mechanism as shown above.

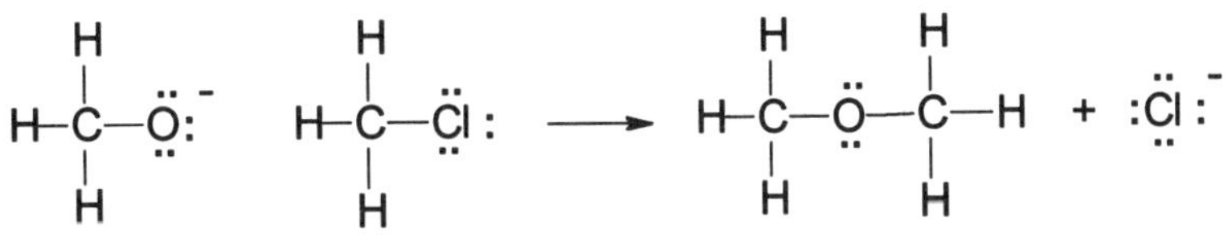

3.16 Factors That Influence Reaction Rates

The factors reviewed in this section are the same four you learned when you studied reaction rates in General Chemistry. Reaction rates are accelerated by higher temperatures. Some reaction rates are accelerated by **catalysts**. Whether or not a change of rate occurs by changing the concentration of reactants depends on the **order** of a reaction with respect to the reactant. The rate of individual reactions depends on the identity of the reactants.

3.17 Reaction Rate Theory

Reactions occur via one or more **transitions states** in which the bonding patterns correspond to neither the reactants or products. The transition state occurs at a maximum in a **reaction coordinate diagram**. Reactions with a high **activation energy** occur at a slower rate than those with a lower activation energy. The reason why reaction rates are accelerated by higher temperature is that a larger fraction of molecules possess an energy equal to or greater than the activation energy and can achieve the transition state structure.

Multistep reactions have more than one transition state, and the lower energy species that forms between transition state is an **intermediate**. Catalysts provide for a mechanism that occurs via a transition state with a lower activation energy.

The **Hammond postulate** relates the structure of the transition state to either the reactant or product depending on the transition state energy and the energy of the reactants and products. Strongly exothermic reactions occur via transition states that more closely resemble the reactant structure. Endothermic reactions occur via transition states that more closely resemble the product.

The Hammond postulate is another of those ideas that are used repeatedly in the text. Transition state structures cannot be determined experimentally. The structure of reactants and products are known, and it is often possible to elucidate the structure of intermediates. Thus, the structure of the transition state is estimated using the Hammond postulate.

3.18 Stability and Reactivity

Stability is a term based on thermodynamic considerations. It refers to the difference in energy relative to a second reference substance. Thus, we always have to question - relative to what? Reactivity is a term based on kinetics and depends on how the substance behaves in the presence of a second substance. A compound may be unreactive in the presence of compound A but be extremely reactive toward compound B.

Solutions to Exercises

3.1 $K = \dfrac{[CH_3CH(OCH_3)_2]\,[H_2O]}{[CH_3CHO]\,[CH_3OH]^2}$

3.2 The conversion of acetylene (C_2H_2) into cyclooctatetraene (C_8H_8) requires four moles of reactant per mole of product. Thus the equilibrium constant expression is:

$K = \dfrac{[C_8H_8]}{[C_2H_2]^4}$

3.3 The hydrolysis reaction written is the reverse of the esterification reaction in Section 3.2. Thus, the equilibrium constant expression for hydrolysis is the reciprocal of the equilibrium constant expression for esterification. The value of the equilibrium constant for hydrolysis is 0.25.

3.4 For an initial concentration of acetone equal to x mole liter^{-1} the theoretical concentration of product for a complete reaction would be 0.5 x mole liter^{-1}. For a 5% yield the actual concentration is 0.025 x mole liter^{-1}. The equilibrium concentration of reactant is 0.95 x mole liter^{-1} because two moles of reactant are require to give one mole of product. The equilibrium constant is approximately 0.028 x^{-1} mole^{-1} liter.

$K = \dfrac{[0.025\,x]}{[0.95\,x]^2}$

3.5 The acid dissociation constants of organic compounds containing atoms within a common group of the periodic table bonded to hydrogen increase down the column. Thus, thiols are more acidic than alcohols. The equilibrium position lies on the side of the equation containing the weaker acid. Thus the position of the equilibrium is to the right where CH_3OH is located and $K > 1$.

3.6 The equilibrium position lies on the side of the equation containing the weaker acid. The acids are methanol located on the right and acetic acid located on the left of the equation. Acetic acid is the stronger acid because its conjugate base is resonance stabilized. Thus, the position of the equilibrium is on the right where the weaker acid, methanol, is found.

3.7 Methanol is the stronger acid by a factor of 10^{33} in K_a. The equilibrium position lies on the left side of the equation, which contains the weaker acid, methane. The equilibrium constant is 10^{-33}.

3.8 Methanol is the stronger acid by a factor of 10^{20} in K_a. The equilibrium position lies on the right side of the equation, which contains the weaker acid, ammonia. The equilibrium constant is 10^{20}.

3.9   The acidity of hydrogen atoms bonded to atoms contained in similarly structured compounds increases from left to right within a period of the periodic table. For example, H_2O is a stronger acid than NH_3. The conjugate acid with a proton located on the oxygen atom of hydroxylamine must be a stronger acid than the conjugate acid with a proton located on the nitrogen atom.

$$
\underset{\underset{H}{|}}{H-N}-\overset{+}{O}: \qquad H-\overset{+}{\underset{\underset{H}{|}}{N}}-\overset{H}{O}:
$$

3.10   The basicity of atoms contained in similarly structured compounds decreases from left to right within a period of the periodic table. For example, NH_2^- is a stronger base than OH^-. The conjugate base with the charge located on the nitrogen atom of hydroxylamine must be a stronger base than the conjugate base with a charge on the oxygen atom.

$$
\overset{-}{H-N}-\overset{H}{\underset{|}{O}}: \qquad H-\underset{|}{N}-\overset{H}{O}:^{-}
$$

3.11   The inductive electron withdrawal by chlorine is larger than that of bromine because chlorine is more electronegative than bromine. As a consequence of flow of electron density away from the O—H group, the acidity of chloroacetic acid exceeds than of bromoacetic acid.

3.12   Dichloroacetic acid is a stronger acid than chloroacetic acid because the two chlorine atoms inductively withdraw more electron density from the O—H group than a single chlorine atom. The pK_a of dichloroacetic acid should therefore be smaller than the pK_a of chloroacetic acid.

3.13   The pK_a of the substituted chlorobutanoic acids increases with increasing distance separating the chlorine atom and the acidic site. Thus, the pK_a of the 4-chloro compound should be greater than 4.02, the pK_a of the 3-chloro compound. It also should be less than the pK_a of butanoic acid, which is 4.82.

3.14   The order of decreasing pK_a values indicate that the groups bonded to the nitrogen atom of the ammonium ions increase in ability to inductively withdraw electron density. Although oxygen is more electronegative than nitrogen, the nitrile compound has a triple bond and is a much more polar group. (See Table 1.4 which shows the average bond moments of C—O and C≡N groups.)

3.15   The conjugate base of propane has its negative charge localized on a single carbon atom. The conjugate base of propene has its negative charge delocalized over two carbon atoms as shown by two contributing resonance structures.

$$
CH_2{=}CH{-}\overset{..}{\underset{}{C}}H_2^{-} \longleftrightarrow {}^{-}\overset{..}{C}H_2{-}CH{=}CH_2
$$

3.16 The conjugate base of ethane has its negative charge localized on a single carbon atom. The conjugate base of ethanonitrile has its negative charge delocalized with some of the charge located on the more electronegative nitrogen atom as shown in one of the two contributing resonance structures.

$$\ ^-:CH_2{-}C{\equiv}N: \longleftrightarrow CH_2{=}C{=}\overset{..}{N}:^-$$

3.17 In amoxicillin, the acidic —CO₂H group is bonded to a carbon atom that is also bonded to a nitrogen atom that inductively withdraws electron density and increases the acidity of the O—H group. In indomethacin, the —CO₂H group is bonded to a carbon atom that is not directly bonded to any electronegative groups. The nitrogen atom in indomethacin is one atom farther removed than that in amoxicillin.

3.18 The greatly increased acidity of the O—H group in phenobarbital reflects the resonance stabilization of the conjugate base, in which the charge is delocalized over two oxygen atoms.

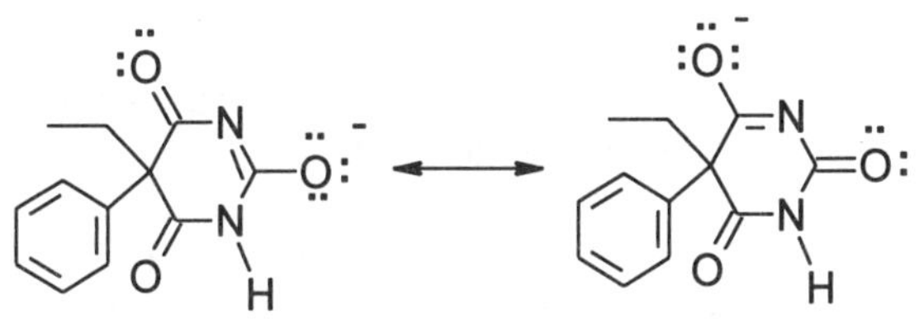

3.19 The nitrogen atom is bonded to a sulfur atom that has two oxygen atoms which are electronegative and hence withdraw electron density from the N—H bond.

3.20 The nitrogen atom is bonded to a carbon atom of a ring that has two nitrogen atoms which are electronegative and hence withdraw electron density from the N—H bond.

3.21 For an equilibrium constant less than 1, the products are less stable than the reactants. Such a reaction is endergonic.

3.22 The reaction with K = 100 is more spontaneous and more exergonic than a reaction with K = 0.01.

3.23 A reaction with K =1 has products and reactants of equal stability—that is, $\Delta G°_{rxn}$ = 0. The free energies of the products and reactants will be shown at the same vertical position in the reaction coordinate diagram.

3.24 Spontaneous reactions are exergonic and have $\Delta G°_{rxn}$ < 0. The reaction with $\Delta G°_{rxn}$ = -15 kJ mole⁻¹ is exergonic.

3.25 Using the relationship $\Delta G°_{rxn}$ = -2.303 RT log K, the equilibrium constant is 0.4. Remember that 25°C is 298 K and that $\Delta G°_{rxn}$ must be expressed in J mole⁻¹.

3.26 Using the relationship $\Delta G°_{rxn}$ = -2.303 RT log K, the $\Delta G°_{rxn}$ = -3.9 kJ mole⁻¹. Remember that 25°C is 298 K and that the $\Delta G°_{rxn}$ obtained is in J mole⁻¹.

3.27   Use the bond dissociation energies listed in Table 3.5.

   (a) -45 kJ mole^{-1} **(b)** -44 kJ mole^{-1} **(c)** +359 kJ mole^{-1}

3.28   Use the bond dissociation energies listed in Table 3.5. Values listed on the back inside cover of the text are required for (b).

   (a) -44 kJ mole^{-1} **(b)** 9 kJ mole^{-1} **(c)** -356 kJ mole^{-1}

3.29   The first step is an addition reaction in which two moles of reactant are converted into one mole of product. The $\Delta S°_{rxn}$ is estimated as 125 J mole^{-1} K^{-1}. The second step is an elimination reaction in which one mole of reactant is converted into two moles of product. The $\Delta S°_{rxn}$ is estimated as -125 J mole^{-1} K^{-1}.

3.30   The first reaction converts two moles of reactant into four moles of product. The $\Delta S°_{rxn}$ is estimated as 250 J mole^{-1} K^{-1}. The second reaction converts three moles of reactant into one mole of product. The $\Delta S°_{rxn}$ is estimated as -250 J mole^{-1} K^{-1}.

3.31   The first reaction is a substitution reaction in which two moles of reactant are converted into two moles of product. The $\Delta S°_{rxn}$ is expected to be 0 J mole^{-1} K^{-1}. The second reaction is an addition reaction in which two moles of reactant are converted into one mole of product. The $\Delta S°_{rxn}$ is estimated as -125 J mole^{-1} K^{-1}.

3.32   In both reactions one mole of reactant is converted into one mole of product. Based solely on the fact that the number of moles of material is neither increased nor decreased, the $\Delta S°_{rxn}$ is estimated as 0 J mole^{-1} K^{-1}. However, the atoms in the cyclic structures are more restricted in their molecular motion compared to the acyclic structures. Thus, $\Delta S°_{rxn}$ is expected to be positive. More molecular freedom is possible in the six–carbon atom structure, so its $\Delta S°_{rxn}$ is more positive.

3.33 $\Delta H°_{rxn}$ = 88.1 kJ mole^{-1} $\Delta S°_{rxn}$ = 147 J mole^{-1} K^{-1} $\Delta G°_{rxn}$ = 14.9 kJ mole^{-1}

3.34 $\Delta H°_{rxn}$ = -56.4 kJ mole^{-1} $\Delta S°_{rxn}$ = -169.5 J mole^{-1} K^{-1} $\Delta G°_{rxn}$ = -5.9 kJ mole^{-1}

3.35 $\Delta H°_{rxn}$ = -18.3 kJ mole^{-1} $\Delta S°_{rxn}$ = -43 J mole^{-1} K^{-1} $\Delta G°_{rxn}$ = 10.6 kJ mole^{-1}

3.36 $\Delta H°_{rxn}$ = -6.3 kJ mole^{-1} $\Delta S°_{rxn}$ = -3 J mole^{-1} K^{-1} $\Delta G°_{rxn}$ = -5.2 kJ mole^{-1}

3.37   In both reactions, the number of moles of product equals the number of moles of reactant. Based on this factor alone, $\Delta S°_{rxn}$ is estimated to be 0 J mole^{-1} K^{-1}. However, in the second equation the atoms in the reactant are converted into a cyclic structure which is more restricted in its molecular motion compared to the acyclic structures. Thus, $\Delta S°_{rxn}$ is expected to be negative and as a result decrease the equilibrium constant relative to the first reaction.

3.38   At 800°C the $\Delta G°_{rxn}$ = 0 because K = 1. Thus, it follows that $\Delta H°_{rxn}$ = T $\Delta S°_{rxn}$. Using $\Delta H°_{rxn}$ = -134,000 J mole^{-1} and T = 1073 K, the calculated $\Delta S°_{rxn}$ = -125 J mole^{-1} K^{-1}.

3.39   The number of moles of reactants is not equal to the number of moles of products in both (a) and (b). Thus $\Delta S°_{rxn}$ is not zero and must be considered in estimating K_{eq}. In the isomerization reaction (c) there are equal numbers of moles of product and reactant. On this basis alone $\Delta S°_{rxn}$ is approximately zero and $\Delta H°_{rxn}$ is approximately equal to $\Delta G°_{rxn}$. However there is a difference in the symmetry of the reactant and product. The product is more symmetrical and has a less positive $S°$ than the reactant (See Section 3.) Thus the $\Delta S°_{rxn}$ is expected to be somewhat negative.

3.40   Assume that $\Delta H°_{rxn} \approx \Delta G°_{rxn}$ and use the relationship $\Delta G°_{rxn} = -2.303\ RT \log K$ to calculate the equilibrium constants at 25°C (298 K). Remember that $\Delta H°_{rxn}$ must be expressed in J mole^{-1}.

(a) 0.00031   (b) 13   (c) 81

3.41   (a) 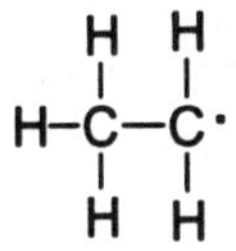   (b) 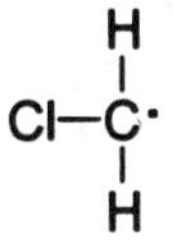   (c)

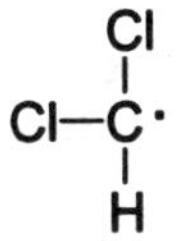

3.42   (a) $CH_3-CH_2-CH_2\cdot$ $CH_3-\overset{\cdot}{C}H-CH_3$

(b) $CH_3-CH_2-CH_2-CH_2\cdot$ $CH_3-CH_2-\overset{\cdot}{C}H-CH_3$

(c) 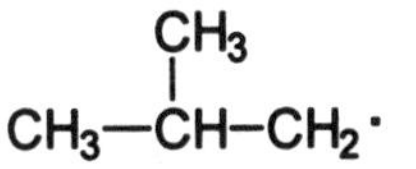 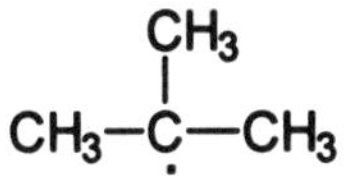 $CH_3-\underset{\underset{CH_3}{|}}{C}H-CH_2\cdot$ $CH_3-\underset{\underset{CH_3}{|}}{\overset{\overset{CH_3}{|}}{C}}-CH_3$

3.43   Oxygen is more electronegative than chlorine. Thus, the electrons of the O—Cl bond should remain with oxygen and the products should be CH_3O^- and Cl^+.

3.44   The $AlCl_3$ combines with Cl^- to give $AlCl_4^-$. Thus, the C—Cl bond cleaves heterolytically. The intermediate is a carbocation $(CH_3)_2CH^+$.

3.45   The heterolytic cleavage placing the O—O bonding electrons on the oxygen atom of water leaves a cation with the positive charge on the oxygen atom of the HO group.

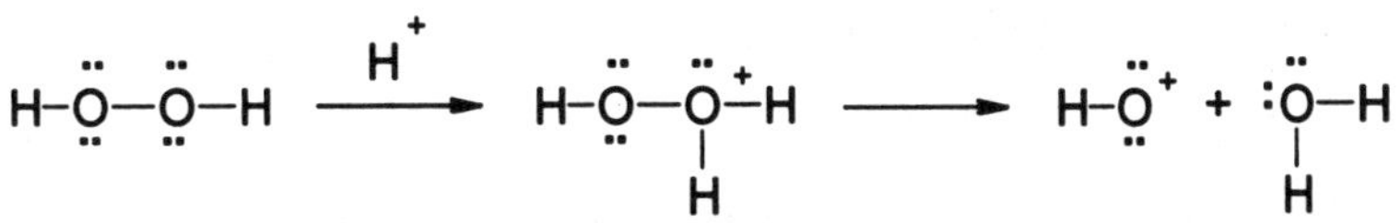

3.46   The homolytic cleavage leaves one electron of the pair of electrons in the O—O bond with each of the two equivalent radical fragments, giving the structure shown on the next page.

$$\underset{\substack{\text{benzene ring} \\}}{\bigcirc}\!\!-\!\!\overset{\displaystyle :\!\overset{..}{O}:}{\underset{\displaystyle \parallel}{C}}\!\!-\!\!\overset{..}{\underset{..}{O}}\cdot$$

3.47 The order of increasing stability is II < I < III, which corresponds to primary < secondary < tertiary.

3.48 The order of increasing stability is I < III < II, which corresponds to primary < secondary < tertiary.

3.49 Heterolytic cleavage of a C—Br bond produces a carbocation and a bromide ion. The secondary carbocation derived from 2-bromopropane is more stable than the primary carbocation derived from 1-bromopropane, so formation of the primary carbocation requires more energy.

3.50 Heterolytic cleavage of a C—Cl bond produces a carbocation and a chloride ion. A resonance stabilized primary carbocation is derived from 3-chloropropene and is thus more stable than the primary carbocation derived from 1-chloropropane, in which the charge is localized.

$$CH_2\!\!=\!\!CH\!\!-\!\!\overset{+}{C}H_2 \quad\longleftrightarrow\quad \overset{+}{C}H_2\!\!-\!\!CH\!\!=\!\!CH_2$$

3.51 The dichlorocarbene is electron deficient; there are only 4 bonding electrons and a lone pair of electrons about the carbon atom. However, either of the chlorine atoms can share one of its lone pairs of electrons in contributing resonance forms. The delocalization of electrons makes CCl_2 more stable than CH_2.

$$\underset{\displaystyle :\overset{..}{C}\!\!-\!\!\overset{..}{\underset{..}{C}l}:}{\overset{\displaystyle :\overset{..}{C}l:}{\big|}} \quad\longleftrightarrow\quad \underset{\displaystyle :C\!\!=\!\!\overset{..}{\underset{..}{C}l}:}{\overset{\displaystyle :\overset{..}{C}l:}{\big|}} \quad\longleftrightarrow\quad \underset{\displaystyle :\overset{..}{C}\!\!-\!\!\overset{..}{\underset{..}{C}l}:}{\overset{\displaystyle :\overset{..}{C}l}{\parallel}}$$

3.52 The cation has 2 fewer electrons than the anion. The cation should act as an electrophile, because it is electron deficient.

$$H\!\!-\!\!\overset{..}{\underset{..}{O}}{}^{+}$$

3.53 The reaction converting A to X has the lower activation energy (E_a), so it proceeds at the faster rate.

3.54 The reaction converting A to X has the more negative $\Delta H°_{rxn}$, so it is more exothermic.

3.55 The reaction of methane with a fluorine atom has the lower activation energy (E_a), so it proceeds at the faster rate.

3.56 The reaction of bromide ion with iodomethane has the lower activation energy (E_a), so it proceeds at the faster rate.

3.57 The rate of reaction is first order in both chloroethane and cyanide ion, so it is second order overall. The rate equation is: rate = k $[CH_3CH_2Cl][CN^-]$.

3.58 The rate of reaction is first order in both *tert*-butyl alcohol and hydrogen ion and is zero order in bromide ion, so it is second order overall.

3.59 (a) homolytic cleavage of a C—H bond and homogenic formation of a H—Br bond.
(b) homolytic cleavage of a Br—Br bond and homogenic formation of a C—Br bond.

3.60 (a) heterogenic formation of a C—O bond.
(b) heterolytic cleavage of a C—Cl bond.

3.61 Heterolytic cleavage of the C—O bond occurs to give a carbocation.

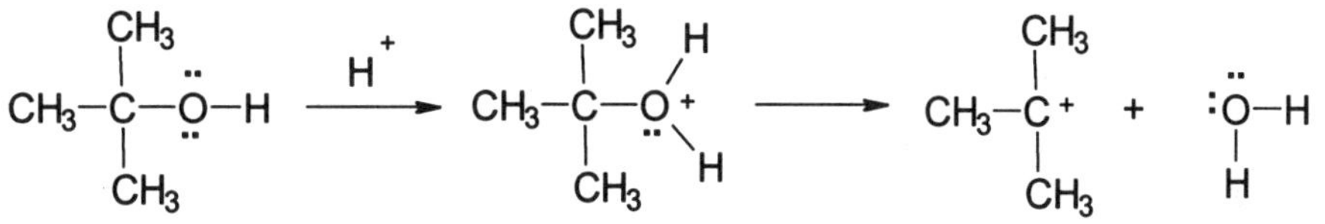

3.62

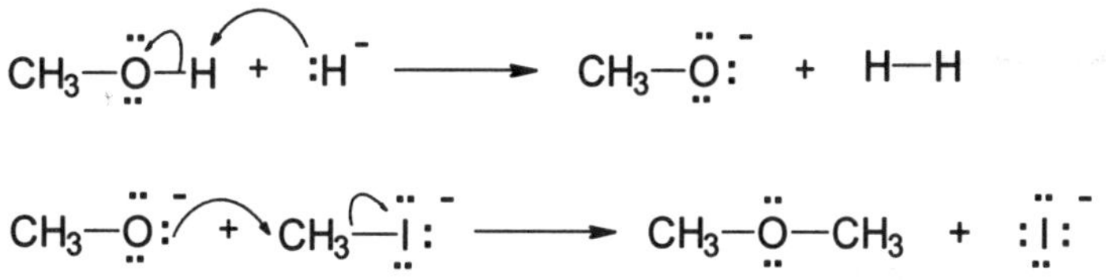

3.63

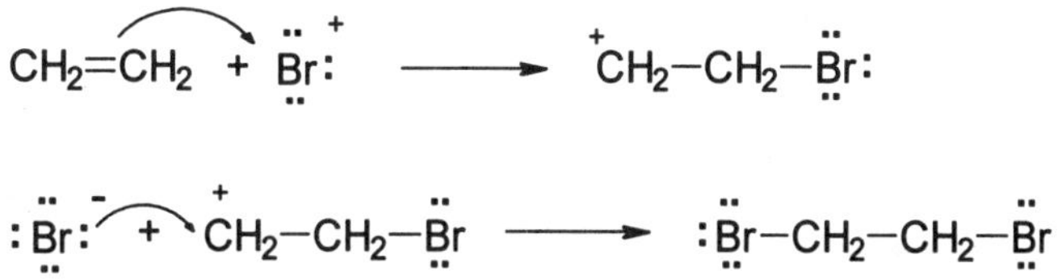

3.64

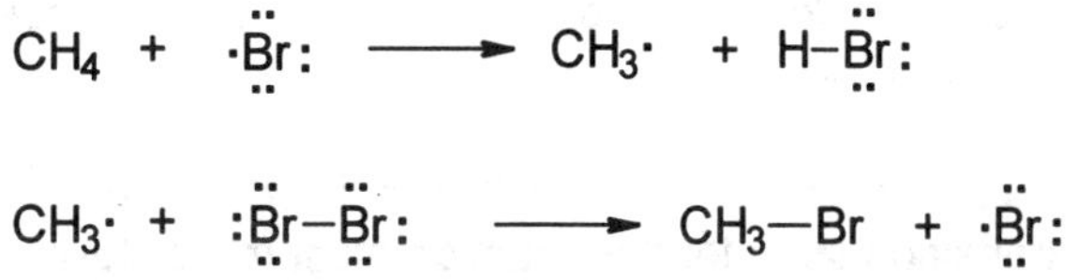

3.65 An intermediate, although short lived, can be detected experimentally. An transition state is a transient species whose structure can only be postulated.

3.66   There are three transitions states—one for each step. There are two intermediates.
One is formed from step 1 and reacts in step 2. The second intermediate is formed from
step 2 and reacts in step 3.

3.67
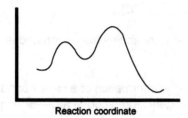

3.68   There are three intermediates and four transition states.

3.69   The activation energy for abstraction of a hydrogen atom from HCl by a methyl radical is
13 kJ mole^{-1}, because it is the reverse of the reaction shown in the reaction coordinate
diagram.

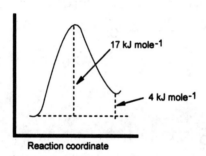

3.70
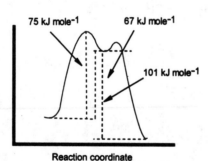

3.71   The energy of activation must be larger for the first reaction listed because it is the more
endothermic reaction. The transition state for this reaction must more closely resemble
the product than for the second reaction.

3.72   The energy of activation must be smaller for the second reaction listed because it is the
more exothermic reaction. The transition state for this reaction must more closely
resemble the reactant than for the first reaction.

4

Alkanes and Cycloalkanes: Structure and Properties

By the title of the chapter, it appears that we are finally starting to consider organic chemistry after a substantial review of General Chemistry. Now it is time to get serious. Wrong—you should have been serious all along. In the first three chapters, you developed tools that will be used to study the structure of organic molecules, their physical properties, and their reactions. These tools won't all be used in any one area of study such as in this chapters, but they will be needed elsewhere. If you encounter problems in understanding material in this or later chapters, you should return to appropriate sections of Chapters 1 through 3 to determine what you missed. In this chapter we depend on an understanding of the sp^3 hybridization of the carbon atom and thermodynamic terms.

Keys to the Chapter

4.1 Classes of Hydrocarbons

In this section we learn a number of new terms that will continue to be used throughout the remainder of the text. **Hydrocarbons** contain only carbon and hydrogen. **Saturated** hydrocarbons contain only single bonds; **unsaturated** hydrocarbons contain carbon-carbon multiple bonds. **Alkanes** have only carbon atoms bonded in chains of atoms. **Cycloalkanes** have only carbon atoms bonded in a ring of atoms. Compounds without rings are **acyclic**; compound with rings are **cyclic**. Other atoms may be found in some rings. Atoms other than carbon within rings are called **heteroatoms**, and the compounds are called **heterocyclic**.

4.2 Normal and Branched Alkanes

Normal alkanes consist of a continuous chain of carbon atoms; branched alkanes have some carbon atoms bonded to more than two other carbon atoms. The general formula for an alkane is C_nH_{2n+2}, whether it is normal or branched. If the number of carbon atoms are known, the number of hydrogen atoms and the molecular formula are known. As you will learn later, other structural features alter the number of hydrogen atoms and affect the molecular formula. Thus, by inspecting the molecular formula and comparing it to the reference molecular formula expected for an alkane, you have a clue about the identity of other structural features. The C_nH_{2n+2} formula is your reference.

A carbon atom is classified as primary (1°), secondary (2°), or tertiary (3°) when it has 1, 2, or 3 alkyl groups, respectively, bonded to it. A carbon atom is quaternary (4°) when it has 4 alkyl groups bonded to it. Learn to classify carbon atoms now because this method will be used in later chapters for the classification of more complex classes of compounds.

4.3 Nomenclature of Alkanes

Some people are not very good at learning other peoples' names. So, what are you to do if you have to discuss even a few dozen of the millions of alkanes possible? Each has a name. However, the name is chosen to make sense and to describe the structure of the molecule. These names are not simply "John", "Frank", "Jane", or "Mary". They are descriptive in the same sense as the terms that you might use to describe someone to another person who can't put a name to a face. A person is described by his or her height or hair color or some other physical characteristics.

A series of nomenclature rules are given in the text. Don't memorize them. Learn them. Get comfortable with the logic of this systematic way of naming compounds, because it is the basis for the names of many other classes of organic compounds containing functional groups. The number of "rules" will increase, but the basic ones for alkanes form the foundation on which all other nomenclature rules are based. The rules can be briefly summarized.

1. Locate the longest carbon chain, called the **parent chain**. Note that the parent chain may not be limited to the carbon atoms written in a horizontal row.

2. Identify the groups that are **substituents** attached to the parent chain. At this time the substituents may only be carbon groups in branches or perhaps halogen atoms.

3. Number the parent chain to give the branching carbon atoms and other substituents the lowest possible numbers.

4. Using a prefix to the name of the parent chain, identify the name and location of all branches and other substituents.

There are a few fine points that cannot be neglected. Remember that each substituent must be assigned a number to indicate its position. Thus, 2-dimethyl is incorrect, because two methyl groups bonded to the C-2 atom must be designated as 2,2-dimethyl. Note that the location of the "first" substituent doesn't serve as a basis for numbering the chain if the same type of substituent is located the same distance from either end of the molecule. In those cases, locate the next substituent with the lowest numbered carbon atom from one of the two ends. In other words, use the point of first difference to select the direction of numbering. Finally, list the substituents alphabetically. Note that the prefixes di, tri, etc. do not affect the alphabetic method of listing alkyl groups. For example, ethyl is listed before dimethyl, because it is the "e" of ethyl that takes precedence over the "m" of methyl.

You should know the common alkyl groups **methyl**, **ethyl**, **propyl**, **isopropyl**, **butyl**, **sec-butyl**, **isobutyl**, and **tert-butyl**. More complex alkyl groups are named by designating the carbon that is "missing" a hydrogen atom as C-1. However, unless your instructor chooses to make up problems specifically designed to test this skill, the names of the more common alkyl groups will be all that you will really have to know to name even quite complex structures.

Drawing the structures of an alkane based on its IUPAC name is much easier than naming a structure. Start by first drawing the chain designated by the parent name. Then number the chain from left to right and place the appropriate branches or substituents at the proper positions. For structural formulas, add enough hydrogen atoms to each carbon atom to give each one four bonds. For bond-line structures, hydrogen atoms are not shown.

4.4 Heats of Formation and Stability of Alkanes

Heats of formation refer to the enthalpy of a reaction forming a compound from the elements. For the majority of organic compounds, ΔH_f is negative, meaning that the reaction is exothermic. Note however, that the direct reaction of elements to form a particular compound may never actually occur. The ΔH_f is most often calculated from the ΔH_{rxn} of several reactions using the Law of Hess. The sign of the heat of formation must always be given, because there are some low molecular weight compounds with multiple bonds that have positive heats of formation.

For the homologous series of normal alkanes, the difference in the heats of formation is approximately a constant (-20.6 kJ mole^{-1}). This value is used as a reference in discussing the stability of other compounds such as cycloalkanes.

The relative stability of isomeric alkanes is determined from the heats of formation, because the value for each alkane is referenced with respect to the same number of carbon and hydrogen atoms in the elemental state. Branching increases the stability of alkanes, which means that the heats of formation of branched alkanes are more negative than for isomeric normal alkanes.

4.5 Cycloalkanes

Cycloalkanes are considered in the same chapter as alkanes, because in essence their bonds and the resulting chemistry of those bonds are the same. The majority of cyclic structures that you encounter in the text contain a single ring. However, there are classes of compounds that have atoms that are shared between two or more rings. These are **spirocyclic**, **bridged-ring**, and **fused-ring** compounds.

The general formula for an alkane is C_nH_{2n+2}. Each ring in a compound reduces the number of hydrogen atoms by 2 relative to an alkane, because a ring contains an extra carbon-carbon bond and therefore two fewer carbon-hydrogen bonds. Thus, the formula for cycloalkanes with one ring is C_nH_{2n}, the formula for compounds with two rings is C_nH_{2n-2}, and so on.

Geometric isomers can result when two or more substituents are attached to the ring at different carbon atoms. If the substituents are on the same side of the ring, the compound is the *cis* isomer. When substituents are on opposite sides of the ring, the compound is the *trans* isomer. You should build models of geometric isomers to convince yourself that they are not interconvertable unless you break the model apart. Geometric isomers are one type of **stereoisomer**. Learn the conventions used to draw geometric isomers so that you can clearly show what you want to convey on a test. You will lose points if it isn't clear what is meant by the structure that you drew on an examination page. Clear communication is vital, because your instructor can't probe what you had in mind!

4.6 Nomenclature of Cycloalkanes

Cycloalkanes are named by prefixing the term cyclo- to a name giving the number of carbon atoms in the ring. The number 1 carbon atom is selected based on the importance of a functional group or alkyl group attached to the ring. The direction of numbering is selected to give the lowest combination of numbers to the remaining substituents at the point of first difference. Geometric isomers are identified with the appropriate *cis-* or *trans-* prefix.

4.7 Stability of Cycloalkanes

As in the case of alkanes, we use heats of formation to discuss the relative stability of cycloalkanes. In a homologous series, the heats of formation become progressively more negative for each additional methylene unit. However, unlike alkanes, the difference is not a constant value (-20.6 kJ mole^{-1}).

Small ring compounds are unstable due to **ring strain** which is the result of the small bond angles required to maintain the structure. The total ring strain is obtained by subtracting the measured heat of formation from the predicted heat of formation, based on the contribution of the number of methylene units in the ring (n x -20.6 kJ mole^{-1}). That **strain energy** is distributed among all of the carbon-carbon bond angles of the ring. The most severely strained compounds are cyclopropane and cyclobutane. (The significance of the strain energy of larger ring compounds is discussed in a later section.)

4.8 Steroids

From time to time in the text you are introduced to the structures of naturally occurring compounds that have a structural feature under discussion at that point in the text. In this case, there is no "chemistry" presented, but more details will be provided later. For now, just get used to seeing complex structures, and examine features such as the number and sizes of ring and the functional groups bonded to those rings.

4.9 Physical Properties of Alkanes and Cycloalkanes

Both alkanes and cycloalkanes have nonpolar covalent bonds. Thus, only London forces control the intermolecular interactions between neighboring molecules and those between solute and solvent molecules in a solvent. Boiling points increase with increasing molecular weight and decrease with branching. Don't worry about the actual values—you don't have to memorize them. Just get familiar with the trends.

4.10 Conformations and Properties

The study of the chemical and physical properties of different conformations of compounds, called **conformational analysis**, formed a basis for understanding the relationships between structure and properties. The energy difference between the conformation of a molecule in its most stable conformation and that required for the molecule in the transition state affects the rates of reactions.

4.11 Conformations of Ethane

The term conformation was introduced in Section 1.16. It refers to different arrangements of atoms in a molecule that result from rotation about carbon-carbon sigma bonds. You should build models of the conformations (conformers) of ethane. The **staggered** conformation is the most stable; the **eclipsed** conformation is the least stable. This difference is related to VSEPR theory (Section 1.10) in that the bonding electron pairs of the carbon-hydrogen bonds of neighboring carbon atoms tend to stay as far apart as possible.

As described in this section, Newman projection formulas give us a method for conveying three-dimensional information in two dimensions. To see how these projection formulas are derived, pick up a molecular model of ethane and look down the axis of the carbon-carbon bond.

The energy difference between one staggered conformation and another is equal to that required to get past an eclipsed conformation. The energy difference is called **torsional strain**. For each carbon-hydrogen bond the contribution to the torsional strain is 4.2 kJ mole^{-1}.

4.12 Conformations of Propane

Rotation about a carbon-carbon bond of propane is similar to that of ethane. However, the eclipsed conformation now has a hydrogen-methyl interaction in addition to two hydrogen-hydrogen interactions, so the barrier to rotation is now larger. The resulting increase in energy is attributed to **van der Waals repulsion** between atoms. Although small in this case, the **steric hindrance** between atoms is larger when atoms are larger or brought closer together in molecules to be discussed later. Examine molecular models to show yourself that atoms designated as 1 and 5 are close to bonding distance when the structure is arranged in a planar conformation.

4.13 Conformations of Butane

With butane, the comparison of conformations become more interesting. In fact the concepts introduced here are important to the understanding of many other phenomena such as the stability of cycloalkanes and many reactions to be studied later. There are two nonequivalent staggered conformations—the **anti** conformation and the **gauche** conformation, which have a **dihedral angle** of 180° and 60°, respectively.

gauche
conformation

anti
conformation

The anti conformation is the more stable because there is van der Waals repulsion between the two methyl groups in the gauche conformation. You can easily view the congestion of atoms if you build a molecular model. Also rotate the model to achieve an eclipsed conformation, and attempt to bring one terminal hydrogen atom of the C-1 atom and one from the C-4 atom as close together as possible. This exercise will illustrate the basis for the Newman "rule of 6", which signals when severe steric repulsion between atoms should occur.

4.14 Conformations of Acyclic Compounds

This section summarizes the features of both staggered and eclipsed conformations of simple alkanes which are used to describe the features of larger molecular weight alkanes. The energies associated with several types of interactions between atoms bonded to adjacent carbon atoms are summarized in Table 4.7. The general conclusions are that eclipsed conformations are less stable than staggered conformations and that gauche conformations are less stable than anti conformations.

4.15 Conformations of Cycloalkanes

If you have avoided spending the money for a molecular model kit, now is the time to consider the price as a worthwhile investment in your grade. The majority of students don't have good spatial perspective and have difficulty seeing the relationships between atoms in two-dimensional representations.

Before attempting to construct a cyclopropane or cyclobutane ring, consult the instructions for your molecular model kit and also determine how flexible the "bonds" are. You should now have an appreciation for the concept of bond angle strain. However, note that the total strain energy of small ring compounds is a combination of bond angle strain and torsional strain because the carbon-hydrogen bonds of adjacent carbon atoms are eclipsed in cyclopropane. In cyclobutane, a little twisting decreases the torsional strain, but the bonds giving rise to the torsional strain are still close. In cyclopentane, the bond angle strain is expected to be small, because you can easily put the structure together using your kit. However, note that torsional interactions still must occur in the molecule.

Representing the structure of cyclohexane in its **chair** conformation on paper is one of the more challenging tasks that you will have in this course. Examine the orientations of the bonds carefully and practice writing out the bonds of the chair conformation. You are going to have to do this many times later in the course. Not only will you lose points for improper drawings of cyclohexane derivatives, but you also won't be able to answer your lecturer's questions about the chemistry of such derivatives, because you won't know where the atoms are in space. There are three **axial** bonds pointed up and three pointed down. The six **equatorial** bonds are pointed out around the ring; three are pointed slightly upward and three slightly downward. Examine each of the following chair cyclohexane conformations and pay attention to the locations of the equatorial and axial bonds at each of the six carbon atoms.

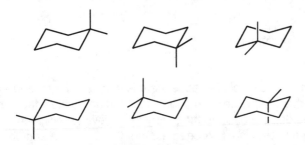

Other representations are incorrect. For example, the following is a common error seen on exams. There is an axial and an equatorial bond but the structure is incorrect.

Now a word to the wise. Learn from molecular models, but don't rely on them as a "crutch". Your instructor may allow you to use molecular models on a test, but the structures are always given on paper and you will have to represent them on paper. So, use the models prior to exams to clarify geometric relationships. However, avoid using them on tests. It takes time to construct them, and they may fall apart and end up on the floor at the most inopportune time!

4.16 Conformational Mobility of Cyclohexane

The chair conformation of cyclohexane can "flip", and this process changes the orientations of all bonds. The equatorial bonds become axial and vice versa. You should examine Figure 4.12 in the text and convince yourself of this fact. It may be necessary to "flip" a substituted cyclohexane ring in order to see a specific spatial interaction between substituents. In addition, the ring "flip" is required in order to determine if the substituents are in their most stable conformation. Boat conformations are relatively unimportant in understanding the chemistry of cyclohexane compounds. Take note of the emphasis (or lack thereof) given to this topic by your instructor.

4.17 Monosubstituted Cyclohexanes

Substituents bonded to the cyclohexane ring have a **conformational preference** for the equatorial position. Substituents in the axial position are sterically hindered to some degree, because they are within the van der Waals radii of the axial hydrogen atoms at the C-3 and C-5 positions, which is called a 1,3-diaxial interaction. Again, molecular models are useful to confirm this fact.

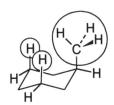

The free energy differences between the axial and equatorial conformations of monosubstituted cyclohexanes are listed in Table 4.8 in the text. These values represent the magnitude of the two 1,3-diaxial interactions, and they depend on the size of the atom, the length of the bond, the polarizability of the atom, and the number of atoms bonded to the atom directly bonded to the cyclohexane ring. These values apply only if the group "encounters" an axial hydrogen atom. If other groups are located in the axial position, then the steric effects are much larger.

4.18 Disubstituted Cyclohexanes

In disubstituted cyclohexanes, not only the stability of the two possible conformations but also the relative stability of the geometric isomers depend on two factors. One is the inherent conformational preference of each substituent, and the other is any possible steric interaction between the two groups themselves. The difference in free energies of formation of dimethylcyclohexanes can be obtained from the data in Table 4.9 in the text. Note that it is the trans-1,2, the cis-1,3, and the trans-1,4 dimethyl compounds that are in diequatorial conformations.

trans-1,2-dimethylcyclohexane   cis-1,3-dimethylcyclohexane   trans-1,4-dimethylcyclohexane

These isomers are more stable than their respective equatorial/axial geometric isomers, because the axial methyl group constitute an unfavorable 1,3-diaxial interaction.

The isomers that are most stable for compounds with two different substituents are also the *trans*-1,2, the cis-1,3, and *trans*-1,4 compounds. The difference in energy between either the isomers or the alternate conformations of each compound can be calculated by appropriate addition of the individual conformational preferences of each group.

4.19 Polycyclic Molecules

The isomeric decalins provide the models for the rings that occur in fused ring compounds such as steroids. From time to time in later sections of the text, decalins will be considered that have substituents on one or both rings. These compounds are useful to elucidate the mechanisms of reactions such as elimination and addition. *trans*-Decalin is a rigid molecule, and neither of its two rings can "flip". Note the location of the hydrogen atoms on the carbon atoms next to a point of fusion, as shown in the following two structures.

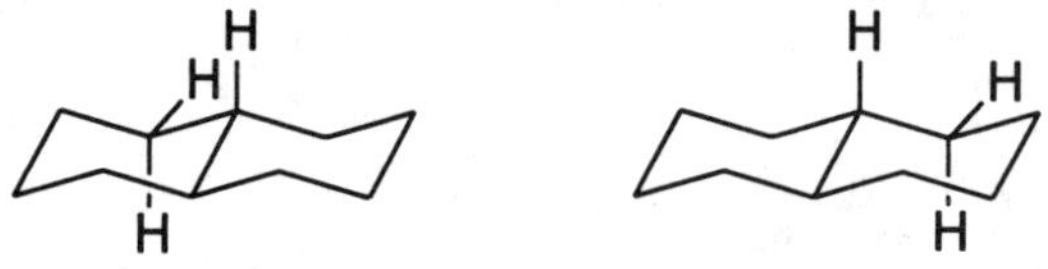

Because the rings are cis-fused in *cis*-decalin, the rings can flip and convert equatorial bonds to axial bonds and vice versa. One representation of this structure is shown below. It takes a bit of concentration to draw the structure properly. The following arrangement of two rings at right angles may help you see how to draw the compound. Just push the two rings together until two sets of carbon atoms are coincident.

4.20 Medium and Large Cycloalkanes

In general it isn't necessary to consider the conformations of medium and large cycloalkanes, because they are very flexible. Constructing a model should convince you of this fact. The main point made in this section is that hydrogen atoms (and other substituents) that are bonded at widely separate carbon atoms may actually be near one another. Steric interactions of this type are **transannular interactions**.

Solutions to Exercises

4.1 Acyclic saturated compounds have the general formula C_nH_{2n+2}. Only (b) and (d) meet this requirement.

4.2 **(a)** This formula is possible because it has $2n + 2$ hydrogen atoms.

(b) This formula is impossible because it has more than $2n + 2$ hydrogen atoms and also has an odd number of hydrogen atoms.

(c) This formula is possible because it has $2n$ hydrogen atoms, which can result from either a cyclic structure or unsaturation.

(d) This formula is impossible because it has more than $2n + 2$ hydrogen atoms.

4.3 The number of carbon atoms represented by n in the general formula for alkanes is 31. The number of hydrogen atoms must be $2n + 2$ or 64. The completely condensed formula for hentriacontane is $CH_3(CH_2)_{29}CH_3$.

4.4 For n = 100, the value of $2n + 2$ is 202. The molecular formula is $C_{100}H_{202}$. The completely condensed formula for this normal alkane is $CH_3(CH_2)_{98}CH_3$.

4.5 **(a)** $CH_3-CH_2-CH_2-CH_3$

(b) 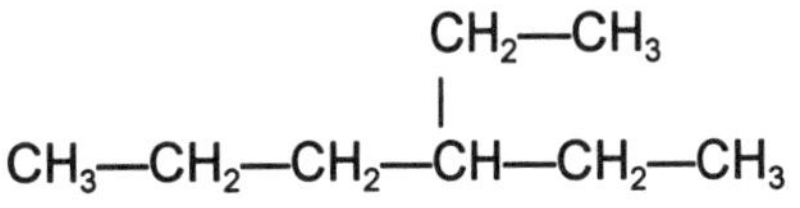
$$CH_3-CH_2-CH_2-\overset{\overset{\displaystyle CH_2-CH_3}{|}}{CH}-CH_2-CH_3$$

(c) 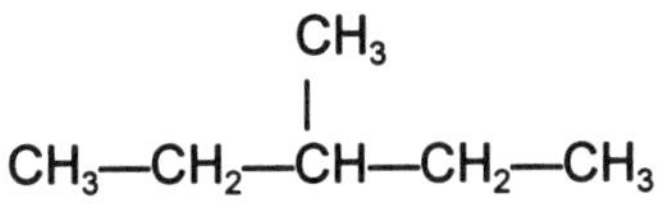
$$CH_3-CH_2-\overset{\overset{\displaystyle CH_3}{|}}{CH}-CH_2-CH_3$$

(d) 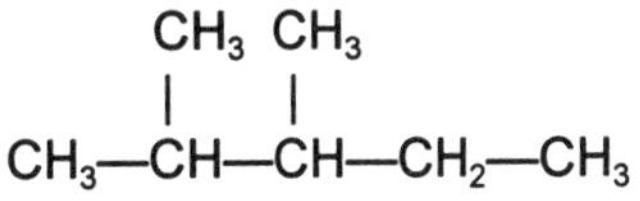
$$CH_3-\overset{\overset{\displaystyle CH_3}{|}}{CH}-\overset{\overset{\displaystyle CH_3}{|}}{CH}-CH_2-CH_3$$

4.6 **(a)** 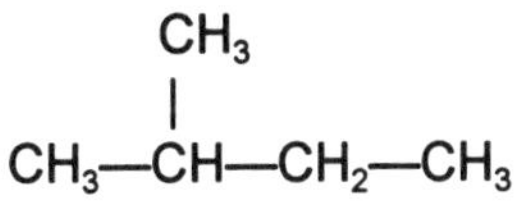
$$CH_3-\overset{\overset{\displaystyle CH_3}{|}}{CH}-CH_2-CH_3$$

(b) 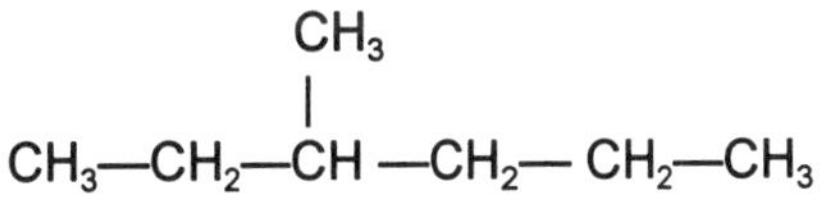
$$CH_3-CH_2-\overset{\overset{\displaystyle CH_3}{|}}{CH}-CH_2-CH_2-CH_3$$

(c) 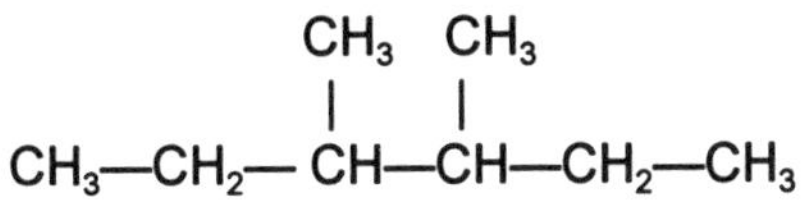
$$CH_3-CH_2-\overset{\overset{\displaystyle CH_3}{|}}{CH}-\overset{\overset{\displaystyle CH_3}{|}}{CH}-CH_2-CH_3$$

(d) 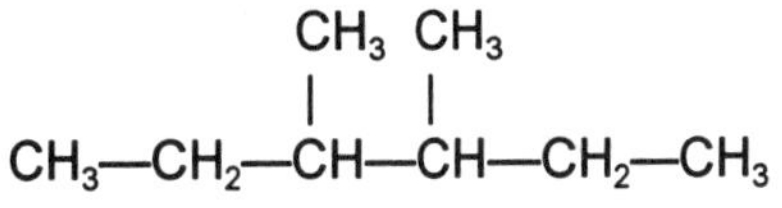
$$CH_3-CH_2-\overset{\overset{\displaystyle CH_3}{|}}{CH}-\overset{\overset{\displaystyle CH_3}{|}}{CH}-CH_2-CH_3$$

4.7 Both I and II have a chain of six carbon atoms with a methyl group at the C-2 atom and an ethyl group at the C-3 atom. Both III and IV have a chain of seven carbon atoms with methyl groups at the C-3 and C-4 atoms.

4.8 Both I and III have a chain of five carbon atoms with methyl groups at the C-2 and C-3 atoms. Both III and IV have a chain of six carbon atoms with a methyl group at the C-3 atom.

4.9   **(a)** methyl   **(b)** propyl   **(c)** *sec*-butyl   **(d)** isobutyl

4.10   **(a)** ethyl   **(b)** isopropyl   **(c)** butyl   **(d)** *tert*-butyl

4.11   **(a)** butyl   **(b)** 2-methylpropyl

        (c) 2-methylbutyl   **(d)** 3-methylbutyl

4.12   **(a)** 2-methylpropyl   **(b)** 2-methylbutyl

        (c) 3-methylpentyl   **(d)** 1,1-dimethylpropyl

4.13   1,1,3,3-tetramethylbutyl

4.14   4,8,12-trimethyltridecyl

4.15   **(a)** 2-methylbutane   **(b)** 3-methylhexane   **(c)** 2-methylpentane

        (d) 3-methylheptane   **(e)** 2-methylpentane   **(f)** 2-methylhexane

4.16   **(a)** 2,3-dimethylhexane   **(b)** 2,5-dimethylhexane   **(c)** 3,4-dimethylheptane

        (d) 3,5-dimethylheptane   **(e)** 3-methylhexane   **(f)** 3-ethylpentane

4.17   5-(1-ethylpropyl)decane

4.18   5-(1,1-dimethylpropyl)nonane

4.19   **(a)** $CH_3-CH_2-\overset{\overset{\displaystyle CH_3}{|}}{CH}-CH_2-CH_3$ **(b)** $CH_3-CH_2-\overset{\overset{\displaystyle CH_3}{|}}{CH}-\overset{\overset{\displaystyle CH_3}{|}}{CH}-CH_2-CH_3$

(c) $CH_3-\overset{\overset{\displaystyle CH_3}{|}}{\underset{\underset{\displaystyle CH_3}{|}}{C}}-\overset{\overset{\displaystyle CH_3}{|}}{CH}-CH_2-CH_3$ **(d)** $CH_3-CH_2-CH_2-\overset{\overset{\displaystyle CH_2CH_3}{|}}{CH}-CH_2-CH_2-CH_3$

(e) $CH_3-\overset{\overset{\displaystyle CH_3}{|}}{CH}-\overset{\overset{\displaystyle CH_3}{|}}{CH}-\overset{\overset{\displaystyle CH_3}{|}}{CH}-\overset{\overset{\displaystyle CH_3}{|}}{CH}-CH_3$

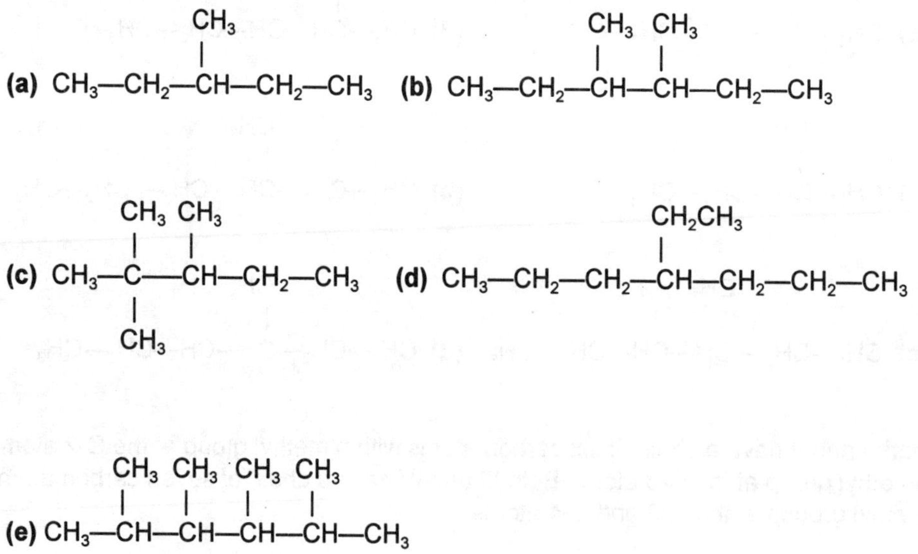

4.20 **(a)** 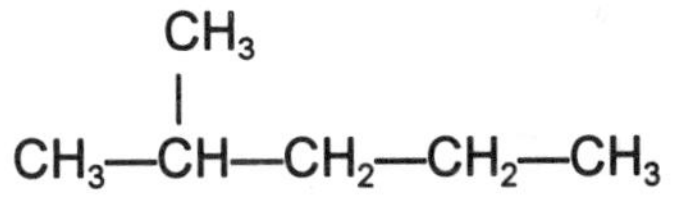
$$CH_3-\overset{\overset{\displaystyle CH_3}{|}}{CH}-CH_2-CH_2-CH_3$$

(b) 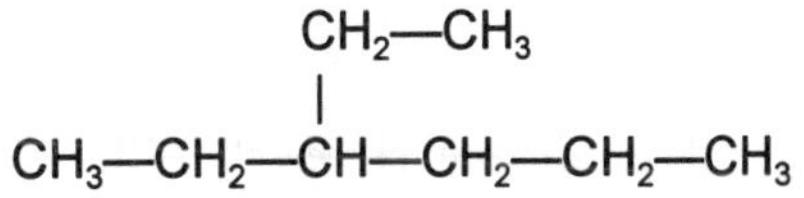
$$CH_3-CH_2-\overset{\overset{\displaystyle CH_2-CH_3}{|}}{CH}-CH_2-CH_2-CH_3$$

(c) 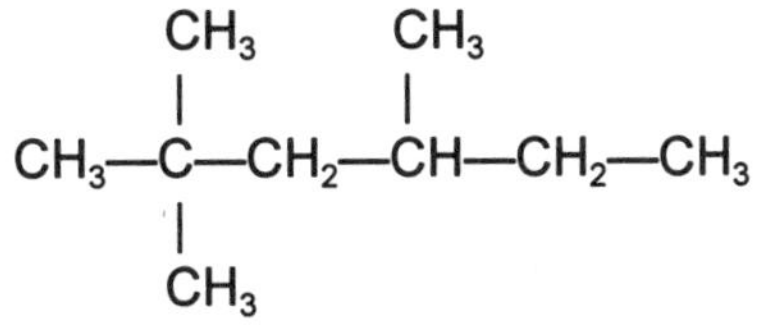
$$CH_3-\overset{\overset{\displaystyle CH_3}{|}}{\underset{\underset{\displaystyle CH_3}{|}}{C}}-CH_2-\overset{\overset{\displaystyle CH_3}{|}}{CH}-CH_2-CH_3$$

(d) 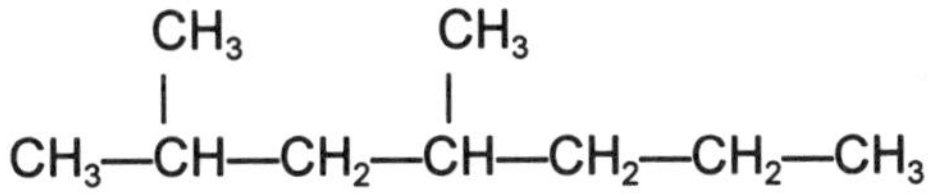
$$CH_3-\overset{\overset{\displaystyle CH_3}{|}}{CH}-CH_2-\overset{\overset{\displaystyle CH_3}{|}}{CH}-CH_2-CH_2-CH_3$$

(e) 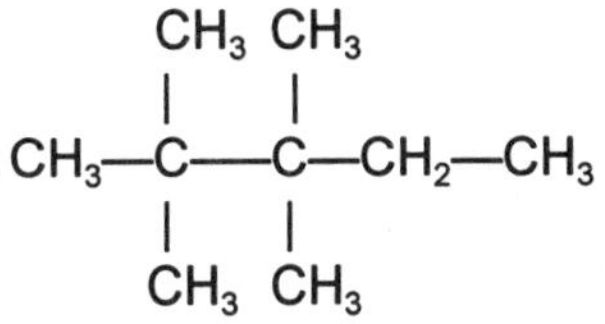
$$CH_3-\overset{\overset{\displaystyle CH_3}{|}}{\underset{\underset{\displaystyle CH_3}{|}}{C}}-\overset{\overset{\displaystyle CH_3}{|}}{\underset{\underset{\displaystyle CH_3}{|}}{C}}-CH_2-CH_3$$

4.21 **(a)** 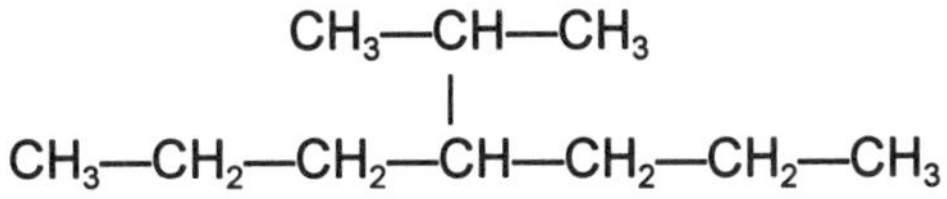
$$CH_3-CH_2-CH_2-\overset{\overset{\displaystyle CH_3-CH-CH_3}{|}}{CH}-CH_2-CH_2-CH_3$$

(b) 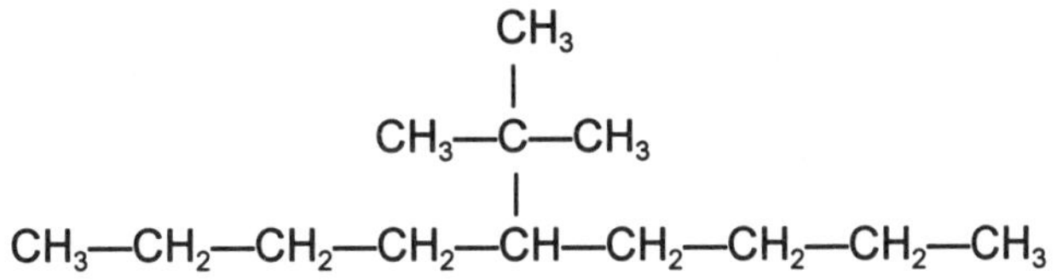
$$CH_3-CH_2-CH_2-CH_2-\overset{\overset{\displaystyle CH_3-\overset{\overset{\displaystyle CH_3}{|}}{C}-CH_3}{|}}{CH}-CH_2-CH_2-CH_2-CH_3$$

(c)
$$CH_3-CH_2-CH_2-CH_2-\overset{\overset{\displaystyle CH-CH_2-CH_3}{\underset{\displaystyle |}{\overset{\displaystyle CH_3}{|}}}}{CH}-CH_2-CH_2-CH_2-CH_2-CH_3$$

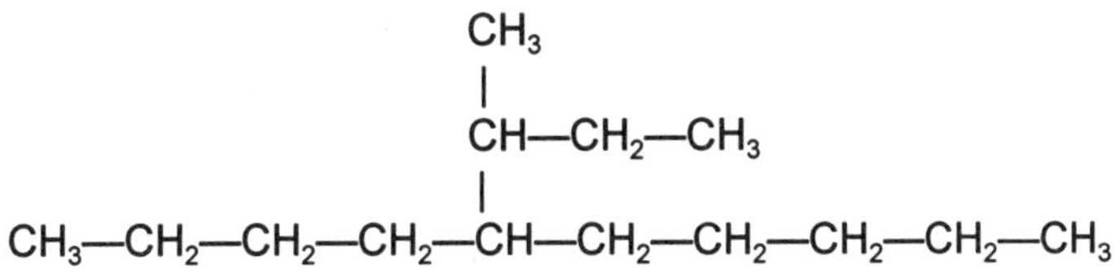

4.22 **(a)** 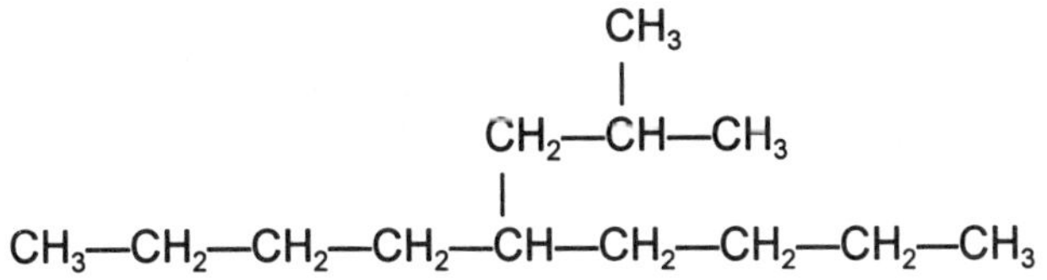
$$CH_3-CH_2-CH_2-CH_2-\overset{\overset{\displaystyle CH_2-CH-CH_3}{\underset{\displaystyle |}{\overset{\displaystyle CH_3}{|}}}}{CH}-CH_2-CH_2-CH_2-CH_3$$

$$CH_2—CH_2—CH_2—CH_3$$

(b) 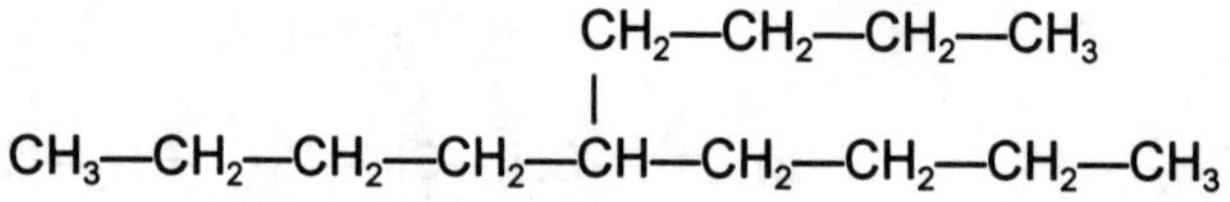 $CH_3—CH_2—CH_2—CH_2—CH—CH_2—CH_2—CH_2—CH_3$

$$CH_3$$

$$CH_3—C—CH_2—CH_3$$

(c) $CH_3—CH_2—CH_2—CH_2—CH—CH_2—CH_2—CH_2—CH_2—CH_3$

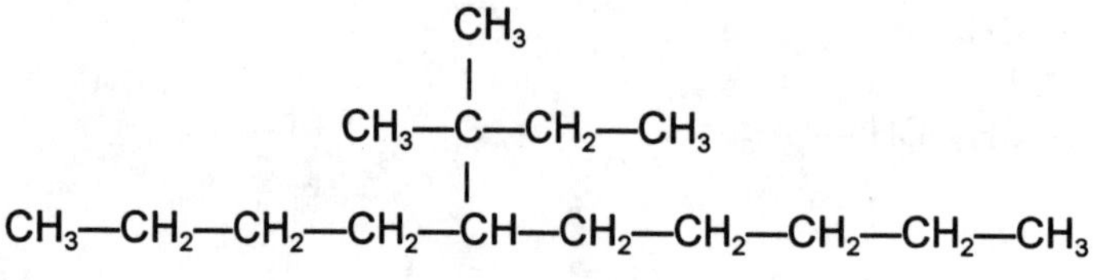

4.23 2-methylhexane and 3-methylhexane

4.24 2,2-dimethylpentane, 3,3-dimethylpentane, 2,3-dimethylpentane, and 2,4-dimethylpentane

4.25 **(a)** 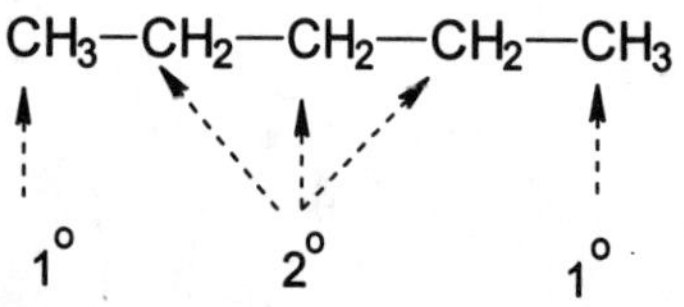 $CH_3—CH_2—CH_2—CH_2—CH_3$
$1°$ $2°$ $1°$

(b) 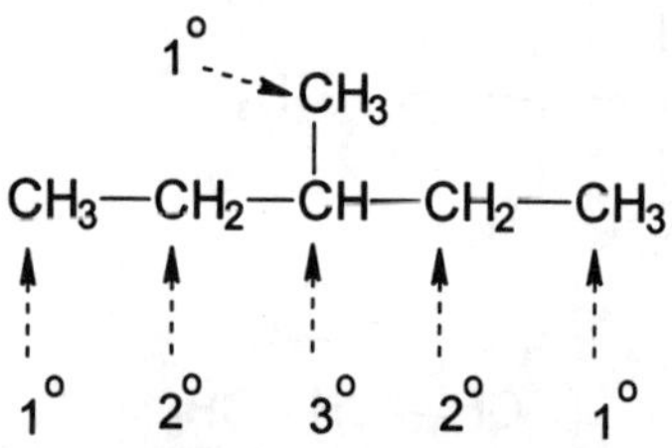 $CH_3—CH_2—CH—CH_2—CH_3$ with $1°$ CH_3 branch
$1°$ $2°$ $3°$ $2°$ $1°$

(c) 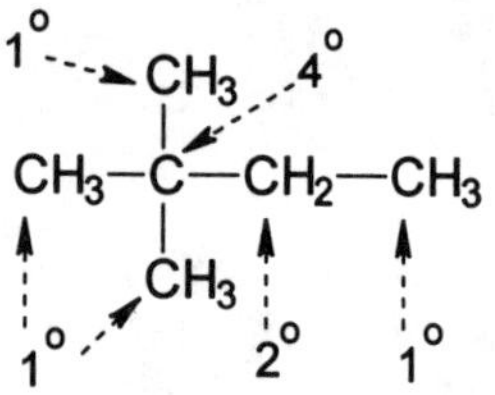 $CH_3—C—CH_2—CH_3$ with $1°$ CH_3 and $4°$, CH_3
$1°$ $2°$ $1°$

(d) 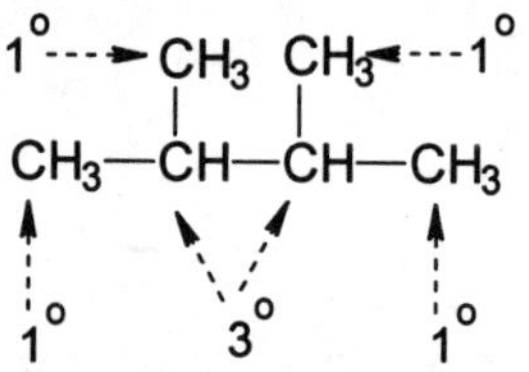 $CH_3—CH—CH—CH_3$ with $1°$ CH_3 CH_3 $1°$
$1°$ $3°$ $1°$

4.26 **(a)** 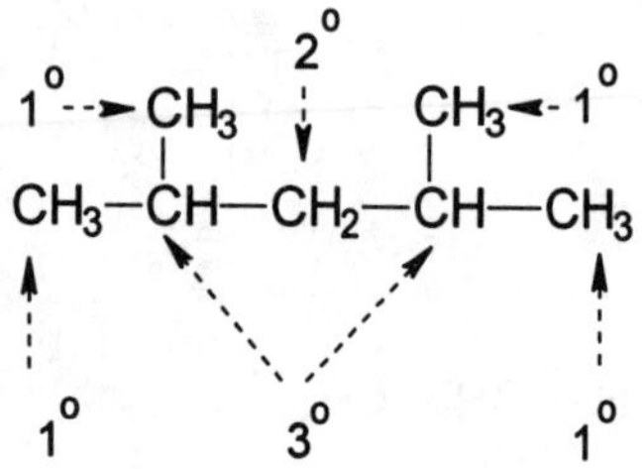 $CH_3—CH—CH_2—CH—CH_3$ with $1°$ CH_3 $2°$ CH_3 $1°$
$1°$ $3°$ $1°$

(b) 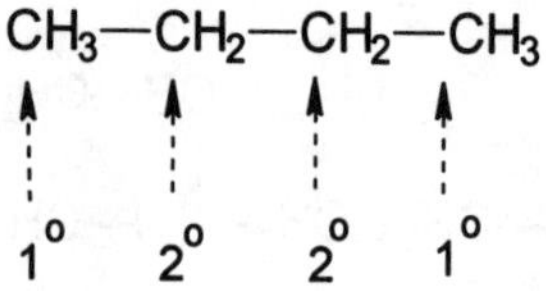 $CH_3—CH_2—CH_2—CH_3$
$1°$ $2°$ $2°$ $1°$

(c) 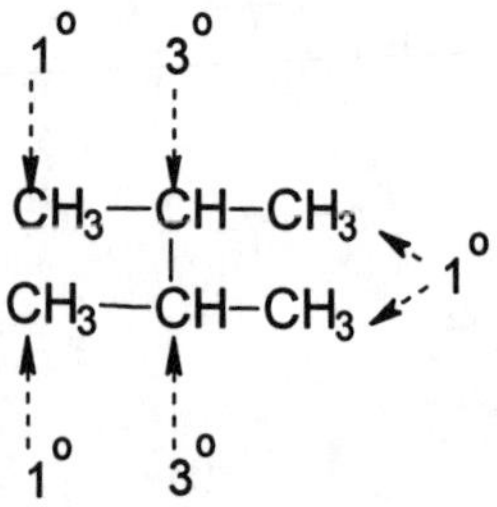 $CH_3—CH—CH_3$ $1°$ $3°$
$CH_3—CH—CH_3$ $1°$
$1°$ $3°$

(d) 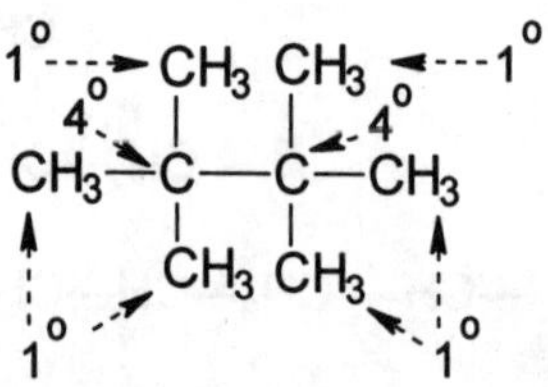 $CH_3—C—C—CH_3$ with $1°$ CH_3 CH_3 $1°$, $4°$ $4°$, CH_3 CH_3
$1°$ $1°$
$1°$ $1°$

4.27

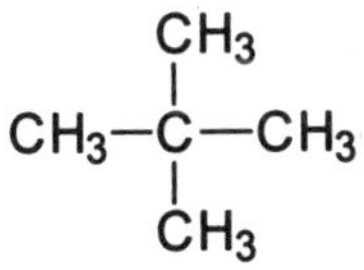

$$CH_3-\underset{\underset{CH_3}{|}}{\overset{\overset{CH_3}{|}}{C}}-CH_3$$

4.28 $CH_3-\underset{\underset{CH_3}{|}}{CH}-\underset{\underset{CH_3}{|}}{CH}-CH_3$

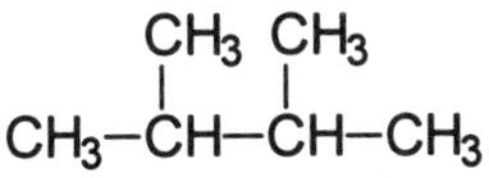

4.29   **(a)** four primary and one quaternary
(b) three primary, one secondary, and one tertiary
(c) three primary, two secondary, and one tertiary
(d) four primary and two tertiary

4.30   **(a)** six primary, one secondary, and two quaternary
(b) four primary, one secondary, and two tertiary
(c) three primary, two secondary, and one tertiary
(d) five primary and three tertiary

4.31   2-Methylhexane has the more negative heat of formation and is thus the more stable of the two isomeric compounds.

4.32   The heats of formation differ by only 0.29 kJ mole^{-1}. Thus, the difference in S° must be considered in determining the ΔG°, and therefore the equilibrium constant, for the isomerization reaction.

4.33   The heat of formation will be 2(20.6) kJ mole^{-1} more negative because the homologous compound has two more methylene units. The value is estimated as -225 kJ mole^{-1}.

4.34   The heat of formation will be 8(20.6) kJ mole^{-1} more negative because the homologous compound has eight additional methylene units. The value is estimated as -473 kJ mole^{-1}.

4.35   **(a)** Use the heat of formation of the isomeric 2-methylheptane as a reference (-212.35 kJ mole^{-1}). The additional branches in 2,5-dimethylhexane should make the compound more stable by approximately 7 kJ mole^{-1}. Thus the heat of formation should be approximately -220 kJ mole^{-1}.

(b) Use the heat of formation of the isomeric 2-methylnonane as a reference (-256.5 kJ mole^{-1}). The additional branch in 2,4-dimethyloctane should make the compound more stable by approximately 7 kJ mole^{-1}. Thus the heat of formation should be approximately -264 kJ mole^{-1}.

(c) Use the heat of formation of the isomeric 2-methylnonane as a reference (-256.5 kJ mole^{-1}). The two additional branch in 2,4,6-trimethylheptane should make the compound more stable by approximately 14 kJ mole^{-1}. Thus the heat of formation should be approximately -270 kJ mole^{-1}.

4.36 The most stable isomer has the more negative heat of formation. Branching tends to increase the stability of isomeric alkanes. The most highly branched isomers, 2,2-dimethylbutane or 2,3-dimethylbutane, are the most likely possibilities.

4.37 (a) 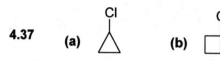 (b) 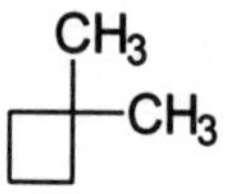 (c)

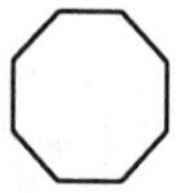

4.38 (a) 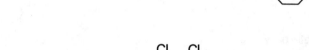 (b) (c)

4.39 (a) 1,1-dimethylcycloheptane (b) cyclodecane

(c) *trans*-1,2-dichlorocyclohexane (d) 1,1-dichlorocyclohexane

4.40 (a) butylcyclooctane (b) isopropylcyclohexane

(c) *cis*-1,3-dimethylcyclohexane (d) 1,1,3-trichlorocyclohexane

4.41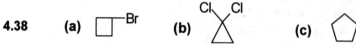

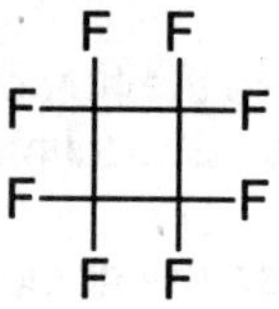

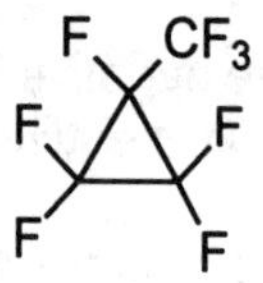

4.42 five: cyclopentane, methylcyclobutane, 1,1-dimethylcyclopropane, *cis*-1,2-dimethylcyclopropane, and *trans*-1,2-dimethylcyclopropane

4.43 (a) $C_{10}H_{18}$ (b) $C_8H_{12}O$ (c) $C_7H_{13}N$ (d) C_6H_6

4.44 (a) $C_{10}H_{18}O$ (b) $C_{10}H_{16}$ (c) $C_{10}H_{14}$ (d) C_8H_{14}

4.45 (a) tricyclic (b) bicyclic (c) pentacyclic (d) tricyclic

4.46 (a) pentacyclic (b) bicyclic (c) bicyclic (d) tricyclic

4.47 The heat of formation will be 20.6 kJ mole^{-1} more negative because the homologous compound has one more methylene unit in the alkyl chain bonded to the cyclohexane ring. The value is estimated as -213.8 kJ mole^{-1}.

4.48   The heat of formation will be 2(20.6) kJ mole^{-1} more negative because the compound has two more methylene units than cyclohexane. The value is estimated as -161.3 kJ mole^{-1}.

4.49   The *trans* isomer has the more negative heat of formation and is thus the more stable of the two isomeric compounds.

4.50   There is a very small difference in the heats of formation. Thus, the difference in S° must be considered in determining the ΔG°, and therefore the equilibrium constant, for the isomerization reaction.

4.51   Octane has the highest boiling point. All of its isomers have branches and should have lower boiling points. 2,2,3,3-Tetramethylbutane has the largest number of branches and should have the lowest boiling point.

4.52   Methylcyclopentane compared to cyclohexane may have a more compact structure and the difference may be due to the same phenomena observed for branched alkanes.

4.53

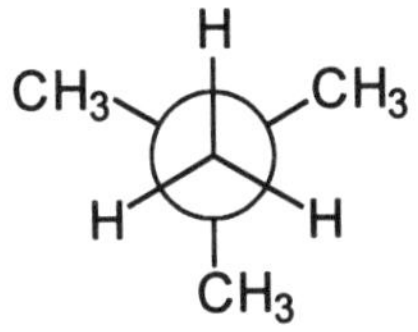

4.54

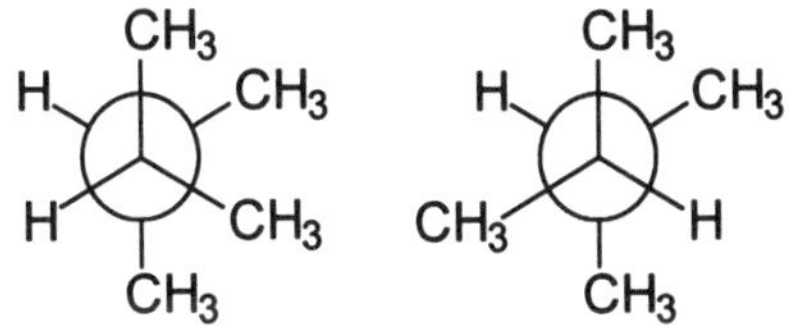

4.55

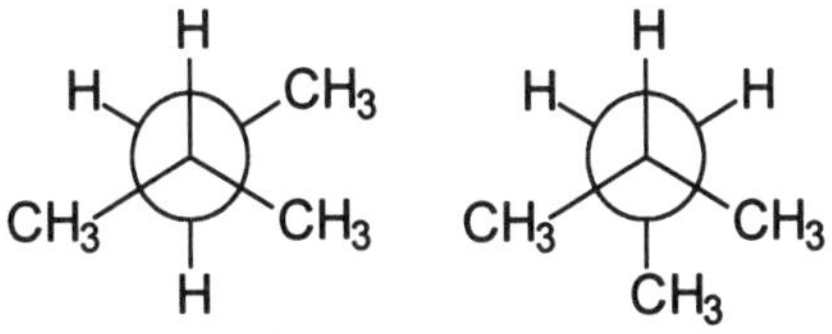

most stable

4.56

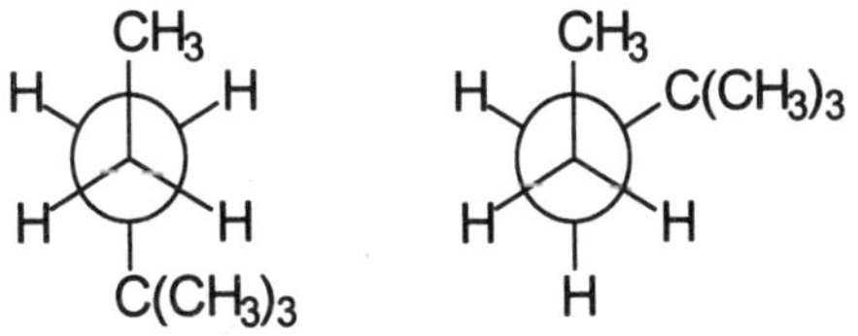

most stable

4.57   Using 5.4 kJ mole^{-1} for the eclipsing interaction of each set of C—H bonds, the remaining 4.1 kJ mole^{-1} can be assigned to the eclipsing interaction of the C—Cl and C—H bonds. Therefore the barrier to rotation for 1,1,1-trichloroethane should be 3(4.1 kJ mole^{-1}) or 12.3 kJ mole^{-1}.

4.58   Using 5.4 kJ mole^{-1} for the eclipsing interaction of each set of C—H bonds, the remaining 4.2 kJ mole^{-1} can be assigned to the eclipsing interaction of the C—Br and C—H bonds.

4.59   The sum of 5.4 kJ mole^{-1} for the eclipsing interaction of one set of C—H bonds and 2(5.4 kJ mole^{-1}) for the eclipsed interactions of the two sets of C—H and C—CH$_3$ bonds is 16.2 kJ mole^{-1}. This predicted value is identical to the observed value.

4.60   The silicon-carbon bond length is longer than a carbon-carbon bond. Thus, the distance between the two methyl groups is larger in the silicon compound than in butane. The barrier to rotation should be lower in the silicon compound because all of the sets of eclipsing bonded pairs of electrons are separated by a larger distance.

4.61

4.62

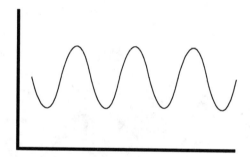

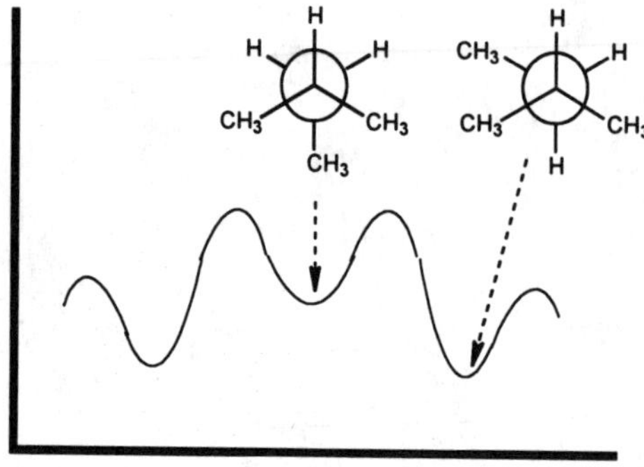

4.63   There is an attractive interaction between chlorine and the hydroxyl group. The interaction may be a hydrogen bond.

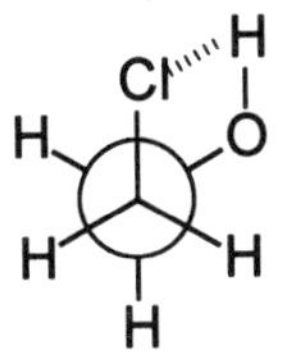

4.64   There is an attractive van der Waals interaction between chlorine and the methyl group which may be the result of the polarizability of the chlorine atom.

4.65   The anti conformation does not have a dipole moment because the bond moments of the C—Cl bonds cancel. The gauche conformation has a dipole moment. The observed dipole moment indicates that some of the compound must exist in the gauche conformation.

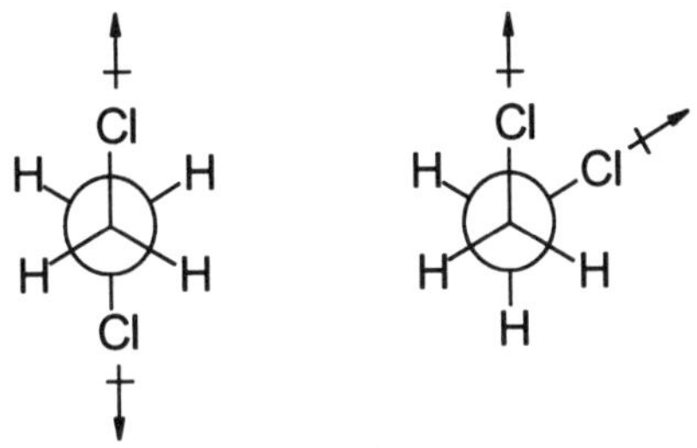

4.66   The anti conformation cannot form intramolecular hydrogen bonds because the hydroxyl groups are widely separated. The hydroxyl groups are sufficiently close in the gauche conformation to form an intramolecular hydrogen bond. The structure has 5 atoms in a hydrogen bonded "ring" counting the atoms starting from the hydroxyl hydrogen atom of one hydroxyl group to the oxygen atom of the other hydroxyl group.

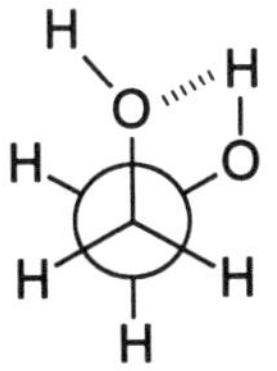

4.67

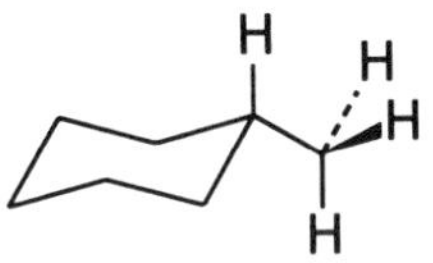

4.68

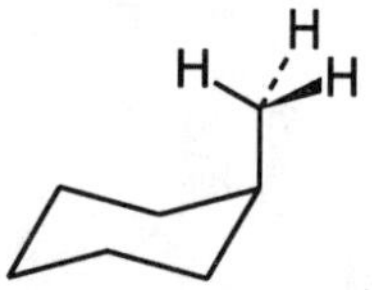

4.69 **(a)** 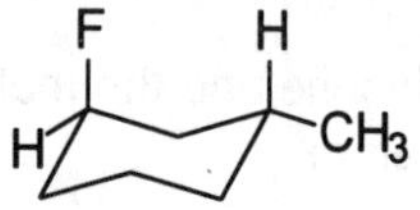   **(b)** 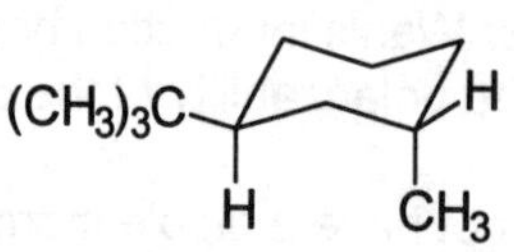   **(c)**

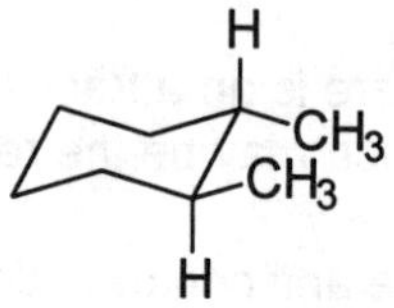

4.70 **(a)** 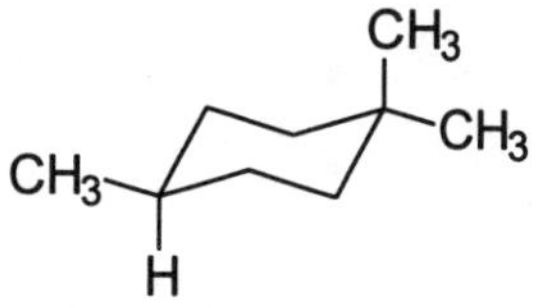   **(b)**

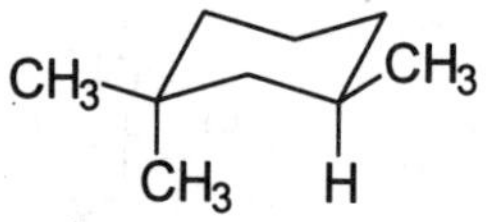

(c)

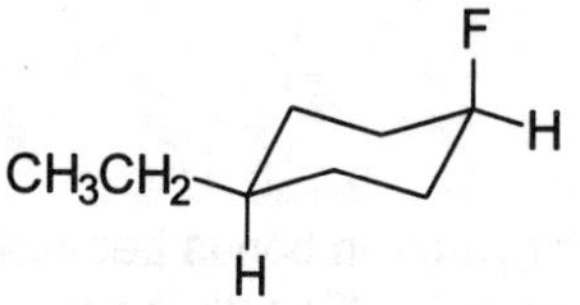

4.71   The group is linear and structurally similar to the —C≡N group, whose conformational preference is 0.8 kJ mole^{-1}.

4.72   The conformational preference of the hydroxyl group is 4.2 kJ mole^{-1}. A CH$_3$O— group differs from a hydroxyl group by a methyl group in much the same way that an ethyl group differs from an methyl group. The conformational preferences of the ethyl and methyl groups are 8.0 and 7.6 kJ mole^{-1}, respectively, a difference of 0.4 kJ mole^{-1}. Thus the conformational preference of the CH$_3$O— group is predicted as approximately 4.6 kJ mole^{-1}.

4.73   In any conformation of the *tert*-butyl compound there is a methyl group located over the cyclohexane ring and there is severe steric repulsion between it and the axial hydrogen atoms of the C-3 and C-5 atoms.

4.74   The C—Br bond is longer than the C—Cl bond, so the distance separating the bromine atom from the C-3 and C-5 hydrogen atoms is larger. In addition, the bromine atom is more polarizable than the chlorine atom and its electrons may be more easily distorted away from the steric congestion in the axial conformation.

4.75   The hydroxyl groups are sufficiently close in the diaxial conformation to form an intramolecular hydrogen bond. The structure has 6 atoms in a hydrogen bonded "ring", counting the atoms starting from the hydroxyl hydrogen atom of one hydroxyl group to the oxygen atom of the other hydroxyl group.

4.76   An axial *tert*-butyl group exists in the trans isomer, which has a steric repulsion of 22 kJ mole^{-1} (Table 4.8). This repulsion is eliminated in the twist boat conformation, even though it is 22 kJ mole^{-1} less stable than the chair conformation, because both *tert*-butyl groups can occupy equatorial-like like positions in this conformation. (See the figure at the bottom of page 178 of the text.).

4.77   The steric strains as listed are all due to repulsion between an axial group and the axial hydrogen atoms at the C-3 and C-5 positions. Repulsion between an axial group and larger atoms at the C-3 and C-5 positions must be substantially larger. In the diaxial conformation of *cis*-1,3-dimethylcyclohexane, that repulsion is between two methyl groups.

4.78   The steric strains as listed are all due to repulsion between an axial group and the axial hydrogen atoms at the C-3 and C-5 positions. Repulsion between an axial methyl group and a larger atom at the C-1 position, such as chlorine must be substantially larger.

4.79   The heat of formation of the trans isomer should be more negative than that of the cis isomer, because the trans isomer with two equatorial methyl groups is the more stable.

4.80   The heat of formation of the trans isomer is more negative than that of the cis isomer, indicating that the trans isomer is the more stable. The methyl groups of the trans isomer do not sterically interfere with one another.

4.81   The cis isomer and trans isomer resemble *cis*- and *trans*-1,2-dimethylcyclohexane in terms of the bridging of one ring to the other ring. Thus, the cis isomer with its axial CH_2 unit should be at least 7.6 kJ mole^{-1} less stable than the trans isomer.

4.82   Compound II should be the less stable because the methyl group is axial. However, the difference should be larger than 7.6 kJ mole^{-1}, because there is a repusive interaction between the methyl group and the CH_2 units of the bridging atoms of the five-membered rings.

steric repulsion

4.83   **(a)**, **(c)**, and **(e)** have their groups in equatorial positions.

(a)

(c)

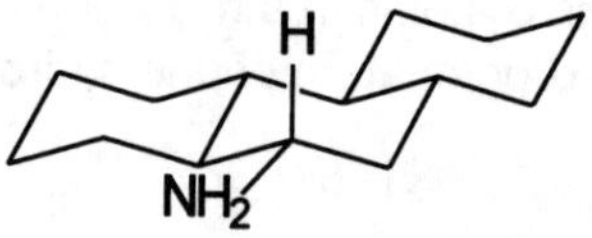

(e)

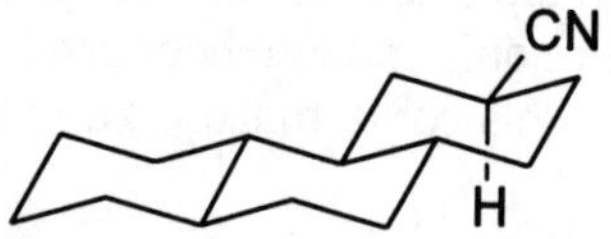

4.84   **(b)** and **(d)** have their groups in equatorial positions.

(b)

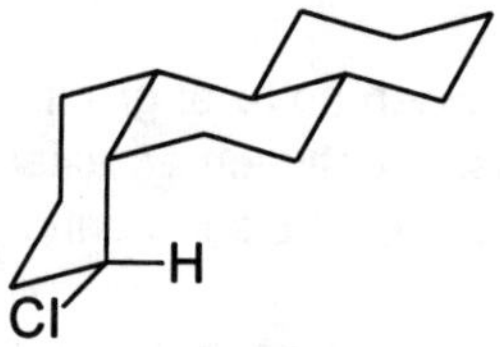

(d)

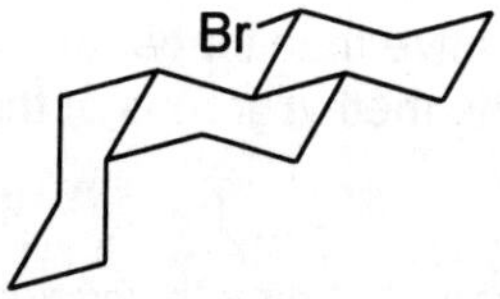

5

Reactions of Alkanes and Cycloalkanes

The focus of this chapter is on both the energy and the thermodynamics of the reactions of the halogens with the carbon-hydrogen bond. The chemistry itself, which involves free radicals, the homolysis of bonds to form those radicals, and the homogenic formation of bonds, is substantially less important than the heterolysis of bonds and the heterogenic formation of bonds encountered for the remainder of the text. However, prior to studying the chemistry of functional groups, it is useful to apply principles that you have learned in earlier chapters to the study of the mechanism of a reaction of the simplest of compounds—the alkanes.

Keys to the Chapter

5.1 Reactivity of Saturated Hydrocarbons

This section informs you of the obvious fact that, because they lack functional groups, alkanes and cycloalkanes are relatively unreactive. Note that the examples cited in the chapter concern alkanes, but the reactions and the interpretation of those reactions apply equally as well to cycloalkanes. The chemistry in both cases is that of σ bonds.

5.2 Bond Dissociation Energies

The term **bond dissociation energy** describes the energy required (an endothermic process) to break a bond and form two atomic or molecular fragments, each with one electron of the original shared pair. Thus, a very stable bond has a large bond dissociation energy—more energy must be added to cleave the bond. A high bond dissociation energy means that the bond (and molecule) is of low energy and is stable.

The bond dissociation energies of the carbon-hydrogen bonds of alkanes depend on the degree of substitution of the carbon atom. Although the values may differ slightly from compound to compound, the general trend is a decrease in the bond dissociation energy in the order primary > secondary > tertiary. In other words it takes less energy to homolytically cleave a carbon-hydrogen bond in compounds with more alkyl substituents. This order is explained based on the electron donating characteristics of alkyl groups, which partially make up for the electron deficiency of the carbon atom of the radical. This characteristic of alkyl groups will be cited time and again for intermediates that are even more electron deficient—the carbocations. This isn't the first time that you have encountered the effect of alkyl groups in stabilizing intermediates. It was first presented in Section 3.13 of the text.

5.3 Combustion of Alkanes and Cycloalkanes

Although commercially important, the combustion of alkanes and cycloalkanes is not a central feature of organic chemistry presented in this text. Many other classes of hydrocarbons burn as well. However, the heat of combustion is used to determine the relative stability of isomeric hydrocarbons. Recall that the heats of formation are most often negative, and they indicate the relative stability of hydrocarbons to the common reference point of the constituent elements.

The values of $\Delta H°_{comb}$ are all negative, and the reference point is the proper number of moles of carbon dioxide and water. The "heat of combustion" is often cited as a positive number, which means that the term refers to $-\Delta H°_{comb}$. In any case, the size of the term indicates how stable the hydrocarbon is relative to its isomers. As shown in the Figure below, formation of isomer B releases more energy (has a larger $\Delta H°_f$) than formation of isomer A, so B is more stable than A and has a lower enthalpy content. Thus, when isomers A and B undergo combustion, B releases less energy (has a lower $\Delta H°_{comb}$) than A in forming the same products. Study this Figure carefully to understand the meaning of both heat of formation and heat of combustion.

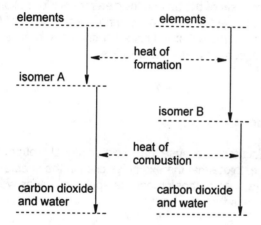

As we determined before when using $\Delta H°_f$ for isomeric hydrocarbons, the $\Delta H°_{comb}$ values establish that branched hydrocarbons are more stable than normal alkanes. Furthermore, the heats of combustion also verify the conclusions regarding the stability of cycloalkanes. For example, cyclopropane is strained and thus is a high energy compound. When it is burned, that strain energy is released and the heat of combustion is larger than predicted based on the number of methylene units in the compound.

5.4 Halogenation of Saturated Hydrocarbons

All alkanes and cycloalkanes can be halogenated to form substituted products. Unfortunately, the reaction doesn't stop after one hydrogen atom is replaced. Moreover the products don't "wait" until each reactant molecule reacts with the halogen. Once formed each product, which also has carbon-hydrogen bonds, can react further with the halogen.

The predominance of a single product in preference to an alternate product is termed **regioselectivity**. In the case of halogenation, there is little regioselectivity in the substitution reaction. Bromination is a more selective reaction than chlorination and leads to larger quantities of the more substituted bromoalkane.

The discussion in this section concerning relative rate of reaction per carbon-hydrogen bond and statistical factors might be difficult to follow at first, but the concept is fundamental to an understanding of the chemistry of molecules that may react at each of several sites. It may help if you consider the analogy of the output of a product by a factory. If all that you see is what comes out the export door, you can't tell whether the number of product units is the result of lots of workers in the factory or whether a small number of workers are extremely productive. The total output is equal to the number of workers times their productivity. Thus, if you want to understand something about the output of two competing factories, you need information about both the relative number of workers and their work habits. In calculating the amount of product that is formed by reaction at one type of carbon-hydrogen bond, you have to know how many bonds of that type there are and the reactivity of that type of bond.

Using the method of statistically correcting for the amount of product formed by reaction at each different type of carbon-hydrogen bond, the reactivity of the various classes of carbon-hydrogen bonds is determined. After dividing the percent of a given isomer by the number of equivalent hydrogen atoms that can give rise to that isomer, the relative reactivity is determined. This order of reactivity is related to the energy required to homolytically cleave a carbon-hydrogen bond. The order of reactivity is tertiary > secondary > primary. The scale of this reactivity variation is larger for bromination than for chlorination. In other word, the bromination reaction is more selective in which hydrogen atoms are replaced.

You can estimate the percent of each isomer formed by halogenation of a compound by first identifying each set of equivalent hydrogen atoms. Then multiply that number by the reactivity of that type of hydrogen atom. This quantity is the relative "output" of that product. Do the same thing for each type of product. To calculate the percentage of a single product, divide the "output" of that product by the total "output", which is obtained by summing the quantities for each isomer produced. Your instructor may expect you to know specific average values for the relative reactivity of the various classes of C—H bonds. However, these values vary somewhat with reaction conditions. Thus, you should also be able to calculate the composition of a reaction mixture given any set of values.

5.5 Enthalpy of Reaction for Halogenation

The overall enthalpy change for a free radical halogenation reaction is the sum of the ΔH_{rxn} values for the two propagation steps. One propagation step forms an alkyl radical and a hydrogen halide as the result of homolytic cleavage of a carbon-hydrogen bond. The second propagation step forms a halogen atom and an alkyl halide as the result of homolytic cleavage of a halogen-halogen bond. Thus, understanding the differences in the reactions of the halogens depends on comparing the bond energies of four types of bonds.

In the "first" propagation step listed, the ΔH_{rxn} depends on the difference in the bond energies of the carbon-hydrogen bond and the hydrogen-halogen bond. In comparisons made between various halogens, the differences in the ΔH_{rxn} lie with the strength of the hydrogen-halogen bond. The ΔH_{rxn} of the "second" propagation step listed depends on the bond energies of both the halogen-halogen bond and the carbon-halogen bond. Because both of these values depend on the halogen, a comparison between halogens is not as easy to make as for the first step.

The "first" propagation step becomes progressively less favorable for the halogens fluorine through iodine. In fact only the fluorination step is exothermic. The reactions of the remaining halogens are progressively more endothermic. The "second" propagation step also becomes progressively less favorable for the halogens fluorine through iodine, but the reaction is exothermic for all four halogens. For the overall reaction obtained by summing the two propagation steps, all except iodination are exothermic. Examining this data, it is concluded that the order of reactivity of halogens is related primarily to the ΔH_{rxn} of the "first" propagation step. This order is related to the bond strength of the hydrogen-halogen bond.

5.6 Activation Energy for Halogenation

(A review of the Hammond postulate in Section 3.17 may be useful in preparation for this section.) The activation energy and the structure of the transition state for the halogenation of an alkane are related to the ΔH_{rxn} of the "first" propagation step. The strongly exothermic reaction for abstraction of a hydrogen atom by a fluorine atom results in a transition state which has a low activation energy and which more closely resembles the reactants. In short, the reaction occurs by an "early" transition state in which the carbon-hydrogen bond is not substantially broken and the hydrogen-fluorine bond is not highly developed. As a result, the reactivity of all types of carbon atoms is similar. Only when the carbon-hydrogen bond is substantially broken and the resulting transition state has developed radical character does the degree of substitution of that carbon atom matter. In chlorination, and even more so in bromination, there is increased radical character in the transition state, which occurs progressively "later" along the reaction coordinate. Thus, the increased selectivity in bromination is the result of forming a transition state which looks more like a free radical, which is the product.

Summary of Reactions

1. Free Radical Bromination

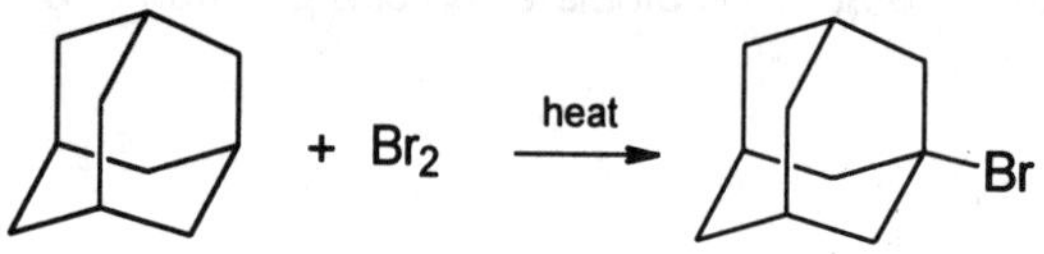

2. Free Radical Chlorination

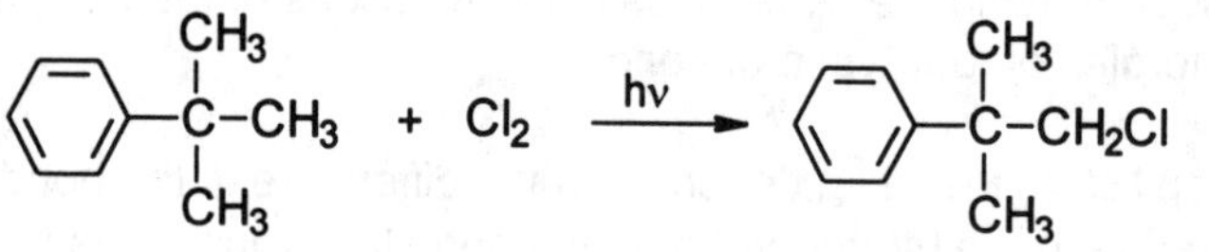

Solutions to Exercises

5.1 (a) 401 kJ mole^{-1} for this tertiary atom
(b) 401 kJ mole^{-1} for this tertiary atom
(c) 410 kJ mole^{-1} for this secondary atom
(d) 422 kJ mole^{-1} for this primary atom

5.2 (a) 401 kJ mole^{-1} for this tertiary atom
(b) 410 kJ mole^{-1} for this secondary atom
(c) 410 kJ mole^{-1} for this secondary atom
(d) 422 kJ mole^{-1} for this primary atom

5.3 (a) $2\ C_8H_{18}\ +\ 25\ O_2\ \longrightarrow\ 16\ CO_2\ +\ 18\ H_2O$

(b) $C_8H_{16}\ +\ 12\ O_2\ \longrightarrow\ 8\ CO_2\ +\ 8\ H_2O$

(c) $C_8H_{16}\ +\ 12\ O_2\ \longrightarrow\ 8\ CO_2\ +\ 8\ H_2O$

(d) $2\ C_8H_{18}\ +\ 25\ O_2\ \longrightarrow\ 16\ CO_2\ +\ 18\ H_2O$

5.4 (a) $C_8H_{16}\ +\ 12\ O_2\ \longrightarrow\ 8\ CO_2\ +\ 8\ H_2O$

(b) $C_8H_{16}\ +\ 12\ O_2\ \longrightarrow\ 8\ CO_2\ +\ 8\ H_2O$

(c) $C_8H_{16}\ +\ 12\ O_2\ \longrightarrow\ 8\ CO_2\ +\ 8\ H_2O$

(d) $C_8H_{16}\ +\ 12\ O_2\ \longrightarrow\ 8\ CO_2\ +\ 8\ H_2O$

5.5 (a) The three compounds in the group are isomeric with the molecular formula C_7H_{16}.
The most stable isomer is 2,4-dimethylpentane, because it is the most branched.
The most stable isomer has the lowest heat of combustion.

(b) The compounds are not isomeric. 2-Methylpentane has the lowest molecular weight
and has the lowest heat of combustion.

(c) All compounds in the group are isomeric with the molecular formula C_6H_{12}. The most
stable isomer is methylcyclopentane, because it is the least strained. The most
stable isomer has the lowest heat of combustion.

5.6 (a) Compound II is the least stable isomer, because its rings have the highest strain
energy. The least stable isomer has the highest heat of combustion.

(b) The compounds are isomeric. II is the least stable isomer, because it has the largest steric hindrance between the two four-membered rings fused to the center ring. Thus, II has the highest heat of combustion.

(c) The compounds are isomeric. II is the least stable isomer, because its methyl group is located in an axial position with respect to the six-membered ring. Thus, I has the highest heat of combustion.

5.7   **(a)** three; 1-chloropentane, 2-chloropentane, and 3-chloropentane
       (b) three; 1-chlorohexane, 2-chlorohexane, and 3-chlorohexane
       (c) four; 1-chloro-2-methylbutane, 2-chloro-2-methylbutane; 2-chloro-3-methylbutane, and 1-chloro-3-methylbutane
       (d) four; 1-chloro-3-methylpentane, 2-chloro-3-methylpentane; 3-chloro-3-methylpentane, and 3-(chloromethyl)-pentane
       (e) one; chlorocyclohexane
       (f) four; 1-(chloromethyl)-1-methylcyclohexane; 2-chloro-1,1-dimethylcyclohexane, 3-chloro-1,1-dimethylcyclohexane, and 4-chloro-1,1-dimethylcyclohexane

5.8   **(a)** three; 1-fluoro-2,2-dimethylbutane, 3-fluoro-2,2-dimethylbutane, and 1-fluoro-3,3-dimethylbutane
       (b) four; 1-fluoro-2,2-dimethylpentane, 3-fluoro-2,2-dimethylpentane, 4-fluoro-2,2-dimethylpentane and 5-fluoro-2,2-dimethylpentane
       (c) two; 1-fluoro-2,3-dimethylbutane and 2-fluoro-2,3-dimethylbutane
       (d) three; 1-fluoro-2,4-dimethylpentane, 2-fluoro-2,4-dimethylpentane, and 3-fluoro-2,4-dimethylpentane
       (e) one; fluorocyclopentane
       (f) six; (fluoromethyl)cyclopentane, 1-fluoro-1-methylcyclopentane, the cis and trans isomers of 1-fluoro-2-methylcyclopentane, and the cis and trans isomers of 1-fluoro-3-methylcyclopentane

5.9

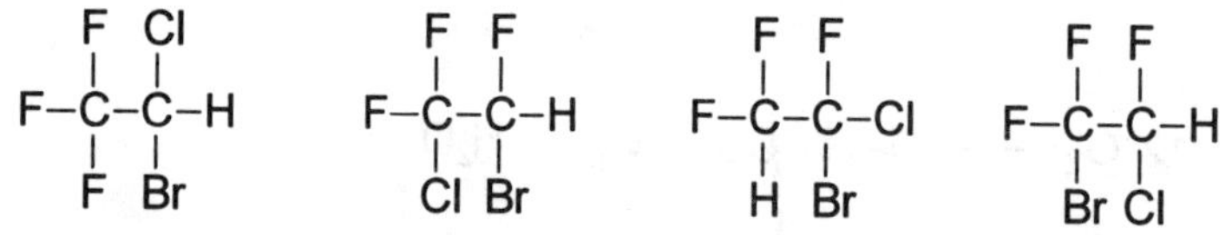

5.10   1,2,2-trichloro-1,1-difluoropropane and 1,2,3-trichloro-1,1-difluoropropane

5.11

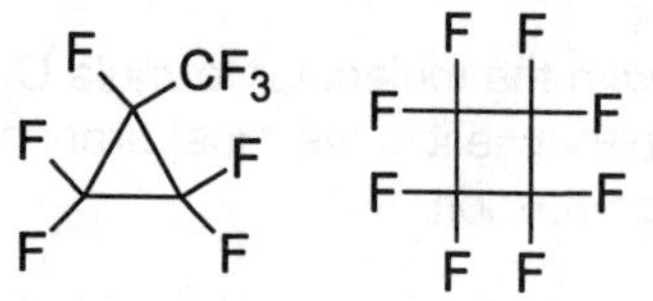

5.12   2-Chloropropane gives two products; 1-bromo-2-chloropropane and 2-bromo-2-chloropropane.

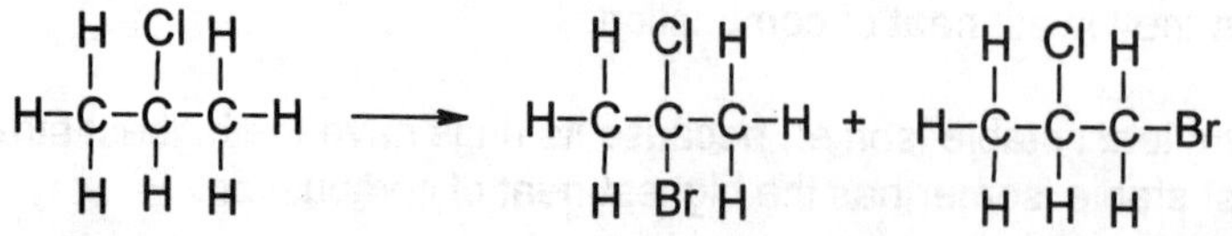

1-Chloropropane gives three products; 1-bromo-1-chloropropane, 2-bromo-1-chloropropane, and 1-bromo-3-chloropropane.

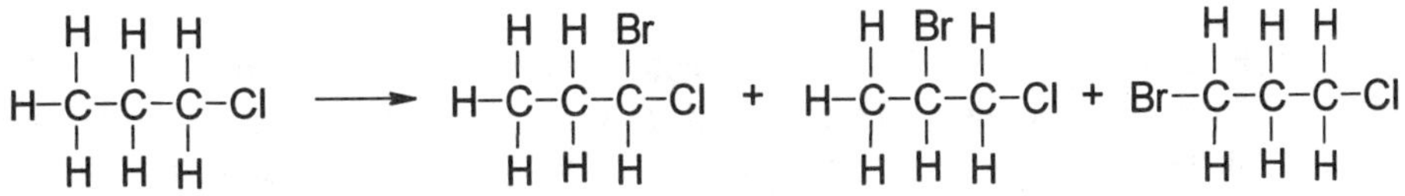

$$
\underset{\substack{| \ | \ |\\H \ H \ H}}{\overset{\substack{H \ H \ H\\| \ | \ |}}{H-C-C-C-Cl}} \longrightarrow
\underset{\substack{| \ | \ |\\H \ H \ H}}{\overset{\substack{H \ H \ Br\\| \ | \ |}}{H-C-C-C-Cl}} +
\underset{\substack{| \ | \ |\\H \ H \ H}}{\overset{\substack{H \ Br \ H\\| \ | \ |}}{H-C-C-C-Cl}} +
\underset{\substack{| \ | \ |\\H \ H \ H}}{\overset{\substack{H \ H \ H\\| \ | \ |}}{Br-C-C-C-Cl}}
$$

5.13 $Cl_3C-\ddot{Br}: \longrightarrow Cl_3C\cdot + \cdot\ddot{Br}:$

$R-H + Cl_3C\cdot \longrightarrow R\cdot + Cl_3C-H$

$R\cdot + Cl_3C-\ddot{Br}: \longrightarrow R-\ddot{Br}: + Cl_3C\cdot$

5.14 $(CH_3)_3C-\ddot{O}-\ddot{Cl}: \longrightarrow (CH_3)_3C-\ddot{O}\cdot + \cdot\ddot{Cl}:$

$R-H + (CH_3)_3C-\ddot{O}\cdot \longrightarrow R\cdot + (CH_3)_3C-\ddot{O}-H$

$R\cdot + (CH_3)_3C-\ddot{O}-\ddot{Cl}: \longrightarrow R-\ddot{Cl}: + (CH_3)_3C-\ddot{O}\cdot$

5.15 $(CH_3)_3C-\ddot{O}-\ddot{O}-C(CH_3)_3 \longrightarrow 2\ (CH_3)_3C-\ddot{O}\cdot$

5.16 $(CH_3CH_2)_3Pb-CH_2CH_3 \longrightarrow (CH_3CH_2)_3Pb\cdot + CH_3CH_2\cdot$

$CH_3CH_2\cdot + :\ddot{Cl}-\ddot{Cl}: \longrightarrow CH_3CH_2-\ddot{Cl}: + \cdot\ddot{Cl}:$

The lower bond energy of the initiating tetraethyllead means that the bond can be cleaved at a lower temperature.

5.17   The ratio of chloroethane to the mixture of deuterium containing isomers relative to chlorethane is 7/93. This ratio reflects both the reactivity of the C—D bond relative to the C—H bond and the ratio of the number of deuterium atoms that may be replaced relative to the number of hydrogen atoms that may be replaced. Chloroethane results from replacement of one deuterium atom. The mixture of deuterium containing isomers results from replacement of any of five primary hydrogen atoms. Thus, the reactivity of a C—D bond relative to the C—H bond is 0.38, which is obtained by solving the expression on the following page.

$$\frac{CH_3CH_2Cl}{CH_3CHDCl + ClCH_2CH_2D} = \frac{7}{93} = \frac{(1 \text{ C—D bond}) \text{ (reactivity of one C—D bond)}}{(5 \text{ C—H bond}) \text{ (reactivity of one C—H bond)}}$$

5.18   Using 438 kJ mole^{-1} for the C—H bond dissociation energy of methane, the $\Delta H°_{rxn}$ is (438 - 497) = -59 kJ mole^{-1}. The OH radical should be more selective than the fluorine radical but less selective than the chlorine radical, because oxygen is less electronegative than fluorine. (See Section 5.6 for $\Delta H°_{rxn}$ for these reactions.)

5.19   Taking into account a statistical factor of 9 primary hydrogen atoms compared to 1 tertiary hydrogen atom, the relative reactivity of a tertiary hydrogen atom compared to a primary hydrogen atom is 1.5, as obtained by solving the following expression.

$$\frac{\text{2-fluoro-2-methylpropane}}{\text{1-fluoro-2-methylpropane}} = \frac{1}{6} = \frac{(\text{one 3° H atom}) \text{ (reactivity of 3° H atoms)}}{(\text{nine 1° H atoms}) \text{ (reactivity of 1° H atoms)}}$$

5.20   Taking into account a statistical factor of 6 primary hydrogen atoms compared to 4 secondary hydrogen atoms, the reactivity of a secondary hydrogen atom relative to a primary hydrogen atom is 1.20, as obtained by solving the following expression.

$$\frac{\text{2-fluorobutane}}{\text{1-fluorobutane}} = \frac{1}{1.25} = \frac{(\text{four 2° H atoms}) \text{ (reactivity of 2° H atoms)}}{(\text{six 1° H atoms}) \text{ (reactivity of 1° H atoms)}}$$

5.21   Divide the decimal equivalent of the percent of each product by the number of hydrogen atoms that can lead to that product. The quotients should stand in order of the relative reactivity of each type of hydrogen atom. Divide each quotient by the smallest quotient to compare to the reactivities cited in the text.

For 2-chloro-3-methylbutane $\dfrac{0.36}{\text{two 2° H atoms}} = 0.18$ 0.18/0.045 = 4.0

For 1-chloro-2-methylbutane $\dfrac{0.27}{\text{six 1° H atoms}} = 0.045$ 0.045/0.045 = 1

For 2-chloro-2-methylbutane $\dfrac{0.23}{\text{one 3° H atom}} = 0.23$ 0.23/0.045 = 5.2

For 1-chloro-3-methylbutane $\dfrac{0.14}{\text{three 1° H atoms}} = 0.047$ 0.047/0.045 = 1

5.22   First calculate the ratio of the two products formed. Then calculate the percent by dividing the relative amount of each product obtained in the ratio by the sum of the two quantities and multiplying by 100% as shown on the following page.

For the chlorination of butane, the products are 2-chlorobutane and 1-chlorobutane.

$$\frac{\text{2-chlorobutane}}{\text{1-chlorobutane}} = \frac{(4 \text{ secondary hydrogen atoms}) (2.5)}{(6 \text{ primary hydrogen atoms}) (1)} = \frac{10}{6}$$

$$\% \text{ 2-chlorobutane} = \frac{10}{16} \times 100\% = 63\% \quad \% \text{ 1-chlorobutane} = \frac{6}{16} \times 100\% = 37\%$$

For the chlorination of 2-methylpropane, the products are 2-chloro-2-methylpropane and 1-chloro-2-methylpropane.

$$\frac{\text{2-chloro-2-methylpropane}}{\text{1-chloro-2-methylpropane}} = \frac{(1 \text{ tertiary hydrogen}) (4)}{(9 \text{ primary hydrogen atoms}) (1)} = \frac{4}{9}$$

$$\% \text{ 2-chloro-2-methylpropane} = \frac{4}{13} \times 100\% = 31\%$$

$$\% \text{ 1-chloro-2-methylpropane} = \frac{9}{13} \times 100\% = 69\%$$

5.23 Fluorination of methane forms a methyl radical as an intermediate whereas fluorination of ethane forms an ethyl radical. The ethyl radical is more stable than the methyl radical because it has a methyl group bonded to the radical center whereas the methyl radical has only hydrogen atoms bonded to it.

The ratio of products depends on the number of sites for attack of the fluorine atom on each alkane and the relative reactivity of those hydrogen atoms.

$$\frac{\text{chloroethane}}{\text{chloromethane}} = \frac{(6 \text{ hydrogen atoms}) (1)}{(4 \text{ hydrogen atoms}) (0.5)} = \frac{6}{2}$$

$$\% \text{ chloroethane} = \frac{6}{8} \times 100\% = 75\% \quad \% \text{ chloromethane} = \frac{2}{8} \times 100\% = 25\%$$

5.24 Bromination of propane should yield essentially only 2-bromopropane by reaction at a secondary carbon atom whose relative reactivity is 80. Bromination of 2-methylpropane should yield essentially only 2-bromo-2-methylpropane by reaction at a tertiary carbon atom whose relative reactivity is 1650. The relative amounts of the two products in a mixture of reactants must take into account the difference in the number of reactive sites in the two compounds.

$$\frac{\text{2-bromopropane}}{\text{2-bromo-2-methylpropane}} = \frac{(2 \text{ hydrogen atoms}) (80)}{(1 \text{ hydrogen atom}) (1650)} = \frac{160}{1650}$$

$$\% \text{ 2-bromopropane} = \frac{160}{1810} \times 100\% = 8.8\%$$

$$\% \text{ 2-bromo-2-methylpropane} = \frac{1650}{1810} \times 100\% = 91\%$$

5.25 Two of the products, 2-chloropentane and 3-chloropentane, are derived from reaction at secondary centers.   The 2-chloropentane should predominate over the 3-chloropentane by a factor of 2:1 because there are twice as many centers that can yield the 2-chloropentane. The third product is 1-chloropentane.

5.26 The remaining 86% of the products are 2-chloroheptane, 3-chloroheptane, and 4-chloroheptane. They can be formed by abstraction of 4, 4 and 2 secondary hydrogen atoms, respectively. Thus, the 86% of product must be distributed in the ratio of these quantities. The approximate percents are 34.4%, 34.4%, and 19.2%, respectively.

5.27 There are three equivalent methyl groups at one end of the molecule and two equivalent methyl groups at the other end. Thus, the products should be formed in a 3:2 ratio. The yields of the two products (I and II) are 33% and 22%, respectively.

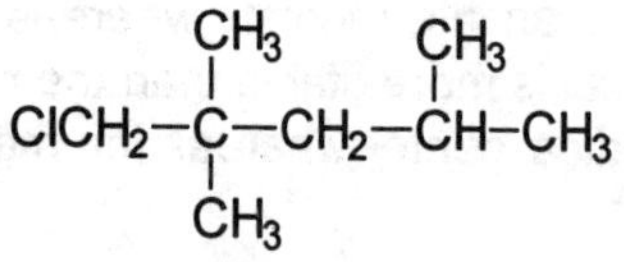

(I) 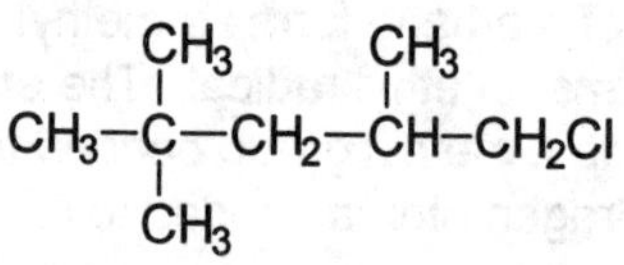   (II)

5.28 There are three equivalent methyl groups at one end of the molecule and two equivalent methyl groups at the other end. Thus, the products (I) and (II) should be formed in a 3:2 ratio.

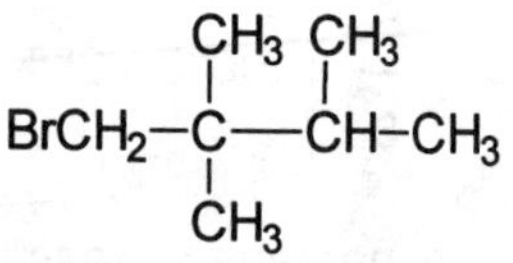

(I) 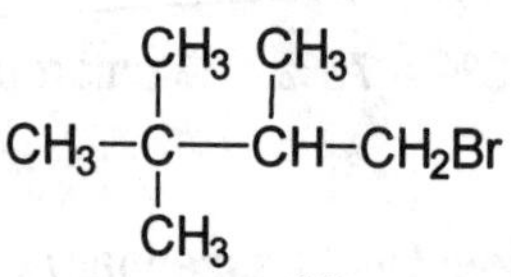   (II)

6

Alkenes—Structure and Properties

Recall from Section 2.1 that the only way to cope with the large numbers of organic compounds and their reactions is by classifying chemical structures based on specific groups of atoms and bonds that are called **functional groups**. This chapter is the first of many dealing with the unique properties of classes of organic compounds characterized by a functional group. As you encounter each new class of functional group, you should should learn what it is that makes the group unique and as a consequence why it behaves as it does. There are no shortcuts. First you have to learn facts much like you had to learn your multiplication tables in your younger years. However, chemistry is more than memorization. Eventually you have to learn how to use the facts and generalize from the specific cases cited to the more general cases of many other molecules sharing those same characteristics. The facts are many and you are going to have to start to put things together in ways that decrease the burden of stuffing more material into your head. For example, in another subject area you might learn that an eagle has wings and an eagle can fly. You might learn this information as two separate facts but of course the two facts are related. At a later time you learn that a pigeon has wings. Of course a pigeon is not an eagle, but there are similarities and you wouldn't be tripped up by a question asking if a pigeon can fly. Sounds a bit ridiculous, but that is exactly the process that is required to succeed in organic chemistry. So keep your focus on the central ideas and learn to generalize. But first, learn the facts which you absolutely have to know. Don't wait a day longer. If you do, you will be swamped and your boat will sink and your dreams of a career along with it.

Keys to the Chapter

6.1 Unsaturated Hydrocarbons

You should review Sections 1.17 and 1.18 to make sure that you understand the models for the formation of double and triple bonds and the difference between **sigma** and **pi** bonds in unsaturated compounds. A double bond in an **alkene** consists of one sigma bond and one pi bond. A triple bond in an **alkyne** consists of one sigma bond and two pi bonds.

On the average, the electrons in pi bonds are more accessible for attack—the subject of Chapter 7. Virtually all of the reactions of alkenes and alkynes (Chapter 11) involve the reactivity of pi bonds. Two or more double bonds separated by one carbon-carbon single bond constitute **conjugated** double bonds. A compound may contain one set or a whole series of conjugated double bonds. Double bonds separated by more than one carbon-carbon single bond are not conjugated. The chemistry of conjugated double bonds is a more specialized subject than reactions of alkenes, and it is discussed in Chapter 12.

6.2 Structure and Classification of Alkenes

Alkenes consist of a sigma bonded framework, which for the most part we can ignore, and two electrons in a π bond that results from the sideways overlap of two 2p orbitals. As a consequence of the π bond, the two carbon atoms and the four atoms bonded to these atoms are in a plane.

The introduction of the concept of the % s character of the hybrid orbitals of an alkene is the first of many times thoughout the study of organic chemistry that you will encounter this terminology. There are chemical consequences to the "tightness" with which electrons are held by a carbon atom in its bonds to other elements. In this section, you find that the % s character affects the bond energy of a σ bond. In order to apply the concept, you have to be careful to understand what is required in the chemical reaction—for example, does the process involve homolytic or heterolytic cleavage of the bond? In the case of a bond dissociation, a homolytic cleavage occurs and it is necessary for each electron in the bond to go its separate way with the two atoms originally bonded. If the electrons are held more tightly by an atom as a consequence of its larger % s character, then it will take more energy to reclaim the electron so that it separate with its own atom. That is why the bond dissociation energy increases with increased % s character.

Bond distances also depend on % s character. If electrons are held more tightly by the carbon atom as a result of its hybridization, then the bond length must be shorter.

Time and time again, we will learn that there are advantages not only to classifying compounds by their functional groups, but also fine tuning the classification by subclasses. Fortunately, the terminology of **degree of substitution** is pretty self-evident. You just count the number of bonds from the double-bonded carbon atoms that go to carbon atoms as opposed to hydrogen atoms. It isn't a big deal now, but you are going to have to recognize the degree of substitution in other contexts without restudying this subject. It is much like knowing that 5 x 7 is 35. Learn the facts now so that you can use them later without searching around for an explanation at points in the text where it assumed that you know the material.

6.3 Unsaturation Number

Before you can consider what functional groups might be part of a structure, you have to limit the possibilities based on the molecular formula. In the case of our biology analogy, recognizing that an animal has wings helps quite a bit. Of course, there are some pitfalls for we know that bats have wings but are not birds. So, in the case of looking at molecular formulas, it is necessary to recognize likely possibilities for functional groups but also to keep in mind that there are some ambiguities as well.

The presence of double or triple bonds in a structure is signaled by the **degree of unsaturation**. Each multiple bond diminishes the maximum number of hydrogen atoms by two. Thus, you really don't have to use the formulas given in this section. Just think 2n+2 and then substract the actual number of hydrogen atoms in the molecular formula from this reference maximum value. Each two "missing" hydrogen atoms can corresponds to a pi bond. However, it may also signal the presence of a ring. Based on this one criterion—like the wings of birds versus bats—you can't say which structural feature is present. But at least you have limited the possibilities. The effect of other atoms on the unsaturation number is summed up as follows:

1. Oxygen and sulfur have no effect.
2. Just count halogen atoms as hydrogen atoms.
3. Add a hydrogen atom to the maximum reference value for each nitrogen atom.

6.4 Geometric Isomerism

As a consequence of the restricted rotation about a carbon-carbon double bond, alkenes can exist as **geometric isomers** if there are two different groups bonded to each of the two carbon atoms of the double bond. If either carbon atom has two identical groups, then only one compound is possible. In later studies of reactions forming carbon-carbon double bonds, both isomers may form or perhaps one will be formed preferentially. However, if you are asked to write the products of a reaction you will always have to keep in mind that geometric isomers may exist. Writing a single structure without recognizing that in fact two compounds exist may mean that you only get half credit for a question on an exam.

Cycloalkenes usually have a cis configuration. You should get out your molecular model kit and convince yourself that bridging a trans configuration in a cycloalkene requires a large number of atoms due to geometric constraints.

6.5 The E,Z Designation of Geometric Isomers

This section deals with the first of many sets of apparently arbitrary rules to give names to compounds. Actually they are arbitrary, but you can't be an individualist and march to your own drummer—at least if you want to earn a reasonable grade. The rules in chemistry aren't any different from those of a language. If the rules aren't applied properly, then we can't communicate effectively and may convey incorrect information.

You probably won't have to be an expert in applying the **sequence rules** and have to decide between "close calls". Most chemistry professors are pleased if you can at least handle the more obvious groups of atoms. After all, there is a whole lot more to chemistry than just naming compounds. However, listen carefully to how much emphasis is placed on the subject by your professor.

The rules are reasonably logical. The most important criterion for deciding on the priority of the groups is atomic number. Furthermore, just one atom of high atomic number wins out over any number of atoms with lower atomic numbers. Thus, —CH_2Br has a higher priority than —CCl_3, because bromine has a higher atomic number than chlorine. Another criterion is the "point of first difference", which means keep going until something important is found in a chain of atoms that makes it more important than a second group of atoms. Thus, the group —$CH_2CH_2CHFCH_2CH_3$ has a higher priority than —$CH_2CH_2CH_2CHBrCH_3$ because the fluorine atom is found sooner on the chain than the bromine atom.

The only part of the rules for E,Z nomenclature that is often not well understood is the treatment of multiple bonds. However, for an atom containing a multiple bond just add as many "phantom" atoms as the number of multiple bonds. So, a C=O counts as a carbon containing two oxygen atoms—one that is really there and a second phantom oxygen atom.

6.6 Nomenclature of Alkenes

This is the second time that you have encountered a lengthy list of rules to name compounds. The first was in Section 4.3 dealing with alkanes. Some of the rules are the same, and others are closely related to those used for alkanes. The double bond takes priority in numbering the longest chain, which must contain the double bond. Only the first number of the two involving the carbon atoms of the double bond is used to name a compound. In cycloalkenes, one of the two carbon atoms of the double bond is given the number 1 and the numbering of the carbon atoms then proceeds through the second carbon atom of the double bond. The name does not include a number locating the double bond, but number are required for the location of substituents such as alkyl groups.

6.7 Physical Properties of Alkenes

As we have observed for alkanes, trends in physical properties are related to the size and structure of the compounds studied. Boiling points for a homologous series of alkenes increase with molecular weight, reflecting the increase in strength of London forces. The identity and geometry of the groups attached to the double-bonded carbon atoms determine whether an alkene is polar or nonpolar. If one isomer is polar and the other is nonpolar, the polar compound is expected to have a higher boiling point.

6.8 Oxidation of Alkenes

The **heats of combustion** of alkenes are used to compare the relative stability of isomeric compounds. For isomers, the same number of moles of carbon dioxide and water are formed. Thus, a comparison of the heats of combustion indicates the difference in the enthalpy content of the isomers. Three generalizations can be made based on the data depicted in Figure 6.4 in the text. These are:

1.   Branched isomers are more stable, so they release less energy when burned.

2.   More highly substituted alkenes are more stable, so they release less energy when burned.

3.   Alkenes with the E configuration are more stable than alkenes with the Z configuration, so they release less energy when burned.

The increased stability of alkenes with increased substitution can be attributed to the release of electron density from the sp^3 hybridized alkyl groups toward the sp^2 hybridized atoms of the carbon-carbon double bond. The electron donating capacity of alkyl groups toward sp^2 hybridized centers is a common feature that "explains" many chemical reactions that you will encounter in later chapters. Thus, it is one of the important facts that you will need to remember when interpreting chemical reactivity.

6.9 Reduction of Alkenes

The reaction of an alkene with hydrogen to give an alkane, called **hydrogenation**, is a reduction reaction. The order of reactivity of alkenes decreases with increased subsitution of the double bond. Thus, the reaction shows some **regioselectivity**, which means that one double bond in a compound with two or more double bonds can often be reduced in preference over another double bond. This phenonomenon of regioselectivity, which is the tendency of a reaction to generate one isomer preferentially over another, is another concept that you wil encounter time and again in your study of organic chemistry.

Transition metal catalysts such as platinum, palladium, **Adams catalyst** (PtO_2), and a special form of nickel called **Raney nickel** can be used. Although the hydrogenation reaction is usually carried out under heterogeneous conditions, one catalyst known as the **Wilkinson catalyst** is used for hydrogenation of alkenes under homogeneous conditions. Neither heterogeneous nor homogeneous catalysts reduce functional groups such as the carbonyl group of ketones, carboxylic acids, or esters under the relatively mild conditions required to hydrogenate a carbon-carbon double bond.

6.10 Mechanism of Catalytic Hydrogenation

Heterogeneous catalytic hydrogenation occurs on the surface of the metal and transfers the two hydrogen atoms to the carbon atoms by a **syn addition**. For many alkenes, there is no difference in the two faces of the double bond, so there is no **stereoselectivity** in the reaction. However, keep your eyes on the environment near the double bond which may make one face less accessible to the transfer of hydrogen. Note that when the hydrogen adds to one face, the atoms bonded to the carbon atoms of the double bond are "pushed" to the opposite side. Because hydrogen adds from the sterically less hindered side, the groups are forced into a more sterically hindered environment.

6.11 Heats of Hydrogenation

The measurement of the heats of hydrogenation ($\Delta H° < 0$) of isomeric alkenes is used to determine the difference in the ΔH_f values of the compounds and hence their relative stabilities. The values of $\Delta H°_{hydrog}$ are all negative, and the reference point is the saturated hydrocarbon. The size of the term indicates how stable the alkene is relative to its isomers. As shown in the Figure below, formation of alkene B releases more energy (has a larger $\Delta H°_f$) than formation of isomeric alkene A, so B is more stable than A and has a lower enthalpy content. Thus, when isomers A and B undergo combustion, B releases less energy (has a lower $\Delta H°_{comb}$) than A in forming the same products. Study this Figure carefully to understand the meaning of both heat of formation and heat of hydrogenation.

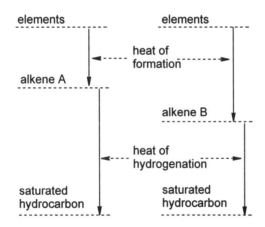

Note that for low molecular weight alkenes the heats of formation are negative meaning that the line for the alkenes would be above that of the elements. However, the conclusions with respect to stability, heats of formation, and heat of hydrogenation are unchanged.

Remember that the heats of hydrogenation can be used to compare the relative stabilities of isomeric alkenes only if the same alkane results from both compounds. For example, isomers with different degrees of branching cannot be compared because they don't produce the same saturated hydrocarbon when hydrogenated.

It is often assumed that the differences in the heats of hydrogenation reflect the differences in the ΔG_f values for the compounds because the S° values are often similar. Based on representative data given in Table 6.3 of the text the following generalizations are made:

1. Branched isomers are more stable and therefore the ΔH_f is smaller.
2. More highly substituted alkenes are more stable and therefore the ΔH_f is smaller.
3. Alkenes with the E configuration are more stable and therefore the ΔH_f is smaller.

Summary of Reactions

1. Heterogenous Catalytic Hydrogenation

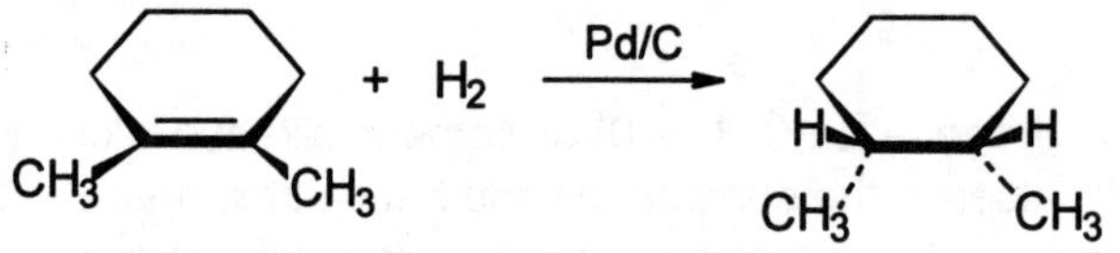

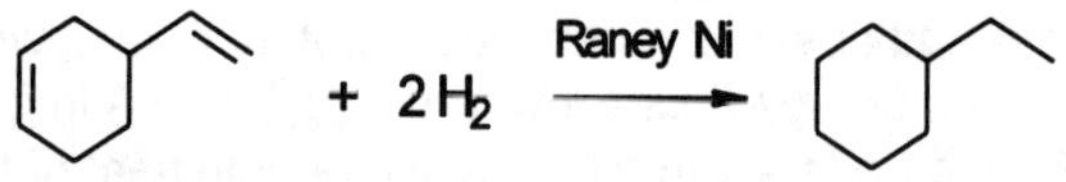

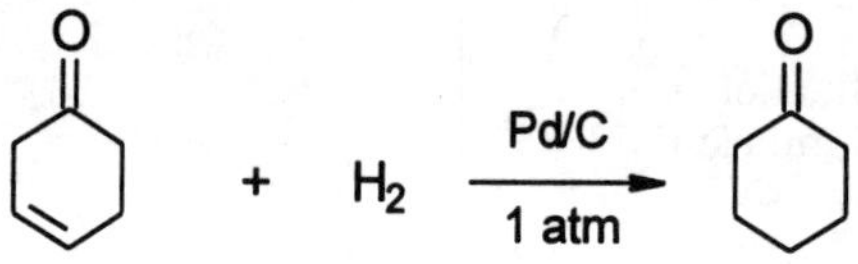

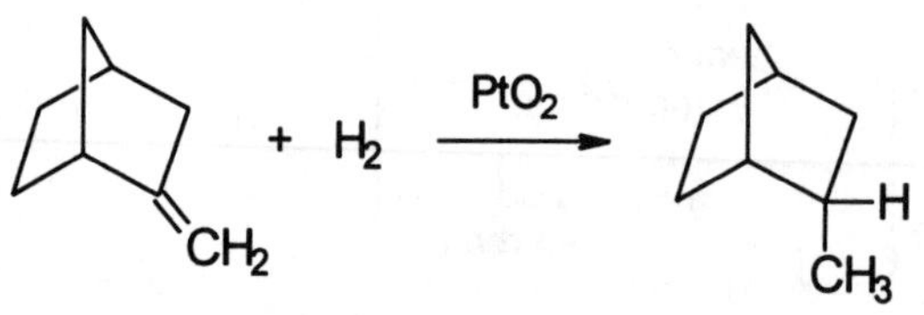

2. Homogeneous Catalytic Hydrogenation

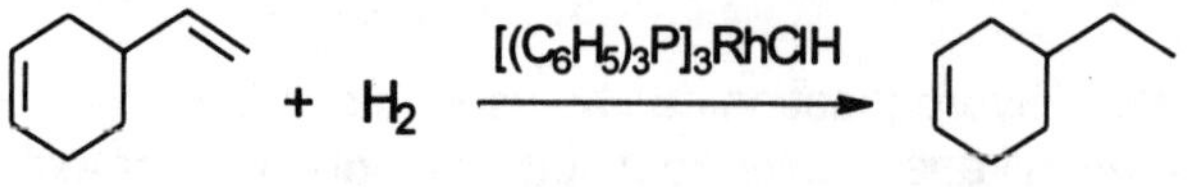

Solutions to Exercises

6.1   **(a)** C_6H_{12} **(b)** C_5H_8 **(c)** C_7H_{12}

6.2   **(a)** C_4H_6 **(b)** $C_{10}H_{18}$ **(c)** $C_{10}H_8$

6.3   **(a)** $C_{10}H_{18}$ **(b)** C_6H_{12} **(c)** $C_{10}H_{16}$ **(d)** C_8H_{12}

6.4   **(a)** C_8H_{12} **(b)** $C_{10}H_{16}$ **(c)** C_7H_{10} **(d)** C_7H_{14}

6.5   **(a)** trisubstituted   **(b)** trisubstituted
           (c) trisubstituted   **(d)** di- and trisubstituted

6.6   **(a)** di- and disubstituted   **(b)** di- and trisubstituted

           (c) disubstituted   **(d)** trisubstituted

6.7   **(a)** trisubstituted   **(b)** disubstituted

           (c) monosubstituted   **(d)** tetrasubstituted

6.8   **(a)** di-, tetra-, and trisubstituted **(b**) di- and disubstituted

           (c) di- and trisubstituted

6.9   **(a)** 3   **(b)** 5   **(c)** 5   **(d)** 8

6.10   **(a)** 13   **(b)** 4   **(c)** 5   **(d)** 6

6.11   **(a)** 6   **(b)** 2   **(c)** 10   **(d)** 6

6.12   **(a)** 5   **(b)** 9   **(c)** 6   **(d)** 8

6.13   only **(a)** and **(c)**

6.14   only **(c)**

6.15   only **(b)** and **(c)**

6.16   only **(d)**

6.17   **(a)** $-CHClCH_3$ **(b)** $-CH_2C\equiv CH$ **(c)** $-OCH_3$

6.18   **(a)** 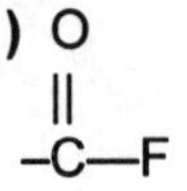
O
‖
–C—F

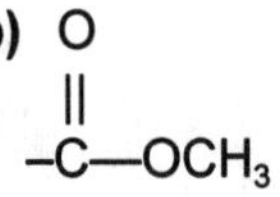

(b)
O
‖
–C—OCH₃

(c)
O
‖
–C—Cl

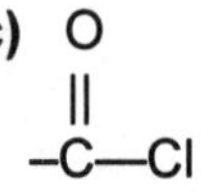

6.19   **(a)** Z   **(b)** Z   **(c)** Z

6.20   **(a)** E   **(b)** Z   **(c)** E

6.21   **(a)**

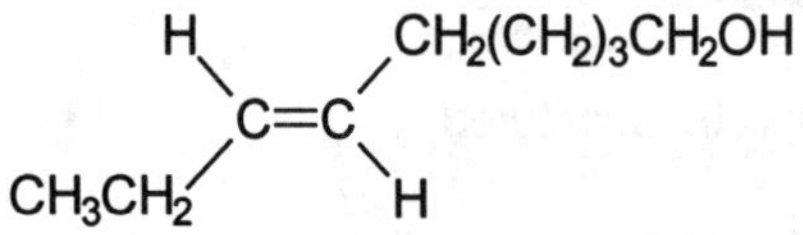

(b)

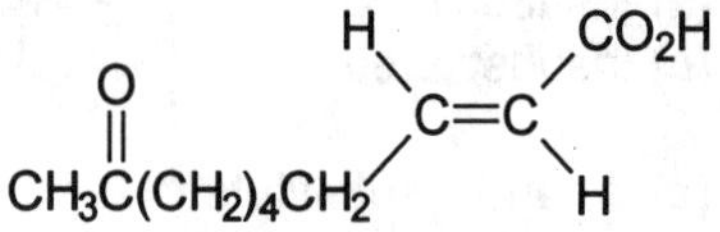

(c)

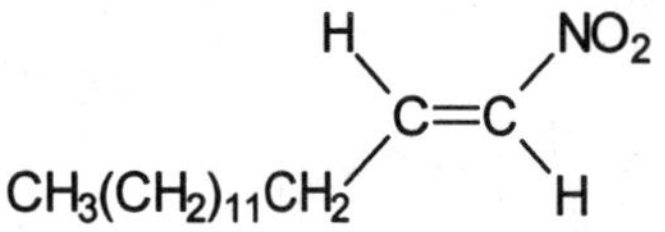

6.22   **(a)** left to right; Z,E   **(b)** left to right; E, Z   **(c)** Z

6.23   **(a)** 2-methyl-1-propene   **(b)** 2,3-dimethyl-2-butene

(c) 2-methyl-2-butene   **(d)** (E)-2,3-dichloro-2-pentene

6.24   **(a)** (E)-1,2-dibromo-1-propene   **(b)** 1-bromo-3-methyl-2-butene

(c) 3-chloro-2-methyl-2-pentene **(d)** (Z)-3,4-difluoro-3-hexene

6.25   **(a)** 1,2-dimethylcyclohexene   **(b)** 1,3-dimethylcylclohexene

(c) cyclodecene   **(d)** 6-ethyl-1-methylcyclohexene

6.26   **(a)** 1-ethylcyclohexene   **(b)** 1,5-dimethylcyclopentene

(c) cyclooctene   **(d)** 1,6-diethylcyclohexene

6.27   **(a)** 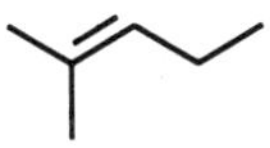   **(b)**

(c) 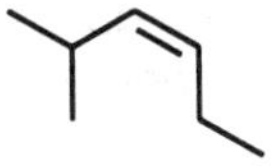   **(d)**

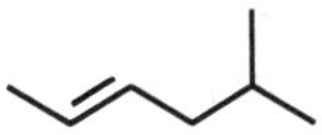

6.28
(a) 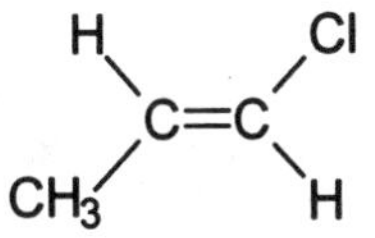 $\underset{CH_3}{\overset{H}{>}}C=C\underset{H}{\overset{Cl}{<}}$ **(b)** $\underset{CH_3}{\overset{Cl}{>}}C=C\underset{CH_3}{\overset{Cl}{<}}$

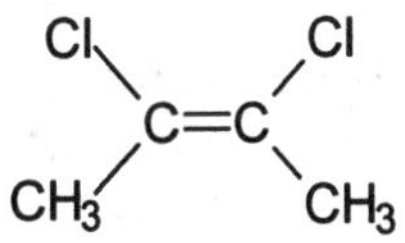

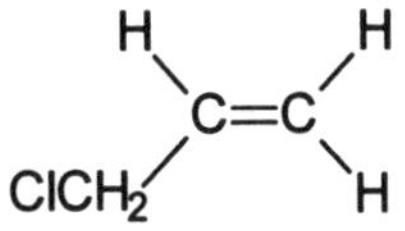

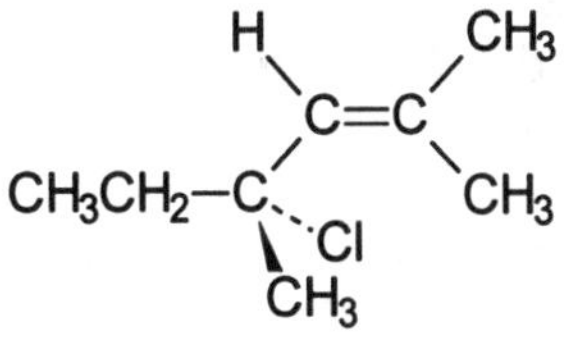

(c) $\underset{ClCH_2}{\overset{H}{>}}C=C\underset{H}{\overset{H}{<}}$ **(d)** $\underset{CH_3CH_2-C\overset{\cdots Cl}{\underset{CH_3}{}}}{\overset{H}{>}}C=C\underset{CH_3}{\overset{CH_3}{<}}$

6.29   **(a)**    **(b)** 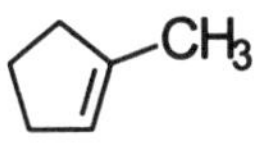 —CH₃   **(c)** 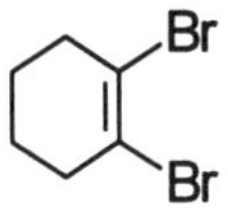 Br ... Br   **(d)** 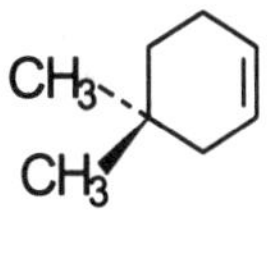 CH₃ ... CH₃

6.30   **(a)**    **(b)** 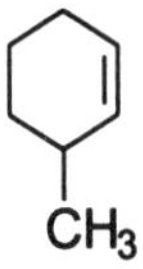 CH₃   **(c)** 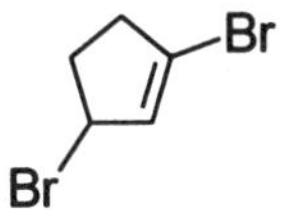 Br ... Br   **(d)**  Cl Cl

6.31   The carbon-carbon double bond of 1-hexene is not symmetrically substituted and the single alkyl group is electron donating to the sp^2-hybridized carbon atom.

6.32   The bond moments of the methyl groups bonded to the sp^2-hybridized carbon atoms of *cis*-2-butene reinforce one another and there is a net dipole moment. In the trans isomer the two bond moments are opposed, so they cancel and there is no dipole moment.

6.33   The two bond moments of the methyl groups bonded to the sp^2-hybridized carbon atom are additive in 2-methylpropene and are larger than the bond moment of the single ethyl group of 1-butene.

6.34   It should be larger than 1.4 D but less than 2.8 D because there will be some partial cancellation of the component of the bond moment along the carbon-carbon bond axis. There will be reinforcement of the component of the bond moment perpendicular to the bond axis.

6.35   The bromine atom is more polarizable and the resulting London forces are stronger than those resulting from the chlorine atom.

6.36   2,3-Dimethyl-2-butene has a small compact shape which may allow close approach of molecules and result in increased London forces. The conformational flexibility of 1-hexene results in a larger effective volume.

6.37   There is substantial van der Waals repulsion of a methyl group and a *tert*-butyl group in *cis*-4,4-dimethyl-2-pentene which makes this isomer much less stable than the trans isomer. The van der Waals repulsion between the two methyl groups in *cis*-2-butene is significantly smaller.

6.38   There is a very large van der Waals repulsion of two *tert*-butyl groups in the cis isomer.

6.39   Compound II has the larger angle strain resulting from the carbon-carbon double bond in the four-membered ring. The compound is less stable (higher energy) and releases more energy in a combustion reaction.

6.40   Both sp^2-hybridized carbon atoms of 1-methylcyclopropene are part of the three-membered ring, giving rise to larger angle strain than the single sp^2-hybridized carbon atom of methylenecyclopropane.

6.41   2-methyl-2-butene < 2-methyl-1-butene < 3-methyl-1-butene, which is the order of decreasing degree of alkyl substitution of the double bond.

6.42   I < III < II, which is the order of decreasing degree of alkyl substitution of the double bond.

6.43   **(a)** 2   **(b)** 2   **(c)** 2   **(d)** 2

6.44   **(a)** 6   **(b)** 3   **(c)** 3

6.45   There are two double bonds because two moles of H_2 are added to the molecular formula. There is one ring because the unsaturation number of the product is 1.

6.46   There are three double bonds because three moles of H_2 are added to the molecular formula. There are no rings because the unsaturation number of the product is 0.

6.47 **(a)** There is no difference in the heats of hydrogenation because the degree of substitution of the double bonds is identical.

(b) The heat of hydrogenation of methylenecyclohexane should be smaller because it is disubstituted whereas the isomeric vinylcyclopentane is monosubstituted.

(c) There is no difference in the heats of hydrogenation because the degree of substitution of the double bonds is identical.

(d) The cyclopropene compound should have the larger heat of hydrogenation because the strain of the double bond in the smaller ring makes the compound less stable.

6.48 1-Methylcyclopentene has the smallest heat of hydrogenation because it has a trisubstituted double bond. The 3-methyl and 4-methyl compounds have disubstituted double bonds.

6.49 I < III < II, which is the order of decreasing degree of substitution of the double bond.

6.50 The compound with a double bond in a six-membered ring is more stable. The compound with a double bond in the four-membered ring is more strained and of higher energy than the isomer, so hydrogenation of the cyclobutene will release a larger amount of energy.

6.51 Both compounds yield the same saturated compound when hydrogenated. Since 1-ethylcyclohexene predominates in an equilibrium mixture, it is more stable. Thus, ethylidenecyclohexane has a larger heat of hydrogenation.

6.52 The degree of substitution of the double bonds is not the same.

6.53 The double bond is present in "two" strained cyclobutene rings, which increases the energy of the compound and makes it unstable.

6.54 There is an increase in the degree of substitution from reactant to product which makes both reactions favorable. However, the four-membered ring compound has more strain in the product as the result of having two sp^2-hybridized carbon atoms in the ring. Thus, the reaction is less favorable.

6.55 The van der Waals repulsion between the methyl and *tert*-butyl groups in (Z)-4,4-dimethyl-2-pentene is much greater than between two methyl groups is (Z)-2-butene, so the difference between the (E) and (Z) isomers is greater also.

6.56 There is a very large van der Waals repulsion of two *tert*-butyl groups in the cis isomer which makes the compound far less stable.

6.57

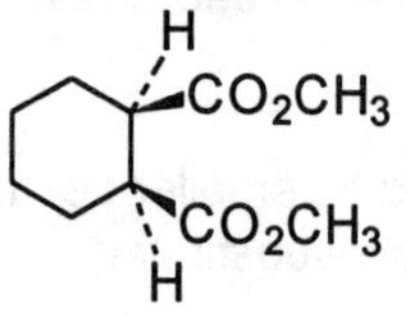

6.58 $CH_3(CH_2)_{10}CH_2-O-\overset{\overset{\displaystyle O}{\|}}{C}-CH_3$

6.59

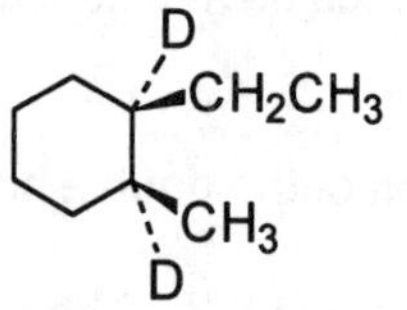

6.60   The second product, with the methyl group and the cyclopropane ring on the same side of the six-membered ring, because hydrogen is added from the sterically less hindered side, which is the face opposite the cyclopropane ring.

6.61   The disubstituted double bond will be hydrogenated somewhat faster than the trisubstituted double bond, but selective hydrogenation is not likely.

6.62   Approach of hydrogen from the "top" of compound I occurs much more easily than for compound II, where the *tert*-butyl group sterically hinders this face of the double bond.

6.63   There are two trisubstituted double bonds and a tetrasubstituted double bond. All should react slowly in a hydrogenation reaction and there should be little selectivity.

6.64   The less substituted double bond is reduced in each case.

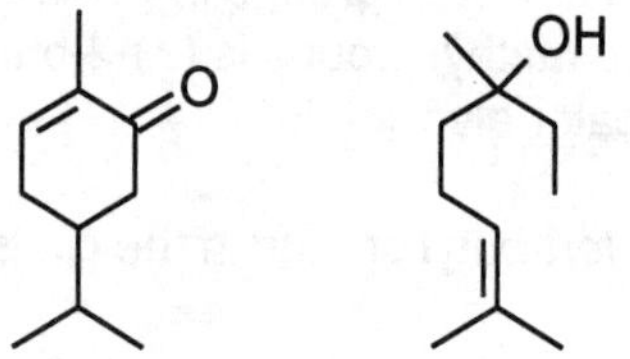

7

Addition Reactions of Alkenes

In this chapter we really commence the study of organic reactions and their mechanisms. From this point on you will have to concentrate on the details of not just a single reaction, as in case of Chapter 5 dealing with the halogenation of alkanes, but on several possible reactions of each type of functional group. It may help to start your collection of flash cards with a "question" on one side and the "answer" on the other side. You need to know what each reactant does and on what compounds it acts.

The reactivity of alkenes is related to the π bond, which on the average has electrons that are larger distances from the associated carbon atoms. Furthermore, the π electrons are more open to attack by reagents than are the σ bonds, which are located in a more protected environment between atoms. The chemistry of the π bond is most easily understood using Lewis acid-base conventions. The bonding electron pair in the π bond can act as a Lewis base and is susceptible to attack by Lewis acids. In organic chemistry these Lewis acids are termed electrophiles.

As you will learn in this chapter there are a variety of electrophiles that can react with double bonds in addition reactions. The electrophile uses the π electrons to form a bond to the carbon atom, and as a result the π bond is "broken". There are minor variations in the way by which various electrophiles react with alkenes. However, the basic theme is essentially the same and as a result you won't have to memorize as many apparently independent facts if you understand the basic mechanism.

Keys to the Chapter

7.1 Characteristics of Addition Reactions

Addition reactions of alkenes result in the incorporation of two atoms or groups of atoms on adjacent carbon atoms which share the double bond. Reagents such as HBr and H_2O add a hydrogen atom and a second species, which in these cases are bromine and the hydroxyl group, respectively. Reagents such as Br_2 can provide two equivalent groups, in which case one bromine atoms adds to each of the two carbon atoms.

The addition reaction of a reagent X—Y to an alkene results in the destruction of one π bond and one σ bond, between two carbon atoms and between X and Y, respectively. However, two σ bonds form—one carbon atom bonds to X and the other carbon atom bonds to Y. Carbon-carbon π bonds are weaker than σ bonds. Thus, the addition reaction is exothermic, because the net result is the replacement of a weak π bond by a σ bond. The magnitude of $\Delta H°_{rxn}$ depends on the reagent, but all the common reagents listed in Table 7.1 give addition products in exothermic reactions.

Addition reactions are also characterized by a negative $\Delta S°_{rxn}$, because two moles of reactant are converted to one mole of product. Recalling that the $\Delta S°_{rxn}$ for such processes is about -125 J mole^{-1} K^{-1} and that the $-T\Delta S°_{rxn}$ term at room temperature is about 37 kJ mole^{-1} gives you some idea about the contributions of $\Delta H°_{rxn}$ and $\Delta S°_{rxn}$ to the $\Delta G°_{rxn}$. The addition of both hydrogen and bromine is strongly favored due to the $\Delta H°_{rxn}$ term. The addition of HBr is only slightly favored as a result of the $\Delta H°_{rxn}$ term, and the hydration reaction is not favorable using these approximate numbers. As a consequence the hydration reaction is reversible (dehydration is the opposite reaction), and the direction of the reaction depends on the specific alkene and the reaction conditions.

Different reactants add to the π in two different ways known as **anti** addition and **syn** addition. Thus, the geometry of the added groups in the final products gives information about the reaction pathway. This type of study is most easily done with cycloalkenes because the products are geometric isomers.

Although several types of reagents are discussed in this chapter, the most common theme is that of attack by electrophiles. As a consequence of the attack of an electrophile on the π electrons to form a bond to the carbon atom, the other carbon atom becomes electron deficient and has a positive charge. The subsequent chemistry of this carbocation is one of the common threads of addition reactions initiated by electrophiles.

7.2 Addition of Hydrogen Halides

The order of reactivity of hydrogen halides with alkenes is related to the acidity of the hydrogen halide. This is a consequence of the availability of the hydrogen ion of the hydrogen halide to serve as an electrophile. Subsequent reaction of the carbocation formed with the nucleophilic halide ion completes the mechanism of the process.

Addition of hydrogen halides is regiospecific, meaning that not only does the hydrogen halide specifically react with the π bond, but it does so to give only one of the two possible products. This result is summarized by Markovnikov's rule—HX compounds add to double bonds so that the hydrogen atom bonds to the carbon atom of the double bond containing the largest number of directly bonded hydrogen atoms. At this point, even without understanding why, you can predict the product of a reaction of a compound of the general formula HX to a double bond.

7.3 Mechanistic Basis of Markovnikov's Rule

The proposed mechanism "explains" Markovnikov's rule. A carbocation is formed by the addition of an electrophile to the π bond. Thus, the direction of the addition depends on the stability of the two possible carbocations that could form. The hydrogen atom adds to the carbon atom of the double bond containing the largest number of directly bonded hydrogen atoms, which means that it is the least substituted carbon atom. As a consequence, the carbocation formed must have the positive charge on the more substituted carbon atom. That in a nutshell is what addition of electrophiles to π bonds is all about. However, to be more precise we really have to consider that the reaction rates for the two possible ways that a hydrogen ion can add are controlled by the energy of the two possible transition states and not by the stability of the intermediate. As shown in Figure 7.3 in the text, the first step in the addition reaction is endothermic, and as a result the structure of the intermediate carbocations should resemble those of the two possible transition states. Thus we say that the transition state reflects the stability of the carbocations formed.

7.4 Rearrangement of Carbocations

Organic chemistry would be a simple subject if all of the atoms of a reactant stayed in place in the conversion to product in reactions such as the addition reaction. However, reactive intermediates such as carbocations react within themselves as well as with other reagents. Either a hydride ion or an alkyl group with a negative charge can move from a center adjacent to the carbocation center to form a new bond and generate a positive charge at that adjacent center. The driving force for such reactions is the formation of a more stable carbocation. Keep your eye open for such possibilities. Hydride ion shifts occur when a secondary or primary carbocation is generated adjacent to a tertiary center. Shifts of alkyl groups most commonly occur when a carbocation is generated at a carbon atom adjacent to a quaternary center. Any of the alkyl groups bonded to that quaternary carbon can migrate. Thus, a mixture of products can result. As shown in Exercise 7.9 in the text, even ring bonds of cycloalkanes can migrate, resulting in rings of different size than the reactant.

7.5 Hydration of Alkenes

The mechanism of the addition of water to a double bond is similar to that of the addition of hydrogen halides. A hydrogen ion first attacks the π bond, followed by capture of the carbocation by the nucleophilic oxygen atom. Note that it is water that reacts, not a hydroxide ion. Remember that the reaction occurs under acidic conditions, in which case the most available nucleophilic material is water. Subsequent loss of a proton from the oxonium ion gives a product which appears to have been the result of addition of a hydrogen ion and a hydroxide ion.

As noted in Section 7.1, the addition reaction is readily reversible because of balancing contributions of $\Delta H°_{rxn}$ and $\Delta S°_{rxn}$ to the $\Delta G°_{rxn}$. The dehydration reaction is much like viewing a film in reverse. The same things happen but in reverse sequence. This in a nutshell is what the principle of microscopic reversibility is all about.

7.6 Addition of Halogens

The addition of halogens such as chlorine or bromine to a double bond has some characteristics that indicate that the mechanism is a little different than for the addition of hydrogen halides. The reaction rate is sensitive to the same features of the alkenes. Thus, compounds that give more stable carbocations as evidenced by the data for hydration reactions are also the more reactive in addition of bromine. Thus, the carbon atom of the double bond bears some positive charge in the transition state. However, rearrangement products are not observed and the stereochemistry of the product of the reaction is the result of a net anti addition.

The data for addition of bromine (and chlorine) are explained by a cyclic bromonium ion (or chloronium ion) in which the positive charge is largely on the bromine atom but is distributed to some extent to the two carbon atoms. Of the two carbon atoms, the more substituted atom has the greater positive charge. The charge is insufficient to induce rearrangement reactions but sufficient to distinguish between the two sites for subsequent attack by a nucleophile. This feature of different charge distribution is shown by data on the formation of halohydrins, where water attacks the more substituted carbon atom because it has the higher positive charge of the two possible carbon atoms.

7.7 Addition of Carbenes

Carbenes (divalent carbon atoms) are electron deficient species and behave as electrophiles toward the π bond of alkenes. However, they are electrically neutral and there is no associated nucleophile. The reaction is a simultaneous formation of two bonds to give a cyclopropane ring. The **Simmons-Smith reagent** is used as a convenient way to form cyclopropane rings. Note that the stereochemistry of the groups of the alkene are unaltered in the product cyclopropane.

7.8 Epoxidation of Alkenes

The addition of an oxygen atom to a double bond results in a cyclic ether known as an **epoxide**. The chemistry of these compounds is discussed in Chapter 17. The reason for showing the reaction at this point is to suggest that there are mechanistic similarities between apparently different attacking reagents. Because oxygen is electronegative you would expect it seek electrons. Moreover, because electrons are "taken" from the π bond to form bonds to oxygen, the reaction rates for various structures reflect the availability of alkyl groups to stabilize the resulting electron deficiency at carbon in transition state.

A number of peroxide reagents are available for the formation of epoxides. In every case the reaction proceeds by a cyclic concerted process in which the stereochemistry of the groups bonded to the original carbon atoms of the double bond are unchanged in the product.

7.9 Dihydroxylation of Alkenes

This section presents two reagents that form **vicinal diols** (glycols) in their reactions with alkenes. Both reagents react by a concerted cyclic mechanism that simultaneously places one oxygen atom on each carbon atom of the original double bond. Subsequent reaction of the cyclic intermediate generates the two hydroxyl groups. The stereochemistry of the addition reaction is syn.

7.10 Ozonolysis of Alkenes

Alkenes react with ozone to give intermediates that subsequently react under work-up conditions to give two products that result from cleaving both bonds of the original carbon-carbon double bond. Those two carbon atoms are identified in the product by the location of oxygen atoms. Under reductive work-up conditions either aldehydes or ketones can result. Under oxidative work-up condition either carboxylic acids or ketones can result. You may be asked to identify an alkene based on the identity of the products. Just "erase" the oxygen atoms and join the carbon atoms of the two products with a double bond.

7.11 Free Radical Addition of Hydrogen Bromide

Radicals are electron deficient but not as much as are carbocations. Nevertheless, certain radicals can add to π bonds by a process in which the radical "takes" one of the two electrons to form a bond to carbon by a homogenic process. As a result the neighboring carbon becomes another radical. As illustrated in Chapter 5, radical reactions often occur by chain propagation steps. An example of this reaction is shown on page 279 of the text for the addition of HBr.

Because the focus of the original Markovnikov addition was the hydrogen atom, the reversed location of hydrogen in the free radical addition is termed anti-Markovnikov addition. Although the direction of the addition of HBr appears to be opposite that of the electrophilic addition of HBr, the reaction is still controlled by the stability of the intermediates formed. In the case of electrophilic addition, it is the hydrogen ion that adds to give the more stable carbocation. In the case of free radical addition, it is the bromine atom that adds to give the more stable radical.

7.12 Polymerization of Alkenes

Addition reactions of alkenes with certain reagents generate reactive intermediates that then react with the original alkene. Continuation of this process generates high molecular weight materials called **polymers**. The intermediates can be carbocations or radicals, and the reaction proceeds as described for electrophilic and free radical addition reactions of alkenes. In each case the appropriate initiating species adds to the double bond to give the most stable intermediate. Subsequent reactions of that intermediate with another molecule of an alkene also give the most stable of two possible intermediates. Addition reactions of alkenes initiated by carbanions are also known. Again, the direction of the addition is controlled by the stability of the intermediate.

Summary of Reactions

1. Electrophilic Addition of Hydrogen Halides

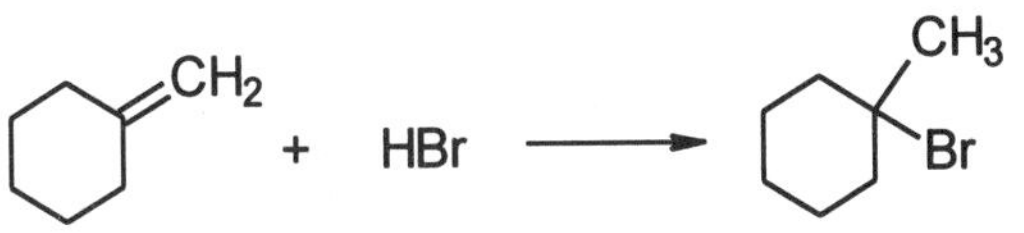

2. Addition of Water (Hydration)

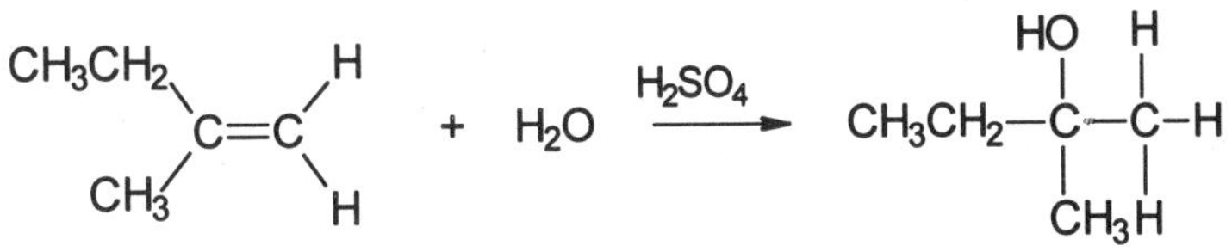

3. Addition of Bromine or Chlorine

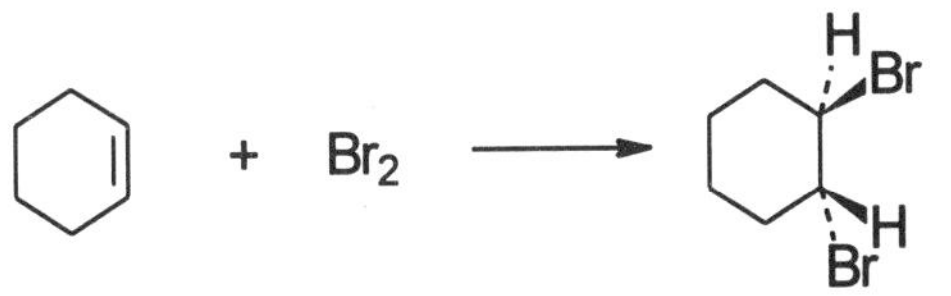

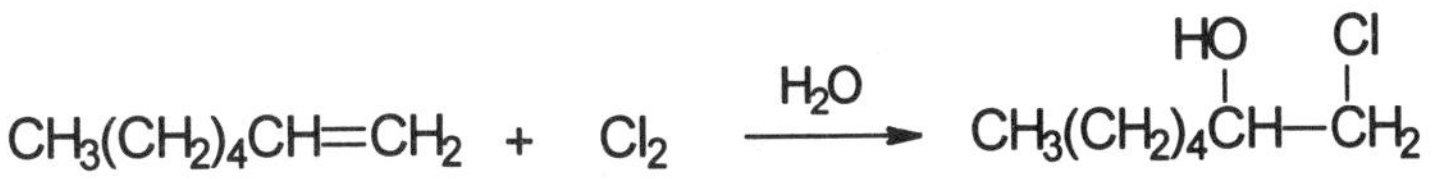

4. Addition of Carbenes

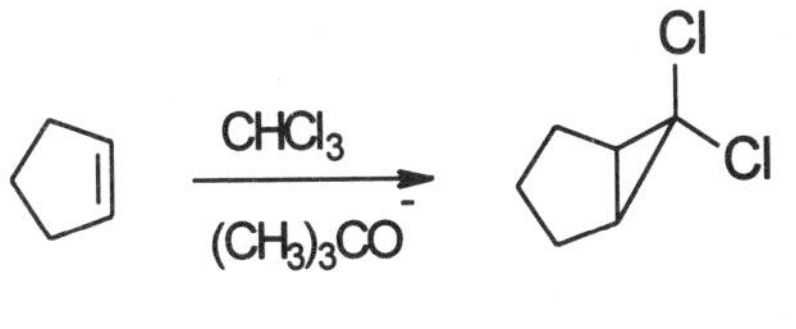

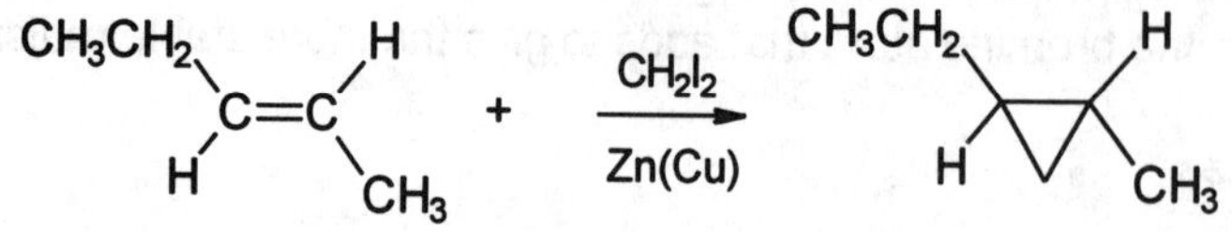

5. Epoxidation of Alkenes

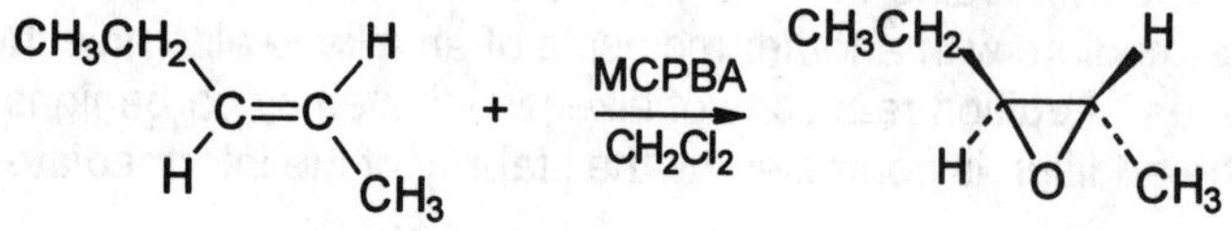

6. Dihydroxylation of Alkenes

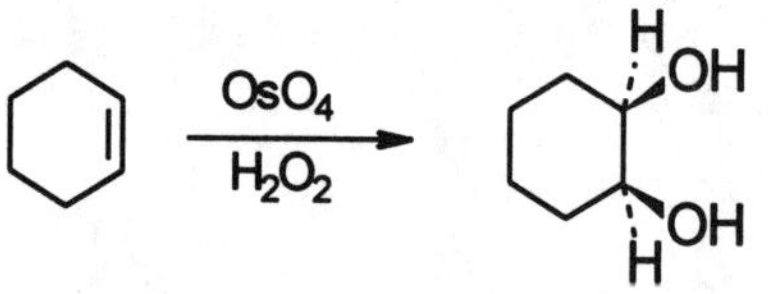

7. Ozonolysis of Alkenes

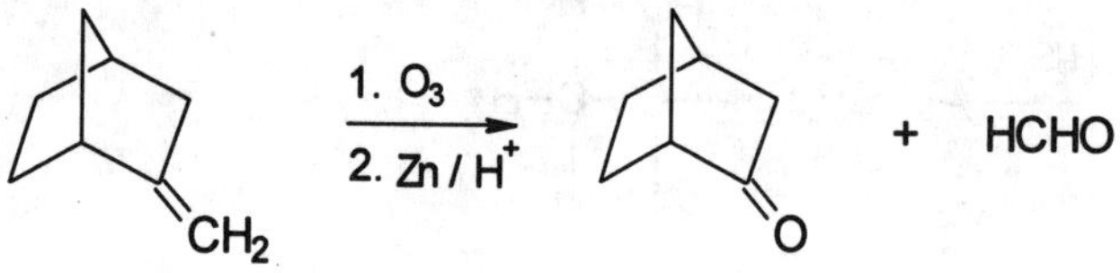

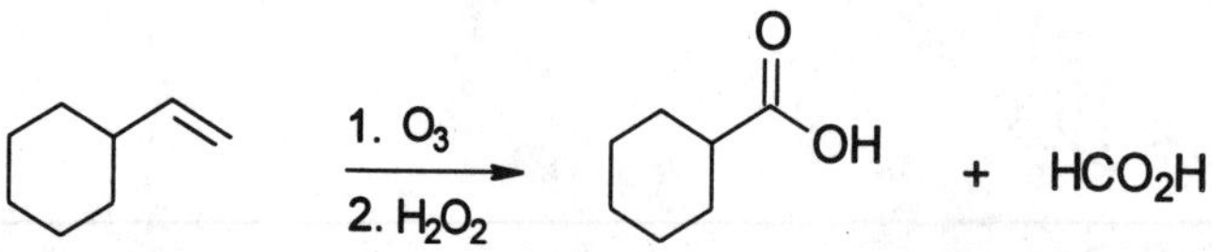

8. Free Radical Addition of Hydrogen Bromide

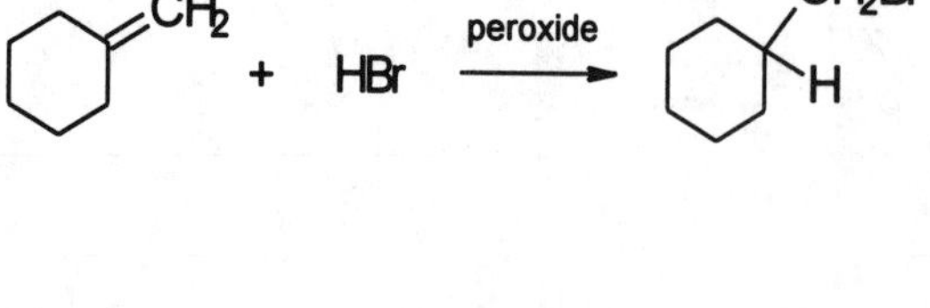

Solutions to Exercises

7.1 Using 240 kJ mole^{-1} as the bond dissociation energy of the π bond (leaving the sigma bond intact), a calculation of the the $\Delta H°$ for addition of IBr also requires the cleavage of an I—Br bond and the formation of C—Br and C—I bonds.

$\Delta H° = 240 + 279 - 289 - 216 = 14$ kJ mole^{-1}

7.2 Using 240 kJ mole^{-1} as the bond dissociation energy of the π bond bond (leaving the sigma bond intact), a calculation of the the $\Delta H°$ for addition of HF also requires the cleavage of an H—F bond and the formation of C—F and C—H bonds.

$\Delta H° = 240 + 567 - 465 - 422 = -80$ kJ mole^{-1}

7.3 The hydroxyl group at the C-1 atom and the hydrogen atom at the C-2 atom are cis as a result of a *syn* addition.

7.4 The hydroxyl groups at the C-1 and C-2 atoms cis as a result of a *syn* addition.

7.5 The heterolytic cleavage of the I—N bond of the IN$_3$ compound gives I$^+$ and N$_3^-$. Addition of the electrophile I$^+$ at C-1 of 1-pentene yields a secondary carbocation at C-2 which is captured by the nucleophile N$_3^-$. The structure of the product is

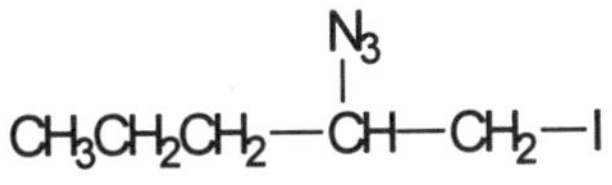

$$CH_3CH_2CH_2-\overset{\overset{\displaystyle N_3}{|}}{CH}-CH_2-I$$

7.6 The heterolytic cleavage of the I—N bond of the INCO compound gives I$^+$ and NCO$^-$. Addition of the electrophile I$^+$ at the double bond from face of the ring opposite the axial methyl group gives a cyclic iodonium ion. In order to achieve a net trans addition, the nucleophilic NCO$^-$ ion must attack at the indicated atom to open the ring of the cyclic iodonium ion via a trans diaxial arrangement. Attack at the other carbon atom would give the diequatorial isomer.

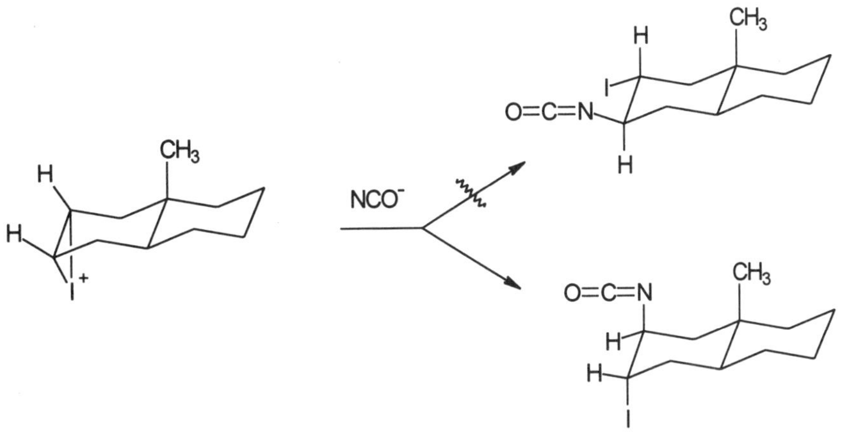

7.7   The heterolytic cleavage of the Se—Cl bond of the NCSeCl compound occurs to give the chloride ion because chlorine is more electronegative than selenium. The NCSe⁺ ion is an electrophile that attacks the C-1 atom to give a tertiary carbocation. In methanol as solvent, the carbocation is captured to give an ether. The structure of the product is

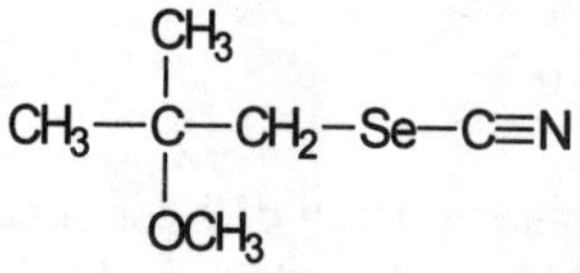

$$CH_3-\underset{\underset{OCH_3}{|}}{\overset{\overset{CH_3}{|}}{C}}-CH_2-Se-C\equiv N$$

7.8   The RSe⁺ ion is an electrophile that attacks the double bond to give a cyclic selenium ion which is then opened by nucleophilic attack of the oxygen atom of the carboxyl group. Note that the oxygen and selenium atoms are trans in the product.

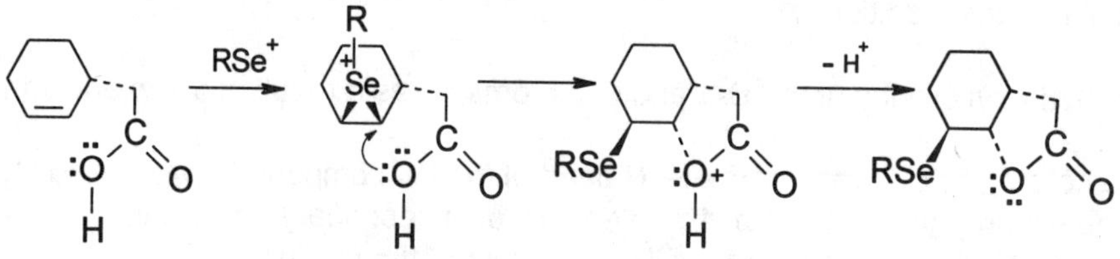

7.9   (a) 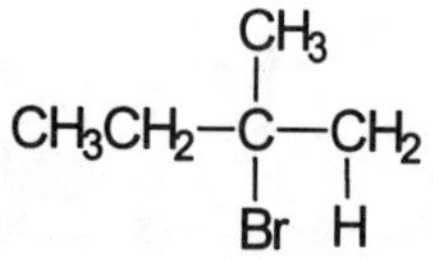 $CH_3CH_2-\underset{\underset{Br}{|}}{\overset{\overset{CH_3}{|}}{C}}-\overset{\overset{}{}}{\underset{\underset{H}{|}}{CH_2}}$ (b) $CH_3CH-\underset{\underset{H}{|}}{\overset{\overset{CH_3}{|}}{C}}-\underset{\underset{Br}{|}}{CH_3}$

(c) $CH_3-\underset{\underset{Br}{|}}{CH}-\underset{\underset{H}{|}}{CH}-CH_2CH_2CH_3$ and $CH_3-\underset{\underset{H}{|}}{CH}-\underset{\underset{Br}{|}}{CH}-CH_2CH_2CH_3$

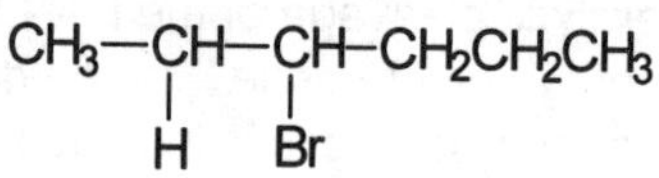

(d) 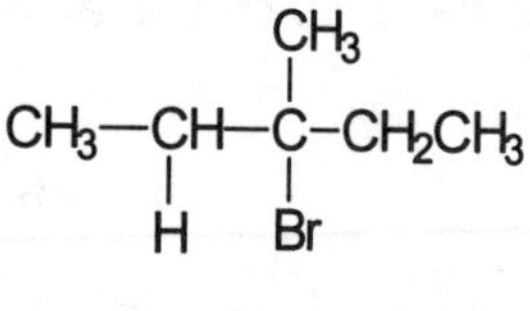 $CH_3-\underset{\underset{H}{|}}{CH}-\underset{\underset{Br}{|}}{\overset{\overset{CH_3}{|}}{C}}-CH_2CH_3$

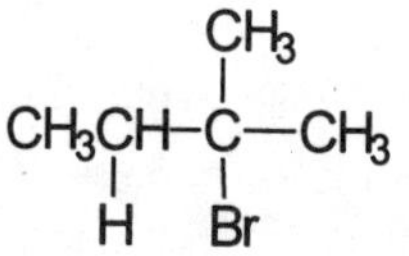

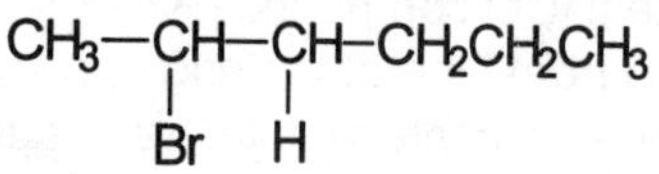

7.10   (a) 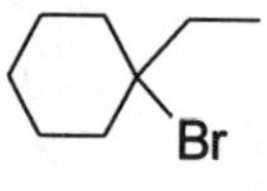   (b)

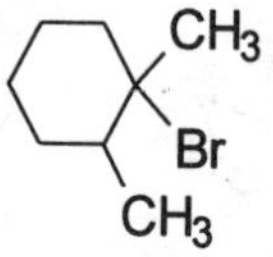

(c) 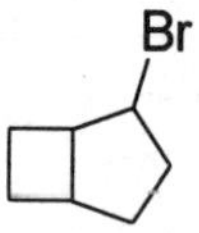   and 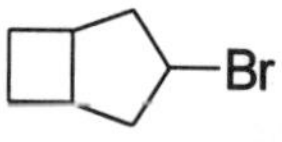   (d)

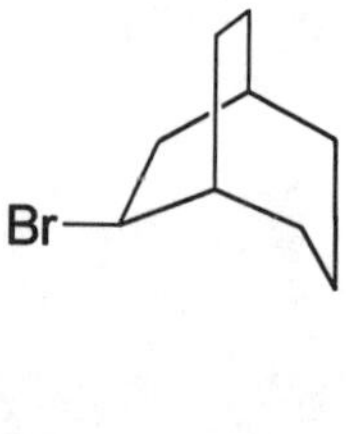

7.11

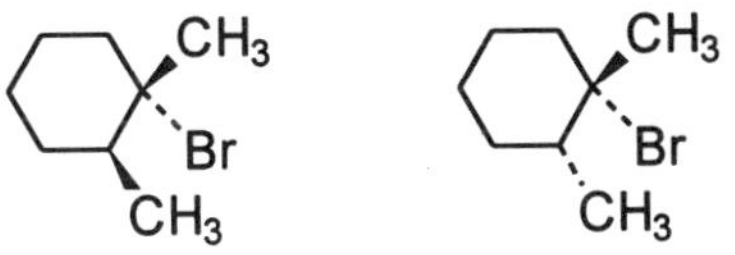

7.12

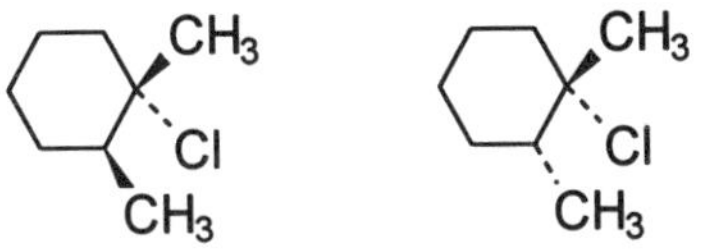

7.13 The addition of a proton to C-1 would give a secondary carbocation that is destabilized by the inductive electron withdrawal of the CF_3 group. Although addition of a proton to C-2 gives a primary carbocation, this intermediate is more stable that the alternative secondary carbocation.

7.14 Addition of a proton to either C-1 or C-2 would give a primary carbocation. In the case of addition to C-2 the positive charge may be delocalized. The lone pair electrons of chlorine may be shared with the carbon atom.

7.15 The amount of rearranged product is smaller in the case of HI. Thus, the iodide ion must capture the carbocation prior to its rearrangement more efficiently than does the chloride ion.

7.16 The major product is 2-bromo-3,3-dimethylbutane, which is expected from a normal addition reaction. The bromide ion can capture the carbocation prior to its rearrangement more efficiently than does the chloride ion but less efficiently than does the iodide ion. The minor product is 2-bromo-2,3-dimethylbutane, which results from rearrangement of a methyl group.

7.17 Assuming that no rearrangement occurs in the hydration reaction the addition products are:

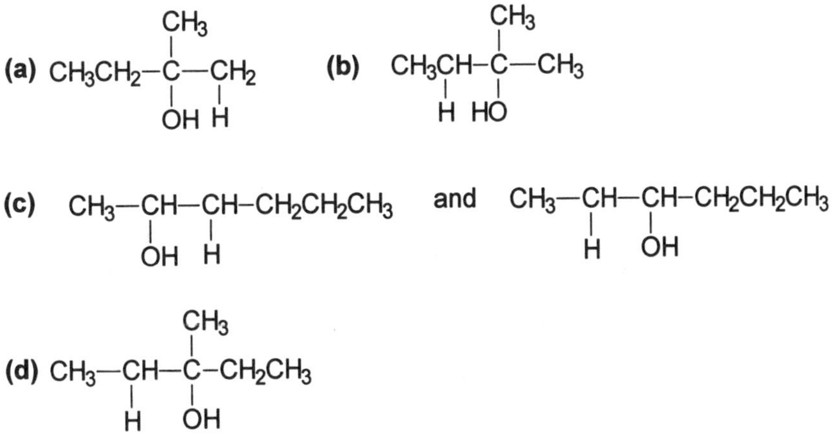

7.18   (a)    (b)

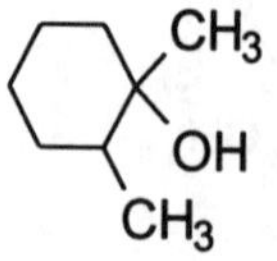

(c) 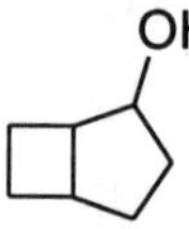 and 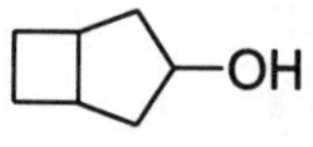   (d)

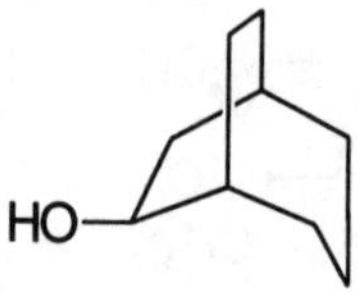

7.19   The same tertiary carbocation is formed from either of the two alkenes. The product is a tertiary alcohol.

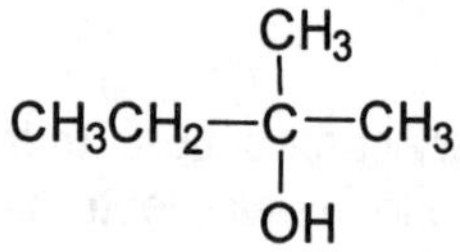

$$CH_3CH_2-\underset{\underset{OH}{|}}{\overset{\overset{CH_3}{|}}{C}}-CH_3$$

7.20   The structure of the tertiary carbocation is the same for both alkenes. However, the rate of the reaction depends on the difference in energy between the reactant and the transition state. The 2,3-dimethyl-2-butene is the more stable isomer because it has the more highly substituted double bond. Thus, the energy required to achieve the transition state is larger for this compound.

7.21   (a) 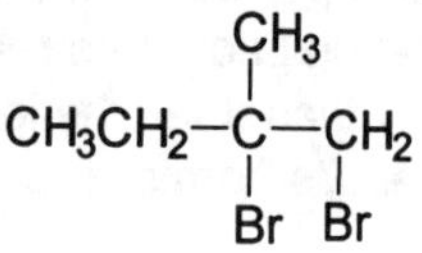 $CH_3CH_2-\underset{\underset{Br}{|}}{\overset{\overset{CH_3}{|}}{C}}-\underset{\underset{Br}{|}}{CH_2}$ (b) $CH_3\underset{\underset{Br}{|}}{CH}-\underset{\underset{Br}{|}}{\overset{\overset{CH_3}{|}}{C}}-CH_3$

(c) $CH_3-\underset{\underset{Br}{|}}{CH}-\underset{\underset{Br}{|}}{CH}-CH_2CH_2CH_3$ (d) 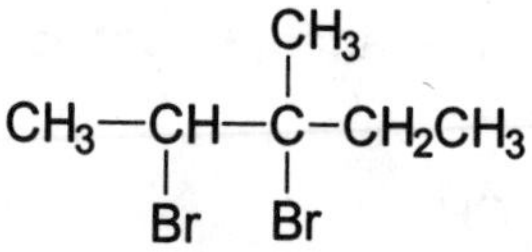 $CH_3-\underset{\underset{Br}{|}}{CH}-\underset{\underset{Br}{|}}{\overset{\overset{CH_3}{|}}{C}}-CH_2CH_3$

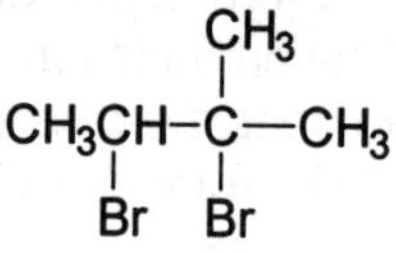

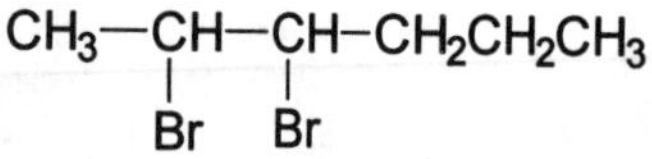

7.22   (a) 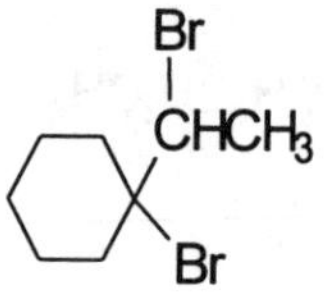   (b) 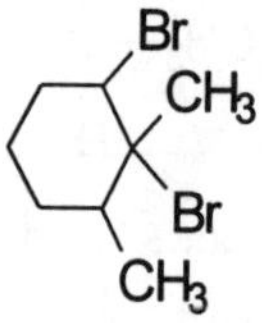   (c) 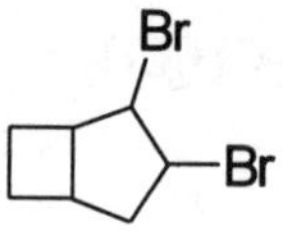   (d)

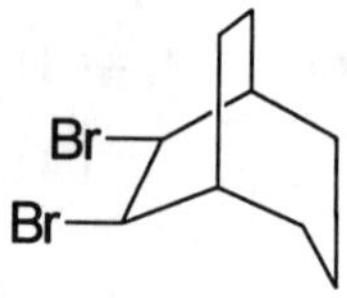

7.23 The bromide ion may attack either of two carbon atoms of the bromonium ion.

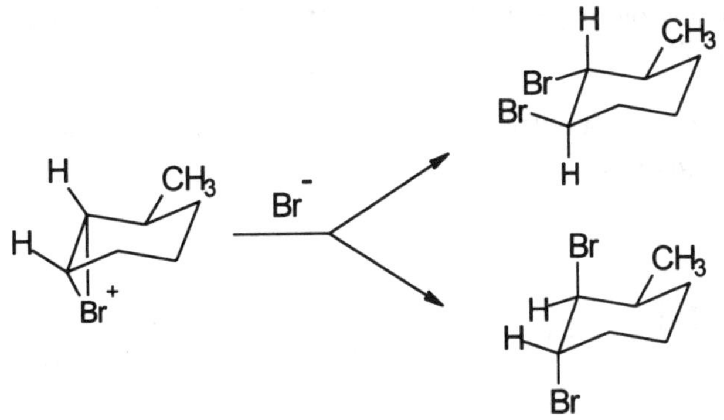

7.24 Addition of a proton occurs at the C-1 atom giving a secondary carbocation at the C-2 atom, which is then temporarily "captured" by the bromine atom at the C-3 atom. The intermediate is thus a bromonium ion that subsequently reacts with bromide ion.

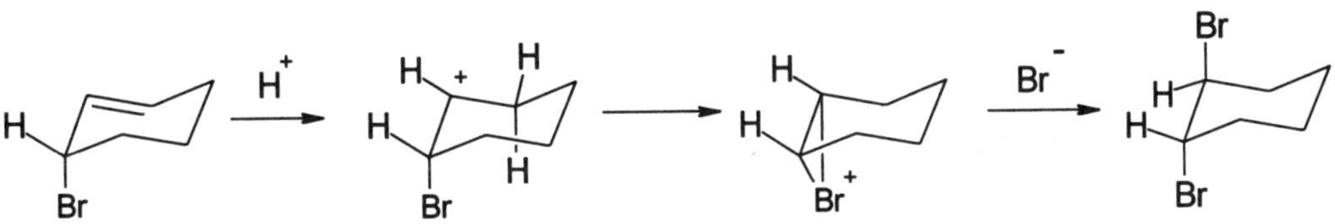

7.25 The reaction of the bromonium ion with water is similar to the reaction of the chloronium ion with water to give a chlorohydrin. The nucleophilic water attacks on the "face" of the original alkene that is opposite the bromine atom of the bromonium ion.

7.26 The reaction of the bromonium ion with chloride ion is similar to that of its reaction with bromide ion in the reaction of bromine. The nucleophilic chloride ion attacks on the "face" of the original alkene that is opposite the bromine atom of the bromonium ion to give *trans*-1-bromo-2-chlorocyclohexane.

7.27 The bromonium ion, which has a bromine atom bridging the C-1 and C-2 atoms, may react with methanol as well as with Br⁻, thus producing a mixture of the dibromo compound and a bromoether. The nucleophilic methanol attacks the C-2 atom which has some tertiary carbocation character rather than the C-1 atom.

7.28 Attack of chlorine forms a chloronium ion which has primary carbocation character at the original methylene carbon atom and tertiary carbocation character at the ring atom. Methanol is expected to attack the tertiary center to give

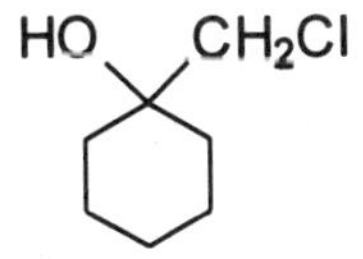

7.29   The bromonium ion has partial positive charges at a secondary and a primary carbon atom. However, the secondary carbocation is destabilized by an electron withdrawal inductive effect of the bromine atom on the adjacent carbon atom. Thus, the nucleophilic water molecule attacks the primary center.

7.30   The bromonium ion undergoes intramolecular attack by the oxygen atom of the hydroxyl group followed by loss of a proton.

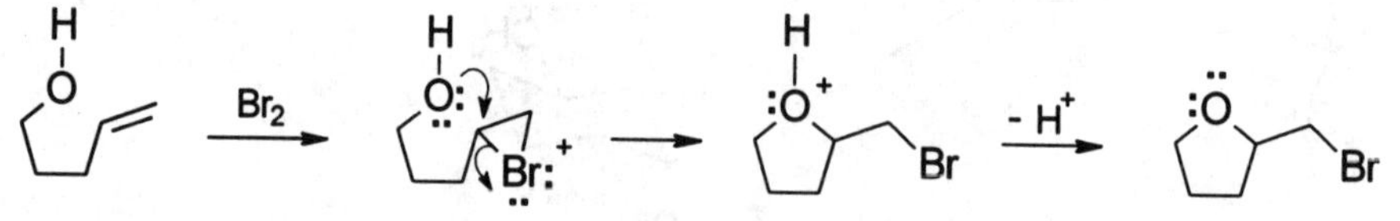

7.31   The *tert*-butoxide ion is not as strong a base as butyl lithium and cannot remove the proton from dichloromethane, which is a weaker acid than trichloromethane.

7.32   The chlorocarbene can add to place the chlorine group either cis or trans to the two methyl groups.

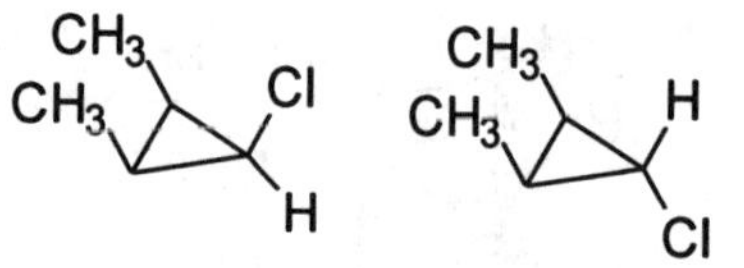

7.33

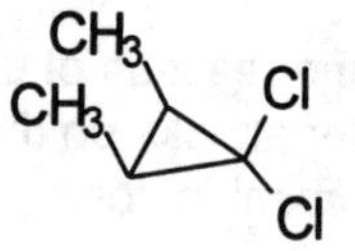

7.34   The product is a cyclopropane obtained by addition of a carbenoid from the Simmons-Smith reagent.

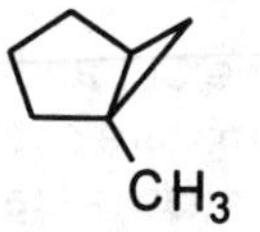

7.35   The methyl groups of the disubstituted *trans*-2-butene should supply electrons to the double bond and increase the availability of electrons to the electrophilic dichlorocarbene. The monosubstituted double bond of 1-butene will be less reactive.

7.36   Based on inductive electron withdrawal of the electronegative chlorine atoms, the dichlorocarbene should be more electrophilic.

7.37   The base can remove a proton from the methyl group and cause an elimination reaction to give vinyl chloride.

7.38   Loss of carbon dioxide yields the trichloromethanide ion, which subsequently loses chloride ion. The by-products are carbon dioxide and sodium chloride.

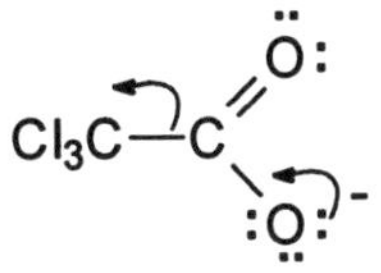

7.39   The trans stereochemistry of the groups about the double bond is retained in the epoxide product.

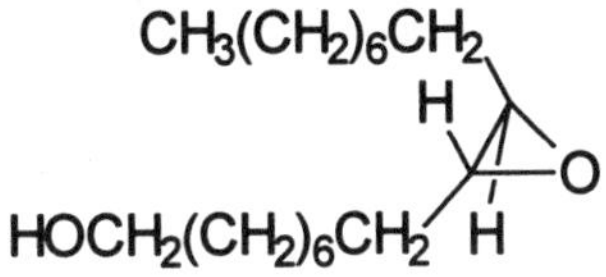

7.40   The cis stereochemistry of the epoxide must exist about the double bond of the unsaturated alcohol.

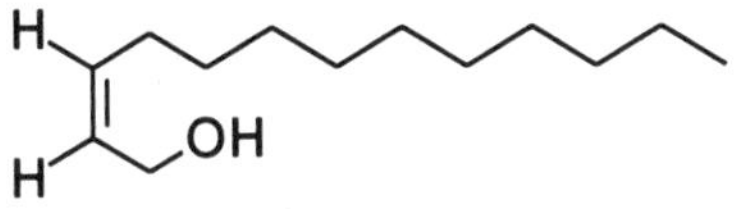

7.41   The first compound because the oxygen of the epoxide is delivered to the double bond from the face that is not sterically hindered by the axial methyl group.

7.42   The isomeric epoxide would place the oxygen atom in a sterically hindered position similar to that of the axial position of a cyclohexane ring.

7.43   The more highly substitued double bond should react with MCPBA. The oxygen atom should be placed on the side of the ring opposite the vinyl group.

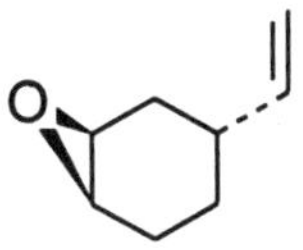

7.44   The order of increasing rate of reactivity is the same as the order of increasing degree of substitution, which is I < III < II < IV.

7.45   The purple color of the permanganate ion is discharged and a brown precipitate of manganese dioxide results. These results are not obtained with cyclopentane.

7.46   A diol is formed when hydroxyl groups are added to the carbon atoms of the vinyl group.

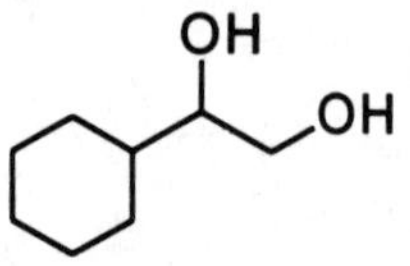

7.47

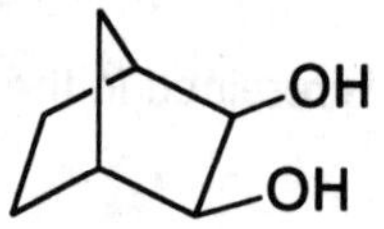

7.48   **(a)** 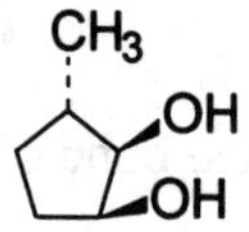   **(b)** 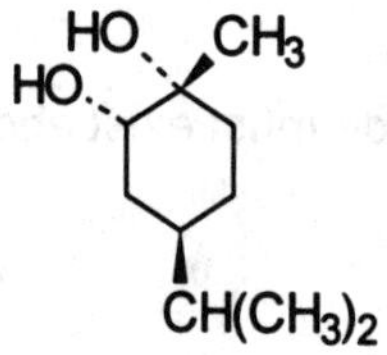   **(c)**

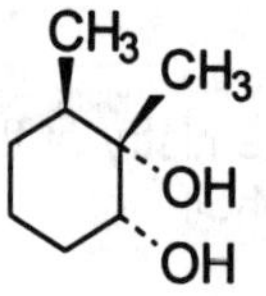

7.49   **(a)** 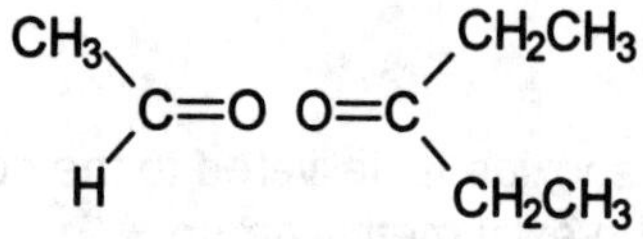   **(b)**

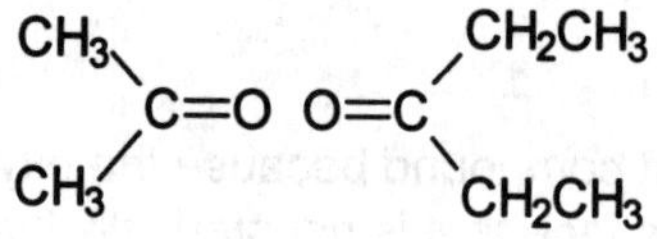

(c) 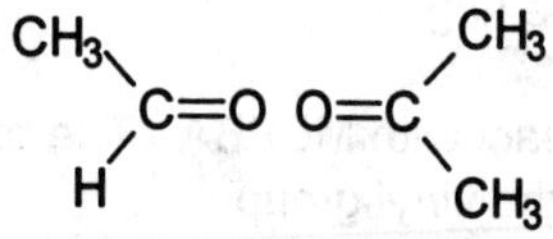   **(d)**

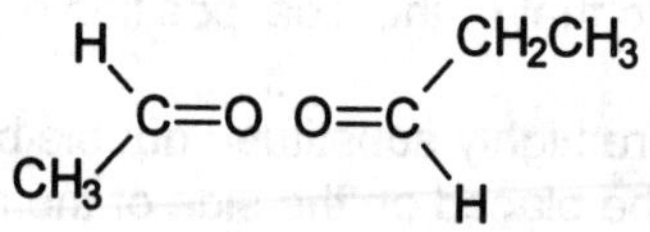

7.50   **(a)** 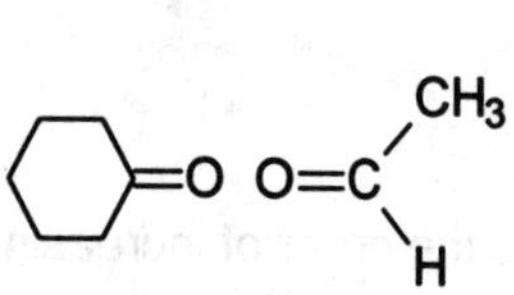   **(b)** 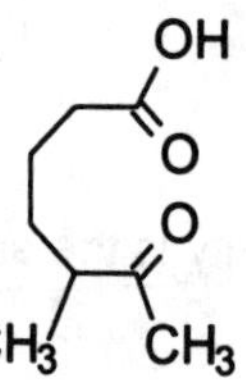   **(c)** 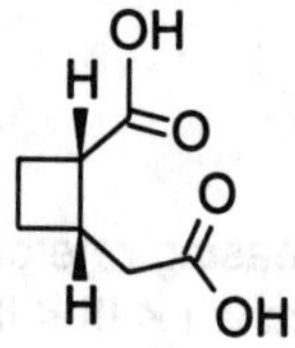   **(d)**

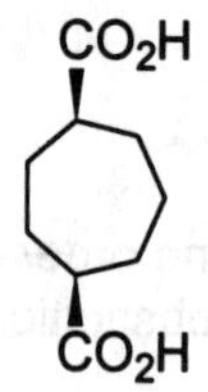

7.51   1,4-Cyclohexadiene gives only one ozonolysis product whereas 1,3-cyclohexadiene gives two.

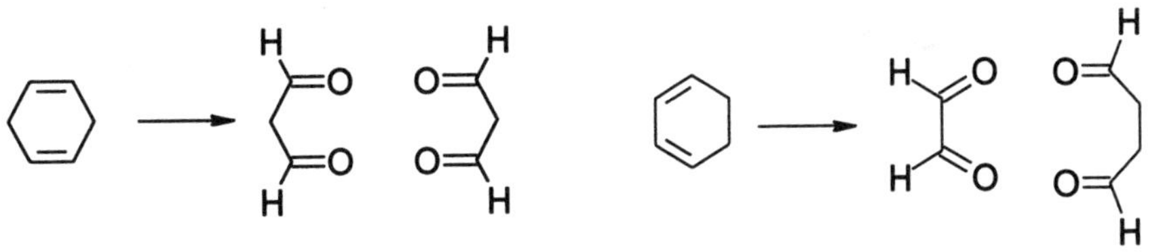

7.52   1-Methylcyclohexene gives a dicarbonyl compound that is an aldehyde and a ketone. Both 3-methylcyclohexene and 4-methylcyclohexene give dialdehydes.

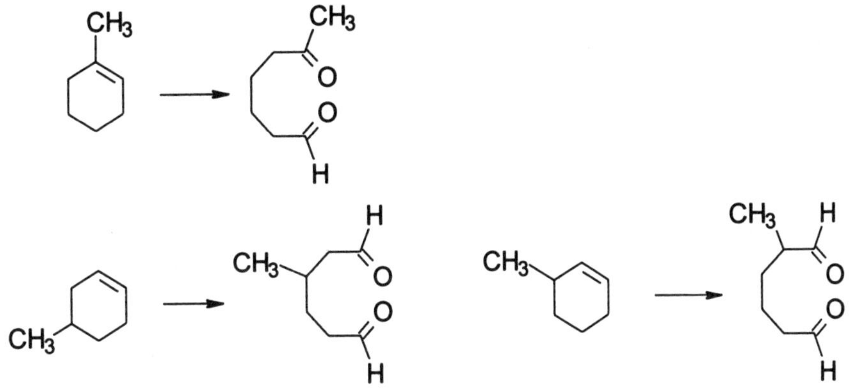

7.53

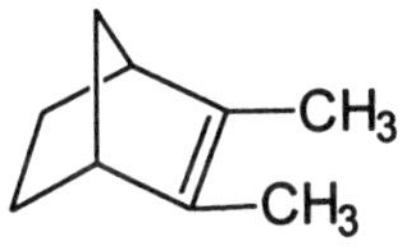

7.54   There are two double bonds in the pheromone and E-Z isomers are possible about each bond, giving rise to four isomers. The 6(E), 9(E) isomer is shown.

7.55   The two compounds are geometric isomers. When the double bond is cleaved, that structural distinction no longer exists and the resulting fragments are identical. The (E) isomer is shown.

7.56   There are three double bonds in the compound and E-Z isomers are possible about each bond giving rise to eight isomers. The 12(E),15(E),18(E) isomer is shown.

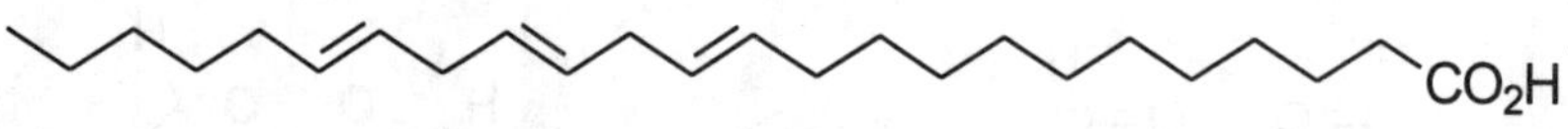

7.57   (a) 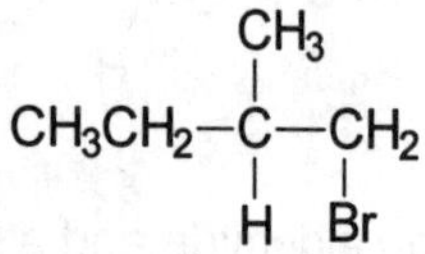 CH$_3$CH$_2$—C—CH$_2$ (b) 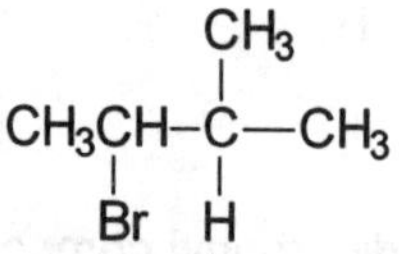   CH$_3$CH—C—CH$_3$

with CH$_3$ above the central C and H, Br below in (a); CH$_3$ above and Br, H below in (b)

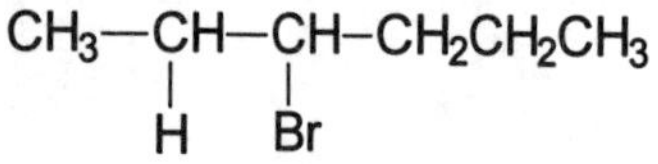

(c)   CH$_3$—CH—CH—CH$_2$CH$_2$CH$_3$ and   CH$_3$—CH—CH—CH$_2$CH$_2$CH$_3$

with H, Br below in the first; Br, H below in the second

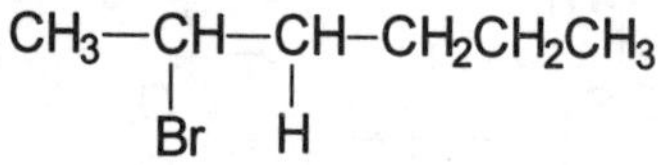

(d) 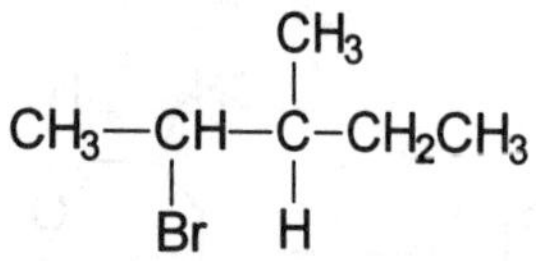 CH$_3$—CH—C—CH$_2$CH$_3$

with CH$_3$ above the central C and Br, H below

7.58   (a)

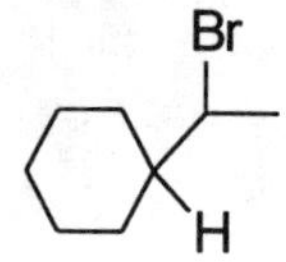

(b)

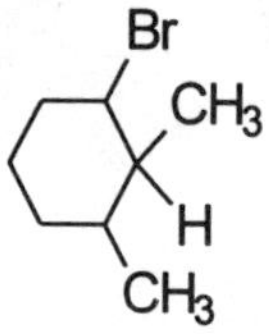

(c)

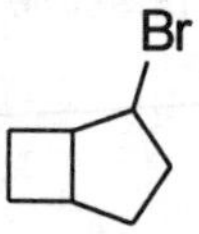

and 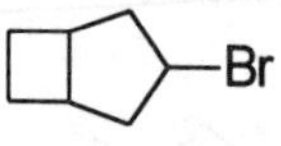   (d) 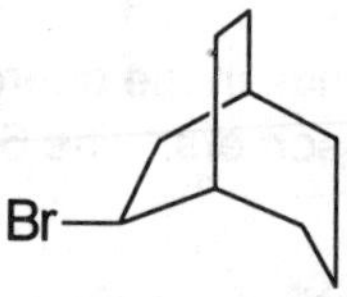 Br—

7.59   The homolytic cleavage of the C—I bond gives a CF$_3$ radical that attacks the terminal carbon atom of 1-butene to give a secondary radical. This radical abstracts an iodine atom from CF$_3$I to give the observed product and regenerates the CF$_3$ radical. The two reactions are propagation steps. The observed regiochemistry is controlled by the stability of the secondary radical compared to a primary radical that would result if the CF$_3$ radical attacked the C-2 atom.

7.60   The tertiary radical formed from 2-methyl-1-pentene is more stable than the secondary radical formed from 3-methyl-1-pentene. However, 2-methyl-1-pentene is also more stable than 3-methyl-1-pentene. Because the energies of both the reactants and their respective transition states differ, it is difficult to predict the relative rates. However, the effect of structure on radical stability is probably greater than the effect of structure on the reactants. Thus, the reaction of 2-methyl-1-pentene is most likely the faster.

7.61 The initiator abstracts a bromine atom to give In—Br and a CCl_3 radical that attacks the terminal carbon atom of 1-hexene to give a secondary radical. This radical abstracts a bromine atom from CCl_3Br to give the observed product and regenerates the CCl_3 radical. The two reactions are propagation steps.

$$In\cdot \; + \; Cl_3C-Br \longrightarrow Cl_3C\cdot \; + \; In-Br$$

$$CH_3(CH_2)_3CH{=}CH_2 \; + \; Cl_3C\cdot \longrightarrow CH_3(CH_2)_3CH-CH_2-CCl_3$$

$$CH_3(CH_2)_3\overset{\displaystyle\cdot}{C}H-CH_2-CCl_3 \; + \; Cl_3C-Br \longrightarrow CH_3(CH_2)_3\overset{\displaystyle Br}{\overset{\displaystyle |}{C}}H-CH_2-CCl_3 \; + \; Cl_3C\cdot$$

7.62 The tertiary radical formed from 2-methyl-1-pentene is more stable than the secondary radical formed from 1-hexene. However, 2-methyl-1-pentene is also more stable than 1-hexene. Because the energies of both the reactants and their respective transition states differ, it is difficult to predict the relative rates. However, the effect of structure on radical stability is probably greater than the effect of structure on the reactants. Thus, the reaction of 2-methyl-1-pentene is most likely the faster. The product is 3-bromo-1,1,1-trichloro-3-methylhexane.

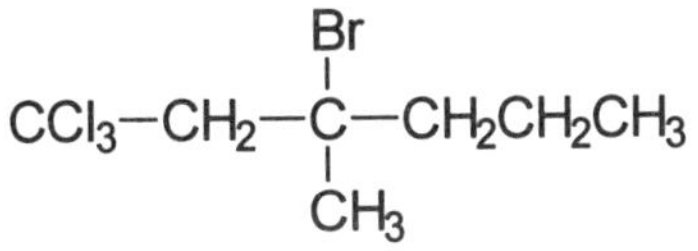

$$CCl_3-CH_2-\overset{\displaystyle Br}{\underset{\displaystyle CH_3}{\overset{\displaystyle |}{\underset{\displaystyle |}{C}}}}-CH_2CH_2CH_3$$

7.63 The bromine radical can add to either the C-1 or C-2 atoms and can attack either from the same side or opposite side of the ring containing the bromine atom already bonded to the ring.

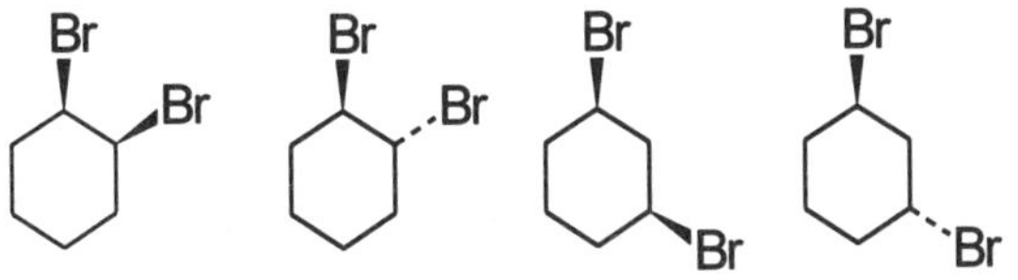

7.64 The sequence of reactions is shown on the following page. In the initiation step, the peroxide abstracts a hydrogen atom to give a sulfur radical. That radical then attacks the terminal carbon atom to give a tertiary radical. This radical abstracts a hydrogen atom from thiol to give the observed product and regenerates the sulfur radical. These two reactions are propagation steps.

$$RO\cdot \ + \ CH_3CH_2S-H \longrightarrow \ CH_3CH_2S\cdot \ + \ RO-H$$

$$(CH_3)_2C=CH_2 \ + \ CH_3CH_2S\cdot \ \longrightarrow \ (CH_3)_2\overset{\cdot}{C}-CH_2-SCH_2CH_3$$

$$(CH_3)_2\overset{\cdot}{C}-CH_2-SCH_2CH_3 + CH_3CH_2S-H \longrightarrow (CH_3)_2CHCH_2SCH_2CH_3 \ + \ CH_3CH_2S\cdot$$

7.65   Addition of a free radical to propene as part of the polymerization process yields a secondary radical that is more stable than the primary radical that results in the polymerization process for ethylene. Assuming that the effect of structure on radical stability is probably greater than the effect of structure on the reactants, the reaction of propene will be the faster.

7.66   The amide ion adds to the terminal carbon atom to give a resonance stabilized anion.

$$NH_2-CH_2-\overset{-}{\underset{..}{C}}H-C\equiv N\colon \ \longleftrightarrow \ NH_2-CH_2-CH=C=\overset{..}{\underset{-}{N}}\colon$$

8

Haloalkanes and Alcohols

In this chapter, the first two of the functional groups containing electronegative atoms are discussed. Both haloalkanes and alcohols have a single bond between sp³ hybridized carbon and an electronegative atom. The result is a polar covalent bond whose properties are responsible for the chemistry of these compounds. Both of the bonds can break heterolytically in reactions with nucleophiles that substitute for the halogen or oxygen atom.

Two exceedingly important related mechanisms are introduced in this chapter. The S_N2 and S_N1 mechanisms are fundamental to understanding a large variety of substitution reactions. Like movie promotions of coming attractions, some of the concepts required to understand and use these mechanisms to predict reactivity of alcohols and haloalkanes are presented to give you a preview of the "big picture" to follow. We will revisit these mechanisms in Chapters 9, 10, 16, and 17. However, this is the chapter where you must learn about these mechanisms because material in later chapters will build on this foundational material.

Keys to the Chapter

8.1 Functionalized Hydrocarbons

The chemistry of compounds containing a halogen or a hydroxyl group is limited in this chapter to compounds with sp³ hybridized carbon atoms. Molecules with sp² and sp hybridized carbon atoms bonded to a halogen or a hydroxyl group have different chemistry. The focus is on chlorine- and bromine-containing haloalkanes.

Both haloalkanes and alcohols are classified as 1°, 2°, and 3° by the same method used to classify carbon atoms in alkanes. If you understand the classification of alkanes, then you are in good shape. If not, then this is the first of the wake up calls that this study guide provides. Don't let your problems build up!

The halogen and hydroxyl groups are interchangeable using certain reagents. Only one reagent is given for each type of reaction. More reagents are introduced later in the chapter and subsequent chapters. Both listed reactions are substitution reactions.

8.2 Uses of Haloalkanes

Haloalkanes are largely the creation of chemical industry and are not common in naturally occurring materials. Nevertheless because of their specialized reactivity they have been used as pesticides.

8.3 Uses of Alcohols

Alcohols are not only common in naturally occurring materials but are produced in very large quantities for a variety of applications. The properties of these compounds and their use in commercial products will be revisited.

8.4 Nomenclature of Haloalkanes

The rules for naming haloalkanes are very similar to the rules for naming alkanes (Section 4.3) and alkenes (Section 6.6). Remember that halogen atoms and branching alkyl groups are treated as "equals". If no other functional groups are present, the chain is numbered from the end closest to the first substituent, whether it is a halogen or an alkyl group. However, the double bond of an alkene or the triple bond of an alkyne takes precedence in numbering a carbon chain, regardless of where the halogen is located or how many halogens there may be. This concept of "priority" of one functional group over another is expanded with each new functional group studied in later chapters. Start learning the order of importance of functional groups in the nomenclature of compounds now, and then add to that base of knowledge.

8.5 Nomenclature of Alcohols

The common names of the simple alcohols are based on the name of the alkyl group bonded to the hydroxyl group. You may wish to review the common names of alkyl groups in Chapter 4. Two additional groups of considerable importance are introduced—the allyl and benzyl groups.

The IUPAC method of naming alcohols is based on the longest continuous chain containing the carbon atom with an attached hydroxyl group. The chain is numbered to give that carbon atom the lowest possible number. The suffix -ol is added to the stem of the name of the alkane which represents the chain. Note that in contrast to halogens, which have lower priority than multiple bonds in assigning names, the hydroxyl group has a higher priority than multiple bonds. Compounds with the hydroxyl group bonded to a ring are numbered from the carbon atom containing the hydroxyl group, and subsequent numbers are assigned in the direction to give the lowest possible numbers for any other structural features.

8.6 Structure and Properties of Haloalkanes

In general we know that the electronegativity of the halogens decreases and the polarizability of the halogens increases from top to bottom in Group VII of the periodic table. The boiling points of homologous R-X compounds increase in the same order, so the intermolecular attractive forces must also increase from top to bottom. This indicates that polarizability is more important than bond polarity in determining the physical properties of haloalkanes. Other factors such as molecular shape and the extent of branching are also likely to influence the intermolecular forces and physical properties of haloalkanes.

8.7 Structure and Properties of Alcohols

This material should look familiar. It was presented initially in Section 2.8. The intermolecular hydrogen bonds of alcohols dominate the physical properties of these compounds. Alcohols have higher boiling points than alkanes of similar molecular weight as a result of intermolecular hydrogen bonding. Hydrogen bonding between alcohol molecules and water also accounts for the solubility of alcohols. Alcohols serve as solvents for polar compounds, especially those that can also form hydrogen bonds with the solvent.

8.8 Organometallic Compounds

This section is your first general introduction to a reactive class of compounds in which the bond from carbon to the substituent has an "unusual" polarity. Most substituents bonded to carbon are electronegative, but metals are electropositive. The **Grignard reagent** is an organomagnesium compound. It is a very reactive compound formed by reacting a haloalkane with magnesium. When a Grignard reagent is exposed to water, an alkane is produced. A method for synthesizing a desired alkane can be devised by "thinking backwards" to identify the necessary Grignard reagent. This method is useful in the synthesis of deuterium substituted compounds. However, the importance of the Grignard reagent will be more obvious in later chapters when its reactivity with several classes of functional groups is discussed. At this point just tuck away the information that the carbon atom of a Grignard reagent has a partial negative charge and behaves as a nucleophile.

Organolithium compounds can also serve as sources of carbon nucleophiles. In addition, these compounds are used as bases.

The **Gilman reagent** is also briefly introduced in this section, and its chemistry will be examined further in some later chapters. The compound is used to "couple" two carbon groups together. One group is provided by an organohalogen compound; the second is provided by the Gilman reagent.

8.9 Reactions of Haloalkanes

The chemistry of haloalkanes is briefly presented in this section. It serves as an advance warning of two reactions that you will encounter throughout the course—substitution and the competing elimination reaction. Nucleophiles can displace a halide ion as a leaving group in a substitution reaction. However, nucleophiles are often sufficiently basic to remove a proton from the carbon atom adjacent to the carbon atom bearing the halogen. The overall result is an elimination reaction. Again, keep in mind throughout this course that substitution and elimination reaction compete with each other. Your instructor may talk about a substitution without mentioning the elimination reaction. It is your responsibility to maintain an active mental file of past information to apply to new information.

8.10 Nucleophilic Substitution Reactions of Haloalkanes

Nucleophilicity refers to the ability of a nucleophile to displace a leaving group in a substitution reaction of a **substrate**. Various trends in nucleophilicity are described in Chapter 10. Nucleophiles most often have a negative charge. However, it is the nonbonding electron pair that is important. Thus, ammonia is a nucleophile even though it is electrically neutral.

Remember to keep track of pairs of electrons in reactions such as the nucleophilic substitution process. In all cases in this section the nucleophile brings a pair of electrons to the site, and the leaving group departs with a pair of electrons that was originally the bonding pair to that group. If the nucleophile has a negative charge and it reacts with a neutral substrate, then the leaving group must also have a negative charge.

Although all of the examples of nucleophiles cited are important because they allow the synthesis of compounds, two are especially interesting. Both the cyanide ion and the alkynide ion result in compounds containing more carbon atoms that the substrate. However, all cited examples will be revisited in later chapters.

8.11 Mechanisms of Nucleophilic Substitution Reactions

In Chapter 7 you learned one of the few central mechanisms that are used to explain a large number of organic reactions. That was the electrophilic addition reaction with unsaturated compounds. Two nucleophilic substitution mechanisms are given in this section. This isn't the last time that you will see them by any means. If you don't take time to understand them now, you are really going to regret it when you keep hearing about them in the future. Don't let mechanisms be your enemy. They really are helpful to organize apparently unrelated information. The essentials of the S_N2 and S_N1 mechanisms are presented in this section. More details and refinements are presented in Chapters 9 and 10, and variations on the theme will be encountered in several other chapters.

The proposed S_N2 reaction is based in part on kinetic experiments. The key points are that both the substrate and the nucleophile are involved in the transition state and that the nucleophile attacks along the same axis that the leaving group departs. (The stereochemical consequences of this process are detailed in Chapters 9 and 10.)

The rate of the S_N2 reaction depends on the structure of the substrate. The order of reactivity is primary > secondary > tertiary. Steric hindrance caused by the groups bonded to the reacting carbon center substantially prevents approach of the nucleophile in tertiary compounds, and it is doubtful that tertiary compounds react by this mechanism. The ability of the nucleophile to displace the leaving group is improved in secondary compounds and still further in primary compounds, so both can react by the S_N2 mechanism.

A different mechanism is postulated as an alternative for substitution reactions at sterically hindered sites. Again, the mechanism is based on kinetic data. The rate determining step is the ionization of the haloalkane to give a carbocation which is subsequently captured by the nucleophile in a faster second step. Thus, the rate of the reaction depends only on the substrate concentration. The order of reactivity is tertiary > secondary > primary. In fact, it is unlikely that a primary compound would react by this mechanism when it has the S_N2 process as an available option. Why do tertiary compounds react so much faster by the S_N1 mechanism compared to secondary compounds, and why don't primary compounds react by this mechanism? It boils down to the stability of the carbocation.

Substrates that react via the S_N1 mechanism may give rearranged products. This process is exactly like the rearrangements described in Section 7.4. The only difference is in the reaction that leads to the carbocation. In the case of alkenes, the carbocation resulted from addition of a proton to a π bond. We now find that the same types of carbocations are formed in a substitution reaction. Either a hydride ion or an alkyl group with a negative charge can move from a center adjacent to the carbocation center to form a new bond and generate a positive charge at that adjacent center. The driving force for such reactions is the formation of a more stable carbocation. Keep your eye open for such possibilities in substitution reactions that occur by the S_N1 mechanism. Hydride ion shifts occur when a secondary or primary carbocation is generated adjacent to a tertiary center. Shifts of alkyl groups most commonly occur when a carbocation is generated at a carbon atom adjacent to a quaternary center. Any of the alkyl groups bonded to that quaternary carbon can migrate. Thus, a mixture of products can result. As shown in Exercise 8.14, even ring bonds of cycloalkanes can migrate, resulting in rings of different size than the reactant.

8.12 Reactions of Alcohols

There are several ways in which alcohols may react, differing in the number and type of bonds cleaved. These are:

1. Cleavage of the oxygen-hydrogen bond
2. Cleavage of the carbon-oxygen bond
3. Cleavage of the carbon-oxygen bond as well as the carbon-hydrogen bond at the carbon atom adjacent to the carbon atom bearing the hydroxyl group.
4. Cleavage of the oxygen-hydrogen bond as well as the carbon-hydrogen bond at the carbon atom bearing the hydroxyl group.

8.13 Acid-Base Reactions of Alcohols

In general, alcohols are somewhat weaker acids than water. Table 8.3 in the text lists K_a and pK_a values for selected alcohols. Ethanol, 2-propanol, and 2-methyl-2-propanol have pK_a values of 15.9, 18.0, and 19.0, respectively. These values indicate a decrease in acidity from 1° to 2° to 3° alcohols. This agrees with our observations in earlier chapters that alkyl substituents are inductively electron-donating; this effect supplies more electron density to the oxygen atom, so it strengthens the O-H bond.

We also see in Table 8.3 that electronegative substituents near the carbon atom bearing the hydroxyl group tend to increase the acidity. Halogen substituents inductively withdraw electron density from the oxygen atom and weaken the O-H bond. The halogens also stabilize the negative charge of the conjugate base—an **alkoxide ion**.

Alcohols, like water, can be protonated to give a conjugate acid known as an **alkyloxonium ion**.

8.14 Substitution Reactions of Alcohols

The hydroxyl group of alcohols can be replaced by a halogen using a hydrogen halide. The order of reactivity is tertiary > secondary > primary. Hydrogen bromide suffices to form bromoalkanes, but zinc chloride is required as a catalyst for the reaction with hydrogen chloride.

The substitution reaction of alcohols parallels that of haloalkanes. However, the hydroxide ion is not the leaving group. Protonation of the hydroxyl group must occur to allow water to become the leaving group. The statement that there is a general correlation between basicity and leaving group ability is an important generalization that you will encounter again in other classes of reactions. The correlation is this: a weaker base is a better leaving group than a stronger base. Tuck it away, but more importantly understand why it is true. In this case, the hydroxide ion is a stronger base than water. Stronger bases have a high affinity for protons and readily bond to them. In the case of their behavior as leaving group, it is necessary that the leaving group (base) depart and take its electron along. That process is counter to the tendency of a strong base to form bonds with a Lewis acid and in this case an electrophilic species—the carbocation. The same argument holds for those reactions that occur by an S_N2 mechanism.

The order of reactivity of alcohols in substitution reactions is tertiary > secondary > primary. Tertiary and secondary alcohols react via an S_N1 mechanism. Primary alcohols react via an S_N2 mechanism.

8.15 Alternate Methods for the Synthesis of Alkyl Halides

In this section two additional reagents are presented that convert alcohols to haloalkanes. They are used for secondary and primary compounds, because these compounds react slowly with hydrogen halides. You will have to adjust to learning several reagents that can accomplish the same synthetic transformation, not only in this reaction, but in other functional group transformations as well. In each case, you will find that different reagents are often necessary to react with the range of possible structures and that each reagent has advantages and disadvantages. Look out for these differences and become skilled in selecting the most appropriate reagent for a reaction.

Thionyl chloride is used to convert alcohols to chloroalkanes. The byproducts are sulfur dioxide and hydrogen chloride. Phosphorus tribromide is used to convert alcohols to bromoalkanes. The byproduct, phosphorous acid, is soluble in water.

8.16 Elimination Reactions

Now we drop the other shoe. As if it isn't hard enough to consider two possible mechanisms for substitution reactions, we now have to cope with the competing elimination reaction that occurs. The elimination reaction removes atoms from adjacent carbon atoms and is called a **1,2-elimination** or a β-**elimination**.

The discussion of the thermodynamics of elimination reactions in this section suggests that elimination reactions are not favored in terms of the $\Delta H°_{rxn}$. However, the $\Delta S°_{rxn}$ is favorable. Remember from Section 3.9 in the text that the $\Delta S°_{rxn}$ for processes in which one additional mole of product is formed is about +125 J mole^{-1} K^{-1}. Thus the $-T\Delta S°_{rxn}$ term at room temperature is about -37 kJ mole^{-1}, which gives you some idea about the possibility of simply reversing an addition reaction to give an elimination reaction. Only dehydration and perhaps dehydrohalogenation are likely possibilities. Both debromination and dehydrogenation have a contribution of $\Delta H°_{rxn}$ that is too positive to be overcome at room temperature by a favorable $\Delta S°_{rxn}$.

8.17 Types of Elimination Reactions

Both dehydrogenation and debromination reactions can occur at sufficiently high temperatures to make the contribution of $-T\Delta S°_{rxn}$ sufficiently negative to overcome the unfavorable $\Delta H°_{rxn}$. However neither of these reactions has wide applicability.

Debromination and more importantly both dehydrohalogenation and dehydration can be made feasible, but not by the microscopic reverse of the comparable addition reactions. In each case a reagent is added that effectively changes the stoichiometry of the reaction. For debromination, the favorable $\Delta H°_{rxn}$ for the formation of zinc bromide contributes to give a net negative $\Delta H°_{rxn}$ for the overall reaction of a dibromide with zinc. For dehydrohalogenation, the reaction of hydroxide ion with a hydrogen halide provides the driving force for the reaction. In the case of dehydration, the reaction of the hydrogen ion with water is the driving force. Note that the analysis presented in no way implies that the reaction occurs by the indicated reactions used in a qualitative analysis of the Law of Hess. In fact the mechanisms are not the two steps listed in each case. However, the thermodynamic conclusions are correct.

8.18 Regioselectivity in Dehydrohalogenation

Dehydrohalogenaton occurs to give a predominance of the most substituted alkene. This regioselectivity results in the so-called **Zaitsev** product. Although no theoretical basis is given at this point, you can nevertheless predict the product of an elimination reaction by simply considering the neighboring carbon atoms of the carbon atom bearing the halogen. Just mentally "lasso" away a bromine atom and an adjacent hydrogen atom and picture the double bond that would result. The one with the most alkyl groups bonded to the carbon atoms of the double bond is favored. If geometric isomers are possible, the more stable trans isomer predominates.

8.19 Mechanisms of Dehydrohalogenation

There are two mechanisms for dehydrohalogenation. The E2 process occurs for primary haloalkanes and usually also for secondary haloalkanes. The E1 process occurs with tertiary haloalkanes. In either case, the reactivity order is iodo > bromo > chloroalkane.

The E2 mechanism is a concerted process in which a base abstracts a proton while a carbon-carbon double bond forms and simultaneously a halide ion leaves. Because a double bond starts to develop in the transition state, the energy of the transition state is lower for more highly substituted compounds. The stereochemistry of the E2 process involves an **anti periplanar** arrangement of the carbon-hydrogen bond and the bond from a carbon atom to the leaving group. This arrangement provides the necessary geometric alignment for the emerging orbitals required to form a π bond.

The E1 mechanism is a two-step process in which the first step—the departure of the leaving group—is rate determining. This is also the first step of the S_N1 reaction. Now the carbocation loses a proton rather than being captured by a nucleophile as in the S_N1 reaction. The loss of a proton or deprotonation results from abstraction by the base. However, note that the base is not included in the rate law because this process occurs after the rate detemining step.

8.20 Regioselectivity in Dehydration of Alcohols

The distribution of product alkenes from dehydration is similar to that for the dehydrohalogenation reaction. The Zaitsev product is the more substituted alkene.

Like substitution reactions of alcohols, both the E2 and E1 reaction require protonation of the oxygen atom before the process can proceed. After that step the separation of compounds into those that react by the E2 or E1 mechanism is similar to that distinction made for S_N2 and S_N1 mechanisms. Tertiary and usually secondary alcohols react via the E1 mechanism; primary alcohols react via the E2 mechanism. In the E1 mechanism, water first leaves in the rate determining step and then abstraction of a proton by a base such as water occurs. In the E2 process, water acts as a base to abstract a proton while water leaves from adjacent carbon atom.

Rearrangement reactions occur in dehydration reactions that proceed by the E1 mechanism because a carbocation is formed. Use the same criteria listed earlier for rearrangement of carbocations formed in S_N1 processes or in the addition reactions of alkenes. There isn't anything really new here. Hydride ion shifts occur when a carbocation is generated adjacent to a tertiary center. Shifts of alkyl groups most commonly occur when a carbocation is generated at a carbon atom adjacent to a quaternary center. Any of the alkyl groups bonded to that quaternary carbon can migrate. After migration occurs, subsequent loss of a proton gives the rearranged alkene product.

Summary of Reactions

1. Formation and Reactivity of Grignard Reagents

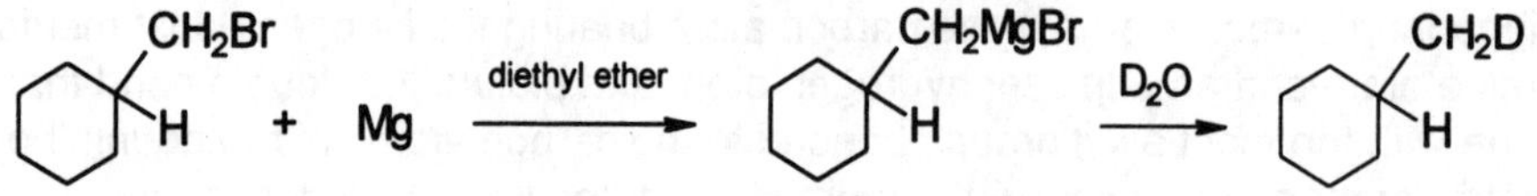

2. Coupling Reaction of Gilman Reagent and Halogen Compounds

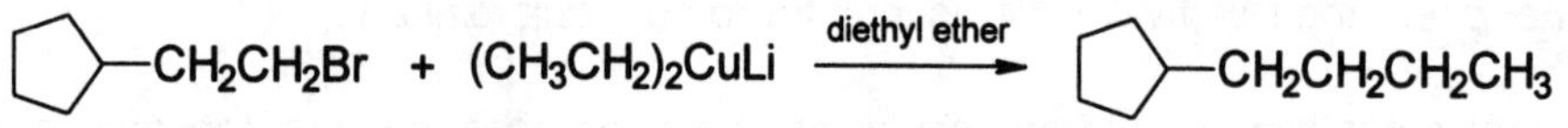

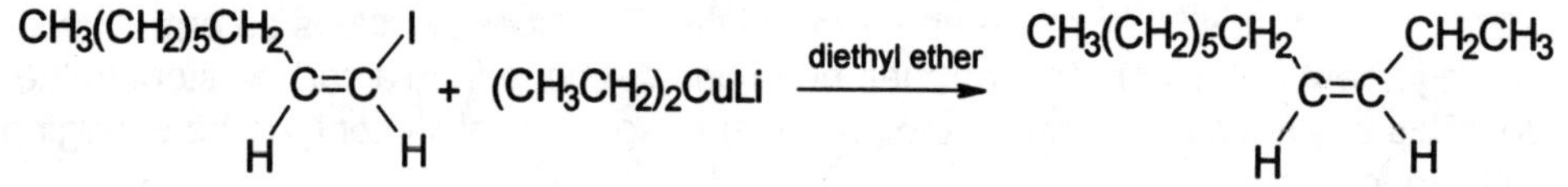

3. Nucleophilic Substitution of Haloalkanes

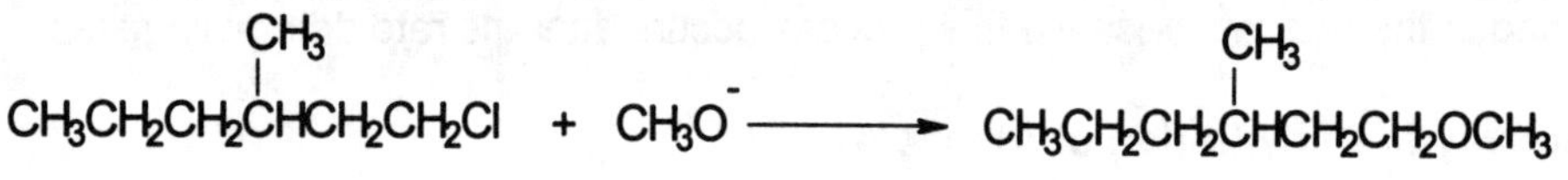

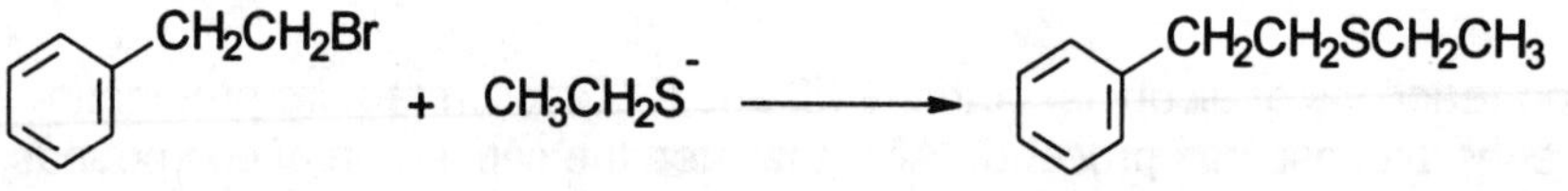

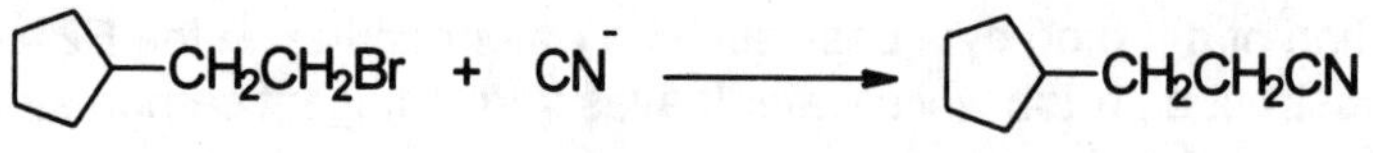

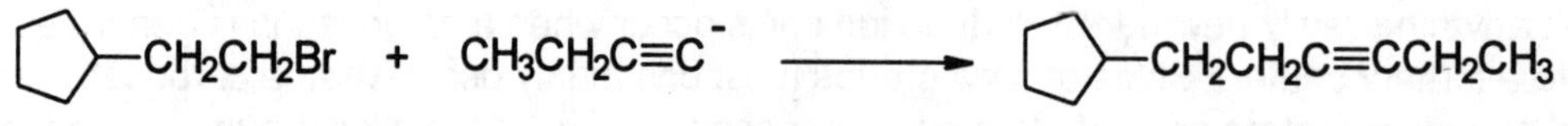

4. Synthesis of Haloalkanes from Alcohols

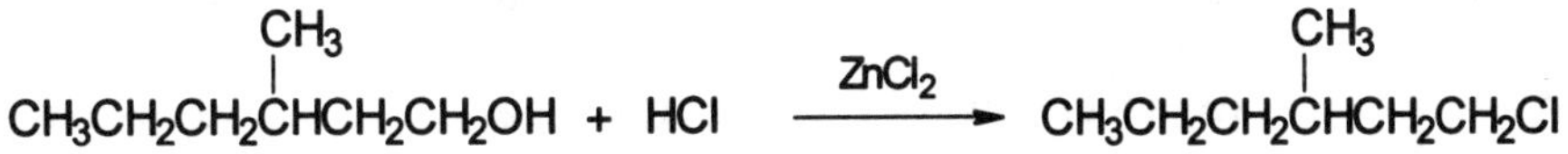

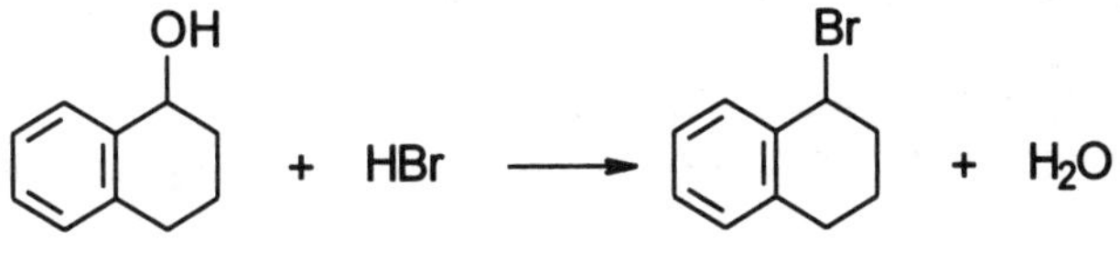

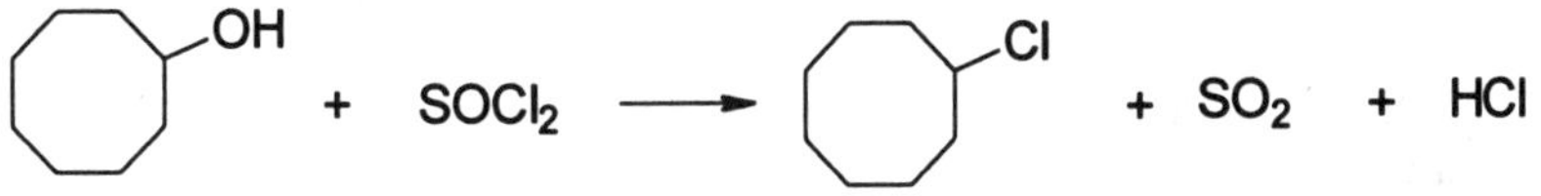

$$CH_3CH_2CH_2\overset{\overset{\displaystyle CH_3}{|}}{C}HCH_2CH_2OH \ + \ HCl \quad \xrightarrow{\quad ZnCl_2 \quad} \quad CH_3CH_2CH_2\overset{\overset{\displaystyle CH_3}{|}}{C}HCH_2CH_2Cl$$

$$3 \ \text{cyclopentyl-}CH_2CH_2OH \ + \ PBr_3 \ \longrightarrow \ 3 \ \text{cyclopentyl-}CH_2CH_2Br \ + \ H_3PO_3$$

5. Dehydrohalogenation of Haloalkanes

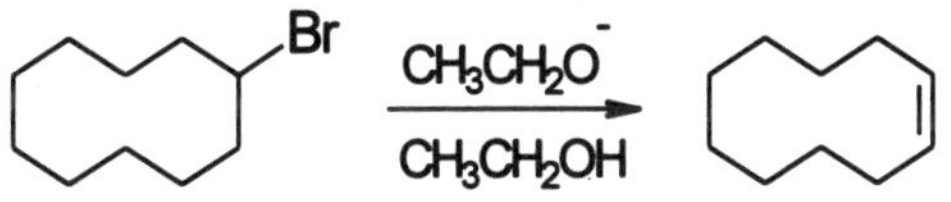

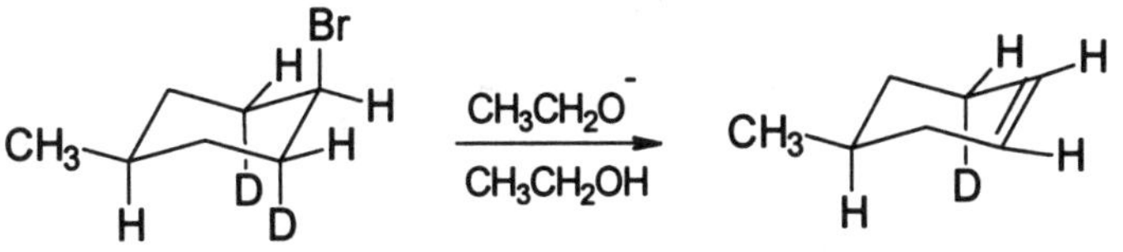

6. Dehydration of Alcohols

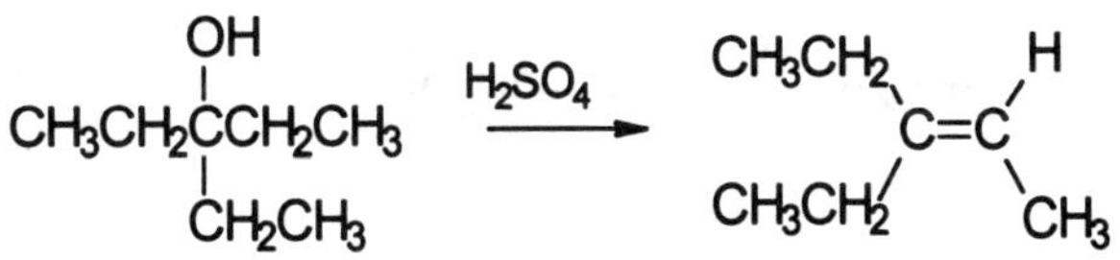

$$CH_3CH_2\overset{\overset{\displaystyle OH}{|}}{\underset{\underset{\displaystyle CH_2CH_3}{|}}{C}}CH_2CH_3 \quad \xrightarrow{\quad H_2SO_4 \quad} \quad \underset{CH_3CH_2}{\overset{CH_3CH_2}{>}}C=C\overset{H}{\underset{CH_3}{<}}$$

Solutions to Exercises

8.1   **(a)** secondary   **(b)** tertiary   **(c)** secondary   **(d)** tertiary

8.2   **(a)** primary   **(b)** secondary   **(c)** secondary   **(d)** primary

8.3   **(a)** phenol and primary alcohol
       (b) primary alcohol
       (c) primary alcohol at top of structure; three secondary alcohol sites

8.4   **(a)** secondary alcohol on left ring; tertiary alcohol at ring juncture
       (b) secondary alcohol on six-membered ring; tertiary on five-membered ring; primary
           alcohol at top of structure
       (c) tertiary alcohol

8.5   **(a)** fluoroethene   **(b)** 3-chloro-1-propene   **(c)** (bromomethyl)benzene

8.6   **(a)** 1-chloro-2,2-dimethylchloropropane
       (b) 1-bromo-3-methylbutane
       (c) 1-fluoro-2-phenylethane

8.7   **(a)** 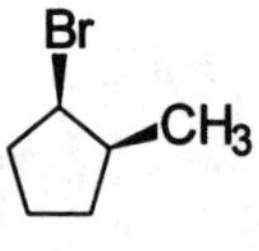 **(b)** 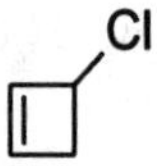 **(c)** 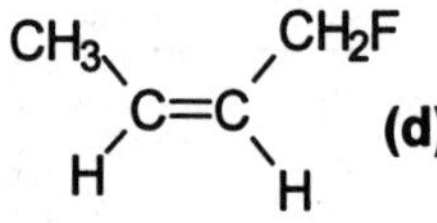 **(d)**

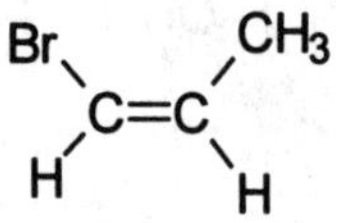

8.8   **(a)** *trans*-1-ethyl-4-iodocycloheptane
       (b) 1-chloro-1-propylcyclohexane
       (c) *trans*-1-bromo-2-cyclobutylcyclodecane
       (d) *trans*-1-bromo-2-cyclopropylcyclooctane

8.9   **(a)** $CH_3CH_2CH_2-\overset{\overset{\displaystyle CH_3}{|}}{\underset{\underset{\displaystyle OH}{|}}{C}}-CH_3$ **(b)** $CH_3CH_2-\overset{\overset{\displaystyle CH_3}{|}}{CH}-CH_2OH$

       (c) $CH_3-\overset{\overset{\displaystyle CH_3}{|}}{CH}-\overset{\overset{\displaystyle CH_3}{|}}{CH}-CH_2OH$ **(d)**    **(e)**

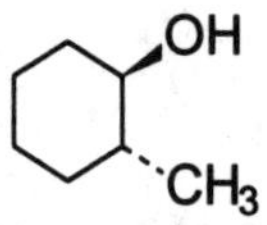

       (f) $HOCH_2-CH_2-CH_2OH$ **(g)** $HOCH_2-\overset{\overset{\displaystyle OH}{|}}{CH}-CH_2-CH_2OH$

8.10 **(a)** 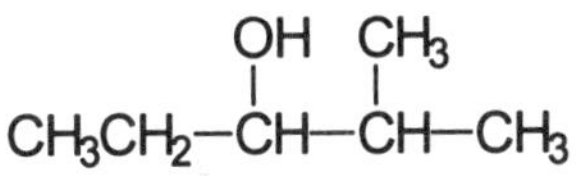 CH$_3$CH$_2$—CH—CH—CH$_3$ (with OH and CH$_3$ substituents)

(b) 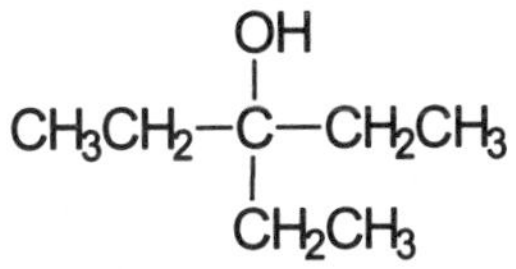 CH$_3$CH$_2$—C—CH$_2$CH$_3$ (with OH and CH$_2$CH$_3$ substituents)

(c) CH$_3$CHCH$_2$—CH—CH$_3$ (with CH$_3$ and OH substituents)

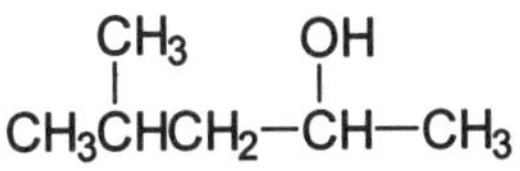

(d) 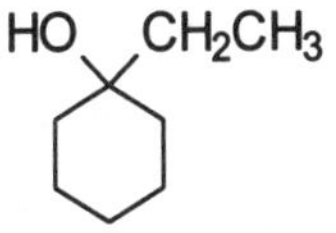 cyclohexane with HO and CH$_2$CH$_3$ substituents

(e) 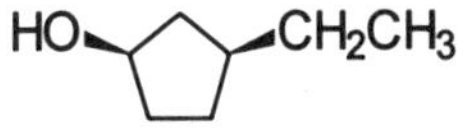 cyclopentane with HO and CH$_2$CH$_3$ substituents

(f) 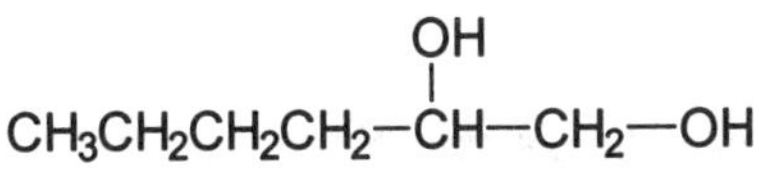 CH$_3$CH$_2$CH$_2$CH$_2$—CH—CH$_2$—OH (with OH substituent)

(g) 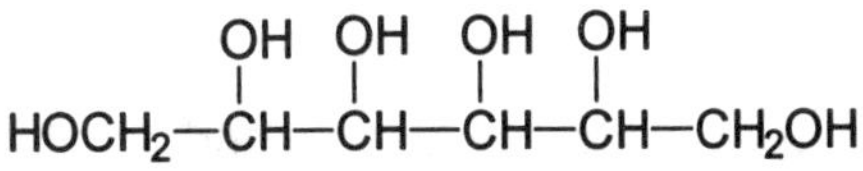 HOCH$_2$—CH—CH—CH—CH—CH$_2$OH (with OH OH OH OH substituents)

8.11 **(a)** 3-ethyl-2-hexanol **(b)** 4,7-dimethyl-5-decanol
 (c) 4-methyl-3-hexanol **(d)** 5,6-dimethyl--3-heptanol

8.12 **(a)** *trans*-4-methylcyclohexanol **(b)** 4-cyclopentyl-1-butanol
 (c) trans-2-bromocyclooctanol **(d)** 2-bromo-3-cyclohexyl-1-propanol

8.13 (E)-6-nonen-1-ol

8.14 2-ethyl-1,3-hexanediol

8.15 Methylene chloride because it has a dipole moment. Carbon tetrachloride has no dipole moment.

8.16 The polar tribromomethane has the lower molecular weight and is less polarizable than the higher molecular weight tetrabromomethane.

8.17 Their molecular weights are similar, 176.4 for chloroiodomethane versus 173.8 for dibromomethane.

8.18 The densities of the two compounds should be similar because they are isomers.

8.19 The (E) isomer has no dipole moment because the bond moments of the two C—Cl groups cancel one another.

8.20 The anti conformation has no dipole moment because the bond moments of the two C—Cl groups cancel one another. Therefore there must be a substantial amount of the gauche conformation in the conformational equilibrium mixture to give a dipole moment of 1.19 D.

8.21 The two hydroxyl groups in 1,2-hexanediol can form more hydrogen bonds with water.

8.22 Ethylene glycol has two hydroxyl groups and can form more hydrogen bonds with water.

8.23 The nonpolar hydrocarbon portion of the molecule is larger in 1-butanol than in 1-propanol.

8.24 The 2-methyl-1-propanol molecule has a more spherical shape, so it interferes less with the network of hydrogen bonded water molecules.

8.25

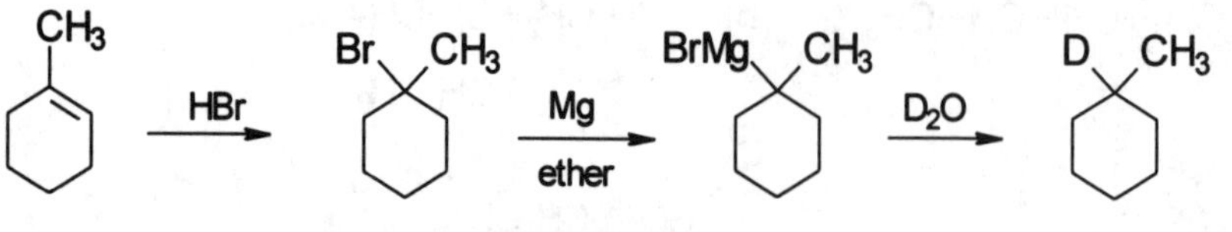

8.26 $Br-CH_2CH_2CH_2CH_2-Br$ $\xrightarrow[\text{ether}]{\text{Mg}}$ $BrMg-CH_2CH_2CH_2CH_2-MgBr$ $\xrightarrow{D_2O}$ $D-CH_2CH_2CH_2CH_2-D$

8.27   Using a Gilman reagent and an alkyl halide, two possible combinations are (1) the Gilman reagent lithium di(3-methyl-1-iodobutyl) cuprate and 1-iodobutane and (2) the Gilman reagent lithium dibutyl cucrate and 3-methyl-1-iodobutane.

8.28   **(a)**

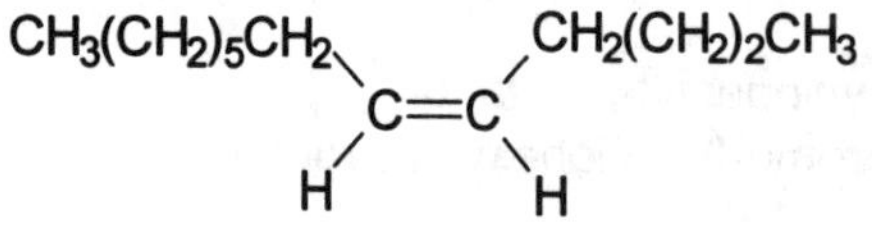

(b)

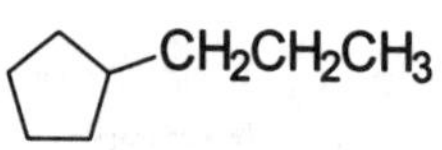

(c)

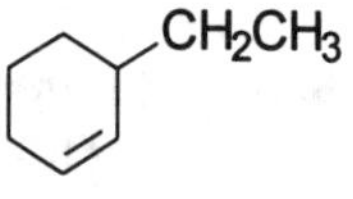

8.29   **(a)** $CH_3CH_2CH_2CH_2CH_2-I$ **(b)** $N\equiv C-CH_2CH_2CH_2-C\equiv N$

(c)

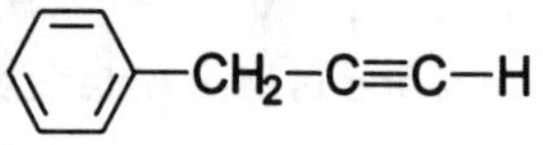

(d)

8.30   **(a)** 1-bromopropane and sodium acetylide
(b) 1-bromo-2-methylpropane and sodium cyanide
(c) 1-bromoethane and sodium ethylthiolate
(d) benzylbromide and sodium hydroxide

8.31   **(a)** 1-chlorohexane because it is a primary halogen compound; 2-chlorohexane is a secondary halogen compound
(b) bromocyclohexane because it is a secondary halogen compound; 1-bromo-1-methylcyclohexane is a tertiary halogen compound
(c) 2-bromo-4-methylpentane because it is a secondary halogen compound; 2-bromo-2-methylpentane is a tertiary halogen compound

8.32   All are primary halogen compounds. The differences in rates are due to steric hindrance of the methyl groups in II and III. The order of reactivity is II < III < I based on the distance of the methyl group from the site of the reaction.

8.33   **(a)** 1-bromo-1-methylcyclohexane is a tertiary halogen compound and is more reactive under S_N1 conditions; bromocyclohexane is a secondary halogen compound
(b) 2-bromobutane is a secondary halogen compound and is more reactive under S_N1

      conditions; 1-bromo-2-methylpropane is a primary halogen compound

    (c) 2-bromo-2-methylbutane is a tertiary halogen compound and is more reactive under S_N1 conditions; 2-bromobutane is a secondary halogen compound

8.34   III < I < II, which is in the order primary < secondary < tertiary

8.35   The product is 3-chloro-1-iodohexane, which results from S_N2 displacement of the primary halogen by iodide ion rather than displacement of the secondary halogen at the C-3 atom.

8.36   Displacement of one chloride ion by sulfide gives an intermediate that can intramolecularly displace a second halide ion.

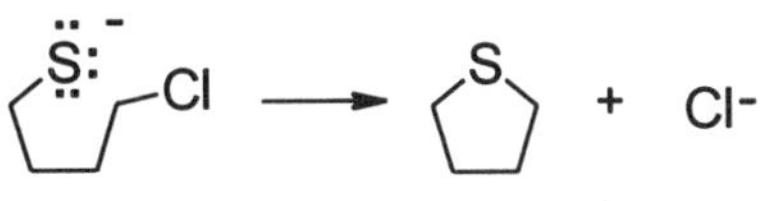

8.37   The compound has a tertiary C—Cl bond which reacts under S_N1 conditions to give a tertiary carbocation. Capture of the carbocation can occur from either face of the planar site, resulting in two geometric isomers.

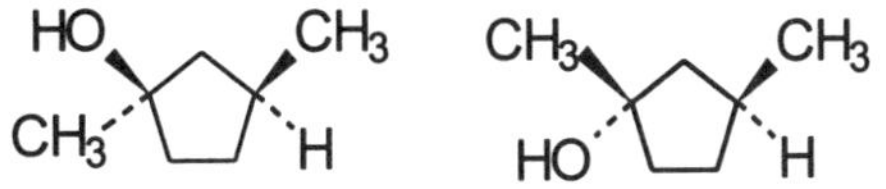

8.38   The same allyl-type carbocation results from either of the two compounds under S_N1 conditions which results in the same product.

8.39   Reaction of the trans isomer requires attack of the nucleophile by a path over the cyclohexane ring to displace the equatorial halogen. This process is more sterically hindered than attack of the nucleophile to displace the axial halogen in the cis isomer via a path in the equatorial plane of the ring.

8.40   Compound I is less sterically hindered and reacts at the faster rate.

8.41   The compound should be more acidic as a result of inductive electron withdrawal by the three chlorine atoms. The K_a should be greater than 10^{-19}, the K_a value of 2-methyl-2-propanol.

8.42   Bromine is less electronegative than chlorine. Thus, the K_a value should be between 1.3×10^{-16} and 5×10^{-15}, the values for ethanol and 2-chloroethanol, respectively.

8.43   Cyclohexanol is a secondary alcohol as is 2-propanol. Its K_a value should be 10^{-18} like that of 2-propanol.

8.44   The *tert*-butoxide ion should be a stronger base than methoxide ion, because *tert*-butyl alcohol is a weaker acid than methanol.

8.45   III < I < II, which is the order primary < secondary < tertiary

8.46   II < III < I, which is the order primary < secondary < tertiary

8.47

(a) $CH_3-\overset{\overset{\displaystyle CH_3}{|}}{\underset{\underset{\displaystyle CH_3}{|}}{C}}-CH_2-CH_2-Br$ (b) $CH_3-\overset{\overset{\displaystyle CH_2-CH_3}{|}}{\underset{\underset{\displaystyle Br}{|}}{C}}-CH_2-CH_3$ (c) $CH_3-\overset{\overset{\displaystyle H}{|}}{\underset{\underset{\displaystyle CH_3}{|}}{C}}-CH_2-\overset{}{\underset{\underset{\displaystyle Br}{|}}{CH}}-CH_3$

8.48   (a)   (b)   (c)

8.49   The reaction proceeds via a carbocation intermediate that has positive charge on two carbon atoms, either of which can react with the bromide ion. The two resonance forms of the carbocation are

$$CH_3-CH=CH-\overset{+}{C}H_2 \longleftrightarrow CH_3-\overset{+}{C}H-CH=CH_2$$

8.50   Both compounds are tertiary alcohols. However, the first compound yields a tertiary allylic carbocation in an S_N1 reaction. As a result, the transition state for formation of this intermediate has a lower activation energy and the reaction occurs at a faster rate.

8.51   Compound I in Exercise 8.45 can rearrange from a secondary carbocation intermediate to give a tertiary carbocation. Compound II in Exercise 8.46 can rearrange from a primary carbocation intermediate to give a tertiary carbocation. Compound III in Exercise 8.46 can rearrange from a secondary carbocation intermediate to give a tertiary carbocation.

8.52   The carbocation with positive charge at the C-2 atom can rearrange via a hydride shift of hydrogen at the C-3 atom to give a carbocation with positive charge at the C-3 atom.

8.53   Addition of -326 kJ mole^{-1} for reaction of zinc with bromine to give zinc bromide to the 105 kJ mole^{-1} for the debromination reaction yields -221 kJ mole^{-1} for the "indirect" debromination reaction using zinc.

8.54   Addition of -55 kJ mole^{-1} for reaction of a proton with a hydroxide ion to give water to the 42 kJ mole^{-1} for the dehydrohalogenation reaction yields -13 kJ mole^{-1} for the dehydrohalogenation reaction using hydroxide ion.

8.55   The reaction of 2-methylpropane would require a lower temperature because the alkene formed is disubstituted and thus more stable than ethylene produced from ethane.

8.56   The reaction of ethylbenzene occurs at a lower temperature because the alkene formed is resonance stabilized as a result of interaction of the vinyl group with the benzene ring.

8.57 At a high temperature, dehydrogenation can occur to give small amounts of various unsaturated compounds. These compounds can then add hydrogen and yield mixtures of isomers. The trans isomer is the more thermodynamically stable isomer and should predominate at equilibrium. One unsaturated compound that can yield both isomers is

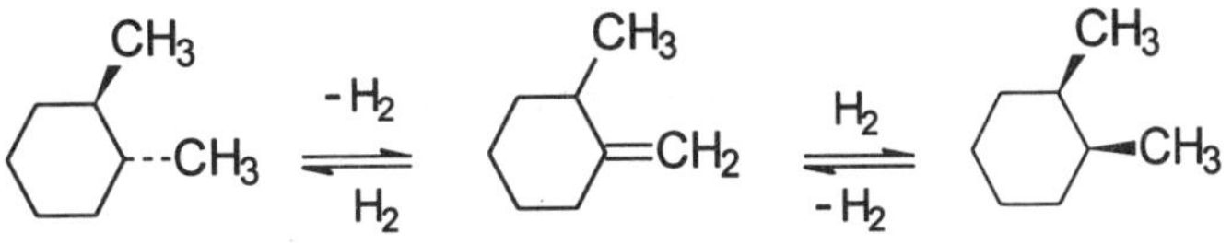

8.58 The endo isomer results from preferential addition of hydrogen from the more sterically accessible face of the double bond, which is the exo side. Under equilibrium conditions, the more thermodynamically stable isomer is formed, which is the exo isomer. The methyl group in this isomer is in an equatorial-like environment as compared to the axial-like environment of the endo isomer.

8.59 The bond dissociation energy of C—Br is smaller than that of the C—Cl bond. Dehalogenation of vicinal dichloroalkanes should require a higher temperature.

8.60 The $\Delta H°_{rxn}$ for the reaction of iodide ion with bromine to give IBr and bromide ion is expected to be negative because the I—Br bond is more polar that the Br—Br bond and as a result has a larger bond dissociation energy. Addition of this reaction to that of "direct" debromination of a vicinal dibromide gives the equation for the "indirect" debromination, which will have a more negative $\Delta H°_{rxn}$.

8.61 A trans diaxial arrangement of the two C—Br bonds is required for debromination. The *tert*-butyl group has a large conformational preference and is in the equatorial position in both compounds. Thus, in II both bromine atoms are in axial positions. In I both bromine atoms are in equatorial positions. Compound II will react, compound I will not react.

8.62 The fused rings of *trans*-decalin are conformationally rigid. As a result, the trans-bromine atoms of the first compound are both in equatorial positions. In the second compound they are in axial positions. Since a trans diaxial arrangement of the C—Br bonds is required for debromination, only the second compound will react.

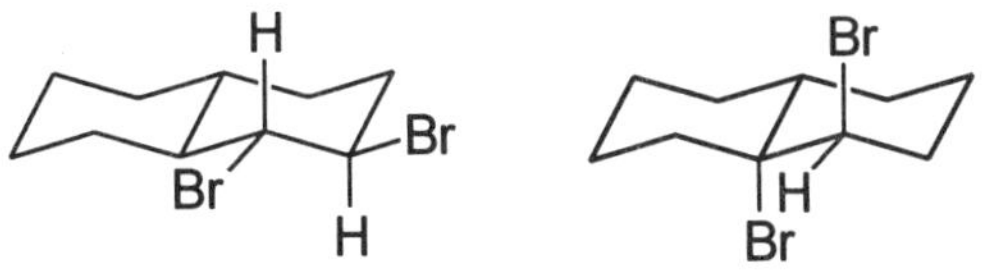

8.63 I, IV, VII, XI, and XII

8.64 XIII only

8.65 IX and XIV

8.66 X only

8.67 II, III, V, VI, VIII, and IX

8.68   VI, VIII, and IX can give trisubstituted alkenes; XIV can give a tetrasubstituted alkene

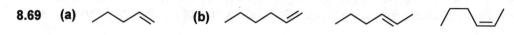

8.69   (a) [structure]   (b) [structures]

(c)  [structures]

(d)  [structures]

8.70   (a) 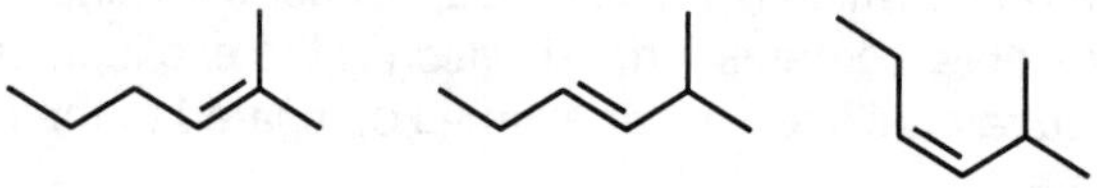 [structures]

(b) 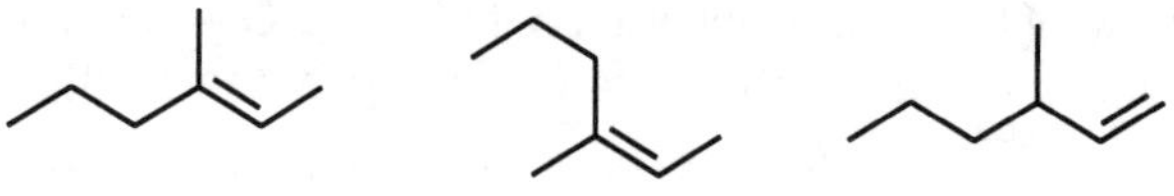 [structures]

(c) 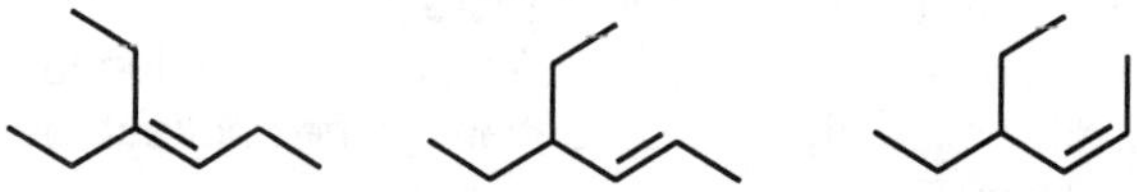 [structures]

(d) 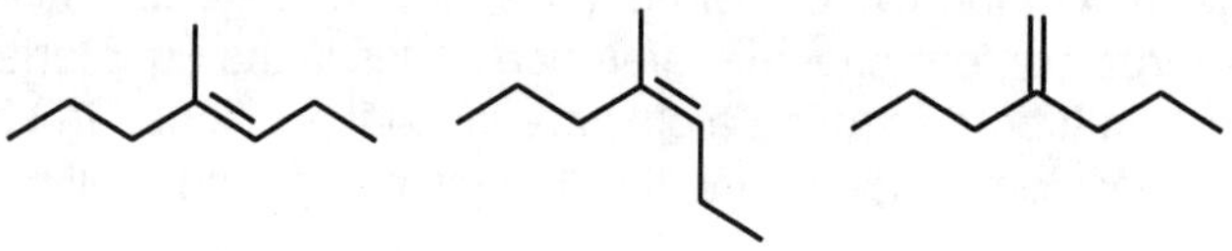 [structures]

8.71

(a) 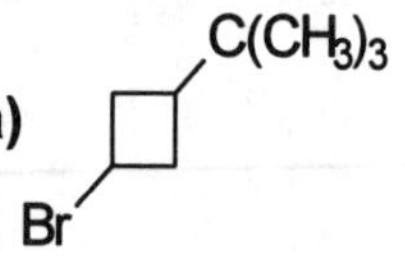 [cyclobutane structure with C(CH₃)₃ and Br]   (b) 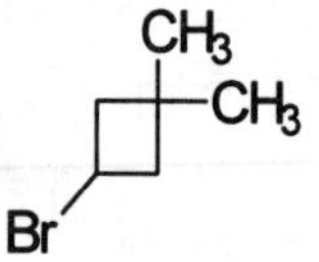 [cyclobutane structure with CH₃, CH₃ and Br]   (c) 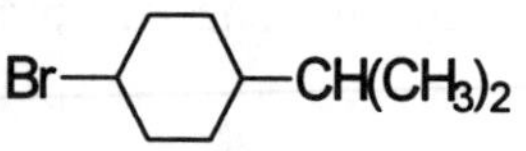 Br–[cyclohexane]–CH(CH₃)₂   (d) 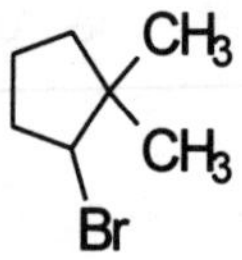 [cyclopentane structure with CH₃, CH₃ and Br]

8.72   Only **(a)** and **(c)** from the following compounds, respectively.

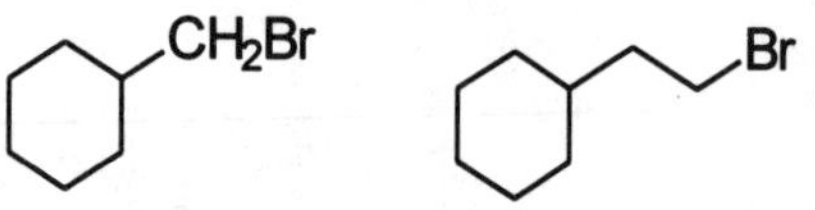

[cyclohexane with CH₂Br]   [cyclohexane with CH₂CH₂Br]

8.73   The required trans-diaxial arrangement of C—H and C—Cl cannot be achieved even by a ring flip. In the alternate conformation, all chorine atoms are in axial positions and all hydrogen atoms are in equatorial positions.

8.74 The fused rings of *trans*-decalin are conformationally rigid. Compound I has an axial C—Br bond that is in a position to undergo trans-diaxial elimination. Compound II does not have the required trans-diaxial arrangement, so the reaction is much slower.

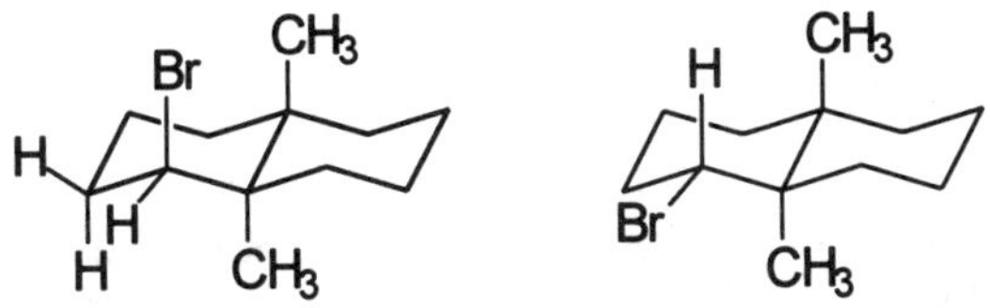

8.75 The anti periplanar arrangement of the C—H and C—Br bonds in the conformation shown gives the following (E) isomer.

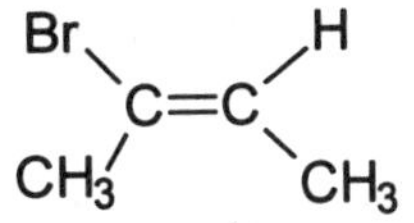

8.76 There would be a larger E/Z ratio of isomers for 1,2-diphenylethene, because the two phenyl groups present a larger steric hindrance in the formation of the (Z) isomer than the methyl and ethyl groups do.

8.77 In the conformation obtained by a ring flip, there is only one axial C—H bond situated to undergo an anti periplanar E2 elimination. That one gives the 2-methene.

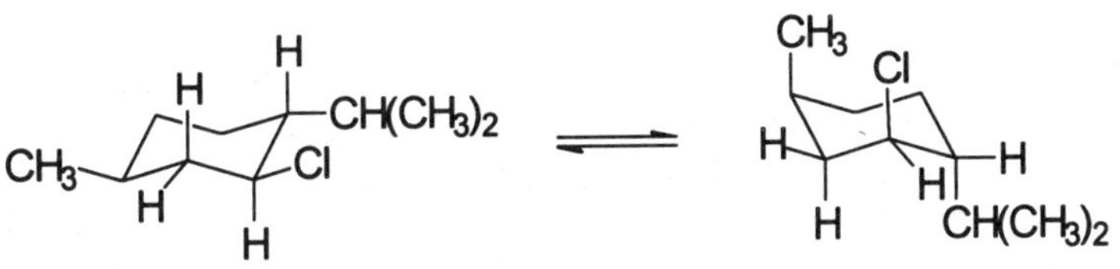

8.78 There are two C—H bonds situated to undergo anti periplanar E2 eliminations. The ratio of products is determined by the degree of substitution of the double bonds. 3-Menthene is the more stable compound because it has a trisubstituted double bond.

8.79 The axial C—H and C—Cl bonds shown in the structure can undergo an anti periplanar E2 elimination reaction.

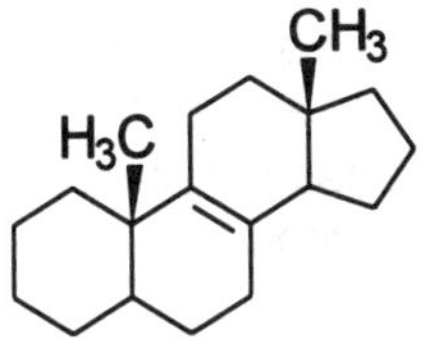

8.80 Compound I has axial C—H and C—Cl bonds as shown in the structure and can undergo an anti periplanar E2 elimination reaction. Compound II has an equatorial C—Cl bond, so it would react at a slower rate.

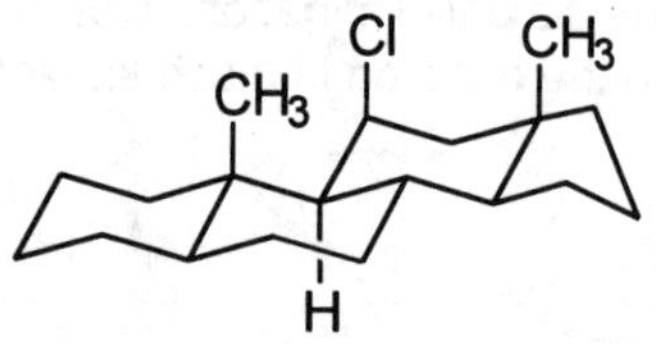

8.81   There is no anti periplanar C—H bond as required for an E2 elimination reaction. The bridgehead carbon cannot form the required planar arrangement for the carbocation that would have to be generated in an E1 reaction.

8.82   The reaction of either the required 1-halo or *trans*-2-halo compound could undergo an E2 elimination because there is a syn periplanar arrangement of the required bonds. However, the resulting compound is very strained and as a consequence a high energy transition state required would be required for elimination.

8.83   The energy required to break a C—D bond is larger than for a C—H bond. This energy difference is reflected in the transition state for the E2 reaction where these bonds are partially broken.

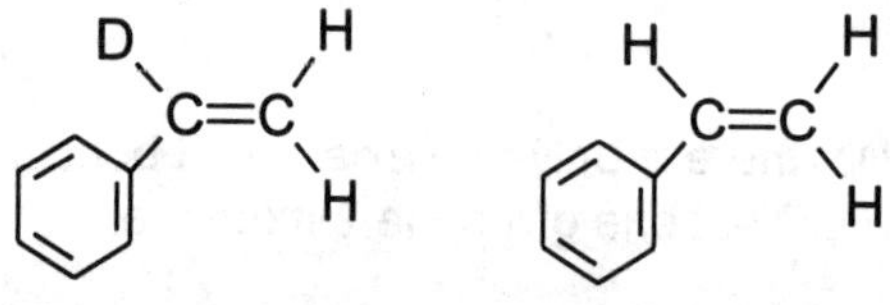

8.84   The second compound has axial C—H and C—Br bonds shown in the structure and can undergo an anti periplanar E2 elimination reaction in which the deuterium is retained. The first compound cannot easily undergo an E2 elimination reaction. However, it can eliminate the deuterium atom situated trans to the bromine atom in a twisted conformation which improves the relationship between the two bonds required for elimination of DBr.

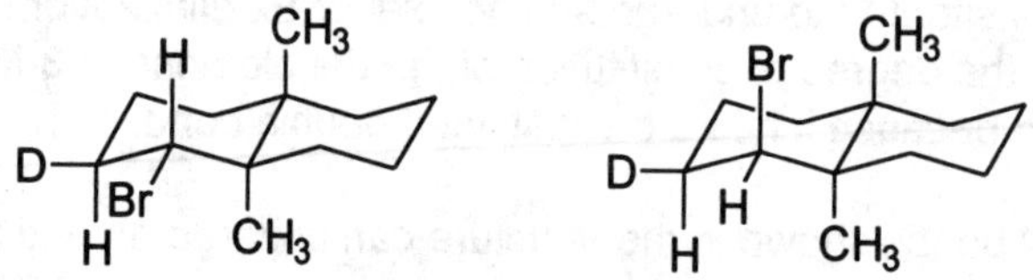

8.85   The E2 reaction occurs via a syn periplanar arrangement of the C—D and C—Br bonds even though the cleavage of a C—D bond requires more energy than the cleavage of a C—H bond. The only C—H bond available for an elimination reaction is at a 120° dihedral angle to the C—Br bond which is unfavorable for an E2 reaction.

8.86   The E2 reaction of the second compound occurs via a syn periplanar arrangement of the C—H and C—Cl bonds in the exo positions. The only C—H bond available for an elimination reaction in the first compound is at a 120° dihedral angle to the C—Cl bond, which is unfavorable for an E2 reaction.

8.87 **(a)**

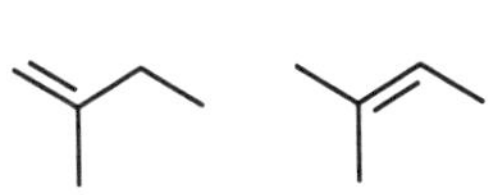

major

(b)

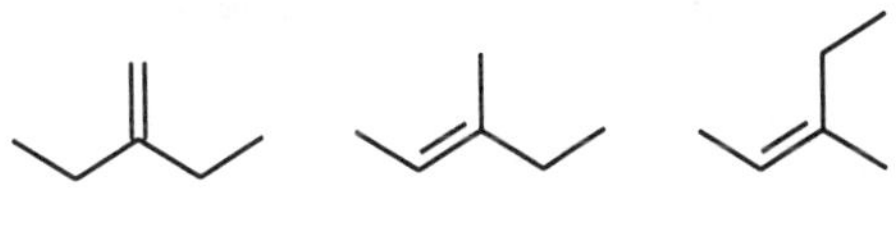

major

(c)

These two trans isomers are of comparable stability.

(d)

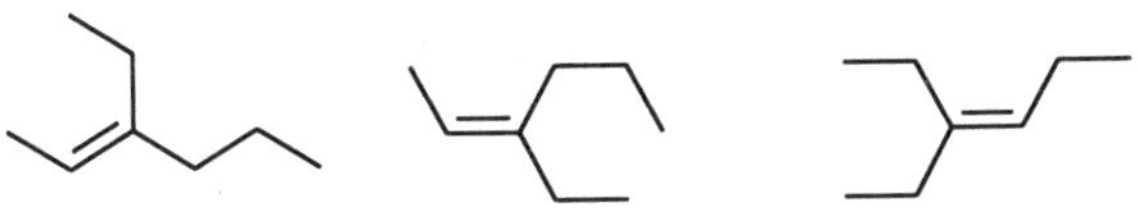

The three isomers are of comparable stability.

8.88 **(a)**

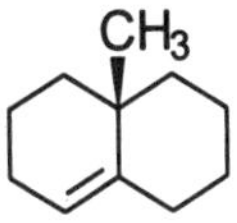

(b)

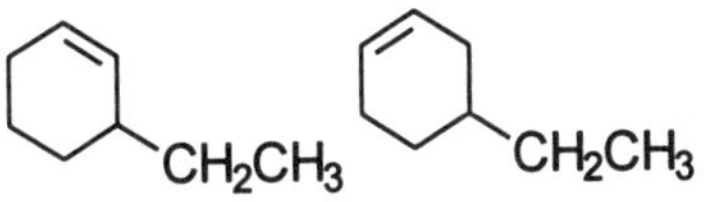

These two isomers are of comparable stability.

(c)

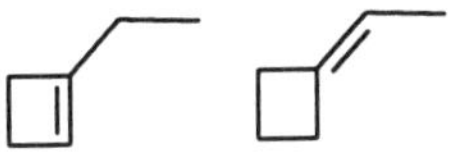

major

8.89 The expected product is the more highly substituted alkene. This product has a double bond that also can interact with the benzene ring giving a resonance stabilized structure, so the dehydration is more rapid than for 2-propanol.

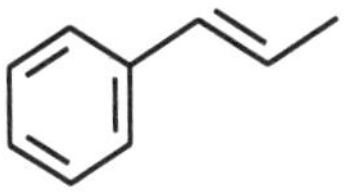

8.90 The alkene has a double bond that can interact with the two benzene rings giving a resonance stabilized structure.

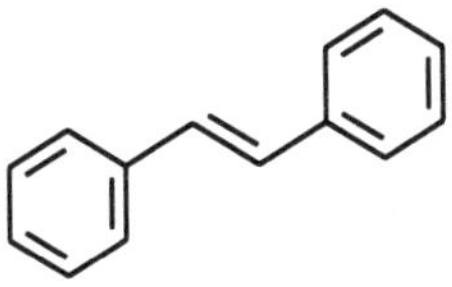

8.91 1-Methylcyclohexene, which has a trisubstituted double bond, predominates over 3-methylcyclohexene, which has a disubstituted double bond.

8.92   The two compounds are geometric isomers (E)-cyclodecene and (Z)-cyclodecene.

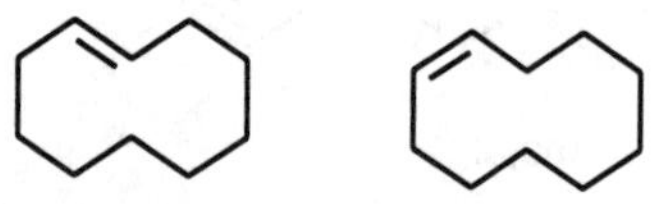

8.93   The first step, which is a migration of a methide group, converts a 2° carbocation into a 1° carbocation and is endothermic. The second step, which is a migration of a hydride ion, converts a 1° carbocation into a 3° carbocation and is exothermic.

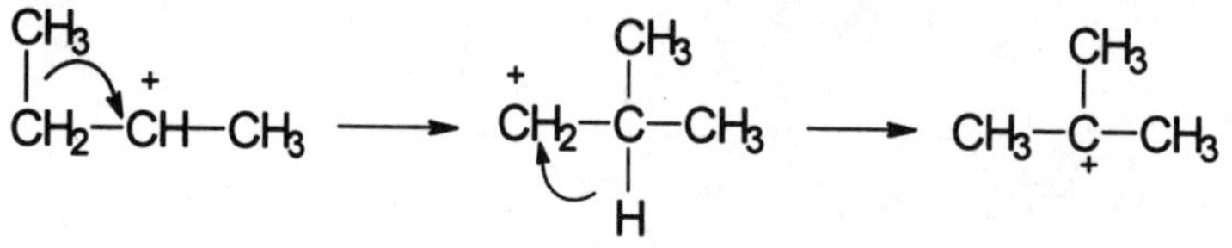

8.94   The first step, which is a migration of a hydride ion, converts a primary carbocation into a tertiary carbocation and is exothermic. The second step, which is a migration of a hydride ion, converts a tertiary carbocation into a secondary carbocation and is endothermic.

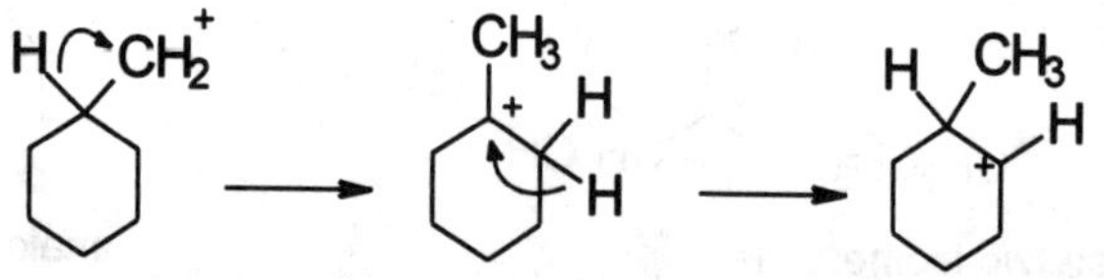

8.95   Protonation of the double bond to give a tertiary carbocation followed by loss of a proton from the C-2 atom of the cyclohexane ring.

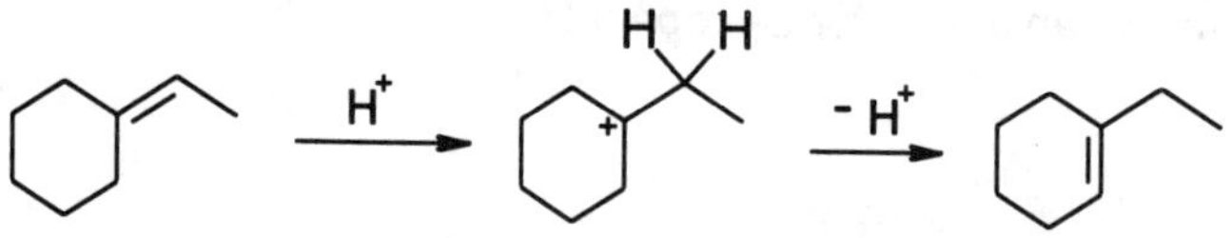

8.96   Protonation of the double bond at C-4 gives a secondary carbocation at C-3. A hydride shift from C-2 to C-3 gives a secondary carbocation at C-2. Finally loss of a proton from the C-1 atom of the cyclohexane ring forms the 1-methylcyclohexene isomer.

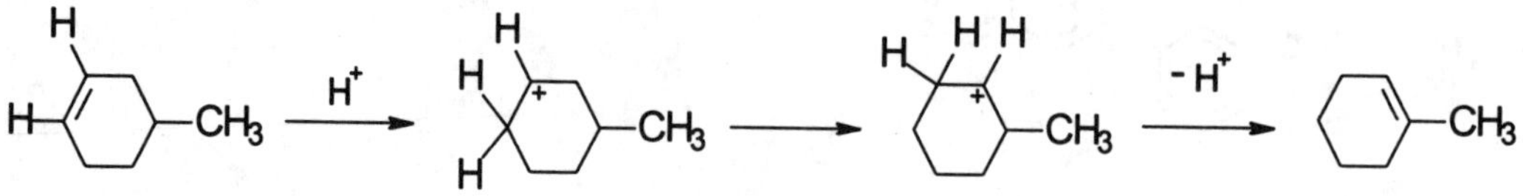

8.97   The initial secondary carbocation with charge at C-3 can rearrange either by a hydride ion shift from C-4 to give a tertiary carbocation or by a methide ion shift from C-2 to give another tertiary carbocation. These three carbocations account for the majority of the alkene products that result from loss of a proton by a carbon atom adjacent to the carbon atom bearing the positive charge. Compounds III and VI are formed from loss of a proton by carbocations that result from further rearrangement of the one of the tertiary carbocations. Note that these compounds are formed in significantly smaller amounts than the other four.

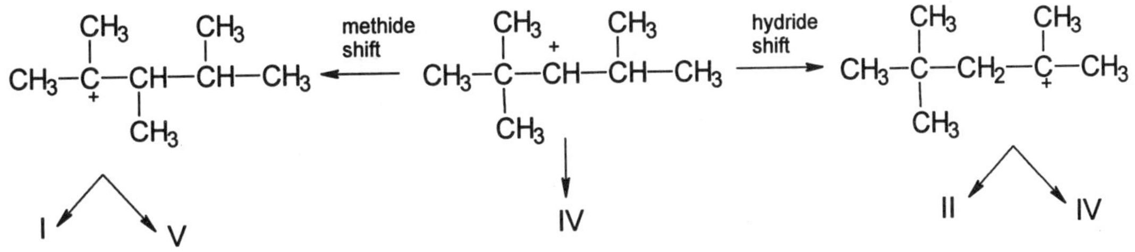

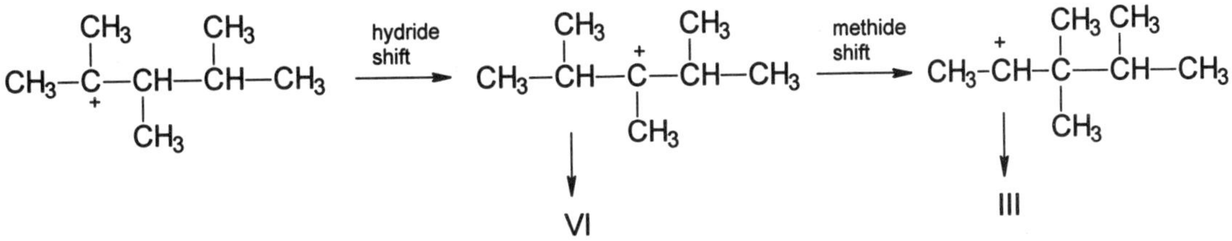

8.98 The initial secondary carbocation can rearrange by transfer of either a methyl group or a methylene group of the cyclohexane ring to form a tertiary carbocation. Subsequent loss of a proton from each of these carbocations give the observed products.

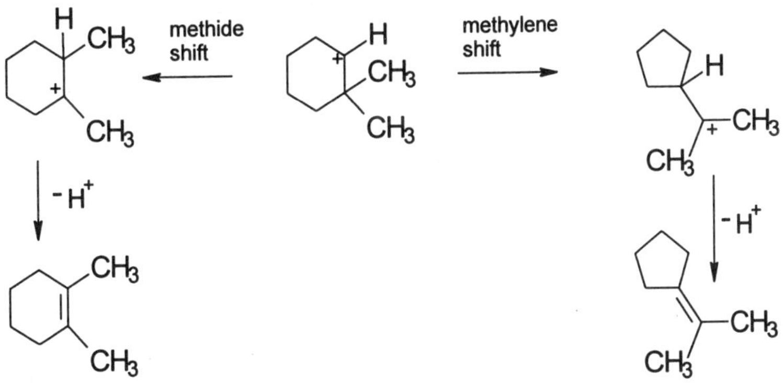

8.99 The initial secondary carbocation can rearrange by transfer of a methylene group of the cyclobutane ring to give another secondary carbocation. The driving force of the reaction is the decrease in ring strain. Subsequent loss of a proton from the cyclopentyl carbocation gives 1-methylcyclopentene.

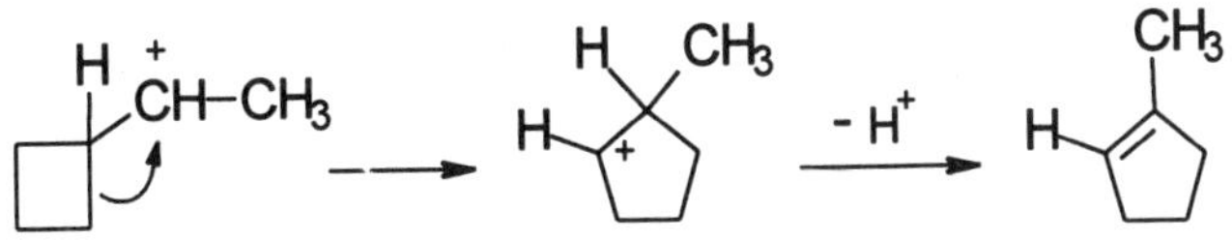

8.100 The initial tertiary carbocation can rearrange by transfer of a methylene group of the cyclobutane ring to give a secondary carbocation. Although the carbocation is less stable, the driving force of the reaction is the decrease in ring strain. Subsequent loss of a proton from the cyclopentyl carbocation gives 3,3-dimethylcyclopentene.

8.101 The initial tertiary carbocation can rearrange by transfer of a carbon atom at the C-3 atom. Another tertiary carbocation results but there is a small reduction of ring strain as a cycloheptane ring is converted into a cyclohexane ring. A subsequent methide shift forms a tertiary butyl group and generates a tertiary carbocation with the charge on the six-membered ring. This carbocation loses a proton to give a trisubstituted double bond.

8.102 The initial tertiary carbocation can rearrange by transfer of a methylene group at the adjacent carbon atom that is part of the other cyclopentane ring. Another tertiary carbocation results, but there is a reduction of ring strain as a cyclopentane ring is converted into a cyclohexane ring. This carbocation loses a proton to give a trisubstituted double bond.

9

Stereochemistry

This chapter introduces a lot of new terms to describe the arrangement of atoms in space about a tetrahedral carbon atom. In order to use these terms properly, you will have to develop skills to visualize structures written in a variety of perspective representations. Sometimes the representations are convenient and other times they are not. Occasionally you may feel like you have to stand on your head to see the necessary relationships to describe the arrangement of the atoms about the tetrahedral carbon atom. Molecular models will help, because then you can turn them around rather than moving yourself into an orientation required to view the structure on the printed page. However, eventually you do have to learn how to deal with the printed page, because that is what you will encounter on an examination. You may be allowed to use models on examinations, but the construction of models takes time. The best grades are earned by those that adapt and develop the necessary visual skills to handle three-dimensional objects as represented on a two-dimensional page.

9.1 Configuration of Molecules

Stereoisomers have different configurations. **Configuration** refers to the arrangement of atoms in space. Geometric isomers, as previously studied in cycloalkanes and alkenes, are stereoisomers. Now we find out that there is another type of stereoisomer.

9.2 Mirror Images and Chirality

Some objects have mirror images that are not **superimposable** on the original object. Such objects are **chiral**. Objects that have a **plane of symmetry** are **achiral**; they are superimposable on their mirror image.

A **stereogenic** center in an organic molecule is a tetrahedral carbon atom bonded to four different atoms or groups of atoms. It is also called a **chiral center**. By inspecting the atoms or groups of atoms bonded to each carbon atom in a molecule you can easily identify any chiral centers. If a carbon atom is bonded to two or more identical atoms or groups, such as two hydrogen atoms or two methyl groups, it is not a chiral center. If a carbon atom is bonded to four different atoms or groups, it is a chiral center, and the molecule has a nonsuperimposable mirror image. The two possible isomers having different configurations about a chiral center are **enantiomers**.

Another way to identify a molecule as chiral or achiral is to look for a plane of symmetry. A plane of symmetry can bisect atoms, groups of atoms, and bonds between atoms. In a molecule with a plane of symmetry, one side of the molecule is the mirror image of the other side. The molecule as a whole is superimposable on its mirror image if the molecule has a plane of symmetry. Thus a molecule with a plane of symmetry is achiral. If a molecule contains chiral carbon atoms and has no plane of symmetry, it is chiral.

Pairs of enantiomers have the same physical properties but behave differently in a chiral environment such as a cell. Most biomolecules are chiral and create a chiral environment that allows distinctions to be made between enantiomers.

9.3 Optical Activity

Each member of a pair of enantiomers rotates the plane of **polarized light** in an instrument called a **polarimeter**. This phenomenon is called **optical activity**. The rotation observed for one enantiomer is equal in magnitude but opposite in direction for the other enantiomer. A chiral substance that rotates plane-polarized light clockwise is **dextrorotatory**; a chiral substance that rotates plane-polarized light counterclockwise is **levorotatory**. The amount of rotation under defined standard experimental conditions is the **specific rotation**. **Optical purity** is a measure of the excess of one enantiomer over another in a mixture.

9.4 Fischer Projection Formulas

Fischer projection formulas for compounds containing one or more chiral centers are two-dimensional representations of a molecule, generated by following a set of rules. First form the eclipsed conformation to arrange all of the carbon atoms along an arc. Then select one end of the molecule termed the "more oxidized" end and place it at the top of a vertical array of atoms. Now "push" the atoms into place to give a vertical lineup of carbon atoms. The two remaining bonds to carbon that would be pointed toward the viewer are also pushed into the plane and are located on the right and left of the line of carbon atoms. For a compound with a chiral center, the location of the appropriate substituent such as a hydroxyl group is used to designate configuration. If the hydroxyl group is on the right in the Fischer projection formula, the compound is D; if it is on the left, the compound is L.

9.5 Absolute Configuration

A method to designate the absolute configuration—the exact position in space of atoms about a chiral center—depends on a set of rules much like those of nomenclature. They are designed to give an unambiguous description of the molecule to all those who know how to use the rules correctly. The method depends on assigning priorities to the four atoms immediately bonded to the chiral center. If the priority is clear, then you don't have to worry about any of the additional atoms bonded farther away. If the priorities of two bonded atoms are equal, then it becomes necessary to proceed to the next atom of each chain to eventually reach the point of first difference.

Priority is assigned to atoms based on atomic number, and additional atoms farther down the chain are ignored even though they may have still higher atomic numbers. Thus, a fluorine atom has a higher priority than a carbon atom even though that carbon atom may be bonded to three chlorine atoms. Those chlorine atoms are irrelevant because the comparison is between the atomic number of fluorine and that of carbon. With a little practice and doing a few problems you should be able to adapt to the "rules". One minor irritation is the funny way that multiple bonds in substituents are handled. However, just obey the rules and double or triple the atoms as if they were really there. In effect they are phantom atoms. Thus, a carbon atom with a triple bond is treated as if that carbon atom were actually bonded to three carbon atoms.

Once the priority order of the atoms or groups of atoms bonded to the chiral carbon atom has been determined, orient the molecule in your mind's eye with the lowest priority group pointing away from you. Now view the three groups pointing toward you and trace a path from the highest priority group to the second highest and then to the third. If you traced a clockwise path, the configuration is R; if you traced a counterclockwise path, the configuration is S. This process is one that requires some visualization skills. Each of us has a way of viewing three-dimensional objects, and there are clear differences in the ease with which this is done. No one can get inside your head and understand what you are seeing. You are going to have to learn to "move" structures or "rotate" about bonds in those structures to give the best orientation to assign the configuration. For example, you may find that you can view the order of priority of the groups from the left better than from the right. In addition you may prefer to have the bond to the lowest priority group in the plane of the page. Thus, given the structure on the left you may choose to rotate by 180° about the C—F bond to obtain the structure on the right. Now view the structure on the right by sighting along the C—H bond and determine that the structure has the S configuration.

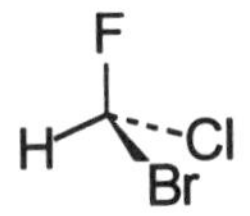

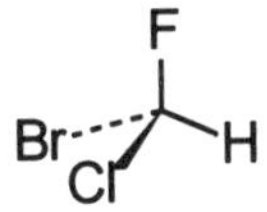

You might find that you can best determine the configuration if the group of lowest priority is placed behind the plane of the page. Using the structure on the left above you can rotate about the C—F bond by 120° to obtain the structure on the right.

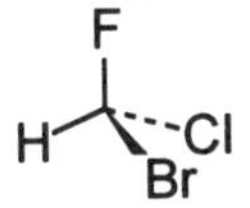

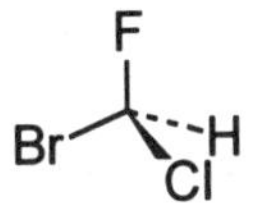

Now you can trace a path from bromine at the "8 o'clock position to chlorine at the "4 o'clock" position to fluorine at the "12 o'clock" position and determine that the configuration is S.

One little "trick" that may help in looking at molecules that are not arranged conveniently but rather have the lowest priority atom in front of the plane of the page. Consider the following structure.

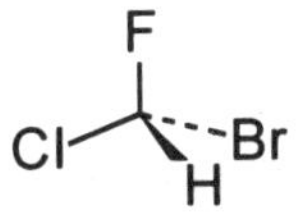

Look the "wrong" way down the bond from the low priority group toward the carbon atom. Then determine the direction of the arrangement of the other three groups. Whatever you see from this perspective is opposite to the perspective you would see if you had been looking from the carbon atom toward the low priority group. Trace a path from bromine at the "4 o'clock" position to chlorine at the "8 o'clock" position to fluorine at the "12 o'clock position. So, the direction corresponds to the R configuration, but it is actually S. Think about how the hands of a clock would appear to move if the clock were transparent and you could look at it from the back.

Note that the assignment of R or S configuration to a compound does not identify its optical rotation as being either (+) or (-). The direction of optical rotation is experimentally determined with a polarimeter. The absolute configuration is experimentally determined by X-ray crystallography.

9.6 Molecules with Two Stereogenic Centers

The consequences of having two (or more) stereogenic centers depend on whether those centers are equivalent or nonequivalent. **Equivalent** centers have identical sets of substituents (which in this case includes hydrogen and carbon atoms). For n nonequivalent centers, there are 2^n stereoisomers. Among those isomers there are pairs of enantiomers. These stereoisomers have opposite configurations at every center and are thus mirror images. All other stereoisomers are termed **diastereomers**. The configurations of some centers may be opposite or not, but all combinations other than the one specific arrangement of enantiomers are diastereomers.

The configuration of each stereogenic center is determined as if it were the only one. Then the configuration of all centers is written as in 2S,3R,4S. The configuration of its enantiomer must be 2R,3S,4R. Any other combination means that the stereoisomer is a diastereomer.

Compounds with two or more equivalent stereogenic centers have fewer stereoisomers than predicted by the 2^n formula. Some of the stereoisomers have a plane of symmetry and are not optically active. Such compounds are **meso**. For two centers, the configurations are R,S, which is the same as S,R because of the plane of symmetry. The isomers R,R and S,S are optically active and are enantiomers.

9.7 Cyclic Compounds with Stereogenic Centers

Cyclic compounds may have one, two, or more stereogenic centers. The same rules apply to assign configuration to cyclic compounds as were used for acyclic compounds. The only difference is that you would eventually end up back at the stereogenic center as you move along the chain of atoms that makes up the ring. However, the point of first difference is usually reached before that time.

Cyclic compounds having two nonequivalent stereogenic centers can exist in four stereoisomeric forms. An interesting feature of these molecules is seen when there are equivalent stereogenic centers. In those cases, there is at least one plane of symmetry. That plane in some cases bisects bonds and in other cases bisects the atoms of the ring. In this latter case it also bisects the atoms bonded to the stereogenic centers.

9.8 Separation of Enantiomers

Enantiomers have the same physical properties and therefore cannot be separated by physical methods. However, diastereomers have different physical properties and can be separated. The process of converting a mixture of enantiomers (**racemic mixture**), by reaction with a single enantiomer of a second compound, into a mixture of diastereomers is illustrated in Figure 9.18 in the text. The diastereomers are separated, after which they are broken down to obtain one enantiomer from one diastereomer and the other enantiomer from the second diastereomer.

9.9 Reactions of Compounds with Stereogenic Centers

The simplest case is considered first. If the reaction does not involve the bonds to the stereogenic center, then the configuration at that center is unchanged. Although it doesn't usually happen, the R,S designation might be different. Usually the priority of the groups is unchanged by the reaction, as in the case of a CH_2OH group being converted into a CH_2OCH_3 group or even a CH_2Cl group. It still may be the highest priority group, especially if all of the other groups are alkyl groups. However, if two of the groups are CH_3 and CH_2F and the reaction involves reaction with chorine to give a CH_2Cl group from the methyl group, then the order of priorities has changed. Initially CH_2F had a higher priority than CH_3. But now the CH_2Cl has a higher priority than CH_2F. This is a little trick that you should watch out for on a test.

Reactions at the stereogenic center affect the configuration of the molecule. Consequently, studies of the optical activity of compounds are used to determine mechanisms. In the case of the S_N2 reaction, the product has a configuration opposite that of the reactant. Thus, we postulate a transition state in which the nucleophile attacks opposite the bond to the leaving group and inverts the configuration as the reaction occurs.

Radical reactions proceed through a planar intermediate which is achiral. Thus, subsequent reaction with another radical can occur with equal probability from either side of the plane of the molecule. The result is a racemic mixture.

We will see in future chapters that retention of configuration, inversion of configuration, or formation of a racemic mixture is an important clue in identifying mechanisms.

9.10 Formation of Compounds with Stereogenic Centers

Formation of compounds with one stereogenic center from achiral compounds using achiral reagents cannot yield a single stereoisomer. However, in cells the reaction of an achiral compound with the chiral reactants within the cell generates a single stereoisomer.

In some reactions where two stereogenic centers are generated from an achiral substrate, some mechanistic information is obtained based on the diastereomers formed. For example the formation of two equivalent centers might give a mixture of the R,R compound and the S,S compound. Although not optically active, that result is different than a process that gives the R,S (meso) compound. Such analysis is used in determining the mechanisms of reactions in later chapters.

9.11 Formation of Diastereomers

If a stereogenic center is generated in a reaction of a substrate that already has a stereogenic center, then a mixture of diastereomers results. The amounts of these isomers are not equal, because the new center is generated in a chiral environment which is the molecule itself. The excess of one diastereomer over another is called the **stereoselectivity** of the reaction.

9.12 Prochiral Centers

In a chiral environment, two apparently equivalent groups can be distinguished and the resulting product of a reaction involving those groups is chiral. The atomic center at which optical activity may result is **prochiral**. The equivalent groups bonded to the prochiral center are **enantiotopic** and are designated pro-R or pro-S to indicate the potential configuration if the group is replaced. Groups at a prochiral center in a molecule that contains a chiral center are **diastereotopic**.

The "faces" of a planar site or functional group that contains a center that can be converted into a stereogenic center are designated as **re** or **si** depending on the priority of the three groups and their arrangement using the R,S conventions.

Solutions to Exercises

9.1   **(a)** none   **(b)** one   **(c)** none

9.2   **(a)** none   **(b)** one   **(c)** one

9.3   2-chloropentane and 2-chloro-3-methylbutane each have one chiral center

9.4   1,2-dichloropropane

9.5   **(a)** one   **(b)** two   **(c)** two   **(d)** none

9.6   **(a)** two   **(b)** one   **(c)** none   **(d)** none

9.7   **(a)** none   **(b)** one   **(c)** one   **(d)** none

9.8   **(a)** none   **(b)** one   **(c)** none   **(d)** two

9.9   **(a)** six   **(b)** seven

9.10   **(a)** six   **(b)** five

9.11   Only (b) and (c) have a plane of symmetry.

9.12   Only (a) and (d) have a plane of symmetry.

9.13   The compounds are enantiomers.

9.14   The specific rotation of the enantiomer is –5.3.

9.15   The optical rotation of a compound is not related to its configuration. Thus, an S isomer can have a positive sign of rotation.

9.16   The R refers to the configuration at the chiral center and is related to the priority sequence of the groups bonded to the stereogenic center. The (–) refers to the sign of the optical rotation of the compound.

9.17   The concentration is 0.06 g/mL. The 10 cm tube is 1 dm long. Using the observed rotation and substituting into the equation on page 353, the specific rotation is 50.

9.18   Using the equation on page 353, the concentration is calculated as 2.4 g/mL.

9.19   The synthetic sample has a majority of the enantiomer of the opposite configuration of the reference (+) isomer. The optical purity of the synthetic sample is calculated as (4.5/13.9) x 100% = 32%, with the majority being the (–) isomer. If there is x% of the (+) isomer, there is (100 – x) % of the (–) isomer. Solving the equation 32% = (100-x) - x, there is 66% of the (–) isomer and 34% of the (+) isomer.

9.20   The synthetic sample has a majority of the reference (S) isomer. The optical purity of the synthetic sample is calculated as (6/24) x 100% = 25%, with the majority being the (S) isomer. If there is x% of the (S) isomer, there is (100 – x) % of the (R) isomer. Solving the equation 25% = x - (100-x), there is 62% of the (S) isomer and 38% of the (R) isomer.

9.21 The structure must first be rotated to obtain an eclipsed conformation of the CH_3 and CO_2H groups and turned to place the most oxidized group at the top. The Fischer projection of the enantiomer given is the first of the series of four stereoisomers depicted below.

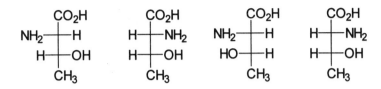

9.22 Compounds I and II are enantiomers, as are compounds III and IV. Any other pairs of stereoisomers are diastereomers.

9.23 **(a)** —OH < —OCH₃ < —SH < —SCH₃

(b) —CH₂Cl < —CH₂Br < —Cl < —Br

(c) —CH₂—CH=CH₂ < —CH₂—C≡C—H < —C≡C—CH₃ < —CH₂—O—CH₃

(d) —CH₂CH₃ < —CH₂CH₂Cl < —CH₂OH < —OCH₃

9.24

(a) $-\overset{\overset{\displaystyle O}{\|}}{C}-CH_3$ < $-\overset{\overset{\displaystyle O}{\|}}{C}-OH$ < $-O-\overset{\overset{\displaystyle O}{\|}}{C}-CH_3$

(b) $-\overset{\overset{\displaystyle O}{\|}}{C}-NH_2$ < $-NH-\overset{\overset{\displaystyle O}{\|}}{C}-CH_3$ < $-O-\overset{\overset{\displaystyle O}{\|}}{C}-CH_3$

(c) $-\overset{\overset{\displaystyle O}{\|}}{C}-Cl$ < $-O-\overset{\overset{\displaystyle O}{\|}}{C}-CH_2Br$ < $-S-\overset{\overset{\displaystyle O}{\|}}{C}-CH_3$

(d) —C≡CH < —C≡N < —N≡C

9.25 **(a)** —CH₂CH₃ < —CH=CHCl < —C≡CH < —OH

(b) 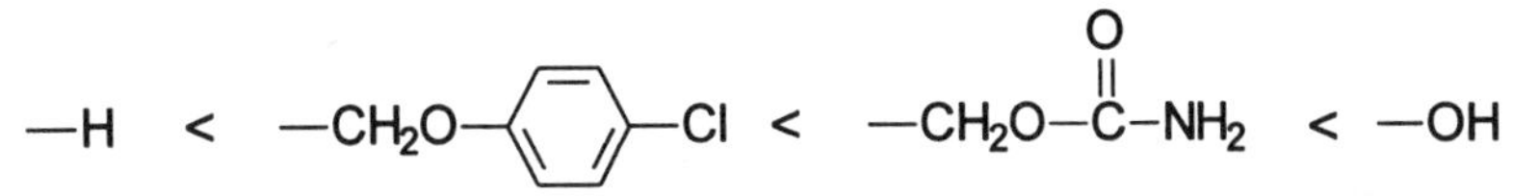 —H < —CH₂O—⟨benzene ring⟩—Cl < $-CH_2O-\overset{\overset{\displaystyle O}{\|}}{C}-NH_2$ < —OH

(c) —H < —CH$_3$ < —CH$_2$O—[2,6-dimethylphenyl] < —NH$_2$

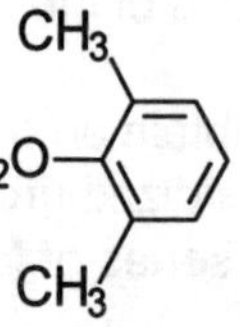

9.26 **(a)** —H < —CH$_2$CH$_2$N(CH$_3$)$_2$ < —[C$_6$H$_4$]—Br < —[pyridyl (N)]

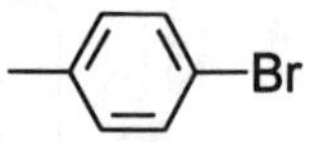

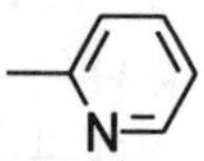

(b) —H < —CH$_2$CH$_2$NHCH$_3$ < —[C$_6$H$_5$] < —[C$_6$H$_4$]—CF$_3$

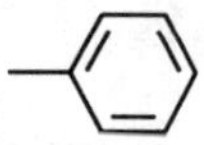

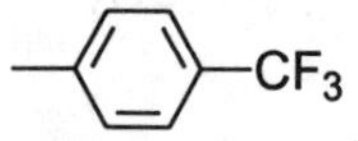

(c) —H < —CH$_2$CO$_2$H < —[C$_6$H$_4$]—Cl < —CH$_2$NH$_2$

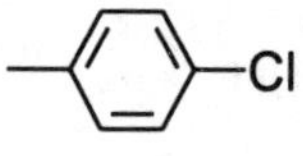

9.27 **(a)** 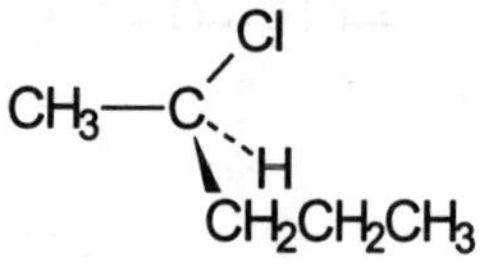 CH$_3$—C(Cl)(H)(CH$_2$CH$_2$CH$_3$)   **(b)** CH$_3$CH$_2$—C(Cl)(H)(CH=CH$_2$)   **(c)** (CH$_3$)$_2$CH—C(Cl)(H)(CH$_2$CH$_3$)

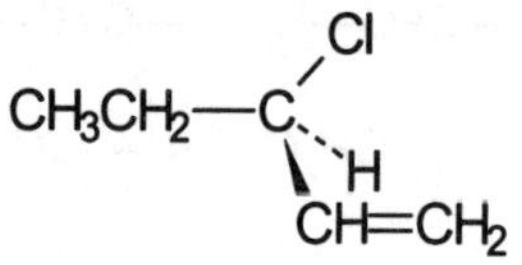

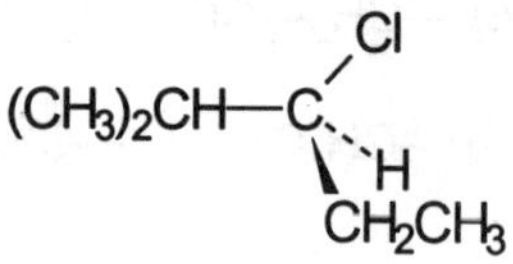

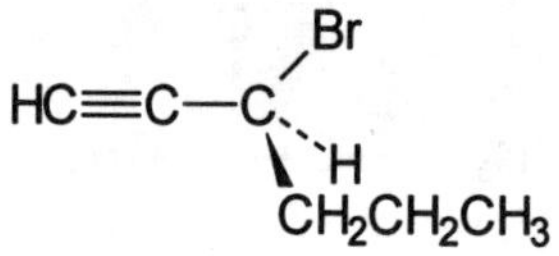

9.28 **(a)** 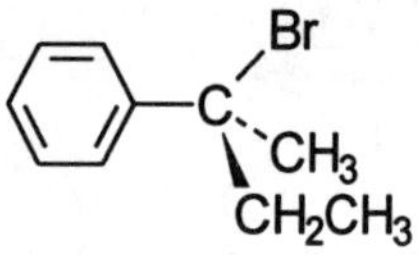 [C$_6$H$_5$]—C(Br)(CH$_3$)(CH$_2$CH$_3$)   **(b)** HC≡C—C(Br)(H)(CH$_2$CH$_2$CH$_3$)   **(c)** Br—C(Cl)(CH$_3$)(CH$_2$CH$_3$)

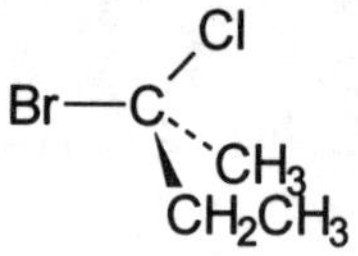

9.29 **(a)** S   **(b)** S   **(c)** R   **(d)** S

9.30 **(a)** S   **(b)** S   **(c)** S   **(d)** S

9.31 R

9.32 S

9.33 Assignments are given for the stereogenic center at the left followed by the one at the right.

(a) R, R   **(b)** S, S   **(c)** R, S   **(d)** S, S

9.34   Assignments are given for the stereogenic center at the left followed by the one at the right.

(a) R, R   **(b)** R, S   **(c)** R, R   **(d)** R, S

9.35   Assignments are given for the stereogenic center at the left followed by the one at the right.

(a) R, R and is not meso   **(b)** R, S and is meso   **(c)** S, S and is not meso

9.36   Assignments are given for the stereogenic center at the left followed by the one at the right.

(a) R, R and is not meso   **(b)** R, S and is meso   **(c)** R, S and is not meso

9.37   Ribitol has a plane of symmetry bisecting the Fischer projection through the C-3 atom and its attached C—H and C—OH bonds. Ribose does not have a plane of symmetry at that point because the "top" and "bottom" groups are not equivalent.

9.38   (a) and (b) are meso compounds. Compound (a) has a plane of symmetry bisecting the Fischer projection through the C-3 atom and its attached carbonyl oxygen atom. Compound (b) has a plane of symmetry bisecting the Fischer projection between the C-3 and C-4 atoms.

9.39   There are two nonequivalent stereogenic centers. One center bears a hydroxyl group. The second center is located at the carbon atom bearing an amino group on the right in the structure. There are four possible stereoisomers.

9.40   There is only one stereogenic center. It is located at the carbon atom bearing a hydroxyl group located next to the carbonyl carbon atom. There are two possible enantiomers.

9.41   The two stereogenic centers in 2,3-dichloropentane are nonequivalent, so there are four possible stereoisomers. The two stereogenic centers in 2,4-dichloropentane are equivalent. A plane of symmetry can be placed through the C-3 atom. Thus, there are two enantiomers and one meso compound possible for this isomer.

9.42   (a) and (c) are meso compounds. Compound (a) has a plane of symmetry bisecting the Fischer projection through the C-3 atom and its attached C—H and C—Cl bonds Compound (c) has a plane of symmetry bisecting the Fischer projection between the C-2 and C-3 atoms.

9.43   (a) has a plane of symmetry bisecting the C-1 to C-2 bond and the C-3 to C-4 bond; (c) and (d) both have planes of symmetry bisecting the C-1 and C-3 atoms as well as the atoms bonded to these atoms.

9.44   (a) has a plane of symmetry bisecting the oxygen atom, the carbon-carbon bond of the epoxide ring and the methylene carbon atom opposite to the epoxide ring. (b) has a plane of symmetry bisecting the C-1 to C-2 bond and the C-3 to C-4 bond.

9.45   **(a)** R at the carbon bearing the bromine atom; S at the carbon bearing the methyl group.
(b) S at the carbon bearing the —OCH$_3$ group; S at the carbon bearing the hydroxyl group.
(c) R at the carbon bearing the —CH$_2$Cl group; S at the carbon bearing the hydroxyl group.

9.46   **(a)** R at the carbon bearing the chlorine atom; R at the carbon bearing the ethyl group.
(b) S at the carbon bearing the fluorine atom; S at the carbon bearing the hydroxyl group.
(c) R at the carbon bearing the —OCH$_3$ group; R at the carbon bearing the bromine atom.

9.47   The X$_S$ compound would give a diastereomeric mixture of A$_R$—X$_S$ and A$_S$—X$_S$ compounds. The A$_R$—X$_S$ compound, which is the enantiomer of the A$_S$—X$_R$ compound, will be the less soluble. Consequently the A$_R$ compound can be obtained optically pure.

9.48   The enantiomer with the larger specific rotation is the pure compound. Thus, the enantiomer with the +44 rotation is pure. The other sample is 75% optically pure.

9.49   The methyl ester also has the R configuration. The bonds to the stereogenic center are not changed, nor is the priority series of the groups bonded to that center. The —CO$_2$H group has the second highest priority in the acid, as does the —CO$_2$CH$_3$ group of the methyl ester. The sign of rotation of the ester cannot be predicted.

9.50   The reaction proceeds with inversion of configuration as the hydride displaces the bromide ion in an S$_N$2 reaction. However, the configuration is not S as it would be if a high priority group replaced the high priority bromine atom. The priority of the groups about the stereogenic center is changed. A high priority bromide ion is substituted by a low priority hydrogen atom. Thus, the configuration is R.

9.51   The reaction proceeds with inversion of configuration as the hydroxide ion displaces the bromide ion in an S$_N$2 reaction. The configuration of the alcohol is R because a high priority hydroxyl group replaces the high priority bromine atom.

9.52   The 2-octanol formed is racemic because the nucleophilic water molecule can capture the carbocation from either face of its plane.

9.53   The products have a chlorine atom bonded to carbon atoms that are not stereogenic centers, but the arrangement of the groups bonded to the C-2 atom is unchanged by the reaction, so the products are still optically active.

9.54   The reaction occurs via a free radical that results from abstraction of hydrogen from the C-2 atom that is the stereogenic center. The radical is achiral and can react with chlorine from either face of its plane. Thus, a racemic mixture results.

9.55   The reaction occurs via a free radical that results from abstraction of hydrogen. The C-2 atom is the stereogenic center and loss of hydrogen from this center yields an achiral intermediate that can react with chlorine from either face of its plane. Thus, a racemic mixture results. The C-3 atom is not a stereogenic center but becomes one once substituted by chlorine. A mixture of 2S,3S and 2S,3R product results, and it has a net optical activity because the optical rotations of the diastereomers do not cancel.

9.56   The C-3 atom is not a stereogenic center but becomes one once substituted by chlorine. The amounts of the 2S,3S and 2S,3R products formed are not equal because the second stereogenic center is generated in the chiral environment of the C-2 atom.

9.57 The C-2 atom is not a stereogenic center but becomes one once bromide ion adds to the carbocation generated at that center. The amounts of the 2R,3R and 2S,3R products formed will be unequal because the second stereogenic center is generated in the chiral environment of the C-3 atom. The 2S,3R compound is meso; the 2R,3R compound is optically active.

9.58 The proton can add to either the C-1 or C-2 atom. When it adds to the C-2 atom, a carbocation results at the C-1 atom. Reaction of that carbocation with bromide ion gives a mixture of cis and trans isomers, neither of which is optically active because both isomers have a plane of symmetry bisecting the C-1 and C-4 atoms. When the proton adds to the C-1 atom, a carbocation results at the C-2 atom. Reaction of that carbocation with bromide ion gives a mixture of cis- and trans-1-bromo-3R-methylcyclohexane. Both diastereomeric compounds are optically active. The configuration of the carbon atom bearing the methyl group remains unchanged.

9.59 The compound would probably not be metabolized because the enzymes responsible for the metabolism of the natural glucose probably do not react with the enantiomer.

9.60 Only part of the racemic mixture is metabolized because the enzymes responsible for the metabolism of one of the stereoisomeric tartaric acids do not react with the enantiomer.

9.61 Biological activity depends on the configuration of molecules that enter the organism. Usually only one member of a pair of enantiomers is biologically active. The lower activity of the enantiomer reflects this difference.

9.62 The receptors responsible for smell are chiral and interact with enantiomeric compounds via diastereomeric complexes. That interaction is perceived differently by the brain for different complexes.

9.63 There are two possible ways to add water across the double bond. Citric acid has two equivalent groups bonded to the carbon atom bearing the hydroxyl group and is achiral. Isocitric acid has two stereogenic centers and is chiral.

$$
\underset{\text{citric acid}}{HO_2CCH_2-\overset{\displaystyle CO_2H}{\underset{\displaystyle OH}{C}}-CH_2CO_2H}
\qquad\qquad
\underset{\text{isocitric acid}}{HO_2CCH-\overset{\displaystyle CO_2H}{\underset{\displaystyle OH\ \ H}{C}}-CH_2CO_2H}
$$

9.64 The enzymes are chiral and, as a result, only one enantiomer is formed. An achiral reagent will give a racemic mixture of two enantiomers of norepinephrine.

9.65 **(a)** The two fluorine atoms are prochiral ,as are the hydrogen atoms of the methylene group. Each of these atoms is enantiotopic.

(b) The two bromine atoms are prochiral, as are the hydrogen atoms of the methylene group. Each of these atoms is enantiotopic.

(c) The two hydrogen atoms of the methylene group are prochiral. They are diastereotopic because the C-1 atom is chiral.

9.66   Draw a three dimensional representation of the Fischer projection and rotate about the C-2 to C-3 bond to obtain a staggered conformation placing the deuterium and the hydroxyl group in an anti arrangement assuming that an anti addition reaction occurred . In this conformation, the two —CO_2H groups are arranged appropriately as predicted if the addition reaction occurred by an anti periplanar transition state. Thus, the addition is indeed anti.

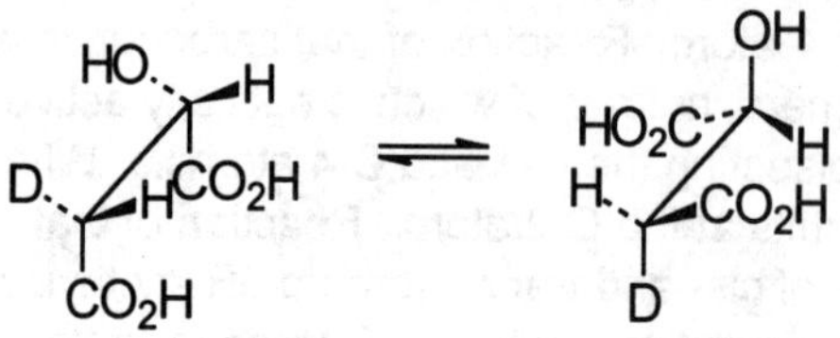

10

Nucleophilic Substitution and Elimination Reactions

In this chapter we complete the information initially presented in Chapter 8 and make use of stereochemical probes that are now possible based on Chapter 9. Because substitution and elimination reactions are two of the limited number of centrally important reactions in organic chemistry, it is important to understand all of the fine details presented in this chapter. In addition, there are underlying principles that "explain" these reactions in terms of their mechanisms. There are a lot of features that vary from reaction to reaction, and your ability to evaluate the relative importance of each feature is critical to your continued success in organic chemistry. You have to learn how to apply these principles logically within the context of the reactions presented in this chapter so that you can extend that knowledge to other reactions in later chapters.

Keys to the Chapter

10.1 Nucleophilicity and Basicity

Nucleophilicity refers to the ability of a nucleophile to displace a leaving group in a substitution reaction. Various trends in nucleophilicity are described in this section. There is a trend within a period, a trend within a group, a trend based on the charge of the nucleophile, and a steric effect of the nucleophile. Nucleophiles are also bases and can abstract protons in elimination reactions. However, in spite of the fact that nucleophilicity and basicity are both related to the availability of the same electron pair, the reactions of a series of nucleophiles do not necessarily parallel those of the same species as bases.

Within a period, nucleophilicity does parallel basicity and decreases from left to right in the periodic table for elements in similarly structured species with the same charge. Thus, hydroxide ion is a better nucleophile than fluoride ion. Nucleophilicity is decreased by hydrogen bonding of protic solvents.

Within a group, the order of nucleophilicity is opposite to the order of basicity. This order of nucleophilicity is related to the polarizability of the group. The order $I^- > Br^- > Cl^-$ is one that you should remember for use over and over again in the study of reaction mechanisms. Another important relationship is $RS^- > RO^-$.

The effect of charge on nucleophilicity is the most obvious of the trends. A species with a negative charge is more nucleophilic that a similarly structured neutral species. Thus, alkoxide ion, RO^-, is a better nucleophile than ROH.

Because nucleophiles must approach a carbon reaction center, there is some steric hindrance that affects the rate of reaction. Larger nucleophiles react at a slower rate than similarly charged smaller nucleophiles containing the same element.

10.2 Stereochemistry of Substitution Reactions

Evidence for the S_N2 and S_N1 mechanisms based on kinetics was presented in Chapter 8. This section provides the second part of the evidence—the stereochemistry of the displacement processes. In the S_N2 process **inversion of configuration** occurs, which is explained by the approach and bonding of the nucleophile in its attack from the back side—the side opposite the leaving group. In the S_N1 process a planar carbocation forms, and it is achiral. Thus, the stereochemical result of the reaction is loss of configuration and the formation of racemic product. The degree of racemization depends on the degree to which the leaving group leaves the site of the carbocation prior to its capture by the nucleophile. If both sides of the plane of the carbocation are symmetrically solvated, then complete racemization occurs. If one side is still shielded by the leaving group, then there some residual net inversion.

10.3 S_N2 Versus S_N1 Reactions

The prediction of which mechanism will prevail for a specific reaction is not always an absolute process. There are four factors to consider.

1. the structure of the substrate
2. the nucleophile
3. the leaving group
4. the solvent

As stated in Chapter 8, the order of reactivity by the S_N1 process is tertiary > secondary > primary, whereas that of the S_N2 process is primary > secondary > tertiary. Thus, at the extremes, tertiary compounds usually react by the S_N1 process and primary compounds usually react by the S_N2 process. The reaction of secondary compounds may proceed by one or the other mechanisms depending on the other factors listed above.

Two features of the structure of the substrate to note are the effect of branching at the β carbon atom and conjugation of allyl and benzyl carbocations. Branching at the β carbon atom decreases the rate of the S_N2 process due to steric hindrance. Conjugation of the allyl and benzyl carbocations makes primary substrates that can form these intermediates behave as if they were secondary substrates. Secondary substrates that can yield the resonance stabilized allylic and benzylic-type carbocations behave like tertiary substrates.

The nucleophile can affect which mechanism prevails in those borderline substrates such as secondary compounds. A charged nucleophile tends to favor the S_N2 process. The structurally related neutral nucleophile tends to favor the S_N1 process.

The leaving group strongly affects the rate of both S_N1 and S_N2 processes. Any feature that stabilizes the negative charge of the leaving group increase the rate of the reaction. The same polarizability order of the halide ions as nucleophiles also results in $I^- > Br^- > Cl^-$ for the effect of rate for the leaving group. The range of rates is larger for the S_N1 process than for the S_N2 process. For leaving groups containing the same element, such as oxygen, the order of leaving group abilities parallels their basicity. Weak bases are better leaving groups. Thus, leaving groups that are derived from strong acids are the best leaving groups.

Polar solvents better stabilize charged intermediates such as carbocations. Thus, the rate of S_N1 process increases with increasing solvent polarity. The effect on S_N2 processes is much smaller. Aprotic solvents increase the effective nucleophilicity of the nucleophile, because decreased solvation of the nonbonding electron pair makes that pair more available for reaction with a substrate.

10.4 Mechanisms of Elimination Reactions

E2 reactions occur when both the basic reagent and the substrate are involved in the rate determining step, which is a concerted process. The E1 reaction occurs in two steps, the first of which is rate determining and does not involve the base. Thus, the formation of a carbocation in an E1 reaction is the same as the first step in the S_N1 process.

Based on the anti periplanar transition state for the E2 process, the stereochemistry of the alkene can be predicted based on the stereochemistry of both stereogenic centers of the substrate. First write the structure with the correct configuration at the centers containing the hydrogen atom to be eliminated and the leaving group. Then rotate about the carbon-carbon bond to the staggered conformation with the hydrogen atom and the leaving group anti to one another. This conformation has the correct geometry for the bonded groups that will remain in the alkene.

The extent of bond breaking of the carbon-hydrogen bond in the transition state is determined by the deuterium isotope effect. The rate of removal of the hydrogen atom from the C-H bond is slower for deuterium in place of hydrogen. If that bond is not broken in the rate determining step as in the E1 process, there is no deuterium isotope effect.

The basicity of the base increases the rate of the E2 process. Thus, the amount of E2 product increases even for substrates that can react by an E1 process. The size of the base affects the regiochemistry. More hindered bases abstract protons from less sterically hindered sites and increase the amount of the less substituted alkene.

10.5 Effect of Structure on Competing Reactions

It will take some review to get all of the many factors that control the S_N1, S_N2, E2, and E1 processes into a single picture. In general tertiary haloalkanes react by S_N1 and E1 mechanisms, and the difference between the amount of each product formed depends on the basicity versus the nucleophilicity of the reagent. Primary haloalkanes react by S_N2 and E2 mechanisms. The amount of elimination increases with the basicity of the nucleophile and with its steric size. Secondary haloalkanes are much more sensitive to the conditions of the reaction and S_N1, S_N2, E2, and E1 processes occur to varying degrees. Highly polarizable nucleophiles favor an S_N2 reaction; neutral nucleophiles are more apt to be seen in an S_N1 reaction. Aprotic solvents favor S_N2 reactions; polar solvents enhance the rate of S_N1 reactions. Effective nucleophiles that are weak bases, such as thiolates, favor substitution over elimination. Strong bases that are less effective nucleophiles, such as *tert*-butoxide, favor elimination.

Solutions to Exercises

10.1   The nucleophilicity for the atoms within a period decreases from left to right. Thus, the nitrogen atom should supply the electrons in nucleophilic substitution reactions.

$$H—\overset{..}{N}—\overset{..}{\underset{..}{O}}—H$$
$$\underset{H}{\overset{|}{}}$$

10.2   The nucleophilicity for the atoms within a family increases from top to bottom. Thus, the sulfur atom of the thiocyanate ion is more nucleophilic than the oxygen atom of the cyanate ion. In the case of the cyanate ion, the nitrogen atom is more basic and a better nucleophile than the oxygen atom, so it forms isocyanate products, R—NCO.

$$:\overset{..}{\underset{..}{O}}—C\equiv N: \longleftrightarrow :\overset{..}{O}=C=\overset{..}{N}:^{-} \qquad ^{-}:\overset{..}{\underset{..}{S}}—C\equiv N: \longleftrightarrow :S=C=\overset{..}{N}:^{-}$$

10.3   The negative charge is localized in the methoxide ion, which makes the ion an effective nucleophile. The negative charge is delocalized in the acetate ion, and the decreased charge on the oxygen atom decreases its nucleophilicity.

10.4   The nucleophilicity for the atoms within a family increases from top to bottom. Thus, the phosphorus atom of the phosphine is more nucleophilic than the nitrogen atom of the aniline.

10.5   Dimethyl sulfide is a smaller nucleophile than diethyl sulfide. Nucleophilicity decreases with increasing size of the nucleophile.

10.6   The nucleophilicity for the atoms within a period decreases from left to right. Thus, if the structures are similar, the nucleophilicity of an arsenic compound should be greater than that of a selenium compound, and the nucleophilicity of a nitrogen compound should be greater than that of an oxygen compound. The sizes of ammonia and water are similar. However, the sizes of triethylarsine and dimethyl selenide are not. The larger size of triethylarsine decreases its nucleophilicity and decreases its rate of reaction.

10.7   The optical purity is 87% - 13% = 74%. Because a mixture of enantiomers results, the mechanism must be S_N1. However, the leaving group (H_2O) substantially shields one face of the carbocation as it leaves and the bromide ion reacts with substantial inversion of configuration.

10.8   The mechanism for a substitution reaction at a primary center is S_N2. The stereogenic center at the C-2 atom is not affected in this reaction. In the process the highest priority group —CH_2OH is changed into the highest priority group —CH_2Br. Thus, the configuration is unchanged and is R.

10.9   Each time a radioactive iodide ion reacts by an inversion process in the S_N2 mechanism, the resulting molecule of inverted product cancels the optical rotation of one molecule of the reactant. Thus, two molecules of optically active 2-iodobutane are effectively "lost" for every one molecule that reacts with radioactive iodide.

10.10 In the substitution reaction with cyanide ion, the highest priority bromine group is replaced by a high priority cyanide group. The reaction occurs by inversion and the net result is a product with the R configuration. In the two step reaction, two inversion steps occur, resulting in net retention of configuration. First the replacement of bromide by iodide ion yields the R product, which subsequently is converted into the S product as cyanide replaces iodide in the second step.

10.11 The stereochemistry at the C-1 atom is inverted when iodide replaces chloride. Thus, the reaction must occur by an S_N2 process.

10.12 The stereochemistry at the C-1 atom is inverted when cyanide ion replaces bromide ion. Thus, the product has the trans configuration.

10.13 The compound is a tertiary halide, which reacts via an S_N1 mechanism and is expected to give a racemic mixture of ethyl ethers.

10.14 The benzene ring stabilizes the secondary carbocation by resonance. The resultant carbocation can be attacked from either side of the plane to give an almost completely racemized product.

10.15 A primary tosylate should tend to react by the S_N2 mechanism. However, the benzene ring can stabilize a primary carbocation by resonance and allow the reaction to occur by the S_N1 mechanism. However, the leaving tosylate group substantially shields one face of the carbocation as it leaves, so the acetate ion reacts with substantial inversion of configuration.

10.16 A secondary chloro compound may react by an S_N1 or S_N2 mechanism depending on reaction conditions and other structural features of the compound. The benzene ring can stabilize a secondary carbocation by resonance and the reaction then occurs by the S_N1 mechanism. The product is largely racemic with some residual inversion of configuration. The slight excess of inverted product results from displacement by trifluoroethoxide ion as the leaving chloride shields one face of the carbocation as it leaves.

10.17 Formation of the tosylate occurs by displacement of a chloride ion from the sulfur atom of the sulfonyl chloride by the oxygen atom of the alcohol. The reaction occurs with retention of configuration because no bonds at the C-2 stereogenic center are affected. The highest priority hydroxyl group in the reactant is transformed into the highest priority tosylate group in the product. Thus, the configuration is still R. Subsequent displacement of the tosylate group by iodide occurs with inversion of configuration. The highest priority tosylate group in the reactant is replaced by a high priority iodide group in the product. Thus, the configuration of the inverted product is S.

10.18   Both compounds are tertiary bromides which react with nucleophiles such as water in aqueous acetone by an S_N1 process. A mixture of compounds with equatorial and axial hydroxyl groups should result. The composition of the mixture is the same for both halogen compounds.

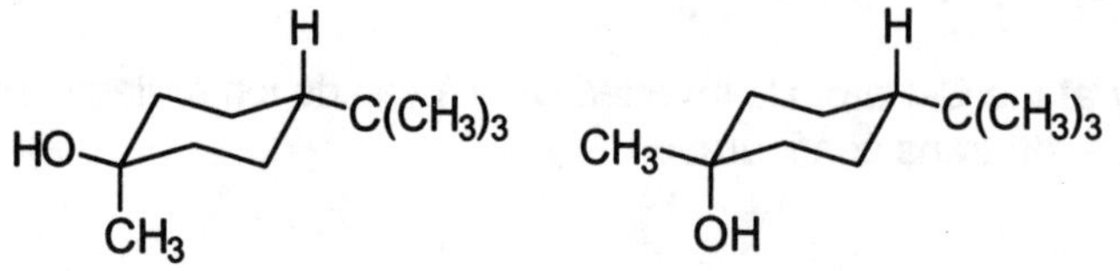

10.19   The bromide ion leaves readily because the resulting carbocation is tertiary and is resonance stabilized by the two benzene rings.

10.20   The tertiary carbocation has two *tert*-butyl groups and a methyl group bonded to the positively charged carbon atom compared to three methyl groups in the *tert*-butyl carbocation. Steric congestion about the C-4 atom of the reactant is decreased in the carbocation compared to the reactant as the chloride ion leaves.

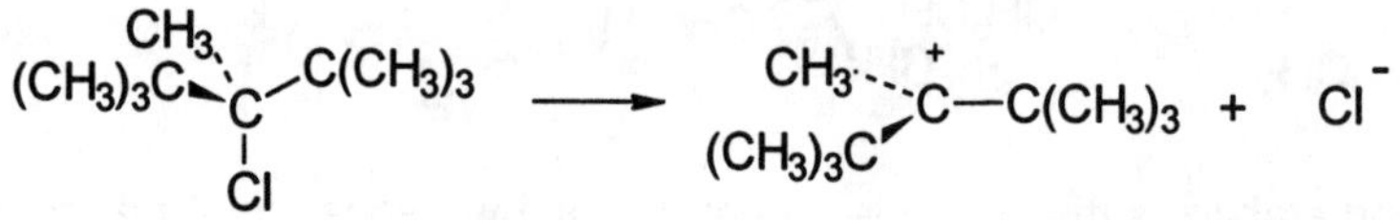

10.21   The same resonance stabilized carbocation results from both compounds. Attack of water at either the C-1 or C-3 atoms gives the same mixture of alcohols.

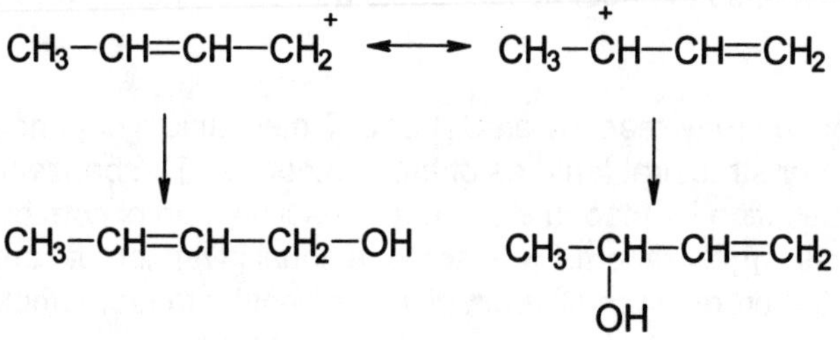

10.22   The bromine atom nearer the benzene ring is replaced because the resulting secondary carbocation is resonance stabilized.

10.23   Dimethyl sulfide is the leaving group and a *tert*-butyl carbocation forms. The substitution products are an alcohol resulting from nucleophilic attack of water on the tertiary carbocation and an ethyl ether resulting from nucleophilic attack of ethanol. The same products are formed from *tert*-butyl chloride because the same carbocation is formed when the chloride ion leaves.

10.24   The methyl compound reacts fastest because the three hydrogen atoms do not present any steric hindrance to the nucleophile in the S_N2 reaction. The other three compounds all have one alkyl group and two hydrogen atoms that are in the path of the nucleophile. Each alkyl group is primary and their steric sizes are similar. The slight decrease in rate suggests that as the length of the chain increases there is some increase in steric hindrance. Note that the change is larger when comparing a methyl group to an ethyl group. Extending the chain to a propyl group does not decrease the rate of reaction by as large a factor.

10.25   The fluorine atoms inductively withdraw electron density and stabilize the trifylate ion compared to the methanesulfonate ion. Thus, the trifluoromethanesulfonate ion is a better leaving group, and therefore the trifylate is more reactive..

10.26   The nitro group inductively withdraws electron density and stabilizes the sulfonate ion compared to the tosylate ion. Thus, the nitro substituted sulfonate ion is a better leaving group, so its esters are more reactive.

10.27   The lone pair electrons of the oxygen atom of both the alcohol and the ether coordinate with cations. The partially positive hydrogen atom of the hydroxyl group of alcohols can help solvate anions. Because the ether cannot solvate anions, the solubility of ionic compounds is much smaller.

10.28   In an aprotic solvent the fluoride ion is sufficiently nucleophilic to react with the iodo compound via an S_N2 mechanism that occurs with inversion of configuration. The product of this reaction is *cis*-1-fluoro-3-methylcyclopentane. In ethanol the fluoride is solvated and is not an effective nucleophile. The substitution product is an ethyl ether.

10.29   Hexamethylphosphoramide is very polar as indicated by its dielectric constant. It has no protic sites and is thus aprotic. The solvent can be used for substitution reactions. Because it has a high dielectric constant, it will favor S_N1 processes. Because it is aprotic, it will accelerate S_N2 processes.

10.30   The fluoride ion is solvated by methanol and its nucleophilicity is much less than it is in an aprotic solvent such as dimethylformamide.

10.31   The reaction rates in water reflect the nucleophilicity of the solvated anion. The degree of solvation decreases with increased size of the ion. Thus, the iodide ion is the best nucleophile of the series. In the aprotic acetone solvent, the order of nucleophilicity is related to the strength of the ions as bases. Chloride ion is the strongest base of the series.

10.32   The position of the equilibrium in water is affected by the degree of solvation of the anion. Bromide ion is more strongly solvated than iodide ion and the reaction tends to proceed to the right as written. In acetone as solvent the position of the equilibrium is controlled by factors other than the stability of the anions, neither of which is solvated.

10.33   Fluoride ion is the strongest base of the halides, because HF is the weakest acid of the
hydrogen halides. In an aprotic solvent the fluoride ion is not solvated and it is an even
stronger base than in protic solvents. Elimination is favored in reactions with strong
bases.

10.34   The *tert*-butoxide ion is not as nucleophilic as the ethoxide ion because it is sterically
larger. Thus, substitution reactions occur less readily with *tert*-butoxide ion. In addition,
the *tert*-butoxide ion is a stronger base than the ethoxide ion and a larger fraction of
elimination product results with a stronger base.

10.35   The unsaturated products are 1-butene, *cis*-2-butene, and *trans*-2-butene. The 2-butenes
are more substituted alkenes and should predominate. The alkenes result in part from an
E1 process derived from loss of a proton from the carbon atom adjacent to the
carbocation center, as well as an E2 process in which either ethanol or ethoxide ion
abstracts a proton from the reactant. Increasing the concentration of ethoxide ion
increases the amount of product of the E2 process.

10.36   The alkoxide derived from 3-ethyl-3-pentanol is a more sterically hindered base than is
the *tert*-butoxide ion. The base strength should be similar. The increased amount of the
1-butene indicates that the more sterically hindered base doesn't abstract the tertiary
hydrogen atom at the C-3 atom as readily, and is far more likely to abstract the primary
hydrogen atom. Thus, the less substituted alkene predominates.

10.37   Arrange the structure in a conformation so that the hydrogen atom at C-3 and the tosylate
group are anti periplanar. The structure of the product has the (E) configuration.

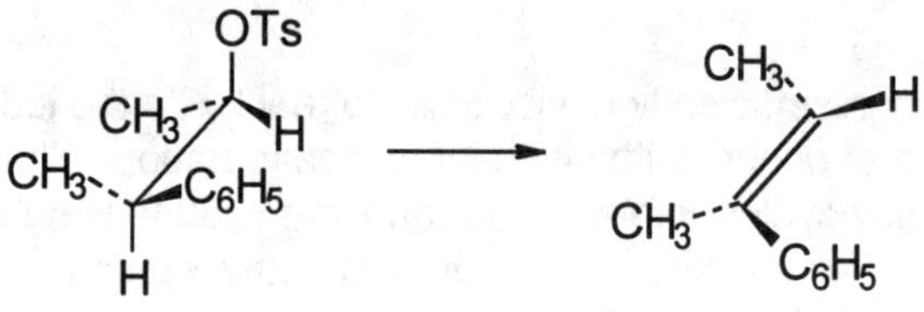

10.38   In the cis isomer, the tosylate group and the hydrogen at the C-2 atom are anti periplanar
and elimination can readily occur.

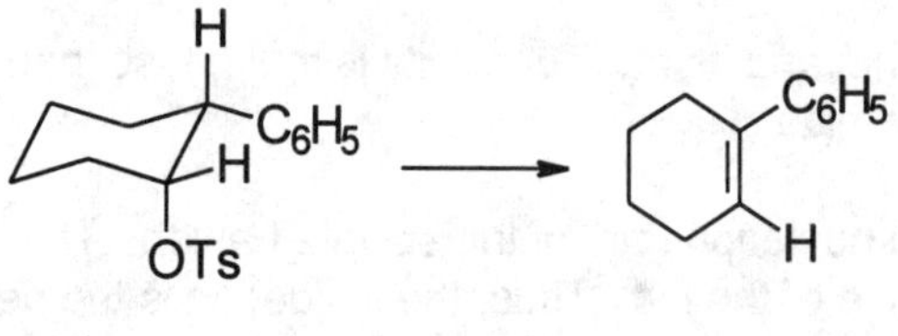

In the trans isomer, elimination cannot occur unless the chair undergoes a ring flip to
place the tosylate in an axial position. However, in this conformation the product will not
be 1-phenylcyclohexene because the phenyl ring will be in an axial position. Elimination
will occur by abstraction of the hydrogen atom at C-6 to give the isomeric 3-phenylcyclo-
hexene, as shown on the following page.

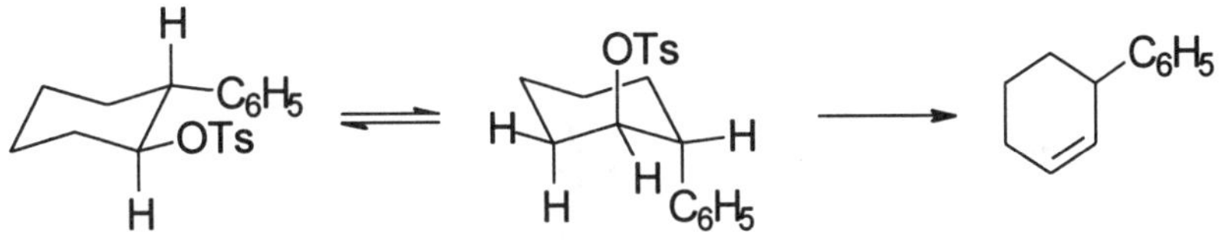

10.39 Arrange the structure in a conformation so that the hydrogen atom at C-3 and the bromide ion are anti periplanar. This conformation gives compound II. An alternate conformation has the deuterium atom at C-3 and the bromide ion in an anti periplanar arrangement. This conformation give compound I.

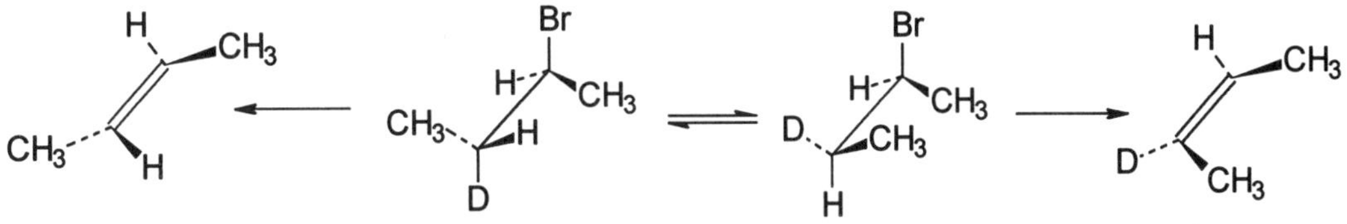

Reversing the configuration at C-3 to S gives a (Z) compound without deuterium and an (E) compound containing deuterium.

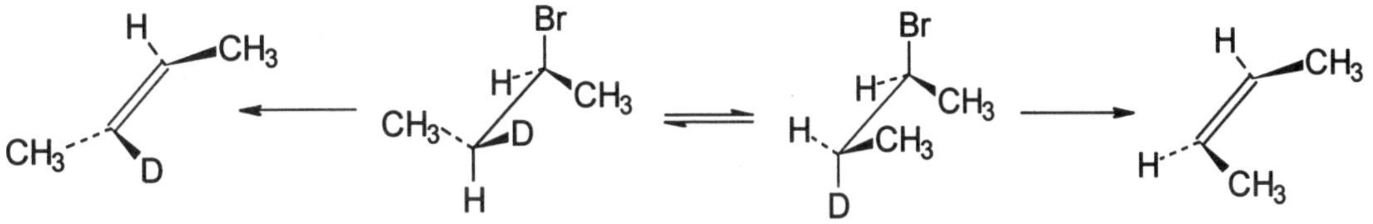

10.40 The carbon deuterium bond is not as easily broken in the E2 reaction. Thus, the major product contains the deuterium atom. It is formed by abstraction of the hydrogen atom from the C-2 atom.

10.41 Compound I has a hydrogen atom at the bridgehead position that may be abstracted to give a tetrasubstituted alkene. The elimination reaction of compound II can only occur by abstraction of a proton from the methyl group.

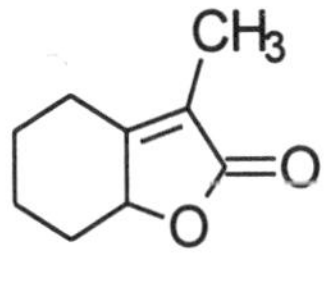

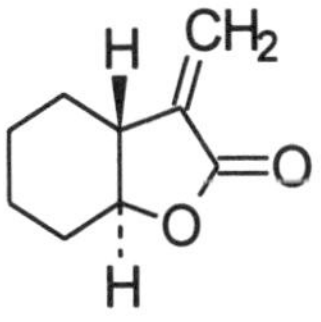

10.42   The second compound has an axial bromine atom that is anti periplanar to the hydrogen atom. This compound reacts at a faster rate than the first compound, which has an equatorial bromine atom. The elimination product of this compound will not contain deuterium.

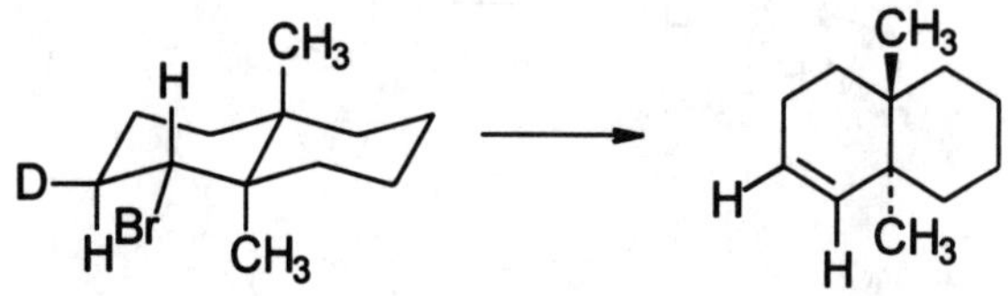

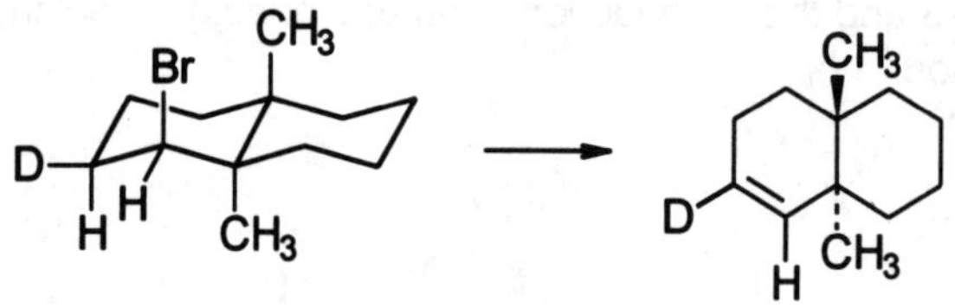

11

Alkynes

If you are comfortable with your understanding of the chemistry of alkenes, then you will find that there isn't as much to learn about the chemistry of alkynes. Both classes of compounds have π bonds which dominate their chemical reactivity, which is largely addition of electrophiles. Moreover at least some of the synthetic methods used to produce alkynes are the same as those used for the synthesis of alkenes, namely elimination reactions. There is one new feature. The C—H bond of terminal alkynes is sufficiently acidic for the proton to be removed by strong bases. As a consequence, the conjugate base formed (a carbanion) is a nucleophile.

Keys to the Chapter

11.1 Occurrence and Uses of Alkynes

Alkynes are less common than alkenes in naturally occurring materials. The few examples cited that are of interest have multiple conjugated triple bonds. Carbon-carbon triple bonds are contained in a limited number of drugs, including oral contraceptives.

11.2 Structure and Properties of Alkynes

A triple bond in an alkyne consists of one sigma bond and two pi bonds. As a result of the geometry of the sp hybrid orbitals, the two carbon atoms of the triple bond and the two atoms directly attached are collinear. There are two classes of alkynes—**monosubstituted** (terminal) and **disubstituted** (internal).

The greater % s character of the sp hybridized carbon atom of alkynes strongly affects the properties of the bond of that carbon atom. The electrons in the bond are held more tightly by the carbon atom, and as a consequence the homolytic cleavage of the C—H bond requires a greater amount of energy. The length of the C—H bond as well as bonds to other atoms is shorter than for sp^2 and sp^3 bonds of the same type.

The bond energy of the carbon-carbon triple bond reflects the less effective bonding of π electrons. The bond energy of the carbon-carbon triple bond is substantially less than three times the bond energy of a carbon-carbon single bond.

The heats of formation of alkynes containing 10 or fewer carbon atoms are positive, because they contain a triple bond that is less stable than carbon-hydrogen and carbon-carbon single bonds. The heats of formation of disubstituted alkynes are less positive than the heats of formation of isomeric monosubstituted alkynes.

Alkynes are relatively nonpolar molecules and their boiling points are controlled by London forces. Terminal alkynes have small dipole moments that are slightly larger than the dipole moments of terminal alkenes. Internal alkynes have no dipole moment if the two alkyl groups bonded to the carbon atoms of the triple bond are identical and have virtually no dipole moment even if the alkyl groups are different. The chemical properties of alkynes are expected to be the same as the properties of alkenes. The only difference is that there are twice as many π bonds to react.

11.3 Nomenclature

Alkynes are named by selecting the longest continuous carbon chain containing the triple bond. The chain is numbered to assign the lowest number to the first carbon atom of the triple bond. Alkyl groups and halogens are disregarded in selecting the direction of numbering unless the same number for the triple bond is obtained from either end of the chain. For compounds containing both double and triple bonds, the chain is numbered from the end nearer the first multiple bond. However, in equivalently placed multiple bonds, the double bond takes precedence over triple bonds in the direction of numbering. Compounds with both double and triple bonds are called enynes, not ynenes.

11.4 Acidity of Terminal Alkynes

Although weakly acidic, terminal alkynes can be converted to their conjugate bases called **alkynide** ions. The pK_a is approximately 25. Thus, a base whose conjugate acid has a pK_a greater than 25 must be used to abstract the hydrogen ion. Hydroxide ion is not sufficiently basic, but the amide ion is. Because the pK_a value of ammonia is approximately 36, the equilibrium constant for the reaction of an alkyne with amide ion is 10^{11}. This reaction is used to produce alkynide ions for use as nucleophiles in displacement of a halide ion from a haloalkane to synthesize alkynes.

11.5 Oxidation of Alkynes

Although a little used reaction, the ozonolysis of alkynes cleaves the triple bond in a manner similar to that of the ozonolysis of alkenes. The products are carboxylic acids. Terminal alkynes give a mole of carbon dioxide.

11.6 Hydrogenation of Alkynes

When the reaction is catalyzed with finely divided platinum, palladium, or nickel, hydrogenation of alkynes is complete and produces alkanes. Hydrogenation requires one mole of hydrogen gas for each π bond in a compound, so triple bonds require two moles of hydrogen gas.

It is also possible to control the hydrogenation of alkynes to stop after adding just one molar equivalent of hydrogen, giving alkenes as the product. This is accomplished by using a specially prepared catalyst. Hydrogenation of alkynes with **Lindlar catalyst** produces cis-alkenes by syn addition, whereas hydrogenation using lithium in liquid ammonia produces trans-alkenes by anti addition.

11.7 Electrophilic Addition Reactions

An unsymmetrical reagent such as HBr adds to a triple bond in a characteristic way given by Markovnikov's rule. The hydrogen atom adds to the less substituted carbon atom of the triple bond. The bromine atom adds to the more substituted carbon atom of the triple bond. The addition product has the two added atoms trans in the resulting alkene, although the stereoselectivity may be low.

Hydrogen bromide adds more slowly to triple bonds than to double bonds. However, after one mole of hydrogen bromide has added, the resulting double bond is less reactive as a result of the electron withdrawing bromine atom. As a consequence it is possible to obtain the product formed from the addition of one mole of HBr. When the second mole of HBr adds, the product has two hydrogen atoms added to the carbon atom that was less substituted originally. Two bromine atoms are located on the other carbon atom.

Bromine will add to compounds with triple bonds. Two moles of Br_2 react with a triple bond. The initial addition product has the two added atoms trans in the resulting alkene. Continued addition yields a tetrabromoalkane.

Hydration of alkynes results in the Markovnikov addition of one mole of water to give an enol that rearranges to give a ketone.

11.8 Synthesis of Alkynes

Alkynes can be prepared from **vicinal** or **geminal** dihalides by a double dehydrohalogenation using a strong base such as $NaNH_2$ in liquid ammonia. Recall that vicinal dihalides are obtained from the addition of a halogen to an alkene. The number of moles of base required for the reaction depends on the type of alkyne produced. A terminal alkyne produced in the synthesis is deprotonated by the reacting base and thus a total of three moles of amide ion is required. Upon work up with water, the terminal alkyne results.

The first step in the dehydrohalogenation give an alkenyl halide which is subsequently dehydrohalogenated. The weaker hydroxide ion gives only the alkenyl halide. That compound can be dehydrohalogenated with the amide ion.

Alkynes can be synthesized by reaction of an alkynide with an alkyl halide. The reaction involves an S_N2 substitution and places an alkyl group at the position of the original hydrogen atom of the terminal alkyne used to produce the alkynide. Because the alkynide ion is a strong base, only primary alkyl halides give this substitution reaction. Other alkyl halides undergo elimination.

11.9 Rearrangement of Alkynes

Alkynes may isomerize as the result of abstraction of a hydrogen ion from the carbon atom bonded to the carbon atom of the triple bond. The process occurs by deprotonation and reprotonation of resonance stabilized intermediates. The equilibrium favors the thermodynamically stable alkyne unless a terminal alkyne can be formed. Although terminal alkynes are thermodynamically less stable than internal alkynes, the formation of the alkynide ion effectively removes that isomer from the equilibrium. Ultimately after aqueous workup the terminal alkyne is the major product.

Summary of Reactions

1. Hydrogenation of Alkynes

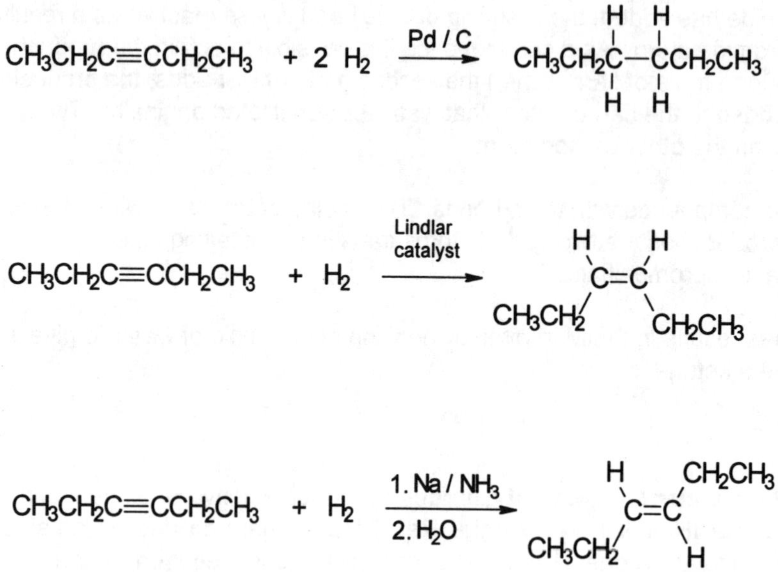

$CH_3CH_2C{\equiv}CCH_2CH_3$ + 2 H_2 $\xrightarrow{Pd\,/\,C}$ $CH_3CH_2\overset{\displaystyle H}{\underset{\displaystyle H}{C}}-\overset{\displaystyle H}{\underset{\displaystyle H}{C}}CH_2CH_3$

$CH_3CH_2C{\equiv}CCH_2CH_3$ + H_2 $\xrightarrow{\text{Lindlar catalyst}}$ $\underset{CH_3CH_2}{\overset{H}{}}C=C\underset{CH_2CH_3}{\overset{H}{}}$

$CH_3CH_2C{\equiv}CCH_2CH_3$ + H_2 $\xrightarrow[\text{2. }H_2O]{\text{1. Na / }NH_3}$ $\underset{CH_3CH_2}{\overset{H}{}}C=C\underset{H}{\overset{CH_2CH_3}{}}$

2. Addition of Hydrogen Halides

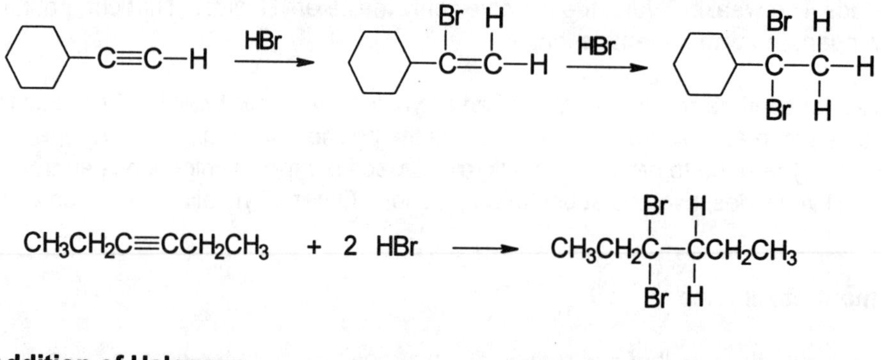

$CH_3CH_2C{\equiv}CCH_2CH_3$ + 2 HBr $\longrightarrow$ $CH_3CH_2\overset{\displaystyle Br}{\underset{\displaystyle Br}{C}}-\overset{\displaystyle H}{\underset{\displaystyle H}{C}}CH_2CH_3$

3. Addition of Halogens

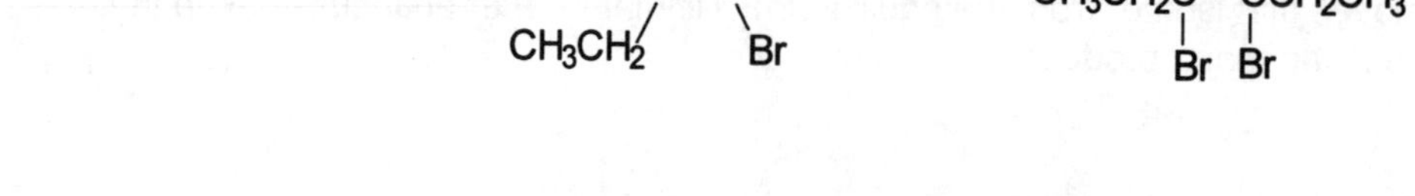

$CH_3CH_2C{\equiv}CCH_2CH_3$ $\xrightarrow{Br_2}$ $\underset{CH_3CH_2}{\overset{Br}{}}C=C\underset{Br}{\overset{CH_2CH_3}{}}$ $\xrightarrow{Br_2}$ $CH_3CH_2\overset{\displaystyle Br}{\underset{\displaystyle Br}{C}}-\overset{\displaystyle Br}{\underset{\displaystyle Br}{C}}CH_2CH_3$

4. Addition of Water (Hydration)

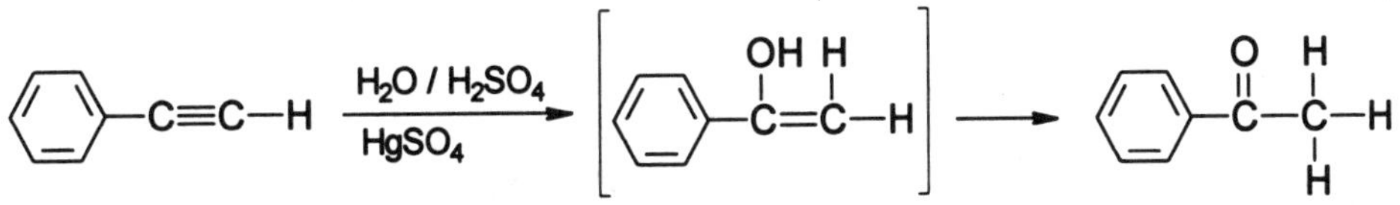

5. Synthesis of Alkynes by Dehydrohalogenation

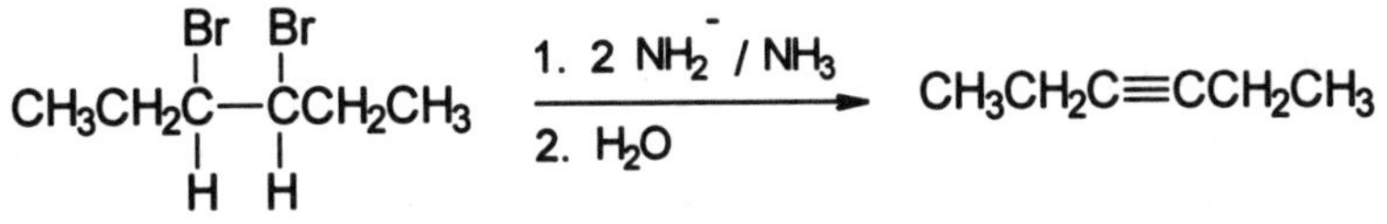

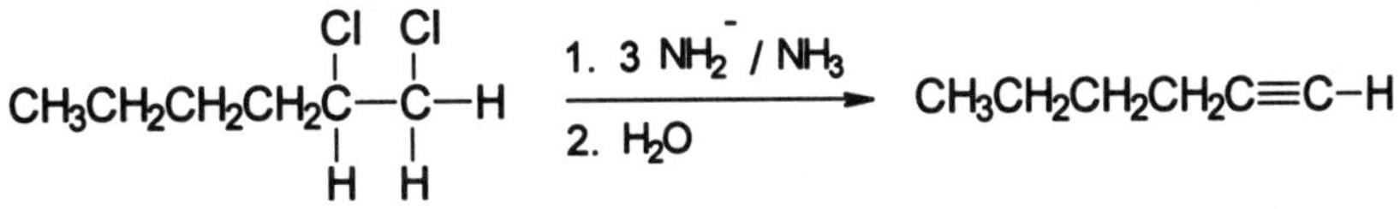

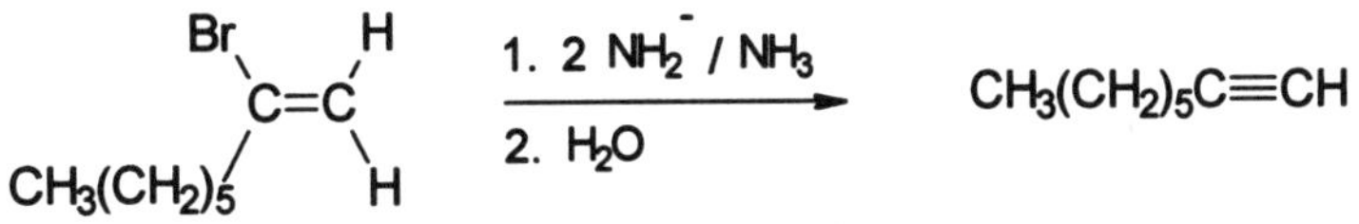

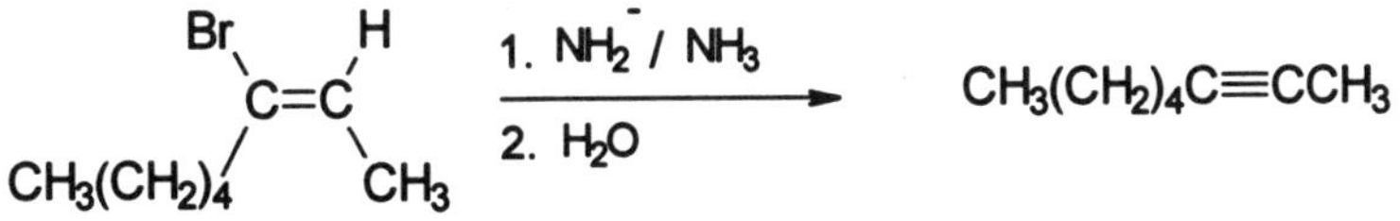

$$CH_3CH_2\underset{\underset{H}{|}}{\overset{\overset{Br}{|}}{C}}-\underset{\underset{H}{|}}{\overset{\overset{Br}{|}}{C}}CH_2CH_3 \quad \xrightarrow[\text{2. } H_2O]{\text{1. 2 } NH_2^- \text{ / } NH_3} \quad CH_3CH_2C\equiv CCH_2CH_3$$

$$CH_3CH_2CH_2CH_2\underset{\underset{H}{|}}{\overset{\overset{Cl}{|}}{C}}-\underset{\underset{H}{|}}{\overset{\overset{Cl}{|}}{C}}-H \quad \xrightarrow[\text{2. } H_2O]{\text{1. 3 } NH_2^- \text{ / } NH_3} \quad CH_3CH_2CH_2CH_2C\equiv C-H$$

$$\underset{CH_3(CH_2)_5}{\overset{Br}{\diagdown}}C=C\underset{H}{\overset{H}{\diagup}} \quad \xrightarrow[\text{2. } H_2O]{\text{1. 2 } NH_2^- \text{ / } NH_3} \quad CH_3(CH_2)_5C\equiv CH$$

$$\underset{CH_3(CH_2)_4}{\overset{Br}{\diagdown}}C=C\underset{CH_3}{\overset{H}{\diagup}} \quad \xrightarrow[\text{2. } H_2O]{\text{1. } NH_2^- \text{ / } NH_3} \quad CH_3(CH_2)_4C\equiv CCH_3$$

6. Synthesis of Alkynes by Alkylation

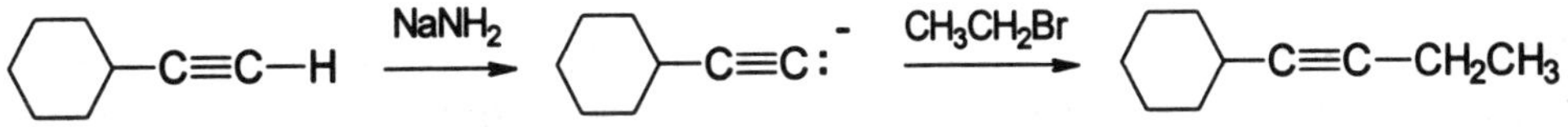

Solutions to Exercises

11.1   **(a)** $C_{13}H_{10}O_2$ **(b)** $C_{12}H_8O$ **(c)** $C_{15}H_{16}O$

11.2   **(a)** monosubstituted   **(b)** disubstituted

11.3   **(a)** C_6H_{12} **(b)** C_5H_8 **(c)** C_7H_{12} **(d)** C_4H_6

11.4   **(a)** C_4H_2 **(b)** C_4H_4 **(c)** $C_{10}H_{18}$ **(d)** $C_{10}H_8$

11.5   One sp hybridized carbon atom decreases the carbon-carbon bond length relative to the sp^3-sp^3 bond of propane from 154 to 146 pm. Changing a second sp^3 hybridized carbon atom to sp should decrease the bond length to 138 pm.

11.6   The bond dissociation energy involves a homolytic cleavage of the C—Cl bond. Because the % s character of the sp^2 orbital is larger than the % s character of the sp^3 orbital, the bonding electrons are drawn closer to the carbon nucleus and more energy is required to "retrieve" the electron required for homolytic cleavage. In the case of the acidity of the C—H bonds, the increased % s character "pulls" electrons away from the bond and allows the proton to leave more readily in the required heterolytic cleavage.

11.7   Note that the heats of formation are positive and that both compounds are unstable with respect to the elements. Thus, the compound with the lower positive heat of formation is more stable assuming that the entropies of formation of the two compounds are approximately equal. In this case 2-pentyne, which has the more substituted triple bond, is the more stable. The heat of combustion measures the heat energy released ($\Delta H° < 0$) when carbon dioxide and water are formed. The less stable (highest energy) isomer releases more energy in the combustion reaction. In this case 1-pentyne has the more negative heat of combustion.

11.8   The heats of formation are positive for both compounds and both are unstable with respect to the elements. Thus, the compound with the lower positive heat of formation is more stable assuming that the entropies of formation of the two compounds are approximately equal. In this case 1,4-pentadiene is the more stable isomer, so two double bonds are more stable than one triple bond.

11.9   The reaction is spontaneous at this temperature in spite of the fact that $\Delta H° > 0$, because $\Delta S° > 0$ for this reaction in which two moles of reactant are converted into four moles of product.

$$2\ CH_4 \longrightarrow C_2H_2\ +\ 3\ H_2$$

11.10 The heats of formation are positive for both compounds and both are unstable with respect to the elements. Thus, the compound with the lower positive heat of formation is more stable assuming that the entropy of formation of the two isomers is the same. In this case 1-propyne is the more stable isomer and should predominate in an equilibrium mixture.

11.11 The positive end of the dipole should be the methyl group, because the sp-hybridized carbon atom in the middle has greater s character and attracts electron density from the sp^3-hybridized carbon of the methyl group. The dipole moment of propene should be smaller than that of propyne because the electrons are less strongly attracted to the sp^2-hybridized carbon atom of propene.

11.12 The 1-alkynes are slightly more polar than 1-alkenes. In addition, the electrons in the two π bonds of an alkyne may be more polarizable than the electrons in the single π bond of an alkene.

11.13 The 3,3-dimethyl-1-butyne compound has a more compact and somewhat spherical structure. The London forces for such compounds are smaller than for cylindrical structures such as 1-hexyne.

11.14 Terminal alkynes have a larger dipole moment than internal alkynes. Thus terminal alkynes should have higher boiling points than isomeric internal alkynes if polarity were the only structural feature determining this physical property. There is decreased freedom of motion of more carbon atoms of internal alkynes compared to terminal alkynes. Thus, the shape of the internal alkyne may allow greater London forces between the more linear chains. The alkyl group of the terminal alkynes may assume conformations that are less amenable to closely packed molecules.

11.15 **(a)** 1-pentyne **(b)** 2,2-dimethyl-3-hexyne **(c)** 4-methyl-2-hexyne

11.16 **(a)** 4,5-dibromo-2-hexyne **(b)** 1-chloro-3-octyne **(c)** 2-chloro-6-methyl-3-octyne

11.17 **(a)** $CH_3CH_2CH_2{-}C{\equiv}C{-}CH_3$

(b) 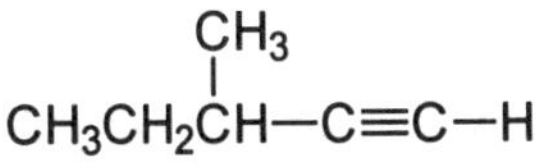
$$CH_3CH_2\overset{\overset{\displaystyle CH_3}{|}}{CH}{-}C{\equiv}C{-}H$$

(c)
$$CH_3CH_2CH_2\overset{\overset{\displaystyle CH_2CH_3}{|}}{CH}{-}C{\equiv}C{-}CH_2CH_3$$

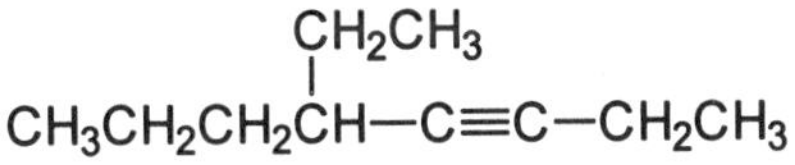

11.18 **(a)** $CH_3CH_2CH_2{-}C{\equiv}C{-}CH_2CH_3$

(b) 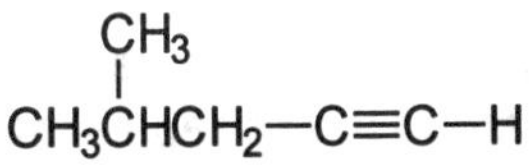
$$CH_3\overset{\overset{\displaystyle CH_3}{|}}{CH}CH_2{-}C{\equiv}C{-}H$$

(c)
$$CH_3CH_2\overset{\overset{\displaystyle CH_3}{|}}{CH}{-}C{\equiv}C{-}CH_2CH_3$$

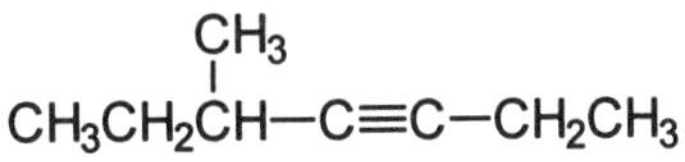

11.19
$$CH_3{-}C{\equiv}C{-}CH{=}CH{-}\overset{\overset{\displaystyle }{|}}{CH}{-}CH_2{-}CH{=}CH_2$$

with substituent:
$$\overset{|}{\underset{|}{C}}\quad \begin{array}{c}C\\ \|||\\ C\\ |\\ H\end{array}$$

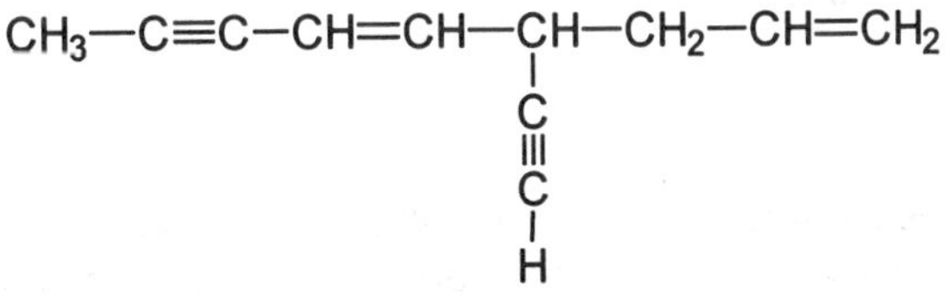

11.20

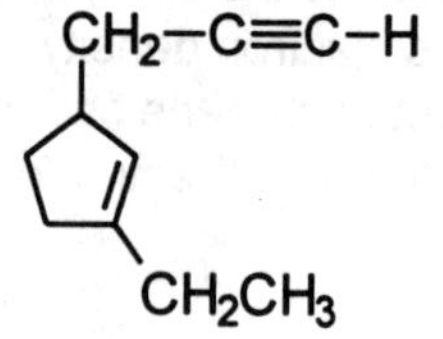

$CH_2-C{\equiv}C-H$

CH_2CH_3

11.21   1-propynyl

11.22   MDL 18962 contains the propargyl group, $-CH_2C{\equiv}CH$.

11.23   The percent conversion is larger using *tert*-butoxide as the base, because it is a stronger base than the methoxide ion.

11.24   The two isopropyl groups are inductively electron donating and should increase the electron density on the nitrogen atom. As a result, the base should be stronger than the amide ion.

11.25   Prepare the conjugate base of 1-propyne using a strong base such as the amide ion. Then add D_2O to the reaction mixture. The conjugate base of 1-propyne is a stronger base than the DO^- ion. As indicated on page 420 of the text, the equilibrium position strongly favors the alkyne, which in this case is 1-deuterio-1-propyne.

11.26   Add a strong base such as amide ion which converts the terminal alkyne into a salt which is expected to be nonvolatile. Distill the internal alkyne from the reaction mixture.

11.27 $CH_3(CH_2)_6CH_2-C{\equiv}C-CH_2(CH_2)_6CO_2H$

11.28 $CH_3(CH_2)_8CH_2-C{\equiv}C-CH_2(CH_2)_3CO_2H$

11.29   Internal alkynes which have identical alkyl groups bonded on each of the two triple-bonded carbon atoms represented as $R-C{\equiv}C-R$ are symmetrical and give two moles of the same carboxylic acid product.

11.30 $HO_2C(CH_2)_3CH_2-C{\equiv}C-CH_2(CH_2)_3CO_2H$

11.31   **(a)** three   **(b)** four   **(c)** four   **(d)** six

11.32   **(a)** eight   **(b)** four   **(c)** seven

11.33   The triple bond of cyclodecyne is less stable than the triple bond of 5-decyne because there is ring strain that results from the location of a linear array of four carbon atoms within the ring. The difference in the heats of hydrogenation of cyclooctyne and 4-octyne will be larger because there will be more ring strain when a linear array of four carbon atoms is located in a smaller ring.

11.34 The heat of formation of 1-pentyne is more positive than the heat of formation of 1,4-pentadiene. In the hydrogenation reaction pentane is the common product. Because heats of hydrogenation are negative, the energy difference between 1-pentyne and pentane is larger than the energy difference between 1,4-pentadiene and pentane.

11.35 Oleic acid is the (Z) isomer and elaidic acid is the (E) isomer.

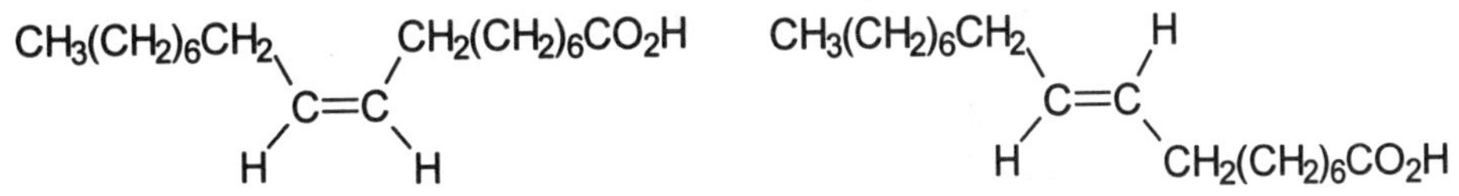

$$CH_3(CH_2)_6CH_2 \qquad CH_2(CH_2)_6CO_2H \qquad CH_3(CH_2)_6CH_2 \qquad H$$
$$\diagdown C=C \diagup \qquad\qquad \diagdown C=C \diagup$$
$$\diagup \qquad \diagdown \qquad\qquad \diagup \qquad \diagdown$$
$$H \qquad\quad H \qquad\qquad H \qquad\quad CH_2(CH_2)_6CO_2H$$

11.36 The required alkene is (Z)-2-methyl-7-octadecene, which can be prepared by catalytic hydrogenation of 2-methyl-7-octadecyne using the Lindlar catalyst.

11.37 The alkene has the (Z) configuration about the double bond. It can be produced by catalytic hydrogenation of a structurally related alkyne using the Lindlar catalyst. Recall that an alkene can be reduced in the presence of a carbonyl group without reduction of the carbon-oxygen double bond. Thus, the carbonyl group of the ester should not be catalytically reduced under conditions that reduce the triple bond but not the resulting double bond.

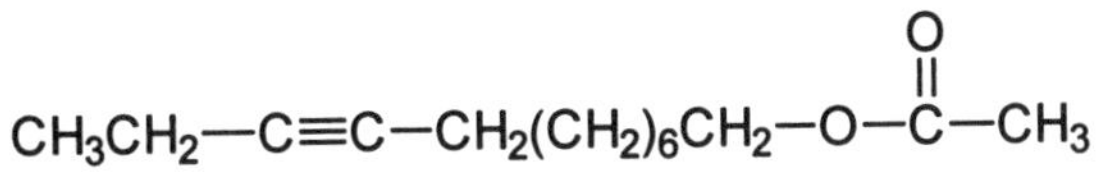

$$\overset{\displaystyle O}{\overset{\displaystyle \|}{CH_3CH_2-C\equiv C-CH_2(CH_2)_6CH_2-O-C-CH_3}}$$

11.38 The (E) configuration can be achieved by reduction of a structurally related alkyne using the sodium in liquid ammonia. The reaction conditions would result in abstraction of a proton from the alcohol, but it would be replaced in the workup of the reaction mixture.

$$CH_3CH_2-C\equiv C-CH_2(CH_2)_8CH_2OH$$

11.39 The terminal triple bond will be reduced and the double bond will be unaffected. The product is

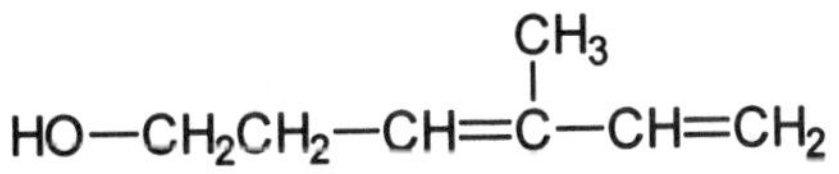

$$\overset{\displaystyle CH_3}{\overset{\displaystyle |}{HO-CH_2CH_2-CH=C-CH=CH_2}}$$

11.40 Hydrogenation using sodium in liquid ammonia gives an alkene with the (E) configuration.

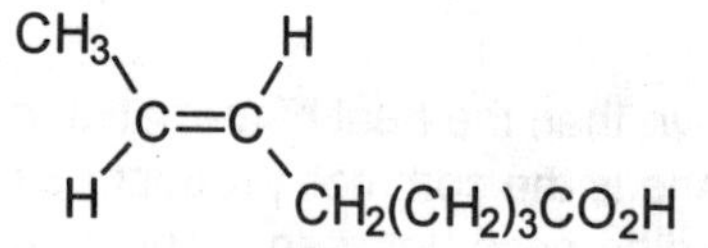

11.41 Hydrogenation with the Lindlar catalyst gives the (Z) isomer. Subsequent reactions in both (a) and (c) give syn addition products which are meso.

11.42 None of them because the initial product is the (E) isomer and all of the reagents listed react with this compound by syn addition processes.

11.43 The C-2 and C-3 carbon atoms each contain an alkyl group. Thus, there is no regioselectivity and either of them could add a proton. The products are

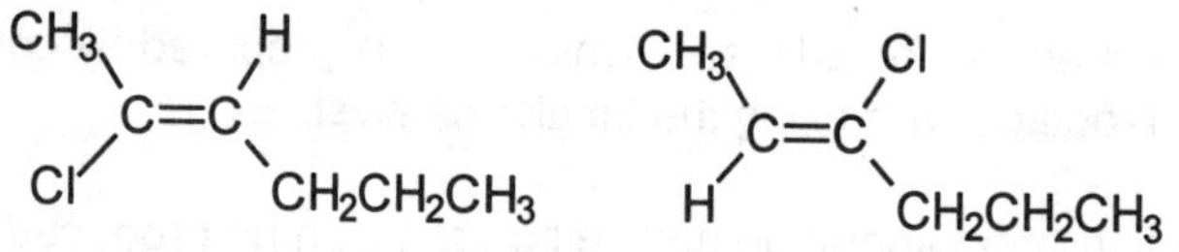

11.44 The trans addition of DBr give the following (E) isomer.

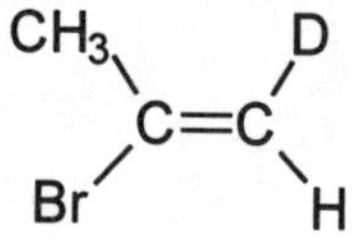

11.45 The addition should occur in an anti-Markovnikov manner similar to that observed for the addition of HBr to alkenes by the free radical mechanism. The stereochemistry of the addition can't be predicted based on information provided in the text but the product shown below involves trans addition.

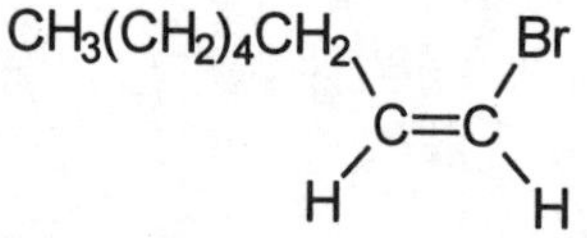

11.46 The C-3 and C-4 atoms each contain an alkyl group. Thus, based on inductive effects alone there should be no regioselectivity and either of them could react with a bromine radical. However, based on the difference in the steric environments of each carbon atom, the bromine can more easily add to the C-4 atom to give the first structure of the two given on the following page.

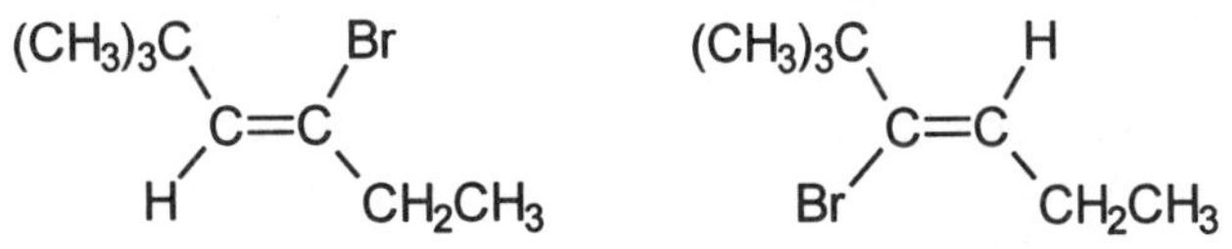

11.47 The bromine should regioselectively add to the triple bond to give the following product.

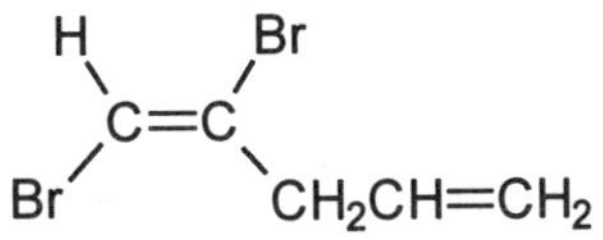

11.48 The bromine should add anti to the triple bond to give the following product which has no dipole moment.

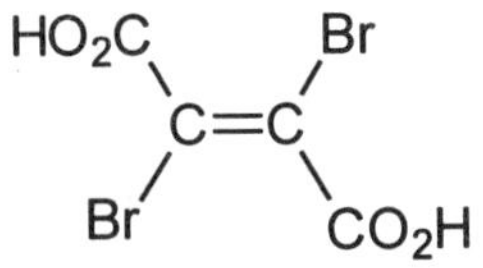

11.49 Compound I is a symmetrical alkyne and the C-3 and C-4 atoms are structurally equivalent. Hydration of this alkyne gives a single product with a carbonyl carbon atom located at C-3. Compound II forms two products. One has a carbonyl carbon atom located at C-3 and the other at C-2.

11.50 The C-2 and C-3 carbon atoms each contain an alkyl group. Thus, based on inductive effects alone there should be no regioselectivity and eiither of them could react to give an intermediate enol in a hydration reaction. However, there is a small difference in the steric environments of each carbon atom. The slight regioselectivity in the reaction may be the result of this difference. The major product has the carbonyl oxygen atom at the least sterically hindered C-2 position.

11.51

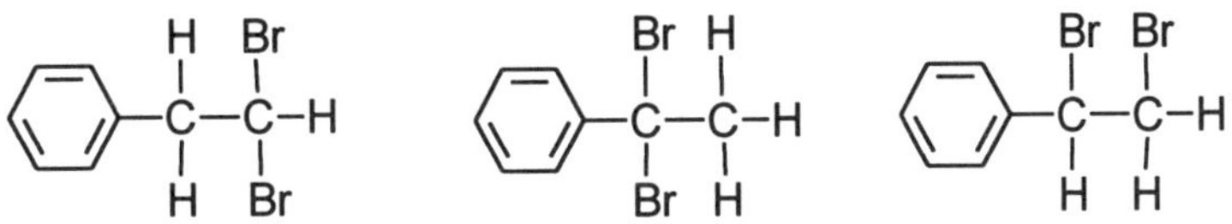

11.52 3,3-Dibromopentane can yield only 2-pentyne because the C-2 and C-4 atoms are equivalent in this symmetrical molecule. 2,2-Dibromopentane can yield 1-pentyne by elimination of hydrogen atoms at C-1 and 2-pentyne by elimination of hydrogen atoms at C-3.

11.53   No, because elimination can result by abstraction of hydrogen atoms at either C-1 or C-3 to give a mixture of 1-pentyne and 2-pentyne.

11.54   Under equilibrium conditions in the presence of a strong base the triple bond isomerizes to the terminal position to give 1-octyne which reacts with the base to give the alkynide salt. This process effectively drives the equilibrium to give the salt of 1-octyne.

11.55   The product is 1,9-decadiyne which forms when the acetylide ion displaces bromide ion from both the 1- and 6- sites of 1,6-dibromohexane.

$$H-C\equiv C-(CH_2)_6-C\equiv C-H$$

11.56   The bromide ion is a better leaving group than the fluoride ion. The product is 8-fluoro-2-octyne.

$$F-(CH_2)_5-C\equiv C-CH_3$$

11.57 $CH_3CH_2CH_2-C\equiv C-CH_2CH_3$

11.58   Prepare the acetylide salt of 3,3-dimethyl-1-butyne and react it with 1-bromobutane. Note that reaction of the acetylide salt of 1-hexyne with 2-bromo-2-methylpropane would give only an elimination product because the alkyl halide is tertiary.

12

Dienes and Allylic Compounds

The focus of this chapter is on the effect of conjugation in resonance stabilized compounds and intermediates. The effect of conjugation is seen in both the physical properties of conjugated compounds and their reactivity. Although an explanation for both features can be provided using conventional Lewis structures, the use of molecular orbitals and the contributing atomic orbitals provides a more "sophisticated" approach. This method will be applied in later chapters as well as for aromatic compounds in Chapter 13 and for pericyclic reactions in Chapter 28.

Keys to the Chapter

12.1 Classes of Dienes

Conjugated dienes and higher polyenes contain a series of alternating single and double bonds. **Isolated dienes** have more than one single bond separating the two double bonds and are regarded as two separate alkenes. **Cumulated dienes** have two double bonds sharing a common carbon atom.

Natural products that can be mentally dissected into isoprene units are **terpenes**. Isoprene is a conjugated butadiene with a methyl branch. The terpenes are named according to the number of five-carbon units that make up their skeleton.

12.2 Stability of Dienes

The effect of the interaction of two double bonds is the first case of the interaction of functional groups that you have seen. Others such as a carbon-carbon double bond and a carbonyl group will be considered in later chapters. The result of conjugation in providing a more stable system is seen in the heats of hydrogenation. Now, as you have heard repeatedly, you can't forget what you learned earlier. If this discussion doesn't make good sense, then you should return to the initial discussion of heats of hydrogenation in Section 6.11 of the text.

Because conjugated dienes are stabilized by resonance interaction of the double bonds, these dienes are of lower energy. Thus, the energy released upon hydrogenation is smaller than predicted based on the heats of hydrogenation of the component double bonds. The difference between the experimental value and the predicted value based on isolated double bonds is the **resonance energy**.

12.3 Molecular Orbital Models of Polyenes

Yes, it has been some time since we examined molecular orbital concepts in Chapter 1. However, this section takes pity on you and reviews the concepts again and then expands on them. Molecular orbitals are pictured as **linear combinations of atomic orbitals**. The mathematics required involves the addition or subtraction of atomic wave functions. Addition of orbitals with the proper sign of the wave function corresponds to **constructive overlap** of atomic orbitals. This combination gives rise to **bonding molecular orbitals**. The subtraction of wave functions or the overlap of atomic orbitals with opposite signs for the wave functions gives **destructive overlap** and produces an **antibonding molecular orbital**.

Molecular orbitals may be **symmetric** or **antisymmetric** based on the sign of the molecular orbital at one point compared to a related point on the other side of a plane that is perpendicular to the molecular axis. The bonding and antibonding molecular orbitals of ethylene are symmetric and antisymmetric, respectively. If the signs of the lobes of the molecular orbital change at the plane used to determine symmetry, that plane is called a vertical **nodal plane**. The electron density at this nodal plane is zero.

Review Table 12.1 to learn the rules of the molecular orbital game. Note especially that you must have the same number of molecular orbitals and contributing atomic orbitals. The relative energies of the molecular orbitals are related to the number of nodal planes.

A group of molecular orbitals for a polyene can be separated into a group of bonding molecular orbitals (corresponding in number to the number of double bonds) and an equal number of antibonding molecular orbitals. There are only sufficient electrons to fill the bonding molecular orbitals. Our picture of the degree of double bond character between adjacent carbon atoms is based on whether the various contributing molecular orbitals are bonding or antibonding. For the C-2 to C-3 bond in butadiene, the π_2 provides no electron density, but the π_1 does provide electron density, so there is some double bond character in this bond.

12.4 Structural Effects on Conjugation

In this section we see the effects of the partial double bond character of the C-2 to C-3 bond of butadiene. First, the bond length is shorter than predicted based on the hybridization of the component sp^2 orbitals that form the sigma bond. This shortening is attributed to the double bond character as a result of the contribution of the π_1 molecular orbital.

In order to maintain the overlap of the constituent p orbitals that make up the π molecular orbitals, the component atoms of the double bonds must be contained in a single plane. Two conformations are observed, consistent with this requirement. The s-trans conformation is more stable than the s-cis conformation, due to steric hindrance in the s-cis conformation. Rotation about the C-2 to C-3 bond in butadiene requires more energy than rotation about a single bond without double bond character. The rotational energy barrier is due to both a torsional component and the loss of resonance energy as a result of rotation. Using an estimation of one energy term gives an estimation of the second energy term.

12.5 Allylic Systems

You have seen the effect of resonance stabilization of charge in the allyl carbocation in Section 10.3 in the text. Again, the same themes keep repeating. If you realize this and review to assemble the big picture, then organic chemistry won't be quite as demanding. Although a substantial part of the course has already been presented, the saying "better late than never" still applies.

The positive charge of an allyl carbocation or any substituted allylic carbocation is distributed between two carbon atoms. (There is no charge on the "center" carbon atom.) However, the charge distribution is equal only in the allyl carbocation itself. Substituted allylic carbocations must of necessity have charge distributions based on the identity of the attached groups. This effect is seen in the product distribution of the S_N1 reactions of substituted allyl chlorides. The major product corresponds to capture of a nucleophile at the more highly substituted center, because the positive charge is better stabilized at that center.

Allylic radicals have an electron deficiency at either end of the radical. (There is no radical character at the "center" carbon atom.) The effect of delocalization in the allyl radical is seen in the lower bond energy of the C—H bond that can give the radical. Selective halogenation at potential allylic sites is accomplished with **N-bromosuccinimide (NBS)**.

Allylic carbanions are also resonance stabilized, with a distribution of negative charge at two carbon atoms. Allylic Grignard reagents consist of two substances because the magnesium atom may be bonded at either of two carbon atoms.

12.6 Molecular Orbital Representation of Allyl Systems

Although you get the "right answer" for the chemical reactivity of allylic system based on resonance stabilization, there is an intellectually more satisfying picture for the same information that uses molecular orbitals. Although not evident at this time, there are instances where only molecular orbitals give a model that is consistent with certain experimental facts.

The molecular orbitals formed from an odd number of p orbitals are arranged somewhat differently that for molecular orbitals from an even number of p orbitals. First of all, there is one molecular orbital, called a nonbonding orbital, in which the nodal plane contains the central carbon atom, which means that there is no electron density contributed to the atom by this molecular orbital. Second, note that the energy of that orbital is the same as a contributing atomic orbital. Hence there is no net stabilization as a result. The symmetry of the molecular orbitals is defined in the same way as for polyenes. Note that as in the case of polyenes, the lowest energy molecular orbital is symmetric and the symmetry alternates with each higher energy molecular orbital. In addition, the increasing number of nodal planes is the same as for the molecular orbitals of polyenes, However, the nodal plane may be at an atom or between atoms.

The distribution of electrons amongst the molecular orbitals follows Hund's rule. The allyl cation has electrons only in π_1, and the "deficiency" of an electron and resulting positive charge is felt at the terminal carbon atoms, corresponding to the π_2 orbital. This orbital contains one electron in the allyl radical and two electrons in the allyl anion.

12.7 Electrophilic Conjugate Addition Reactions

Conjugate dienes undergo addition by electrophiles. However, rather than forming a localized carbocation as in alkenes, an allylic carbocation is formed. The nucleophile can then add to either of two carbon atoms that bear the positive charge of the allylic carbocation. Addition to the "end" of the allylic system that is bonded to the carbon atom attacked by the electrophile gives a net **1,2-addition**. Addition to the other "end" gives a net **1,4-addition**. The amounts of these two products depend on the groups bonded to the two "end" carbon atoms of the allylic system.

If the product of the reaction is stable and does not have sufficient time to revert to an allyl cation by ionization of the original nucleophile, then the result is termed **kinetic control**. This simply means that whatever is formed fastest is the major product. However, if the product of the reaction is not stable and has sufficient time to ionize, then the repetition of this process will eventually give the product that is the most thermodynamically stable, hence the term **thermodynamic control**.

The product of kinetic control may be influenced by the stability of the intermediate and the partial positive charge at the two "ends" of the allylic carbocation. However, the product of thermodynamic control is influenced only by structural features of the product, such as the stability of the remaining double bond. Products with the more highly substituted double bond are favored.

12.8 Cumulated Dienes

Cumulated dienes are also called **allenes**. These compounds are less stable than both conjugated dienes and isolated dienes. Part of the reason for this lack of stability is explained by the sp hybridization of the central carbon atom. How is the order of stability of the different types of dienes determined? Well, of course by heats of hydrogenation. (You'll encounter this technique in the next chapter as well and in some other chapters. So, again make sure that you really understand what is going on in the hydrogenation process and how the stability of the reactants and products affects this experimental quantity).

The two π bonds of allene are perpendicular to one another, as shown in Figure 12.10 in the text. Note that the central carbon atom contributes one p orbital to each π bond. Therefore that atom must be sp hybridized. An interesting consequence of the arrangement of the two double bonds of allene is the fact that suitably substituted allenes are chiral in spite of the absence of a chiral carbon atom. The molecule itself is chiral and that is what counts. If two groups bonded to one of the carbon atoms are equivalent, then the molecule is achiral because it has a plane of symmetry.

Summary of Reactions

1. Allylic Bromination

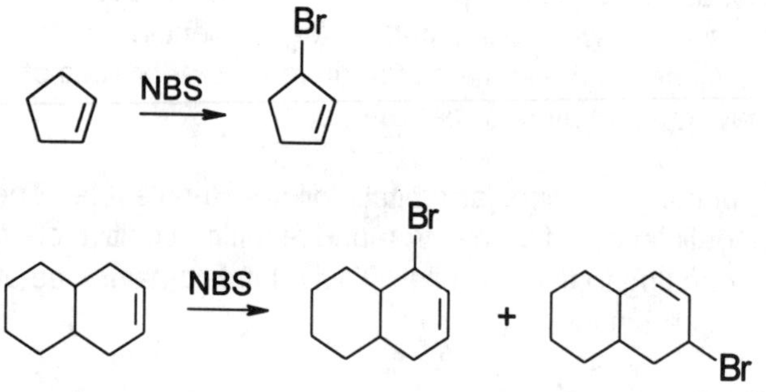

2. Electrophilic Conjugate Addition

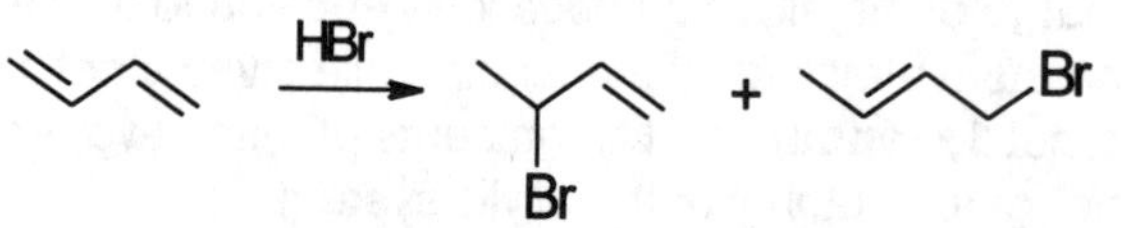

Solutions to Exercises

12.1 Conjugated compounds have only one single bond separating the double bonds. Only (a) has conjugated double bonds. In (b) the double bonds are separated by two single bonds. In (c) four single bonds separate the two double bonds, although as drawn the double bonds appear close to one another. In (d) each of the three double bonds is separated by three single bonds.

12.2 The compounds in (a), (b), and (c) all have conjugated double bonds. In (d) the closest double bonds are the one on the left and the vinyl group depicted as a branch. However, there are two intervening single bonds, so the double bonds are not conjugated.

12.3 **(a)** Only the middle compound of the three has conjugated double bonds.
(b) The first and second compounds listed have conjugated double bonds.

12.4 **(a)** The two double bonds to a common carbon atom near the middle of the structure are cumulated double bonds. The two double bonds near the right of the structure are conjugated double bonds.
(b) A series of six double bonds are separated by one single bond between each pair of double bonds. The entire series is conjugated.
(c) All three double bonds are isolated or nonconjugated.

12.5 The series of five carbon-carbon triple bonds and the carbon-nitrogen triple bond are all separated by single bonds. Recall that one set of p orbitals making one π bond of a triple bond is perpendicular to a second set of p orbitals making the other π bond. Thus, there is one set of six conjugated π bonds that is perpendicular to a second set of six conjugated π bonds.

12.6 All 16 double bonds are conjugated.

12.7 **(a)** monoterpene   **(b)** sesquiterpene   **(c)** sesquiterpene

12.8 **(a)** sesquiterpene   **(b)** monoterpene   **(c)** sesquiterpene

12.9   Conjugated dienes are more stable than isomeric compounds with isolated double bonds by an amount of energy called the resonance energy. Because they are of lower energy, they release a smaller amount of energy upon hydrogenation. Compound III is conjugated and has a smaller heat of hydrogenation than I and II.

12.10   Use -125 kJ mole^{-1} and -116 kJ mole^{-1} for the heats of hydrogenation of terminal monosubstituted and trans disubstituted double bonds, respectively.

(a) Add -125 and -116 kJ mole^{-1} for the terminal monosubstituted and trans disubstituted double bonds, but decrease the quantity by 15 kJ mole^{-1} resonance energy because the bonds are conjugated. The predicted value is -226 kJ mole^{-1}.

(b) Add -116 and -116 kJ mole^{-1} for the two trans disubstituted double bonds, but decrease the quantity by 15 kJ mole^{-1} resonance energy because the bonds are conjugated. The predicted value is -217 kJ mole^{-1}.

(c) Add -125 and -116 kJ mole^{-1} for the isolated terminal monosubstituted and trans disubstituted double bonds. The predicted value is -241 kJ mole^{-1}.

(d) Add -125 and -125 kJ mole^{-1} for the two terminal monosubstituted double bonds which are isolated from one another. The predicted value is -250 kJ mole^{-1}.

12.11   Protonate the C-1 atom of the π bond to give a carbocation at the C-2 atom of the original double bond. Deprotonate at the saturated C-3 atom of the original compound to give the conjugated 1,3-cyclohexadiene. This compound is the more stable due to resonance stabilization and should be the major component of the equilibrium mixture.

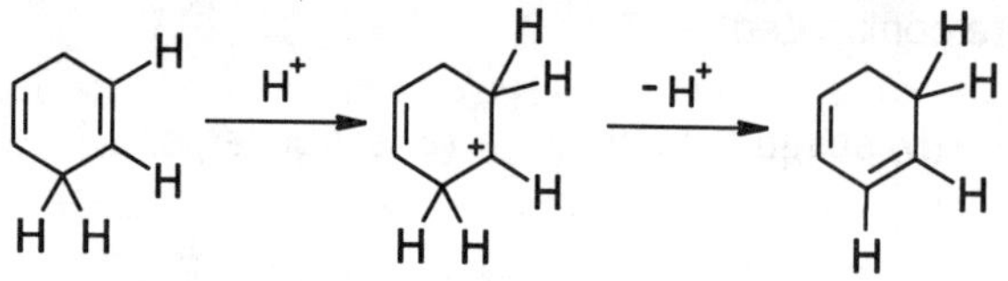

12.12   The two dienes are similarly substituted, so equal amounts of the two compounds should be formed in an equilibrium mixture. To achieve equilibrium, protonate the terminal carbon atom of the conjugated diene in the leftmost six-membered ring. Write an alternate resonance form of the allylic carbocation. Deprotonate at a methylene carbon atom adjacent to the carbocation center in this resonance form.

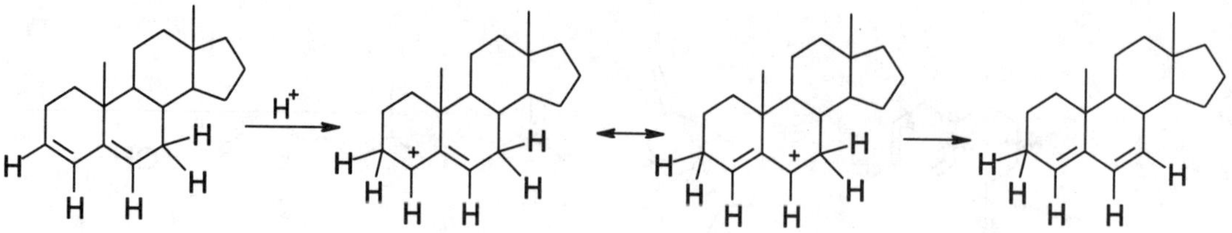

12.13 The highest occupied bonding molecular orbital, which is π_4. It has three nodal planes separating the system into four "isolated" π bonds, which is the conventional Lewis structure.

12.14 The symmetry of π_1 through π_4 is symmetric, antisymmetric, symmetric, and antisymmetric, respectively. The highest occupied bonding molecular orbital of a conjugated polyene with n double bonds, which is π_n, has n - 1 vertical nodal planes separating the system into n "isolated" π bonds resembling the conventional Lewis structure.

12.15 **(a)** 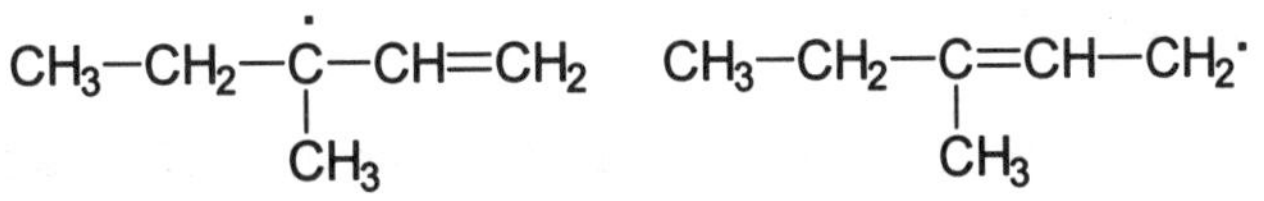 $CH_3-CH_2-\overset{\displaystyle .}{C}-CH=CH_2$ $CH_3-CH_2-C=CH-CH_2\cdot$
$\qquad\qquad\qquad CH_3 \qquad\qquad\qquad\qquad\qquad CH_3$

(b) $\cdot CH_2-\overset{CH_3}{\underset{|}{C}}=CH-CH_2-CH_3$ $CH_2=\overset{CH_3}{\underset{|}{C}}-\overset{\displaystyle .}{C}H-CH_2-CH_3$

(c) $CH_3-CH_2-\overset{}{\underset{|}{C}}=CH-CH_2\cdot$ $CH_3-CH_2-\overset{\displaystyle .}{C}-CH=CH_2$
$\qquad\qquad\qquad CH_3 \qquad\qquad\qquad\qquad\qquad CH_3$

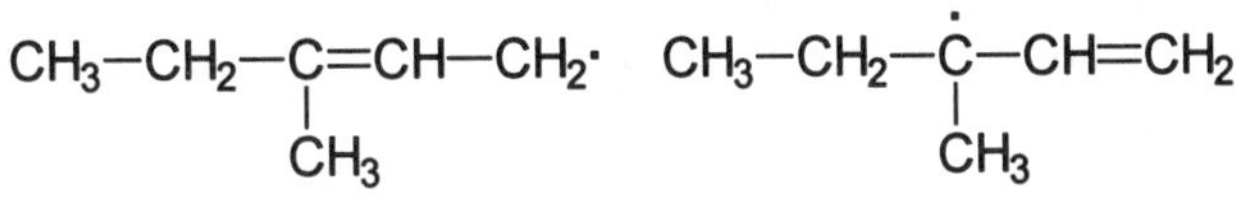

(d)  $CH_3-\overset{CH_3}{\underset{|}{C}}=CH-\overset{\displaystyle .}{C}H-CH_3$ $CH_3-\overset{CH_3}{\underset{|}{\underset{.}{C}}}-CH=CH-CH_3$

12.16 **(a)** 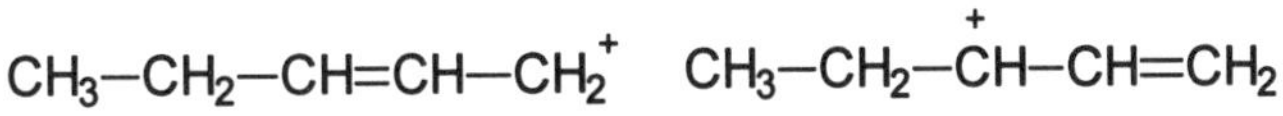 $CH_3-CH_2-CH=CH-\overset{+}{C}H_2$ $CH_3-CH_2-\overset{+}{C}H-CH=CH_2$

(b) $CH_2-\overset{CH_3}{\underset{|}{C}}=CH-\overset{+}{C}H_2$ $CH_3-\overset{CH_3}{\underset{|}{\underset{+}{C}}}-CH=CH_2$

(c) $CH_3-CH_2-\overset{}{\underset{|}{C}}=CH-\overset{+}{C}H_2$ $CH_3-CH_2-\overset{+}{\underset{|}{C}}-CH=CH_2$
$\qquad\qquad\qquad CH_3 \qquad\qquad\qquad\qquad\qquad CH_3$

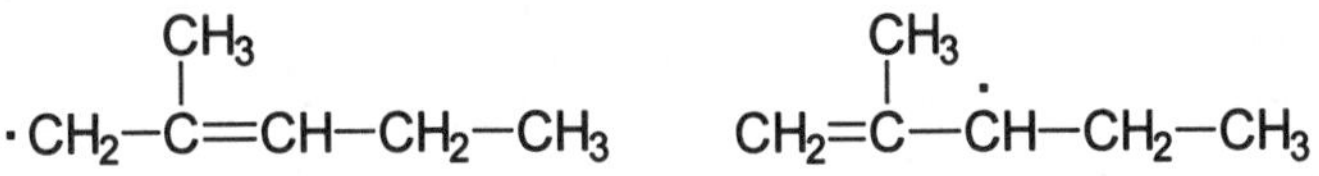

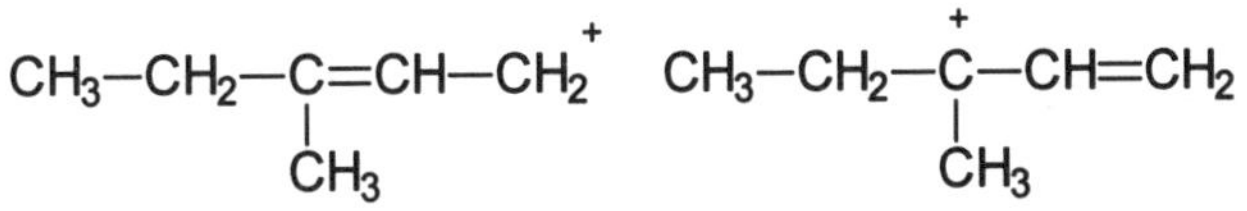

(d) $CH_2=\overset{CH_3}{\underset{|}{C}}-\overset{+}{C}H-CH_3$ $\overset{+}{C}H_2-\overset{CH_3}{\underset{|}{C}}=CH-CH_3$

12.17 The positively charged carbon atoms of the allyl carbocation are both primary.

$CH_2=CH-\overset{+}{C}H_2 \longleftrightarrow \overset{+}{C}H_2-CH=CH_2$

The positively charged carbon atoms of the carbocation derived from 1-chloro-3-methyl-2-butene are primary and tertiary. Hence this carbocation is more stable than the allyl carbocation.

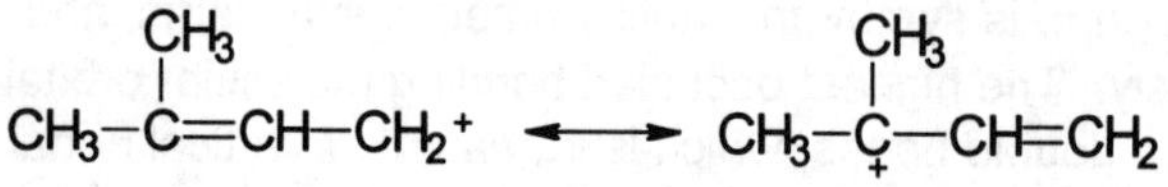

$$CH_3-\overset{\overset{\displaystyle CH_3}{|}}{C}=CH-CH_2{}^+ \quad \longleftrightarrow \quad CH_3-\overset{\overset{\displaystyle CH_3}{|}}{\underset{+}{C}}-CH=CH_2$$

The stabilization of this intermediate is reflected in the transition state that generates it. As a consequence the activation energy required for the reaction is smaller and the reaction occurs at a faster rate.

12.18 Allylic carbocations are generated as the protonated oxygen atom of the alcohol leaves as water in an S_N1 reaction. The positively charged carbon atoms of the carbocation derived from compound I are primary and secondary. The positively charged carbon atoms of the carbocation derived from compound II are primary and tertiary. Thus, the carbocation derived from II is more stable than that derived from I and the activation energy required for the reaction of II is smaller and the reaction occurs at a faster rate.

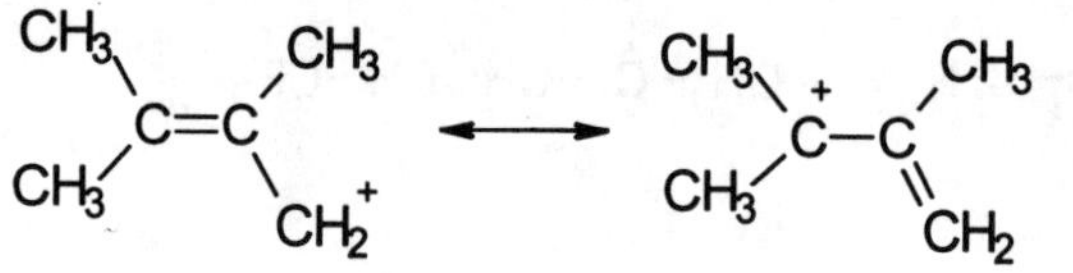

12.19 The carbon atom bearing the pyrophosphate group in both compounds is primary. However, that carbon atom in 2-isopentenyl pyrophosphate is allylic. Thus, a carbocation forms more readily for this compound, and the reaction rate is greater than for the reaction of 3-isopentenyl pyrophosphate.

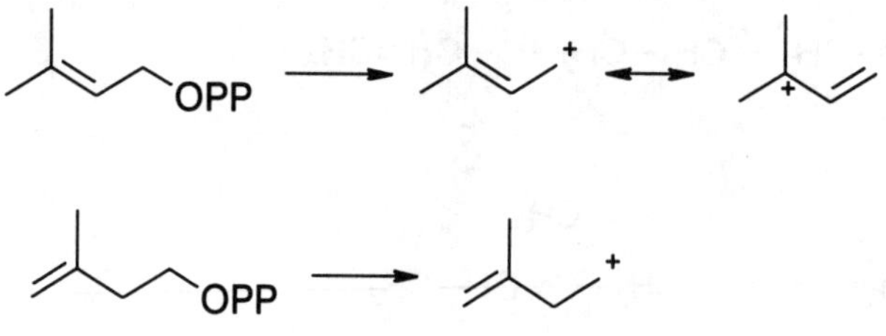

12.20

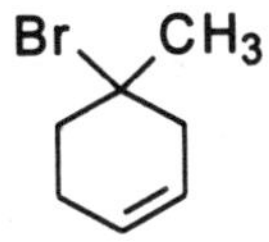

12.21 Abstraction of an allylic hydrogen atom gives a resonance stabilized allylic radical which can abstract a bromine atom at either end of the radical.

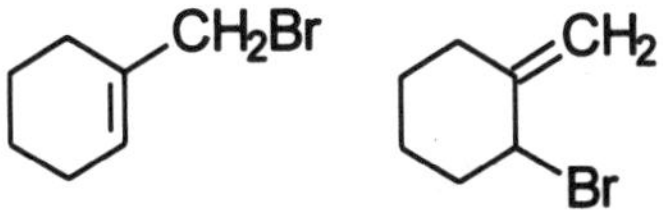

12.22 Both the (E) and (Z) isomers are derived from abstraction of bromine by the radical at its primary C-1 atom, shown in one of the two contributing resonance forms. These compounds are the major products because they have the more substituted double bond and the reaction of the radical at the primary center is less sterically hindered. The isomeric 3-bromo-1-octene is derived from abstraction of bromine by the radical at its secondary C-3 atom.

$$CH_3CH_2CH_2CH_2CH_2-\overset{\bullet}{C}H-CH=CH_2 \longleftrightarrow CH_3CH_2CH_2CH_2CH_2-CH=CH-CH_2\overset{\bullet}{}$$

12.23 1-Chloro-4,4-dimethyl-2-pentene is the major isomer because the double bond is disubstituted and the chlorine atom is abstracted from the tert-butyl hypochlorite by a primary radical. The isomeric product 3-chloro-4,4-dimethyl-1-pentene has a monosubstituted double bond and the chlorine atom is abstracted by a secondary radical.

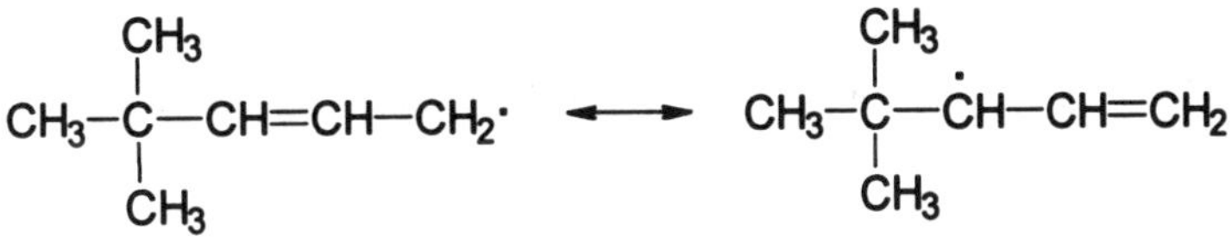

12.24 The same resonance stabilized radical is generated by abstraction of a hydrogen atom at C-5 of 1,3-pentadiene or abstraction of a hydrogen atom at C-3 of 1,4-pentadiene. Thus, the mixture of brominated products is the same for both compounds.

$$CH_2=CH-CH=CH-CH_2\overset{\bullet}{} \longleftrightarrow CH_2=CH-\overset{\bullet}{C}H-CH=CH_2$$

1,3-Pentadiene is a conjugated diene and is more stable than 1,4-pentadiene. Thus, the activation energy for abstraction of its hydrogen atom is larger because both isomers yield the same intermediate and the transition states are of similar energy. 1,4-Pentadiene would undergo allylic bromination at a faster rate.

12.25 In order to stabilize an allylic system there must be proper alignment of 2p orbitals. In compound II the orbital of the radical at the bridgehead carbon cannot overlap with the 2p orbitals making up the π bond, so no resonance stabilization occurs.

12.26 The bond between C-3 and C-4 is the weakest because cleavage yields two resonance stabilized allyl radicals.

$$CH_2=CH-CH_2-CH_2-CH=CH_2 \longrightarrow CH_2=CH-CH_2\cdot \ + \ \cdot CH_2-CH=CH_2$$

12.27 There are two isomeric allylic Grignard structures. Reaction of D⁺ from D_2O can occur at either the C-1 or C-3 atoms.

$$CH_3-CH=CH-CH_2-MgBr \ \rightleftharpoons \ CH_3-\overset{\overset{\displaystyle MgBr}{|}}{CH}-CH=CH_2$$

$\Big\downarrow D_2O$ $\Big\downarrow D_2O$

$$CH_3-CH=CH-CH_2-D \qquad\qquad CH_3-\overset{\overset{\displaystyle D}{|}}{CH}-CH=CH_2$$

12.28 There are two isomeric allylic Grignard reagents. One is a primary Grignard reagent with a trisubstituted double bond. The other is a tertiary Grignard reagent with a monosubstituted double bond.

$$CH_3-\overset{\overset{\displaystyle CH_3}{|}}{C}=CH-CH_2-MgBr \qquad CH_3-\overset{\overset{\displaystyle CH_3}{|}}{\underset{\underset{\displaystyle MgBr}{|}}{C}}-CH=CH_2$$

12.29 There are two isomeric allylic Grignard reagents. The most stable is the one where the Grignard reagent is primary and the double bond is disubstituted. The less stable is a secondary Grignard reagent with a monosubstituted double bond. In addition, the reactive site of this reagent is near a quaternary carbon atom and is sterically hindered. The major product has the deuterium atom at the C-1 atom.

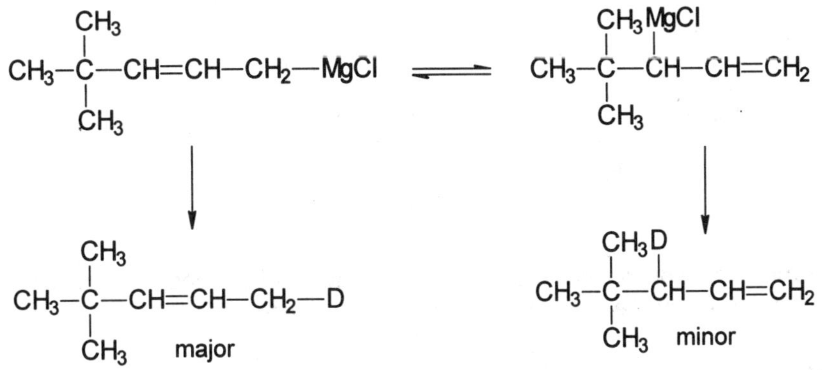

$$CH_3-\underset{\underset{CH_3}{|}}{\overset{\overset{CH_3}{|}}{C}}-CH=CH-CH_2-MgCl \;\rightleftharpoons\; CH_3-\underset{\underset{CH_3}{|}}{\overset{\overset{CH_3\,MgCl}{|}}{C}}-CH-CH=CH_2$$

$$CH_3-\underset{\underset{CH_3}{|}}{\overset{\overset{CH_3}{|}}{C}}-CH=CH-CH_2-D \qquad\qquad CH_3-\underset{\underset{CH_3}{|}}{\overset{\overset{CH_3\,D}{|}}{C}}-CH-CH=CH_2$$

major   minor

12.30 There are two isomeric allylic Grignard reagents. The most stable is the one where the Grignard reagent is secondary and the double bond is tetrasubstituted. The less stable is a tertiary Grignard reagent with a trisubstituted double bond. The major product has the deuterium atom at a secondary carbon atom.

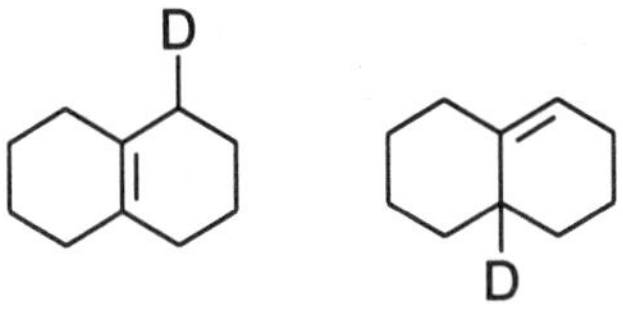

major   minor

12.31 The π_3 molecular orbital has two vertical nodal planes. They are located that the C-2 and C-4 atoms. As a consequence there is electron density only at the C-1, C-3, and C-5 atoms. The negative charge is distributed at these three atoms.

12.32 A nucleophile has to bond to the lowest energy orbital that does not have any electrons. In the cation the four electrons are located in the π_1 and π_2 molecular orbitals. Thus, the reaction occurs at the π_3 molecular orbital, which is depicted above in the answer for Exercise 12.31. The nucleophile can bond only at the C-1, C-3, and C-5 atoms. These atoms correspond to the atoms having a positive charge in the three contributing resonance forms.

$$CH_2=CH-CH=CH-\overset{+}{C}H_2 \longleftrightarrow CH_2=CH-\overset{+}{C}H-CH=CH_2 \longleftrightarrow \overset{+}{C}H_2-CH=CH-CH=CH_2$$

12.33 Unlike the addition reactions to 1,3-butadiene where the double bond may be located in two different places, there is only one position for the double bond in the product cyclohexene. In both 1,2 and 1,4 addition, the bromine is located at a C-3 position using the double bond to number the ring. The compound is 3-bromocyclohexene.

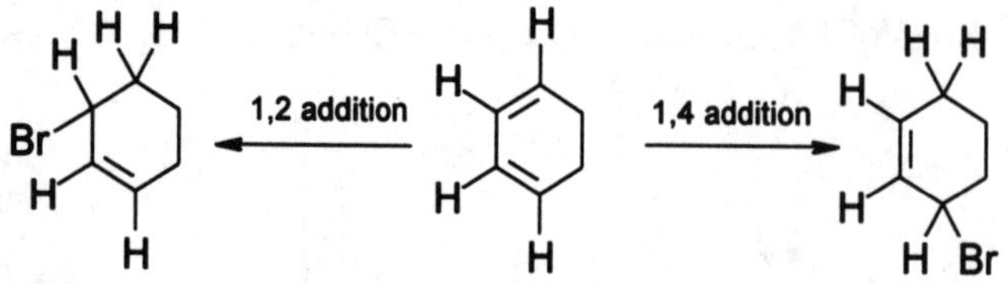

12.34 Unlike the similar reaction in Exercise 12.33, the products of 1,2 and 1,4 addition of DBr are different, because the deuterium positions can be distinguished.

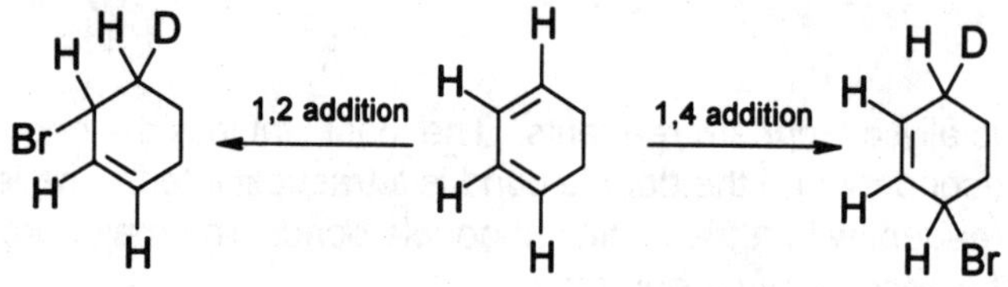

12.35 Addition of a proton to the C-1 carbon atom gives a resonance stabilized allylic carbocation with equal electron density at the C-2 and C-4 atoms, which are also structurally equivalent. Capture of the carbocation by chloride ion at the C-2 and C-4 atoms gives the same product, 4-chloro-2-pentene.

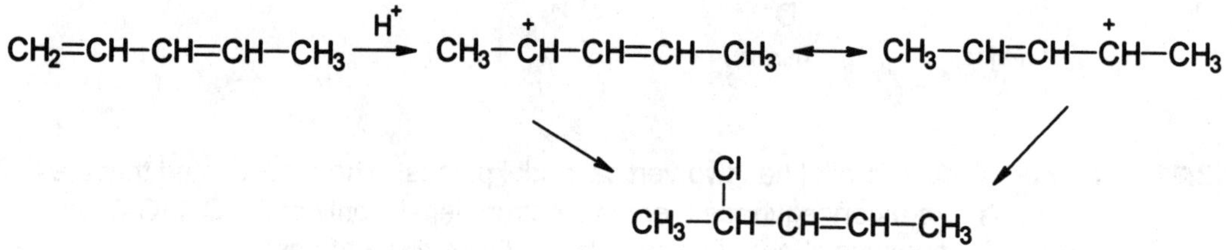

12.36 Addition of a deuterium to the C-1 carbon atom gives a resonance stabilized allylic carbocation with equal electron density at the C-2 and C-4 atoms. However, unlike the reaction with HCl given in Exercise 12.35, the C-2 and C-4 atoms are not structurally equivalent with respect to the two terminal carbon atoms, one of which contains a deuterium atom. Capture of the carbocation by chloride ion at the C-2 atom of the original compound gives a 1,2-addition product that is isomeric with the product of reaction at the C-4 atom, which corresponds to a 1,4-addition reaction. Thus the amount of each addition process can be determined.

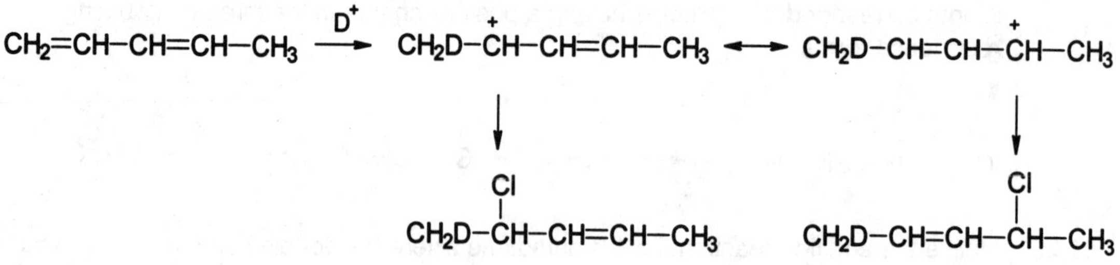

12.37 Both the 1,2 and 1,6 addition products have conjugated double bonds. These products are more stable the product of 1,4 addition, which has isolated double bonds.

1,2-addition product CH_2=CH–CH=CH–CH–CH$_2$ (with Br Br on the 5th and 6th carbons)

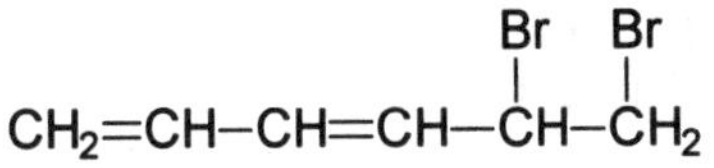

1,6-addition product CH_2–CH=CH–CH=CH–CH$_2$ (with Br on first and last carbon)

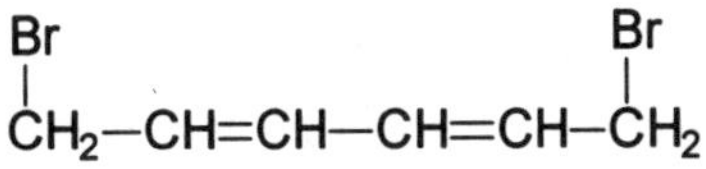

1,4-addition product CH_2=CH–CH–CH=CH–CH$_2$ (with Br Br)

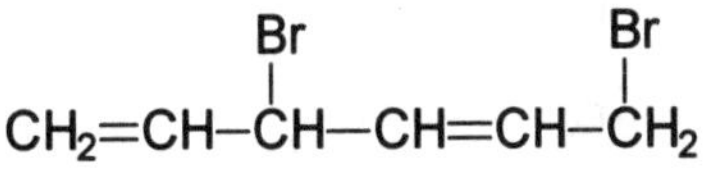

12.38 At the higher temperature, the more stable isomer is favored. This compound, which contains a disubstituted double bond, is the result of 1,4-addition and is the isomer obtained in 90% yield. The 1,2- addition product is less stable because it has a monosubstituted double bond. At the lower temperature the amounts of the two products are kinetically controlled.

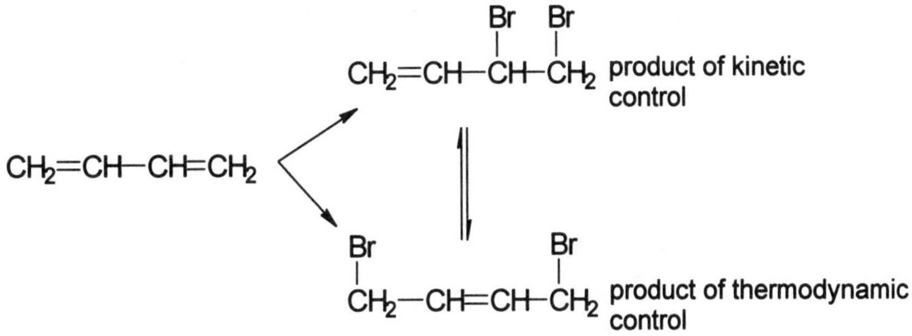

CH_2=CH–CH=CH$_2$

CH_2=CH–CH–CH$_2$ (Br Br) product of kinetic control

CH_2–CH=CH–CH$_2$ (Br Br) product of thermodynamic control

12.39 The product is 1-bromo-2,3-dimethyl-2-butene, which is the result of a 1,4 addition reaction. The double bond is tetrasubstituted, compared to a disubstituted double bond in the 1,2-addition product, which is not observed.

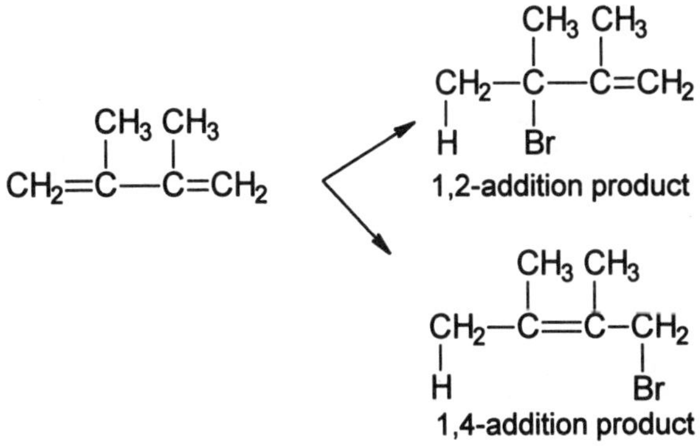

CH_2=C–C=CH$_2$ (CH$_3$ CH$_3$)

CH_2–C–C=CH$_2$ (CH$_3$ CH$_3$, H Br) 1,2-addition product

CH_2–C=C–CH$_2$ (CH$_3$ CH$_3$, H Br) 1,4-addition product

12.40 The product is 1,2-dichloro-4-phenyl-3-butene and is formed by 1,2-addition. This product is favored because the double bond is in conjugation with the benzene ring.

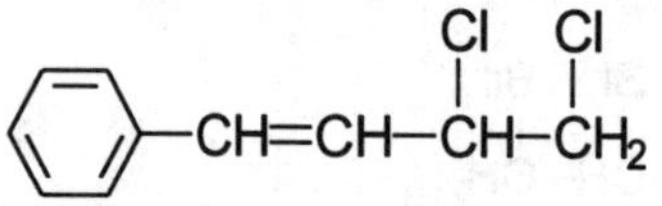

The initial electrophilic attack is at the terminal position of the diene because the resulting allylic carbocation is resonance stabilized by the benzene ring.

Attack at the carbon atom bonded to the benzene ring would give an allylic carbocation lacking such stabilization.

12.41

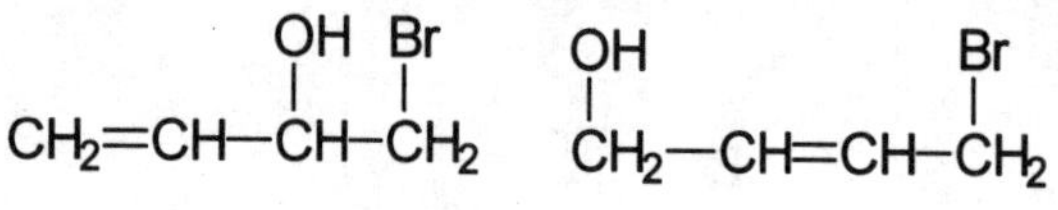

12.42 Attack of the electrophilic chlorine cation at the C-4 position would give an allylic carbocation that has positive charge distributed between a secondary and a primary carbon atom.

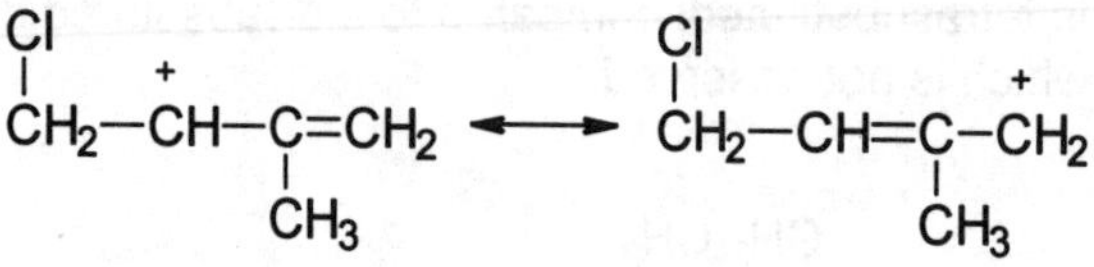

However, the electrophilic chlorine cation adds to the C-1 atom, giving a more stable allylic carbocation that has positive charge distributed between a tertiary and a primary carbon atom.

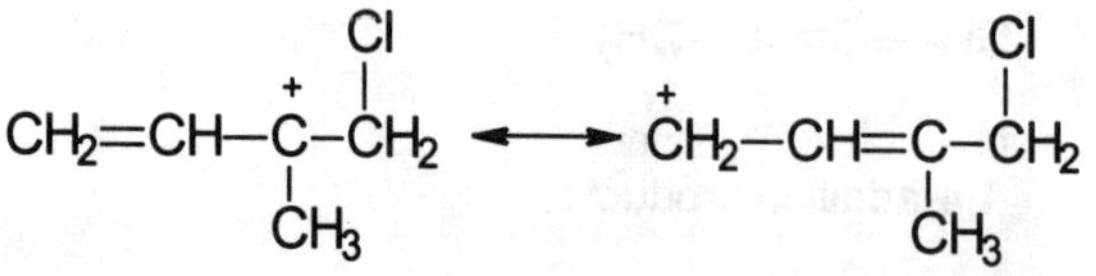

Reaction of this more stable carbocation with water acting as a nucleophile occurs at the C-4 atom of the original diene, because that 1,4-addition product has a trisubstituted double bond.

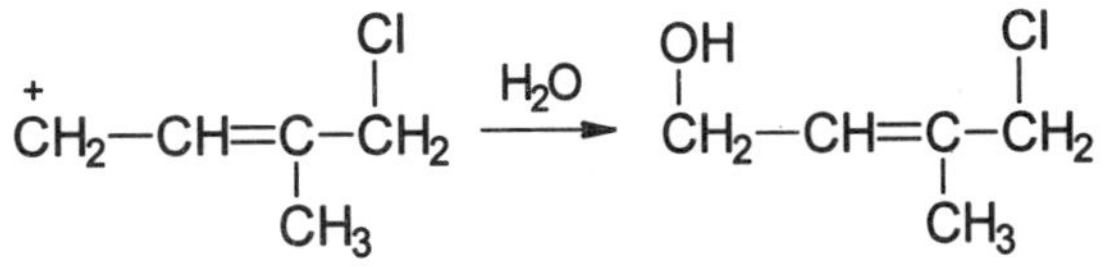

Reaction at the C-2 atom would give a less stable 1,2-addition product with a monosubstituted double bond.

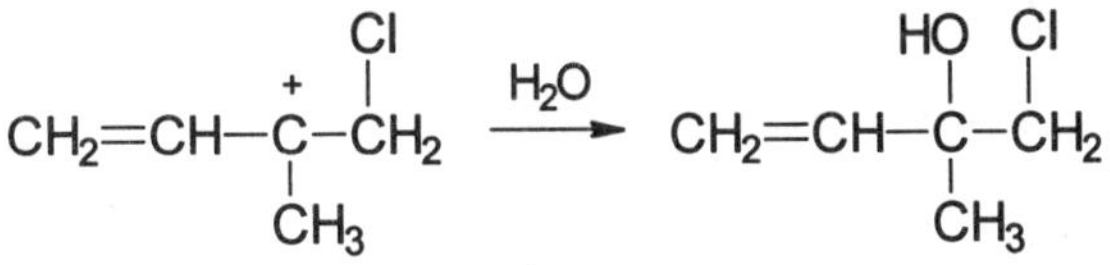

12.43 The σ bond of the double bond of propene is formed by overlap of two sp^2 orbitals. The central carbon atom of allene is sp-hybridized; its σ bond of the double bond overlaps with an sp^2-hybridized orbital of the terminal carbon atoms. Bonds formed by orbitals with increased % s character are shorter.

12.44 The two *tert*-butyl groups are located at the sp^2 hybridized terminal carbon atoms of the allene system in planes that are perpendicular to one another. Thus, the distance separating the groups is larger than the distance separating the two *tert*-butyl groups in the alkene, where they are located in a common plane.

12.45 There is an allene system contained within the molecule. Each of the sp^2 hybridized terminal carbon atoms of the allene system has two nonequivalent atoms or groups of atoms bonded to it. Thus, the compound can exist in two enantiomeric forms. The naturally occurring mycomycin is one of these enantiomers and is optically active.

12.46 Each of the sp^2 hybridized terminal carbon atoms of the allene system within the molecule has two nonequivalent atoms or groups of atoms bonded to it. Thus, the compound can exist in two optically active enantiomeric forms.

12.47 Each compound has two nonequivalent atoms or groups of atoms bonded to each of the sp^2 hybridized terminal carbon atoms of the allene system and can exist as pairs of enantiomers.

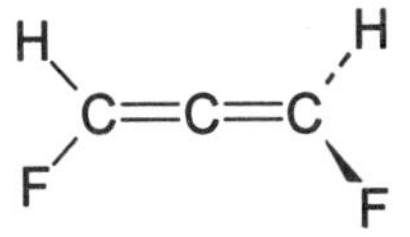

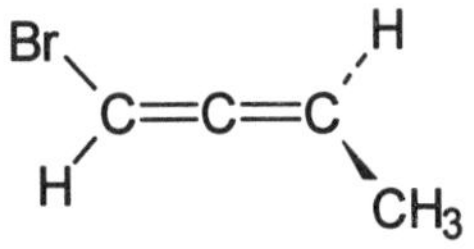

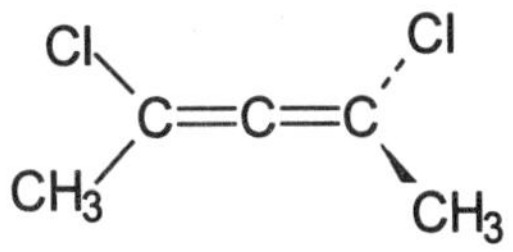

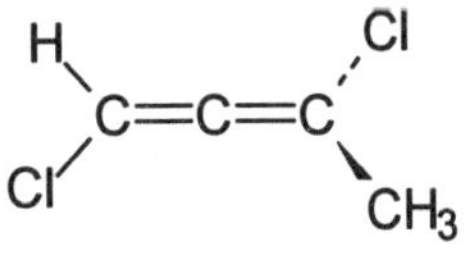

12.48   The addition of a third multiple bond to form an extended cumulene places the sigma bonds at each end of the system in a common plane. Thus, the molecule has a plane of symmetry and cannot be optically active. Geometric isomers are possible.

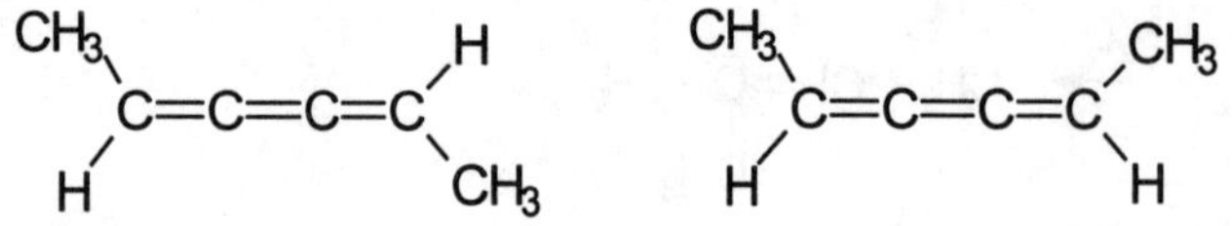

12.49

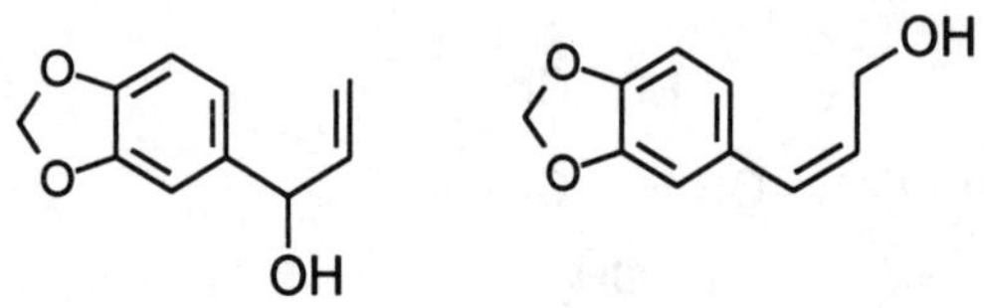

12.50

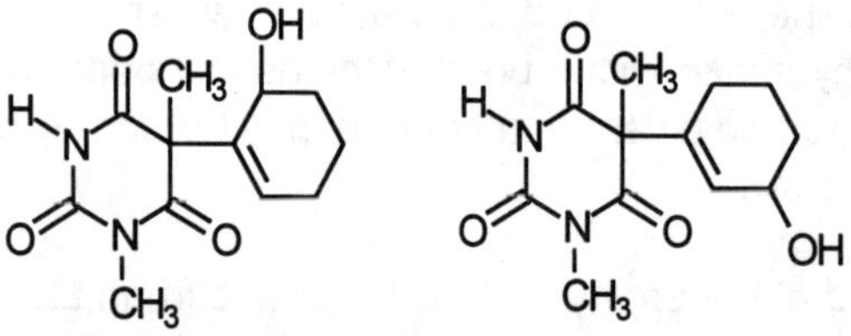

13

Arenes and Aromaticity

In this chapter we encounter molecular structures that are used to represent unusually stable and quite unreactive molecules, but the characteristics of those structures don't look "right" for molecules of such stability. The problem lies with the method of representation and the limitations of conventional Lewis structures. So you will have to get used to the fact that the representations are used in spite of the fact that they aren't really representative!

Aromatic compounds have certain physical and chemical properties that you should always compare with those classes of compounds studied in earlier chapters. Which properties are similar and which are distinctly different? This chapter deals with the structure of aromatic compounds as well as ways to quantitatively evaluate their special stability which accounts for their very different reactivity.

Keys to the Chapter

13.1 Aromatic Compounds

The most common aromatic compounds contain a benzene ring that may have one or more of its hydrogen atoms replaced by substituents or carbon groups. Although many of these compounds are termed "aromatic" because of their odor, that property is a human physiological response and is not the criterion used to classify these compounds as a separate class.

Several benzene rings can be fused to include two common carbon atoms between rings. The possibilities are endless, as more and more rings can be fused. The common feature in all compounds is the alternating series of single and double bonds within the rings. You should know the structures of benzene, naphthalene, anthracene, and phenanthrene, which are the simplest **arenes**.

13.2 Aromaticity

The chemical criterion for aromaticity is the lack of reactivity toward reagents, such as bromine, that are normally reactive toward carbon-carbon double bonds. The stability of benzene is due to the resonance stabilization of the cyclic conjugated arrangement of the π bonds. In spite of the fact that we know that molecular orbitals distribute the electron density over the entire benzene ring, it is still convenient to write just one of the two conventional Lewis structures (called Kekulé structures in aromatic compounds) to represent benzene.

The resonance energy of benzene is determined by that old reliable method called the heat of hydrogenation. The difference between the experimental quantity and the estimated quantity based on the heat of hydrogenation of three isolated double bonds is called the

resonance energy. Note that the resonance energy for benzene is substantially larger than for butadiene, an acyclic diene.

13.3 The Hückel Rule

The special stability of benzene and other aromatic compounds is related to the energy of the molecular orbitals. You can predict this stability without knowing anything about those molecular orbitals by using the Hückel rule. Before you even consider using the rule to determine whether or not a compound is aromatic, first establish that the molecule is cyclic and planar. Then make sure that it contains only sp^2 hybridized atoms in the ring, so the p orbital from each ring atom can be used to form a delocalized system. If these criteria are met, then count the number of π electrons in the ring and compare that number to the Hückel formula of 4n+2. The Hückel rule states that a compound is aromatic if the number of π electrons equals 4n+2, where n is an integer.

Both cyclobutadiene and cyclooctatetraene have an alternating series of single and double bonds but yet are not aromatic. The Hückel formula of 4n+2 "explains" why. There is no integer value of n that can give 4 or 8 upon substituting into the 4n +2 expression.

Some ions are aromatic because they do have 4n +2 π electrons. Note that the Hückel rule applies to contributing electrons—not the number of carbon atoms and not the number of 2p orbitals that make up the cyclic delocalized system. It is the number of electrons contained within the molecular orbitals and distributed over the entire molecule that determines aromaticity.

13.4 Molecular Orbitals and the 4n +2 Rule

This section provides a rationale for why the molecular orbitals containing a certain number of electrons give a stable molecule. No support is given for the indicated arrangement of the molecular orbitals, but you can "fill" them just as you did the atomic orbitals in General Chemistry. Note that there is one lowest energy orbital and one highest energy orbital for benzene. The remaining ones occur as pairs of equal energy orbitals said to be **degenerate**.

You have to accept the order of molecular orbitals for cyclobutadiene and cyclooctatetraene, also without proof. Each has a set of degenerate nonbonding molecular orbitals that are at the same energy level as the contributing p orbitals. In each case a diradical would result. However that conclusion is correct only if the molecule is planar and each side of the polygon is equal. In fact, these two molecules do not exist in this form because there are lower energy alternatives. In the case of cyclooctatetraene that alternative is a tub conformation.

One way to remember the number of molecular orbitals and their energy is to inscribe the corresponding polygon in a circle, placing one of the vertices at the six o'clock position. The points of contact of the polygon with the circle give the relative energies of the molecular orbitals. Thus for benzene, cyclobutadiene, and cyclooctatetraene the pictures are as follows.

The order of the energies of the cyclopentadienyl anion and the cycloheptatrienyl cation follows a similar pattern to those of benzene. There is a single lowest energy molecular orbital and all others are degenerate pairs. In each case there are three bonding molecular orbitals and thus six electrons can be distributed in each system. Using the same device as above the pictures are as follows.

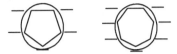

13.5 Heterocyclic Aromatic Compounds

The criteria described above for establishing aromaticity apply to heterocyclic compounds as well. Heteroatoms in the ring must be sp^2 hybridized and the Hückel rule of $4n+2$ p electrons must be obeyed. The most common heteroatoms are nitrogen, oxygen, and sulfur. The point to remember is that some aromatic rings require the contribution of one electron of the heteroatom for the aromatic ring and others require two. If none of the "double bonds" in the ring is bonded to the heteroatom, then it must contribute two electrons in a p orbital. This is the case in pyrrole, furan, and thiophene. If there is a "double bond" in the ring bonded to a heteroatom, as in the case of pyridine, then that heteroatom contributes only one electron to the aromatic system. Note that there is one sp^2 hybridized orbital that either forms a bond to an atom attached to the ring or contains a nonbonded pair of electrons. This pair is in an orbital that is perpendicular to the π system and cannot be used to write resonance forms.

13.6 Polycyclic Aromatic Compounds

Polycyclic aromatic compounds containing fused benzene rings contain $4n + 2$ electrons, the same criterion used to predict aromaticity in monocyclic compounds. Thus, naphthalene and anthracene contain 10 and 14 π electrons, respectively.

    Make sure that you don't inadvertently end up writing resonance structures that have too many bonds to carbon. It's easy to forget that there are some bonds to hydrogen atoms that are not shown. In addition, remember that no matter what resonance form is written, there are no bonds to hydrogen at any carbon atoms at points of fusion.

13.7 Nomenclature of Benzene Compounds

Some monosubstituted benzene compounds have common names such as toluene, phenol, and aniline. You should know the common names for the nine compounds given on the bottom of page 485 and top of page 486 in the text, simply because these names are used so commonly, even when IUPAC rules are observed.

    Disubstituted compounds may be named using a numbering system or the prefixes o-, m-, and p- for ortho, meta, and para, respectively. However, compounds with three or more substituents must use numbers to locate the substituents. Substituents are named in alphabetical order.

    Make sure that you know the difference between phenyl and benzyl. Students often forget this distinction and use the phenyl group rather than benzyl when confronted with the term on an exam covering material in later chapters. Don't let this happen to you.

13.8 Physical Properties of Substituted Benzene Compounds

The physical properties of substituted benzene compounds depend on the substituents themselves. Thus, you should check whether the substituents have individual bond moments and determine whether they are additive or not. Also remember that hydrogen bonding of substituents strongly affects the solubility and boiling points of compounds.

13.9 Reactions of Side Chains

A group of atoms bonded to an aromatic ring is a side chain. Substituents bonded to atoms other than the atom directly bonded to the aromatic ring are unaffected by the ring. However, any substituent bonded to the **benzyl carbon atom** has substantially different reactivity. Furthermore, intermediates that involve the benzyl carbon atom are stabilized by resonance. Examples include carbocations and free radicals. Electrophiles attack double bonds in conjugation with the aromatic ring to give a benzyl carbocation. Free radical abstraction of a hydrogen atom at the benzyl carbon atom requires less energy than for other sp^3 hybridized carbon atoms.

13.10 Oxidation of Side Chains

Reaction of potassium permanganate with an aromatic ring that has a side chain of carbon atoms completely oxidizes the side chain and gives a carboxylic acid. Under these vigorous oxidation conditions the aromatic ring itself is unaffected. The oxidation of a methyl group is a convenient way to introduce a carboxyl group on a benzene ring.

13.11 Reduction of Aromatic Compounds

Aromatic rings cannot be easily reduced. The reaction requires high pressures of hydrogen gas and often high temperatures. Carbon-carbon double bonds in side chains are easily reduced without affecting the aromatic ring. The special reactivity of functional groups bonded directly to the aromatic ring is exemplified by the reduction of ketones directly to saturated carbon atoms. The reduction of a carbonyl group in a nonaromatic compound requires substantially higher pressures of hydrogen gas.

Summary of Reactions

1. Reactions of Side Chains of Aromatic Compounds

2. Oxidation of Side Chains of Aromatic Compounds

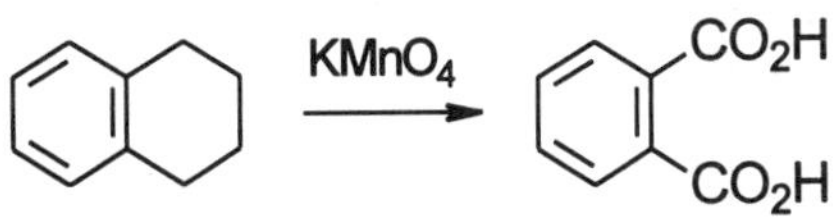

3. Reduction of Substituted Aromatic Compounds

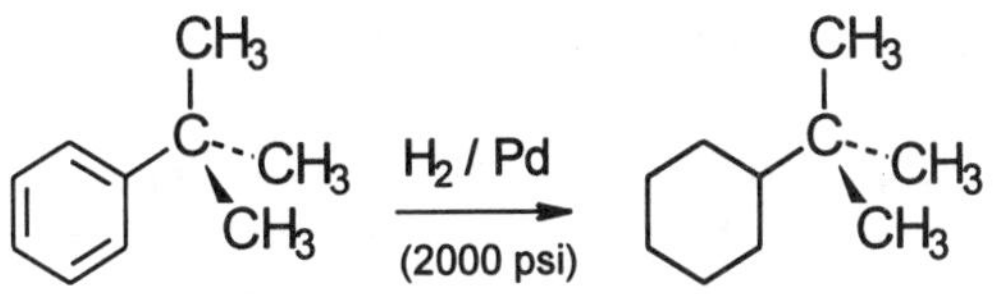

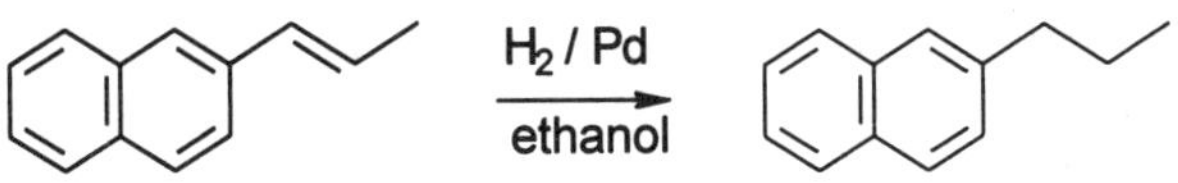

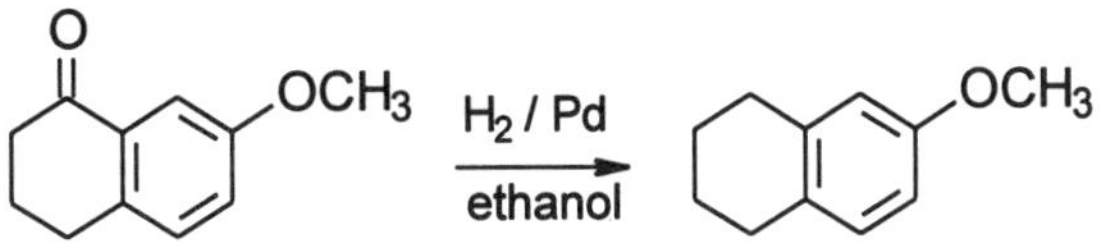

Solutions to Exercises

13.1   **(a)** Although there are six π electrons in the triene, the compound is not aromatic because the π bonds are not in a cyclic arrangement.

(b) There are six π electrons in the cyclic triene. However, there is an sp^3-hybridized carbon atom interrupting the conjugation of the π bonds, so the compound is not aromatic.

(c) There are six π electrons in the bicyclic triene. However, only two of the π bonds are conjugated and each end of that system is separated from the third π bond by an sp^3-hybridized carbon atom. The compound is not aromatic.

(d) There are eight π electrons, so the compound does not have 4n+2 π electrons. Two other facts that tell us that the compound is not aromatic are that the π bonds are not in a continuous cyclic arrangement and there are intervening sp^3-hybridized carbon atoms.

13.2   **(a)** There are ten π electrons in the bicyclic compound arranged in a continuous array of alternating single and double bonds. The number of π electrons is consistent with the Hückel rule for n = 2. The compound is aromatic.

(b) The π bonds are in a cyclic arrangement without any intervening sp^3-hybridized carbon atoms. However there are 8 π electrons, a number that is not consistent with the Hückel rule. The compound is not aromatic.

(c) The π bonds are in a cyclic arrangement without any intervening sp^3-hybridized carbon atoms. However, there are 12 π electrons, a number that is not consistent with the Hückel rule. The compound is not aromatic.

(d) There are 14 π electrons, a number that is consistent with the Hückel rule for n = 3. However, there are two intervening sp^3-hybridized carbon atoms. The compound is not aromatic.

13.3   There are two isolated double bonds in the ring that are hydrogenated in forming cyclohexane. Hydrogenation of one of the two double bonds to form cyclohexene should require one-half the amount required for total hydrogenation, so $\Delta H° = -120$ kJ mole^{-1}.

13.4   Based on the heats of hydrogenation to form the same compound, benzene is more stable than 1,3-cyclohexadiene. Thus, the hydrogenation of benzene to form 1,3-cyclohexadiene is an endothermic reaction, and $\Delta H° = +22$ kJ mole^{-1}.

13.5   Using $\Delta H° = -120$ kJ mole^{-1} for the hydrogenation of one double bond in cyclohexene, the calculated heat of hydrogenation of seven double bonds is -840 kJ mole^{-1}. However, the stated resonance energy means that the actual heat of hydrogenation will be smaller by 351 kJ mole^{-1}. Thus, the heat of hydrogenation for the aromatic compound is -489 kJ mole^{-1}.

13.6   The heats of formation are both positive, meaning that the compounds are less stable than the elements under standard conditions. Phenanthrene is the more stable of the two aromatic compounds because it has the smallest positive heat of formation. The greater stability is the result of a larger resonance energy.

13.7   Using the heats of formation of furan and tetrahydrofuran, the calculated heat of hydrogenation of furan is -149 kJ mole^{-1}. There are two double bonds in furan, so the heat of hydrogenation would be twice that of cyclopentene or -220 kJ mole^{-1} if the compound were not aromatic. The difference between this value and the actual heat of hydrogenation is the resonance energy, which is 71 kJ mole^{-1}.

13.8   Using the heats of formation of pyridine and piperidine, the calculated heat of hydrogenation of pyridine is -193 kJ mole^{-1}. The sum of the heats of hydrogenation of two carbon-carbon double bonds, using -120 kJ mole^{-1} from cyclohexene as a reference value, and -87.6 kJ mole^{-1} as the heat of hydrogenation of a carbon-nitrogen double bond, gives -328 kJ mole^{-1} of pyridine if it were not aromatic. The difference between this value and the actual heat of hydrogenation is the resonance energy, 135 kJ mole^{-1}.

13.9   If benzene were not aromatic, more energy would be released upon combustion of this hypothetical compound. The calculated heat of combustion would be more negative by 152 kJ mole^{-1}, which is the resonance energy of benzene. The value would be -3452 kJ mole^{-1}.

13.10   The same number and types of bonds are made and broken in each reaction. However, there are differences such as the strain energies of the two compounds. In addition, there is a significant difference in energy due to the "loss" of the benzene ring in the second compound. Thus, the reaction of this compound should be less exothermic than the first by an amount equal to the resonance energy of benzene. Some of that difference is counterbalanced by the stability of the conjugated tetraene product and its associated resonance energy.

13.11   The experimental heat of hydrogenation of benzene is -208 kJ mole^{-1}. The experimental value for the heat of hydrogenation of biphenyl, -415 kJ mole^{-1}, is twice that value, within experimental error. Thus, there is no additional stability associated with the joining of two benzene rings in biphenyl, so the rings must not be conjugated. Therefore, the rings must not be coplanar. Steric effects between two sets of C—H bonds ortho to the juncture of the two rings twists the two rings at an angle that does not allow conjugation.

13.12   If the benzene ring were in conjugation with the double bond, the resulting resonance stabilization would cause the heat of hydrogenation to be less than that of a comparably substituted double bond. However, it is likely that the heat of hydrogenation is unaffected by the benzene ring because steric effects between the ortho hydrogen atom on the benzene ring and the vinyl hydrogen of the cyclobutene will rotate the benzene out of conjugation.

13.13 Boron is a member of group III and has only six bonding electrons in its trivalent compounds. Thus, it is sp^2-hybridized and has a vacant p orbital. Nitrogen has an unshared pair of electrons in its trivalent compounds. The six electrons of nitrogen can be delocalized over the six atom ring using the p orbitals of both nitrogen and boron. Borazole is isoelectronic with benzene.

13.14 Boron is sp^2-hybridized and has a vacant p orbital. There is a network of 2p orbitals of carbon and boron that can delocalize electrons. However, there are only four π electrons indicated by the two carbon-carbon π bonds. This number does not fit the Hückel rule and the compound is not aromatic.

13.15 **(a)** No, because 12 π electrons does not fit the Hückel rule of 4n+2..

(b) Yes, because 10 π electrons does fit the Hückel rule.

(c) Yes, because oxygen contributes one of its two unshared electron pairs to the π system. The four electrons of the carbon-carbon π bond and the carbon-nitrogen π bond, and one unshared electron pair of electrons from oxygen give a total of 6 π electrons, which does fit the Hückel rule. Note that the other electron pair of oxygen and that of nitrogen are in sp^2 orbitals perpendicular to the π system.

(d) No, because oxygen would contribute one of its two unshared electron pairs to the π system. The six electrons of the three carbon-carbon π bonds and one unshared electron pair from oxygen give a total of 8 π electrons, which does not fit the Hückel rule. Note that the other electron pair of oxygen cannot be included to give a total of 10 electrons, because it is in an sp^2 hybrid orbital perpendicular to the π system.

13.16 **(a)** No, because 16 π electrons does not fit the Hückel rule.

(b) No, because 8 π electrons does not fit the Hückel rule.

(c) Yes, because the nitrogen atom bonded to a hydrogen atom contributes one of its unshared electron pairs to the π system. The four electrons of the two carbon-carbon π bonds and this one unshared electron pair from nitrogen give a total of 6 π electrons, which does fit the Hückel rule. Note that each of the other two nitrogen atoms contribute one electron to the π system as part of their respective double bonds. The remaining unshared electron pair of each nitrogen atom is in an sp^2 orbital perpendicular to the π system.

(d) No, because each nitrogen atom would contribute its unshared electron pair to the π system. The four electrons of the two carbon-carbon π bonds and the two unshared electron pairs give a total of 8 π electrons, which does not fit the Hückel rule.

13.17 There are four resonance forms and the positive charge is located at the C-1, C-3, C-5, and C-7 atoms of the carbocation.

13.18 A resonance stabilized carbocation results from the acyclic iodo compound, leading to a mixture of 3-methoxy-1,4-pentadiene and 1-methoxy-2,4-pentadiene.

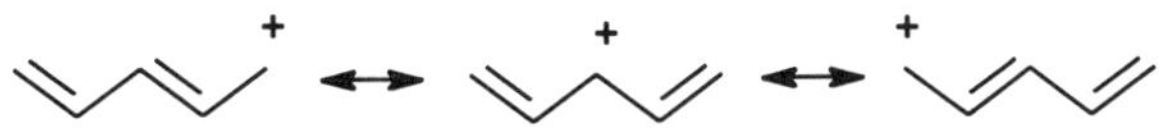

The cyclopentadienyl cation derived from the cyclic iodo compound has only four π electrons, so it does not fit the Hückel rule. The carbocation would be a diradical having one electron each in the π_2 and π_3 molecular orbitals. The formation of this high energy intermediate requires a large activation energy and the reaction rate should be much slower than for the acyclic compound.

13.19 The carbocation that results from abstraction of a chloride ion by the Lewis acid $SbCl_5$ is a cyclopropenium ion. It has two π electrons and fits the Hückel rule (with $n = 0$). The electrons can be delocalized over the three carbon atoms of the ring.

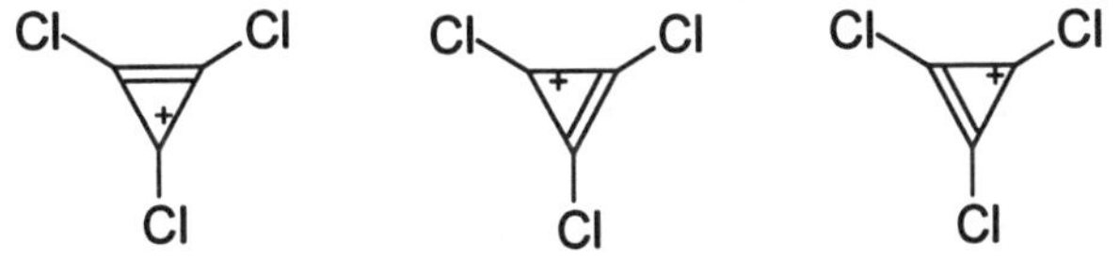

13.20 The dipolar resonance form of the carbonyl group places a negative charge on oxygen and a positive charge on one of the carbon atoms of the cyclopropene ring. The resulting cyclopropenium ion has two π electrons and fits the Hückel rule. The positive charge can be delocalized over the three carbon atoms of the ring. The increased polarity of the carbonyl group as a result of resonance increases the dipole moment.

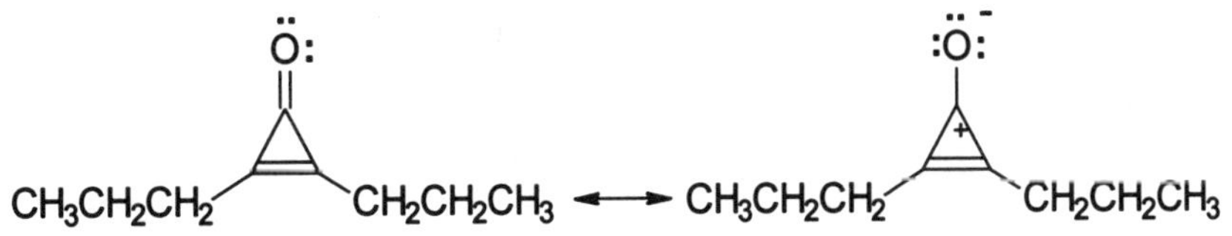

13.21 Loss of a proton from the methylene group gives a carbanion that has 10 π electrons, which fits the Hückel rule. The cyclononatetraenyl anion is aromatic.

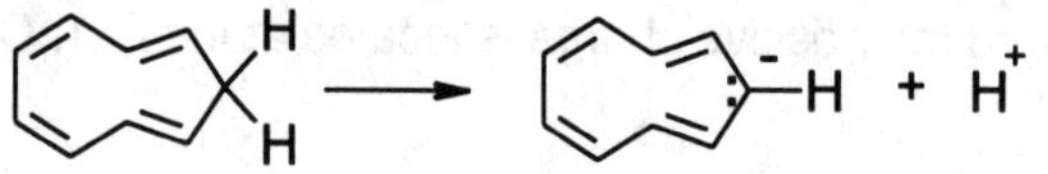

13.22 The dianion has 10 π electrons, which fits the Hückel rule. The cyclooctatetraenyl dianion is aromatic. The most stable resonance form places the negative charges apart as shown in the second resonance form.

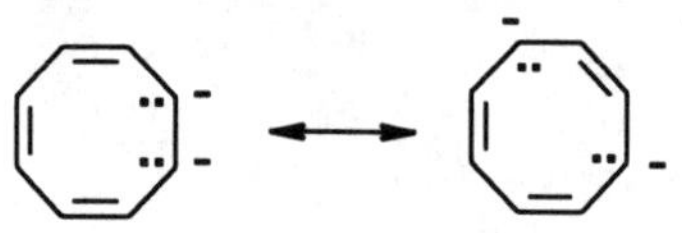

13.23 Loss of one proton from each of the methylene groups gives a dianion that has a total of 10 π electrons, which fits the Hückel rule. The dianion is aromatic.

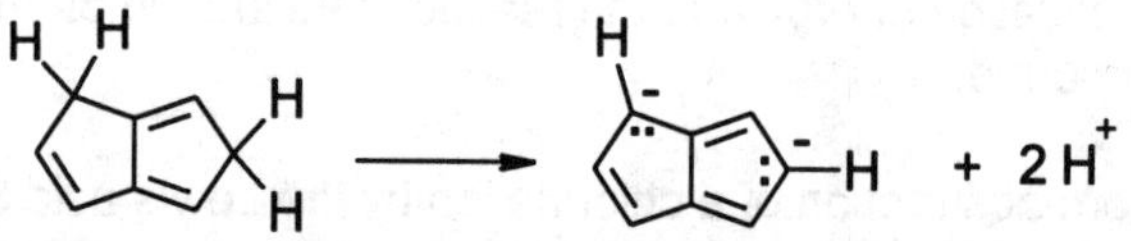

13.24 The cyclopropenium cation product is aromatic. Thus, its resonance stabilization is larger than the stabilization of the allyl cation. The reaction should be exothermic.

13.25 **(a)** $C_{20}H_{12}$ **(b)** $C_{16}H_{12}$ **(c)** $C_{14}H_{10}$ **(d)** $C_{18}H_{12}$

13.26 **(a)** $C_{22}H_{14}$ **(b)** $C_{16}H_{10}$ **(c)** $C_{14}H_{14}$ **(d)** $C_{14}H_{10}$

13.27

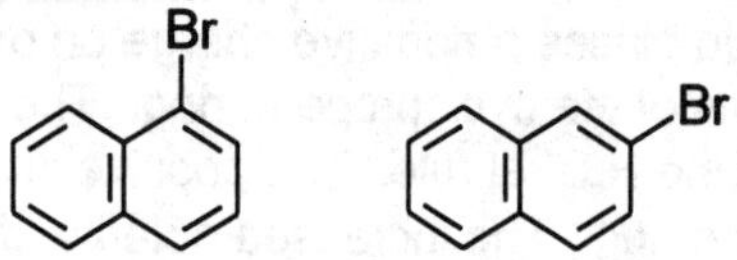

13.28

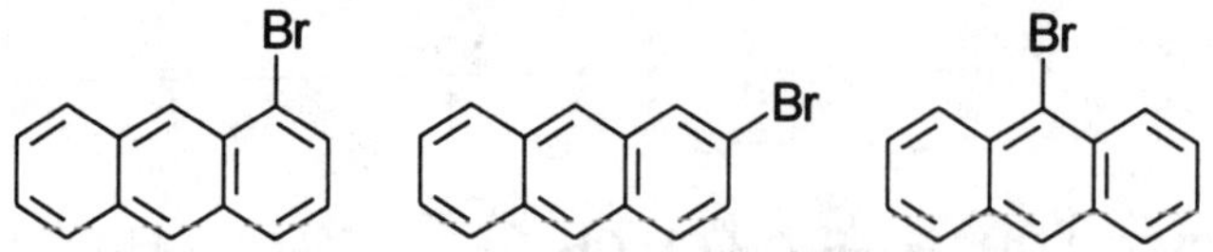

13.29 The most stable resonance form has four "benzene-type" rings. Note that the points of fusion all have double bonds that are shared by two rings.

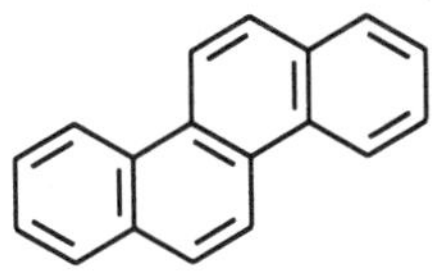

13.30 The resonance form shown only has one "benzene-type" ring. Two more stable resonance forms can be written that each have two "benzene-type" rings. In the form on the left, a double bond at the point of fusion of the first and second rings is shared between both. In the form on the right, a double bond at the point of fusion of the second and third rings is shared between both.

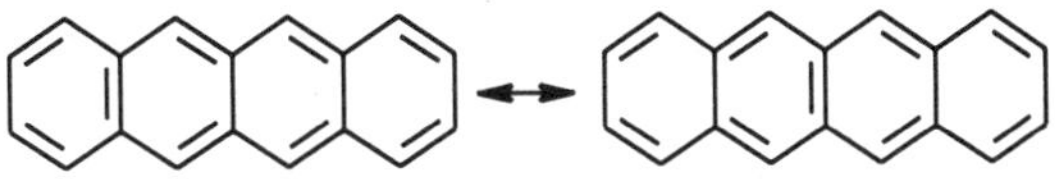

13.31 The first of the three isomers shown below has no dipole moment.

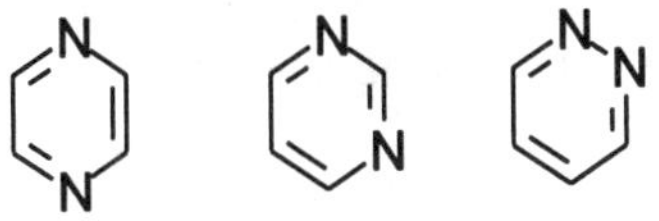

13.32 The first of the three isomers shown below has no dipole moment.

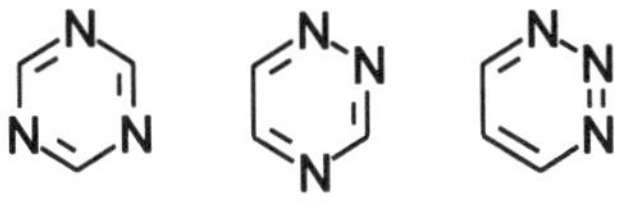

13.33 **(a)** nitrogen contributes one and oxygen contributes two

(b) each nitrogen atom contributes one

(c) nitrogen contributes one and sulfur contributes two

13.34 **(a)** the nitrogen atom bonded to hydrogen contributes two; the other nitrogen atom contributes one

(b) the nitrogen atom bonded to hydrogen in the five-membered ring contributes one; the other three nitrogen atoms each contribute two

(c) each nitrogen atom contributes one

13.35 **(a)** furan   **(b)** pyrrole   **(c)** thiophene

13.36 **(a)** pyrrole   **(b)** thiophene   **(c)** furan   **(d)** pyridine

13.37 The third of the three isomers shown below has no dipole moment.

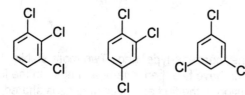

13.38 The middle structure of the three isomers shown below has no dipole moment.

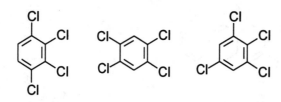

13.39 The compounds are isomeric and have the same general shape. Thus, the London forces should be similar. Both compounds are polar and should have similar dipole-dipole forces. However, benzyl alcohol can form hydrogen bonds whereas anisole cannot.

13.40 The boiling point of the ortho isomer is lower because the hydroxyl hydrogen atom forms an intramolecular hydrogen bond. The other two isomers form intermolecular hydrogens bond between hydroxyl groups and thus have higher boiling points.

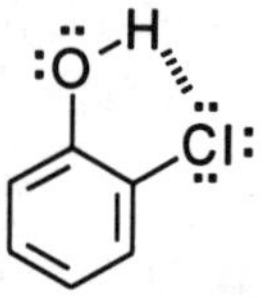

13.41 **(a)** para   **(b)** ortho   **(c)** meta

13.42 **(a)** meta   **(b)** para   **(c)** ortho

13.43 **(a)** ethylbenzene
          (b) isopropylbenzene
          (c) 1,4-diethylbenzene
          (d) 1,3,5-triethylbenzene

13.44 **(a)** 1,2,4-trichlorobenzene
          (b) 2,4-dibromophenol
          (c) 3-bromoaniline
          (d) 3,4-dichlorotoluene

13.45 (a) 4-chloro-3,5-dimethylphenol
(b) 2,6-dimethylaniline
(c) 4-chloro-2-phenylbenzene

13.46 (a) (b) 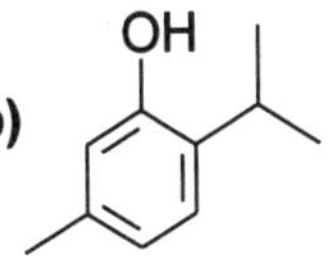 (c)

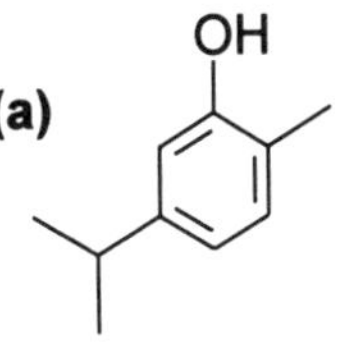

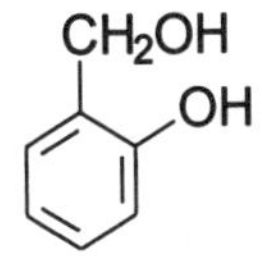

13.47

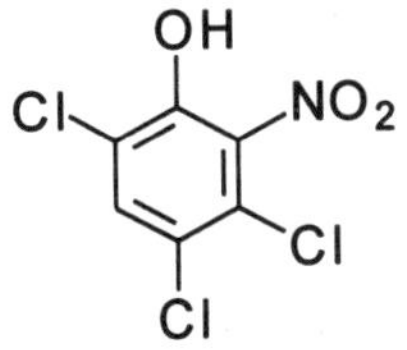

13.48

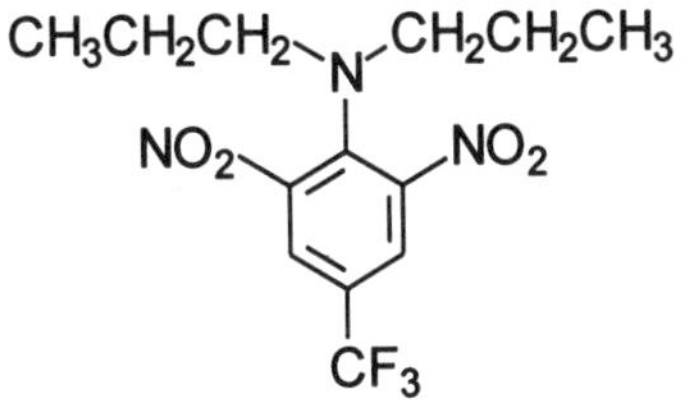

13.49 (a) 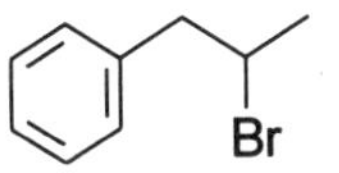 (b) 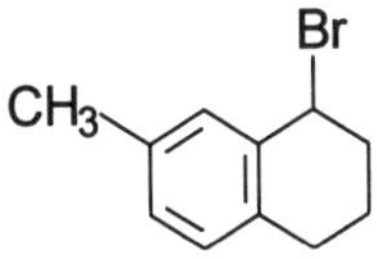 (c)

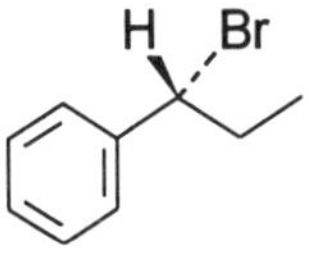

13.50 They are different. The first compound regioselectively adds HBr via an intermediate benzylic resonance-stabilized carbocation that then captures a bromide ion. The two carbon atoms of the double bond of the second compound are equivalent. Addition of HBr can give only one product.

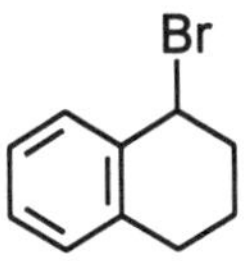

I 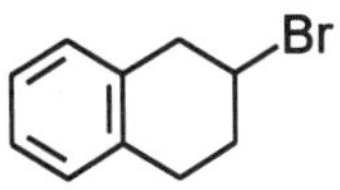   II

13.51   Under free radical conditions, a benzyl radical is formed by abstraction of a hydrogen atom by a bromine atom. Reaction of the benzyl radical with bromine gives a benzyl bromide.

(a) NO_2—⟨ ⟩—CH_2Br (b) [bicyclic structure with Br]   (c) CH_3—⟨ ⟩—$\overset{Br}{\underset{|}{C}}(CH_3)_2$

13.52   (a) [benzene ring with Br substituent and $\overset{Br}{\underset{}{C}}CH_3$ group]   (b) [phenanthrene-type structure with Br and H]   (c) $BrCH_2$—⟨ ⟩—$C(CH_3)_3$

14

Electrophilic Aromatic Substitution

The fact that benzene compounds have special stability as a consequence of resonance stabilization should suggest that the reactivity of these compounds should also be substantially different than that of other unsaturated hydrocarbons. In fact this chapter deals exclusively with essentially a single reaction that is characteristic of aromatic compounds. Only one reaction? Well, only one type of reaction! It may look like there are many reactions, but they are all characterized as electrophilic aromatic substitution. The rest of the story is simply variations on a theme. In this chapter we consider the types of electrophiles that react and the mechanism of the reaction. Then the discussion turns to the fine points which include explanation for differences in the rate of the reaction and the reason why certain compounds are formed in preference to isomeric compounds.

Keys to the Chapter

14.1 Reactivity of Aromatic Rings

Aromatic rings are attacked by electrophiles represented by E^+ to give substituted aromatic compounds represented by Ar—E. The common reactions of aromatic compounds are designated by the type of group substituted on the aromatic ring. Thus, halogenation means than a halogen has substituted for hydrogen on the aromatic ring to give a product represented by Ar—X. In the case of chlorination the product is Ar—Cl.

The mechanism of aromatic substitution consists of two steps. The first is electrophilic attack to give a carbocation intermediate. The second is loss of a proton to regenerate the aromatic system.

14.2 Typical Electrophilic Substitution Reactions

Five types of substitution are listed in this section. At the very least, you have to know the reagents that are required for each reaction. The reactions and reagents are as follows:

1. halogenation	iron trihalide and the halogen
2. nitration	nitric acid and sulfuric acid
3. sulfonation	sulfur trioxide and sulfuric acid
4. alkylation	alkyl halide (usually chloride or bromide) and AlX_3
5. acylation	acyl halide (usually chloride) and $AlCl_3$

On a more sophisticated level, you should know how the electrophilic reagent is generated from the reagents. Only at this level of understanding can you predict what might occur with other combinations of reagents. It is this predictive power that makes organic chemistry a more manageable area of study Note that in each case, one of the two materials is a Lewis acid, and it reacts with the other material to give the electrophile. You should be able to write the mechanisms of the reactions that form the electrophiles.

14.3 Limitations of Friedel-Crafts Reactions

Except in the case of alkylation, the electrophile that forms then attacks the aromatic ring. However, in alkylation the electrophile is a carbocation that tends to rearrange to the most stable carbocation, which then alkylates the aromatic ring. These rearrangements shouldn't be unexpected, because you have already encountered them in the addition reactions of alkenes and in S_N1 reactions. An alternate approach to obtaining the desired alkyl group on the aromatic ring is to acylate first and then reduce the carbonyl group using the **Clemmensen** method. Keep this method in mind, because a common question on exams is how to synthesize alkylbenzenes in which the alkyl group cannot be directly introduced by the Friedel-Crafts alkylation.

A second limitation is the result of substituents already on the aromatic ring. Many of these substituents make the ring less reactive toward alkylation. In addition, these same substituents preclude Friedel-Crafts acylation.

14.4 Substituent Effects on Reactivity of Aromatic Rings

Any substituents already on the aromatic ring affect the rate of substitution by electrophiles. That observation should make sense, because the electrophile seeks electrons of the π system and the substituents on the ring should affect their availability. Groups that increase the reactivity are **activating groups**; groups that decrease the reactivity are **deactivating groups**. The degree to which the groups affect the reactivity is qualified by the adjectives "strongly" and "weakly". You need to learn the properties of the groups listed in Table 14.1 in the text.

The substituents already on the aromatic ring determine the position that the electrophile attacks. All substituents fall into two classes termed **ortho,para directors** and **meta directors**. All activating groups are ortho,para directors; with the exception of the halogens, the deactivating groups listed are meta-directors. Halogens are a little different because they are ortho,para directors in spite of being deactivating groups.

14.5 Interpretation of Rate Effects

Any group that can increase the electron density of the aromatic ring makes it more inviting to an electrophile, so the rate of reaction is faster. Conversely, groups that reduce the electron density decrease the rate of reaction. Groups can affect the electron density by either an inductive effect or a resonance effect or a combination of both. In some cases the inductive and resonance effects have opposite effects on the rate.

With the exception of alkyl groups, the inductive effect of all common groups is withdrawal of electron density. The degree of electron withdrawal is largest for groups with a formal positive charge (such as the nitro group), followed by groups with a partial positive charge (such as the carbonyl group). In addition, electron withdrawal is related to the electronegativity of the atom (such as a halogen) bonded to the aromatic ring.

Resonance effects are seen for groups that either have lone pair electrons or have multiple bonds to an electronegative atom. Groups with lone pair electrons (such as hydroxyl and amino) increase the electron density of the ring. Groups with multiple bonds to an electronegative atom (such as the carbonyl group) withdraw electron density from the ring.

Halogens are "unusual" in that they deactivate the ring toward electrophilic substitution but are ortho-para directors. This feature is related to the fact that the halogens are inductively electron withdrawing but can weakly supply electrons by resonance. Substituents such as the hydroxyl and amino groups donate electrons to the ring by resonance efficiently enough to oppose and overcome the inductive electron withdrawal by their electronegative atoms.

Some groups such as the nitro group and cyano group withdraw electrons by a combination of both an inductive effect and a resonance effect. As a consequence such groups are strongly deactivating.

14.6 Interpretation of Directing Effects

In order to understand the directing effect of a group, it is first necessary to know where the positive charge occurs in resonance forms when an electrophile attacks any of the positions of a substituted benzene ring. The charge is always ortho and para to the position at which the electrophile enters. Why is that important? The answer is that a substituent already on the ring can strongly affect the rate of the reaction if the positive charge ends up there in one of the resonance forms. If the substituent can stabilize a positive charge by donation of electrons, then substitution at the original site is favored. The terminology used for the substituent already on the ring implies that it "directs" where the electrophile goes. More correctly, it facilitates the reaction if the electrophile attacks at that site. Ortho,para directors can stabilize the positive charge when an electrophile attacks ortho and para to them. Meta directors destabilize positive charge when an electrophile attacks ortho and para to them. As a consequence the electrophile enters the meta position "by default".

14.7 Multiple Substituent Effects

Multiple substituents on a benzene ring all affect the reactivity and site of attack of an electrophile. Often several isomers result. However, if the directing properties of the groups reinforce each other by enhancing the reactivity at the same position, then a single isomer will result. If the directing influence of the two groups affect different positions and can give rise to two isomers, then the isomer that is formed under the direction of the more activating substituent dominates. Regardless of the activity of a position toward electrophilic attack, substitution between two groups situated meta to each other does not occur because of steric hindrance.

14.8 Functional Group Modification

Some functional groups can only be introduced on the benzene ring indirectly. To introduce such groups, a functional group already on the ring is transformed by a chemical reaction. Several of these reactions do not change the attachment of the atom bonded to the aromatic ring. These are

1. oxidation of an alkyl group to a carboxylic acid
2. reduction of an acyl group to an alkyl group
3. reduction of a nitro group to an amino group

The amino group can be converted to other functional groups by replacement of the nitrogen atom after formation of a diazonium ion. The groups that can replace the diazonium ion are

1. halogens, using copper(I) halide
2. nitrile, using copper(I) cyanide
3. hydroxyl, using aqueous acid
4. hydrogen, using hypophosphorous acid

14.9 Synthesis of Substituted Aromatic Compounds

Devising a synthesis is like assembling a jigsaw puzzle: you go step by step as you discover which pieces fit together. An important difference is that the order of the steps may be very significant in the synthesis. Start your synthesis plan by evaluating each substituent on the ring and how it might be placed there. Consider the ortho,para or meta directing effects of various substituents and the possibility of modifying functional groups once they have served their purpose. Then put all these pieces together in the appropriate order to accomplish the desired synthesis.

If two groups can be introduced by direct reaction and their location is consistent with the directing characteristic of one of them, then the synthesis is straightforward. Put the group on first that will direct the proper position of the second group. More sophisticated syntheses require that groups be introduced that can be modified later to give another group. Then the problem is to determine whether the second group should be introduced before or after the chemical modification. Synthesis problems are very common on organic chemistry exams, so it is important to work many sample problems to hone your skills.

14.10 Polycyclic and Heterocyclic Aromatic Compounds

This subject area is often not treated extensively in the first organic course. However, your lecturer could decide to expand the coverage. This shouldn't worry you if you understand the theory of aromatic substitution established for benzene compounds. Substitution in other polycyclic aromatic rings proceeds to give products that are most favored by the stabilization of the intermediate carbocation.

Substitution in heterocyclic aromatic compounds is a bit more complicated because it is necessary to evaluate the effect of the heteroatom on charge distribution. The five-membered compounds pyrrole, thiophene, and furan are more reactive toward electrophilic aromatic substitution and are regarded as electron rich compounds as a result of the availability of electrons from the heteroatom in contributing resonance forms. Pyridine is less reactive toward electrophilic aromatic substitution.

Summary of Reactions

1. Halogenation

2. Nitration

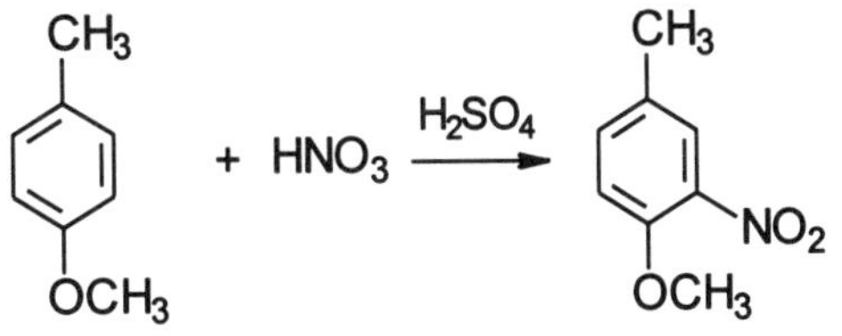

3. Sulfonation

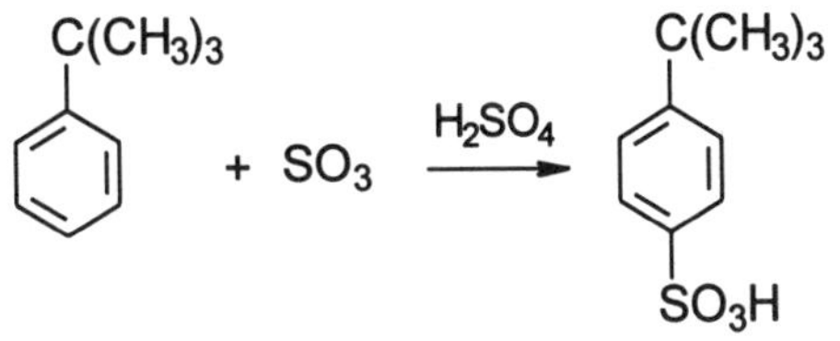

4. Alkylation (Friedel-Crafts)

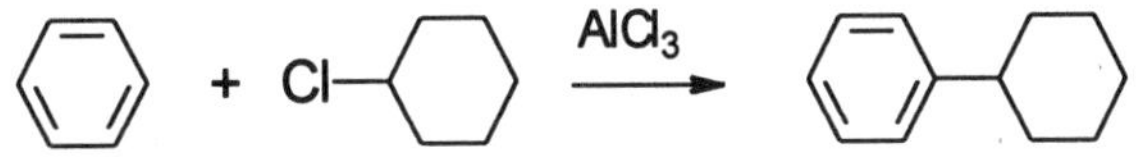

5. Acylation (Friedel-Crafts)

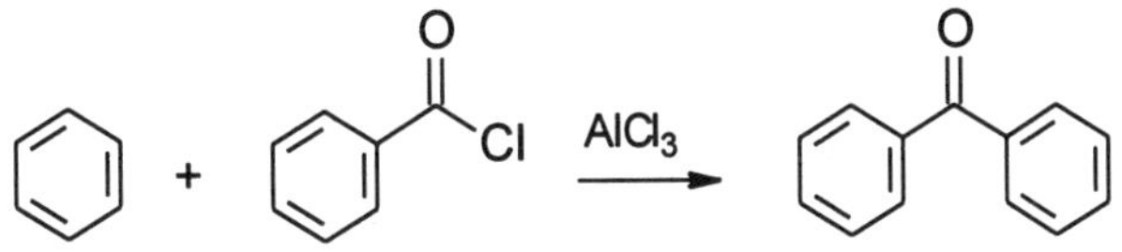

6. Oxidation of Side Chains

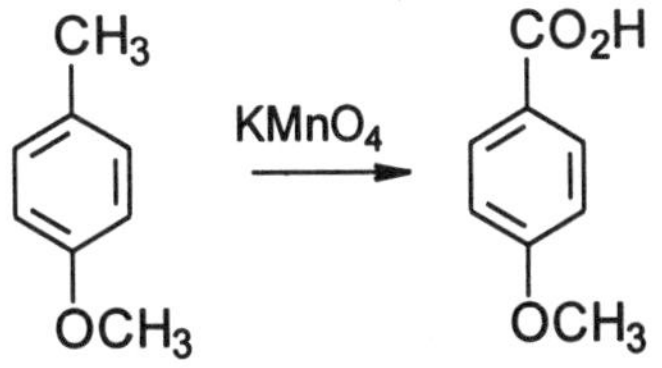

7. Reduction of Acyl Side Chain

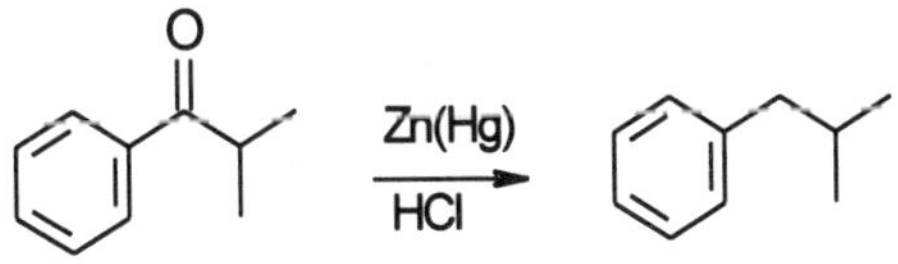

8. Reduction of Nitro Groups

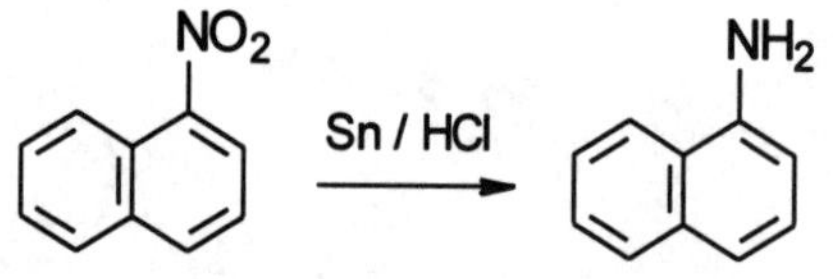

9. Reactions of Amino Groups via Diazonium Ion

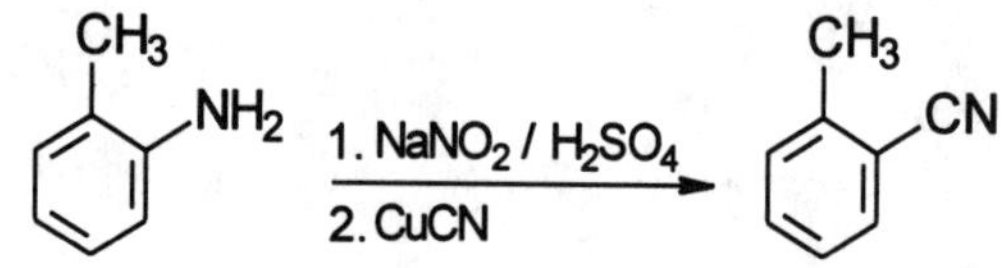

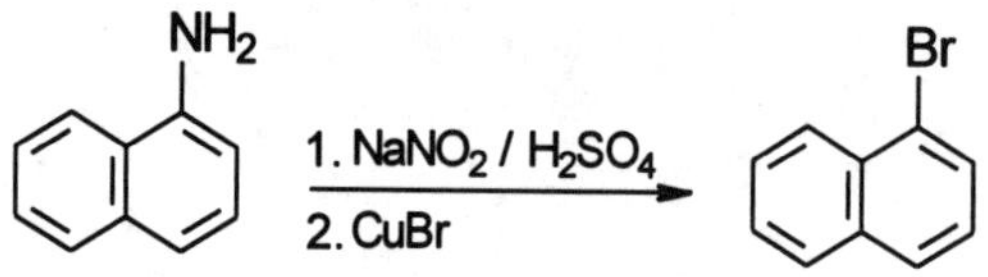

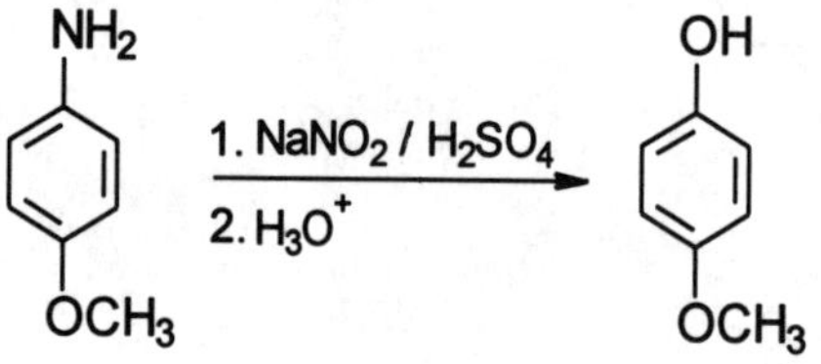

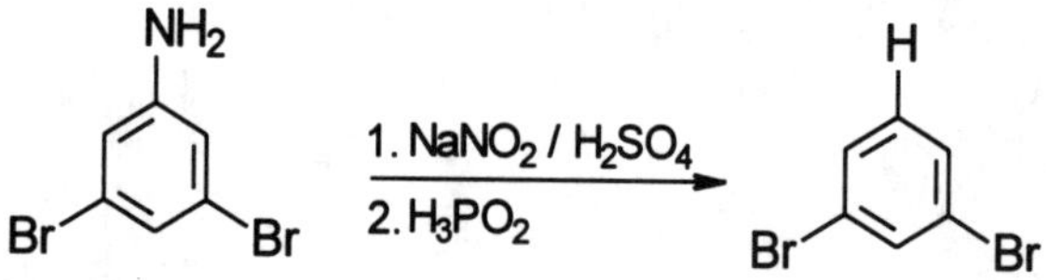

Solutions to Exercises

14.1   The oxygen-oxygen bond of the conjugate acid of hydrogen peroxide cleaves heterolytically to give water and a hydroxyl cation, which is the electrophile.

$$H-\overset{\cdot\cdot}{\underset{\cdot\cdot}{O}}-\overset{\cdot\cdot}{\underset{\underset{H}{|}}{O}}{}^{+}\!-H \longrightarrow H-\overset{\cdot\cdot}{\underset{\cdot\cdot}{O}}{}^{+} \ + \ \overset{\cdot\cdot}{:}\underset{\underset{H}{|}}{O}-H$$

14.2   The electrophile is I^+, which is produced by the heterolytic cleavage of the I—Cl bond.

14.3   The electrophile is $AcOHg^+$, which is formed by protonation of one of the acetate groups of the covalent $Hg(OAc)_2$ followed by heterolytic cleavage of the Hg—O bond. Acetic acid is the other product.

14.4   The electrophile is the *tert*-butyl carbocation, $(CH_3)_3C^+$, which is formed by protonation of the oxygen atom followed by heterolytic cleavage of the C—O bond. Water is the leaving group.

$$(CH_3)_3C-\overset{\cdot\cdot}{\underset{\underset{H}{|}}{O}}{}^{+}\!-H \longrightarrow (CH_3)_3\overset{+}{C} \ + \ \overset{\cdot\cdot}{:}\underset{\underset{H}{|}}{O}-H$$

14.5   The properties of the —SCH_3 group are similar to those of the chloro group because both chlorine and sulfur are third row elements. Because sulfur is less electronegative than chlorine, the ring is less deactivated by inductive electron withdrawal. Although third row elements are not effective electron donors by resonance, sulfur is a better donor of electrons because it is less electronegative. As a consequence, the —SCH_3 group will be an ortho,para directing group.

14.6   The sulfur atom is bonded to three electronegative atoms. As a result the sulfonamide group will inductively withdraw electron density from the aromatic ring and deactivate benzene toward electrophilic substitution. There are no lone pair electrons on sulfur to be donated by resonance. Thus the group is a meta director.

14.7   An ammonium ion is formed in acidic solution because the nitrogen atom is protonated. The conjugate acid of N,N-dimethylaniline has a positive charge on the nitrogen atom and as a result the group withdraws electron density from the aromatic ring and deactivates benzene toward electrophilic substitution. Because there are no lone pair electrons on the protonated nitrogen atom to be donated by resonance, the group is a meta director.

14.8   As the number of chlorine atoms bonded to the carbon atom increases, the inductive withdrawal of electron density from that carbon atom and the adjacent benzene ring increases. Because there are no lone pair electrons on the carbon atom to be donated by resonance, the groups become increasingly better meta directors by increased electron withdrawal.

14.9   The reagents are as follows. The structures of the products are listed after the reagents.

(a) bromine with iron(III) bromide
(b) fuming sulfuric acid (sulfur trioxide and sulfuric acid)
(c) nitric acid with sulfuric acid
(d) acetyl chloride with aluminum trichloride

(a)

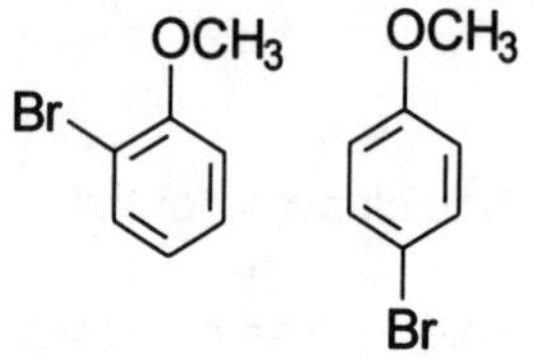

(b)

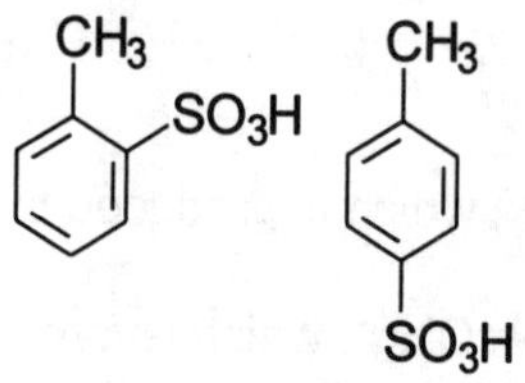

(c)

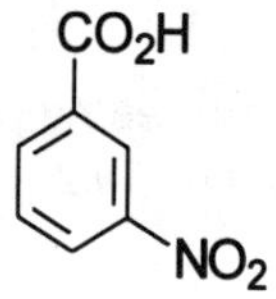

(d)

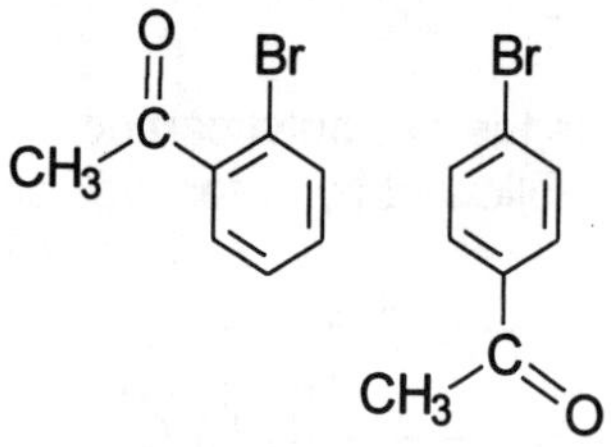

14.10   The reagents are as follows. The products are listed after the reagents.

(a) chlorine with iron(III) chloride
(b) methyl chloride with aluminum trichloride
(c) acetyl chloride with aluminum trichloride
(d) nitric acid with sulfuric acid

(a)

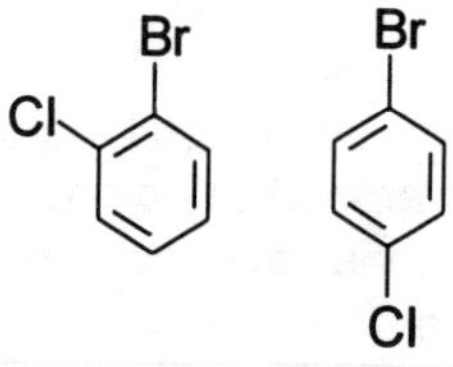

(b)

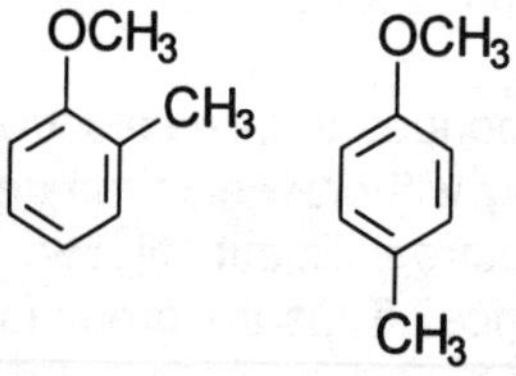

(c)

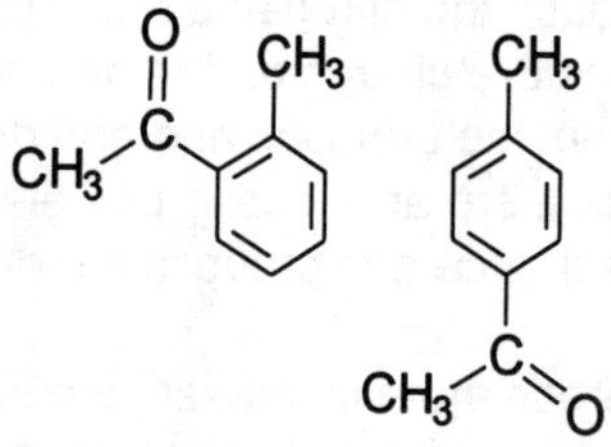

(d)

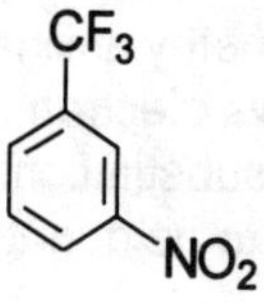

14.11   Reaction of chlorocyclohexane with aluminum trichloride gives the cyclohexyl carbocation, which alkylates the benzene ring to give cyclohexylbenzene.

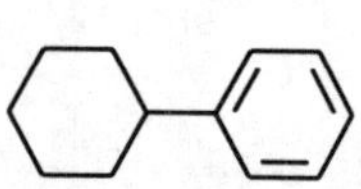

14.12 Reaction of this primary chloroalkane with aluminum trichloride to give a primary isobutyl carbocation will lead to the rearranged *tert*-butyl carbocation. Thus, alkylation of benzene will yield *tert*-butylbenzene.

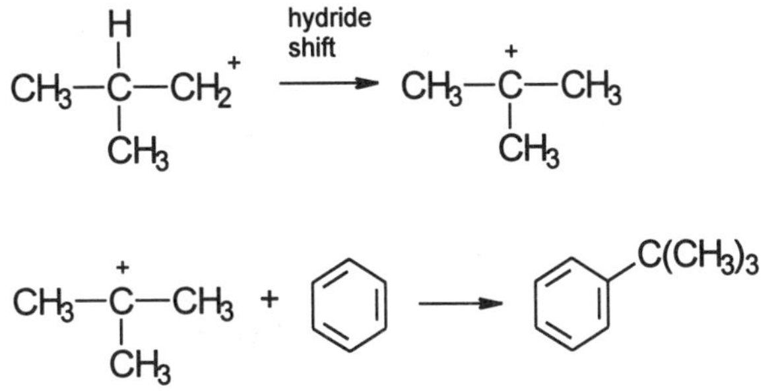

14.13 Addition of a proton to the C-1 atom of propene gives a secondary carbocation which is the electrophile. The product will be isopropylbenzene.

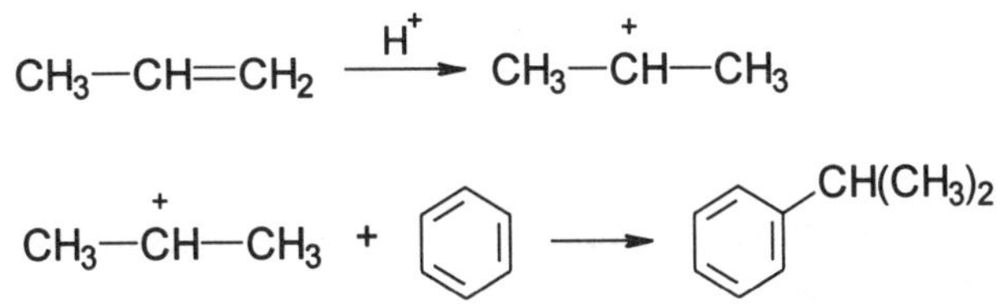

14.14 Addition of a proton to the C-1 atom of 2-methylpropene gives a tertiary carbocation which is the electrophile. Although both the methoxy and the methyl groups are activating and are ortho,para directing, the *tert*-butyl carbocation will alkylate ortho to the methoxy group because it is the stronger activating group.

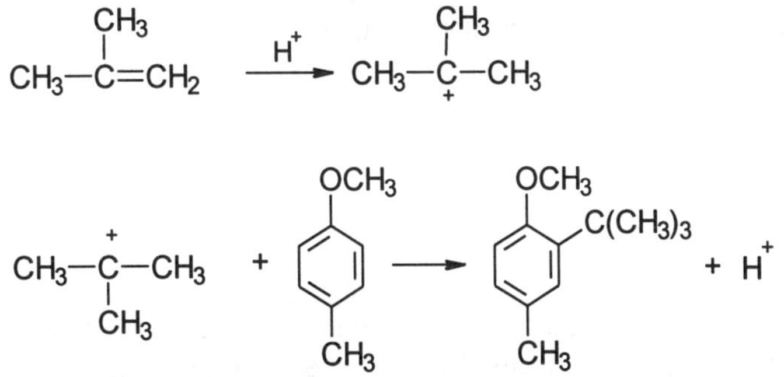

14.15 The electrophile is the isopropyl carbocation, $(CH_3)_2CH^+$, which is formed by protonation of the oxygen atom followed by heterolytic cleavage of the C—O bond. Water is the leaving group. The isopropyl carbocation will alkylate ortho or para to the methyl group, as shown on the following page. The major product will be the para isomer due to steric hindrance in the attack at the ortho position.

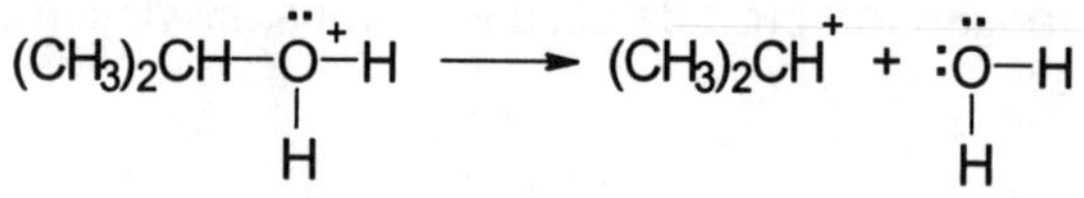

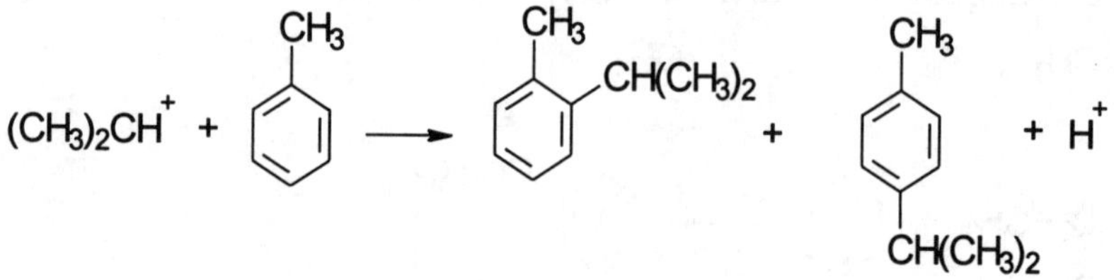

$$(CH_3)_2CH-\overset{\overset{..}{+}}{\underset{H}{O}}-H \longrightarrow (CH_3)_2CH^+ + \overset{..}{\underset{H}{O}}-H$$

$$(CH_3)_2CH^+ + \text{[toluene]} \longrightarrow \text{[2-isopropyltoluene]} + \text{[4-isopropyltoluene]} + H^+$$

14.16

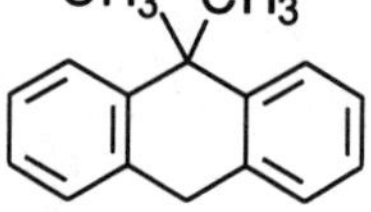

14.17 An intramolecular Friedel-Crafts acylation reaction occurs at the position ortho to the butanoyl chain to give a cyclic ketone.

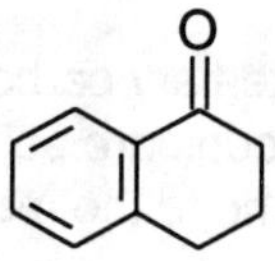

14.18 An intramolecular Friedel-Crafts acylation reaction occurs on the ring containing the methoxy group because it is the more activated aromatic ring.

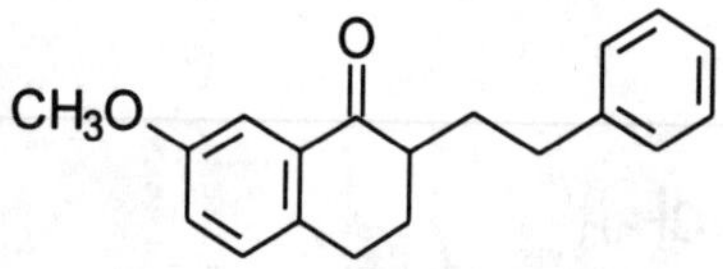

14.19 **(a)** Reaction occurs in the ring bonded to the nitrogen atom because that ring is activated. The other ring is deactivated by the carbonyl group. Reaction will occur at the ortho or para position.

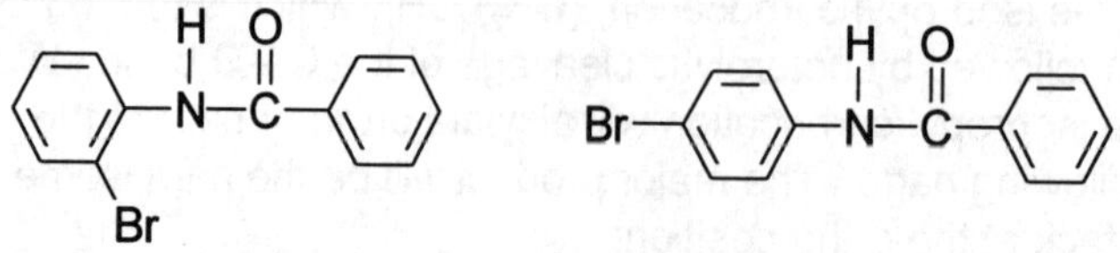

(b) Reaction occurs in the ring bonded to the methylene group because that ring is slightly activated. The other ring is deactivated by the carbonyl group. Reaction will occur at the ortho or para position.

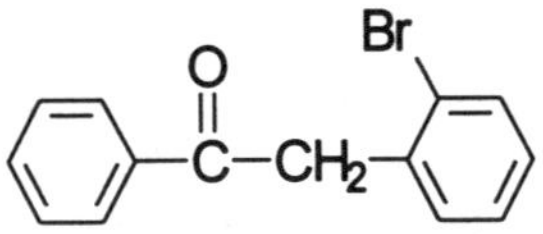

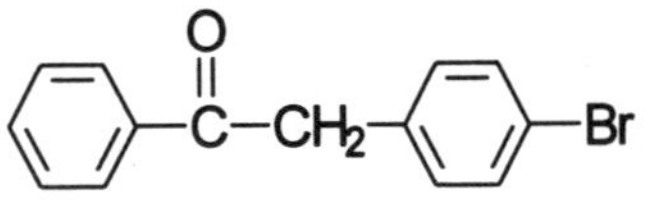

14.20 **(a)** Both rings are activated by the oxygen atom. However, the ring on the right is deactivated by the nitro group. Nitration will occur in the other ring at the ortho or para position.

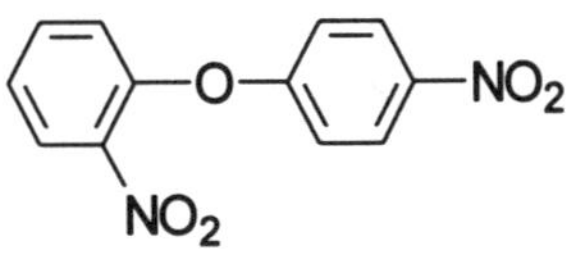

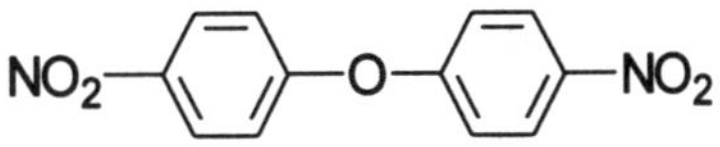

(b) Both rings are slightly activated by the methylene group. However, the ring on the right is also activated by the hydroxyl group. Nitration will tend to occur in that ring ortho to the activating group. A smaller amount of product with the nitro group ortho to the methylene group may form.

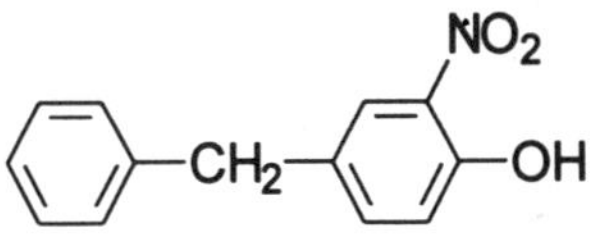

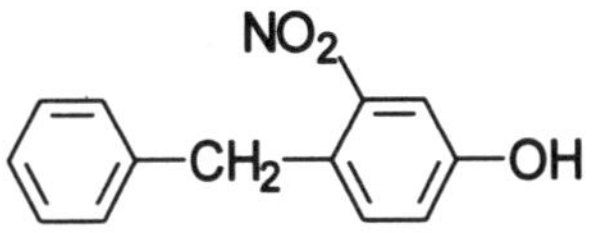

14.21 **(a)** The alkyl group is slightly activating and the bromo group is slightly deactivating. Bromination occurs ortho to the alkyl group as shown below.

(b) The bromo groups are deactivating but are ortho,para directing groups. Nitration at the C-2 position between the two bromine atoms is sterically hindered. The other ortho positions at C-4- and C-6 are structurally equivalent and only one product results.

(c) The dimethylamino group is strongly activating and the isopropyl group is slightly activating. Acetylation occurs ortho to the dimethylamino group.

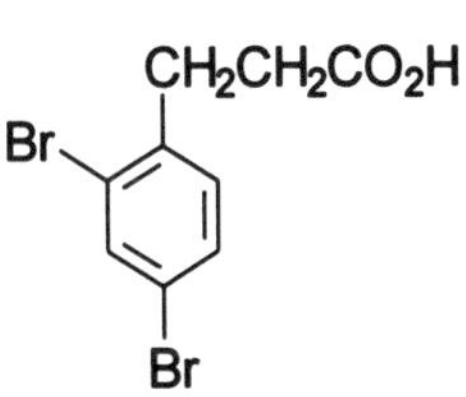

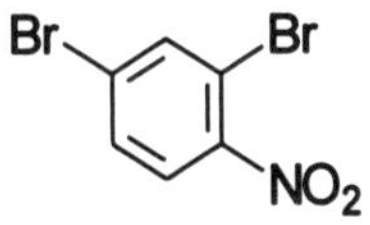

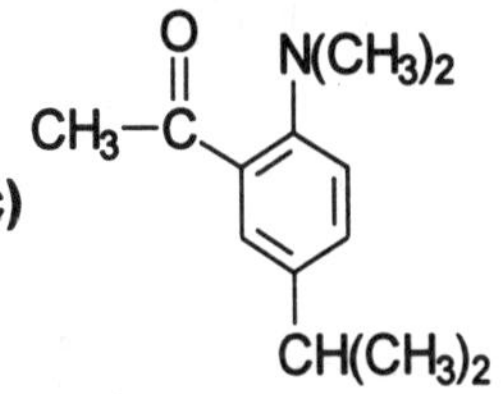

14.22 **(a)** The amino group is strongly activating and the nitro group is strongly deactivating. Bromination occurs ortho to the amino group as shown below.

(b) The two carboxylic acid groups are deactivating and are meta directing groups. Nitration occurs at the C-5 position, which is meta to both groups.

(c) The isopropyl group is slightly activating and the bromo group is slightly deactivating. Acetylation may occur ortho or para to the diisopropyl group.   The ortho position between the isopropyl and bromo group is sterically hindered. The other ortho position is substantially hindered by the isopropyl group. The para position with respect to the isopropyl group is also ortho with respect to the bromine. This is the most likely site for attack.

(a) 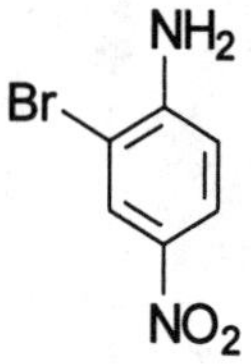   (b) 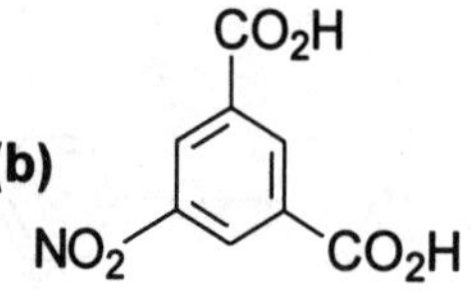   (c)

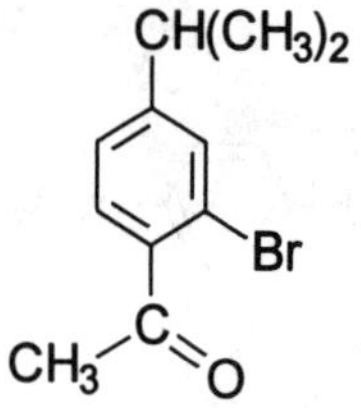

14.23 The reagent oxidizes alkyl groups to —CO_2H groups.

(a) 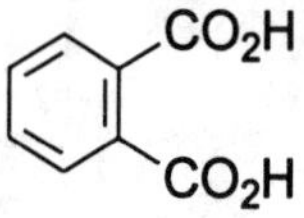   (b) 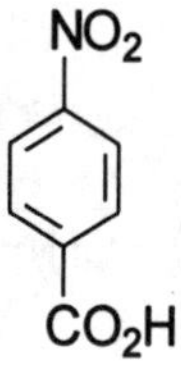   (c)

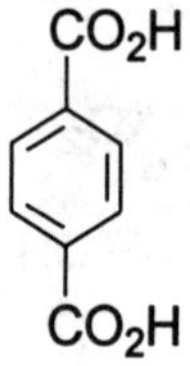

14.24 The reagent reduces carbonyl groups to methylene groups.

(a) 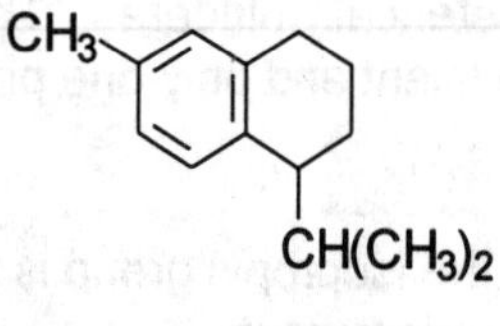   (b) 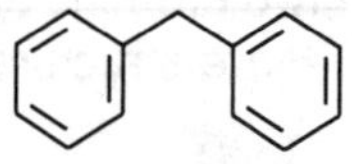   (c)

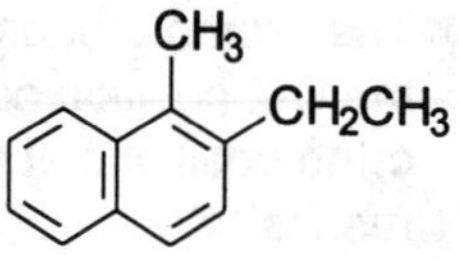

14.25 A Grignard reagent forms which has some carbanionic character. Reaction with D^+ from D_2O replaces the MgX group with deuterium.

(a) 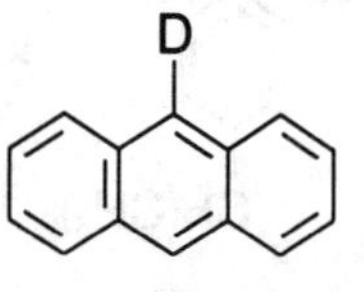   (b) 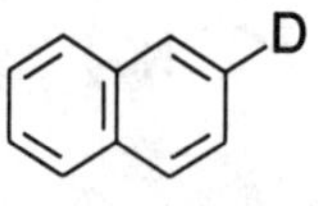   (c)

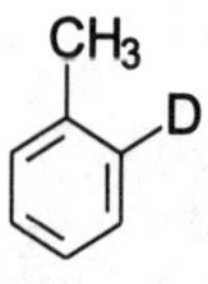

14.26   The nitro group is reduced to an amino group by the reagent.

(a)

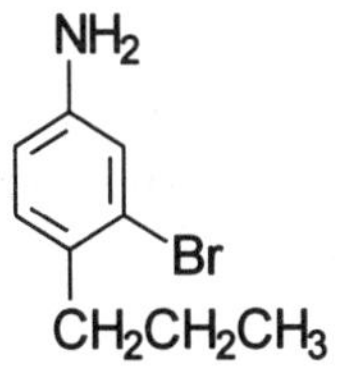

(b)

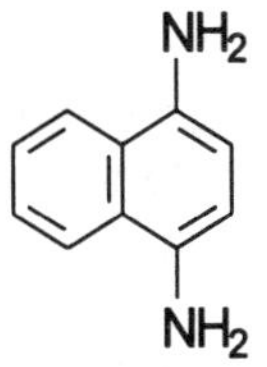

(c)

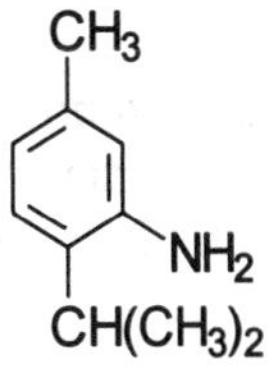

14.27   (a)

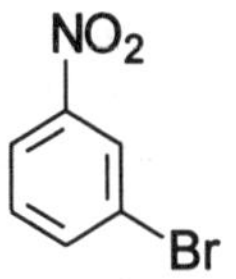

(b)

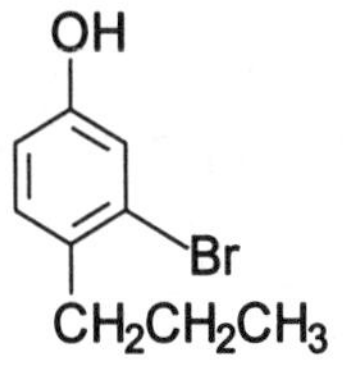

(c)

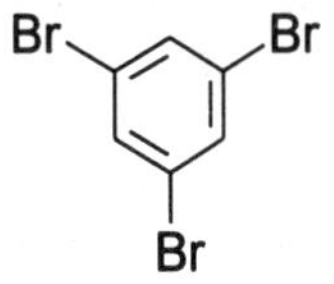

14.28   (a)

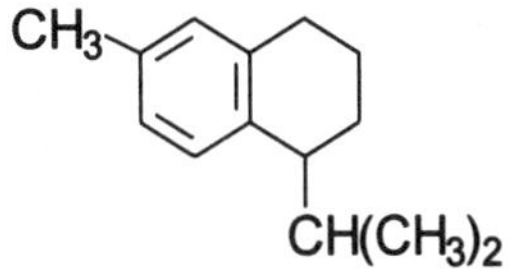

(b)

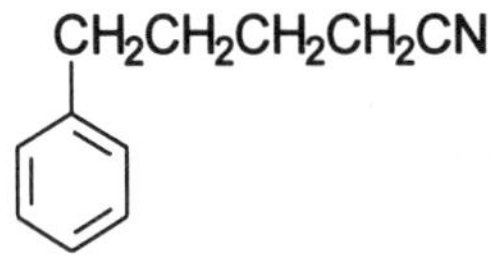

(c)

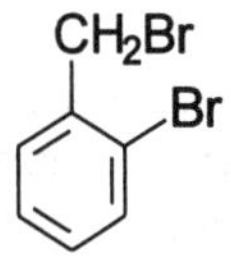

14.29   **(a)** Nitric acid with sulfuric acid is required. The products are a mixture of ortho and para isomers of bromonitrobenzene, because the bromo group is ortho,para directing.

(b) Fuming sulfuric acid is the reagent. m-Nitrobenzenesulfonic acid is the product, because the nitro group is meta directing.

(c) Bromine and iron(III) bromide are required. The products are a mixture of ortho and para isomers of bromoethylbenzene, because the ethyl group is ortho,para directing.

(d) Chloromethane and aluminum trichloride are required. The products are a mixture of ortho and para isomers of methylanisole, because the methoxy group is ortho,para directing.

14.30   **(a)** Bromine and iron(III) bromide are required. The product is m-bromobenzoic acid, because the carboxylic acid group is meta directing.

(b) Acetyl chloride and aluminum trichloride are the reagents. The products are a mixture of ortho and para isomers of isopropylacetophenone, because the isopropyl group is ortho,para directing.

(c) Nitric acid with sulfuric acid is required. The product is m-nitroacetophenone, because the acetyl group is meta directing.

(d) Nitric acid alone may be sufficient without sulfuric acid, because the hydroxy group is strongly activating. The products are a mixture of ortho and para isomers of nitrophenol, because the hydroxy group is ortho,para directing.

14.31 **(a)** The bromo group is an ortho,para director, whereas the nitro group is a meta director, so bromination must occur first. Brominate benzene using bromine and iron(III) bromide. Nitrate the bromobenzene using nitric acid with sulfuric acid.

(b) The nitro group is a meta director, whereas the bromo group is an ortho,para director, so nitration must occur first. Nitrate benzene using nitric acid with sulfuric acid. Brominate the nitrobenzene using bromine and iron(III) bromide.

(c) The ethyl group is weakly activating, whereas the bromo group is weakly deactivating, so alkylation is done first. Alkylate benzene using bromoethane and aluminum tribromide. Brominate ethylbenzene using bromine and iron(III) bromide.

(d) Both an ethyl and a bromo group are ortho,para directors, but a meta director is required. The acetyl group is a meta director, so acylation must occur first. Acetylate benzene using acetyl chloride and aluminum trichloride. Then brominate acetophenone using bromine and iron(III) bromide. Finally, reduce the m-bromoacetophenone using zinc/mercury amalgam and HCl.

14.32 **(a)** The bromo group is an ortho,para director. The sulfonic acid group is a meta director. Thus, sulfonation must occur first. Sulfonate benzene using fuming sulfuric acid. Then brominate the benzenesulfonic acid using bromine and iron(III) bromide.

(b) Bromine is an ortho,para director, whereas the sulfonic acid group is a meta director. Thus, bromination must occur first. Brominate benzene using bromine and iron(III) bromide. Sulfonate bromobenzene using fuming sulfuric acid to give a mixture of isomers, and separate them by a physical method.

(c) The methyl group is an ortho,para director, whereas the nitro group is a meta director. Therefore, alkylation must occur first. Alkylate benzene using bromomethane and aluminum tribromide. Nitrate the toluene using nitric acid with sulfuric acid to give a mixture of isomers, and separate them by a physical method.

(d) Both the nitro group and the carbonyl group are deactivating groups and meta directors. A methyl group is an ortho,para director. Therefore, alkylate benzene using bromomethane and aluminum tribromide. Then nitrate the toluene using nitric acid with sulfuric acid to give a mixture of isomers, and separate them by a physical method. Finally, oxidize the para isomer using potassium permanganate to convert the methyl group to —CO_2H.

14.33 **(a)** The nitro group is a meta director, whereas the chloro group is an ortho,para director. Therefore, nitrate benzene twice using nitric acid and sulfuric acid to give m-dinitrobenzene. Then chlorinate the m-dinitrobenzene using chlorine and iron(III) chloride. The chlorine is introduced at a position that is meta to each of the nitro groups.

(b) The methyl group of toluene is an ortho,para director, whereas the nitro group is a meta director. Nitrate toluene three times using nitric acid with sulfuric acid. The reaction will become progressively more difficult as each nitro group further deactivates the products. The nitro groups are placed in positions that are meta to each other as well as being ortho and para to the methyl group.

(c) The methyl group of toluene is an ortho,para director. The bromo group is an ortho,para director also, whereas the nitro group is a meta director. Nitrate toluene using nitric acid with sulfuric acid to obtain a mixture of the ortho and para isomers of nitrotoluene. Separate the para isomer by a physical method. Brominate twice using bromine and iron(III) bromide. The bromine atoms are introduced at the two equivalent positions that are ortho to the methyl group and meta to the nitro group.

14.34 **(a)** The bromo group is an ortho,para director and the nitro group is a meta director, and both are deactivating. The methyl group is an ortho,para director. Nitrate toluene three times using nitric acid with sulfuric acid to give 2,4,6-tribromotoluene. Oxidize this compound using potassium permanganate to convert the methyl group into a carboxylic acid group.

(b) The methyl group of toluene is an ortho,para director. Nitrate toluene using nitric acid with sulfuric acid. Separate the para isomer by a physical method. Brominate this compound using bromine and iron(III) bromide. The bromine is placed in a position that is ortho to the methyl group and meta to the nitro group.

(c) The bromo group is an ortho,para director and the nitro group is a meta director. Thus, nitration must occur first. Nitrate benzene twice using nitric acid with sulfuric acid to give m-dinitrobenzene. Then brominate the m-dinitrobenzene using bromine and iron(III) bromide. The bromine is introduced at a position that is meta to each of the nitro groups.

14.35 **(a)** Both the bromo and hydroxyl groups are ortho,para directors. The nitro group is a meta director that can be converted into another functional group. Nitrate benzene to give nitrobenzene. Then brominate nitrobenzene to give m-bromonitrobenzene. Reduce the nitro group to an amino group using tin and HCl. Convert the amino group into a diazonium group using nitrous acid, and react with hot aqueous acid to give m-bromophenol.

(b) Both the bromo and amino groups are ortho,para directors. The nitro group is a meta director. Nitrate benzene to give nitrobenzene. Then brominate nitrobenzene to give m-bromonitrobenzene. Reduce the nitro group to an amino group using Sn and HCl.

(c) The methyl group of toluene is an ortho,para director. Nitrate toluene to give a mixture of the ortho and para isomers of nitrotoluene. Separate the para isomer by a physical method. Reduce the nitro group to an amino group using tin and HCl. Convert the amino group into a diazonium group using nitrous acid, and react with hot aqueous acid to replace the diazonium group with a hydroxyl group.

14.36 **(a)** The bromo and chloro groups are ortho,para directors, but they are meta to each other in the product. The nitro group is a meta director. Nitrate benzene using nitric acid with sulfuric acid. Brominate the nitrobenzene using bromine and iron(III) bromide to give the meta isomer. Reduce the nitro group to an amino group using tin and HCl. Convert the amino group into a diazonium group using nitrous acid, and react with copper(I) chloride.

(b) Nitrate toluene to give a mixture of the ortho and para isomers of nitrotoluene. Separate the para isomer by a physical method. Reduce the nitro group to an amino group using tin and HCl. Convert the amino group into a diazonium group using nitrous acid, and react with react with copper(I) cyanide to give the nitrile.

(c) The bromo and methyl groups are ortho,para directors, but they are meta to each other in the product. The nitro group is a versatile group that can be converted into other groups. Nitrate toluene to give a mixture of the ortho and para isomers of nitrotoluene. Separate the para isomer by a physical method. The nitro group is a meta director, but the bromo groups must be ortho to it and meta to the methyl group. Reduce the nitro group to an ortho,para directing amino group using tin and HCl. The amino group is strongly activating, whereas the methyl group is only weakly activating. Brominate this compound twice using bromine and iron(III) bromide. The bromine atoms are placed in the equivalent positions ortho to the amino group, since the para position is already occupied by —CH$_3$. Convert the amino group into a diazonium group using nitrous acid, and then react with hypophosphorous acid to replace the diazonium group by hydrogen.

14.37 **(a)** Nitration tends to occur at atoms adjacent to the sulfur atom of thiophene. These are the C-2 and C-5 positions. However, the C-2 position is also activated by the ortho methoxy group. The C-5 position is less favored because it is meta to the methoxy group.

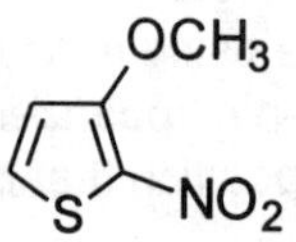

(b) Both the C-2 and C-5 positions have substituents. Nitration will occur at the position that is ortho to the activating methyl group rather than ortho to the deactivating carbonyl group.

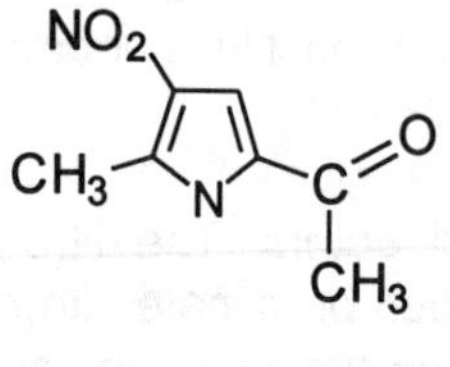

(c) Nitration tends to occur at the C-3 and C-5 positions. The ortho,para directing chlorine atom at C-2 will also direct the nitro group into those positions. A mixture of isomers results.

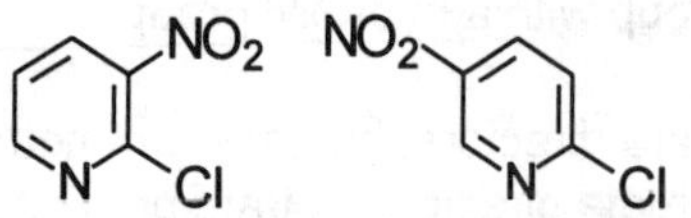

14.38 **(a)** There is one set of four equivalent C—H bonds that are ortho to a carbonyl group. These positions are deactivated. Reaction should occur at one of the other four equivalent sites that are meta to a carbonyl group.

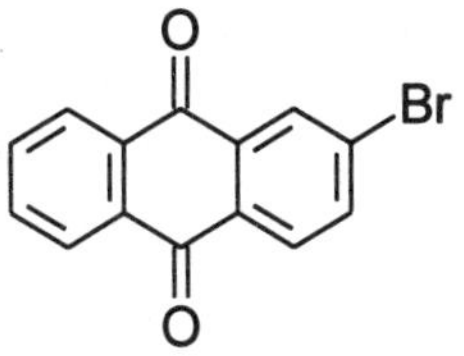

(b) Reaction should occur in the ring activated by the hydroxyl group. The naphthalene ring tends to react at a C-1 type site rather than a C-2 site. Thus, reaction should occur at C-4 which is also activated by the hydroxy group. Note that the C-2 position is also activated by the hydroxy group, but the C-3 position is not.

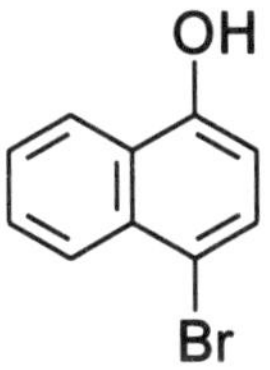

(c) Reaction should occur in the ring that is not deactivated by the nitro group. The naphthalene ring tends to react at a C-1 type site rather than a C-2 site. Thus, reaction should occur at either C-5 or C-8 of this compound. However, attack is less likely at the C-8 position due to steric hindrance.

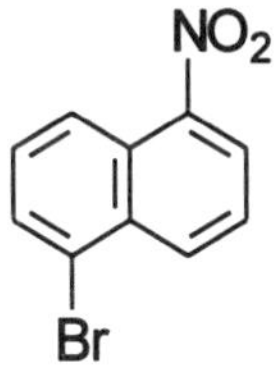

14.39 **(a)** 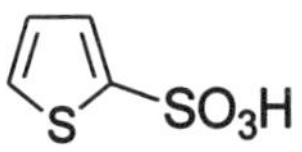 **(b)** 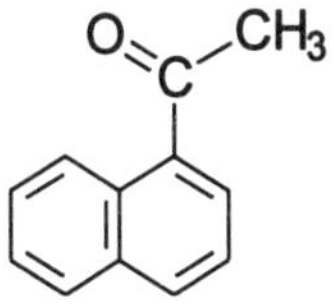 **(c)**

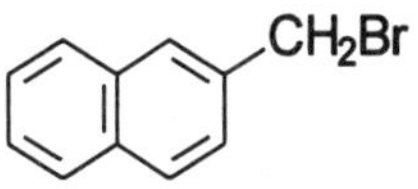

14.40 Sulfonation and desulfonation are reversible reactions. At a higher temperature the yield of the thermodynamically more stable isomer will increase. The para isomer is more stable than the more sterically hindered ortho isomer.

14.41   The right ring is somewhat deactivated compared to the left ring due to the chloro group. The indicated position is para to the activating nitrogen atom of the substituted amine. The position ortho to the nitrogen atom is also activated but may not be as reactive due to steric hindrance to attack at that site from the alkyl group bonded to nitrogen.

14.42   There are two deactivating groups which are in positions to affect all available sites for oxidation. Oxidation requires removal or transfer of electrons by an oxidizing agent. The electron withdrawing substituents decrease the electron density of the ring and hence its susceptibility to oxidation.

15

Spectroscopy

It might be said that "chemistry" isn't discussed in this chapter because there are no reactions. In fact, spectroscopy looks a lot like physics. But, spectroscopy is very much a part of chemistry in spite of the absence of reactions. The experimental techniques presented probe the very heart of chemistry—the structure of molecules themselves. Without an understanding of structure, then it isn't possible to understand chemical transformations.

Keys to the Chapter

15.1 Structure Determination

Although the identity of a molecule and its structure can be determined indirectly based on its reactions, such methods destroy some portion of the sample of the compound. The various spectroscopic methods determine certain aspects of structure by physical methods. As such there is no alteration of structure and the original molecule remains.

To gain some idea of the number and variety of possible isomers for a given molecular structure, you may want to write a few examples of each member of the classes of compounds mentioned in this section that have the molecular formula $C_5H_{10}O$. Then consider the functional groups and keep them in mind for subsequent discussions. For now the question is "how would I know that this or that functional group is present?"

15.2 Spectroscopy

The energy of light is directly proportional to its **frequency**. The **wavelength** of light is inversely proportional to the frequency, as given by $\lambda = c/\nu$. The **wavenumber** is the reciprocal of the wavelength. Thus, you should be able to understand whether the energy increases or decreases for a change in any of these quantities.

Spectroscopy is used to probe the physical changes in a molecule as the result of absorption of energy. The energy absorbed may change the extent to which a bond is stretched or the location of electrons in orbitals, and may even cause certain changes in the very nuclei of molecules. Ultraviolet, infrared, and nuclear magnetic resonance spectroscopy are three areas of use to organic chemists.

15.3 Ultraviolet Spectroscopy

The portions of the electromagnetic spectrum known as the ultraviolet and the visible regions are associated with energies sufficient to cause electronic transitions of conjugated systems. The position of the absorption peak is referred to as λ_{max}. Each transition corresponds to a promotion of an electron from the **highest occupied molecular orbital (HOMO)** to the **lowest unoccupied molecular orbital (LUMO)**.

The λ_{max} of a series of molecules increases with the length of the conjugated system. Thus, a triene absorbs at longer wavelengths than a diene. This difference is related to the decrease in the energy difference between the HOMO and LUMO as illustrated in Figure 15.5 in the text.

15.4 Woodward-Fieser Rules

These rules as listed in Table 15.1 are empirical. Their use requires only that you recognize all contributing structural features in the vicinity of the unsaturated carbon atoms. Counting the number of double bonds is straightforward. Then make sure not to overlook any **homoannular** component (conjugated double bonds in a common ring) of the system, because its effect on λ_{max} is substantial.

Counting alkyl groups isn't too hard, but remember that this terminology requires that bonds to saturated carbon atoms of a ring are considered as alkyl groups. Finally don't forget those **exocyclic** double bonds (double bonds containing one ring atom and one atom outside the ring). Check not only the "ends" of the π system but also the internal double bonds.

The color of polyunsaturated compounds is related to the λ_{max}. The color we see is the complement of the color corresponding to the light absorbed. Table 15.2 in the text gives the relationship between absorbed and reflected light.

15.5 Infrared Spectroscopy

For a science organized by functional group, it is evident why infrared spectroscopy is so useful. It confirms the presence (and sometimes more importantly the absence) of functional groups. The infrared spectrum is displayed so that absorptions of energy are related to wavelength or wavenumber. The "position" of the absorption is indicated by an inverted "peak" pointed down from a baseline.

Infrared absorptions correspond to the stretching of a bond or the bending of a bond angle. The strength of the bond as given by a force constant. Multiple bonds have higher force constants and require higher energy to stretch them. The energy required to stretch a bond is also related to the atomic mass of the bonded atoms. Bonds to hydrogen such as C—H, O—H, and N—H require higher energy than bonds such as C—C, O—C, and N—C.

The amount of energy required to stretch a specific bond in an organic molecule depends on the nature of the bonded atoms and the type of bond between them. The energy required to stretch bonds falls in a region of the electromagnetic spectrum called the infrared (IR). The full interpretation of the IR spectrum of a molecule is difficult, but certain functional groups have characteristic absorptions which can be used to propose a structure for an "unknown" compound. Table 15.4 in the text lists some of the characteristic frequencies for certain functional groups.

15.6 Structure Identification Using Infrared Spectroscopy

The spectrum of an "unknown" can be established by comparison to the spectrum of a known compound. If the spectrum of the "unknown" has all of the same absorption peaks as a compound of known structure, then the two samples are identical. If the "unknown" has one or more peaks that differ from the spectrum of a known, then the two compounds are not identical or some impurity in the "unknown" sample is causing the extra absorptions. If the "unknown" lacks even one absorption peak that is present in the known structure, the "unknown" has a different structure than the known.

15.7 Identifying Hydrocarbons

In addition to the atoms forming the bond and the number of those bonds, there is one additional structural feature that determines the position of an IR absorption. It is our old friend the % s character of the bond. With increased % s character, the electrons are more tightly held by an atom, so a bond to that atom requires higher energy to stretch. This difference is used to detect alkene and alkynes providing they have C—H bonds as well as C=C or C≡C bonds.

15.8 Identifying Oxygen-Containing Compounds

The presence of a carbon-oxygen double bond is easily detected by its characteristic strong absorption near the middle of an IR spectrum. The exact location is controlled by the extent to which a dipolar resonance form contributes to the structure. If the dipolar resonance form is stabilized by atoms bonded to the carbonyl group, then there is more single bond character and the energy required to stretch the bond is smaller.

Alcohols and ethers both contain C—O bonds which are difficult to unambiguously confirm in IR spectra. However the presence of an O—H bond in an alcohol is easily detected by a strong absorption on the left of the spectrum.

15.9 Bending Deformations

The absorption due to bending of one C—H bond or a group of C—H bonds with respect to a plane in compounds such as alkenes and aromatic compounds is observed on the right of the IR spectrum. Tables 15.5 and 15.5 in the text provide data that can be used to determine structure.

15.10 Nuclear Magnetic Resonance

Certain nuclei have a nuclear spin, a concept that is not familiar to you. A spinning nucleus generates a magnetic field, and its energy depends on the direction of spin in the presence of an applied magnetic field. The NMR method depends on detecting the absorption of energy required to change the direction of the spin of a nucleus.

Two nuclei that are important in the determination of the structures of organic compounds are ^{1}H and ^{13}C. The magnetic field strength required to "flip" the spin of various hydrogen atoms (or carbon atoms) within a molecule differs. The local magnetic fields differ throughout a molecule because the bonding characteristics differ. Thus, each hydrogen (or carbon) nucleus is unique, and distinct resonances are obtained for each structurally nonequivalent atom in a molecule.

15.11 Chemical Shift

Nuclei that have a local magnetic field that opposes the applied magnetic field are **shielded**. A relative scale called the **delta scale**, in which one delta unit (δ) is 1 ppm of the applied magnetic field, is used to measure the chemical shift of hydrogen atoms. The resonance for the hydrogen atoms of tetramethylsilane, $(CH_3)_4Si$, is defined as 0 δ. The delta scale is independent of the applied magnetic field. Shielded nuclei are found at high field and have small δ values. A summary of the relationships that you will use is illustrated as follows.

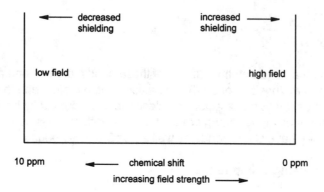

15.12 Detecting Sets of Nonequivalent Hydrogen Atoms

To understand the relationship between an NMR spectrum and the structure of a compound, you have to be able to recognize the equivalence of nuclei in the structure. The simplest examples, such as the six equivalent protons on the two methyl groups of 2-bromopropane, should be clear. However, some nuclei that might look equivalent at first glance are actually nonequivalent. Thus, both hydrogen atoms in 1-bromo-1-chloroethene are bonded to the same carbon atom, but they are not equivalent. One hydrogen atom is cis to the chlorine atom and the other is cis to the bromine atom. Replacing one hydrogen atom or the other by deuterium gives a set of diastereomers, and the hydrogen atoms are **diastereotopic**. Such hydrogen atoms usually have different chemical shifts.

Enantiotopic hydrogen atoms as defined in Section 9.12 of the text have the same chemical shift. Replacement of such hydrogen atoms by deuterium atoms gives enantiomers, and their physical properties are identical including the chemical shift.

15.13 Structural Effects on Chemical Shift

The chemical shifts of hydrogen atoms depend on the local electron density, which in turn affects the local magnetic field. Electronegative atoms deshield hydrogen atoms by an effect analogous to the inductive effect with which you are already familiar. Note that deshielding results in a shift to lower field and larger δ values.

Electrons in π bonds are easily polarized and strongly affect local magnetic fields. In addition these effects extend to some distance. Of particular importance is the deshielding of hydrogen atoms bonded to an aromatic ring. The chemical shifts of such atoms are large, meaning that they are low field.

The chemical shifts of hydrogen atoms in certain types of structural environments are listed in Table 15.7 in the text. You should be thoroughly familiar with these values.

15.14 Relative Peak Areas and Proton Counting

By examining the NMR spectrum, you can tell how many sets of structurally nonequivalent hydrogen atoms are contained in a molecule. Each set causes resonance absorptions in its own characteristic region. Another method is used to confirm the assignment of the type of hydrogen atoms. It is called "proton counting", and it relies on the fact that the area of each resonance peak is proportional to the relative number of hydrogen atoms of each kind. Information about the relative peak areas may be provided in a problem or right on the spectrum.

15.15 Spin-Spin Splitting

Multiple absorptions for a set of equivalent hydrogen atoms is known as **spin-spin splitting**. It results from the interaction of the nuclear spin(s) of the hydrogen atom(s) with the set of hydrogen atoms on an adjacent carbon atom. In general, sets of hydrogen atoms on nonequivalent neighboring carbon atoms **couple** with each other. If hydrogen atom A couples and causes splitting of the resonance for hydrogen atom B, then the resonance for hydrogen atom B is also split by hydrogen atom A. The separation of the components of the multiplet is the **coupling constant**, which is designated as J.

A set of one or more hydrogen atoms that has n equivalent neighboring hydrogen atoms has n+1 peaks in the NMR spectrum. Common multiplets include **doublets**, **triplets**, and **quartets**. The appearance of several sets of multiplets resulting from n = 1 to n = 4 is shown in Figure 15.16 of the text. The areas of the component peaks of a doublet are equal; the areas of the component peaks of other multiplets are not equal but are listed in Table 15.8 in the text.

You should recognize the characteristics of certain common structures such as the ethyl and isopropyl groups. The component multiplets are recognized not only by their multiplicity but by their intensity as well. The vinyl group is another common structure that is characterized by its NMR spectrum.

One caveat about the appearance of multiplets. All of the "rules" are observed if the chemicals shift difference between the coupled protons is large compared to the magnitude of the coupling constant. If the two quantities are similar, then the intensities of the components of the multiplet do not correspond to those listed in Table 15.8 and the actual number of components of the multiplet may be fewer or larger than predicted by the n+1 rule.

15.16 Structural Effects on Coupling Constants

There are orientational contributions to the magnitude of the coupling constant of vicinal hydrogen atoms. The coupling constant is largest for anti periplanar arrangements in saturated H—C—C—H compounds. The coupling constant for hydrogen atoms in an E arrangement is larger than for hydrogen atoms in an isomer with a Z arrangement.

Coupling constants over more than three bonds are termed **long range**. These coupling constants can be used to determine the structures of isomeric aromatic compounds.

15.17 Effect of Dynamic Processes

Not all hydrogen atoms "stay" in one place from the viewpoint of an observer. In those cases, due to dynamic processes such as rotation about a single bond, the resulting chemical shifts of two seemingly nonequivalent hydrogen atoms are actually equivalent. One such circumstance is found in cyclohexane, because the ring flip interchanges the axial and equatorial hydrogen atoms. A second example is found in alcohols due to the rapid exchange of protons between oxygen atoms of various alcohol molecules in a sample.

15.18 ^{13}C NMR Spectroscopy

This method of structural determination is a more recent development than that of hydrogen NMR spectroscopy. Its power lies in the direct determination of the number of nonequivalent carbon atoms in a structure. If you can see that the carbon atoms are different, then you know that each will give a distinct resonance. More importantly, the count of the number of resonances gives the number of sets of equivalent carbon atoms that must be in the structure. Some idea about the identity of each carbon atom responsible for a resonance is provided by the list of chemical shifts in Table 15.9 of the text.

Additional information about the identity of a carbon atom is provided by the multiplicity of the resonance as a result of coupling to directly bonded hydrogen atoms. The splitting of the resonances can be eliminated by **proton decoupling**.

The intensity of the resonances of ^{13}C atoms in a structure is usually not proportional to the number of equivalent carbon atoms. Thus, it is necessary to reconcile the number of resonances and the number of carbon atoms in a molecule based on the possibility of equivalent carbon atoms in the structure.

Solutions to Exercises

15.1 The increased number of aromatic rings gives a more conjugated system, which decreases the energy difference between the highest occupied molecular orbital and the lowest unoccupied molecular orbital. The 380 nm value for anthracene is slightly outside the wavelength of light of the visible spectrum. However, the 480 value for tetracene indicates that light in the blue region is absorbed. As a result, the compound will have an orange color. See Table 15.2 on page 543 of the text.

15.2 Both compounds have 11 conjugated double bonds. There are two more double bonds in lycopene than in β-carotene, but they are not part of the conjugated system. Thus, lycopene should have the same color as β-carotene, which is yellow-orange. See page 543 of the text.

15.3 2,4-Hexadiyne has two conjugated triple bonds and should absorb in the ultraviolet region. The triple bonds in 2,5-hexadiyne are isolated and should not absorb in the ultraviolet region.

15.4 The absorption of the conjugated system consisting of a carbon-carbon double bond and a carbonyl group should be affected by alkyl group substitution similar to that observed for conjugated double bonds. Compound I has three alkyl groups located at the carbon atoms of the carbon-carbon double bond whereas II has only one. Thus, compound I should absorb at the longer wavelength.

15.5 Compound IV has two equivalent sets of two conjugated double bonds and should absorb at the shortest wavelength. Compound III has three conjugated double bonds and an isolated double bond and absorbs at longer wavelength. Compounds I and II both have five conjugated double bonds and absorb at still longer wavelength. However, II also has two additional methyl groups and absorbs at the longest wavelength.

15.6 The tetrayne has four conjugated triple bonds and the triyne has three conjugated triple bonds. The difference of 27 nm in the absorbed light is the result of the increased conjugation of the tetrayne. The difference is similar to that observed with increased conjugation of double bonds.

15.7 **(a)** There are two double bonds that are homoannular, and there are two alkyl groups that are part of the ring system. The value is $(217 + 39 + 2 \times 5) = 266$ nm.
(b) There are two conjugated double bonds and one is exocyclic. There are four alkyl groups. The value is $(217 + 5 + 4 \times 5) = 242$ nm.
(c) There are three double bonds. Two are homoannular and two are exocyclic. There are four alkyl groups. The value is $(217 + 30 + 39 + 2 \times 5 + 4 \times 5) = 316$ nm.
(d) There are three double bonds. Two are homoannular and one is exocyclic. There are three alkyl groups. The value is $(217 + 30 + 39 + 5 + 3 \times 5) = 306$ nm.

15.8 **(a)** There are two double bonds that are homoannular, and there are three alkyl groups, two of which are part of the ring system. The value is $(217 + 39 + 3 \times 5) = 271$ nm.
(b) There are three conjugated double bonds and two are exocyclic. There are three alkyl groups. The value is $(217 + 30 + 2 \times 5 + 3 \times 5) = 272$ nm.
(c) There are two double bonds and one is exocyclic. There are four alkyl groups. The value is $(217 + 5 + 4 \times 5) = 242$ nm.
(d) There are three double bonds. Two are homoannular and two are exocyclic. There

are four alkyl groups. The value is (217 + 30 + 39 + 2 x 5 + 4 x 5) = 316 nm.

15.9   The longer wavelength for the trans isomer indicates that there is more extensive conjugation than in the cis isomer. The proximity of the ortho hydrogen atoms of the cis isomer causes severe steric hindrance and prevents the two aromatic rings from being planar, thus decreasing the amount of conjugation.

15.10   The extra dimethylamino group of the first compound increases the degree of conjugation and stabilization of the positive charge over the rings of the compound. This compound should absorb at a longer wavelength. The violet and blue-green colors correspond to absorption of light near 510 and 640 nm, respectively. Thus, the first compound is blue green and the second is violet.

15.11   Abstraction of a proton gives the phenoxide ion, in which the electrons of the oxygen atom are more readily delocalized into the aromatic ring.

15.12   The amino group has electrons that are delocalized into the aromatic ring of p-methyl-aniline. As a result this compound absorbs light at a longer wavelength. When HCl is added, the amino group is protonated to give an ammonium ion that has no lone pair electrons. Thus the absorption of the conjugate acid is similar to that of benzene.

15.13   The carbonyl group of propanone has a strong absorption in the 1700-1780 cm^{-1} region. 2-Propen-1-ol has an absorption for the carbon-carbon double bond in the 1630-1670 cm^{-1} region and an absorption for the oxygen-hydrogen bond in the 3400-3600 cm^{-1} region.

15.14   1-Pentyne is a terminal alkyne, so its sp-hybridized C—H bond has an absorption in the 3300-3320 cm^{-1} region, whereas 2-pentyne, which is an internal alkyne, does not have an absorption in this region.

15.15   The longer wavelength absorption (smaller wavenumber) corresponds to a lower energy vibration. The dipolar resonance form of a ketone is more stable than that of an aldehyde, because the extra alkyl group donates electron density. The increased contribution of the resonance form with a carbon-oxygen single bond means that the ketone carbonyl bond requires less energy to stretch it.

15.16   Consider the inductive and resonance effects of oxygen and nitrogen based on these properties as discussed in the chapter on aromatic substitution. Both oxygen and nitrogen are inductively electron withdrawing and destabilize the dipolar resonance form of the carbonyl group. Based on their electronegativities, this effect is larger for oxygen than for nitrogen, so the dipolar resonance form of an ester should be less stable that of an amide. In addition the relative ability of the two atoms to donate electrons by resonance is important. Because nitrogen is more effective in donation of electrons by resonance, there is an increased contribution of a dipolar resonance form for the amide.

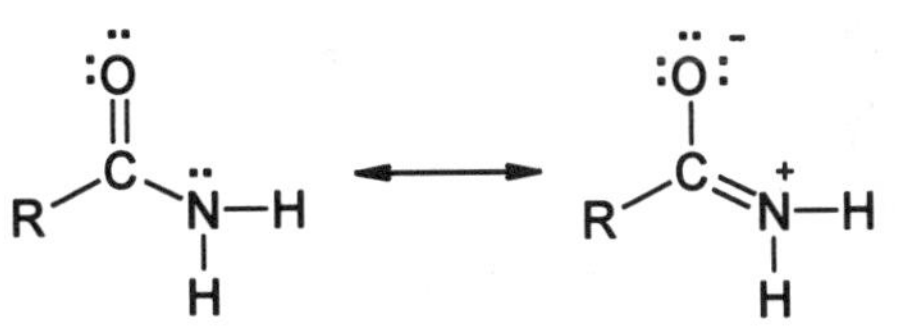

The greater stability and increased contribution of the dipolar resonance form of the amide means the carbonyl bond of the amide requires less energy to stretch it, so its stretching vibration occurs a lower wavenumber.

15.17 The absorptions correspond to an O—H and a carbonyl group, respectively. Only the carboxylic acid group of III has both of these two structural features. The other two compounds are esters that would have an absorption corresponding to a carbonyl group but, because they have no O—H group, would have no absorption in the 3500-3000 cm^{-1} region.

15.18 The carbonyl group of the second compound is conjugated with a double bond. As a result there is some contribution of a resonance form in which the carbon-oxygen bond has single bond character. Thus, the functional group requires less energy to stretch, which corresponds to absorption at a smaller wavenumber.

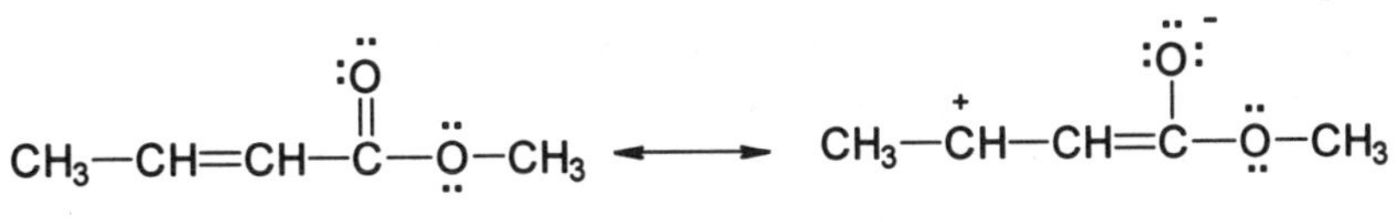

15.19 The ortho nitro isomer has four adjacent C—H bonds, and the out-of plane bending of these bonds should occur in the 770-735 cm^{-1} region. The para nitro isomer has two sets of two adjacent C—H bonds, and the out-of plane bending should occur in the 860-800 cm^{-1} region.

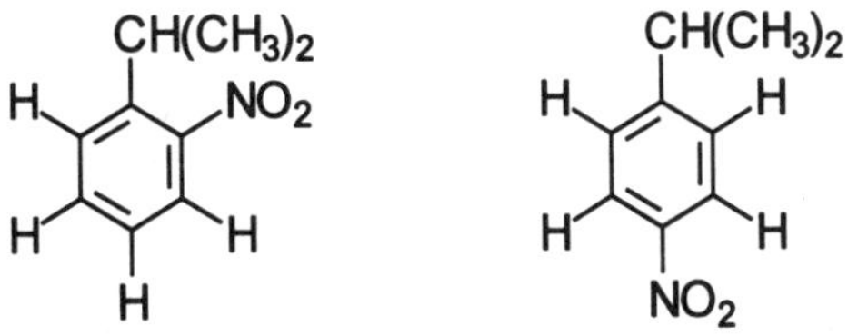

15.20 The 1,2,3 isomer has three adjacent C—H bonds, and the out-of plane bending absorptions of these bonds should occur in the 810-750 cm^{-1} region. In addition there is another absorption that occurs in the 745-690 cm^{-1} region. The 1,2,4 isomer has two adjacent C—H bonds and one lone C—H bond. The absorptions for these bonds should occur in the 860-800 and 900-860 cm^{-1} regions, respectively. The 1,3,5 isomer has three lone C—H bonds. The absorptions for these bonds should occur in the 900-860 cm^{-1} region.

15.21 Divide 437 Hz by 60 x 10^6 and multiply by 10^6 to obtain 7.28 δ.

15.22 Multiply 5.37 δ by 360 Hz because each ppm or δ unit equals 360 Hz. The chemical shift is 1933 Hz downfield.

15.23 **(a)** Only one, because all four methyl groups are equivalent. The C-2 atom has no C—H bonds.
  (b) Two, because there are two equivalent methyl groups and two equivalent sp^2-hybridized C—H bonds.
  (c) Two, because there are three equivalent methyl groups and three equivalent sp^2-hybridized C—H bonds on the benzene ring.
  (d) There are four. The C-1 methyl group and the branching methyl group at C-2 are not equivalent! In addition there is an sp^2-hybridized C—H bond at C-3, and the C-4 methyl group.

15.24 **(a)** Only one, because there are two equivalent sp^2-hybridized C—H bonds at C-2.
  (b) The sp^2-hybridized C—H bonds at C-2 are not equivalent. Thus, these two hydrogen atoms and the hydrogen at C-1 give three resonances.
  (c) The sp^2-hybridized C—H bonds at C-3 are not equivalent. Thus, these two hydrogen atoms, the hydrogen atom of the sp^2-hybridized C—H bond at C-2, and the C-1 methylene hydrogen atoms give four resonances.
  (d) The sp^2-hybridized C—H bonds at C-2 are not equivalent, so each hydrogen atom has a different resonance.

15.25 **(a)** Each compound has resonances integrating as three hydrogen atoms in the 3.3-4.0 δ region associated with ethers. However, this signal for *tert*-butyl methyl ether is a singlet due to the three hydrogen atoms of the methyl group. Isopropyl ethyl ether has two resonances in this region. The resonance arising from the two hydrogen atoms of the methylene group is a quartet due to coupling of the methyl group. The other hydrogen resonance in this region is due to the methine hydrogen of the isopropyl group and is a heptet because it is split by the two methyl groups.
  (b) All twelve of the hydrogen atoms of cyclohexane are equivalent, and the resonance appears as a singlet at high field. The isomeric alkene has a multiplet in the 5.0-6.5 δ region due to the hydrogen atoms of the sp^2-hybridized C—H bonds, as well as peaks due to the methylene and methyl hydrogen atoms.
  (c) The resonance due to the hydrogen atoms of the equivalent methyl groups of the 2,2-dimethyloxirane is a singlet. The hydrogen atoms of the equivalent methyl groups of the isomeric compound are split by the hydrogen atom of the ring.

15.26 **(a)** The equivalent C-1 and C-3 methylene groups of 1,3-dibromopropane give a triplet due to splitting by the hydrogen atoms at C-2. The methylene hydrogen atoms at C-2 give a quintet. The equivalent methyl groups of 2,2-dibromopropane give one peak, a singlet.

(b) The low-field portion of the spectrum of 1,1-dichlorobutane has an absorption due to the C-1 hydrogen atom. It is a triplet. The low-field portion of the spectrum of the isomeric 1,4-dichlorobutane has an absorption with an integrated intensity of four hydrogen atoms due to the C-1 and C-4 methylene groups. The signal is also a triplet.

(c) Both compounds have resonances in the 5.0-6.5 δ region due to the two hydrogen atoms of the sp²-hybridized C—H bonds. However, those of 2-methyl-1-propene have no nearest neighbor hydrogen atoms to split the signal. The signal for the isomeric cis-2-butene is split by the hydrogen atoms of the methyl groups.

15.27 **(a)**  **(b)**

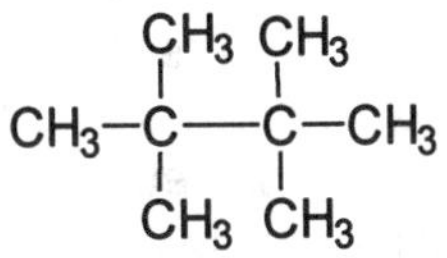

(c) 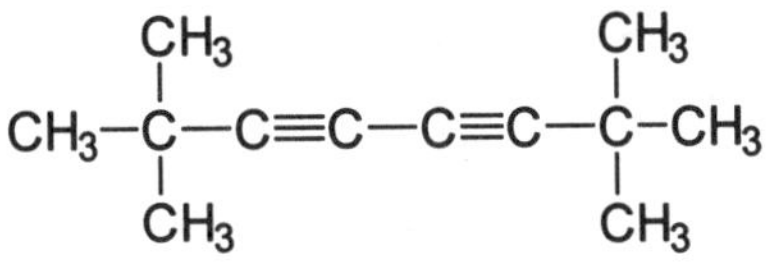 **(d)**

15.28 **(a)** 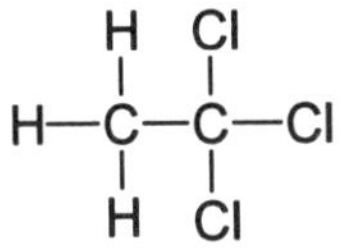 **(b)** 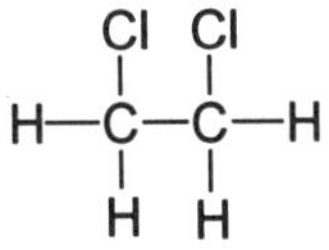 **(c)** 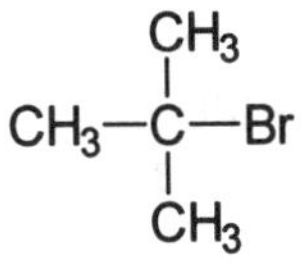 **(d)**

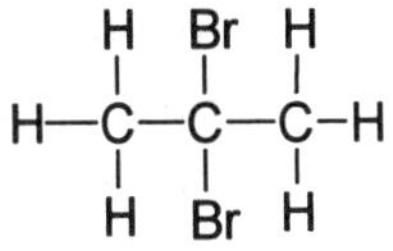

15.29   The local magnetic field generated by the π electrons of the ring is similar to that shown for benzene in Figure 15.13 on page 559 of the text. The location of the hydrogen atoms of [18] annulene is shown. Twelve hydrogen atoms "outside" the ring are deshielded and their resonance occurs at 8.8 δ. There are also 6 hydrogen atoms on the "inside" of the ring. The magnetic field experienced by these hydrogen atoms is reversed and they are shielded. Thus, the resonance occur at high field. The negative value indicates that the resonance is at higher field than TMS.

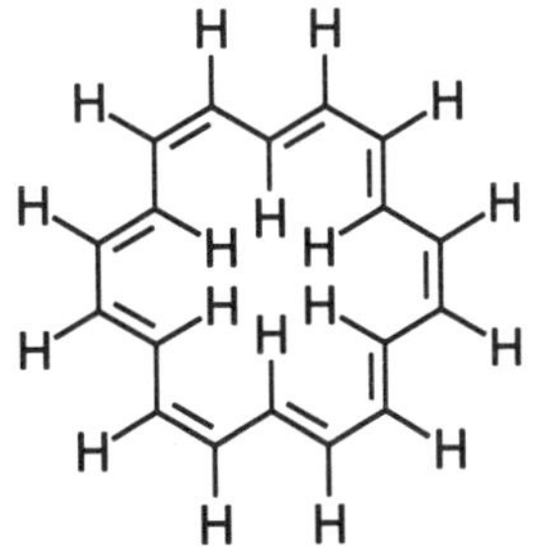

15.30   Ten of the hydrogen atoms are "outside" the ring and are deshielded. Their resonance occurs at 7.8 δ. There are also 4 hydrogen atoms on the "inside" of the ring. The magnetic field experienced by these hydrogen atoms is reversed. They are shielded and the resonance occurs at -0.6 δ. The ratio of the low-field to high-field resonances is 5:2.

15.31 **(a)** The resonance of the six hydrogen atoms of the two equivalent methyl groups is a triplet, and the resonance of the four hydrogen atoms of the two equivalent methylene groups is a quartet.

   (b) The resonance of the hydrogen atoms of the methyl group bonded to the oxygen atom is a singlet. The resonance of the hydrogen atom of the other carbon atom bonded to the oxygen atom is a heptet. The resonance of the hydrogen atoms of the two equivalent methyl groups is a doublet.

   (c) The resonance of the hydrogen atoms of the methylene group bonded to a chlorine atom is a singlet. The resonance of the hydrogen atom of the other carbon atom bearing a chlorine atom is a quartet. The resonance of the hydrogen atoms of the methyl group is a doublet.

   (d) The resonance of the hydrogen atom at the carbon atom bearing two chlorine atoms and an oxygen atom is a singlet. The resonance of the hydrogen atom of the carbon atom bearing an oxygen atom and one chlorine atom is a doublet. The resonance of the third hydrogen atom is a doublet.

15.32 The lowest field resonance in each case is for hydrogen atom(s) at the carbon atom that is bonded to the chlorine atom.

   (a) triplet, split by the two hydrogen atoms on C-2
   (b) singlet, because there are no hydrogen atoms on C-2
   (c) quintet, split by the four equivalent hydrogen atoms at C-2 and C-4
   (d) singlet, because there are no hydrogen atoms on C-2

15.33 The resonance for the hydrogen atom at C-2 is split into a quartet by the three hydrogen atoms at C-3 and is further split into doublets by the hydrogen atom at C-1.

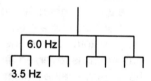

15.34 Because H_b is split by both H_a and H_c, its resonance should be a doublet of doublets. However, the coupling constants $J_{a,b}$ and $J_{b,c}$ are both 6 Hz. As a result, the inner lines of each doublet in the doublet of doublets overlap, so the resonance of H_b appears as a triplet.

15.35 The spectrum is consistent with 1,3-dibromopropane. The resonance of the C-2 hydrogen atoms appears as a quintet. The resonance of the hydrogen atoms at C-1 and C-3 is a triplet.

15.36 1,1,2,3,3-pentachloropropane has equivalent hydrogen atoms at C-1 and C-3 that give the low-field resonance which is split into a doublet by the hydrogen atom at C-2. The higher field resonance is due to the hydrogen atom at C-2, which is split by the hydrogen atoms at C-1 and C-3.

$$\begin{array}{ccccccc} & \text{Cl} & & \text{Cl} & & \text{Cl} & \\ & | & & | & & | & \\ \text{Cl} & - & \text{C} & - & \text{C} & - & \text{C} & - \text{Cl} \\ & | & & | & & | & \\ & \text{H} & & \text{H} & & \text{H} & \end{array}$$

Note that 1,1,1,2,3-pentachloropropane also has two equivalent hydrogen atoms and another nonequivalent hydrogen atom. However, the low-field resonance would be for a single hydrogen atom at C-2 and it would be a triplet. The high-field resonance would be due to two hydrogen atoms at C-1 and it would be a doublet. These resonances are just the opposite of those in the spectrum described.

$$\begin{array}{ccccccc} & \text{Cl} & & \text{Cl} & & \text{H} & \\ & | & & | & & | & \\ \text{Cl} & - & \text{C} & - & \text{C} & - & \text{C} & - \text{Cl} \\ & | & & | & & | & \\ & \text{Cl} & & \text{H} & & \text{H} & \end{array}$$

15.37 **(a)** The triplet corresponds to the four hydrogen atoms at C-1 and C-3. The quintet corresponds to the two hydrogen atoms at C-2.
(b) The low-field resonance corresponds to the two hydrogen atoms at C-1. The apparent heptet corresponds to the hydrogen atom at C-2, which is split by eight hydrogen atoms and should appear as nine lines. However, the intensity of two of the "outer" lines is too small to be seen relative to the other seven lines. The high-field doublet corresponds to the six hydrogen atoms in the two methyl groups.
(c) The low-field resonance corresponds to the four hydrogen atoms on the carbon atoms bonded to the oxygen atom. The high-field resonance corresponds to the four hydrogen atoms in the other two methylene groups. These two sets of atoms apparently have a small coupling constant, because the resonances appear to be unsplit.

(a) $\begin{array}{ccccccc} & \text{H} & & \text{H} & & \text{H} & \\ & | & & | & & | & \\ \text{Cl} & - & \text{C} & - & \text{C} & - & \text{C} & - \text{Cl} \\ & | & & | & & | & \\ & \text{H} & & \text{H} & & \text{H} & \end{array}$
(b) $\text{CH}_3 - \overset{\displaystyle\text{CH}_3}{\underset{\displaystyle\text{H}}{\text{C}}} - \text{CH}_2 - \text{Cl}$
(c)

15.38 **(a)** The low-field heptet corresponds to the hydrogen atom at C-2. The high-field doublet corresponds to the six hydrogen atoms of the two equivalent methyl groups of C-1 and C-3. The resonance at 3.4 δ is due to the hydroxyl hydrogen atom.
(b) The low-field quintet corresponds to the hydrogen atom at C-2. The doublet corresponds to the four hydrogen atoms at C-1 and C-3.
(c) The multiplet at 5.8 δ is due to the hydrogen atom at C-2. The peaks at 5.0-5.2 δ are due to the two hydrogen atoms at C-1, which are not equivalent. The resonance at 3.4 δ is due to the two hydrogen atoms at C-4, which also bears a bromine atom. The resonance at 2.6 δ is due to the two hydrogen atoms at C-3.

(a)
```
    H  OH H
    |  |  |
H—C—C—C—H
    |  |  |
    H  H  H
```

(b)
```
    Cl Cl Cl
    |  |  |
H—C—C—C—H
    |  |  |
    H  H  H
```

(c)
```
  H        CH₂CH₂Br
   \       /
    C=C
   /       \
  H         H
```

15.39 (a) There are three sets of equivalent carbon atoms. One set is the C-1, C-4, C-5, and C-8 atoms. The second set is the C-2, C-3, C-6, and C-7 atoms. The third is the two carbon atoms at the points of fusion of the rings.

(b) There are six signals, because there are two sets of nonequivalent methyl groups and four sets of nonequivalent ring carbon atoms. The carbon atom of the methyl groups at C-1 and C-3 are equivalent and have a different resonance than the carbon atom of the C-2 methyl group. The C-1 and C-3 ring atoms are equivalent. The C-4 and C-5 ring atoms are equivalent. The C-2 and C-5 ring atoms each have their own resonance.

(c) There are three signals, because there are a set of three equivalent methyl groups and two sets of nonequivalent ring carbon atoms. The C-1, C-3, and C-5 ring atoms are equivalent. The C-2, C-4, and C-6 ring atoms are equivalent.

(d) There are three signals, because there is a set of two equivalent methyl groups and two sets of nonequivalent ring carbon atoms. The C-1 and C-4 ring atoms are equivalent. The C-2, C-3, C-5, and C-6 ring atoms are equivalent.

15.40 (a) There are three signals, because there is one set of two equivalent methyl groups and one set of two equivalent methylene groups, as well as the carbonyl carbon atom.

(b) There are five signals, because all carbon atoms are nonequivalent. The methyl carbon atoms are situated differently with respect to the carbonyl carbon atom, as are the two methylene carbon atoms.

(c) There are five signals. Only the two methyl carbon atoms bonded to the common carbon atom on the right of the structure are equivalent.

15.41 Since the spectrum has two signals, the hydrocarbon has two sets of equivalent carbon atoms. The molecular formula and the signals described correspond to cyclobutene. The resonance of the carbon atom of the methylene groups is at high field and is split into a triplet by the two hydrogen atoms. The resonance of the carbon atoms of the double bond is at low field and is split into a doublet by the single hydrogen atoms.

```
   H       H
    \     /
     ▢═▢
    /     \
   H       H
   :       :
   H       H
```

15.42 The spectrum has two signals, so the hydrocarbon has two sets of equivalent carbon atoms. The molecular formula and the signals correspond to 2,3-dimethylbutane. The resonance of the carbon atoms of the four methyl groups is at high field and is split into a quartet by the three hydrogen atoms. The signal for the tertiary carbon atoms at C-3 and C-4 is at lower field and is split into a doublet by their respective single hydrogen atoms.

$$CH_3-\overset{\overset{\displaystyle CH_3}{|}}{\underset{\underset{\displaystyle H}{|}}{C}}-\overset{\overset{\displaystyle CH_3}{|}}{\underset{\underset{\displaystyle H}{|}}{C}}-CH_3$$

15.43 There is a singlet in the spectrum for 2,2-dimethylbutane which corresponds to the C-2 atom. All of the resonances of 3-methylpentane are split, because, each carbon atom has at least one hydrogen atom bonded to it.

15.44 The lowest field resonance in each compound is for the carbon atom bonded to the oxygen atom. That signal for 1-butanol is a triplet because there are two hydrogen atoms bonded to the C-1 atom. The signal for 2-butanol is a doublet because there is only one hydrogen atom bonded to the C-2 atom.

15.45 The (E)- and (Z)-4-octenes as well as the (E)- and (Z)-3,4-dimethyl-3-hexenes each have four sets of equivalent carbon atoms. For the (E)- and (Z)-4-octenes, the low-field resonance is for the C-4 and C-5 atoms. The highest field resonance is for the C-1 and C-8 atoms. The next highest is for the C-2 and C-7 atoms, and the next to lowest is for the C-3 and C-6 atoms. For the (E)- and (Z)-3,4-dimethyl-3-hexenes, the low-field resonance is for the C-3 and C-4 atoms. The highest field resonance is for the C-1 and C-6 atoms. The next highest is for the carbon atoms of the methyl groups at the C-3 and C-4 atoms. The next to lowest is for the C-2 and C-5 atoms. The (E) and (Z) isomers cannot be distinguished by a proton coupled ^{13}C spectrum, but the different carbon skeletons of the two compounds can be distinguished. The low-field resonance of the (E)- and (Z)-4-octenes is a doublet whereas that for the (E)- and (Z)-3,4-dimethyl-3-hexenes is a singlet.

15.46 There are four isomeric compounds for the molecular formula. However there are three equivalent carbon atoms in the methyl groups of 2-methyl-2-chloropropane and two equivalent carbon atoms in the methyl groups of 1-chloro-2-methylpropane. Thus, these isomers would have only two and three resonances, respectively. Only 1-chlorobutane and 2-chlorobutane have four nonequivalent carbon atoms as needed to provide a spectrum with four resonances. The lowest field resonance is for the carbon atom bonded to the chlorine atom. In the proton coupled ^{13}C spectrum, that resonance for 1-chlorobutane is a triplet, but it is a doublet for 2-chlorobutane.

15.47 The multiplicity rules for the coupling of fluorine to another nucleus are the same as the rules for hydrogen atoms, because each atom has a nuclear spin of ½.
   (a) It is a quartet because there are three hydrogen atoms bonded to the carbon atom which can couple to fluorine through two bonds.
   (b) It is a triplet because there are two hydrogen atoms bonded to the benzylic carbon atom which can couple to fluorine through two bonds.
   (c) It is a doublet because the three equivalent fluorine atoms can couple to one hydrogen atom.

15.48 The multiplicity rules for the coupling of fluorine to another nucleus are the same as the rules for hydrogen atoms, because each atom has a nuclear spin of ½.
   (a) It is a doublet because the three equivalent hydrogen atoms can couple to one fluorine atom.
   (b) It is a triplet because the two equivalent hydrogen atoms couple to two fluorine atoms.

(c) It is a doublet of doublets because the hydrogen atom is coupled to one fluorine atom via two bonds and the other through three bonds.

15.49   The multiplicity rules for the coupling of phosphorus to another nucleus are the same as the rules for hydrogen atoms, because each atom has a nuclear spin of ½. The resonance is a doublet, because there is just one phosphorus atom coupled to the hydrogen atoms of the three equivalent methyl groups via three bonds.

15.50   The nine hydrogen atoms of the three equivalent methyl groups will split the phosphorus resonance into a decet (10 lines).

15.51   Both methyl groups are equivalent, and they are in a diequatorial conformation in the trans isomer. A ring flip to give the diaxial conformation also has the methyl groups in equivalent environments.

15.52   The reference compounds are "locked" into a single conformation because the conformational preference of the *tert*-butyl group is very large. Thus, the signals for the indicated hydrogen atoms are typical of axial and equatorial hydrogen atoms. In bromocyclohexane, the conformation with the equatorial bromine predominates, and hence the axial hydrogen atom at C-1 contribute more to the time-averaged resonance. The value 3.95 δ is closer to the 3.81 δ value of the trans isomer with its equatorial bromine atom than to the 4.62 δ value of the cis isomer with its axial bromine atom.

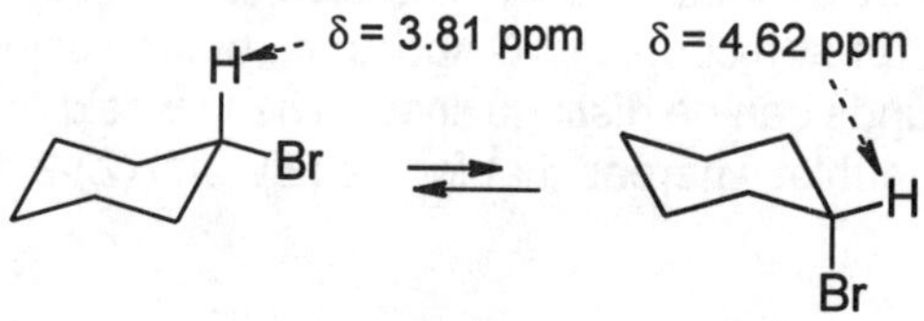

15.53   At 25°C the rate of rotation about the carbon-carbon bond is rapid and the NMR spectrum reflects the time average of two conformations. At the lower temperature the spectrum is that of two conformations. Each conformation has one set of equivalent hydrogen atoms and each of their resonances is a singlet.

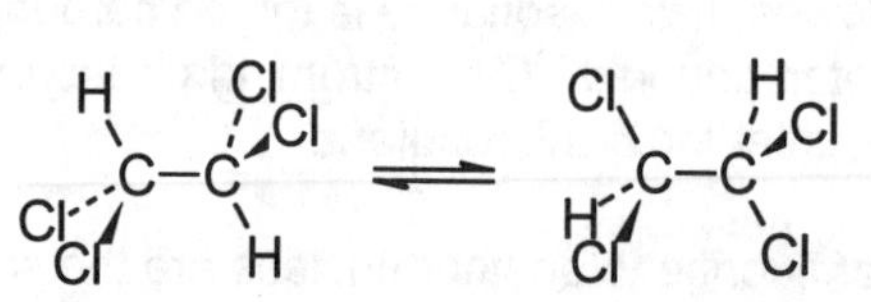

15.54   The conformation with the hydrogen atoms anti to one another will have a large coupling constant and the one with gauche hydrogen atoms will have a small coupling constant. Thus, the time average coupling constant will be intermediate and be closer to the value for the gauche conformation, because it is present in the larger amount.

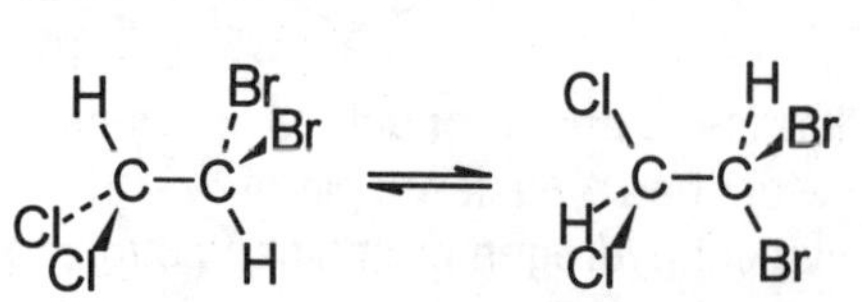

16

Alcohols—Reactions and Synthesis

Some of the chemistry of alcohols was presented in Chapter 8. In this chapter the remaining chemistry of alcohols is considered in detail. Then, another important part of the story is considered—the synthesis of alcohols. In future chapters, the same approach of considering reactions and synthesis is presented for other functional groups. For each functional group you should always be asking two questions: how does the functional group react and how can it be formed?

Keys to the Chapter

16.1 Overview of Alcohol Reactions

The focus of the chemistry of functional groups such as the hydroxyl group is on the bonds of the group and any nonbonded pairs of electrons that could be sites of reactivity. The two sites of the hydroxyl group at which reaction may occur are the C—O and the O—H bonds. You should review the chemistry at the C—O bond as given in Chapter 8 to compare with extensions of related substitution reactions given in this chapter. The only chemistry of the O—H bond presented in Chapter 8 was its reaction as an acid. In this chapter the replacement of the hydrogen atom by carbon groups is considered.

There are two types of reactions that involve C—H bonds in the vicinity of the hydroxyl group. The elimination reaction which occurs by elimination of a hydrogen atom at the carbon atom β to the hydroxyl group and the hydroxyl group itself was considered in Chapter 8 and will not be reviewed in this chapter. However, that doesn't mean that you can forget the reaction. It is one that you have to tuck away and recognize as one of the steps in synthetic sequences. The oxidation of an alcohol occurs by removal of a hydrogen atom at the carbon atom α to the hydroxyl group and the hydrogen atom of the hydroxyl group, a reaction considered in this chapter.

16.2 Conversion of Alcohols into Esters

Esters formed by the reaction of an alcohol with either an inorganic acid or a carboxylic acid have one feature in common. In both reactions, the oxygen atom bridging the alcohol and acid fragments is derived from the alcohol. The "reason" for this result is pictured by a substitution reaction mechanism in which a nucleophile attacks the carbonyl carbon atom to give a tetrahedral intermediate which subsequently ejects a leaving group. This process, known as nucleophilic acyl substitution, is another of the limited number of important mechanisms that dominate organic chemistry. You will meet it again in later chapters on the chemistry of acids and acid derivatives.

This section also emphasizes the point that replacement of a hydroxide ion as a leaving group is not as favorable a process as replacement of a better leaving group such as the chloride ion. Thus, the formation of esters by reaction of alcohols with acid chlorides is a more favorable process than reaction of alcohols with acids themselves.

16.3 Conversion of Alcohols into Alkyl Halides

The conversion of an alcohol into its conjugate acid allows the substitution of water as a leaving group and replacement of oxygen by a halogen. The same conversion can be accomplished using thionyl chloride or phosphorus tribromide. In both cases an intermediate is formed that converts the hydroxyl group into a better leaving group.

The reaction mechanism of the reaction of an alcohol with thionyl chloride involves a chlorosulfite which is formally an ester. Subsequent displacement by chloride ion can occur with retention or inversion of configuration depending on the solvent. In pyridine as solvent, inversion occurs. In dioxane, retention occurs via a solvated **ion pair**, in which the chloride attacks by an **internal return mechanism**.

The reaction mechanism of the reaction of an alcohol with phosphorus tribromide involves a phosphite ester. The oxygen atom is now bound to phosphorus and as such is a better leaving group that the hydroxide ion. Nucleophilic substitution by bromide gives an alkyl bromide. Note that the stoichiometry is one mole of PBr_3 to three moles of alcohol.

At this point, you should start to make lists of reagents that accomplish essentially the same transformation. You can prepare alkyl halides from alcohols by using a hydrogen halide, thionyl chloride, or phosphorus tribromide. The choice of reagents may be dictated by the structure of the substrate or by the presence of other functional groups that may react competitively with the selected reagent.

16.4 Oxidation of Alcohols

In this section we learn that alcohols are oxidized to carbonyl compounds by chromium(VI) compounds. The products depend on the structure of the substrate as well as the specific chromium reagent.

Alcohols are oxidized by the Jones reagent (CrO_3 in H_2SO_4). Primary alcohols are oxidized to aldehydes, which are further oxidized to carboxylic acids under the reaction conditions. The aldehyde usually cannot be isolated. Secondary alcohols are oxidized to ketones. Tertiary alcohols are not oxidized. Pyridinium chlorochromate (PCC), a reagent generated from CrO_3 and pyridine in CH_2Cl_2 as solvent, oxidizes alcohols only to carbonyl compounds. Secondary alcohols are oxidized to ketones, but primary alcohols are converted into aldehydes without being further oxidized to carboxylic acids.

16.5 Reactions of Vicinal Diols

Vicinal diols can be produced by direct oxidation with potassium permanganate to give a cis compound. This stereochemistry is the result of formation of a cyclic manganate ester (Section 7.9), which then is hydrolyzed to the diol. Vicinal diols can also be prepared by ring opening of epoxides to give a product with trans stereochemistry (Section 17.9).

Vicinal diols undergo a specialized reaction called the **pinacol rearrangement**. The mechanism should be recognized as a combination of steps, all of which you have seen before. (As is often the case, a new reaction isn't really new in a mechanistic sense. Apparently different mechanisms are just the result of a mix and match of steps that you already know.) First protonation occurs so that water can be the leaving group (you know that). The resulting carbocation rearranges to a more stable carbocation (you are aware of that idea). The only new feature is the type of stabilized carbocation. It is an oxocarbocation which is stabilized by donation of lone pair electrons from oxygen (another fact that you know based on the effect of oxygen in electrophilic aromatic substitution). Finally the loss of a proton is a simple reaction of an acid.

The details of the mechanism of the oxidative cleavage of a diol by periodic acid is not given, but the results are clear. Break the bond between the carbon atoms bearing the hydroxyl groups and convert the carbon-oxygen single bond into a carbonyl group.

16.6 Synthesis of Alcohols

An effective synthesis of any type of functional group depends on the type of substrates available and on the specificity of several possible reagents that could be used. As you go through the study of any functional group, you will have to recall what occurred earlier in your study. For an earlier reaction, the focus was on the reactivity of functional group A, which coincidentally resulted in a product with functional group B. That reaction must also be "learned" as a method to synthesize compounds with functional group B. You will also encounter reactions in the future that give additional examples of transformations to give a functional group encountered earlier in the text. This section gives you past history as well as previews of coming attractions.

Reactions that you know can be used to synthesize alcohols include nucleophilic substitution of an alkyl halide or hydration of an alkene. The former reaction occurs in competition with an elimination reaction. The second reaction is reversible and may also give rearranged products. That is the past history. Do you know it?

Alcohols can be prepared by reduction of carbonyl compounds. Aldehydes yield primary alcohols; ketones yield secondary alcohols. If you look ahead, you will find that reduction of alcohols by hydrogen gas with a transition metal catalyst occurs more slowly that the reduction of alkenes, so both carbonyl groups and double bonds are reduced in compounds that contain both functional groups. In addition you will find that lithium aluminum hydride and sodium borohydride both reduce carbonyl compounds to alcohols without affecting carbon-carbon double bonds. In fact, your instructor may choose to introduce that information at this point.

Alcohols with more complex hydrocarbon structures can be made by alkylation methods using a carbonyl compound and a Grignard reagent, which is carbanion-like reagent. This material is discussed in Section 16.10 and Chapter 18.

16.7 Synthesis of Alcohols from Haloalkanes

The substitution of a halide ion by a hydroxide ion occurs in competition with an elimination reaction. This problem can be circumvented by using an oxygen-containing derivative that is a weak base as the nucleophile. The acetate ion is such a species. The product is not an alcohol but rather an ester. Subsequent hydrolysis of the ester gives the alcohol and acetate ion. Note that the hydrolysis occurs by a substitution reaction at an acyl carbon atom. Sound familiar? Didn't you encounter the reaction of a nucleophile with an acyl derivative earlier in the chapter?

You have to be alert to these types of relationships as you encounter each apparently new piece of information. The ultimate discussion of acyl substitution occurs in Chapter 22. By that time you should be armed with lots of facts and experience.

16.8 Indirect Hydration Methods

Direct hydration of an alkene to give an alcohol is limited by both rearrangement and the reversibility of the process. Two indirect methods for hydration are given in this section, each with different regioselectivity.

The first step of the oxymercuration-demercuration method should look like the addition of bromine in a mechanistic sense. Both involve cyclic intermediates that are not prone to rearrangement, because much of the positive charge is on the heteroatom. Subsequent attack by a nucleophile gives the addition product. The new step, known as demercuration, results in replacement of mercury by hydrogen using sodium borohydride. The entire sequence, known as **oxymercuration-demercuration**, results in addition of water with the same regiospecificity as the direct hydration reaction. Thus, the process is a Markovnikov addition reaction.

The hydroboration-oxidation method gives an alcohol in a two step sequence that is equivalent to an anti-Markovnikov addition of water. The first step is the addition of an H—B bond across the double bond by a cyclic four-center mechanism. The regiochemistry is controlled by the fact that the hydrogen atom that adds is not a proton but rather hydridic in character, and the boron is the "positive" part of the reagent. Thus, in a sense the boron adds to the same carbon atom that adds a proton in the addition of reagents such as HBr.

The stereochemistry of the addition of an H—B bond across the double bond is syn. The subsequent replacement of the boron atom by oxygen using basic hydrogen peroxide occurs with retention of configuration.

16.9 Reduction of Carbonyl Compounds

The reduction of alkenes by hydrogen in the presence of certain metals occurs readily at atmospheric pressure, but the reduction of carbonyl compounds requires substantially higher pressures. Thus, unsaturated carbonyl compounds can be reduced to saturated carbonyl compounds. The opposite reaction, namely the reduction of a carbonyl group without affecting a carbon-carbon double bond, is also possible by using a different reagent. Both lithium aluminum hydride and sodium borohydride can reduce carbonyl compounds to alcohols without affecting carbon-carbon double bonds.

The mechanism of reduction of carbonyl compounds with hydride reagents occurs by attack of a nucleophilic hydrogen atom in the form of a hydridic hydrogen atom bonded to a metal atom. As such, the mechanism is very different from that of catalytic hydrogenation, so it should be reasonable to expect different regiochemistry for the two reducing agents. Note that lithium aluminum hydride must be used in an aprotic solvent such as ether to avoid its reaction with protons. Thus, this reagent shouldn't be used for reduction of carbonyl compounds that have hydroxyl groups or other acidic protons. Sodium borohydride can be used in protic solvents such as alcohols, and in fact it is necessary to do so. The proton of the alcohol solvent is transferred to the oxygen atom, giving the alcohol product. The hydrogen atom bonded to the carbon atom bearing the hydroxyl group comes from the hydride reagent.

The difference in the reactivity of lithium aluminum hydride and sodium borohydride is shown by their reactivity with esters. Lithium aluminum hydride reduces esters to alcohols; sodium borohydride does not.

The stereoselectivity of metal hydride reagents is not high. As a result, even in carbonyl compounds where the "faces" of the carbonyl group are different as a consequence of different steric environments, mixtures of isomeric alcohols are formed.

16.10 Synthesis of Alcohols Using Grignard Reagents

Interconverting functional groups is a somewhat pedestrian, albeit important, technique. Putting together two "small" groups of atoms to form a larger molecule is very different and is what much of the art of organic chemistry is all about. The Grignard reagent is a versatile material that can be used to stitch pieces together. The reagent behaves as a nucleophile, and it attacks the carbonyl carbon atom to give an alkoxide which has as its counter ion $MgBr^+$. Subsequent hydrolysis gives the alcohol.

How do you know what components to put together to give an alcohol? Examine the atoms or groups of atoms bonded to the carbon atom bearing the hydroxyl group. One of the groups of atoms can come from the Grignard reagent. The other two atoms or groups of atoms must have been present in the carbonyl compound. Thus, primary alcohols can come from only one Grignard reagent and formaldehyde. There are two possible combinations for a secondary alcohol. One of the alkyl groups can be from the Grignard reagent, and the other must be bonded to the carbonyl carbon atom of an aldehyde. As a result, there are two possible ways to make the compound if the alkyl groups bonded to the carbon atom bearing the hydroxyl group are different. Tertiary alcohols can be made from three different combinations of a Grignard reagent and a ketone if the three alkyl groups are different.

The use of the Grignard reagent is precluded if there is an acidic hydrogen in the substrate selected to react with the Grignard reagent. A reaction occurs, but it is one that destroys the Grignard reagent before it adds to the carbonyl group.

Alkynides formed directly from reaction of terminal alkynes with sodium amide, or Grignard reagents of alkynes can be added to carbonyl compounds to give acetylenic alcohols. The Grignard of an alkyne is prepared by reacting it with methyl Grignard reagent.

16.11 Sulfur Compounds

Thiols contain the -SH group, whereas alcohols contain the -OH group. Thiols, also called **mercaptans**, have significantly different physical and chemical properties than alcohols. Thiols are lower boiling compounds, because the -SH group does not form hydrogen bonds. In addition, the thiols are stronger acids than alcohols. Like alcohols, thiols can be synthesized by displacement of a halide ion from haloalkanes. The SH⁻ ion is an excellent nucleophile. Oxidation of thiols produces disulfide bonds rather than analogs of aldehydes and ketones.

Summary of Reactions

1. Formation of Esters

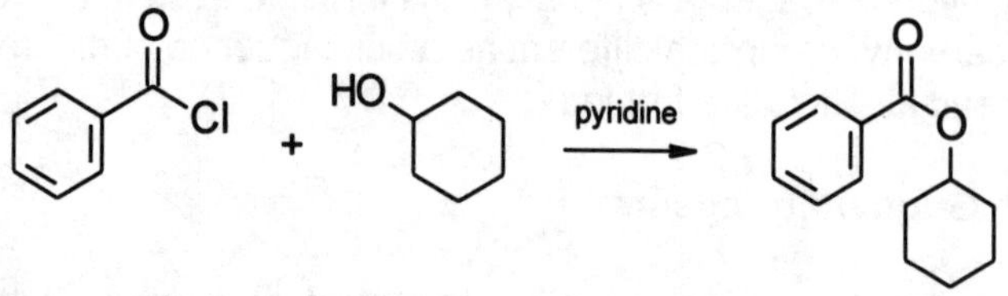

2. Synthesis of Alkyl Halides

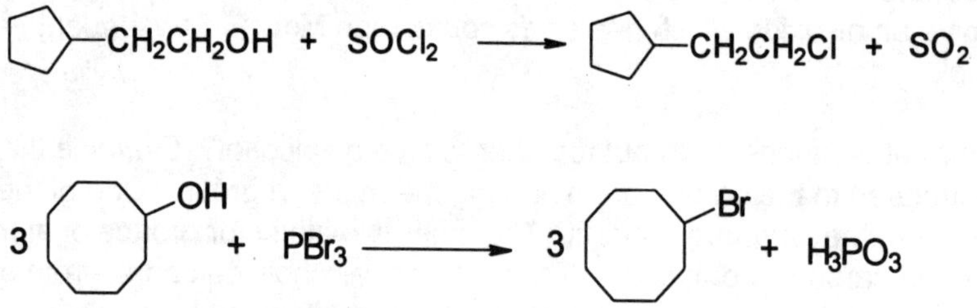

$$\text{cyclopentyl–CH}_2\text{CH}_2\text{OH} + \text{SOCl}_2 \longrightarrow \text{cyclopentyl–CH}_2\text{CH}_2\text{Cl} + \text{SO}_2$$

3. Oxidation of Alcohols

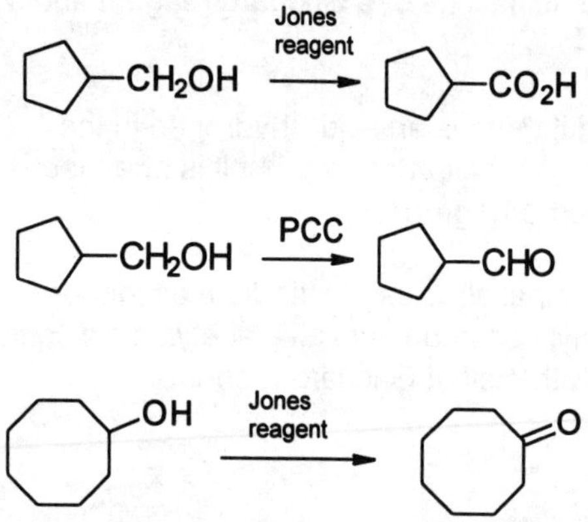

4. Pinacol Rearrangement

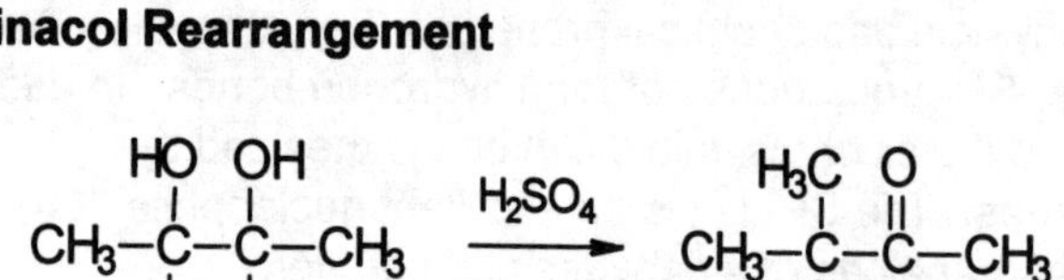

5. Oxidative Cleavage of Vicinal Diols

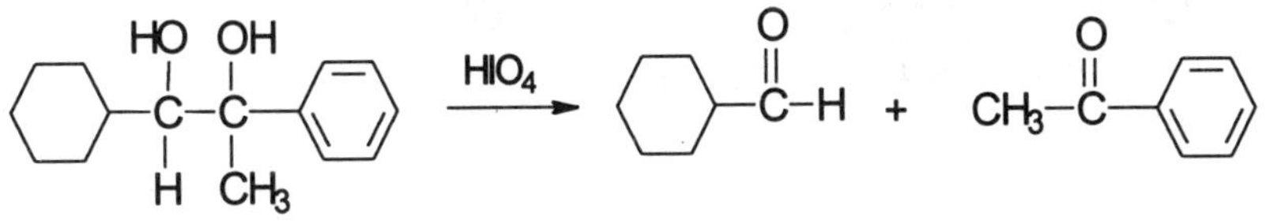

6. Synthesis of Alcohols from Haloalkanes

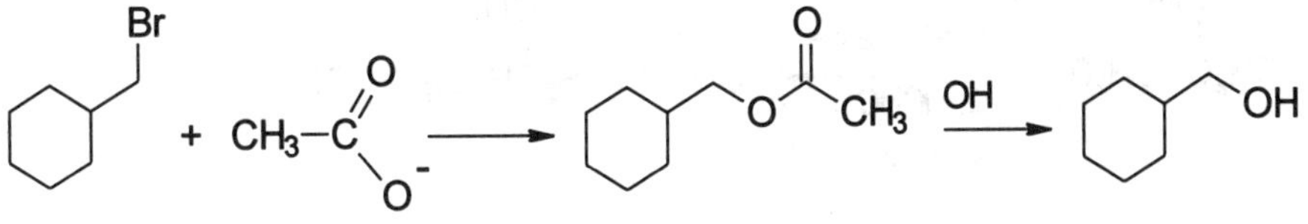

7. Synthesis of Alcohols from Alkenes

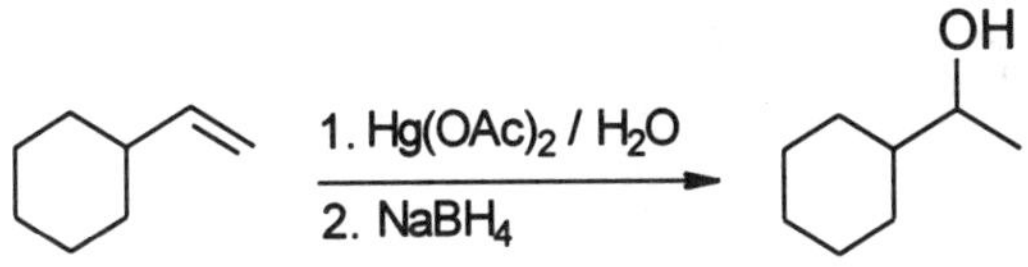

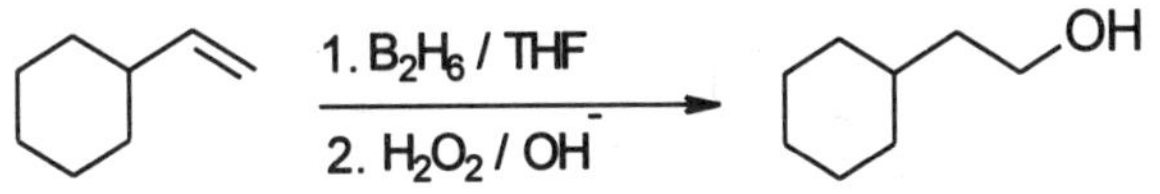

1. Hg(OAc)$_2$ / H$_2$O
2. NaBH$_4$

1. B$_2$H$_6$ / THF
2. H$_2$O$_2$ / OH$^-$

8. Reduction of Carbonyl Compounds

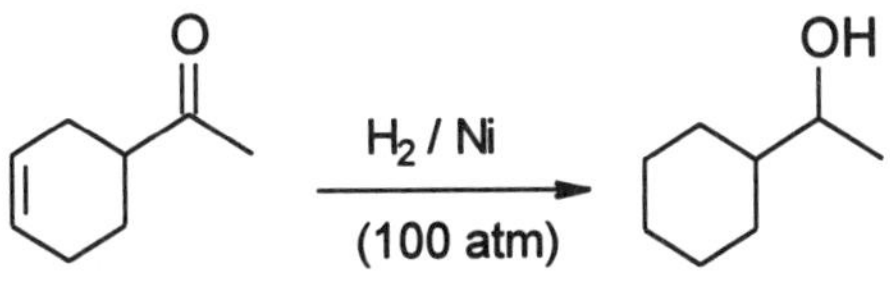

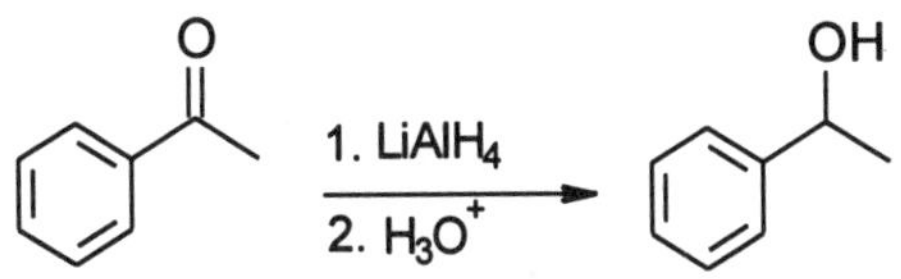

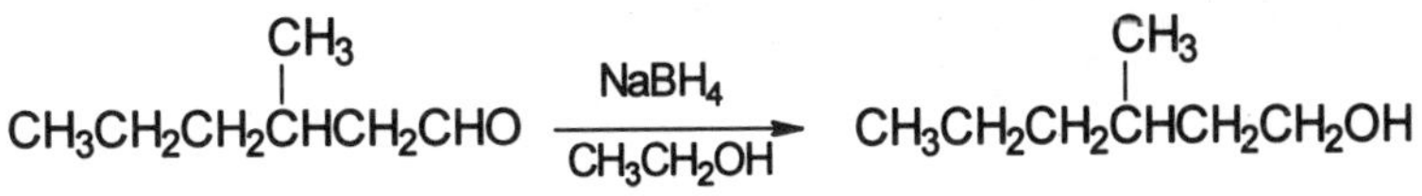

H$_2$ / Ni
(100 atm)

1. LiAlH$_4$
2. H$_3$O$^+$

$$CH_3CH_2CH_2\overset{\overset{\displaystyle CH_3}{|}}{C}HCH_2CHO \xrightarrow[CH_3CH_2OH]{NaBH_4} CH_3CH_2CH_2\overset{\overset{\displaystyle CH_3}{|}}{C}HCH_2CH_2OH$$

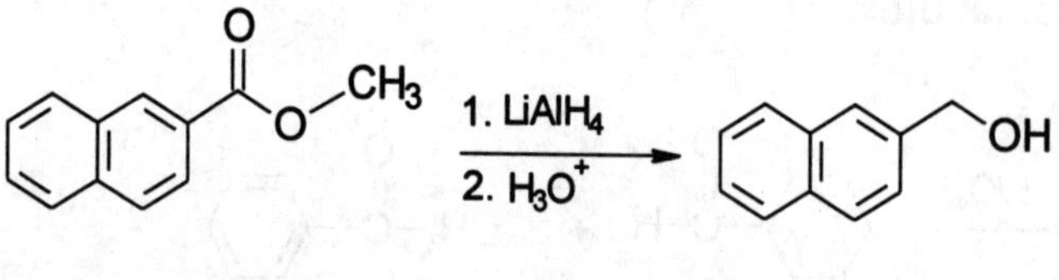

9. Synthesis of Alcohols Using Grignard Reagents

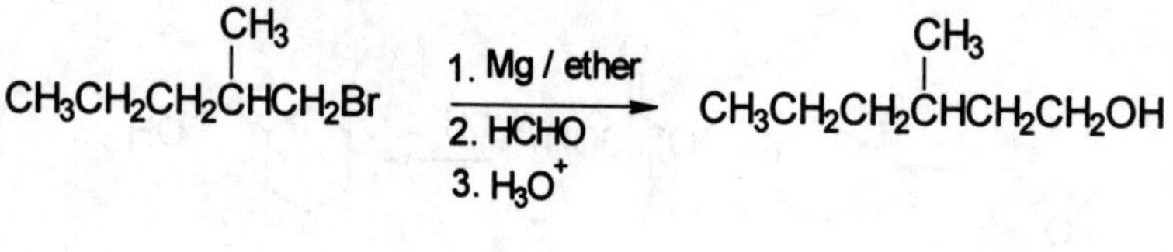

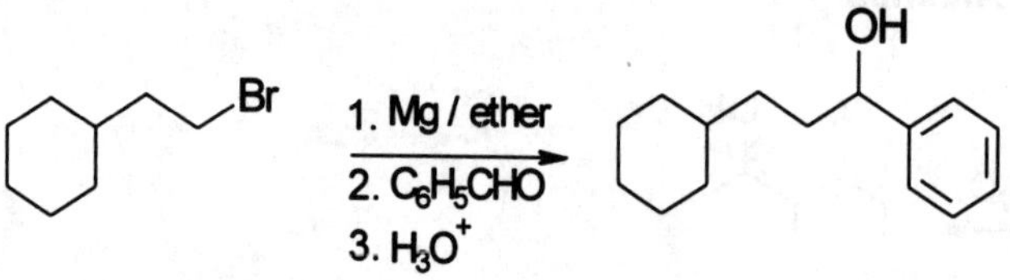

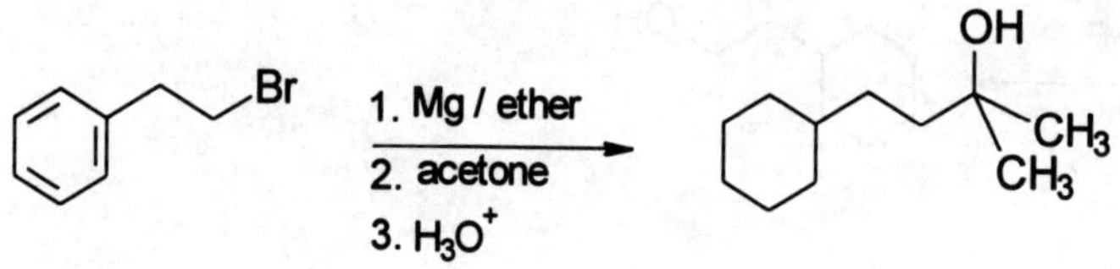

Solutions to Exercises

16.1 The $\Delta S°_{rxn}$ must be zero, as is expected for a gas phase reaction in which the number of moles of reactant and product are equal.

16.2 Solving $\Delta G° = -2.303\ RT \log K_{eq}$ at 25°C and with $K_{eq} = 4$, the $\Delta G°_{rxn}$ is -0.8 kJ mole^{-1}.

16.3 **(a)** 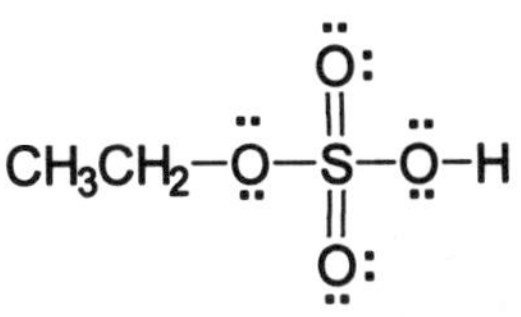 CH$_3$CH$_2$–O–S(=O)(=O)–O–H

(b) 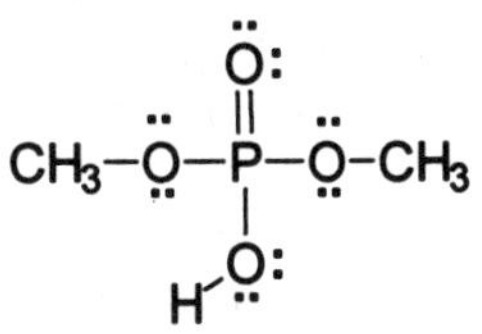 CH$_3$–O–P(=O)–O–CH$_3$ with H–O

(c) 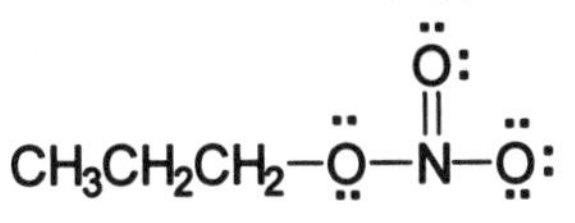 CH$_3$CH$_2$CH$_2$–O–N(=O)–O:

(d) 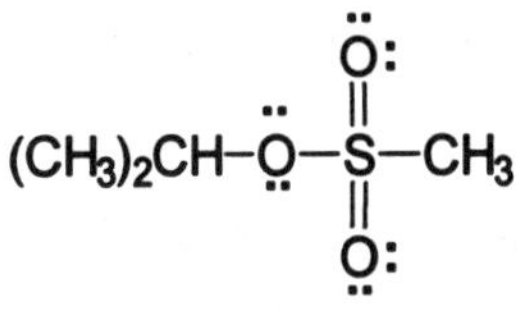 (CH$_3$)$_2$CH–O–S(=O)(=O)–CH$_3$

16.4 **(a)** 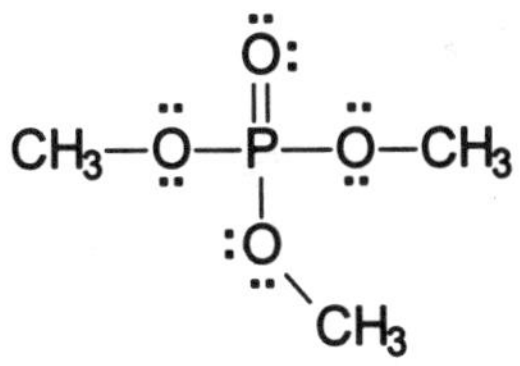 CH$_3$–O–P(=O)–O–CH$_3$ with :O–CH$_3$

(b) 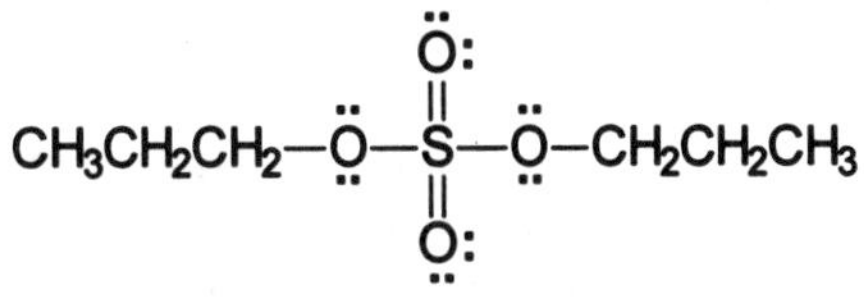 CH$_3$CH$_2$CH$_2$–O–S(=O)(=O)–O–CH$_2$CH$_2$CH$_3$

(c) 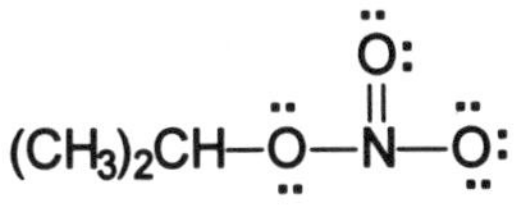 (CH$_3$)$_2$CH–O–N(=O)–O:

(d) 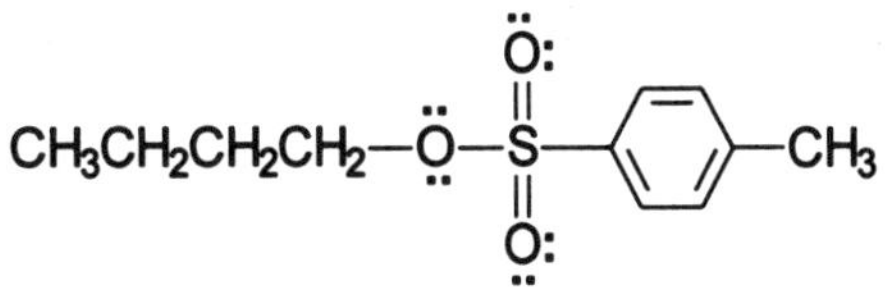 CH$_3$CH$_2$CH$_2$CH$_2$–O–S(=O)(=O)–C$_6$H$_4$–CH$_3$

16.5 H–O–C(=O)–C(=O)–O–H   C$_6$H$_5$–CH$_2$–O–C(=O)–C(=O)–Cl: CH$_3$–O–C(=O)–C(=O)–O–CH$_3$

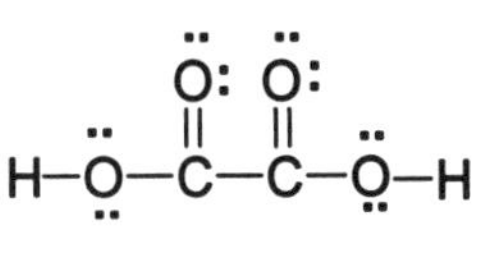

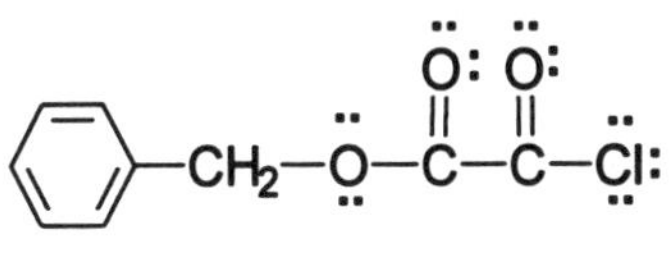

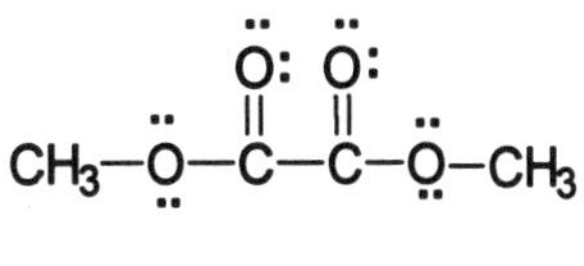

16.6 The compound is a dimethyl ester of carbonic acid, which is an unstable acid. The acid chloride is phosgene.

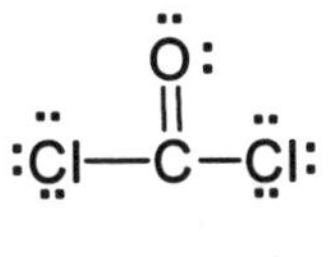

:Cl–C(=O)–Cl:

16.7 The ester is formed with the secondary alcohol, because its oxygen atom is a better nucleophile than the oxygen atom of the tertiary alcohol, which is more sterically hindered. The tertiary hydroxyl group does not react because only one equivalent of tosyl chloride is used. The ester is shown on the following page.

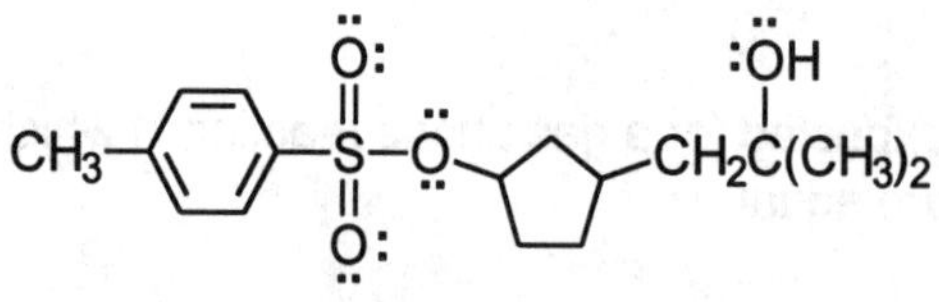

$$CH_3-\text{(ring)}-\overset{\overset{\ddot{O}:}{\|}}{\underset{\underset{:\ddot{O}:}{\|}}{S}}-\ddot{O}-\text{(cyclopentane)}-CH_2\overset{:\ddot{O}H}{\underset{|}{C}}(CH_3)_2$$

16.8   The equatorial hydroxyl group in the A ring of the steroid is less sterically hindered than the axial hydroxyl group of the C ring. Esterification occurs at the oxygen of the equatorial hydroxyl group because it is a more effective nucleophile.

16.9   The leaving group of the trifluoromethanesulfonic acid is a weak conjugate base of the more acidic trifluoromethanesulfonic acid. Weak bases are better leaving groups. Thus, alkyl esters of trifluoromethanesulfonic acid are more reactive than those of methanesulfonic acid, because trifluoromethanesulfonic acid is the stronger acid and its conjugate base is the weaker base.

16.10   The leaving groups are the perchlorate and nitrite ions, respectively. Perchloric acid is a strong acid and nitrous acid is a weak acid. Methanesulfonic acid is a strong acid. Thus, the perchlorate ion should be comparable to the methanesulfonate ion as a leaving group. The nitrite ion should be a much poorer leaving group.

16.11   Benzenesulfonic acids are strong acids, and benzoic acids are weak acids. Thus, esters of benzenesulfonic acids are more reactive because the benzenesulfonate ion is a weaker base than the benzoate ion.

16.12   Bromine is an electron withdrawing group and makes the benzenesulfonate ion a weaker base. Thus, the bromine substituted ion is a better leaving group than the tosylate ion.

16.13   Protonation of the oxygen atom gives an oxonium ion, and water can leave the C-2 atom while water of the solvent acts as a nucleophile. The result is net inversion of configuration. Eventually total racemization will occur.

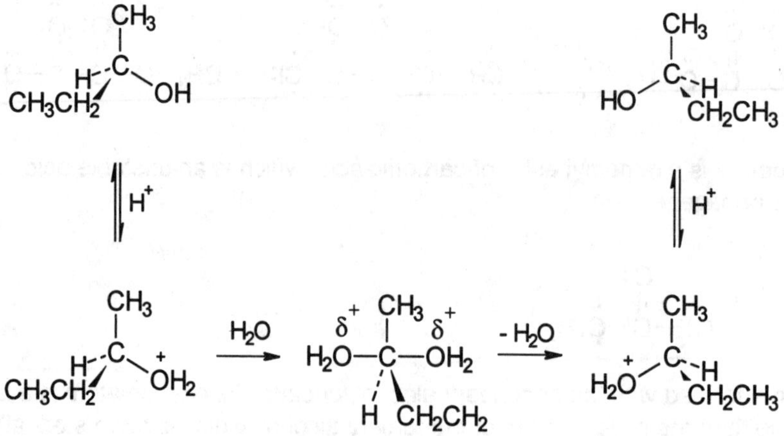

16.14   Protonation of the oxygen atom gives an oxonium ion which loses water from the C-1 atom to give a resonance stabilized allyl carbocation. Reaction of water at the original C-3 atom gives an alcohol in which the double bond is conjugated with the aromatic ring.

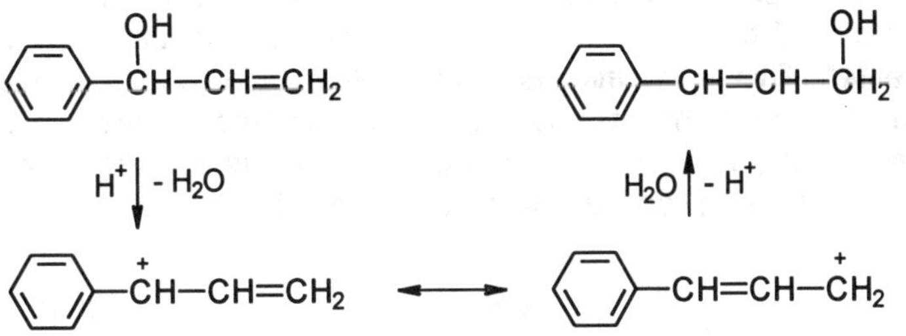

OH
⬡—CH—CH=CH₂

OH
⬡—CH=CH—CH₂

H⁺ | - H₂O

H₂O↑ - H⁺

⬡—ĊH—CH=CH₂ ⟷ ⬡—CH=CH—ĊH₂

16.15 Protonation of the oxygen atom gives an oxonium ion, and water can leave the C-1 atom giving a resonance stabilized allyl carbocation. Reaction of water at the original C-3 atom gives 3-buten-2-ol. Reaction of water at the original C-3 atom gives a mixture of the original alcohol and its trans isomer.

16.16 Protonation of the tertiary alcohol at the C-4 atom and loss of water gives a tertiary carbocation. This carbocation is achiral and subsequent attack of the oxygen atom of the hydroxyl group at C-1 can occur at either side of the plane to give the observed product, which is racemic.

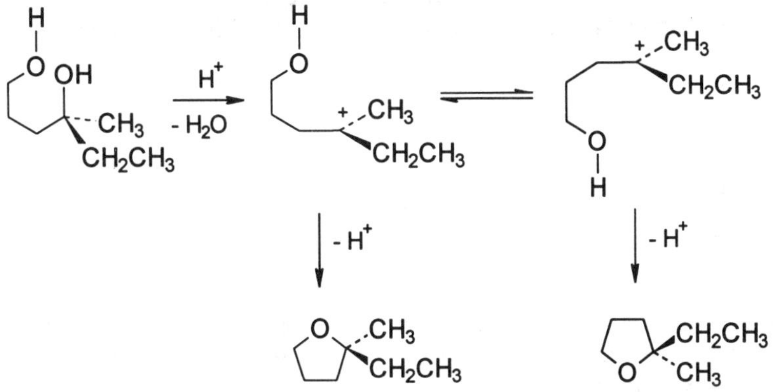

16.17 (a)

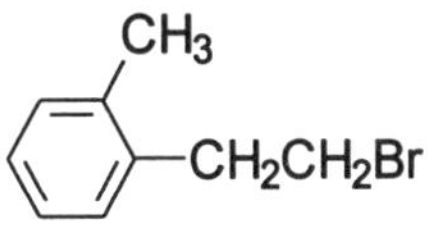

CH₃
⬡—CH₂CH₂Br

(b)

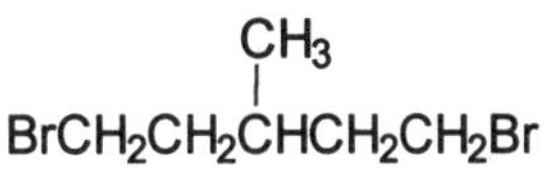

CH₃
BrCH₂CH₂CHCH₂CH₂Br

(c)

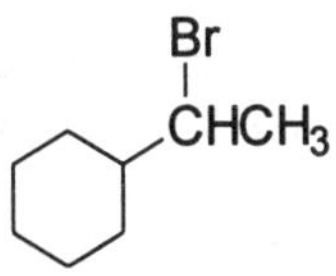

Br
⬡—CHCH₃

(d)

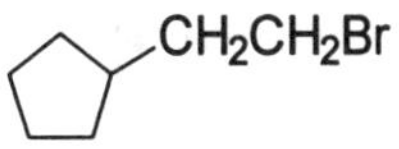

⬠—CH₂CH₂Br

16.18 (a)

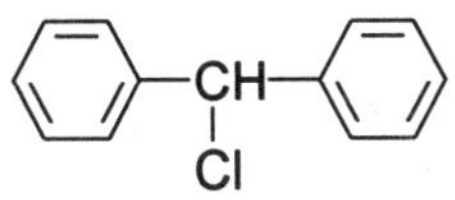

⬡—CH—⬡
|
Cl

(b) 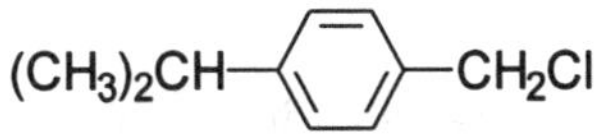 (CH₃)₂CH—⬡—CH₂Cl

(c)

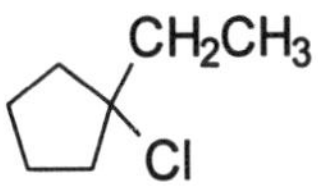

CH₂CH₃
⬠
Cl

(d)

CH₃
[decalin structure]
Cl

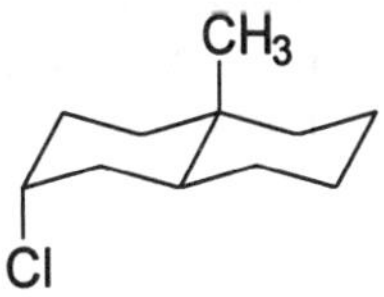

16.19 Both compounds give resonance stabilized carbocations. However, the carbocation derived from 3-methyl-2-buten-1-ol has its positive charge distributed between a primary and a tertiary center. This carbocation is more stable than the carbocation derived from 2-methyl-2-buten-1-ol, which has its positive charge distributed between a primary and a secondary center. Thus, 3-methyl-2-buten-1-ol reacts at a faster rate because a lower energy transition state that resembles the carbocation is formed.

$$CH_3-\underset{\underset{CH_3}{|}}{C}=CH-CH_2OH \xrightarrow[- H_2O]{H^+} CH_3-\underset{\underset{CH_3}{|}}{C}=CH-\overset{+}{C}H_2 \longleftrightarrow CH_3-\underset{\underset{+}{|}}{\overset{\overset{CH_3}{|}}{C}}-CH=CH_2$$

$$CH_3-CH=\underset{\underset{CH_3}{|}}{C}-CH_2OH \xrightarrow[- H_2O]{H^+} CH_3-CH=\underset{\underset{+}{\overset{\overset{CH_3}{|}}{C}}}{}-CH_2 \longleftrightarrow CH_3-\overset{+}{C}H-\underset{\underset{CH_3}{|}}{C}=CH_2$$

16.20 The allylic carbocation has its positive charge distributed between a secondary center at the original C-1 atom and a tertiary center at the C-3 atom. Under conditions of kinetic control, the bromine atom attacks the more positive center and gives 3-bromo-3-methyl-cyclopentene. If the products can equilibrate, the compound with the more substituted double bond would be favored. That isomer is 1-methyl-3-bromo-cyclopentene.

16.21 If the HBr escapes then the nucleophilic bromide ion required to displace the substituted phosphite ion as the leaving group is lost. Thus, the phosphite ester remains and will regenerate the original alcohol when the reaction mixture is treated with water.

16.22 The extra HBr provides an added source of nucleophilic bromide ion to displace the substituted phosphite ion as the leaving group.

16.23 Use thionyl chloride in dioxane, which gives a substitution product with retention of configuration by an internal return mechanism, to prepare the trans isomer. Use thionyl chloride in pyridine to achieve inversion of configuration to prepare the cis isomer.

16.24 Thionyl chloride in pyridine gives inversion and yields the (S) isomer. In diethyl ether, as in dioxane, the (R) isomer results by retention of configuration via an internal return mechanism.

16.25 Formation of a methanesulfonate followed by reaction with the nucleophilic chloride ion in an aprotic solvent tends to give substitution at the original C-1 atom. If HCl is used, a resonance stabilized carbocation forms which can react with chloride ion at the C-1 atom to give the same product as with the listed reaction or at the C-3 atom to give an isomeric compound.

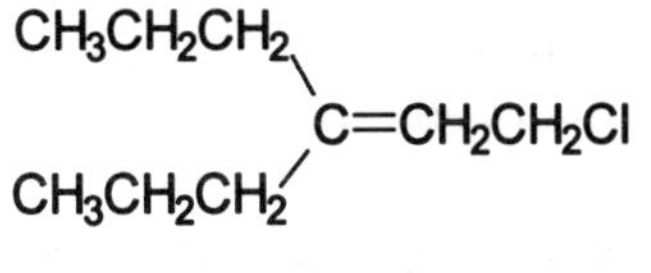

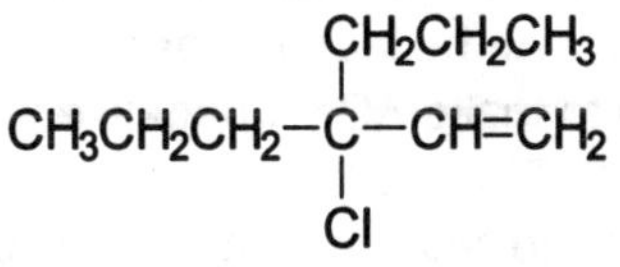

$$CH_3CH_2CH_2 \diagdown$$
$$C=CH_2CH_2Cl$$
$$CH_3CH_2CH_2 \diagup$$

$$\overset{\displaystyle CH_2CH_2CH_3}{CH_3CH_2CH_2-\underset{\displaystyle Cl}{\overset{\displaystyle |}{C}}-CH=CH_2}$$

16.26 Formation of a tosylate followed by reaction with the nucleophilic bromide ion in an aprotic solvent tends to give substitution at the methylene atom.

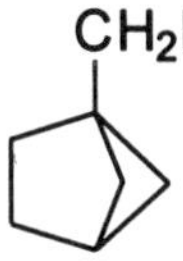

If HBr is used, a carbocation forms which rearranges to give a tertiary carbocation which has an expanded ring system that has lower strain energy.

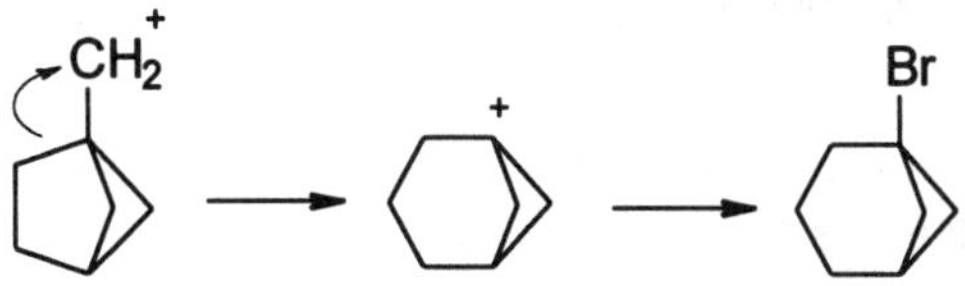

16.27 Sterically hindered alcohols are less likely to undergo S_N2 reactions. The competing S_N1 reaction gives a primary carbocation that can undergo a methide shift, which gives the 2-bromo-2-methylbutane product. Although the resulting carbocation is less stable, a hydride shift of the carbocation resulting from the methide shift gives a secondary carbocation that leads to the 2-bromo-3-methylbutane product.

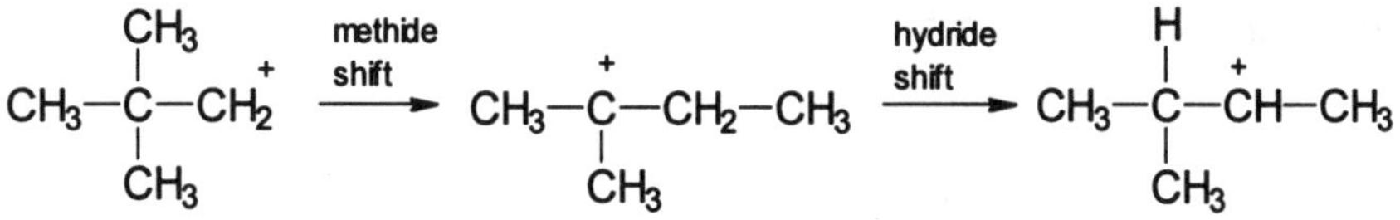

16.28 The zinc chloride serves as a Lewis acid to abstract a chloride ion and form a carbocation.

A hydride shift interconverts secondary carbocations with positive charge at the C-2 and C-3 atoms. The larger amount of the product at C-2 is due to the equivalence of two possible ions compared to the single ion with charge at C-3.

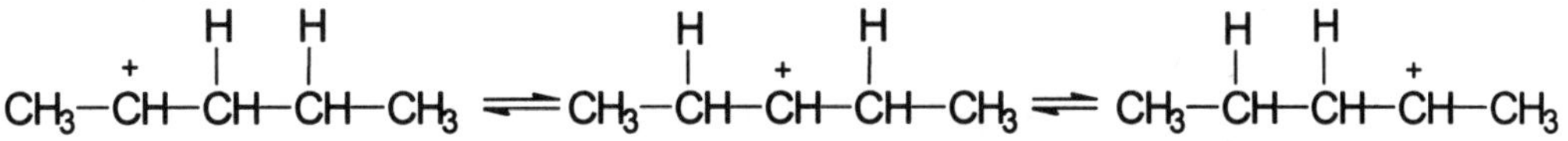

16.29   For this gas phase reaction, the $\Delta S°_{rxn}$ is positive because one mole of reactant gives two moles of product. At the high temperature of the reaction the $-T\Delta S°_{rxn}$ term is sufficiently negative to make $\Delta G°_{rxn}$ negative in spite of the positive $\Delta H°_{rxn}$ term.

16.30   The net difference in the two reactions is the combination of hydrogen and oxygen to give water, which is a reaction with $\Delta H°_{rxn} < 0$. Addition of the following two equations and their associated $\Delta H°_{rxn}$ values gives the net $\Delta H°_{rxn}$ for the reaction of ethanol and oxygen even though the mechanism for the actual reaction may not involve these two steps.

$$CH_3CH_2OH \longrightarrow CH_3CHO + H_2 \qquad \Delta H° > 0$$

$$H_2 + 1/2\ O_2 \longrightarrow H_2O \qquad \Delta H° < 0$$

$$CH_3CH_2OH + 1/2\ O_2 \longrightarrow CH_3CHO + H_2O \quad \Delta H° < 0$$

16.31   The product from 2-octanol is a ketone, which is insoluble in water because it is a nonpolar compound even though there is a carbonyl group. The product from 1-octanol is a carboxylic acid, which reacts with base to form a carboxylate ion which is soluble in water.

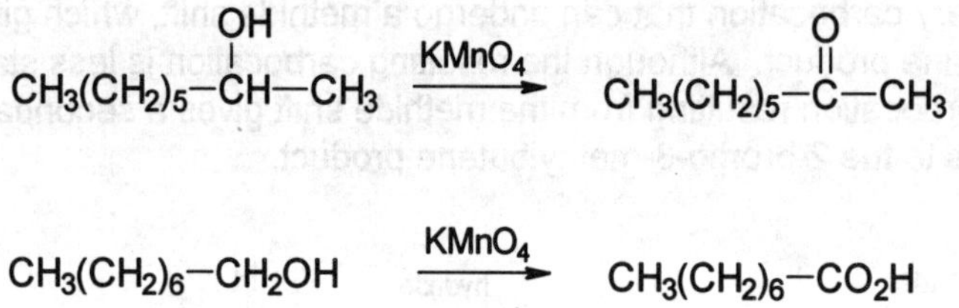

$$CH_3(CH_2)_5\overset{\underset{|}{OH}}{CH}-CH_3 \xrightarrow{KMnO_4} CH_3(CH_2)_5\overset{\overset{O}{\|}}{C}-CH_3$$

$$CH_3(CH_2)_6-CH_2OH \xrightarrow{KMnO_4} CH_3(CH_2)_6-CO_2H$$

16.32   **(a)** $CH_3(CH_2)_2-C\equiv C-CH_2\overset{\overset{O}{\|}}{C}CH_3$ **(b)** $CH_2=CH-CH=CH-CH_2-\overset{\overset{O}{\|}}{C}-H$

(c) $CH_3O-\bigcirc-\overset{\overset{O}{\|}}{C}-H$ **(d)**

16.33   **(a)**   **(b)** $H-\overset{\overset{O}{\|}}{C}-CH_2\overset{\underset{|}{CH_3}}{C}HCH_2-\overset{\overset{O}{\|}}{C}-H$

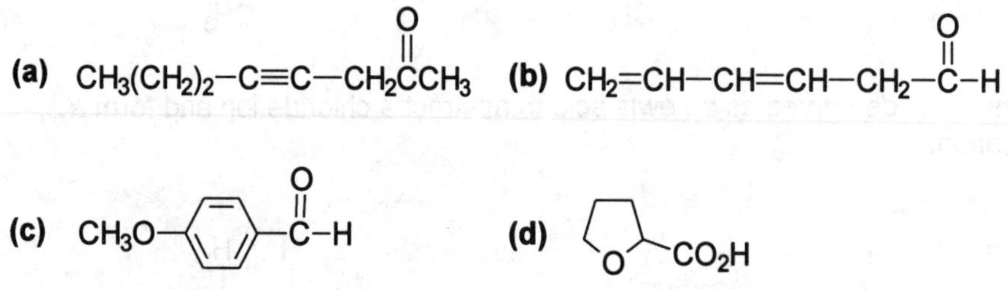

(c)   **(d)**

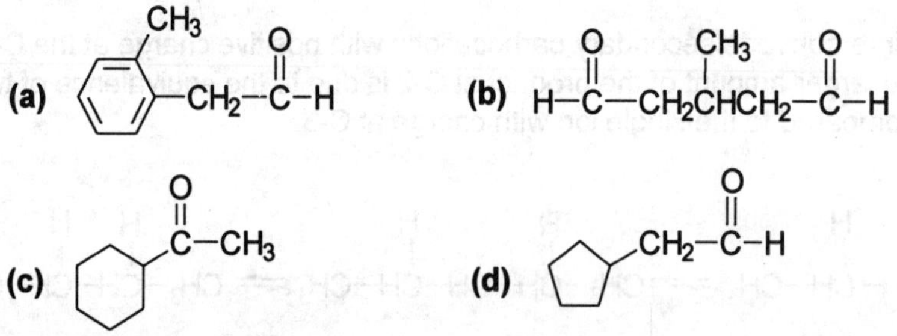

16.34 **(a)**

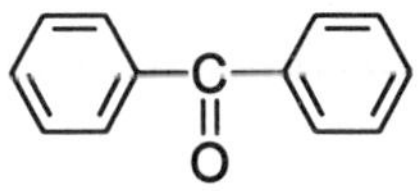

(b) 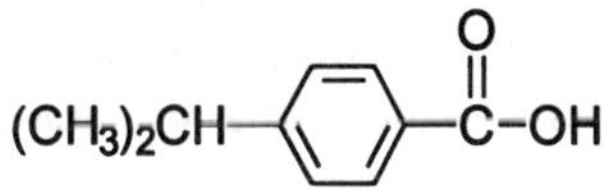 $(CH_3)_2CH$—⟨benzene ring⟩—$\overset{\overset{\displaystyle O}{\|}}{C}$—OH

(c)

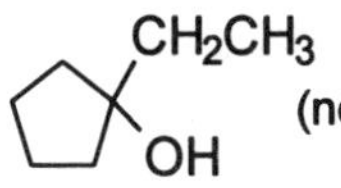

CH₂CH₃
OH
(no reaction)

(d)

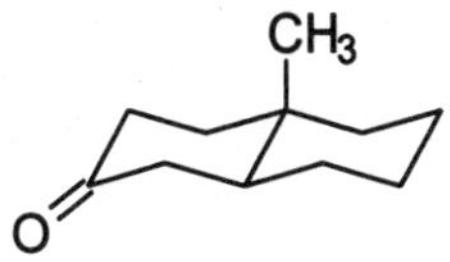

CH₃
O

16.35 PCC converts the primary alcohol into an aldehyde without further oxidation to a carboxylic acid.

CH_3CH_2 H
  C=C
 H $CH_2CH_2CH_2CH_2$—$\overset{\overset{\displaystyle O}{\|}}{C}$—H

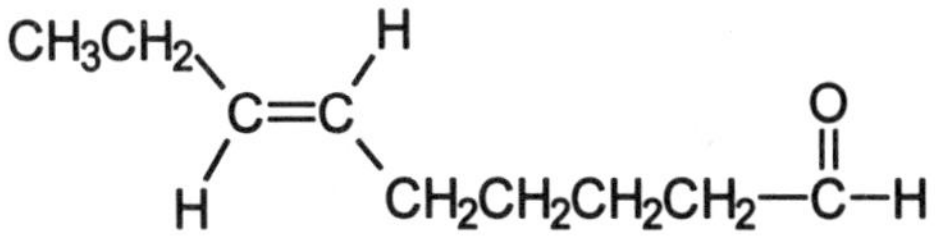

16.36 PCC converts the primary alcohol into an aldehyde without further oxidation to a carboxylic acid. It also converts the secondary alcohol into a ketone.

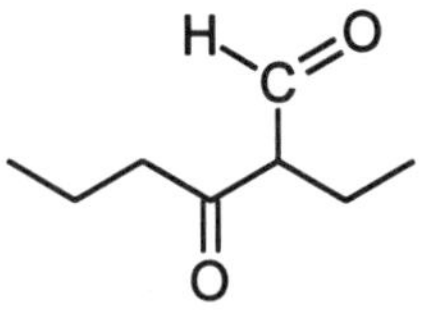

H—C=O

O

16.37 The slower rate for the compound with deuterium on the C-2 atom indicates that the C—D bond is cleaved in the rate determining step.

16.38 The rate determining step is not the formation of the chromium ester because the more hindered endo alcohol should react at the slower rate. The rate determining step involves cleavage of the C—H bond, which is exo in the endo alcohol and is sterically more accessible than the endo C—H bond in the exo alcohol.

16.39

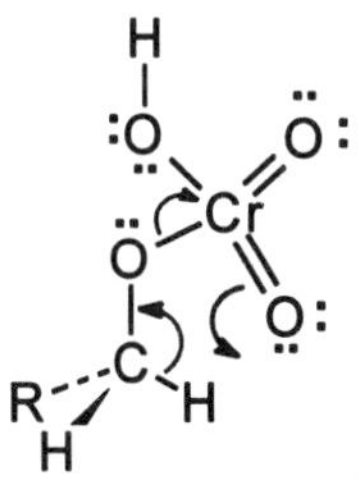

H
:O: :O:
   Cr
:O: :O:
 R—C—H
    H

The reaction involves five atoms in an intramolecular process, which is both strain free and highly probable.

16.40 The rate determining step involves cleavage of the C—H bond, as seen in Exercise 16.38 above. That bond is equatorial for the alcohol of the C ring of the steroid. It is oxidized faster than the equatorial alcohol of the A ring which has an axial C—H bond.

16.41 A pinacol-type rearrangement occurs. Protonation occurs at either of the two equivalent oxygen atoms and water readily leaves to give a tertiary carbocation. Migration of either a methyl or an ethyl group from the adjacent carbon atom gives a protonated carbonyl compound that then loses a proton to give a mixture of two products..

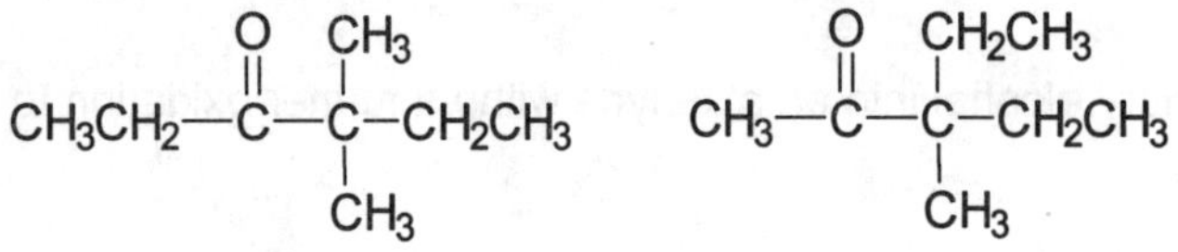

$$CH_3CH_2-\overset{\overset{O}{\|}}{C}-\overset{\overset{CH_3}{|}}{\underset{\underset{CH_3}{|}}{C}}-CH_2CH_3 \qquad CH_3-\overset{\overset{O}{\|}}{C}-\overset{\overset{CH_2CH_3}{|}}{\underset{\underset{CH_3}{|}}{C}}-CH_2CH_3$$

16.42 A pinacol-type rearrangement occurs. Protonation occurs at the hydroxyl group of the carbon atom bonded to the two phenyl groups, and water readily leaves to give a tertiary carbocation that is stabilized by the phenyl groups. Migration of a methylene group of the cyclopentane ring gives a hydroxy carbocation that then loses a proton.

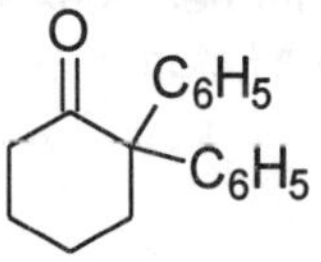

16.43 **(a)** 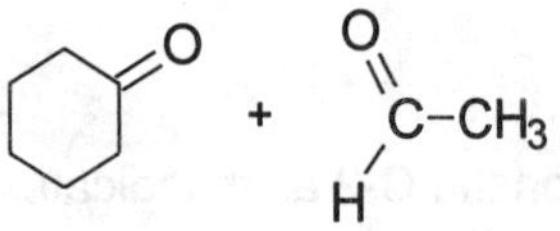   +   **(b)** 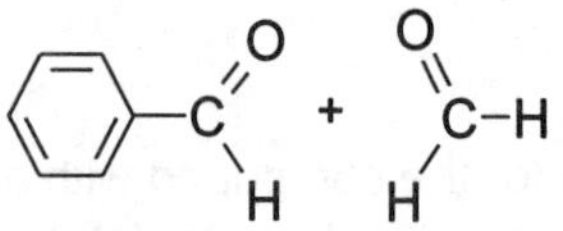   +

(c) 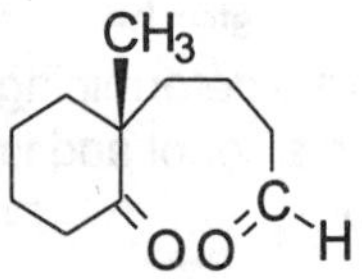   **(d)**

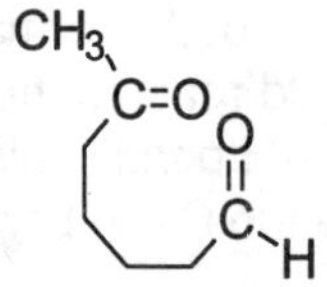

16.44 Compounds with two hydroxl groups in a syn periplanar arrangement or a cis arrangement can form the cyclic iodate ester more easily and then react to cleave the carbon-carbon bond. In (a), (b), and (c), the first compound reacts at the faster rate.

16.45 Osmium tetroxide reacts with a double bond to give a vicinal diol. Subsequent reaction with periodate cleaves the carbon-carbon bond that was originally a double bond. The compound has a double bond between C-9 and C-10 of an unsaturated carboxylic acid and could be either cis or trans. The trans compound is shown but the natural product is actually cis.

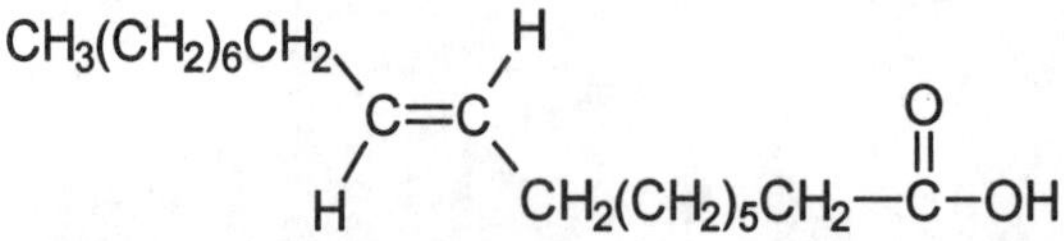

$$CH_3(CH_2)_6CH_2\underset{H}{\overset{}{\diagdown}}C=C\underset{CH_2(CH_2)_5CH_2-\overset{\overset{O}{\|}}{C}-OH}{\overset{H}{\diagup}}$$

16.46 Osmium tetroxide reacts with a double bond to give a vicinal diol. Subsequent reaction with periodate cleaves the carbon-carbon bond that was originally a double bond. The product is a dicarbonyl compound which indicates that the original double bond was contained in a ring that has been cleaved.

16.47 **(a)** The iodide ion is a better leaving group that the bromide ion. Thus, 1-iodohexane reacts faster than 1-bromohexane with the same nucleophile.

(b) A secondary benzylic carbocation results from 1-bromo-1-phenylethane which reacts faster than 1-bromo-2-phenylethane, which is a primary halogen compound.

(c) Both compounds are primary but 1-bromo-2,2-dimethylpropane is sterically hindered and reacts via an S_N2 mechanism at a lower rate.

16.48 DMF would be a better solvent because it is aprotic and does not decrease the nucleophilicity of the ethanoate ion as does the ethanol, which forms hydrogen bonds to the nucleophile.

16.49 Reaction via an S_N2 mechanism is impossible because the nucleophilic ethanoate ion cannot approach from the back side of C—Br bond. Although the center is tertiary, the carbocation required for the S_N1 mechanism cannot form because it cannot be planar due to restriction of the bicyclic ring. As a result the substitution reaction cannot occur.

16.50 Displacement of the axial bromo group by attack of the ethanoate occurs with inversion of configuration to give the trans ester which has an equatorial C—O bond. Hydrolysis of the ester occurs with nucleophilic attack at the carbonyl carbon atom and releases the alkoxy portion of the ester. The product is the *trans*-4-*tert*-butylcyclohexanol.

16.51 2-Butanol can be prepared by acid-catalyzed hydration of 1-butene or either of the isomeric 2-butenes. 2-Methyl-2-propanol can be prepared from 2-methyl-1-propene. 1-Butanol cannot be prepared by acid-catalyzed hydration of an alkene.

16.52 2-Propanol is the product of the acid-catalyzed hydration reaction. Synthesis of 1-propanol requires the anti-Markovnikov addition of water by a hydroboration-oxidation procedure that is not as easily adapted to industrial scale.

16.53 Addition of a proton to the C-1 atom gives a secondary carbocation with an adjacent quaternary center. A shift of either a methyl or ethyl group can occur to give a tertiary carbocation. The products are 2,3-dimethyl-3-pentanol and 2,3-dimethyl-2-pentanol.

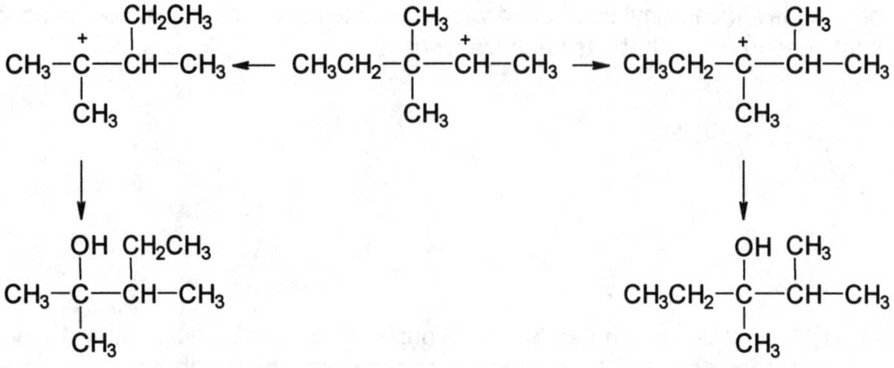

$$\underset{\underset{\textstyle CH_3}{|}}{\overset{\overset{\textstyle CH_2CH_3}{|}}{CH_3-\overset{+}{C}-CH-CH_3}} \longleftarrow \underset{\underset{\textstyle CH_3}{|}}{\overset{\overset{\textstyle CH_3}{|}}{CH_3CH_2-\overset{+}{C}-CH-CH_3}} \longrightarrow \underset{\underset{\textstyle CH_3}{|}}{\overset{\overset{\textstyle CH_3}{|}}{CH_3CH_2-\overset{+}{C}-CH-CH_3}}$$

$$\underset{\underset{\textstyle CH_3}{|}}{\overset{\overset{\textstyle OH}{|}\overset{\textstyle CH_2CH_3}{|}}{CH_3-C-CH-CH_3}} \qquad\qquad \underset{\underset{\textstyle CH_3}{|}}{\overset{\overset{\textstyle OH}{|}\overset{\textstyle CH_3}{|}}{CH_3CH_2-C-CH-CH_3}}$$

16.54 Addition of a proton can occur at either the C-1 or C-2 atoms to give secondary carbocations in both cases.

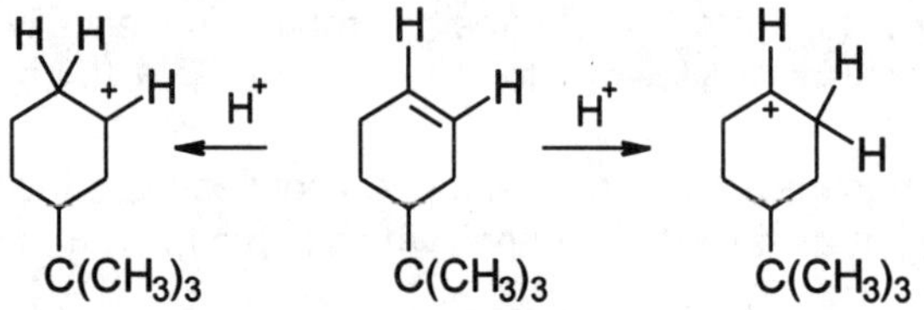

Addition of water then occurs at either the C-2 or C-1 center of the original compound. Attack from either side of each carbocation generates a mixture of cis and trans isomers. The products are *cis*- and *trans*-4-tert-butylcyclohexanol, as well as *cis*- and *trans*-3-tert-butylcyclohexanol.

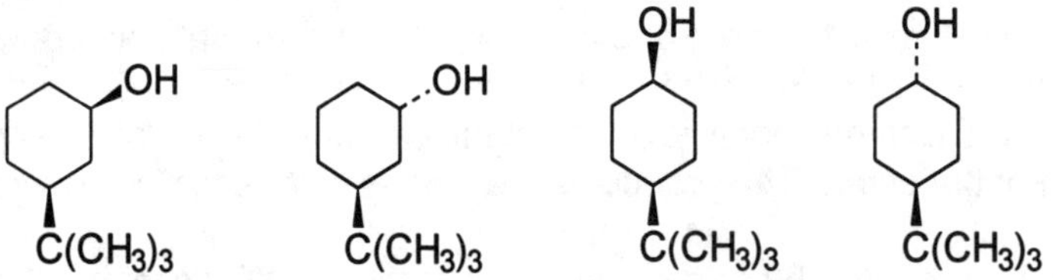

16.55 (a) 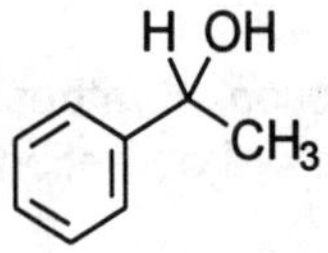 (b) 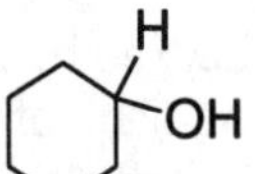 (c)

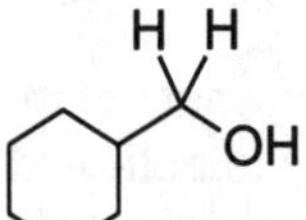

16.56 (a) 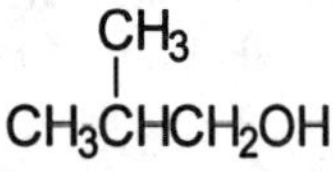 $CH_3\overset{\underset{\textstyle CH_3}{|}}{CH}CH_2OH$ (b) 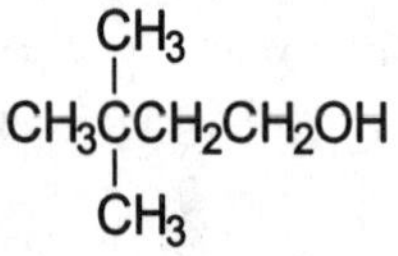 $CH_3\overset{\underset{\textstyle CH_3}{|}}{\overset{\overset{\textstyle CH_3}{|}}{C}}CH_2CH_2OH$ (c) $CH_3-\overset{\underset{\textstyle }{}}{\overset{\overset{\textstyle OH}{|}}{CH}}-\overset{\underset{\textstyle CH_3}{|}}{\overset{\overset{\textstyle CH_3}{|}}{C}}-CH_3$

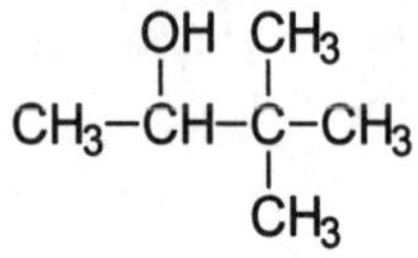

16.57 Attack of hydride ion can occur from either side of the plane of the carbonyl group although not with equal probability. A mixture of geometric isomers occurs in both (a) and (b).

16.58 Each compound is symmetrical with respect to the plane of the carbonyl group. Only one alcohol can result from reduction which occurs via attack of a hydride ion on the carbonyl carbon atom.

16.59 Attack of the hydride ion should occur at the more sterically accessible position from "below" the average planes of the rings, which avoids the axial methyl groups. In the product from (a) the hydroxyl group is equatorial, and in the product from (b) it is axial.

16.60 Sodium borohydride reduces only aldehydes and ketones. The ester group is unaffected.

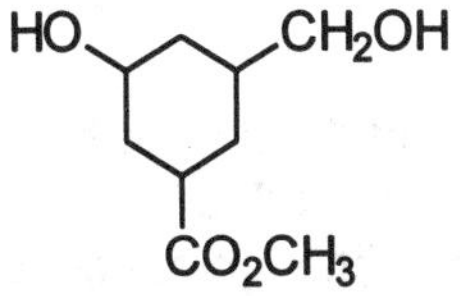

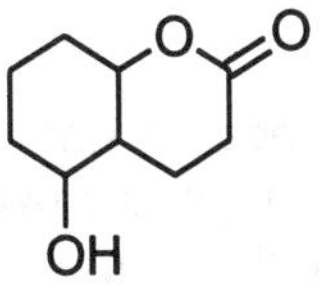

Lithium aluminum hydride reduces esters as well as aldehydes and ketones.

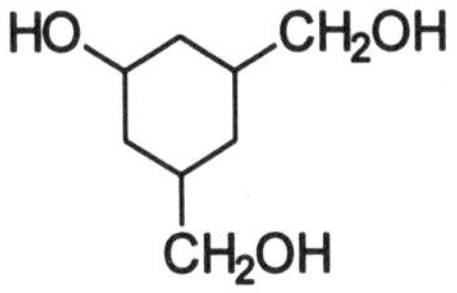

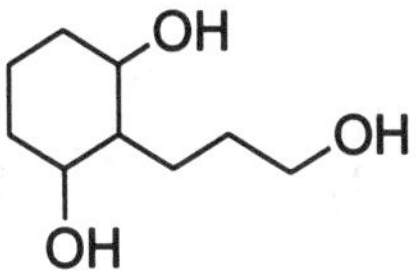

16.61 The product of each reaction is that which corresponds to Markovnikov addition of water.

(a) 2-methyl-2-propanol   **(b)** 2,3-dimethyl-2-butanol
(c) 2-butanol   **(d)** 2-propanol

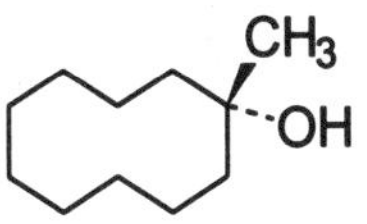

16.62 The product of the reaction is that which corresponds to Markovnikov addition of water.

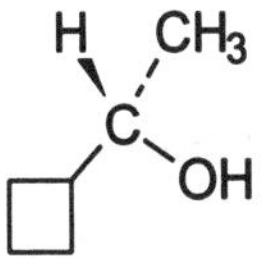

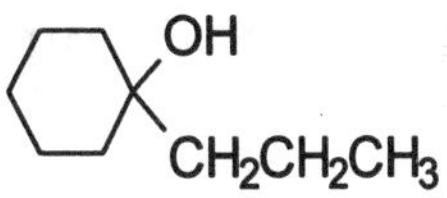

16.63 The product of the reaction for both (a) and (b) is that which corresponds to Markovnikov addition of water.

(a) 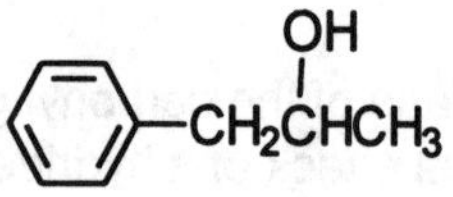 C$_6$H$_5$—CH$_2$CHCH$_3$ with OH

(b) 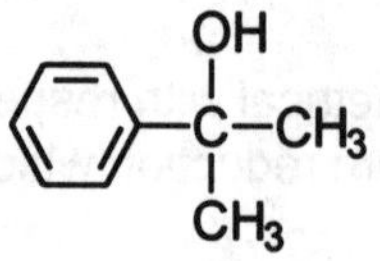 C$_6$H$_5$—C(CH$_3$)(CH$_3$)—CH$_3$ with OH

For (c) the regioselectivity is controlled by the stability of the positive charge in the mercurinium ion at the benzylic position. The product is

(c) 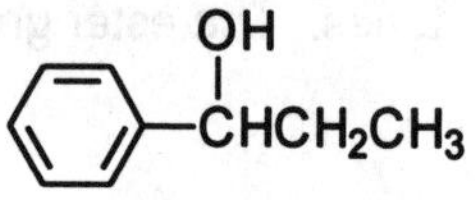 C$_6$H$_5$—CHCH$_2$CH$_3$ with OH

16.64 There is no regioselectively because each carbon atom of the double bond has a single alkyl group that is part of the ring and a hydrogen atom bonded to it. In addition, the addition could occur from either side of the double bond although not necessarily in equal amounts. Thus, four products are possible.

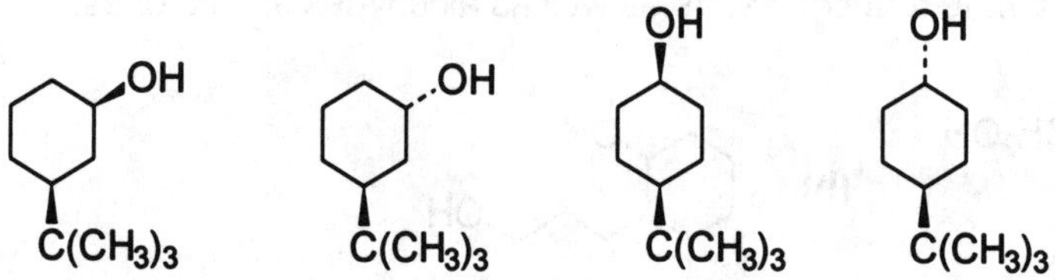

16.65 The product of each reaction is that which corresponds to anti-Markovnikov addition of water.

(a) 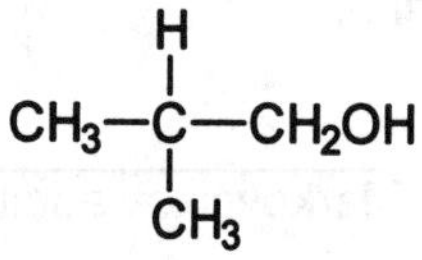 CH$_3$—C(CH$_3$)(H)—CH$_2$OH

(b) CH$_3$—C(CH$_3$)(H)—C(CH$_3$)(OH)—CH$_3$

(c) CH$_3$CH$_2$—C(H)(H)—CH$_2$OH

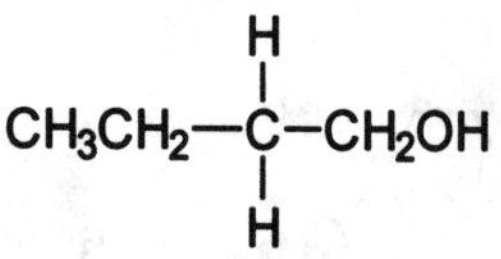

(d) 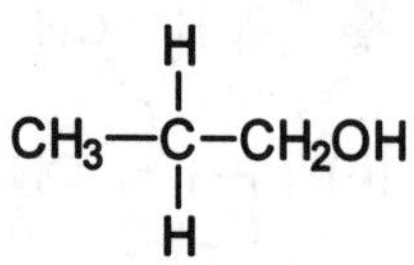 CH$_3$—C(H)(H)—CH$_2$OH

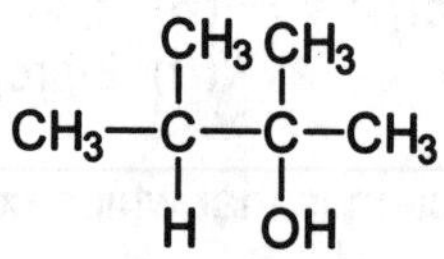

16.66 The product of each reaction is that which corresponds to anti-Markovnikov addition of water.

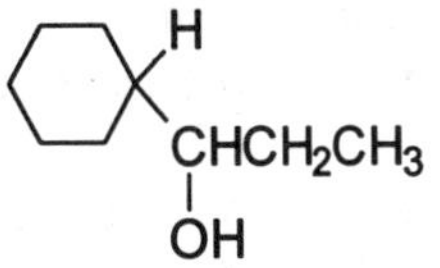

16.67 The product of each reaction is that which corresponds to anti-Markovnikov addition of water.

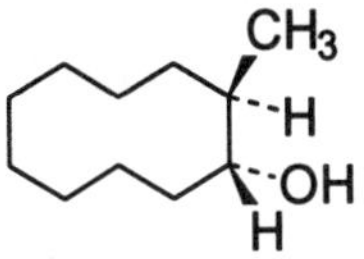

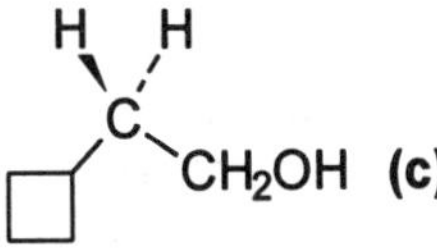

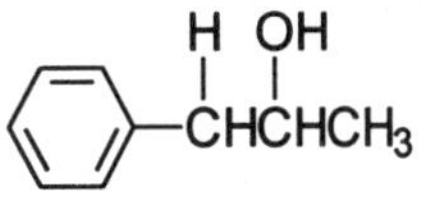

16.68 The product of the reaction is that which corresponds to anti-Markovnikov addition of water.

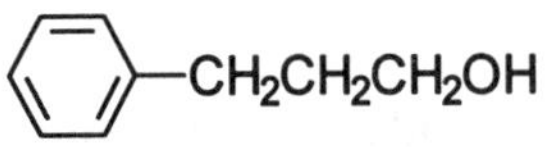

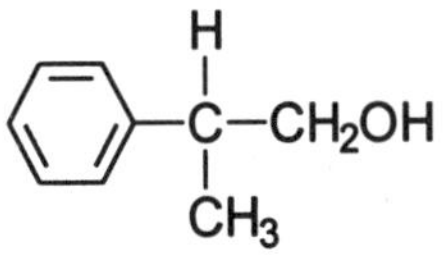

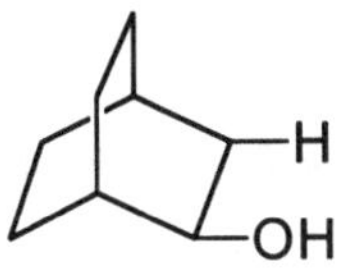

16.69

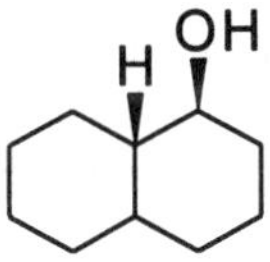

16.70 No, because the indicated alcohol requires the Markovnikov addition of water which is opposite that obtained from the hydroboration-oxidation method.

16.71 (a) Use the Grignard reagent of bromoethane and propanal.
    (b) Use the Grignard reagent of iodomethane and 2-methylpropanal or the Grignard reagent of 2-bromopropane and ethanal.
    (c) Use the Grignard reagent of bromoethane and 2-methylpropanal or the Grignard reagent of 2-bromopropane and propanal.
    (d) Use the Grignard reagent of bromoethane and 3-methylbutanal or the Grignard reagent of 1-bromo-2-methylpropane and propanal.

16.72 **(a)** Use the Grignard reagent of bromocyclopentane and formaldehyde.
(b) Use the Grignard reagent of iodomethane and 4,4-dimethylcyclohexanone.
(c) Use the Grignard reagent of bromoethane and benzaldehyde or the Grignard reagent of bromobenzene and propanal.
(d) Use the Grignard reagent of bromocyclohexane and 2-butanone or the Grignard reagent of iodomethane and 1-cyclohexyl-1-propanone or the Grignard reagent of iodoethane and 1-cyclohexyl-1-ethanone.

16.73 **(a)** Prepare the Grignard reagent of bromocyclopentane and add it to propanone, followed by hydrolysis.
(b) Prepare the Grignard reagent of bromocyclopentane and add it to ethanal, followed by hydrolysis.
(c) Prepare the Grignard reagent of 1-bromooctane and add it to formaldehyde, followed by hydrolysis.
(d) Prepare the Grignard reagent of bromoethane and add it to pentanal, followed by hydrolysis.

16.74 **(a)** Prepare the Grignard reagent of bromobenzene and add it to cyclopentanone, followed by hydrolysis.
(b) Prepare the Grignard reagent of bromobenzene and add it to 3-hexanone, followed by hydrolysis.
(c) Prepare the Grignard reagent of bromobenzene and add it to octanal, followed by hydrolysis.
(d) Prepare the Grignard reagent of bromobenzene and add it to formaldehyde, followed by hydrolysis.

16.75 There are three possible combinations of Grignard reagent and a carbonyl compound.

(1) Prepare the Grignard reagent of an ethynyl group by reacting acetylene with a Grignard reagent such as methyl magnesium bromide. Then react it with 2-butanone, followed by hydrolysis.
(2) Prepare the Grignard reagent of iodoethane and react with it with 3-butyn-2-one, followed by hydrolysis.
(3) Prepare the Grignard reagent of iodomethane and react with it with 1-pentyn-3-one, followed by hydrolysis .

16.76 The ethynyl Grignard reagent adds to the carbonyl carbon atom from the least hindered side of the plane of the carbonyl group. Attack from the "top" of the of the ring is more hindered due to the axial methyl group at the ring juncture.

16.77 **(a)** 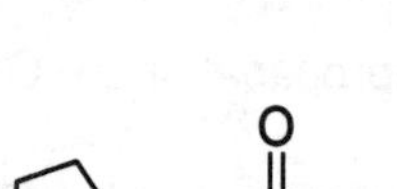 ⬠—CH$_2$CCH$_3$ **(b)** ⬡—CH$_2$—C—H **(c)** ⬜ CH$_2$CH$_2$—C—OH

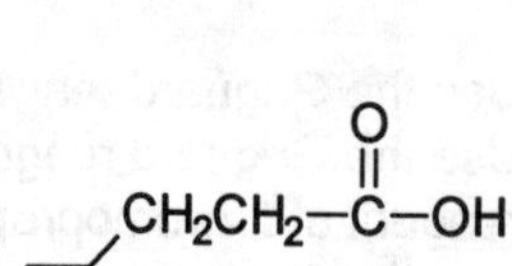

16.78 **(a)** 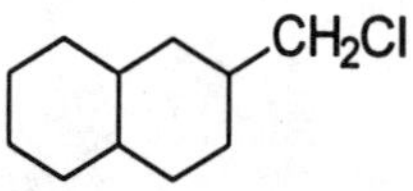 (decalin)—CH$_2$Cl **(b)** 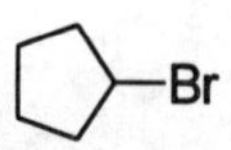 ⬠—Br **(c)** 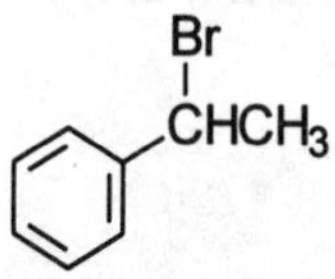 ⬡—CHCH$_3$ with Br

16.79 Reduce the carbon-carbon double bond using hydrogen and a platinum catalyst. Then oxidize the secondary alcohol using PCC or the Jones reagent.

16.80 Reduce the carbon-carbon double bond using hydrogen and a platinum catalyst. Then reduce the ketone with sodium borohydride.

16.81 React the ketone with the methyl Grignard reagent to obtain 1-methylcyclohexanol. Dehydration of the alcohol using an acid catalyst yields 1-methylcyclohexene as the major product. Methylenecyclohexane is the minor product.

16.82 React the ketone with the ethyl Grignard reagent to obtain cyclohexanol with an ethyl group bonded to the carbon bearing the hydroxyl group.. Dehydration of the alcohol using an acid catalyst yields an alkene which, when hydrogenated using hydrogen and a platinum catalyst, gives the desired product.

16.83

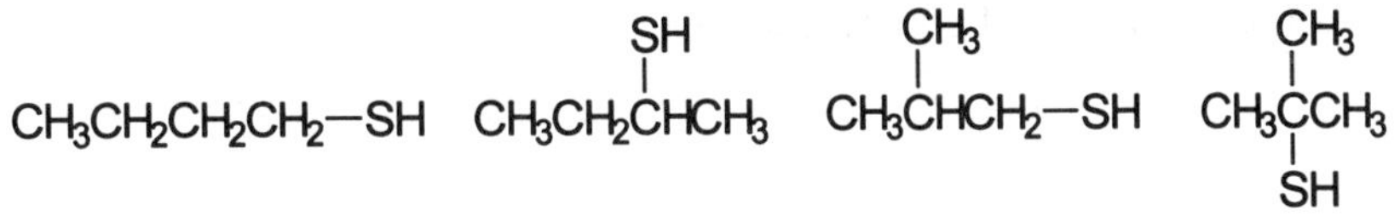

$CH_3CH_2CH_2CH_2{-}SH$ $CH_3CH_2\overset{\overset{\displaystyle CH_3}{|}}{C}HCH_3$ $CH_3\overset{\overset{\displaystyle CH_3}{|}}{C}HCH_2{-}SH$ $CH_3\overset{\overset{\displaystyle CH_3}{|}}{\underset{\underset{\displaystyle SH}{|}}{C}}CH_3$

16.84 $CH_3CH_2CH_2{-}SH$ $CH_3\overset{\overset{\displaystyle SH}{|}}{C}HCH_3$ $CH_3CH_2{-}S{-}CH_3$

16.85 (a) $CH_3CH_2CH_2{-}SH$ (b) $CH_3CH_2\overset{\overset{\displaystyle SH}{|}}{\underset{\underset{\displaystyle CH_3}{|}}{C}}HCHCH_3$ (c) ⬠—SH

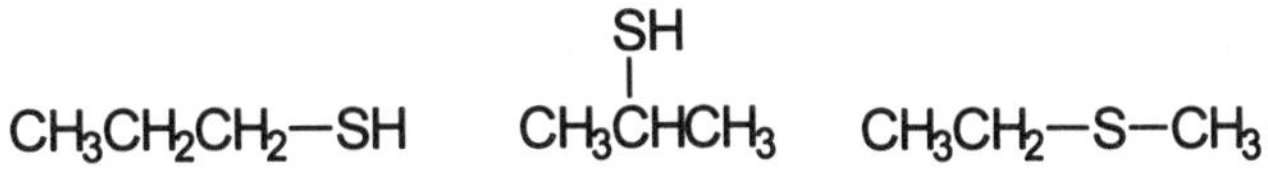

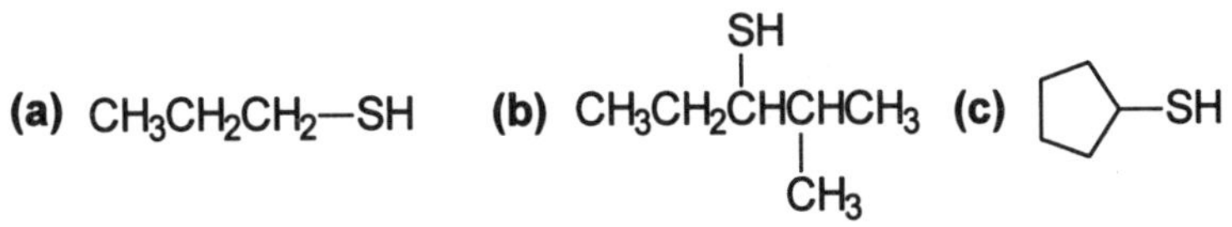

16.86 (a) $CH_3\overset{\overset{\displaystyle SH}{|}}{C}HCH_3$ (b) $CH_3\overset{\overset{\displaystyle CH_3}{|}}{C}HCH_2{-}SH$ (c) ⬜$\overset{\displaystyle SH}{}$

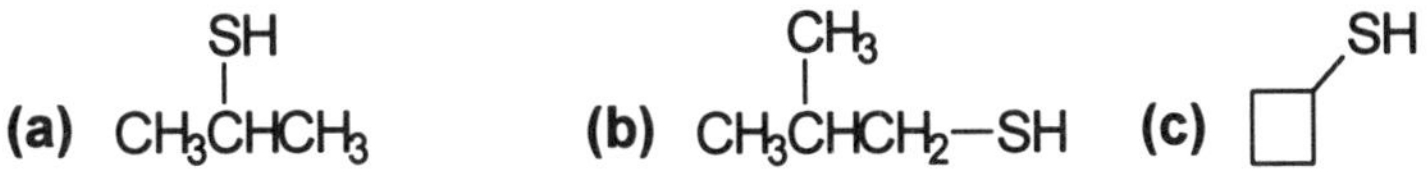

16.87 The strong base reacts with the thiol to give a thiolate ($CH_3CH_2CH_2S^-$) that is nonvolatile.

16.88 The S—H group does not form hydrogen bonds. As a result, the intermolecular attractive forces of both the thioether and the thiol are similar. There are similar London forces in each of the compounds because they are isomers.

16.89 Use either of the following combinations of a thiol and a haloalkane in the presence of base.

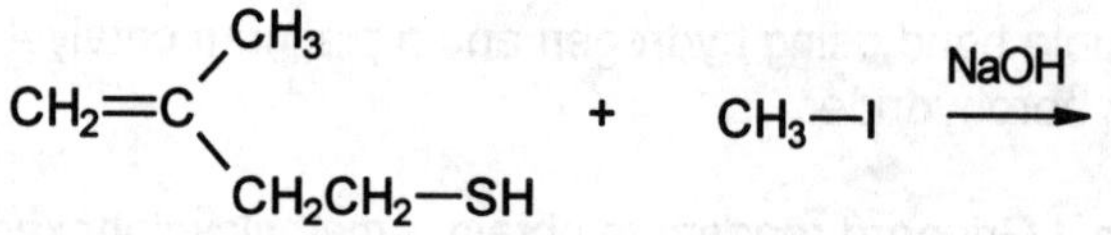

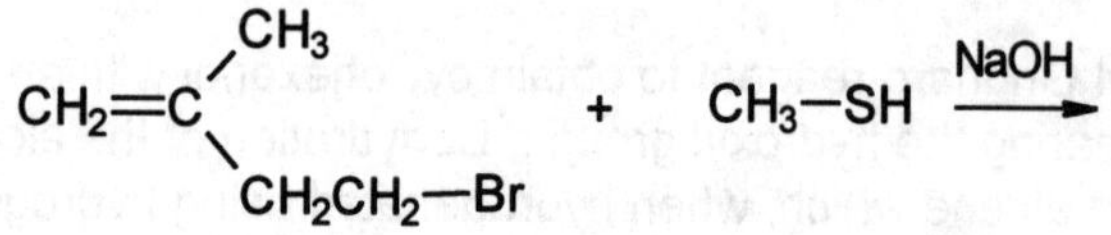

$$CH_2\!\!=\!\!C\!\!<^{CH_3}_{CH_2CH_2-SH} \quad + \quad CH_3-I \quad \xrightarrow{\ NaOH\ }$$

$$CH_2\!\!=\!\!C\!\!<^{CH_3}_{CH_2CH_2-Br} \quad + \quad CH_3-SH \quad \xrightarrow{\ NaOH\ }$$

16.90 Add HBr to the double bond of 3-methyl-1-butene in the presence of peroxide to achieve the anti-Markovnikov addition. React the product 1-bromo-3-methylbutane with thiourea to give an isothiouronium salt and hydrolyze it.

17

Ethers and Epoxides

The chemistry of ethers has substantially less variety than the chemistry of alcohols, because several of the reactions of alcohols involve the O—H bond, namely dehydration and oxidation reactions. Thus, ethers are said to be less reactive than alcohols. However, if comparable reactions are considered, such as substitution, then alcohols and ethers have similar reactivities. The specific set of reactions that occur with epoxides are the result of ring strain, which is manifested by formation of ring-opened products.

Keys to the Chapter

17.1 Structure of Ethers

Alcohols and ethers both contain oxygen atoms bonded to two other atoms by single bonds. In the case of alcohols, one bond is to a carbon atom which is part of an alkyl group or a cycloalkyl group, and the second bond is to a hydrogen atom. Ethers contain an oxygen atom bonded to two alkyl groups, two aryl groups, or one of each. The geometry of ethers resembles that of alkanes, with the substitution of a methylene carbon atom by an sp³-hybridized oxygen atom. Conformations of ethers resemble those of alkanes. The two nonbonding electron pairs, which are directed to the corners of a tetrahedron, will be important to depict in reaction mechanisms.

17.2 Nomenclature of Ethers

The common names of simple ethers are based on the names of the alkyl or aryl groups bonded to the oxygen atom. The name results from listing the alkyl (or aryl) groups in alphabetical order and appending the name ether.

The IUPAC name is based on the longest carbon chain bonded to the oxygen atom. The smaller group bonded to the oxygen atom is identified as an **alkoxy group** and is regarded as a substituent on the longer chain.

The three-, five-, and six-membered cyclic ethers have common names. Three-membered ring compounds are called **epoxides** of the corresponding alkene from which they may be synthesized. The common names of five- and six-membered ring compounds are called tetrahydrofurans and tetrahydropyrans, respectively. In the IUPAC system, each ring size has a specific name. The names for cyclic ethers having three-, four-, five-, and six-membered rings are oxirane, oxetane, oxolane, and oxane, respectively. The oxygen atom in each ring is assigned the number 1, and the ring is numbered in the direction that gives the lowest numbers to substituents.

17.3 Physical Properties of Ethers

Ethers have two polar C—O bonds and have substantial dipole moments. They are more polar than alkanes, but less polar than alcohols. Ethers do not have an O—H bond and cannot serve as hydrogen bond donors. They can, however, serve as hydrogen bond acceptors, which makes the low molecular weight ethers soluble in water. Ethers have boiling points substantially lower than those of alcohols of comparable molecular weight, because they cannot form intermolecular hydrogen bonds. The boiling points of ethers are very close to the boiling points of alkanes of similar molecular weight.

Ethers are excellent solvents for both nonpolar and nonpolar solutes. They are aprotic, so they are used as solvents for reagents that react with acidic protons, as is the case for the Grignard reagent. Polyethers readily dissolve polar compounds and hydrogen bond donors.

17.4 Industrial Synthesis of Ethers

Symmetrical ethers can be synthesized by dehydration of alcohols. Review the reaction mechanism. It isn't much different from other nucleophilic substitution reactions. First protonate the oxygen atom of one molecule to prepare oxygen to leave as water. Then attack that protonated alcohol in a nucleophilic substitution reaction with a second molecule of the alcohol.

A second industrial synthesis involves the addition of an alcohol to an alkene. This reaction resembles that of hydration. First protonate the alkene and then have the nucleophilic alcohol react with the carbocation. The regiospecifity of the reaction is described by the term Markovnikov addition.

17.5 Alkoxymercuration-Demercuration of Alkenes

Alkoxymercuration occurs by the same mechanism as oxymercuration. The only difference is the alkyl group bonded to the oxygen atom of an alcohol compared to the hydrogen atom bonded to the oxygen atom of water. The regiospecificity is described by the term Markovnikov addition.

17.6 The Williamson Ether Synthesis

The Williamson ether synthesis is yet another example of a parallel reaction of an alcohol compared to water. Hydroxide ion is a nucleophile that can displace a halide ion from an alkyl halide to give an alcohol. Similarly an alkoxide ion is a nucleophile that can displace a halide ion from an alkyl halide to give an ether. Based on that fact we expect the reaction to occur with inversion of configuration and to be limited by possible competing elimination reactions.

Intramolecular Williamson ether synthesis occurs at rates that depend on the number of atoms in the transition state. The rates are affected by the probability of the alkoxide approaching the carbon atom bearing the halide ion, as well as the strain of the resulting ring compound. The resulting order of reactivity given in terms of the ring size is 3 > 5 > 6 > 4.

17.7 Reactions of Ethers

Both ethers and alcohols can act as bases, because they have two lone pairs of electrons on the oxygen atom. They are both very weak bases and can only be protonated to form the conjugate acid, an oxonium ion, by a strong acid. The formation of an oxonium ion is analogous to the reaction of water with a strong acid to give the hydronium ion. Oxonium ions are intermediates in many reactions catalyzed by strong acids in the reactions of both ethers and alcohols.

Ethers react with strong acids such as HBr and HI to give cleavage products. The reaction proceeds by a two-step process in which first the oxygen atom is protonated and then the halide ion attacks one of the alkyl groups to displace an alcohol by an S_N2 process. The alkyl group attacked by the halide ion is controlled by the reactivity order 1° > 2° > 3°. The product alcohol can react with additional HX to give a second mole of a haloalkane.

17.8 Ethers as Protecting Groups

Protecting groups are provided by reagents that easily form derivatives of the functional group to be "protected" but can also be easily removed when required. A protecting group is used to render a functional group unreactive toward specific reagents that are required to transform a second functional group in the molecule.

Alcohols are protected by preparing silyl ethers of the general formula R'—O—SiR_3. The compound is obtained by reacting an alcohol R'—OH with a chlorosilane Cl—SiR_3. The silyl ether is cleaved by fluoride ion to liberate the alcohol after other transformation are completed.

17.9 Synthesis of Epoxides

Epoxides are synthesized by one of two methods. Epoxidation is the reaction of an alkene with certain peroxyacids such as MCPBA or MMPP. The stereochemistry of the groups of the alkene is retained in the epoxide. Halohydrins undergo an intramolecular Williamson ether synthesis. Not that the reaction involves inversion of configuration at the center where the halide ion is displaced. Recall that halohydrins are formed by addition of halogen to a double bond in aqueous solution.

17.10 Reactions of Epoxides

The ring strain of the three-membered epoxide ring results in reactions based on "displacement" of the oxygen atom from one of the two carbon atoms. Of course, the oxygen atom doesn't leave completely, because it is still bonded to the other carbon atom. The regiochemistry of the ring opening and the stereochemistry of the product depend on whether the reaction occurs under basic or acidic conditions.

Nucleophiles displace an alkoxide ion, a reaction not observed in acyclic compounds. As in the case of other S_N2 reactions, the order of reactivity for a substrate with a nucleophile is primary > secondary > tertiary. The resulting product has the nucleophile and the hydroxyl group on adjacent carbon atoms. The nucleophile is on the less substituted carbon atom. Grignard reagents react with epoxides to give an alcohol containing two additional carbon atoms between the alkyl group and the carbon atom bearing the hydroxyl group in the product. Thus, for the reaction of a Grignard reagent, RMgBr, with ethylene oxide, the product is RCH_2CH_2OH.

Under acidic conditions, the oxygen atom is protonated to give a cyclic intermediate that resembles the bromonium ion and the mercurinium ion in other mechanisms. Subsequently, the nucleophilic reagent attacks the more substituted carbon atom because it has the greater partial positive charge. Thus, the nucleophile is bonded to the more substituted carbon atom in the product.

The stereochemistry of the ring opening of an epoxide is easily predicted. Nucleophilic attack of a reagent such as methoxide ion occurs with inversion of configuration at the least substituted carbon atom. The stereochemistry of the other carbon atom of the original epoxide is unchanged, because no bond to that center is broken in the reaction. In a substituted epoxide, the ring opening reaction under acidic conditions occurs by inversion of configuration at the more substituted center where the nucleophile attacks. The configuration at the least substituted center is unchanged.

17.11 Sulfides

The sulfur analogs of ethers are **sulfides**. They are named using alkylthio groups as substituents to the parent chain. Cyclic sulfides have common names.

Sulfides are prepared by reaction of a thiolate, the conjugate base of a thiol, with an alkyl halide. Because the thiolate ion is less basic and is a better nucleophile than an alkoxide. The sulfur analog of a Williamson synthesis has fewer complications due to elimination reactions.

Sulfide are oxidized to **sulfoxides** and then to **sulfones**.

17.12 Spectroscopy of Compounds with C—O and C—S Bonds

Infrared spectroscopy is usually not used to confirm the presence of either C—O or C—S bonds, because the stretching vibrations occur in a region complicated by other absorptions. The O—H stretching vibration of alcohols is easily seen as a strong broad absorption on the "left" of the spectrum at high wavenumbers.

The chemical shift of hydrogen atoms bonded to the carbon atom bearing the oxygen atom of either alcohols or ethers occurs in the 3—4 δ region. The chemical shift of hydrogen atoms bonded to the carbon atom bearing the sulfur atom of either thiols or sulfides is less deshielded and occurs in the 2.5 δ region. Both O—H and S—H hydrogen atoms have variable chemical shifts depending on concentration.

The α carbon atom of an alcohol or an ether has a chemical shift that reflects the deshielding of the electronegative oxygen atom. The deshielding due to a sulfur atom is less.

Summary of Reactions

1. Synthesis of Ethers by Addition of an Alcohol to an Alkene

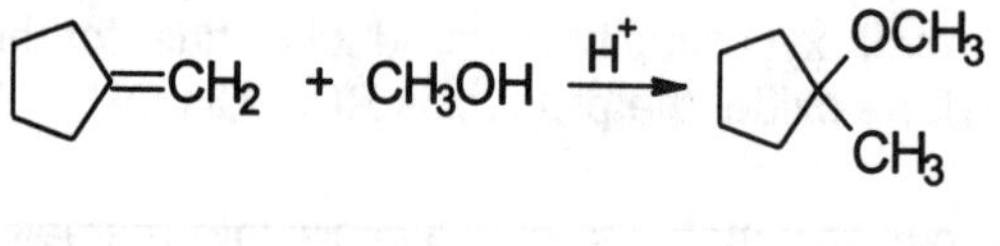

2. Synthesis of Ethers by Alkoxymercuration-Demercuration of Alkenes

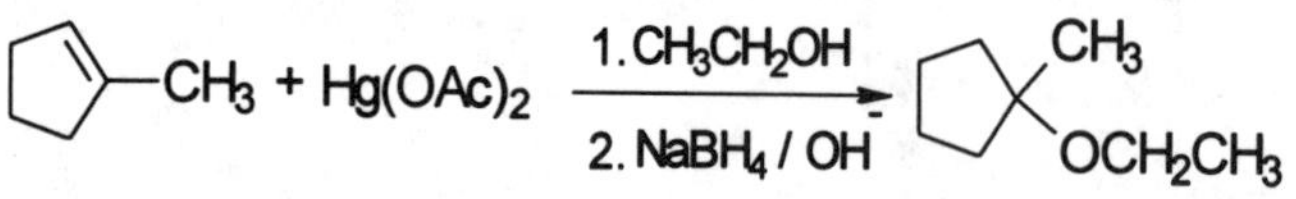

3. Williamson Ether Synthesis

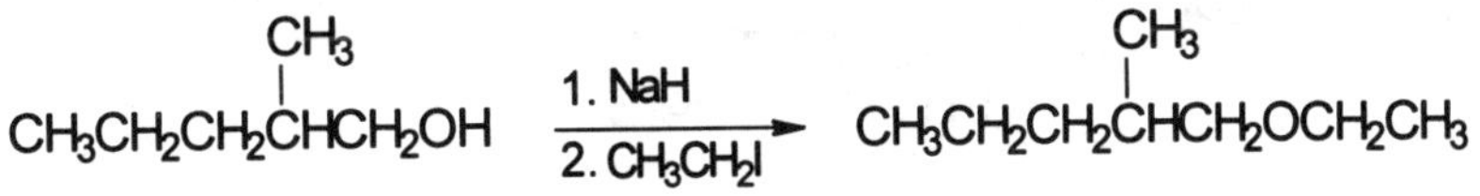

$$CH_3CH_2CH_2\overset{\underset{\textstyle CH_3}{|}}{C}HCH_2OH \quad \xrightarrow[\text{2. CH}_3\text{CH}_2\text{I}]{\text{1. NaH}} \quad CH_3CH_2CH_2\overset{\underset{\textstyle CH_3}{|}}{C}HCH_2OCH_2CH_3$$

4. Cleavage of Ethers

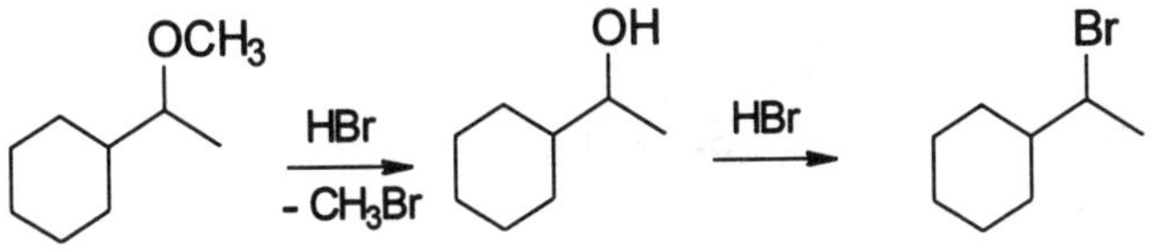

5. Use of Silyl Ethers as Protecting Groups

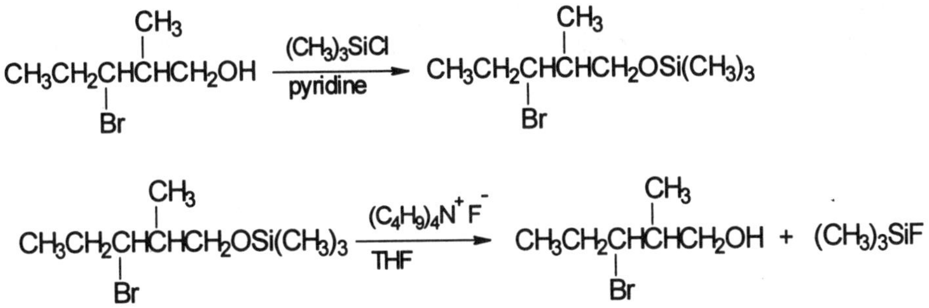

$$CH_3CH_2\overset{\underset{\textstyle Br}{|}}{\overset{\overset{\textstyle CH_3}{|}}{C}}HCHCH_2OH \quad \xrightarrow[\text{pyridine}]{(CH_3)_3SiCl} \quad CH_3CH_2\overset{\underset{\textstyle Br}{|}}{\overset{\overset{\textstyle CH_3}{|}}{C}}HCHCH_2OSi(CH_3)_3$$

$$CH_3CH_2\overset{\underset{\textstyle Br}{|}}{\overset{\overset{\textstyle CH_3}{|}}{C}}HCHCH_2OSi(CH_3)_3 \quad \xrightarrow[\text{THF}]{(C_4H_9)_4N^+F^-} \quad CH_3CH_2\overset{\underset{\textstyle Br}{|}}{\overset{\overset{\textstyle CH_3}{|}}{C}}HCHCH_2OH \; + \; (CH_3)_3SiF$$

6. Synthesis of Epoxides

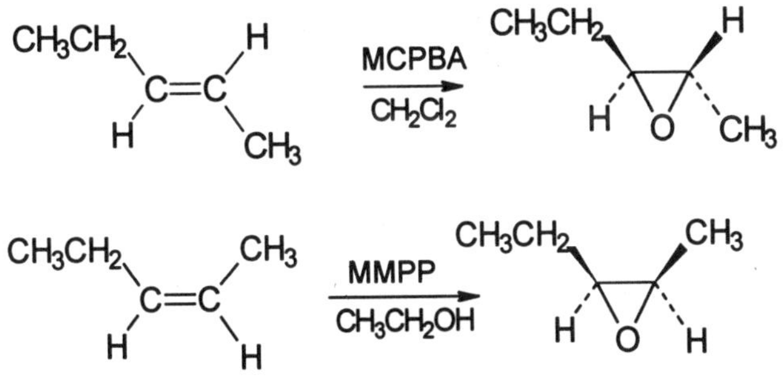

7. Ring Cleavage of Epoxides

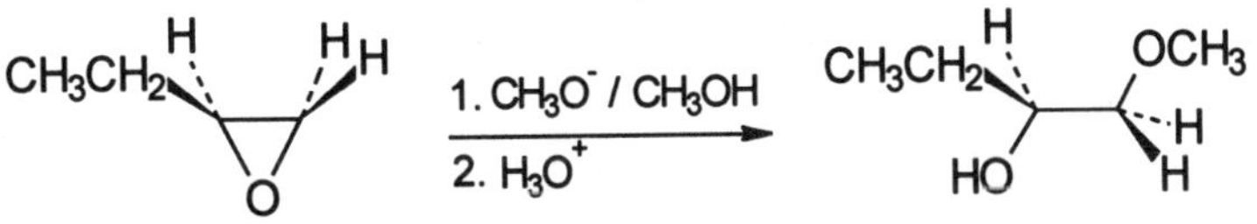

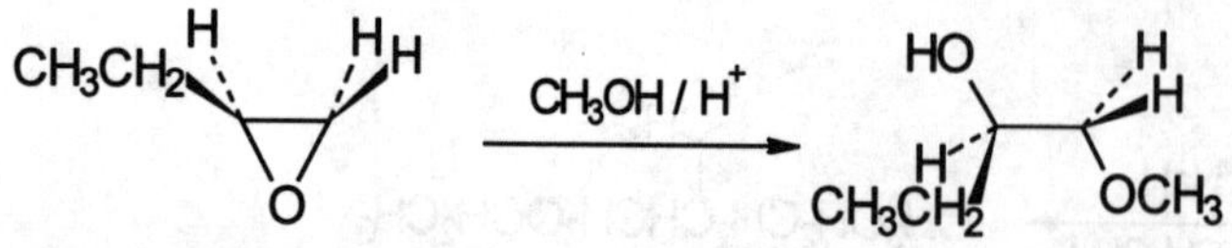

8. Synthesis of Sulfides

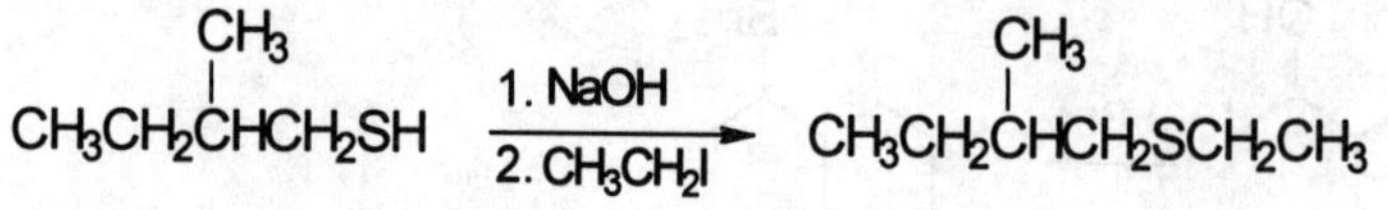

9. Oxidation of Sulfides

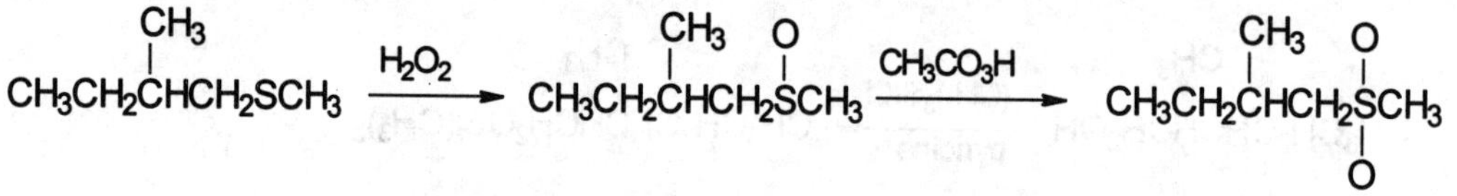

Solutions to Exercises

17.1 **(a)** CH₃CH₂CH₂—O—CH₃   CH₃CH₂—O—CH₂CH₃   CH₃CH—O—CH₃ (with CH₃ substituent above)

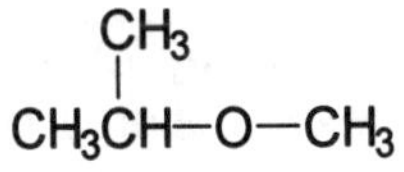

(b) CH₃CH₂CH₂CH₂—OCH₃   CH₃CH₂CHCH₃ (OCH₃ above)   CH₃CHCH₂—OCH₃ (CH₃ above)   CH₃CCH₃ (CH₃ above, OCH₃ below)

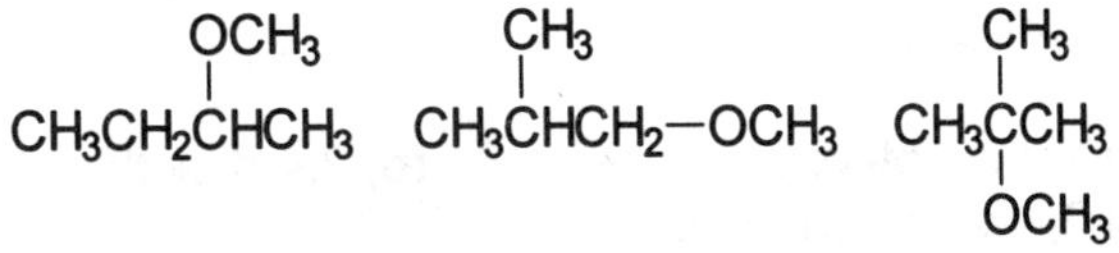

(c)

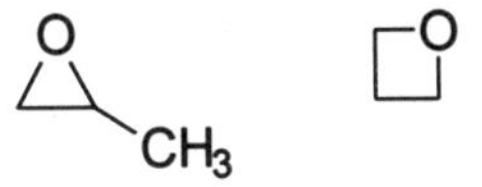

17.2 **(a)**

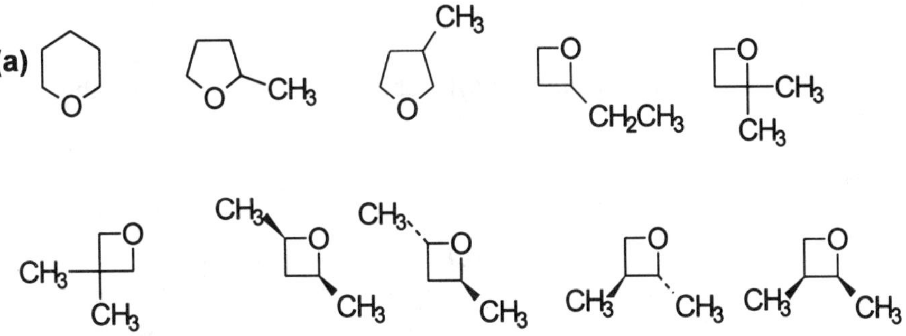

(b)

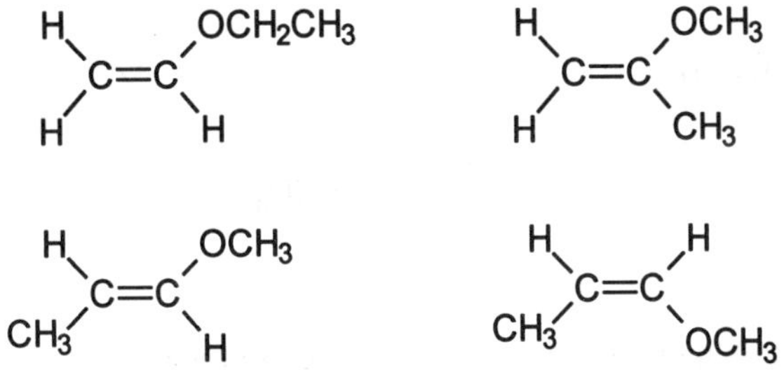

(c)

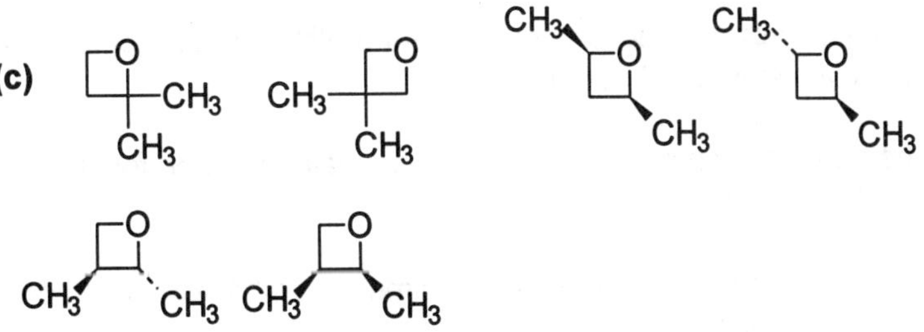

17.3   (a) dicyclopentyl ether   (b) phenyl propyl ether
(c) cyclopentyl propyl ether

17.4   (a) *tert*-butyl phenyl ether   (b) benzyl *tert*-butyl ether
(c) cyclohexyl propargyl ether

17.5   (a) 2-methoxypentane   (b) 2-methoxy-4-methylpentane
(c) 3-ethoxyhexane

17.6   (a) 1,2-dimethoxypentane   (b) 2,4-dimethoxypentane
(c) 2,4-diethoxyhexane

17.7   (a)   (b)

17.8   (a) divinyl ether   (b) difluoromethyl 1-chloro-2,2,2-trifluoroethyl ether

17.9   (a)   (b)   (c)   (d)

17.10   (a)   (b)   (c)   (d)

17.11   Diethyl ether is somewhat soluble in ether. The added oxygen atom of 1,4-dioxane increases the extent of hydrogen bonding with water, leading to increased solubility. The ratio of carbon to oxygen atoms decreases from 4:1 in diethyl ether to 2:1 in 1,4-dioxane.

17.12   Ethoxybenzene is only a hydrogen bond acceptor, but p-ethylphenol is both a hydrogen bond acceptor and a hydrogen bond donor.

17.13   Diisopropyl ether has a more compact shape and has smaller London forces than the cylindrically shaped dipropyl ether.

17.14   Although they have similar molecular weights, 1,2-dimethoxyethane has an additional oxygen atom and is more polar.

17.15 The oxygen atom of dipropyl ether is sufficiently basic to accept a proton from the strong acid, sulfuric acid. The conjugate base is a cation and is soluble in the polar medium. Heptane cannot be protonated.

17.16 Aluminum trichloride is a Lewis base. It forms a coordinate covalent bond with one set of lone pair electrons on the oxygen atom of tetrahydrofuran. The process of bond formation is exothermic.

17.17 The atomic radius of the rubidium cation is larger than the atomic radius of the potassium cation. 18-Crown-6 has a cavity suitable to solvate the potassium cation, which means that it would be too small for the rubidium cation.

17.18 The atomic radius of the lithium cation is smaller than the atomic radius of the sodium cation. 15-Crown-5 has a cavity suitable to solvate the sodium cation, which means that a smaller ring such as 12-crown-4 would be suitable for the lithium cation.

17.19

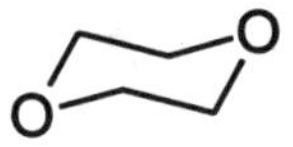

17.20 The methoxy group should occupy the equatorial position of 1,3-dioxane much like it does in cyclohexane. The hydroxyl group occupies the axial position of 1,3-dioxane because it forms hydrogen bonds with the ring oxygen atoms.

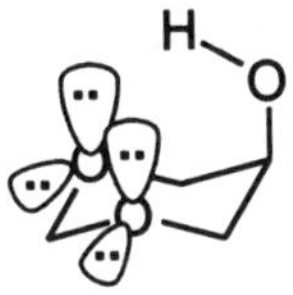

17.21

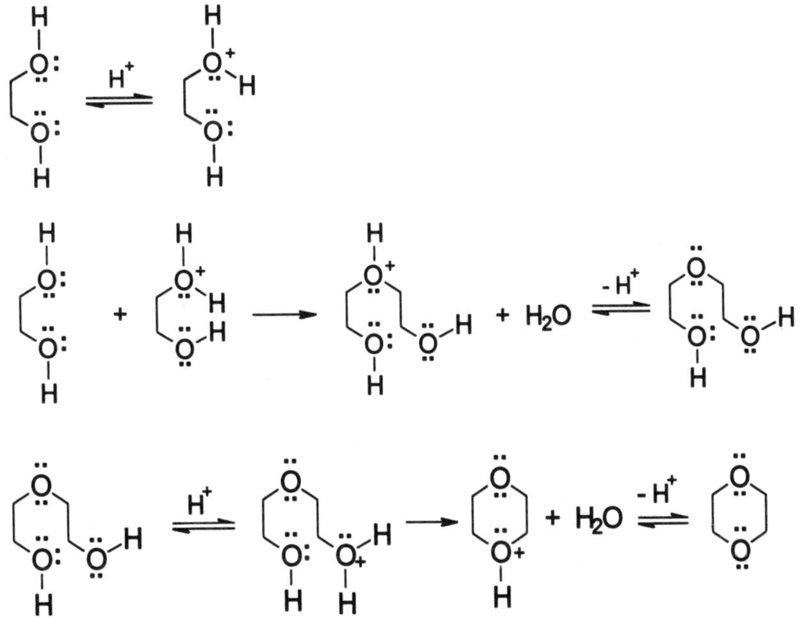

17.22

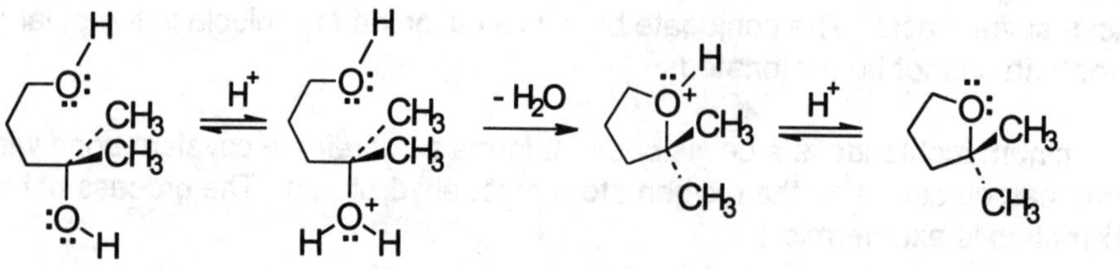

The oxygen atom bonded to the C-1 atom remains in the ether.

17.23 Electrophilic attack of $AcOHg^+$ gives a mercurinium ion that then reacts with the nucleophilic methanol at the carbon atom with the larger positive charge, which is the more substituted carbon atom.

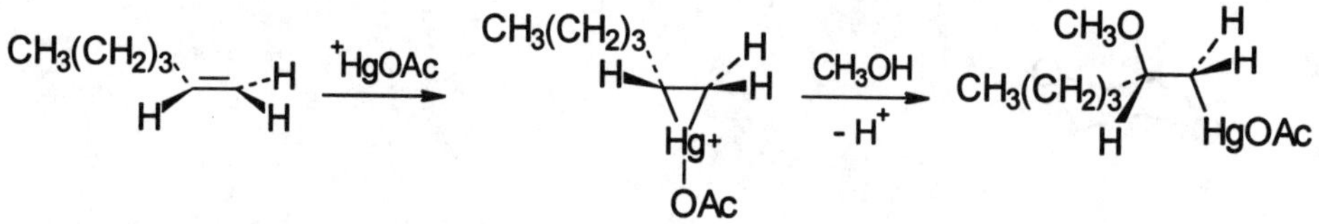

17.24 Electrophilic attack of $AcOHg^+$ gives a mercurinium ion that then reacts with the nucleophilic oxygen atom of the alcohol at C-1. Attack occurs at the carbon atom with the larger positive charge, which is the more substituted carbon atom, C-4.

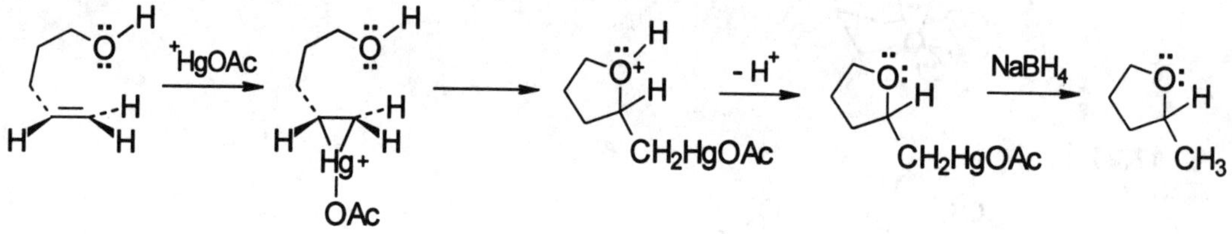

17.25 An alkoxide ion, formed by reaction with the hydride ion, intramolecularly displaces a chloride ion.

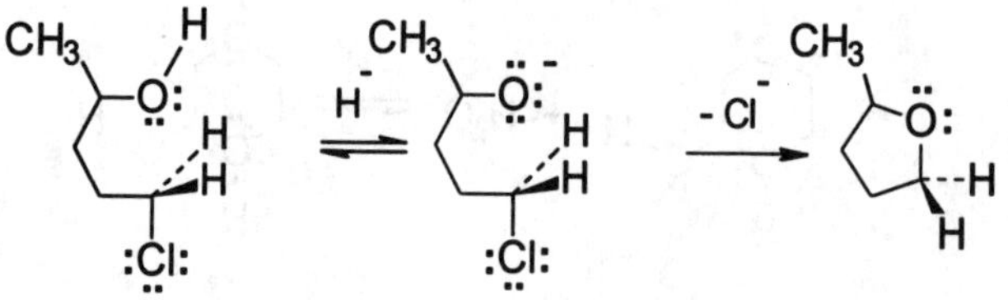

17.26 The alkoxide ion displaces a bromide ion from the C-4 atom, giving a five-membered cyclic ether. Displacement of a bromide ion from the C-3 atom would give a more highly strained four-membered cyclic ether.

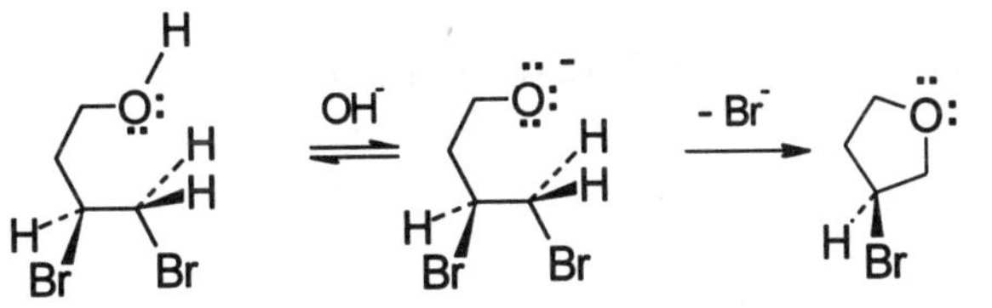

17.27 **(a)** Reaction of the alkoxide of cyclopentanol with iodoethane gives a good yield because the S_N2 reaction occurs at a primary center.

(b) Although the alkoxide of 1-methylcyclohexanol is sterically hindered, the reaction of this nucleophile with iodomethane give a good yield because the alkyl halide is unhindered and cannot undergo an elimination reaction.

(c) The reaction of the alkoxide of cyclohexanol with 2-bromo-2-methylpropane will give only elimination product. The reaction of *tert*-butoxide with bromocyclohexane may give some ether but the major product will be cyclohexene, an elimination product.

(d) The reaction of the alkoxide of 2-butanol with 2-bromobutane will give largely elimination products.

(e) The reaction of the phenoxide with 3-bromo-2-methylhexane will give largely elimination products.

17.28 Select a primary alkyl halide in order to avoid a competing elimination product. The combination of alkyl halide and the alcohol required to form the alkoxide ion are given.

(a)

(b)

(c)

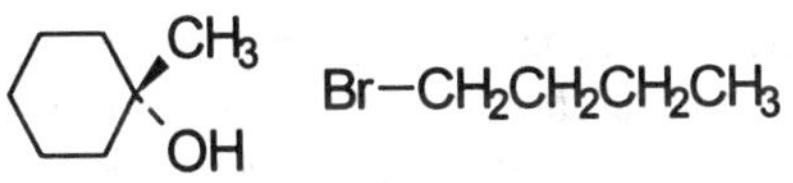

17.29 The displacement reaction occurs at the carbon atom bearing the iodine atom. The C—O bond of the alkoxide of 2-octanol is unaffected, so the configuration at the C-2 chiral center is unchanged.

17.30 The tosylate is formed by reaction of the oxygen atom of the alcohol with the sulfur atom of the toluenesulfonyl chloride. Hence, the tosylate has the same configuration as the original alcohol. The subsequent displacement reaction occurs at the carbon atom bearing the tosylate group. The C—O bond is broken in an S_N2 reaction when ethoxide ion displaces the tosyl group. Thus, the ether is formed with inversion of configuration. The ether is the enantiomer of the compound formed by the reaction of Exercise 17.29 and has the opposite sign of rotation.

17.31 Displacement of one halide ion by hydroxide ion gives an alcohol which can exchange a proton with hydroxide ion to give an alkoxide ion. This ion then intramolecularly displaces a halide ion to give dioxane.

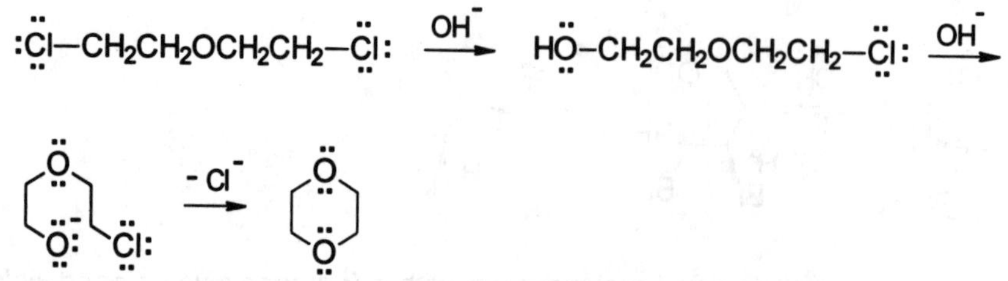

:Cl—CH₂CH₂OCH₂CH₂—Cl: $\xrightarrow{OH^-}$ HO—CH₂CH₂OCH₂CH₂—Cl: $\xrightarrow{OH^-}$

$\xrightarrow{-Cl^-}$

17.32 Each compound can be made by an intramolecular Williamson synthesis using a halogen substituted alcohol and a strong base such as sodium hydride. In both (a) and (b) there two possible isomeric reactants that can be considered. However, the isomer that has a primary halide is the best choice to preclude a competing elimination reaction.

(a) 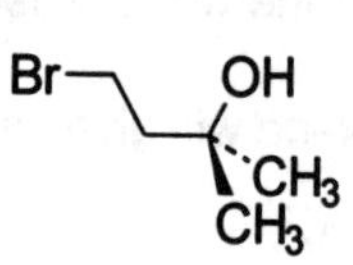 Br—OH ‚CH₃ CH₃   **(b)** 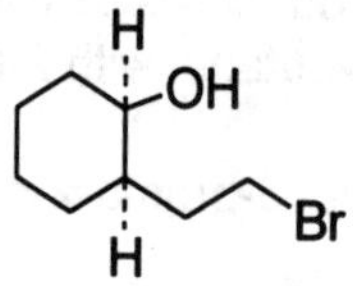 H OH H Br   **(c)** 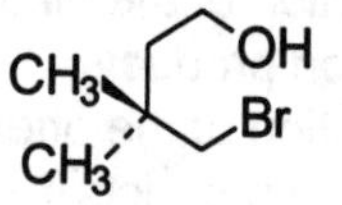 CH₃ CH₃ OH Br

17.33 React the conjugate base of the following phenol with 1-bromobutane.

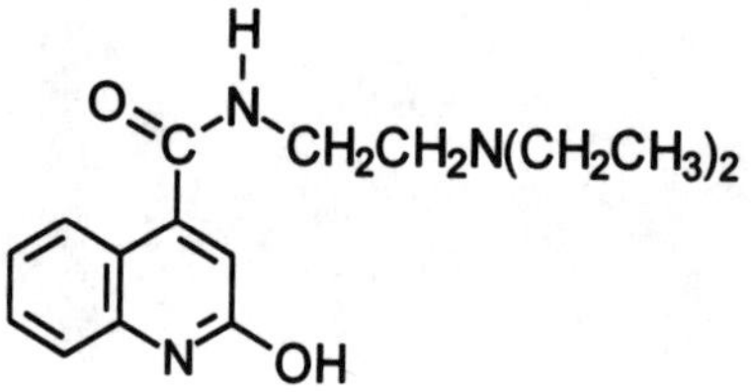

O=C—N(H)—CH₂CH₂N(CH₂CH₃)₂

N—OH

17.34 The second of the following two combinations of reagents would give only an elimination reaction. The combination can only give a substitution reaction. However, the halogen compound is secondary and hindered by the two phenyl groups bonded to the site where substitution occurs.

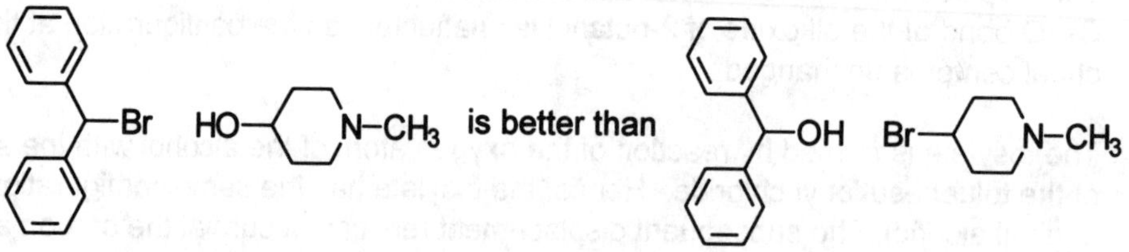

—Br   HO—N—CH₃   is better than   —OH   Br—N—CH₃

17.35 **(a)** Addition of mercuric acetate followed by reduction with sodium borohydride gives the product of "indirect" addition of water in a Markovnikov sense. Formation of the alkoxide using sodium hydride followed by displacement of iodide from methyl iodide is a Williamson synthesis of an ether. The product is shown on the next page.

(b) Addition of borane followed by oxidation with hydrogen peroxide gives the product of "indirect" addition of water in an anti-Markovnikov sense. Formation of the alkoxide using sodium hydride followed by displacement of bromide from ethyl bromide is a Williamson synthesis of an ether. The product is shown below.

(c) Addition of borane followed by oxidation with hydrogen peroxide gives the product of "indirect" addition of water in an anti-Markovnikov sense. Formation of the alkoxide using sodium hydride followed by nucleophilic attack on ethylene oxide opens the ring to give an ether product that is also an alcohol. The product is shown below.

(a)

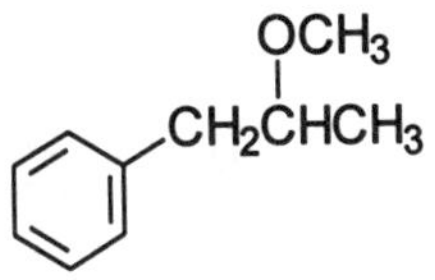

$$OCH_3$$
$$CH_2CHCH_3$$

(benzene ring structure)

(b) 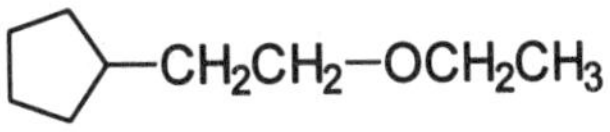 (cyclopentane ring)$-CH_2CH_2-OCH_2CH_3$

(c) 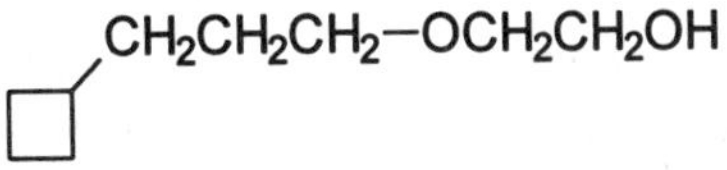 (cyclobutane ring)$CH_2CH_2CH_2-OCH_2CH_2OH$

17.36 **(a)** Reduction with sodium borohydride gives a primary alcohol. Reaction of the alcohol with the active metal potassium gives an alkoxide that displaces iodide ion from methyl iodide in a Williamson ether synthesis.

(b) Reduction with lithium aluminum hydride gives a secondary alcohol. Reaction of the alcohol with the strong base sodium hydride gives an alkoxide that displaces iodide ion from ethyl iodide in a Williamson ether synthesis.

(c) Reduction with sodium borohydride gives a primary alcohol. Reaction of the alcohol with the active metal sodium gives an alkoxide that attacks the primary carbon atom of the oxirane to give an ether that is also an alcohol.

(a) 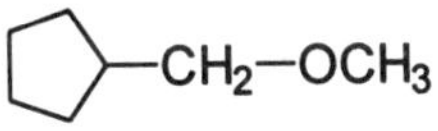 (cyclopentane ring)$-CH_2-OCH_3$ **(b)** 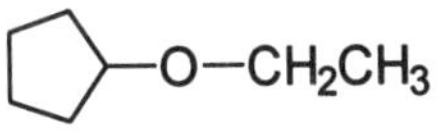 (cyclopentane ring)$-O-CH_2CH_3$

(c) 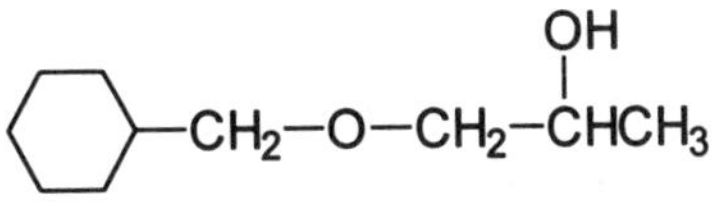 (cyclohexane ring)$-CH_2-O-CH_2-\overset{\overset{\displaystyle OH}{|}}{C}HCH_3$

17.37   **(a)** $CH_3O-CH_2CH_2-OCH_2CH_3$ **(b)** $CH_3O-CH_2CH_2CH_2-OCH_3$

(c) 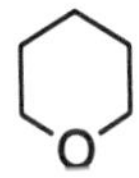 (tetrahydropyran ring with O)

(d) 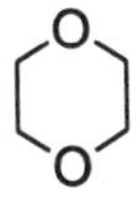 (dioxane ring with two O)

17.38 Iodide ion in the aprotic solvent dimethylformamide is an excellent nucleophile. Attack of iodide ion at the methyl group occurs because the phenoxide ion is a good leaving group. Phenol is a stronger acid than an alcohol and therefore phenoxide ion is a weaker base than an alkoxide ion. Recall that weaker bases are better leaving groups.

17.39 Protonation of the ether followed by an S$_N$2 displacement by bromide ion gives a bromo alcohol. The stereochemistry of the product is the result of inversion at the center attacked by the bromide ion.

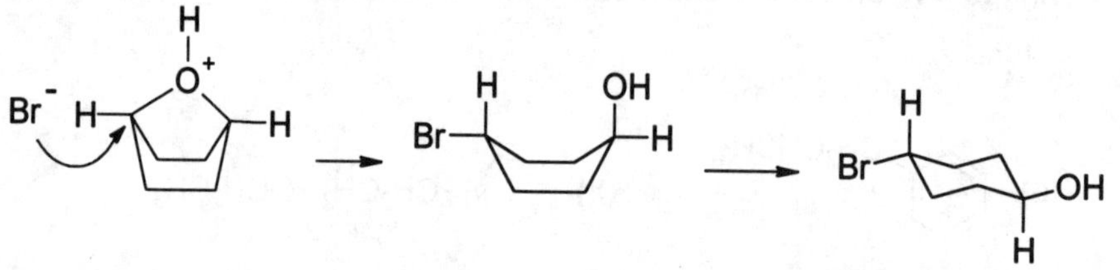

17.40 The reactant is a bicyclic ether that gives the product as the result of protonation of the ether. The cleavage of either of the two C—O bond occurs by an S$_N$1 mechanism because both centers are tertiary. Capture of the carbocation by chloride gives a chloro alcohol with the OH group on a tertiary carbon atom. Reaction of that alcohol with HCl gives the dichloro product.

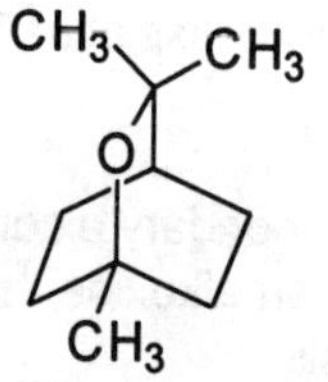

17.41 The alkoxide ion derived from the alcohol is situated trans with respect to the C—Br bond, and a Williamson synthesis reaction occurs to give an epoxide.

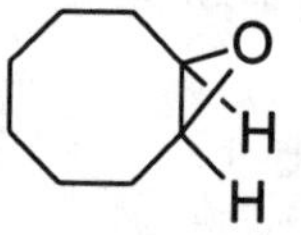

17.42 The alkoxide ion derived from the alcohol is situated trans with respect to the C—Cl bond in a diaxial conformation, which is formed by a ring flip of the more stable diequatorial conformation. A Williamson synthesis reaction readily occurs to give an epoxide. In the cis isomer, a trans arrangement of the C—Cl and C—O⁻ bonds is not possible.

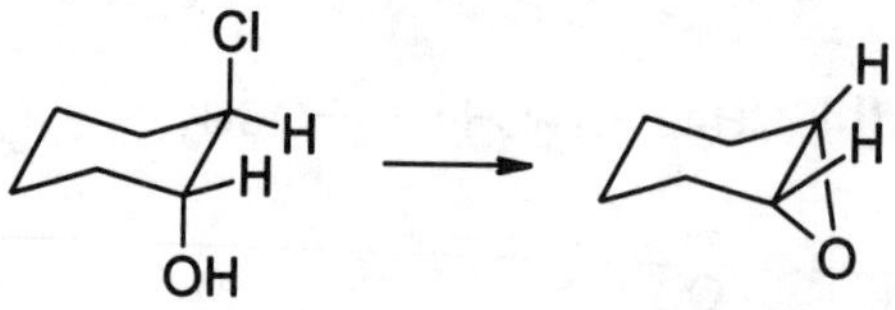

17.43 The anti addition reaction of Br⁺ and water followed by loss of a proton gives a bromohydrin. Intramolecular displacement of bromide by an alkoxide in a base-catalyzed reaction gives an epoxide. Note that the trans groups in the (E)-2-butene are also trans in the epoxide as shown on the next page.

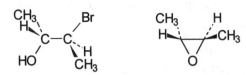

17.44 The displacement of bromide by an alkoxide derived from the alcohol inverts the configuration at the C-3 center, so the product is (2S,3R), which is meso.

17.45 The epoxide ring could form on the same "side" of the ring as the methyl group or on the opposite side.

17.46 The alkyl groups of the epoxide are cis, so the alkyl groups in the alkene must also be cis.

17.47 The required compound is (2R,3S)-3-bromo-2-butanol.

17.48 The first of the two compounds should give the higher yield, because it will react fastest in a displacement of a halide ion from a primary center. Possible competing reactions with the second compound include intermolecular elimination reactions.

17.49 **(a)** A ring opening reaction of ethylene oxide with ethanol under either acidic or basic conditions is required.

(b) React the product of (a) with a second mole of ethylene oxide in acid or base. The hydroxyl group of ethyl cellosolve is the nucleophile that opens the epoxide ring.

(c) First, a ring opening reaction of ethylene oxide with ammonia gives 2-aminoethanol. Reaction of this compound with ethylene oxide leads to ring opening as the result of nucleophilic attack of the nitrogen atom rather than the oxygen atom.

17.50   React ethylene glycol with hydroxide to obtain the conjugate base of ethylene glycol. React this conjugate base with ethylene glycol. Convert the diol to a dichloro compound by reacting it with thionyl chloride. Dehydrohalogenate the dichloro compound using sodium hydroxide.

17.51

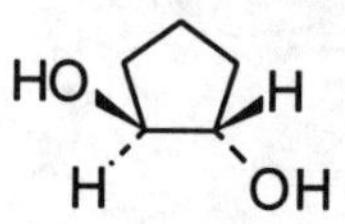

17.52   Reaction of the methyl carbanion inverts the configuration at the center where it attacks. Thus, the methyl group and the hydroxyl group are trans in the product. The most stable conformation is diequatorial.

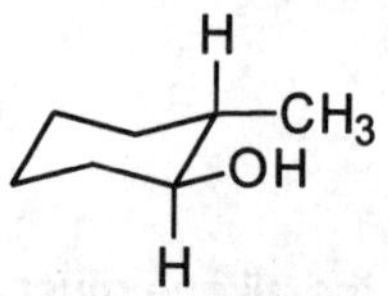

17.53   Ammonia is a better nucleophile than water. The reaction occurs with inversion of configuration at the center attacked by the nucleophilic ammonia. Attack at one carbon atom gives the (2R,3R) product and attack at the other carbon atom gives the (2S,3S) products. Thus a racemic mixture of enantiomers results.

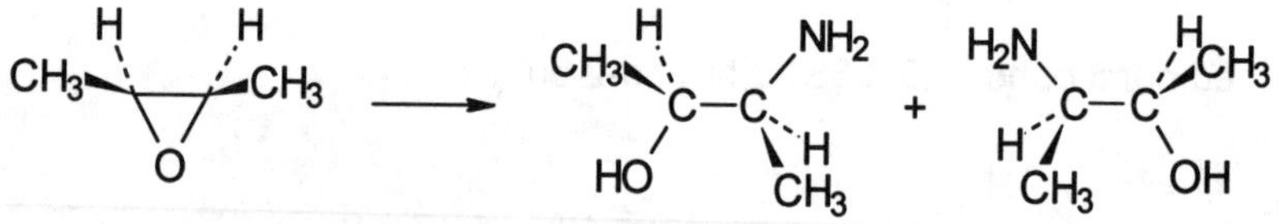

17.54   A ring opening reaction of the following epoxide and the conjugate base of the phenol will occur by nucleophilic attack at the primary carbon atom of the epoxide.

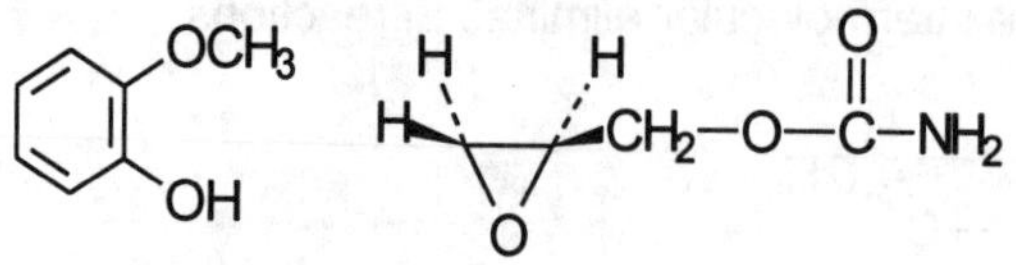

17.55   The ring opening reaction of the epoxide by the nucleophilic thiolate ion occurs at the primary carbon atom of the epoxide. The product is shown on the next page.

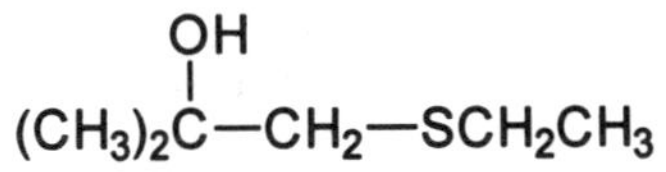

$$\underset{\displaystyle (CH_3)_2\overset{\displaystyle OH}{\overset{\displaystyle |}{C}}-CH_2-SCH_2CH_3}{}$$

17.56   The deuteride ion attacks at the less hindered secondary carbon atom rather than the tertiary carbon atom. Thus, the deuteride ion and the hydroxyl group formed will be trans. The most stable conformation of the product has the methyl group in the equatorial position because the conformational preference of a methyl group is larger than for a hydroxyl group.

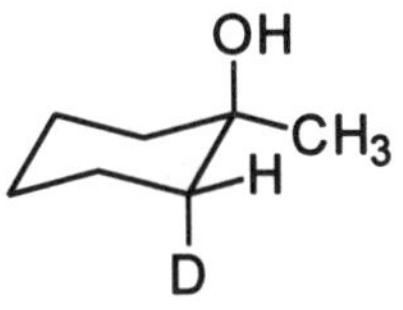

17.57   The reaction occurs with inversion of configuration at the center of the conjugate acid of the epoxide attacked by the nucleophilic water. Attack at one of the secondary carbon atoms gives the (2R,3R) product and attack at the other secondary carbon atom gives the (2S,3S) products. Thus a racemic mixture of enantiomers results.

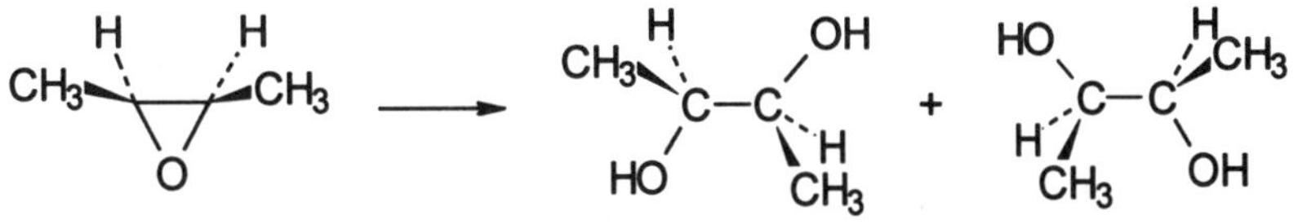

17.58   The reaction occurs with inversion of configuration at the center of the conjugate acid of the epoxide attacked by the nucleophilic water. Attack at one carbon atom gives a product designated (2S,3R) and attack at the other carbon atom gives a product designated (2R,3S). However, the two centers in the diol product are equivalent and thus only a single product which is meso is obtained.

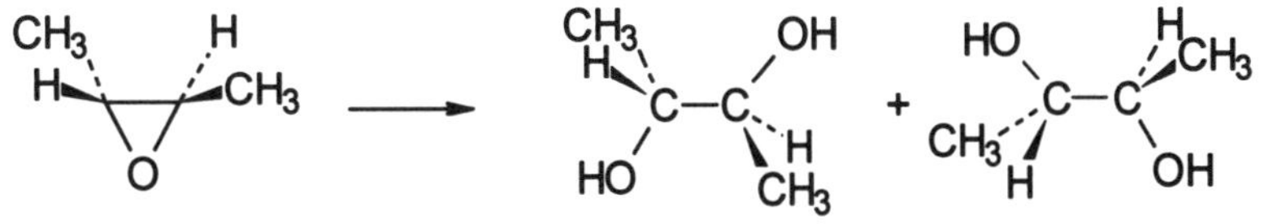

17.59   Ring opening of the epoxide as the result of nucleophilic attack by ethoxide ion yields an alkoxide that can react to displace the primary chloride and form an epoxide ring. The mechanism is shown on the next page.

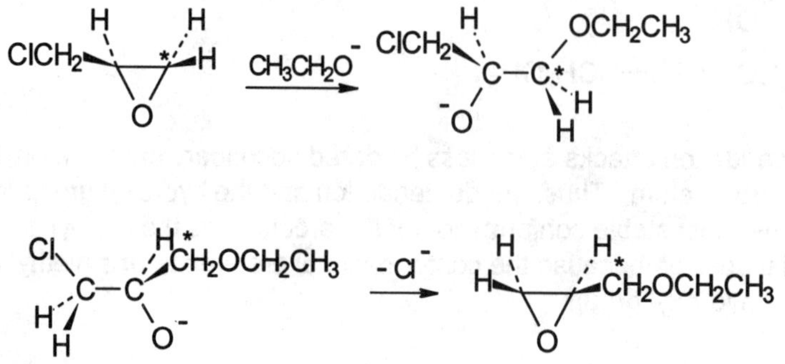

17.60 The alkoxide ion formed by proton exchange with base is a nucleophile and it attacks the primary carbon atom of the epoxide to open the three-membered ring. The result of the intramolecular displacement reaction is a six-membered ring.

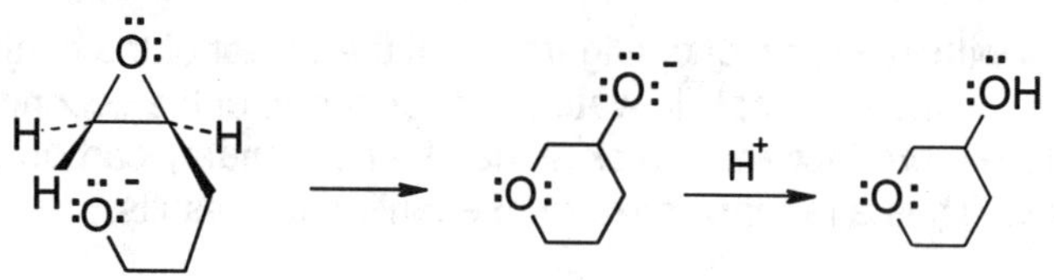

17.61 There are no unsaturated carbon atoms as evidence by the absence of low field resonances in the C-13 NMR spectrum. Thus, the compound must be cyclic to account for the unsaturation number of one. The compound is not an alcohol, based on the absence of a signal of intensity one in the hydrogen NMR. There are two sets of equivalent carbon atoms which means that the compound must be tetrahydrofuran.

17.62 There are two sp^2-hybridized carbon atoms based on the two low field peaks in the C-13 NMR spectrum. There are two protons on those carbon atoms based on the integration of the split signal at 5.5 δ in the hydrogen NMR spectrum. The signal near 4.5 δ is due to a hydroxyl group. The singlet at 3.9 δ is due to two hydrogen atoms bonded to the carbon atom bearing the hydroxyl group. The singlet at 1.6 δ is due to three hydrogen atoms of a methyl group. Neither the —CH_2OH nor the —CH_3 groups have neighboring hydrogen atoms because the signals are not split. The compound must be 2-methyl-2-propen-1-ol.

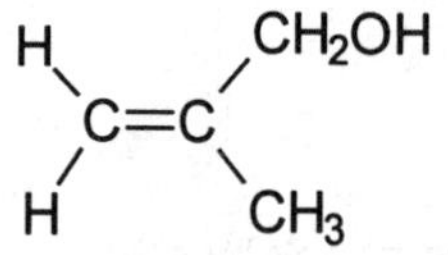

18

Aldehydes and Ketones

At this point, we start an adventure of studying the reactions of the first of many functional groups that contain a carbonyl group. There is some common chemistry but also considerable differences in the chemistry of the various carbonyl-containing compounds. As such we separate the chemistry of those compounds with electronegative groups bonded to the carbonyl atom from those that do not. Those containing an electronegative group are acids and acid derivatives which are the subject of later chapters. However, remember that their chemistry is not an entirely separate subject. Much of what you learn in this and the next chapter dealing with aldehydes and ketones will be used in the study of acids and acid derivatives.

The chemistry of the carbonyl group itself is presented in this and the next chapter. Additional chemistry that results from the effect of the carbonyl group on the adjacent carbon atom is presented in Chapter 23.

Keys to the Chapter

18.1 The Carbonyl Group

The **carbonyl group** is C=O with the carbon atom bonded to two other atoms. Carbonyl compounds with only hydrogen, alkyl, or aryl groups bonded to the carbonyl carbon atom are aldehydes or ketones. **Aldehydes** have one hydrogen atom and one alkyl or aryl group bonded to the carbonyl carbon atom. **Ketones** have only alkyl or aryl groups bonded to the carbonyl carbon atom.

Examine the structure of formaldehyde and the electronic structure of the carbonyl group shown in Figure 18.1 in the text. The reactivity of the carbonyl group is interpreted based on its π electrons and the two sets of nonbonded electrons. In addition, pay particular attention to the dipolar structure that is a contributing resonance structure for the carbonyl group. This structure is considered in discussing both the stability of carbonyl compounds and their reactivity.

18.2 Occurrence of Aldehydes and Ketones

The importance of carbonyl compounds in naturally occurring products is illustrated by the cited compounds. Even though the compounds have complex structures, their chemistry is predictable based on the chemistry presented in this and the subsequent chapter. After all, that is the organizational basis for organic chemistry.

18.3 Nomenclature of Aldehydes and Ketones

Aldehydes and ketones are named based on the longest continuous chain containing the carbonyl carbon atom. In aldehydes, the location of the carbonyl carbon atom is not designated because it is the number 1 carbon atom. The chain is numbered starting from this carbon atom. In ketones, the chain is numbered by starting from the end of the chain which gives the lowest possible number for the carbonyl carbon atom. Cyclic ketones are numbered from the carbonyl carbon atom, with subsequent numbers assigned in the direction which gives the lowest possible numbers for any other structural features. Aldehyde groups appended to a ring are named as carbaldehydes.

Note that the priority of carbonyl groups exceeds that of either double or triple bonds. Among carbonyl groups, the priority order is carboxylic acid > aldehyde > ketone.

18.4 Physical Properties of Aldehydes and Ketones

Carbonyl compounds are polar compounds and as a result have higher boiling points than alkanes of similar molecular weight. The lower molecular weight carbonyl compounds are soluble in water, because water can form hydrogen bonds to the carbonyl oxygen atom. Alkanes are insoluble in water.

Carbonyl compounds cannot form intermolecular hydrogen bonds like alcohols can, because they lack a hydrogen atom bonded to an electronegative atom such as oxygen. Thus, they have lower boiling points than alcohols of similar molecular weight.

18.5 Redox Reactions of Carbonyl Compounds

Aldehydes are easily oxidized by a variety of oxidizing agents, such as Tollens' reagent, Benedict's solution, and Fehling's solution. A reaction is detected by the formation of metallic silver with Tollens's reagent or a red precipitate of Cu_2O with Benedict's or Fehling's solution. Ketones are not oxidized by these reagents. The product of oxidation is a carboxylate ion which yields the carboxylic acid when neutralized. Recall that stronger oxidizing agents such as the Jones reagent (Chapter 16) oxidize primary alcohols to carboxylic acids. Thus, aldehydes are also oxidized by this reagent.

Aldehydes and ketones are reduced to primary and secondary alcohols, respectively. Hydrogen gas with a platinum catalyst may be used as a reducing agent, but high pressures are required and any carbon-carbon double bond would be reduced first. Lithium aluminum hydride and sodium borohydride reduce carbonyl groups without affecting carbon-carbon double bonds. Aldehydes yield primary alcohols; ketones yield secondary alcohols.

The carbonyl group in both aldehydes and ketones can be reduced to a methylene group with Zn/Hg and HCl or with NH_2NH_2 and KOH. The two reactions are termed the Clemmensen reduction and the Wolff-Kishner reduction, respectively.

18.6 Synthesis of Carbonyl Compounds

In Chapter 16 of this study guide, it was stated that an effective synthesis of any type of functional group depends on the type of substrates available and on the specificity of several possible reagents that could be used. In addition you were cautioned that as you go through the study of any functional group, you must recall what occurred earlier in your study. For an earlier reaction the focus might have been on the reactivity of functional group A which coincidently resulted in a product with functional group B. That reaction must also be "learned" as a method to synthesize compounds with functional group B. In addition you will also have to be prepared to encounter future reactions that give additional examples of transformations to give a functional group encountered earlier in the text. This section in the text gives you a summary of the past history of reactions to prepare carbonyl group. Section 18.7 gives the previews of coming attractions.

There are four general methods to prepare carbonyl compounds that have already been presented. These are:

1. oxidation of alcohols
2. Friedel-Crafts acylation of aromatic compounds
3. ozonolysis of alkenes
4. oxidative cleavage of diols

In addition, although not specifically covered earlier, hydration by the hydroboration-oxidation method discussed for alkenes in Section 16.8 of the text is a reaction that also occurs with alkynes as substrates. The product is a carbonyl compound rather than an alcohol.

Primary alcohols are oxidized by PCC to give aldehydes. Secondary alcohols are oxidized by the same reagent to give ketones. The Jones reagent (CrO_3 in acetone/H_2SO_4) further oxidizes primary alcohols beyond the aldehyde state to give carboxylic acids. The Jones reagent reacts with secondary alcohols to give ketones, which are not further oxidized.

Friedel Crafts acylation gives a carbonyl compound with the carbonyl group directly attached to an aromatic ring. The review is the reaction between an acid chloride and the aromatic compound. The reaction is limited to aromatic compounds lacking strongly deactivating groups. Two variations on the acylation reaction are given. One is the intramolecular cyclization reaction of carboxylic acids using HF. The second is the **Gatterman-Koch** reaction using CO and HCl, which behave together like formyl chloride to give an aldehyde.

Ozonolysis of an alkene gives two carbonyl fragments. If the fragments are identical or if one of the two fragments can be "thrown away", then the method may be useful. For example, the ozonolysis of a methylenecycloalkane give formaldehyde and a cycloalkanone. This method depends on the availability of the alkene, which in turn must usually be prepared by an elimination reaction of another substrate.

The oxidative cleavage of diols also gives two carbonyl fragments. Use of this method as a synthesis of carbonyl compounds is thus limited, because usually a synthetic method focuses on formation of a single product. Furthermore the diol must be prepared by oxidation of an alkene.

Indirect hydration of alkynes can be controlled to add one mole of a borane. The resulting "alcohol" is actually an enol—it has the hydroxyl group bonded to an unsaturated carbon atom. The enol rapidly rearranges to the isomeric ketone. Because an alkyne can react twice with any reagent, the addition reaction of a borane is controlled by using disiamylborane, which has two sterically hindering 1,2-dimethylpropyl groups. Only one mole of the borane compound adds to the double bond.

18.7 Synthesis of Carbonyl Compounds—A Preview

Each of the synthetic methods given in this section is based on the chemistry of functional groups yet to be encountered. Thus, you don't know yet how to make the necessary starting material nor do you understand the typical reactivity of those functional groups. However, assuming that you have the compounds, then the given reactions do occur. Memorize the methods now, and you will meet them again later.

Acid chlorides can be prepared by the reaction of carboxylic acids with thionyl chloride. They are reduced to aldehydes by either of two methods. The **Rosenmund reduction** uses hydrogen gas and a modified palladium catalyst. The newer preferred reagent is a modified metal hydride reagent—lithium tri(tert-butoxy)aluminum hydride. This reagent replaces the chloride of the acid chloride to give an aldehyde. Note that the aldehyde is not further reduced by this hydride reagent as it would be by $LiAlH_4$.

Esters are reduced to aldehydes using diisobutylaluminum hydride (**DIBAL**). The aldehyde itself is not formed in the reaction but is formed in the aqueous workup reaction to give a hemiacetal which decomposes to the aldehyde. The details of these latter reactions are given in Chapter 19.

Carboxylic acids and acid chlorides react with certain organometallic compounds to give ketones. Addition of an organolithium compound to a carboxylic acid gives a salt of a geminal diol that gives a diol upon hydrolysis, which in turn eliminates water to give a ketone. Note that the two groups bonded to the carbonyl carbon atom can come from two possible combinations of reactants. Reaction of an acid chloride with a Gilman reagent gives a ketone directly. The product does not react with the Gilman reagent.

Nitriles are reduced by modified hydride reagents to give imine anion intermediates. Upon aqueous workup, the imine formed is hydrolyzed to give the more stable carbonyl compound, which is an aldehyde. Grignard reagents react with nitriles to give a salt of an imine, which is converted to an imine and then to a ketone in the aqueous workup.

18.8 Spectroscopy of Aldehydes and Ketones

Aldehydes and ketones are characterized by the strong absorption due to the carbonyl group in the IR spectra of these compounds. Aldehydes absorb at slightly higher wavenumber than ketones, and in addition aldehydes have a characteristic absorption at 2710 cm^{-1} due to the aldehydic C—H bond. The position of the carbonyl absorption occurs at lower wavenumber with conjugation due to a greater contribution of the dipolar resonance form, because it decreases the double bond character of the carbonyl group. The position of the carbonyl absorption of cycloalkanones depends on ring size.

The proton NMR of aldehydes show a characteristic absorption near 10 δ due to the aldehydic C—H bond. The α hydrogen atoms of both aldehydes and ketones have NMR absorptions in the 2.0-2.5 δ region

The α carbon atoms of both aldehydes and ketones have ^{13}C NMR absorptions in the 30-50 δ region. The carbonyl carbon atom is easily identified by its absorption in the 190-220 δ region.

Summary of Reactions

1. Oxidation of Aldehydes

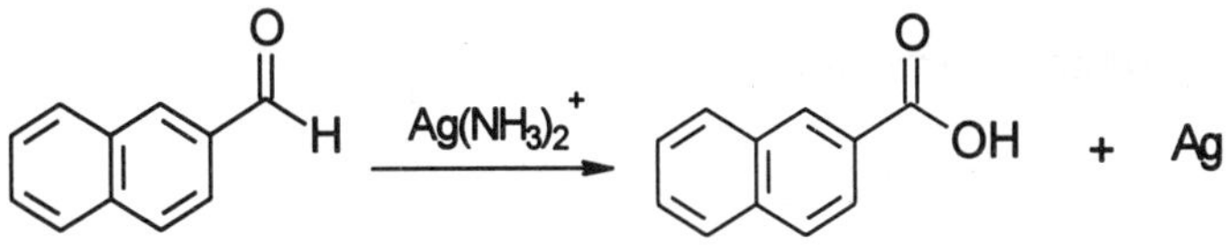

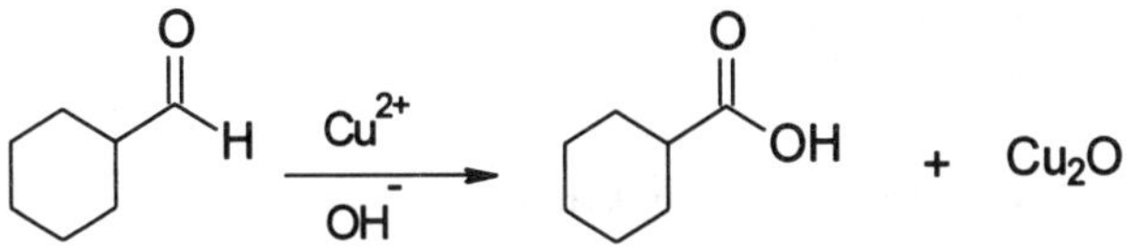

2. Reduction of Aldehydes and Ketones to Alcohols

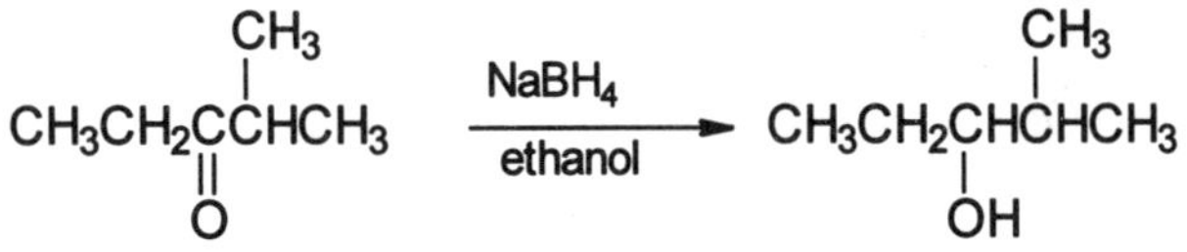

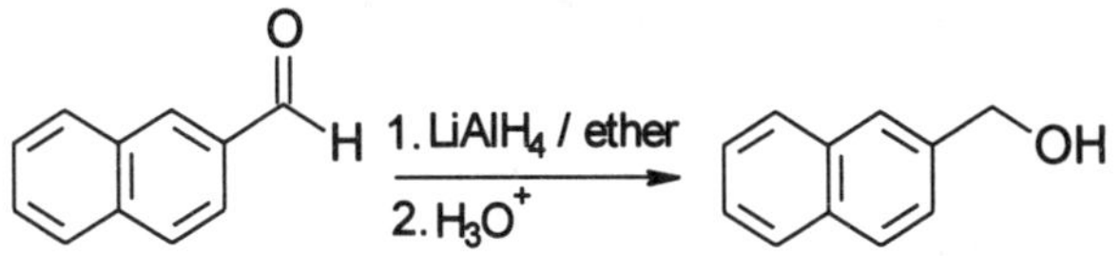

3. Reduction of Aldehydes and Ketones to Methylene Groups

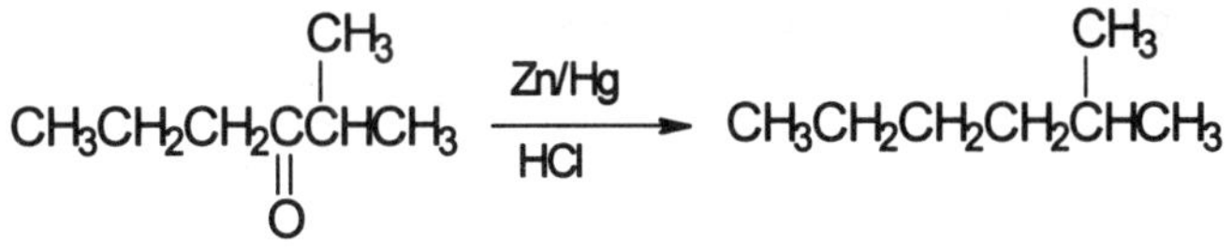

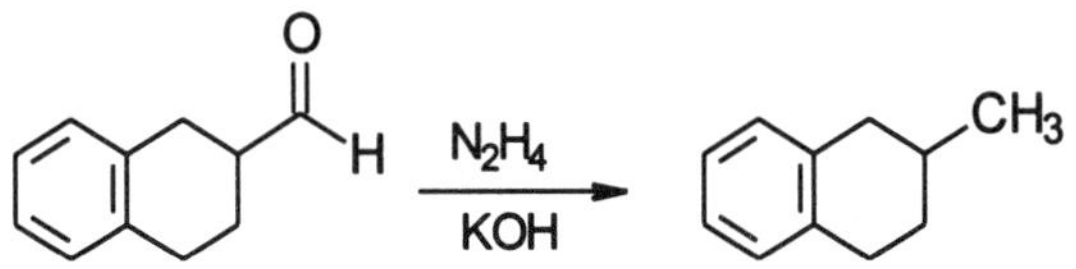

4. Synthesis of Aldehydes and Ketones by Oxidation of Alcohols

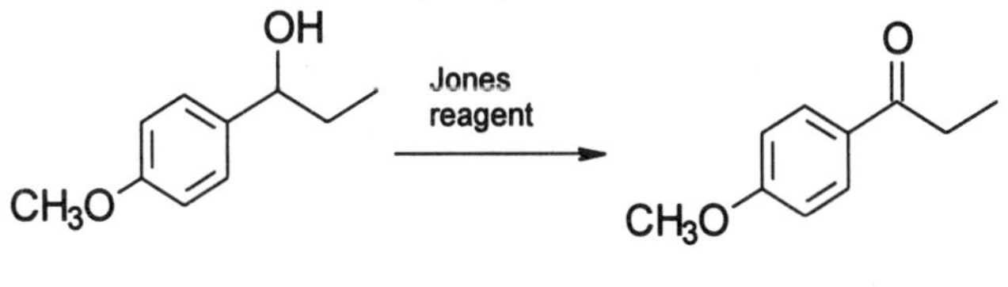

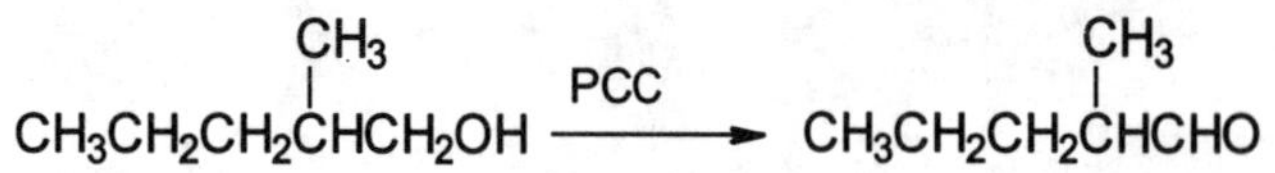

$$\underset{\substack{| \\ \text{CH}_3}}{\text{CH}_3\text{CH}_2\text{CH}_2\text{CHCH}_2\text{OH}} \xrightarrow{\text{PCC}} \underset{\substack{| \\ \text{CH}_3}}{\text{CH}_3\text{CH}_2\text{CH}_2\text{CHCHO}}$$

5. Synthesis of Aryl Ketones by Friedel-Crafts Acylation

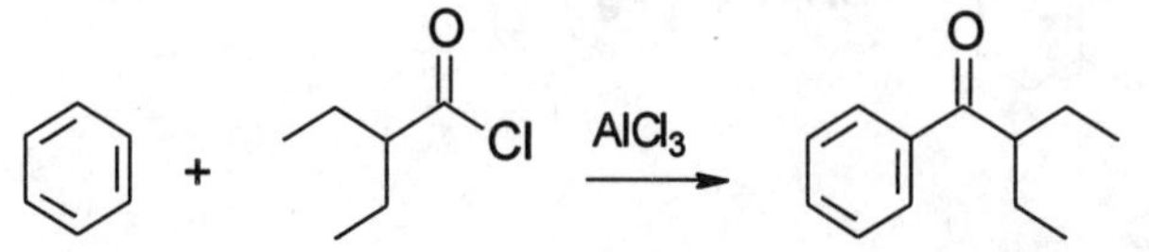

6. Synthesis of Carbonyl Compounds by Ozonolysis of Alkenes

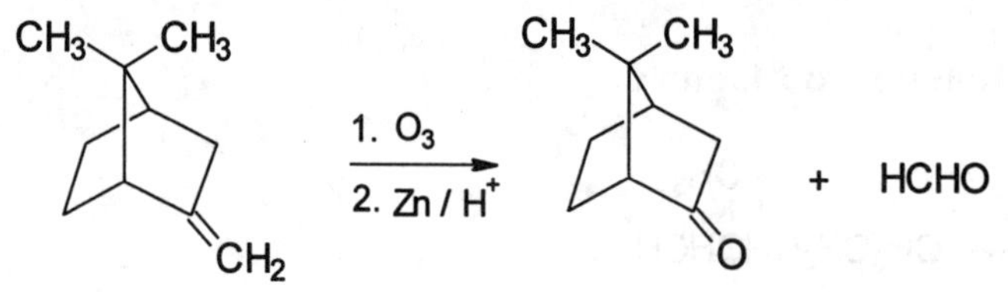

7. Synthesis of Carbonyl Compounds by Oxidative Cleavage of Vicinal Diols

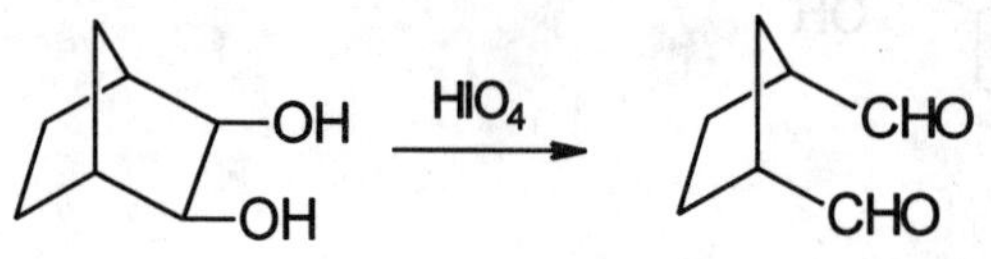

8. Synthesis of Carbonyl Compounds by Hydration of Alkynes

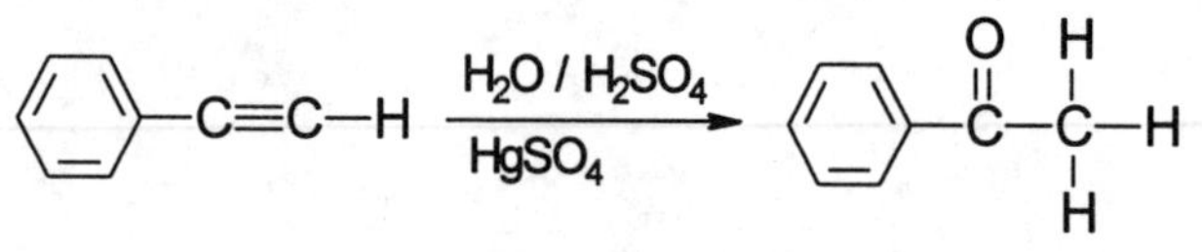

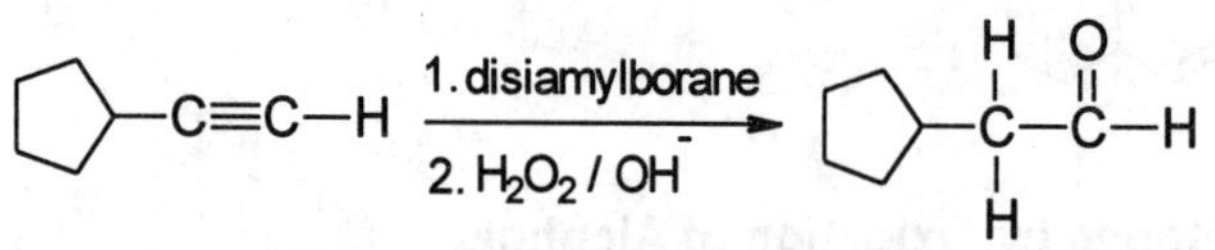

9. Synthesis of Carbonyl Compounds by Reduction of Acid Derivatives

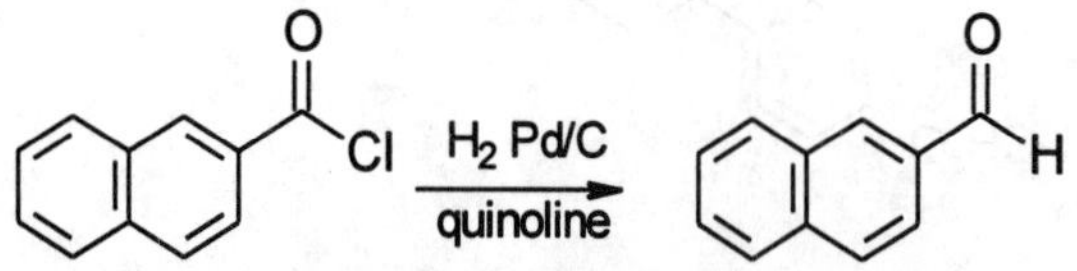

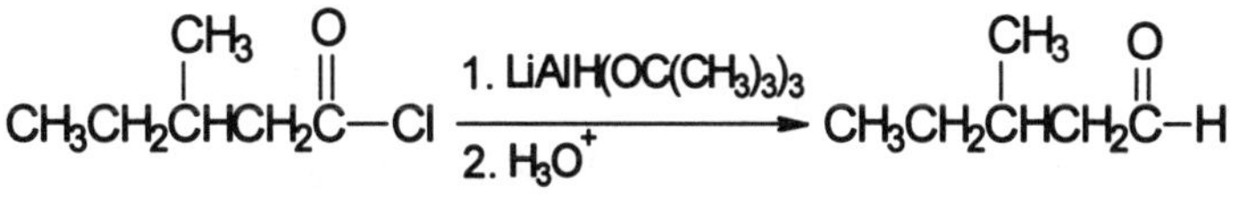

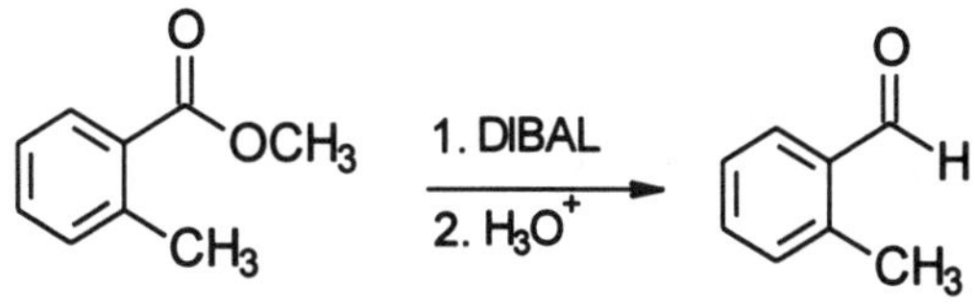

$$CH_3CH_2\overset{\underset{\displaystyle CH_3}{|}}{C}HCH_2\overset{\underset{\displaystyle ||}{O}}{C}-Cl \quad \xrightarrow[\text{2. } H_3O^+]{\text{1. LiAlH(OC(CH}_3)_3)_3} \quad CH_3CH_2\overset{\underset{\displaystyle CH_3}{|}}{C}HCH_2\overset{\underset{\displaystyle ||}{O}}{C}-H$$

$$\xrightarrow[\text{2. } H_3O^+]{\text{1. DIBAL}}$$

10. Reactions of Organometallic Reagents with Acid Derivatives

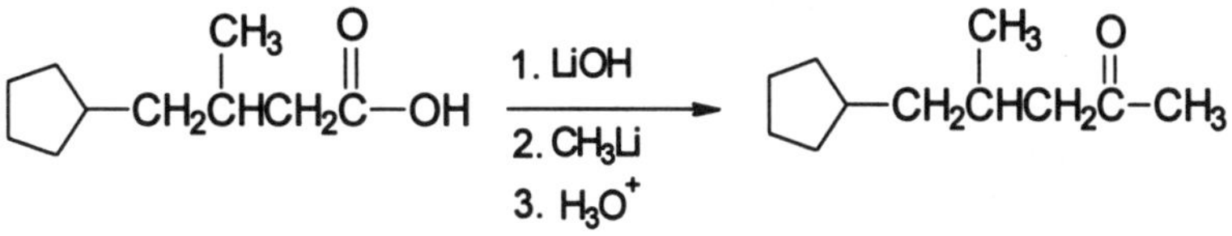

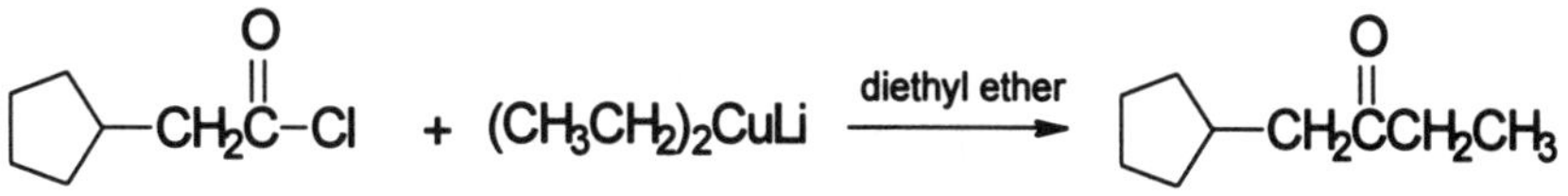

$$\xrightarrow[\substack{\text{2. CH}_3\text{Li} \\ \text{3. } H_3O^+}]{\text{1. LiOH}}$$

$$\xrightarrow[]{\text{diethyl ether}}$$

11. Formation of Carbonyl Compounds from Nitriles

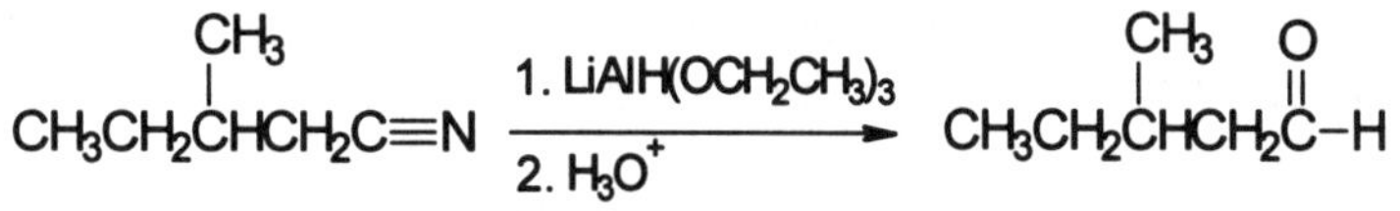

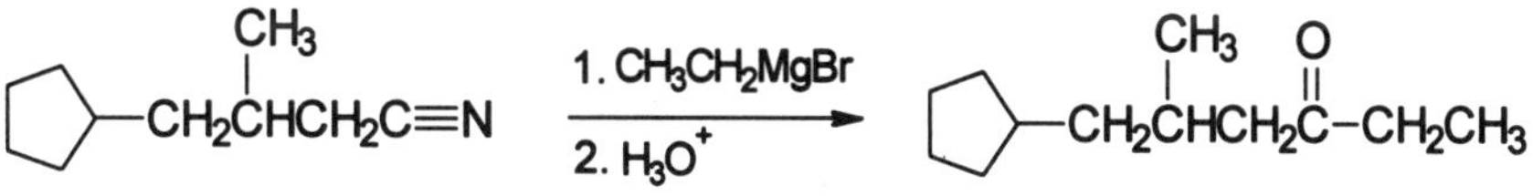

$$\xrightarrow[\text{2. } H_3O^+]{\text{1. LiAlH(OCH}_2\text{CH}_3)_3}$$

$$\xrightarrow[\text{2. } H_3O^+]{\text{1. CH}_3\text{CH}_2\text{MgBr}}$$

Solutions to Exercises

18.1 **(a)** CH$_3$CH$_2$CHCHO with CH$_3$ on the CH

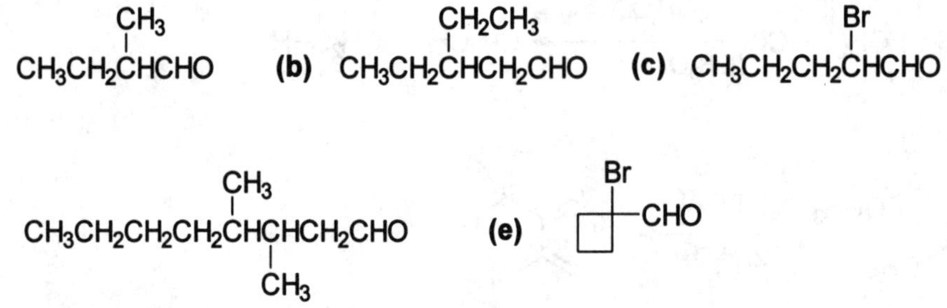

(b) CH$_3$CH$_2$CHCH$_2$CHO with CH$_2$CH$_3$ on the CH

(c) CH$_3$CH$_2$CH$_2$CHCHO with Br on the CH

(d) CH$_3$CH$_2$CH$_2$CH$_2$CHCHCH$_2$CHO with CH$_3$ above and CH$_3$ below

(e) cyclobutane ring with Br and CHO substituents

18.2 **(a)** CH$_3$CH$_2$CHCCH$_3$ with =O on the C and Br below

(b) CH$_3$CHCCHCH$_3$ with =O on the C and CH$_3$, CH$_3$ below

(c) CH$_3$CHCH$_2$CCH$_3$ with =O on the C and CH$_3$ below

(d) CH$_3$CHCHCCH$_3$ with CH$_3$ and O above, CH$_3$ below

(e) cyclohexane-1,3-dione ring with CH$_3$ substituent

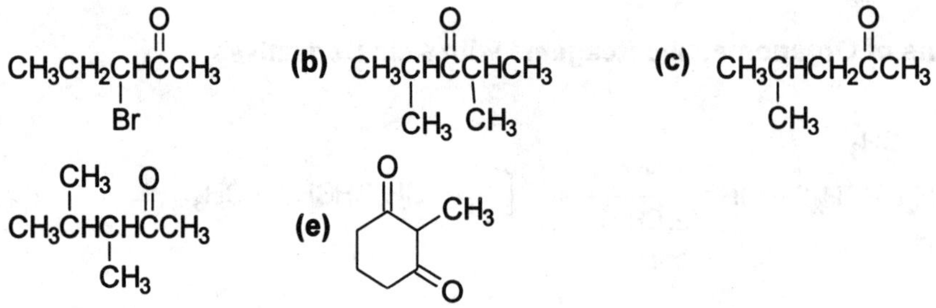

18.3 **(a)** butanal   **(b)** 3,3-dimethylbutanal
(c) 2-methylpropanal   **(d)** 2-ethyl-3-methylpentanal

18.4 **(a)** 3-pentanone   **(b)** 3,3-dimethyl-2-butanone
(c) 2-methyl-3-pentanone   **(d)** 5-methyl-3-hexanone

18.5 **(a)** 4-chloro-2,3-dimethylheptanal   **(b)** 6-ethyl-3-methyl-2-nonanone
(c) 4-ethyl-2,3,5-trimethylheptanal   **(d)** 8-methyl-4-nonanone

18.6 **(a)** 2-methylcyclobutanone   **(b)** 5-chlorocyclodecanone
(c) 2-ethylcyclohexanone   **(d)** 1-cyclohexyl-1-pentanone

18.7 **(a)** CH$_3$CCHO with CH$_3$ above and CH$_3$ below

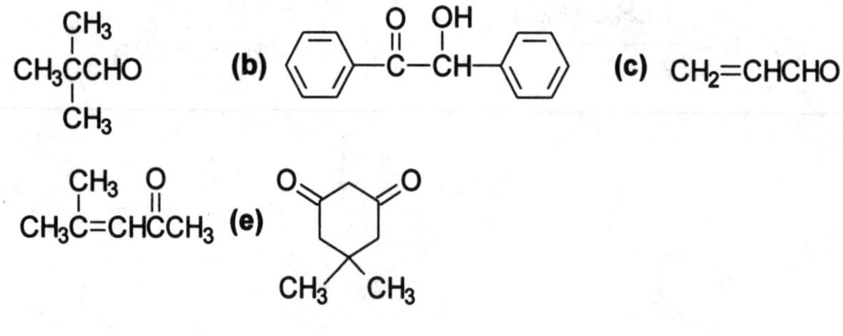

(b) phenyl-C(=O)-CH(OH)-phenyl

(c) CH$_2$=CHCHO

(d) CH$_3$C=CHCCH$_3$ with CH$_3$ above and O above

(e) cyclohexane-1,3-dione ring with CH$_3$ and CH$_3$ substituents

18.8 **(a)** CH$_3$—C—C—CH$_3$ with CH$_3$ above, O above, CH$_3$ below

(b) CH$_3$CCH$_2$CCH$_3$ with OH above, O above, CH$_3$ below

(c) C=C with CH$_3$ and H on left carbon, H and CHO on right carbon

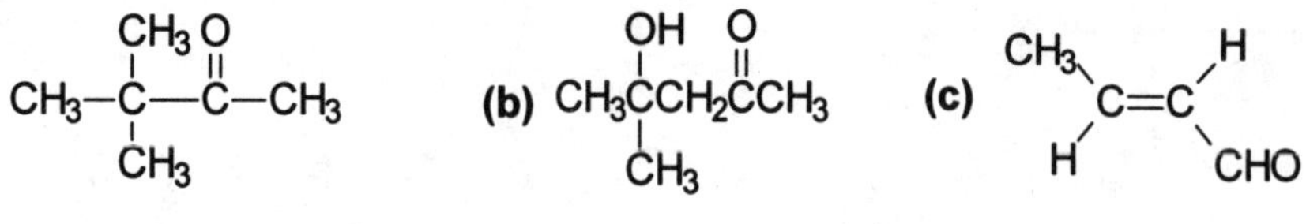

(d) 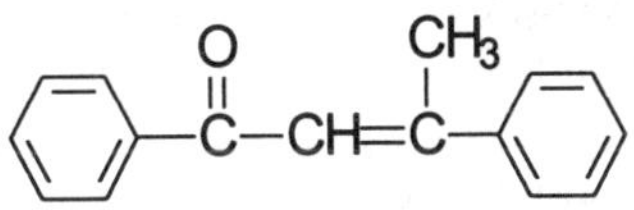   **(e)** 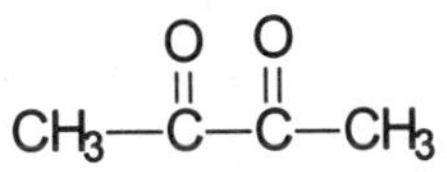 $CH_3-\overset{\overset{O}{\|}}{C}-\overset{\overset{O}{\|}}{C}-CH_3$

18.9   Although sterically small, the methyl and hydrogen atoms bonded to the carbonyl carbon atom occupy more space than two hydrogen atoms of formaldehyde and the bond angle widens to accommodate the methyl group.

18.10   The atomic radius of oxygen is smaller than the atomic radius of carbon, and the bond length between two atoms reflects their respective atomic radii.

18.11   The oxygen atom of the "two" carbon-oxygen bonds of the carbonyl group "pulls" electron density away from the carbonyl carbon atom more so than does the single oxygen atom of the carbon-oxygen bond of the alcohol.

18.12   An additional dipolar resonance form of propenal gives added stabilization of the positive charge and increases the polarity of the molecule.

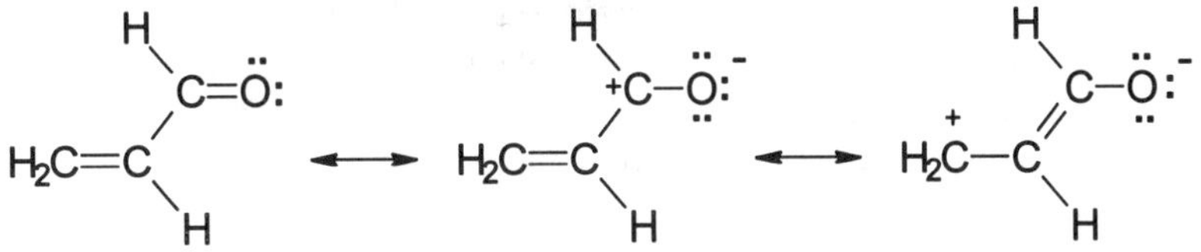

18.13   2-Methylpropanal is a more compact molecule that has a more spherical shape than the cylindrically shaped butanal, so its London forces are smaller.

18.14   As the polar carbonyl group moves to the interior of the molecule, the compound more closely resembles an alkane structure and the intermolecular attractive forces are smaller.

18.15   The 2-hydroxybenzaldehyde can form intramolecular hydrogen bonds, which decrease the number of intermolecular hydrogen bonds that would cause a higher boiling point like that of 3-hydroxy-benzaldehyde.

18.16   The 2-hydroxyacetophenone can form intramolecular hydrogen bonds, which decrease the number of intermolecular hydrogen bonds that would cause a higher boiling point like that of 3-hydroxyacetophenone.

18.17   Butanal is a hydrogen bond acceptor of water molecules but 1-butanol is both a hydrogen bond acceptor and a hydrogen bond donor, so it is somewhat more soluble.

18.18   2-Methylpropanal is a more compact molecule that has a more spherical shape than the cylindrically shaped butanal, so it interferes less with the hydrogen bonding arrays of water.

18.19   A red precipitate of Cu_2O forms when an aldehyde reacts with Benedict's solution. A precipitate of silver usually seen as a silver mirror forms when an aldehyde reacts with Tollens's reagent.

18.20 Ketones yield secondary alcohols when reduced using sodium borohydride. Aldehydes yield primary alcohols when reduced using lithium aluminum hydride.

18.21 **(a)** Oxidation of an aldehyde by Tollens's reagent yields a carboxylate ion in solution. Upon neutralization, the isolated product is the carboxylic acid. The product is shown below.
(b) Oxidation of an aldehyde by Benedict's solution or Fehling's solution yields a carboxylate ion in solution. Upon neutralization, the isolated product is the carboxylic acid.
(c) Oxidation of a primary alcohol by PCC yields an aldehyde.

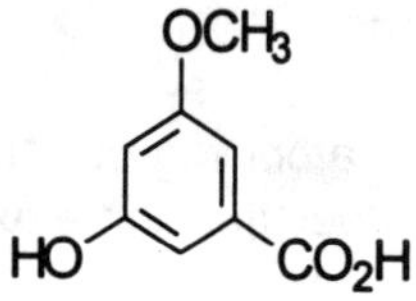

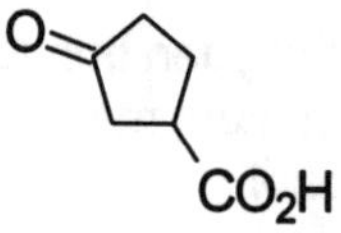

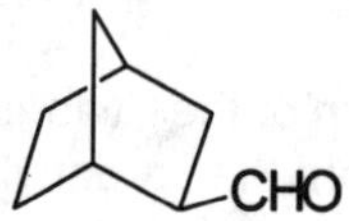

18.22 **(a)** Oxidation of an aldehyde by Fehling's or Benedict's solution yields a carboxylate ion in solution. Upon neutralization, the isolated product is the carboxylic acid. The product is shown below. Note that the ketone is unaffected by the reagent.
(b) Oxidation of an aldehyde by Tollens's reagent yields a carboxylate ion in solution. Upon neutralization, the isolated product is the carboxylic acid.
(c) Oxidation of an aldehyde by Jones reagent yields a carboxylic acid.

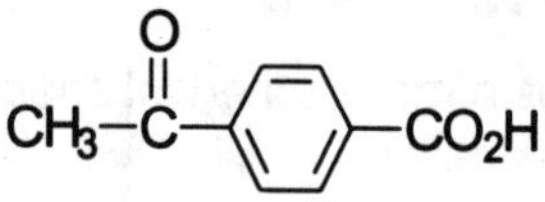

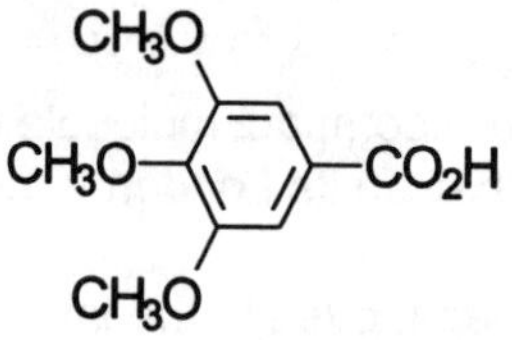

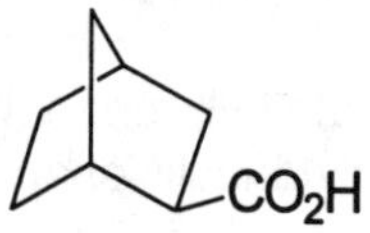

18.23 Lithium aluminum hydride reduces aldehydes to primary alcohols and ketones to secondary alcohols. Carbon-carbon double bonds and aromatic rings are unaffected.

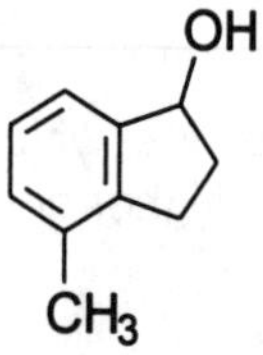

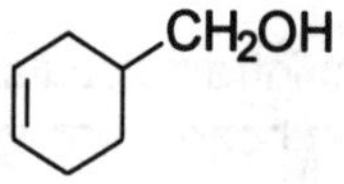

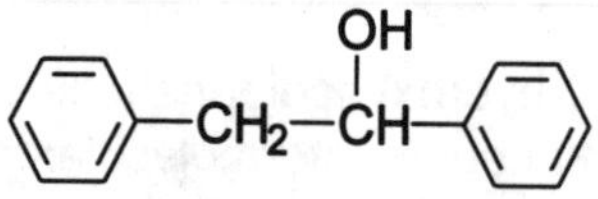

18.24 Sodium borohydride reduces aldehydes to primary alcohols and ketones to secondary alcohols. Carbon-carbon double bonds and aromatic rings are unaffected.

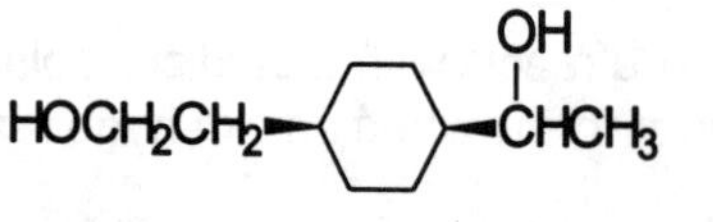

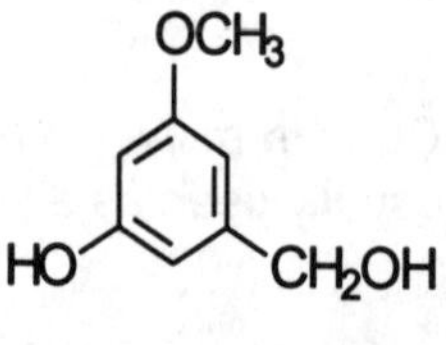

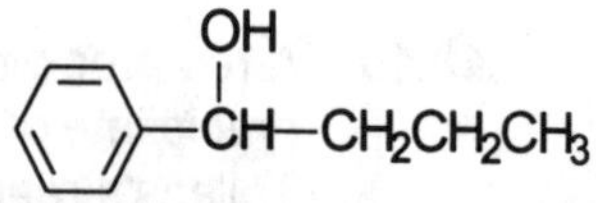

18.25 The hydride reagent can attack from either face of the carbonyl group, so both cis and trans isomers may form.

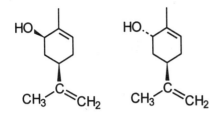

18.26 The hydride reagent can attack from either face of the carbonyl group, so both cis and trans isomers may form.

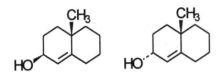

18.27 **(a)** The Wolff-Kishner reduction converts the ketone into a methylene group. The product is shown below.
   (b) The Clemmensen reduction converts the ketone into a methylene group and the aldehyde group into a methyl group.
   (c) At atmospheric pressure the carbon-carbon double bond is reduced.

18.28 **(a)** The Wolff-Kishner reduction converts the ketone into a methylene group. The carbon-carbon double bond is unaffected.
   (b) The Clemmensen reduction converts the ketone into a methylene group.
   (c) At atmospheric pressure the carbon-carbon double bond is reduced. The carbonyl group is unaffected.

(a) ⬡⬠ **(b)** ⬚ **(c)** ⟋⟍⟋⟍CHO

18.29 **(a)** PCC oxidizes primary alcohols to aldehydes. The product is shown on the next page.
   (b) Carboxylic acids are acylating agents of aromatic rings in the presence of HF. The cyclization occurs at the C-1 atom, which is more reactive the C-3 atom.
   (c) Ozonolysis of the alkene yields formaldehyde and a ketone.
   (d) Reaction with osmium tetroxide gives a diol that is subsequently cleaved by periodic acid.
   (e) Reaction of a terminal alkyne with disiamylborane followed by hydrogen peroxide yields an aldehyde.

(a) 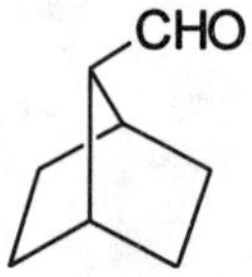 [structure: norbornane with CHO group]

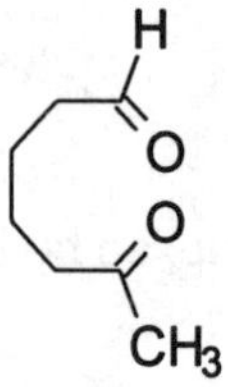

(b) 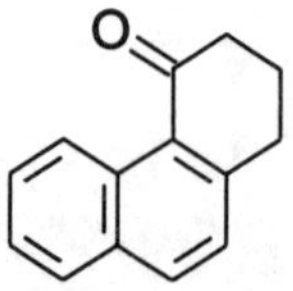 [structure: tricyclic ketone]

(c) 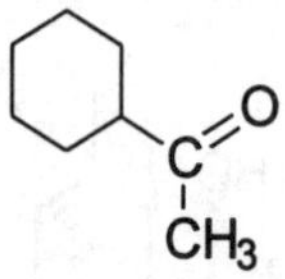 [structure: cyclohexyl C=O with CH₃]

(d) [structure: cyclic peroxide with H, O, O, CH₃]

(e) CH₃CH₂CHCH₂CH₂CHO with CH₃ substituent

18.30 (a) PCC oxidizes primary alcohols to aldehydes. The product is shown below.

(b) Carboxylic acids are acylating agents of aromatic rings in the presence of HF. The cyclization occurs at either of two ortho positions with respect to the side chain containing the carboxyl group.

(c) Ozonolysis of the alkene contained in a ring gives a dicarbonyl group which in this case is a dialdehyde.

(d) Reaction with osmium tetroxide gives a diol that is subsequently cleaved by periodic acid.

(e) Reaction of an internal alkyne with disiamylborane followed by hydrogen peroxide yields a ketone with the carbonyl group at the least hindered carbon atom of the original alkyne.

(a) 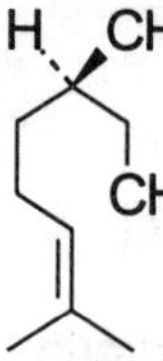 [structure: chain with H, CH₃, CHO and isopropenyl end]

(b) 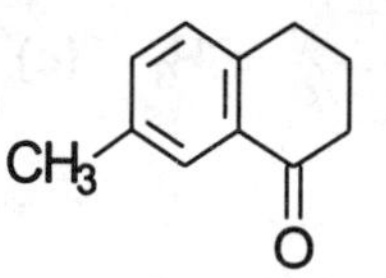 [structure: tetralone with CH₃]

(c) 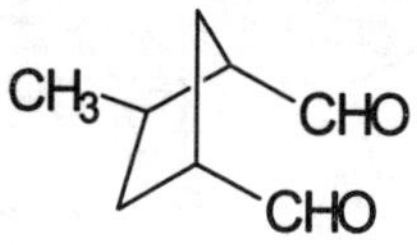 [structure: bicyclic with CH₃ and two CHO groups]

(d) 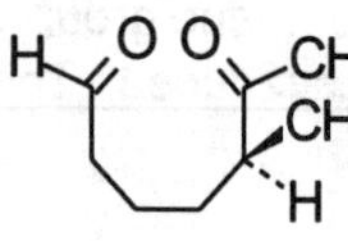 [structure: open chain with H, O, O, CH₃, CH₃, H]

(e) 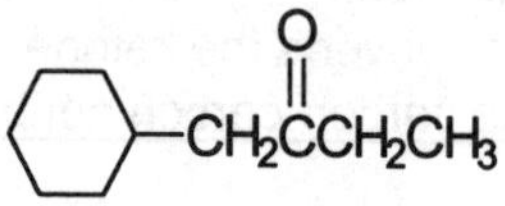 [structure: cyclohexyl—CH₂CCH₂CH₃ with O]

18.31 (a) The Rosenmund reaction reduces acid chlorides to form aldehydes. The product is shown on the next page.

(b) DIBAL reduces esters to form aldehydes. The product is shown on the next page.

(c) Methyl lithium reacts with carboxylic acids to give ketones by adding a methyl group to the carbonyl carbon atom of the carboxylic acid. The product is shown on the next page.

(d) Reduction of nitriles by complex hydride reagents containing alkoxy groups gives aldehydes. The product is shown on the next page.

(a) 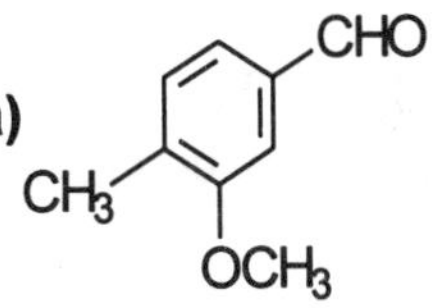   (b) 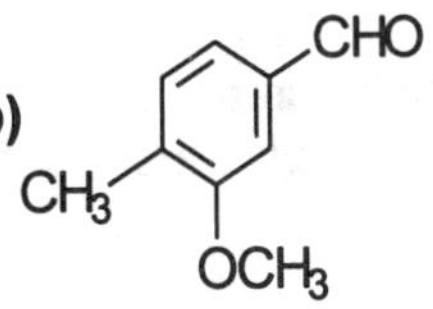   (c) 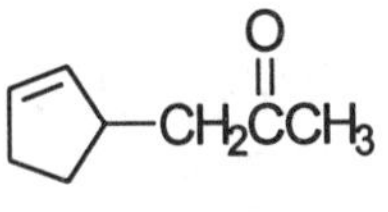   (d)

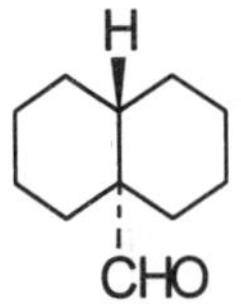

18.32 **(a)** The Rosenmund reaction reduces acid chlorides to form aldehydes. The product is shown below.
(b) DIBAL reduces esters to form aldehydes. The product is shown below.
(c) Phenyl lithium reacts with carboxylic acids to give ketones by adding a phenyl group to the carbonyl carbon atom of the carboxylic acid. The product is shown below.
(d) Addition of a Grignard reagent to nitriles gives ketones containing the group bonded to the triple bonded carbon atom of the original nitrile plus the alkyl group of the Grignard reagent.

(a) 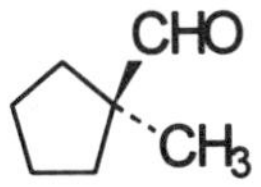   (b) 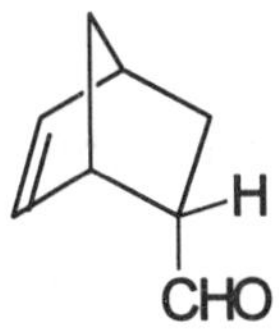   (c) 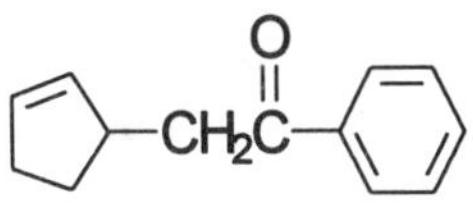   (d)

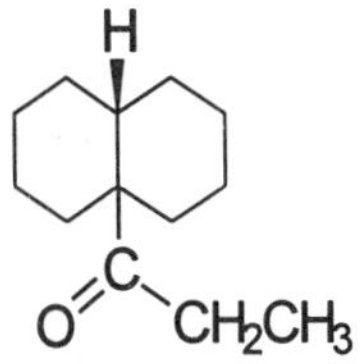

18.33 The reducing agent must transfer a hydride ion from behind the plane of the molecule as written below in order to give the (S) alcohol. The attack is on the *re* face. (Refer to Section 9.12 in the text if you need to review the concept of prochiral centers.)

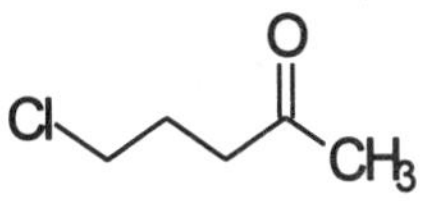

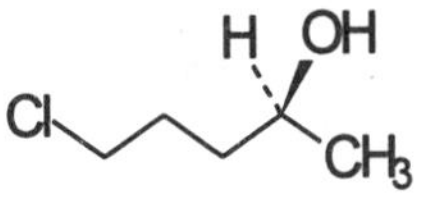

18.34 The reducing agent transfers a hydride ion from above the plane of the molecule as written below and gives the (S) alcohol.

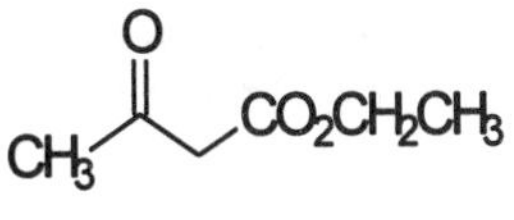

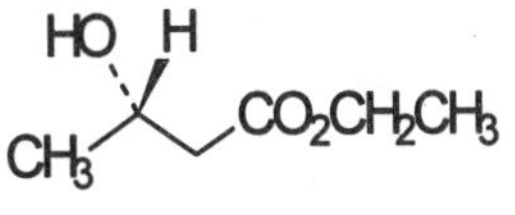

18.35 The double bond of 3-buten-2-one is conjugated with the carbonyl group. One of the resonance forms of the compound has a single bond between the C-3 and C-4 atoms. As a consequence there is some decrease in the double bond character, which makes stretching easier.

18.36   The bromine is inductively electron withdrawing. The partial positive charge of the C-2 atom destabilizes the dipolar resonance form of the carbonyl group. As a result the carbonyl group has increased double bond character, and the absorption occurs at higher wavenumber.

18.37   Structures with a formula of C_4H_8O must contain a ring or a double bond. Structures to consider include cyclic ethers, cyclic alcohols, unsaturated alcohols, unsaturated ethers, aldehydes, and ketones.

(a) A compound with only two signals must have two sets of nonequivalent carbon atoms. The signals correspond to an internal methylene group and a methylene group bonded to an oxygen atom. The compound is tetrahydrofuran.

(b) The compound has three sets of nonequivalent carbon atoms, with signals corresponding to methyl, methylene, and aldehyde carbon atoms. The compound is 2-methylpropanal.

(c) The compound has four nonequivalent carbon atoms, with signals corresponding to methyl, methylene, and carbonyl carbon atoms. The signals at 7.9, 29.4, and 36.9 ppm indicate the likelihood of two α carbon atoms and one further away. The compound is 2-butanone.

(d) The compound has four nonequivalent carbon atoms, with signals corresponding to methyl, methylene, and carbonlyl carbon atoms. The signals at 13.7, 15.7, and 45.8 ppm indicate the likelihood of one α carbon atom and two further away. The compound is butanal.

18.38   Structures with a formula of $C_5H_{10}O$ must contain a ring or a double bond. Structures to consider include cyclic ethers, cyclic alcohols, unsaturated alcohols, unsaturated ethers, aldehydes, and ketones.

(a) The compound has four nonequivalent carbon atoms, with signals corresponding to methyl, methylene, and carbonyl carbon atoms. The signals at 27.3 and 41.5 ppm indicate the likelihood of two α carbon atoms, with one being more highly substituted. The compound is 3-methyl-2-butanone.

(b) The compound has five nonequivalent carbon atoms, with signals corresponding to methyl, methylene, and carbonyl carbon atoms. The signals at 29.3 and 45.2 ppm indicate the likelihood of two α carbon atoms, with different degrees of substitution. The compound is 2-pentanone.

(c) The compound has three sets of nonequivalent carbon atoms, with signals corresponding to methyl, methylene, and carbonyl carbon atoms. The compound is 3-pentanone.

18.39   Isomeric ketones with the formula $C_{10}H_{12}O$ must have a total of four rings and/or double bonds to account for the formula. Keep this in mind as you try to match the hydrogen NMR signals to an appropriate structure.

(a) The signal of intensity 5 at 7.2 δ identifies the five hydrogen atoms on a phenyl group. The signals of intensity 2 at 3.6 and 2.3 δ are due to protons on two carbon atoms adjacent to a carbonyl group. The triplet at 1.0 δ is due to a methyl group adjacent to a methylene group. The compound is 1-phenyl-2-butanone.

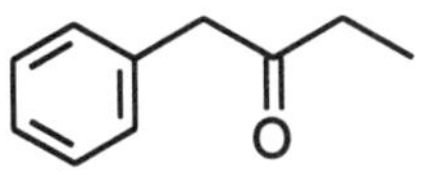

(b) The doublet of doublets with intensity 4 at 7.5 δ is due to two sets of equivalent protons on a benzene ring. The singlet of intensity 3 at 2.5 δ is due to a methyl group adjacent to a carbonyl group or carbon-carbon double bond, so it is likely due to a para-methyl substituent on the ring already identified. The signal of intensity 2 at 2.7 δ is due to a methylene group adjacent to a carbonyl group, but note that the signal is split by neighboring protons. The triplet at 1.2 δ is due to a methyl group adjacent to a methylene group. The compound is p-methylpropiophenone.

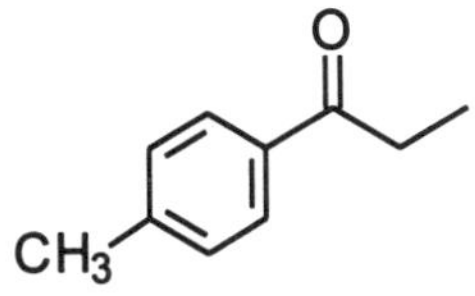

(c) The signal of intensity 5 at 7.2 δ is due to an unsubstituted phenyl group. The multiplet of intensity 4 at 2.8 δ is due to protons on carbon atoms adjacent to a carbonyl group or carbon-carbon double bond. The singlet of intensity 3 is due to a methyl group with no neighboring protons. The compound is 4-phenyl-2-butanone.

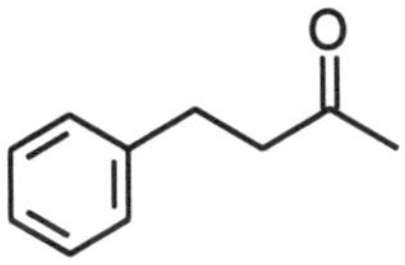

18.40 Isomeric ketones with the formula $C_9H_{10}O$ must have a total of five rings and/or double bonds to account for the formula. Keep this in mind as you try to match the hydrogen NMR signals to an appropriate structure.

(a) The doublets with intensity 2 at 6.9 and 7.9 δ are due to two sets of equivalent protons on a benzene ring. The singlet of intensity 3 at 2.6 δ is due to a methyl group adjacent to a carbonyl group or a carbon-carbon double bond, with no neighboring protons. The singlet of intensity 3 at 3.9 δ is due to a methyl group bonded to an oxygen atom. The compound is p-methoxyacetophenone.

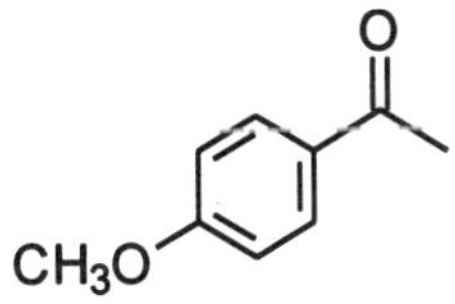

(b) The multiplet with intensity 5 at 7.0 δ is due to a phenyl group. The singlet of intensity 2 at 4.4 δ is due to a methylene group bonded to an oxygen atom. The singlet of intensity 3 at 2.6 δ is due to a methyl group bonded to a carbonyl carbon atom. The compound is phenoxyacetone.

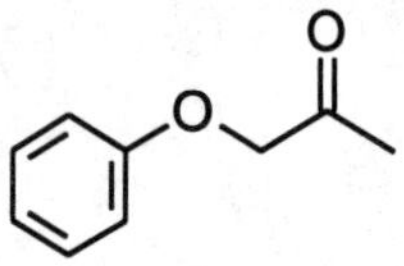

(c) The singlet of intensity 1 at 9.9 δ is due to an aldehydic hydrogen atom. The doublets of intensity 2 at 7.0 and 8.8 δ are due to two sets of equivalent protons on a benzene ring. The quartet of intensity 2 at 4.1 δ is due to a methylene group bonded to an oxygen atom and adjacent to a methyl group. The triplet of intensity 3 at 1.4 δ is due to a methyl group adjacent to a methylene group. The compound is p-ethoxybenzaldehyde.

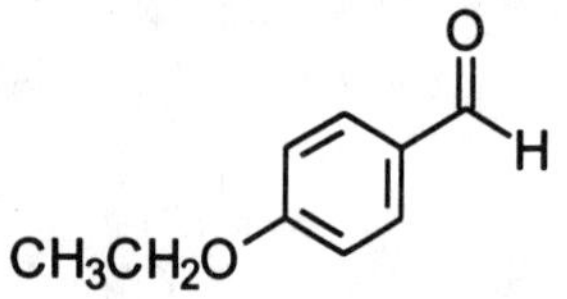

19

Aldehydes and Ketones: Nucleophilic Addition Reactions

Now that you know how to synthesize carbonyl compounds, we consider their use to synthesize other classes of compounds. The chemistry depends on the bonding characteristics of the carbonyl group. Recall that there is a dipolar resonance form that contributes to the carbonyl group. Much of the chemistry of carbonyl compounds involves addition of a nucleophile to the partially positive carbon atom of the carbonyl group. Thus, the substituents bonded to this center affect the reactivity of carbonyl compounds. Those same groups also affect the reactivity of the α carbon atom, which is the subject of Chapter 23. This chapter concentrates on the reactions that involve only the carbonyl group itself.

Aldehydes and ketones undergo addition reactions with unsymmetrical reagents such as HNu. The nucleophilic part of the reagent adds to the carbon atom of the carbonyl group, and the electrophilic part adds to the oxygen atom. These products are obtained in both acid- and base-catalyzed reactions.

Keys to the Chapter

19.1 Thermodynamic Considerations

In this section the ΔH°_{rxn} for the addition reactions of alkenes is contrasted with the ΔH°_{rxn} for the addition reactions of carbonyl compounds. Because the bond energy of the π bond of carbonyl groups is larger than the bond energy of the π bond of alkenes, there is a smaller driving force for addition reactions of carbonyl compounds. Only the hydrogenation reaction of carbonyl compounds is exothermic. The addition reaction of water is nearly enthalpically neutral.

The addition reactions of an aldehyde are more favorable in a thermodynamic sense than the addition reactions of ketones. This observation is consistent with the stabilization of the dipolar resonance form of the carbonyl group. In a ketone, two alkyl groups stabilize the positive charge on the carbon atom in this resonance form. The stabilization of a reactant makes a reaction that eliminates that feature less favorable.

19.2 Irreversible and Reversible Addition Reactions

Unlike the reactions of other functional groups such as alkenes, alkynes, alkyl halides, and alcohols, the reactions of the carbonyl group may be reversible. The consequence of this reversibility must be considered in the choice of conditions to drive the reaction toward the product side of the equilibrium. The reversibility of the reaction depends on the characteristics of the group originally added as a nucleophile to the carbonyl carbon atom. Both the alkyl group of a Grignard reagent and the hydride ion of a metal hydride are strong bases and hence poor leaving groups. Thus, the addition of a Grignard to a carbonyl compound to give an alcohol and the reduction of a carbonyl compound using hydride ion are both irreversible reactions.

If a nucleophilic group that adds to a carbonyl carbon atom is a good leaving group, then the reaction is reversible. Such groups are generally weak bases. The cyanide ion is an example. The addition product of HCN to a carbonyl group is a cyanohydrin and it may reversibly eliminate HCN to give the original carbonyl group.

The products of addition reactions of carbonyl compounds may be determined by **kinetic control** or **thermodynamic control**. In the case of irreversible reactions, the possible products formed are controlled by the rate of the reaction. In reversible reactions, the possible products formed are controlled by their thermodynamic stability.

Based on the equilibrium constants for several types of reactions, several conclusions can be made about the effect of structure on the % conversion to product. These conclusions can be used to estimate the effect of structure on any carbonyl addition reaction. Each of the three statements listed on page 704 of the text are the consequence of the stabilization or destabilization of the dipolar resonance form of the carbonyl group. Make sure that you understand that any feature that stabilizes a positive charge then stabilizes the carbonyl group and hence decreases its tendency to undergo addition reactions.

19.3 Hydration of Carbonyl Compounds

The equilibrium constant for the addition of water to a carbonyl compound is greater than one for only a few compounds such as formaldehyde. Although not a useful synthetic reaction, the equilibrium constant provides a basis for evaluating the effect of structure on carbonyl addition reactions in general.

Aldehydes are more hydrated than ketones in these addition reactions as a result of their structural difference of a hydrogen atom versus an alkyl group. Nucleophiles are attracted to a carbonyl carbon atom because of its partial positive charge. The partial positive charge is larger in an aldehyde, which has only one alkyl group bonded to the carbonyl carbon atom, than in a ketone, which has two alkyl groups. There is a steric effect also. The carbonyl carbon atom has a larger bond angle than the tetrahedral product. As a result of increased steric hindrance, the addition product is less stable for ketones than for aldehydes.

Groups bonded to the α carbon atom can affect the electron density of the carbonyl carbon atom by either resonance or inductive effects. Resonance stabilization of the carbonyl group as illustrated in Section 18.8 of the text decreases the equilibrium constant for addition reactions. Inductive electron withdrawal by electronegative groups such as halogen atoms destabilizes the carbonyl group and increases the equilibrium constant for addition reactions.

19.4 Mechanisms of Addition Reactions of Carbonyl Compounds

Addition reactions to carbonyl groups can occur by either of two mechanisms depending on whether the reaction conditions involve acid or base. In both cases, addition of a nucleophile occurs at the carbonyl carbon atom and addition of an electrophile occurs at the oxygen carbon atom. The difference between the mechanisms for addition under acidic or basic conditions is in the sequence of two possible reactions. Under acidic conditions, a proton adds to the carbonyl oxygen atom to give a conjugate acid that has a greater positive charge on the carbonyl carbon atom, which then reacts with a nucleophile. Under basic conditions, the nucleophilicity of the nucleophile is sufficient to attack the carbonyl carbon atom to give an alkoxide ion, which subsequently reacts with the electrophile—usually a proton.

19.5 Kinetic Effects in Addition Reactions

The same structural features that control the equilibrium constant for the addition reaction of carbonyl compounds also affect the rate of the reaction. In the case of the rate of a reaction, the stabilization of the partial positive charge on the carbonyl carbon diminishes its attraction to a nucleophile. Thus aldehydes react faster than ketones. The steric features of the alkyl groups in aldehydes and to a greater degree in ketones affect the reaction by decreasing the rate of reaction with increasing size of the groups. The groups sterically hinder the approach of the nucleophile in a manner reminiscent of the same effect observed in S_N2 reactions.

19.6 Addition of Alcohols to Carbonyl Compounds

Alcohols add to aldehydes and ketones to give **acetals** and **ketals**, respectively. The equilibrium constant is less than one. However, intramolecular reactions are favored. Analysis of the bond energies of the bonds broken in the reactant and formed in the product suggests that the $\Delta H°_{rxn}$ for intermolecular as well as intramolecular reactions are only slightly negative. As a result the unfavorable $\Delta S°_{rxn}$ for the intermolecular reaction makes the $\Delta G°_{rxn}$ positive. The $\Delta S°_{rxn}$ for the intramolecular reaction is much less negative because there is no difference in the number of moles of reactant and product.

You should learn to recognize the hemiacetal and hemiketal structures because they are present in more complicated structures where other functional groups may obscure the issue. You will have to recognize these functional groups in compounds such as carbohydrates. Look at the structure for glucose at the top of page 710 in the text. Could you pick out and identify the hemiacetal without the label pointing to that center? The clue is the two oxygen atoms bonded to the same carbon atom, one in an —OH group and the other in an —OR group..

Hemiacetals result from the reaction of an aldehyde, RCHO, with an alcohol, R'OH. Thus a hemiacetal is characterized by a hydroxyl group, an -OR' group from the alcohol, an -R group from the aldehyde, and a hydrogen atom all bonded to the same carbon atom. Given the structural formula of a hemiacetal, we can mentally reverse the reaction to determine the aldehyde and alcohol from which it was formed.

Hemiketals result from the reaction of a ketone with an alcohol. A hemiketal is characterized by an -OH group, an -OR' group from the alcohol, and two -R groups from the ketone all attached to the same carbon atom. The -R groups may be alkyl or aryl groups in either hemiacetals or hemiketals.

19.7 Formation of Acetals and Ketals

Hemiacetals and hemiketals can be converted into **acetals** and **ketals**, respectively. An acetal results from the reaction of a hemiacetal with an alcohol in acidic solution. An acetal is characterized by two -OR' groups from the alcohol, an -R group from the original aldehyde, and a hydrogen atom all bonded to the same carbon atom. A ketal results from the reaction of a hemiketal with an alcohol in acidic solution.   A ketal has two -OR' groups from the alcohol and two -R groups from the ketone attached to the same carbon atom.   The -R groups may be alkyl or aryl groups.

Given the structural formula of an acetal or a ketal, we can determine the aldehyde or ketone and the alcohol that led to its formation.   The two -OR' groups identify the alcohol reactants as R'OH, and the remainder of the molecule identifies the aldehyde or ketone which produced the hemiacetal or hemiketal which was then converted to an acetal or a ketal, respectively.

The formation of acetals from aldehydes and alcohols is essentially enthalpically neutral because the same number and types of bonds are made and broken.   Furthermore, the $\Delta S°_{rxn}$ is essentially zero.   Thus, the reaction is driven by the removal of the water formed.   Remember, of course, that the reaction to form the hemiacetal is not favored either.   In the case of ketones, the initial reaction forming the hemiketal is disfavored and, as a result, so is the formation of ketals.

The mechanism of conversion of a hemiacetal into an acetal (or a hemiketal into a ketal) proceeds by protonation of the alcohol to convert it into a better leaving group.   (This isn't the first time that you have seen this step in other reactions where a C—O bond is heterolytically cleaved.)   The resulting oxocarbocation is resonance stabilized. (Remember that such stabilization was the driving force for the rearrangement of an alkyl group in the pinacol rearrangement discussed in Section 16.5.)   It subsequently reacts with the nucleophilic oxygen atom of the alcohol, and the last step is proton transfer from the oxonium ion.   Note that the reaction requires acid conditions.   As a result, acetals (or ketals) can regenerate the carbonyl compound under acidic conditions in the presence of water.   However, both acetals and ketals are stable under basic conditions.

19.8 Acetals as Protecting Groups

Protecting groups are provided by reagents that easily form a derivative of the functional group to be "protected" but can also be easily removed when required.   A protecting group is used to render a functional group unreactive toward specific reagents that are required to transform a second functional group in the molecule.   Acetals and ketals are ideal protecting groups because they are easily formed by driving the reaction to completion, and as a result they spontaneously react to liberate the carbonyl group in the reverse of that reaction.

Cyclic acetals derived from ethylene glycol are used as protecting groups for carbonyl compounds, because their formation is not entropically disfavored as are the formation of acetals using two moles of an alcohol.   As a consequence of the elimination of an unfavorable entropy of reaction term, ketals are also easily formed.   In dicarbonyl compounds it may be possible to selectively form a derivative of one carbonyl group if the structural features are sufficiently different.

Alcohols can also be protected by their incorporation into an acetal or ketal.   Both 1,2-diols and 1,3-diols react with acetone to give a ketal.   Alcohols react under acid-catalyzed conditions with dihydropyran to give a THP derivative which is easily hydrolyzed.

19.9 Thioacetals and Thioketals

The sulfur analogs of acetals and ketals are synthesized using the dithiols 1,2-ethanedithiol or 1,3-ethanedithiol.   The thioacetals and thioketals are stable not only under basic conditions, but under acidic conditions as well.   Unprotecting the carbonyl group requires mercury(II) chloride.   Thioacetals and thioketals can be reduced to methylene compounds using Raney nickel.   Thus, the sequence of thioacetal formation followed by reduction is another method to reduce carbonyl compounds to hydrocarbons.

19.10 Addition of Nitrogen Compounds

Amines and other nitrogen derivatives that are sufficiently nucleophilic attack the carbonyl carbon atom to give a tetrahedral addition product, called a **hemiaminal**, but it is unstable with respect to dehydration. As a result, an imine forms. The overall reaction is termed an **addition-elimination reaction**. Keep this reaction in mind whenever a compound with an —NH_2 group is in the presence of a carbonyl group. Examine the several reagents on page 721 of the text. In every case the net result of the reaction can be obtained by mentally stripping away an oxygen atom from the carbonyl compound and two hydrogen atoms from the amino group. Then join the two fragments by a double bond. You should be able to write such products without resorting to writing out each step of the mechanism.

$$\underset{R}{\overset{R}{>}}C=O \ + \ NH_2-G \ \longrightarrow \ \underset{R}{\overset{R}{>}}C=N \overset{G}{} \ + \ H_2O$$

19.11 The Wittig Reaction

Alkenes can be synthesized from carbonyl compounds by reaction with a phosphorus ylide in the **Wittig reaction**. First you have to know how to make the ylide. React an alkyl halide with triphenylphosphine to obtain a phosphonium ion. Then remove a proton by using a strong base such as butyllithium. The negatively charged carbon atom of the ylide reacts as a nucleophile and attacks the carbonyl carbon atom to give an intermediate that eliminates oxygen and a phosphorus moiety, leaving the desired alkene.

Summary of Reactions

1. Formation of Cyanohydrins

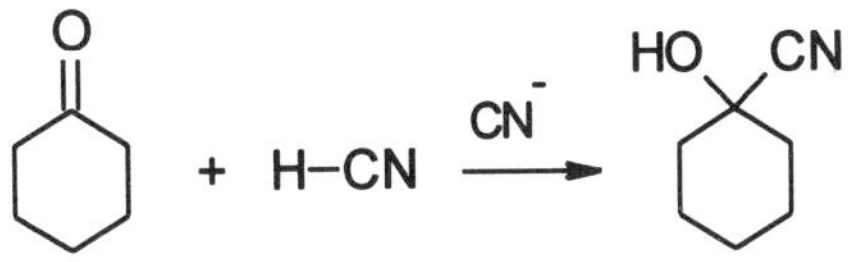

2. Formation of Acetals and Ketals

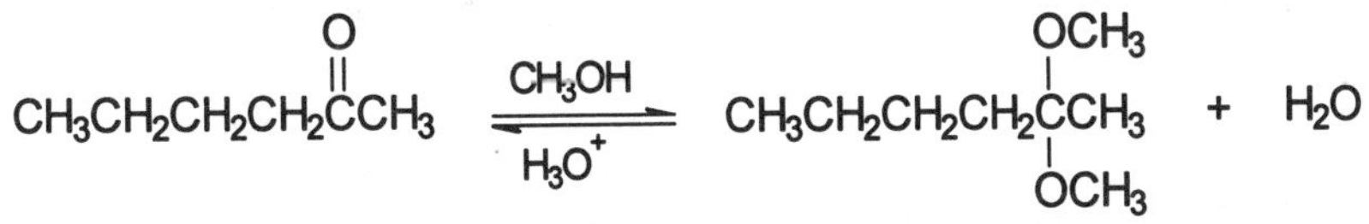

$$CH_3CH_2CH_2CH_2\overset{O}{\overset{\|}{C}}CH_3 \ \underset{H_3O^+}{\overset{CH_3OH}{\rightleftharpoons}} \ CH_3CH_2CH_2CH_2\underset{OCH_3}{\overset{OCH_3}{\underset{|}{\overset{|}{C}}}}CH_3 \ + \ H_2O$$

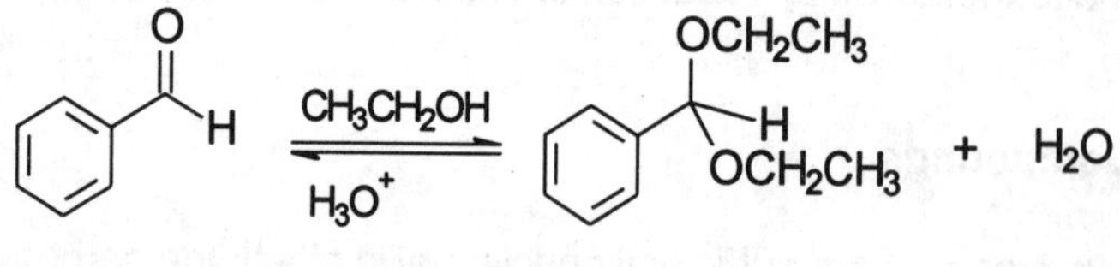

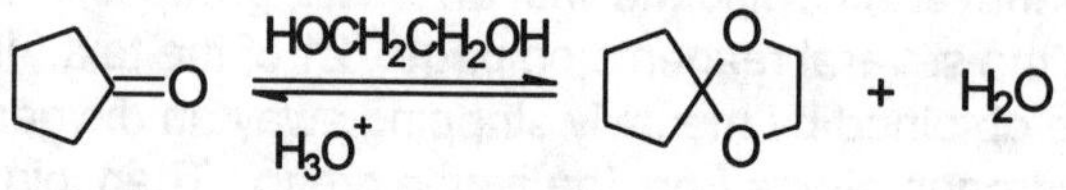

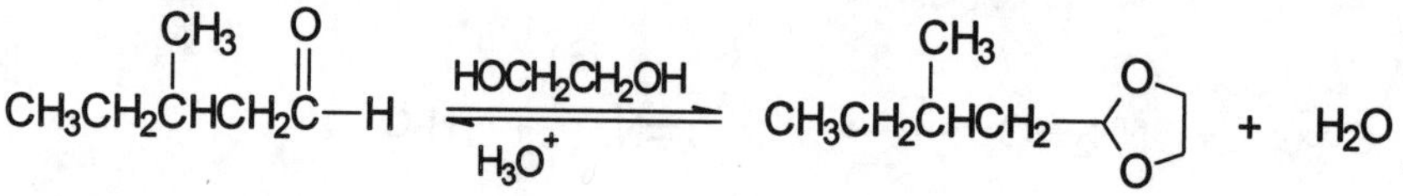

3. Protection of Alcohols by Acetal Formation

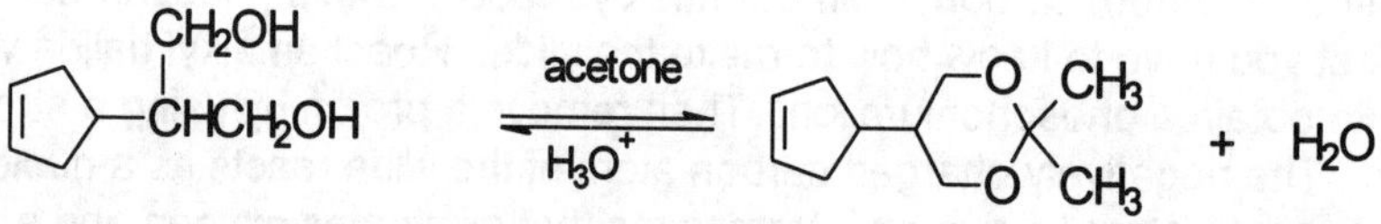

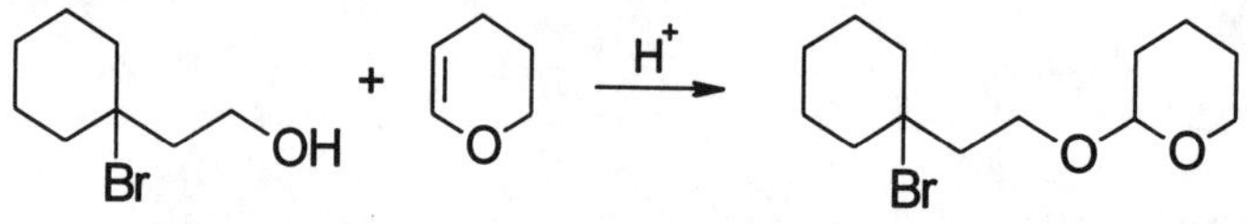

4. Use of Thioacetals and Thioketals as Protecting Groups

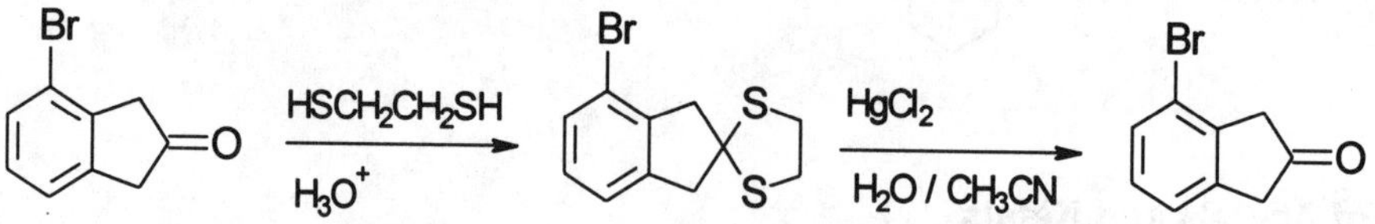

5. Use of Thioacetals and Thioketals to Reduce Carbonyl Groups to Methylene Groups

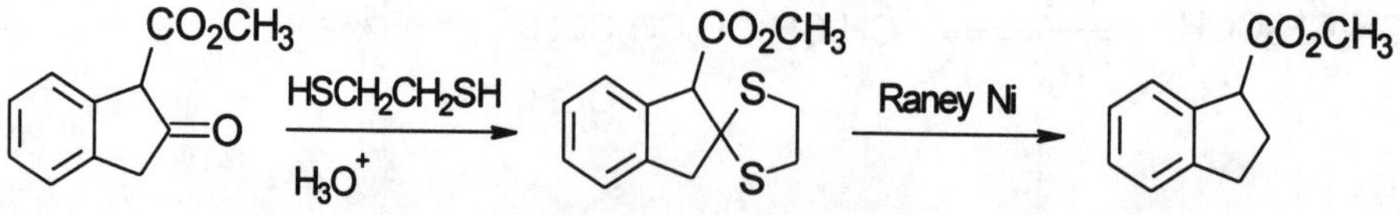

6. Addition of Nitrogen Compounds to Carbonyl Compounds

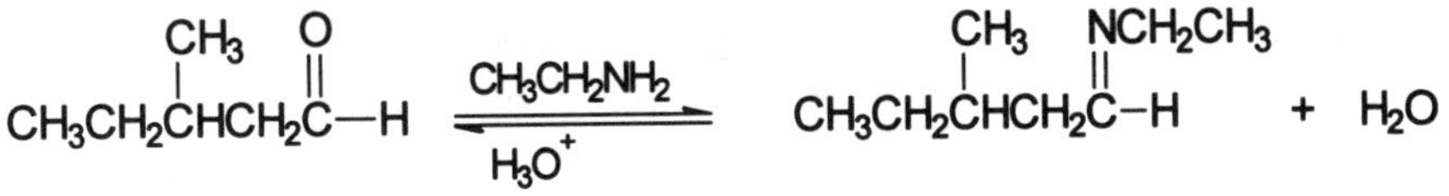

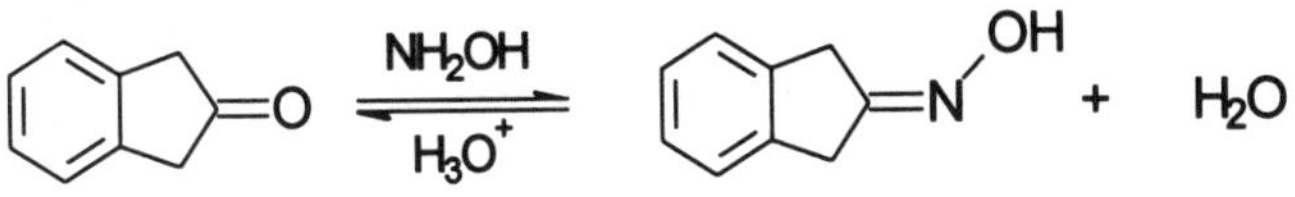

$$CH_3CH_2CHCH_2\overset{\overset{\displaystyle O}{\|}}{C}-H \quad \underset{H_3O^+}{\overset{CH_3CH_2NH_2}{\rightleftharpoons}} \quad CH_3CH_2CHCH_2\overset{\overset{\displaystyle NCH_2CH_3}{\|}}{C}-H \quad + \ H_2O$$

(with CH_3 substituents on the third carbon)

$$\underset{}{\text{indanone}}=O \quad \underset{H_3O^+}{\overset{NH_2OH}{\rightleftharpoons}} \quad =N^{\diagup OH} \quad + \ H_2O$$

7. Synthesis of Alkenes using the Wittig Reaction

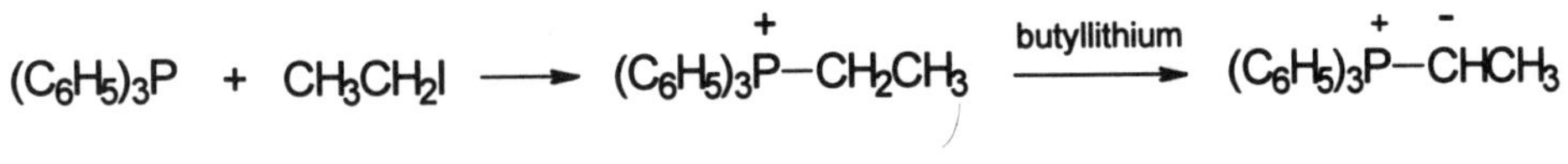

$$(C_6H_5)_3P \ + \ CH_3CH_2I \ \longrightarrow \ (C_6H_5)_3\overset{+}{P}-CH_2CH_3 \ \underset{}{\overset{butyllithium}{\longrightarrow}} \ (C_6H_5)_3\overset{+}{P}-\overset{-}{C}HCH_3$$

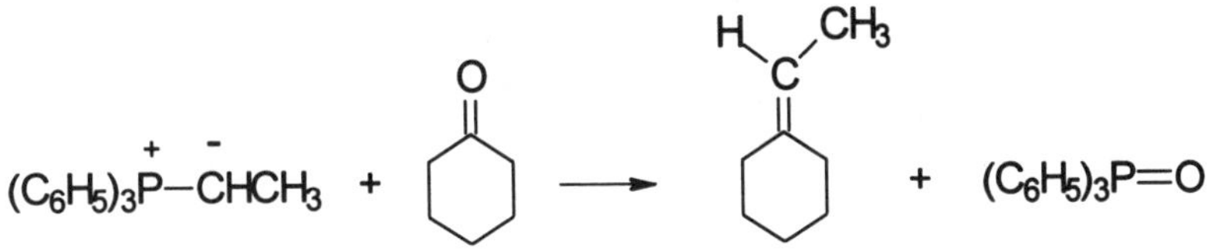

$$(C_6H_5)_3\overset{+}{P}-\overset{-}{C}HCH_3 \ + \ \underset{}{\overset{O}{\bigcirc}} \ \longrightarrow \ \underset{}{\overset{H\diagup C \diagdown CH_3}{\bigcirc}} \ + \ (C_6H_5)_3P{=}O$$

Solutions to Exercises

19.1   The total of the energy released for formation of an O—H bond and a C—CN bond is 425 + 510 = 935 kJ mole^{-1}. The total energy required to replace a carbon-oxygen double bond by a carbon-oxygen single bond and to break an H—CN bond is 365 + 545 = 910 kJ mole^{-1}. Thus, the overall reaction is exothermic by 25 kJ mole^{-1} and $\Delta H° = -25$ kJ mole^{-1}.

19.2   The total of the energy released for formation of an O—H bond and a C—S bond is 425 + 270 = 695 kJ mole^{-1}. The total energy required to replace a carbon-oxygen double bond by a carbon-oxygen single bond and to break an H—S bond is 365 + 345 = 710 kJ mole^{-1}. Thus, the overall reaction is endothermic by 15 kJ mole^{-1} and $\Delta H° = + 15$ kJ mole^{-1}.

19.3   In the reaction as written, a carbon-carbon double bond is replaced by a carbon-carbon single bond ($\Delta H° = +250$ kJ mole^{-1}), and a carbon-oxygen single bond is replaced by a carbon-oxygen double bond ($\Delta H° = -365$ kJ mole^{-1}), for a net release of 115 kJ mole^{-1}. In addition one C—H bond is broken and two C—H bonds are formed (net $\Delta H° = -420$ kJ mole^{-1}), and one O—H bond is broken ($\Delta H° = +425$ kJ mole^{-1}). Thus the overall reaction has $\Delta H°_{rxn}$ 250 - 365 - 420 + 425 = -120 kJ mole^{-1}.

19.4   The heats of formation of 2-propanone and 2-propanol are -217 and -272 kJ mole^{-1}, respectively (from page 703 of the text). Therefore, $\Delta H° = +55$ kJ mole^{-1} for the reaction. For a reaction converting one mole of reactant into two moles of product, the $\Delta S° = +125$ J mole^{-1} deg^{-1} (from page 115 in the text). The minimum temperature at which the reaction becomes favorable occurs when $K_{eq} = 1$, which is when $\Delta G° = 0$, which occurs when $\Delta H° = T\Delta S°$. Substituting for $\Delta H°$ and $\Delta S°$, that temperature is 440 K or 167°C.

19.5   Formaldehyde has the most reactive carbonyl group because it does not have substituents to stabilize the dipolar resonance form. It is also the least sterically hindered carbonyl compound and is readily attacked by nucleophiles.

19.6   The compound is a liquid and is relatively easy to handle. The aldehyde functional group is less sterically hindered than a ketone and is more readily attacked by nucleophiles. This particular dialdehyde is "twice" as effective in capturing nucleophilic sites, because once the first carbonyl group reacts, the second aldehyde is contained within the biomolecule and can react with another nucleophilic site in the molecule. As a result the biological effectiveness of the biomolecule is severely compromised.

19.7   **(a)** Cyclopropanone is the more reactive because the strain associated with the incorporation of an sp^2-hybridized carbon atom in a ring is diminished by the lower bond angle requirements of an sp^3-hybridized carbon atom in a ring. Thus, cyclopropanone reacts faster with sodium borohydride.

(b) Both compounds have resonance stabilized carbonyl groups. However, acetophenone also has a methyl group bonded to the carbonyl carbon atom and is a ketone. Thus, acetophenone has a more stabilized carbonyl group and is less reactive. The carbonyl group of acetophenone is also more sterically hindered than the carbonyl group of benzaldehyde. Thus, benzaldehyde reacts faster with sodium borohydride.

(c) Both compounds are ketones and the inductive effects of the alkyl groups are similar in both. However, one of the alkyl groups of 3,3-dimethyl-2-butanone is a *tert*-butyl group. Thus, the carbonyl group of this compound is more sterically hindered than the carbonyl group of acetone. Thus, acetone reacts faster with sodium borohydride.

19.8 (a) Benzaldehyde has a resonance stabilized carbonyl group and is less reactive than an aldehyde such as acetaldehyde that does not have such stabilization. The carbonyl group of benzaldehyde is also more sterically hindered than the carbonyl group of acetaldehyde. Thus, acetaldehyde reacts faster with sodium borohydride.

(b) Reduction of cyclopentanone gives an alcohol with additional torsional strain. In the case of cyclohexanone the product does not have added torsional strain because the C—H and C—OH bonds are each staggered with respect to bonds on adjacent carbon atoms. Thus, cyclohexanone reacts faster with sodium borohydride.

(c) The trifluoromethyl group is inductively electron withdrawing and destabilizes the dipolar resonance form of the carbonyl group. As a result the carbonyl group of p-trifluoromethylbenzaldehyde is more reactive than that of benzaldehyde.

19.9 Each additional chlorine atom may increase the equilibrium constant by a factor of 37. The estimated equilibrium constant is $(37)^3$, which is approximately 5×10^4.

19.10 Each of the two chlorine atoms may increase the equilibrium constant by a factor of 37. The estimated equilibrium constant is $0.0014 \times (37)^2$, which is approximately 1.9.

19.11 The p-methoxy group stabilizes the carbonyl group by a resonance interaction in which the methoxy group releases electrons. The m-methoxy group destabilizes the carbonyl group somewhat by an inductive effect in which the methoxy group withdraws electron density. No donation of electrons by resonance is possible for two groups located in meta positions.

19.12 One phenyl group can stabilize the carbonyl group because it can exist in a coplanar conformation with the carbonyl group, allowing the π electrons to overlap. Two phenyl groups cannot both be coplanar with the carbonyl group, because there is a steric repulsion between the ortho hydrogen atoms of the two rings. One of those hydrogen atoms is shown.

The ortho hydrogen atom of the other ring would have to occupy the same space as the indicated hydrogen atom if both rings were coplanar. Rotation of the ring about the bond to the carbonyl carbon atom relieves this repulsion but eliminates the resonance interaction with the carbonyl group. The resonance interaction between the carbonyl group and the first phenyl group stabilizes acetophenone compared to acetaldehyde, but the change in stability is small as the second phenyl group is considered.

19.13   Cyclopropanone has a larger equilibrium constant because the strain associated with the incorporation of an sp²-hybridized carbon atom in a ring is diminished by the lower bond angle requirements of an sp³-hybridized carbon atom in a ring. The product is less strained, so the reaction is favored.

19.14   Hydration of cyclopentanone gives a hydrate with additional torsional strain between the C—OH bonds and the adjacent C—H bonds. In the case of cyclohexanone the hydrate does not have added torsional strain because the C—OH bonds are each staggered with respect to adjacent C—H bonds. Thus, cyclohexanone has the larger equilibrium constant.

19.15   The carbonyl carbon atom of 2,2,4,4-tetramethyl-3-pentanone has two *tert*-butyl groups bonded to it at 120° angles. Formation of the cyanohydrin decreases the bond angle between the two *tert*-butyl groups to 109°, resulting in increased steric congestion. Thus, the reaction is less favored than the reaction with 2-propanone.

19.16   Addition of HCN to cyclopentanone gives a cyanohydrin with additional torsional strain between the C—OH and C—CN bonds and the adjacent C—H bonds. In the case of cyclohexanone the cyanohydrin does not have added torsional strain because the C—OH and C—CN bonds are each staggered with respect to adjacent C—H bonds. Thus, cyclohexanone has the larger equilibrium constant.

19.17   The carbonyl carbon atom of 3,3-dimethyl-2-butanone has a *tert*-butyl and a methyl group bonded to it at 120° angles. Formation of the cyanohydrin decreases the bond angle between the two groups to 109°, resulting in increased steric congestion. The alkyl groups in 2-butanone are methyl and ethyl, so the steric repulsion is far less. Thus, the reaction with 3,3-dimethylbutanone is less favored than the reaction with butanone.

19.18   The p-methoxy group stabilizes the carbonyl group by a resonance interaction in which the methoxy group releases electrons. Thus, its conversion to a cyanohydrin in which this resonance stabilization is lost is less favorable than for benzaldehyde.

19.19   The p-methyl group donates electron density to aromatic rings and hence stabilizes the carbonyl group. Thus, its conversion to a cyanohydrin in which this stabilization is lost is less favorable than for benzaldehyde. Its equilibrium constant should be smaller.

19.20   Nitrogen is a better donor of electrons by resonance than oxygen. As a consequence the dimethylamino group should more strongly stabilize the carbonyl group by a resonance interaction. Thus, the conversion of p-dimethylaminobenzaldehyde to a cyanohydrin in which this stabilization is lost is less favorable than for p-ethoxybenzaldehyde.

19.21   Conversion into a cyanohydrin must place either a hydroxyl or a cyano group in an axial position, so there will be a 1,3-diaxial interaction with the axial methyl group at the C-3 atom. This steric repulsion between groups decreases the stability of the product and therefore decreases the equilibrium constant compared to that of cyclohexanone.

19.22 Attack of the cyanide occurs fastest in the equatorial direction to form the cyanohydrin with the hydroxyl group in the axial position. This is the kinetic product. However, the hydroxyl group has a larger conformational preference for the equatorial position than the cyano group. As a consequence the reversible formation of the cyanohydrin under equilibrium conditions will favor formation of the compound with the cyano group axial and the hydroxyl group equatorial.

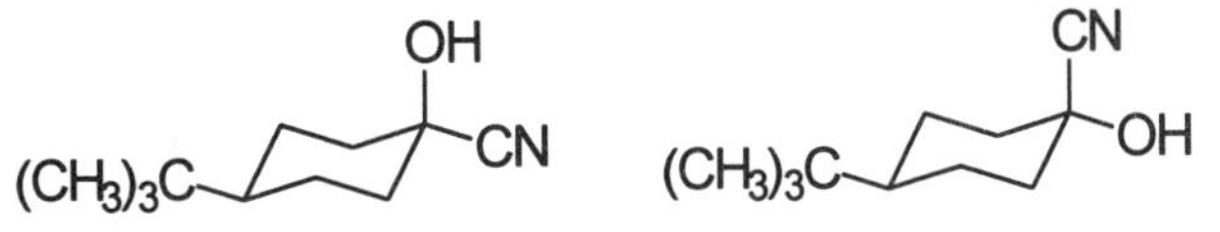

19.23 There is no direct reaction between the two reactants. Each carbonyl compound can exist in equilibrium with its cyanohydrin and HCN. Thus, the HCN is simply distributed between the two carbonyl compounds. The mechanism is the standard one for addition of HCN to a carbonyl compound. The equilibrium constants for formation of the cyanohydrins for benzaldehyde and propanone are 210 and 30, respectively. Thus, the reaction as written has $K = 7$.

19.24 **(a)** Both carbonyl groups are equivalent and only one reacts with one equivalent of HCN.

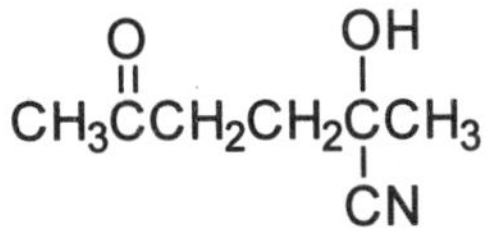

(b) The equilibrium constant for the saturated aldehyde side chain is larger than for benzaldehyde, so reaction occurs at that site.

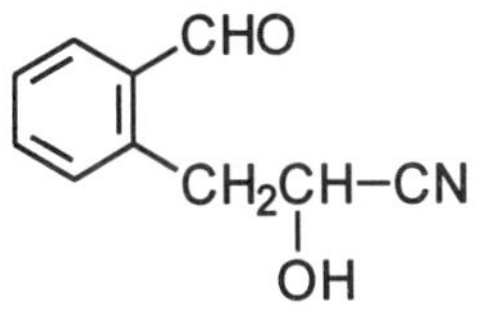

(c) The ketone that is para to the methoxy group is resonance stabilized, so the equilibrium constant for its reaction with HCN is smaller than for the ketone that is meta to the methoxy group. The methoxy group is inductively electron withdrawing and should destabilize the dipolar resonance form of the carbonyl group meta to it, making it more reactive.

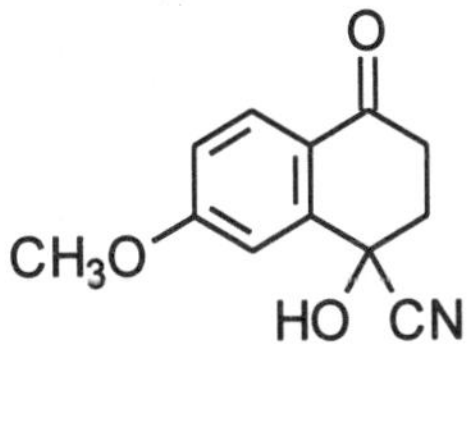

(d) The cyclic ketone is sterically hindered by the two methyl groups at the adjacent carbon atom. Thus, the cyanohydrin formed occurs at the methyl ketone of the chain bonded to the ring.

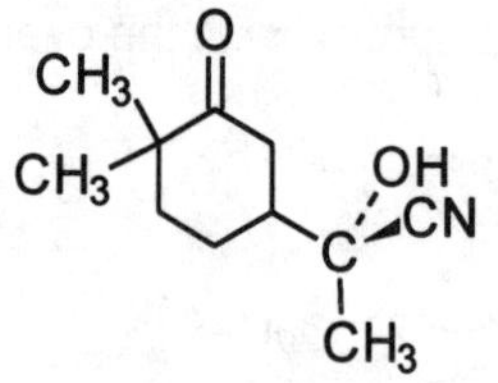

19.25 **(a)** acetal   **(b)** ketal   **(c)** acetal   **(d)** hemiacetal

19.26 **(a)** ketal   **(b)** diacetal   **(c)** hemiketal   **(d)** ketal

19.27 **(a)** acetal   **(b)** hemiacetal **(c)** hemiketal   **(d)** ketal

19.28 **(a)** hemiacetal **(b)** ketal   **(c)** acetal   **(d)** acetal

19.29 There is a primary alcohol, a secondary alcohol, and a ketal.

19.30 There is a secondary alcohol, an amino group, and a hemiacetal.

19.31 The order of the formation constants for acetals should parallel those for the formation of hydrates. The fluoro groups should favor formation of the acetal of trifluoroethanal compared to ethanal. Thus, the equilibrium constant should be greater than one.

19.32 Formation of a five-membered hemiacetal ring of 4-hydroxybutanal is less favored than formation of a six-membered hemiacetal ring of 5-hydroxypentanal because there is torsional strain in the five-membered ring. The six-membered ring has no torsional strain.

19.33 The benzene ring and the methyl group can be cis or trans in the cyclic acetal.

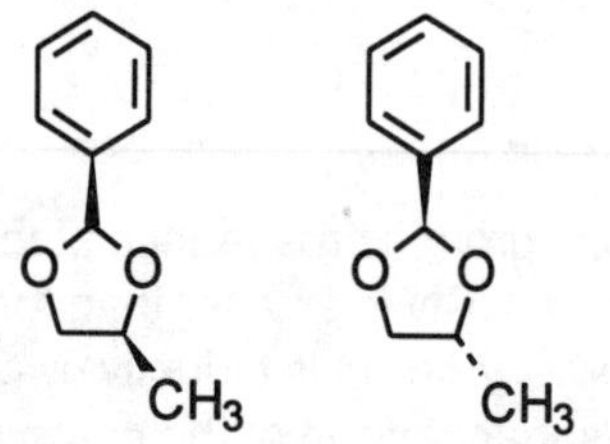

19.34 Acetone can form a ketal with a six-membered ring by reacting with the C-1 and C-3 hydroxyl groups or a ketal with a five-membered ring by reacting with the C-1 and C-2 hydroxyl groups.

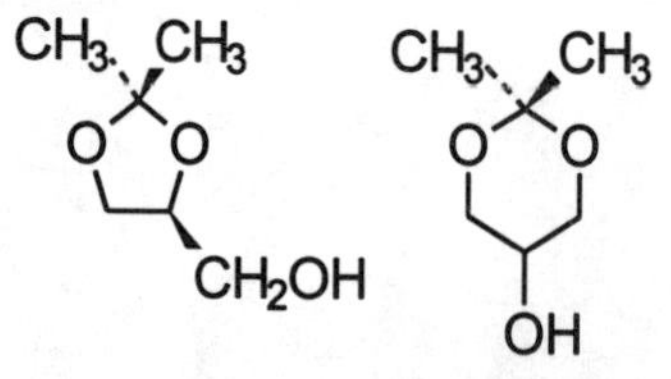

19.35 An acetal forms. The ketone is less reactive, so ketal formation is not favored.

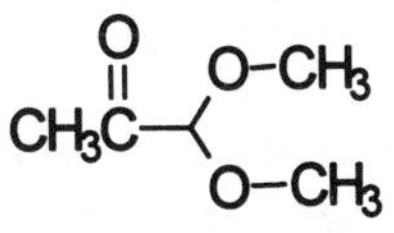

19.36 Both an acetal and a ketal form (at the original C-1 and C-2 atoms, respectively), because the equilibrium constants for formation of cyclic derivatives are more favorable than those for reaction with two moles of an alcohol to form acyclic derivatives.

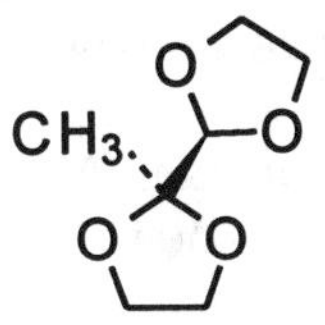

19.37 **(a)**

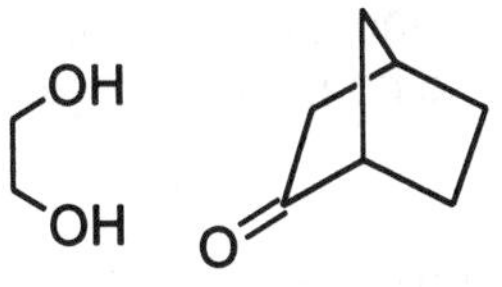

(b)

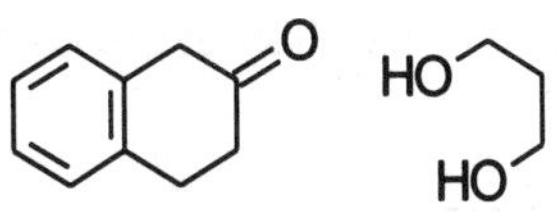

(c)

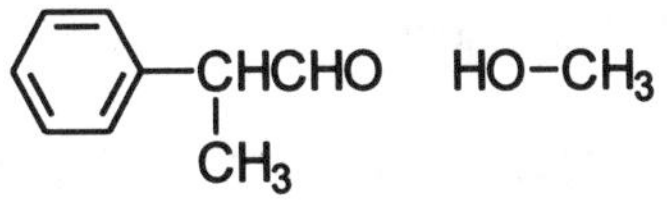

(d)

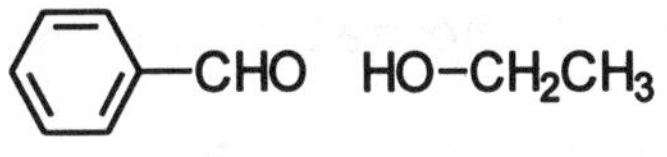

19.38 **(a)**

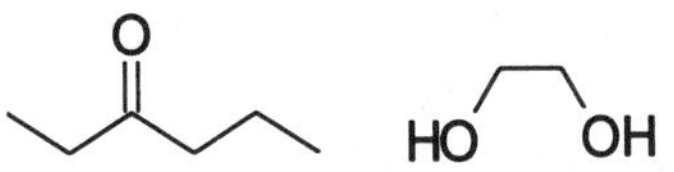

(b)

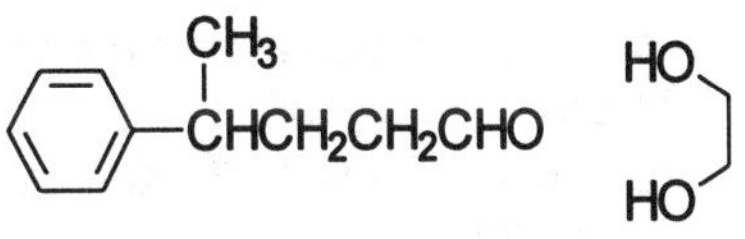

(c)

(d) OHCCH₂CH₂CHO   HO–CH₃

19.39 **(a)** Prepare a cyclic ketal of the ketone using ethylene glycol. Reduce the ester with lithium aluminum hydride to give the primary alcohol. Hydrolyze the ketal with dilute acid.

(b) Prepare a cyclic acetal of the aldehyde using ethylene glycol. Oxidize the secondary alcohol to a ketone using PCC. Add the Grignard reagent prepared from iodomethane. Hydrolyze the acetal with dilute acid, which will also convert the magnesium alkoxide to a tertiary alcohol.

(c) Prepare a cyclic ketal of the ketone using ethylene glycol. Reduce the ester with lithium aluminum hydride to give the primary alcohol. Hydrolyze the ketal with dilute acid.

19.40 (a) Prepare either an acetal of the aldehyde with methanol or a cyclic acetal using one mole of ethylene glycol. The ketone is less reactive and it remains unprotected. Reduce the ketone with sodium borohydride to give a secondary alcohol. Hydrolyze the acetal with dilute acid.

(b) Prepare a THP derivative of the alcohol using dihydropyran. Add the Grignard reagent prepared from iodomethane. Hydrolyze the THP derivative with dilute acid, which will also convert the magnesium alkoxide to a secondary alcohol.

(c) Prepare a THP derivative of the alcohol using dihydropyran. Add sodium amide to make the acetylide ion of the acetylene. Add bromoethane to alkylate the acetylide ion. Hydrolyze the THP derivative with dilute acid. Oxidize the secondary alcohol to a ketone using PCC.

19.41 (a) Prepare a cyclic ketal using ethylene glycol. Obtain the secondary alcohol by hydroboration-oxidation of the carbon-carbon double bond. Hydrolyze the ketal with dilute acid.

(b) Prepare a THP derivative of the alcohol using dihydropyran. Form the Grignard reagent of this protected bromo compound. React the Grignard reagent with ethanal. Hydrolyze the THP derivative with dilute acid.

(c) Oxidize the secondary alcohol using PCC. Prepare a ketal using ethylene glycol. Add sodium amide to make the acetylide ion of the acetylene. React the acetylide ion with ethanal to give the acetylenic alcohol. Hydrolyze the ketal derivative with dilute acid.

19.42 (a) Prepare either an acetal of the aldehyde with methanol or a cyclic acetal using one mole of ethylene glycol. The ketone is less reactive and it remains unprotected. Reduce the ketone with sodium borohydride to give a secondary alcohol. Hydrolyze the acetal with dilute acid.

(b) Prepare a cyclic ketal of the diol using acetone. Form the Grignard reagent of this protected bromo compound. React the Grignard reagent with ethanal. Hydrolyze the cyclic ketal with dilute acid.

(c) Prepare a THP derivative of the alcohol using dihydropyran. Add sodium amide to make the acetylide ion of the acetylene. React the acetylide ion with ethanal to give the acetylenic alcohol. Hydrolyze the THP derivative with dilute acid.

19.43 Under the basic conditions of the Wolff-Kishner reduction, the ketone is reduced to a methylene group and the cyclic ketal is not affected. Under the acidic conditions of the Clemmensen reduction, the ketal hydrolyzes and both ketones of the product are reduced. The structures of these two products are shown on the following page.

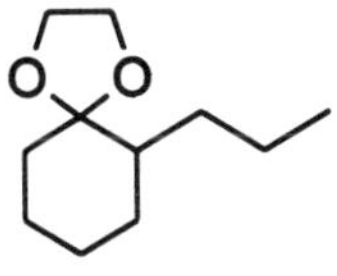

Wolff-Kishner product 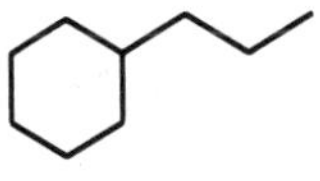   Clemmensen product

19.44 Under the basic conditions of the Wolff-Kishner reduction, the ketone is reduced to a methylene group and the cyclic ketal is not affected. Under the acidic conditions of the Clemmensen reduction, the ketal hydrolyzes and both ketones of the product are reduced. The products of the reaction are shown on the next page.

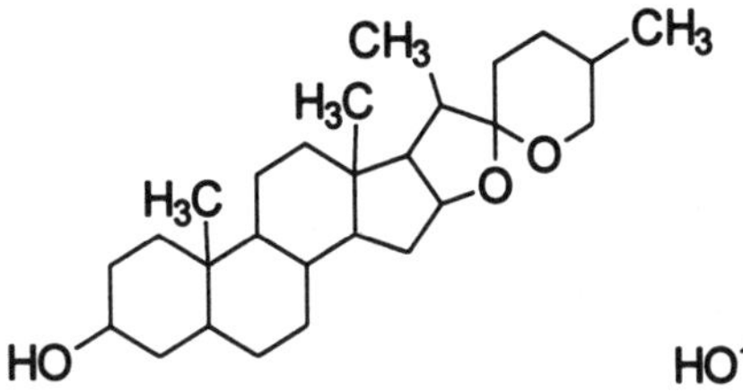

Wolff-Kishner product 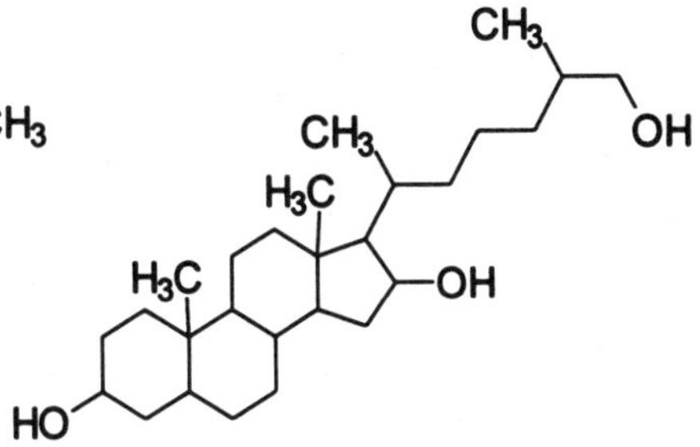   Clemmensen product

19.45 **(a)** The aldehyde reacts with two moles of the thiol to form a thioacetal.
      (b) The ketone reacts with one mole of the 1,2-dithiol to form a cyclic thioketal.
      (c) The ketone reacts with one mole of the 1,3-dithiol to form a cyclic thioketal.

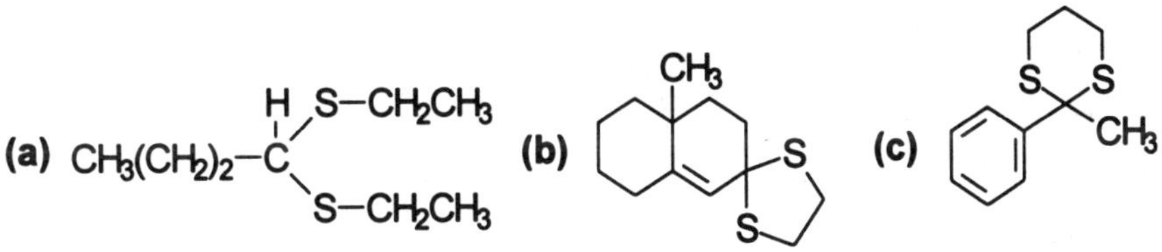

19.46

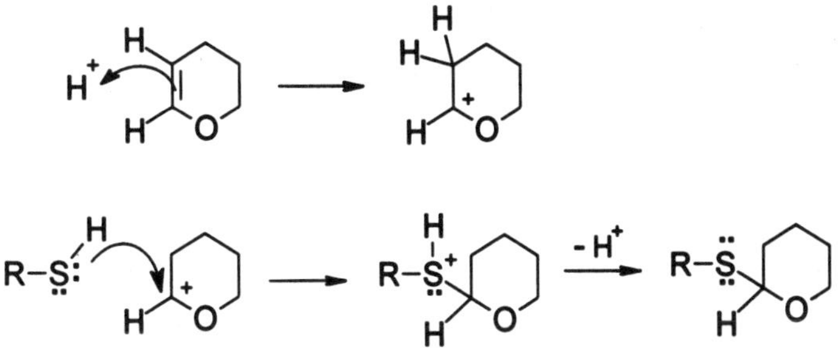

19.47

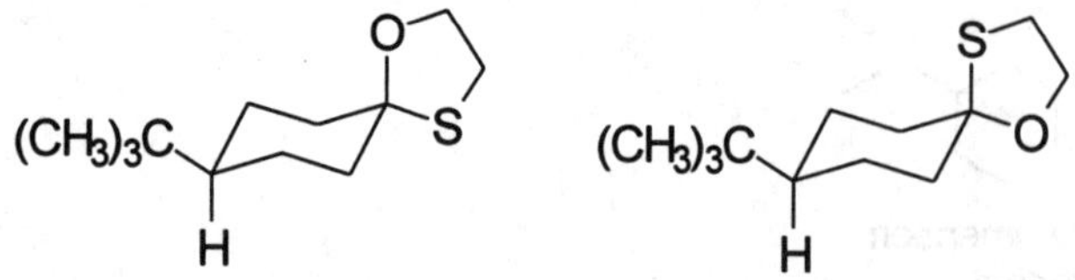

19.48   The cyclohexanone ring reacts because it produces a derivative in which no torsional strain is introduced. The cyclic thioketal of the cyclopentanone would have torsional interactions between the C—S and C—H bonds on adjacent carbon atoms. The product of the reaction is shown below.

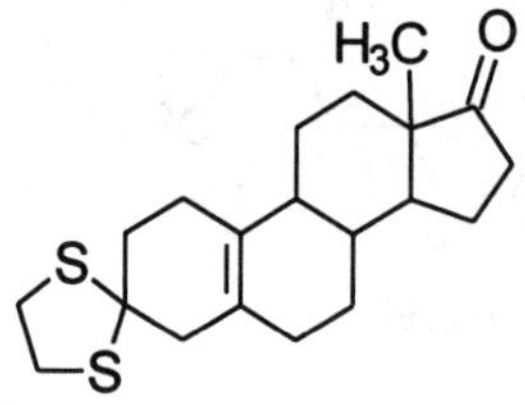

19.49   (a) 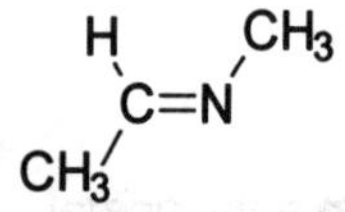   (b) 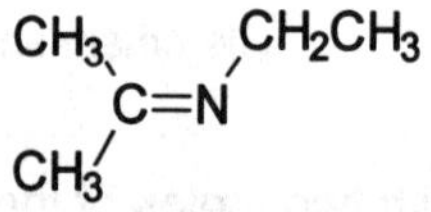   (c)

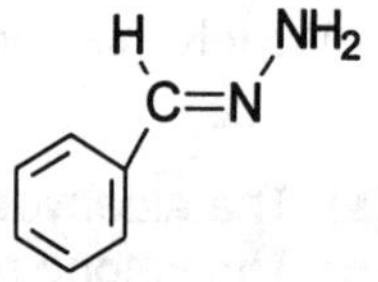

(d) 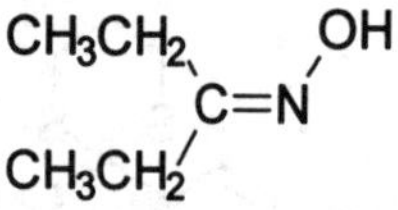   (e)

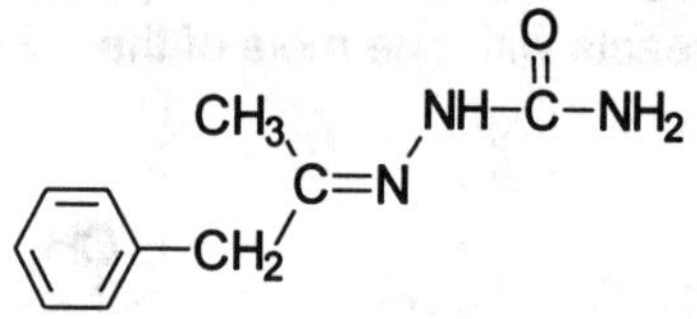

19.50   (a) cyclopentanone and hydroxylamine   (b) benzaldehyde and phenylhydrazine
(c) acetylcyclohexane and semicarbazide

19.51   (a) 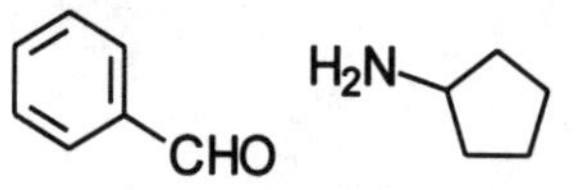   (b)

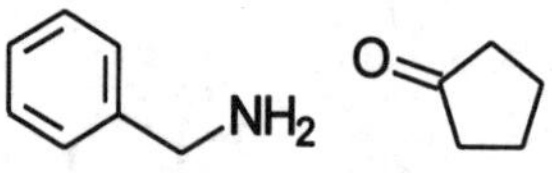

19.52   (a) 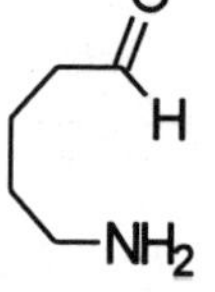   (b)

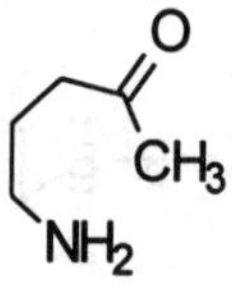

19.53   Geometric isomers are possible about the carbon-nitrogen double bond if the two groups bonded to the sp²-hybridized carbon atom are nonequivalent. The two methylene groups of the ring in cyclohexanone oxime are equivalent. In the oxime of cyclopentanecarbaldehyde there is a hydrogen atom and a cyclopentane ring bonded to the sp^2-hybridized carbon atom and geometric isomers result.

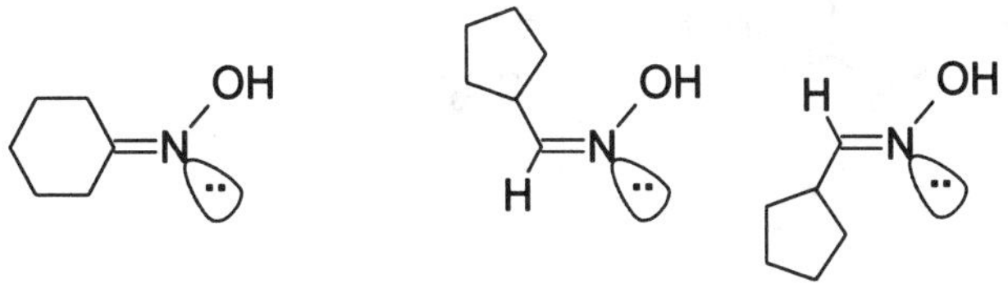

19.54

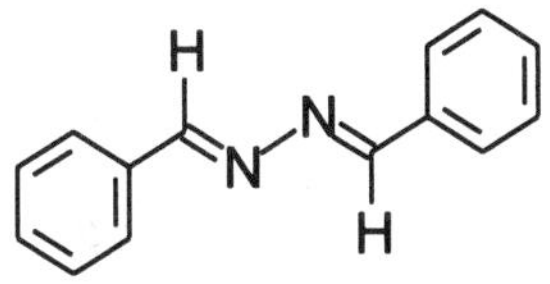

19.55   An oxime forms with one of the two equivalent carbonyl groups. The hydroxyl group of the oxime adds to the other carbonyl carbon atom, giving a hemiacetal which subsequently dehydrates to give an heterocyclic aromatic compound.

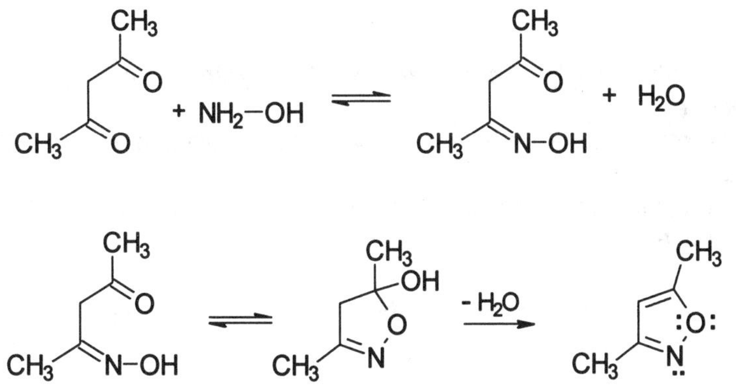

19.56   A hydrazine forms with one of the two equivalent carbonyl groups. The amino group of the hydrazine adds to the other carbonyl carbon atom, giving a hemiaminal. Hemiaminals normally dehydrate. However, in this case dehydration involving the carbon atom adjacent to the carbon atom bearing the hydroxyl group gives a more stable aromatic compound as shown below and on the next page..

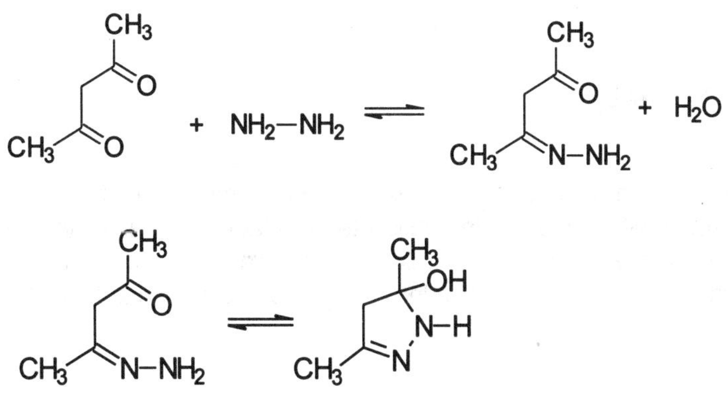

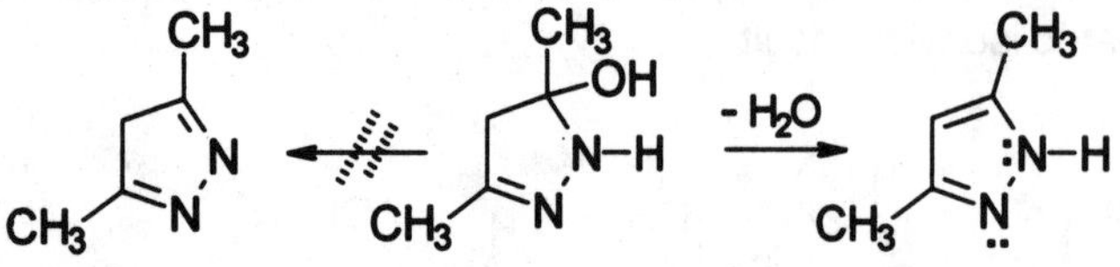

19.57 **(a)** 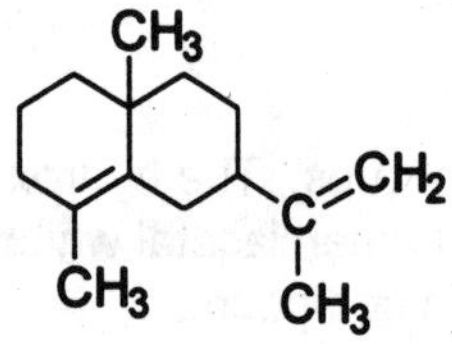 **(b)** 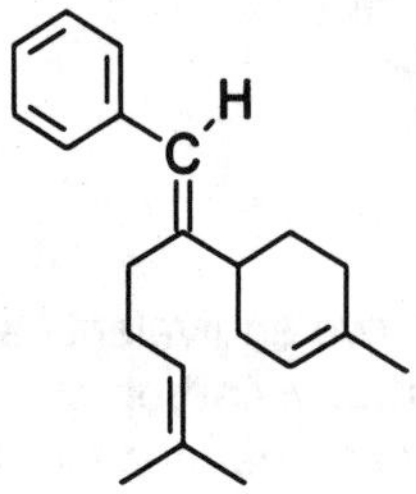 **(c)**

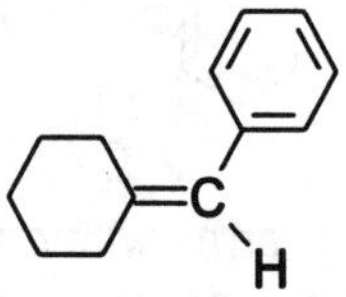

19.58 **(a)** Prepare the ethyltriphenylphosphonium salt, add butyllithium and react the ylide with cyclopentanone.
(b) Prepare the methyltriphenylphosphonium salt, add butyllithium and react the ylide with 3-hexanone.
(c) Prepare the propyltriphenylphosphonium salt, add butyllithium and react the ylide with 4-heptanone.

19.59 Two geometric isomers can result from this combination of reagents.

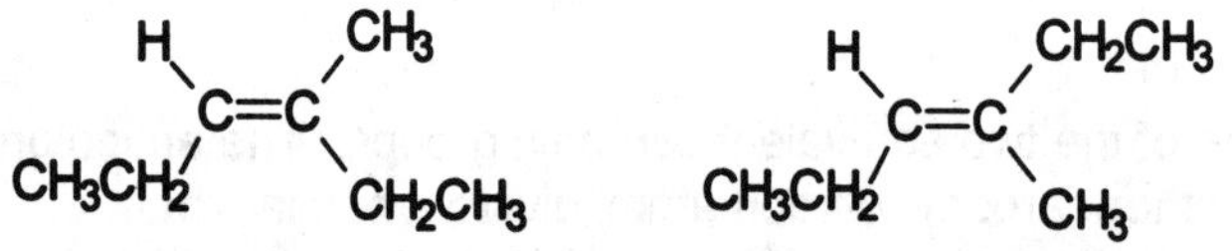

19.60 The negative charge on the C-1 atom of the original halogen compound may be delocalized by resonance with the triple bond. Thus, a weaker base than butyllithium is sufficient to abstract the required proton .

19.61 The dipolar resonance form of dimethylsulfonium methylide has a negative charge on the methylene carbon atom, which is a nucleophilic site. It attacks the carbonyl carbon atom giving an alkoxide which then forms the epoxide via an intramolecular Williamson synthesis as shown on the next page. The leaving group is dimethyl sulfide.

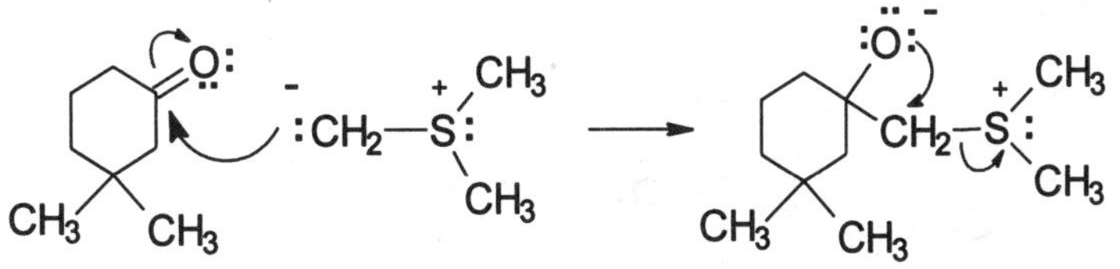

19.62 The dipolar resonance form of diazomethane has a negative charge on the methylene carbon atom, which is a nucleophilic site. It attacks the carbonyl carbon atom, giving an alkoxide which then forms the epoxide via an intramolecular Williamson synthesis. The leaving group is nitrogen.

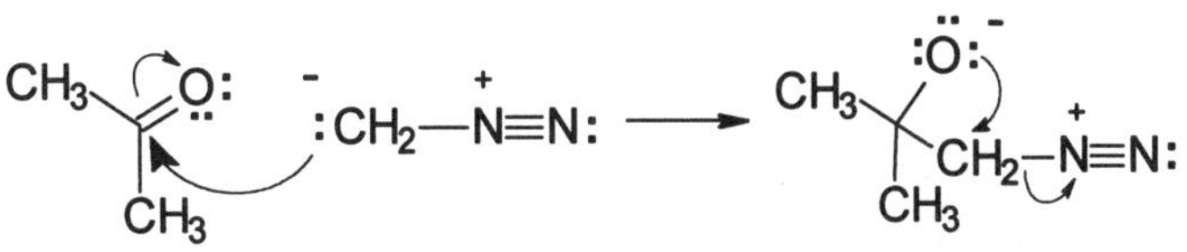

19.63 The initial attack of the negatively charged methylene group would be affected by the steric environment of the carbonyl carbon atom. Attack along an equatorial pathway is less sterically hindered that attack from an axial direction. Thus, the methylene group of the more favored epoxide product is in an equatorial-like position.

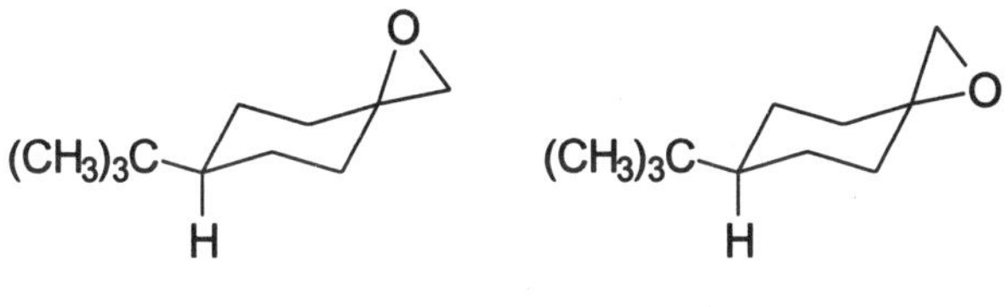

19.64 Sodium hydride abstracts a proton from the central methylene group. The resulting negative charge is resonance stabilized by conjugation with the carbonyl group. Note that a dipolar resonance form of the phosphorus grouping has a positive charge on the phosphorus atom. Thus, the conjugate base of the compound resembles the ylide used in the Wittig reaction.

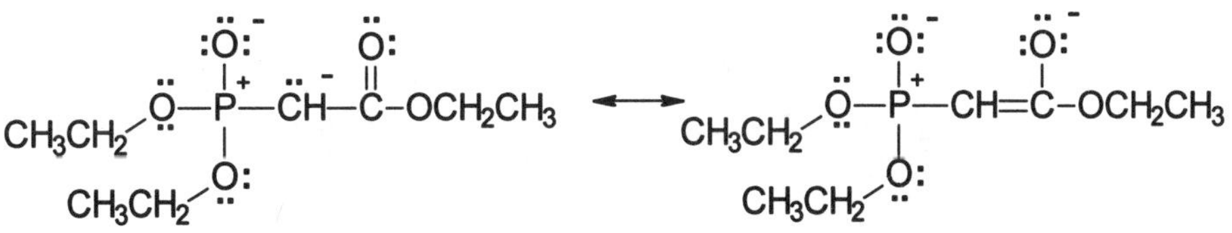

Based on the unbalanced equation given, the carbonyl compound must transfer an oxygen atom to the phosphorus atom. The by product is

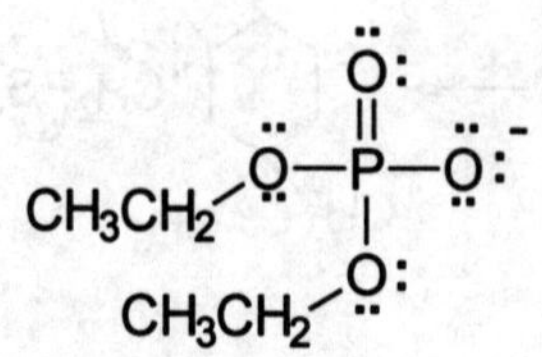

$$CH_3CH_2 \overset{\displaystyle \ddot{O}:}{\underset{\displaystyle \overset{|}{\underset{CH_3CH_2}{\ddot{O}:}}}{\overset{..}{O}-P-\ddot{O}:^-}}$$

20

Carbohydrates

The molecules displayed in this chapter have functional groups on every carbon atom! However, their chemistry is essentially everything that you would expect. Carbohydrates have alcohol and carbonyl functional groups, and you know the chemistry of both functional groups. The only additional skill required is the ability to focus on the reaction site while ignoring all those other functional groups that clutter the picture. In order to do so you will also have to use and recognize stereochemical aspects of both structures and reactions.

Keys to the Chapter

20.1 Carbohydrates and Energy

Carbohydrates are polyhydroxy aldehydes and ketones or compounds that produce them in hydrolysis reactions. Carbohydrates are a major source of stored chemical energy. That energy is incorporated into carbohydrates in their formation from carbon dioxide and water in photosynthesis. The energy is liberated in metabolic reactions in animals.

20.2 Classification of Carbohydrates

Monosaccharides are carbohydrates that cannot be hydrolyzed to simpler carbohydrates. These compounds may be aldehydes or ketones, or acetals or ketals which yield an alcohol and a mono-saccharide when hydrolyzed. **Oligosaccharides** contain **glycosidic linkages** (acetal or ketal) which release two or more monosaccharide units upon hydrolysis. One monosaccharide serves as the acetal or ketal center that reacts with the hydroxyl group of the next monosaccharide. **Disaccharides** are a class of oligosaccharides containing two monosaccharide units. **Polysaccharides** contain a large number of monosaccharide units bonded to each other by a series of glycosidic bonds. Each unit serves both as the acetal (or ketal) center to form one glyco-sidic bond and as an alcohol to form the other glycosidic bond to neighboring monosaccharides.

A monosaccharide is classified as an **aldose** if the carbonyl group is an aldehyde and as a **ketose** if the carbonyl group is a ketone. The number of carbon atoms is commonly between three and six. Both pieces of information are used in classifying monosaccharides. The term aldohexose indicates that the compound is an aldehyde and contains six carbon atoms. The term ketopentose indicates that the compound is a ketone and contains five carbon atoms.

20.3 Chirality of Monosaccharides

The structures of the D series of monosaccharides are given as Fischer projections in Figures 20.1 and 20.3 in the text for aldoses and ketoses, respectively. The terms D and L refer to the configuration at the chiral center farthest away from the carbonyl group, which is an aldehyde group in aldoses and a ketone group in ketoses. The nomenclature used for each of these monosaccharides is a name that conveys the internal configuration at each center with respect to the carbon atom establishing the D configuration assigned by Fischer. Thus, you have to have a program to recognize the players. Your instructor will probably indicate which structures you should know. (One of the author's instructors made the class memorize all of them!) Because the relative configuration of all chiral centers is given by the name, the configuration of every center of an L-monosaccharide with the same name is reversed compared to the D-monosaccharide.

Each isomeric monosaccharide of the D-series is a diastereomer of every other isomer. The enantiomer of a particular D compound is the L compound. Remember that diastereomers have different physical properties.

20.4 Isomerization of Monosaccharides

Inversion of one or more of the stereogenic centers of a monosaccharide gives diastereomers. Such reactions generally require specialized chemistry that involves protection of one or more of the functional groups. Compounds with multiple stereogenic centers that differ in configuration at one stereogenic center are **epimers**. Epimerization at selected centers, as in galactose and glucose, occurs in biochemical reactions catalyzed by enzymes. Epimerization at the α carbon atom occurs readily via an **enediol** intermediate and can occur with an enzyme. Note that not only are aldoses interconverted, but ketoses can be transformed into aldoses and vice-versa in this equilibrium reaction.

20.5 Hemiacetals and Hemiketals

Monosaccharides form intramolecular hemiacetals or hemiketals containing five-membered and six-membered rings called **furanoses** and **pyranoses**, respectively. These structures are the predominant form of monosaccharides—there is very little of the open chain form depicted by the Fischer projection.

The Haworth projection formulas depict the cyclic form of carbohydrates in a flat five- or six-membered ring drawn as if perpendicular to the plane of the page. The ring oxygen atom is drawn in the back of the ring in a furanose (five-membered ring). In a pyranose (six-membered ring), the oxygen atom is shown in the back right position in the plane of the ring. The hemiacetal or hemiketal carbon atom is on the right and lies on the line where the plane of the ring and the plane of the page intersect. The hydroxyl group at this center may be below or above the plane of the ring, designated as α or β, respectively. The hydroxyl groups on the right in the open chain form are drawn below the plane of the ring in the Haworth projection formula; hydroxyl groups on the left are drawn above the plane of the ring.

You should work through the method of obtaining the Haworth projection formula. However, in the end you have to know that the —CH$_2$OH group is "up". In addition you have to know that hydroxyl groups on the right in the Fischer projection formula are "down" in the Haworth projection formula.

As a result of forming a cyclic hemiacetal or hemiketal, an additional stereogenic center is created. The configuration at this center is the only difference between the α or β forms of a monosaccharide, and the structures are diastereomers. These special types of diastereomers are called **anomers**. At equilibrium in aqueous solution, mixtures of anomers of both pyranoses and furanoses are formed from most pentoses and hexoses. The composition of such mixtures is given in Table 20.1 in the text.

Haworth projection formulas convey stereochemical information but do not realistically picture the location of the functional groups in space. For that you have to draw cyclohexane-like conformations. There is a standard way of orienting the ring. Note that the foot of the chair is on the right and the back of the chair is on the left. Of course such a ring can be "flipped", but this is the preferred way to represent the location of the hydroxyl groups. Remember that up and down for substituents in cyclohexoses does not mean axial or equatorial. In fact, that relationship alternates with each carbon atom.

Mutarotation involves the reversible opening and closing of a hemiacetal or hemiketal center in the cyclic forms of a carbohydrate. When the ring is open, rotation about the bond between C-1 and C-2 can occur, changing the position of the hydroxyl group. When the ring closes again, α may have changed to β or vice versa. Eventually a point of equilibrium is attained between the α and β forms. The α and β anomers of a monosaccharide are diastereomers, so the optical rotation values of the two anomers are not related. The optical rotation of a mixture of anomers has an intermediate value determined by the percent composition of the two anomers.

20.6 Reduction and Oxidation of Monosaccharides

Any hemiacetal or hemiketal of a carbohydrate can exist in equilibrium with its open-chain form. The resultant aldehyde or ketone is reduced to an **alditol** by reaction with $NaBH_4$. An alditol is a polyalcohol obtained when the carbonyl group is reduced to a —CH_2OH group. Thus, to write the structure of the reduction product, simply write the open-chain structure of the monosaccharide, replacing —CHO with —CH_2OH.

As a result of the equilibrium of an hemiacetal of a monosaccharide with its open-chain form, the aldehyde of an aldose will reduce Tollens's or Benedict's reagent. The ketone of a ketose will undergo the same reactions, because it undergoes tautomerization to the aldehyde form. The products, which have a carboxylic acid at C-1, are **aldonic acids**. All monosaccharides that are oxidized to aldonic acids are called reducing sugars because they reduce the reagent.

Aldaric acids result from oxidation at both terminal positions of an aldose to give a dicarboxylic acid. Special enzyme-catalyzed reactions convert the highest numbered carbon atom of an aldose into a carboxylic acid without affecting the aldehydic carbon atom. These compounds are **uronic acids**.

20.7 Glycosides

The hemiacetal and hemiketal forms of monosaccharides react with alcohols to form acetals and ketals called **glycosides**. The C—O bond formed is a **glycosidic bond**, and the —OR from the alcohol is called an **aglycone**. The formation of glycosides occurs by a mechanism identical to that described in Section 19.7 of the text for simple hemiacetals and hemiketals. A carbocation is formed that is stabilized by the lone pair electrons of a directly bonded oxygen atom. As a result, a mixture of anomeric glycosides results.

Any carbohydrate that exists as an acetal or ketal cannot revert to an aldehyde or ketone in the presence of the basic solution of Tollens' or Benedict's reagent. Thus, no reaction occurs and the compound is called a non-reducing sugar.

20.8 Disaccharides

Disaccharides are glycosides formed from two monosaccharides which can be either aldoses or ketoses. One of the —OR groups is provided by the original cyclization to give the hemiacetal or hemiketal. The second —OR group is derived from an aglycone, which is a second monosaccharide that provides the alcohol functional group. In order to identify a disaccharide it is necessary to:

1. Know the constituent monosaccharides.
2. Know the ring form of each monosaccharide.
3. Know which monosaccharide serves as the acetal center or ketal center and its configuration.
4. Know the configuration of the hemiacetal center of this monosaccharide.
5. Know which hydroxyl group of the monosaccharide serving as the aglycone is bonded to the acetal center.

The common disaccharides are maltose, cellobiose, lactose, and sucrose. How much you have to learn about these structures will be indicated by your instructor. Use the above scheme to organize the information. You might have to identify some disaccharide other than those given in this chapter. Again, go through the list. The first two steps are straightforward. The third step requires that you locate the acetal center. It is the one with two oxygen atoms bonded to a common carbon atom and each of them is bonded to a second carbon atom. For step 4, check the hemiacetal center at the carbon atom bearing both an —OH and —OR group. Finally step 5 is straightforward, because it is derived from the —OR group that is not part of the furanose or pyranose ring of the monosaccharide that is not providing the acetal/ketal center.

20.9 Polysaccharides

Polysaccharides that consist of only one type of monosaccharide are **homopolysaccharides**. Starch and cellulose are both homopolysaccharides of glucose. They differ only in the stereochemistry of the glycosidic linkage, which is bonded to the C-4 atom in both cases. In starch the linkage is α; in cellulose it is β. The ability of organisms to hydrolyze α or β-glycosidic bonds controls what they can use as food sources. Most animals can only digest α-linked polysaccharides of glucose. Cattle and other herbivores can digest β-linked polysaccharides, because they harbor microorganisms which have the necessary enzymes.

20.10 Proof of Structure of Monosaccharides

A monosaccharide can be shown to be an aldose or a ketose based on the products of oxidation by periodate. Aldoses give an equivalent of formic acid, whereas ketoses give an equivalent of carbon dioxide. (Note that each secondary alcohol site also gives an equivalent of formic acid.)

Oxidation by nitric acid may provide information about the relationship of the stereogenic centers in an aldose. If the secondary hydroxyl groups are symmetrically arranged, the aldaric acid will be meso. That fact decreases the number of possible arrangements to consider for the stereogenic centers.

Formation of an osazone identical with that of a second monosaccharide establishes the identity of the unknown monosaccharide. All stereogenic centers other than the one α to the carbonyl group have the same configuration.

Both the chain extension and chain degradation methods provide information about the configuration of the stereogenic centers of an unknown monosaccharide. Chain degradation indicates the configuration of all but one of the original stereogenic centers, and thus the unknown may be one of two possible compounds. Chain elongation gives two diastereomeric products whose configuration at the C-3 atom and beyond is the same as for the original monosaccharide.

20.11 Determination of Ring Size

The size of a monosaccharide ring is determined by first preparing a glycoside with an aglycone such as a methyl group. Subsequent oxidation by periodate gives a dialdehyde whose structure can be used to deduce the ring size. See page 768 in the text to discover these differences for a furanose and a pyranose.

An alternate method of determining the ring size involves complete methylation of the monosaccharide, followed by hydrolysis of the acetal. Subsequent treatment using a strong oxidizing agent oxidizes the aldehyde group, but also cleaves a carbon-carbon bond at the carbon atom bearing the oxygen atom of the original ring and gives a second carboxylic acid site.

20.12 Structure of Disaccharides

The structure of a disaccharide can be determined by complete methylation followed by acid-catalyzed hydrolysis of the acetal (or ketal) center. Then, the structures of the methylated monosaccharides must be determined. Based on these structures, the identity of the original disaccharide can usually be deduced. The "free" hydroxyl group in one of the methylated monosaccharides was the one linking the two rings.

Summary of Reactions

1. Reduction of Monosaccharides

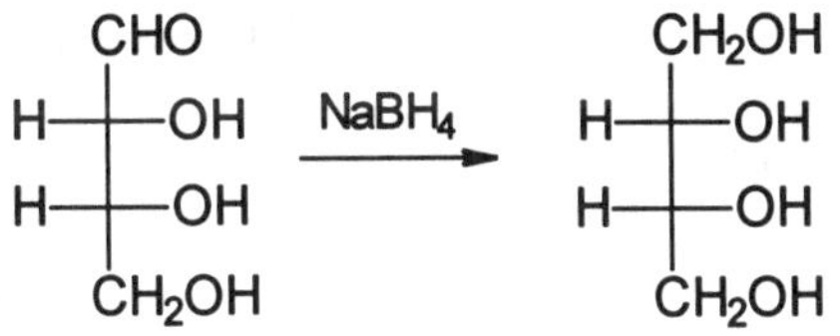

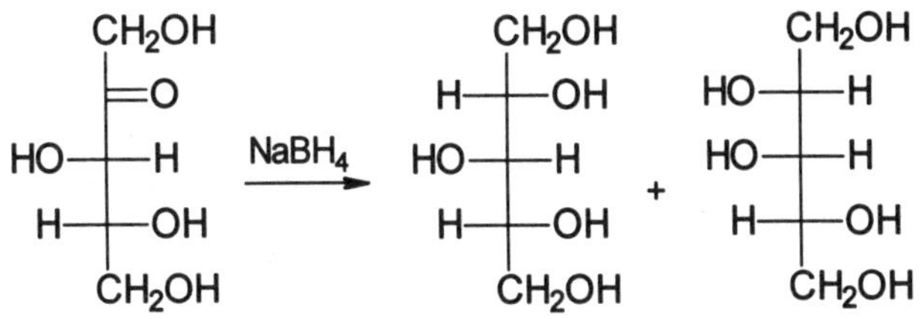

2. Oxidation of Monosaccharides

$$
\begin{array}{c}
\text{CHO} \\
\text{H}\!-\!\!-\!\text{OH} \\
\text{H}\!-\!\!-\!\text{OH} \\
\text{CH}_2\text{OH}
\end{array}
\xrightarrow{\text{Ag(NH}_3\text{)}_2{}^+}
\begin{array}{c}
\text{CO}_2\text{H} \\
\text{H}\!-\!\!-\!\text{OH} \\
\text{H}\!-\!\!-\!\text{OH} \\
\text{CH}_2\text{OH}
\end{array}
$$

$$
\begin{array}{c}
\text{CH}_2\text{OH} \\
\text{H}\!-\!\!-\!\text{OH} \\
\text{HO}\!-\!\!-\!\text{H} \\
\text{H}\!-\!\!-\!\text{OH} \\
\text{CH}_2\text{OH}
\end{array}
\xrightarrow{\text{HNO}_3}
\begin{array}{c}
\text{CO}_2\text{H} \\
\text{H}\!-\!\!-\!\text{OH} \\
\text{HO}\!-\!\!-\!\text{H} \\
\text{H}\!-\!\!-\!\text{OH} \\
\text{CO}_2\text{H}
\end{array}
$$

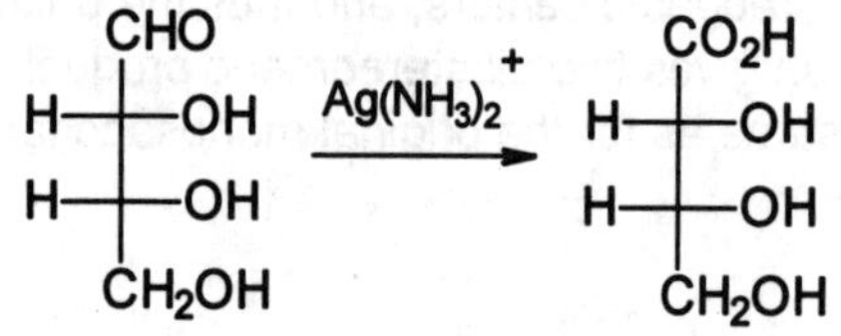

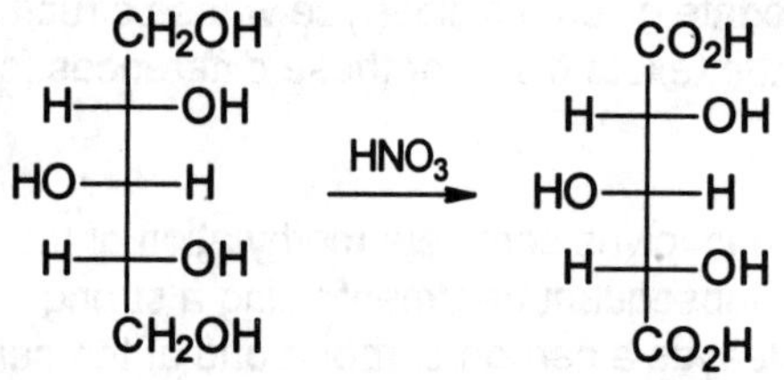

3. Isomerization of Monosaccharides

$$
\begin{array}{c}
\text{CHO} \\
\text{H}\!-\!\!-\!\text{OH} \\
\text{H}\!-\!\!-\!\text{OH} \\
\text{H}\!-\!\!-\!\text{OH} \\
\text{CH}_2\text{OH}
\end{array}
\rightleftharpoons
\begin{array}{c}
\text{CH}_2\text{OH} \\
=\!\text{O} \\
\text{H}\!-\!\!-\!\text{OH} \\
\text{H}\!-\!\!-\!\text{OH} \\
\text{CH}_2\text{OH}
\end{array}
\rightleftharpoons
\begin{array}{c}
\text{CHO} \\
\text{HO}\!-\!\!-\!\text{H} \\
\text{H}\!-\!\!-\!\text{OH} \\
\text{H}\!-\!\!-\!\text{OH} \\
\text{CH}_2\text{OH}
\end{array}
$$

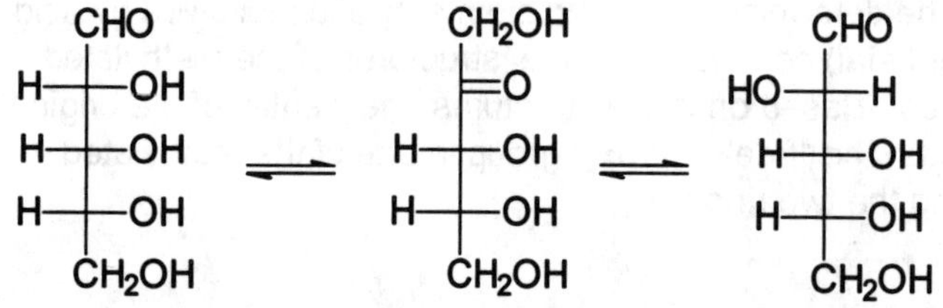

4. Formation of Glycosides

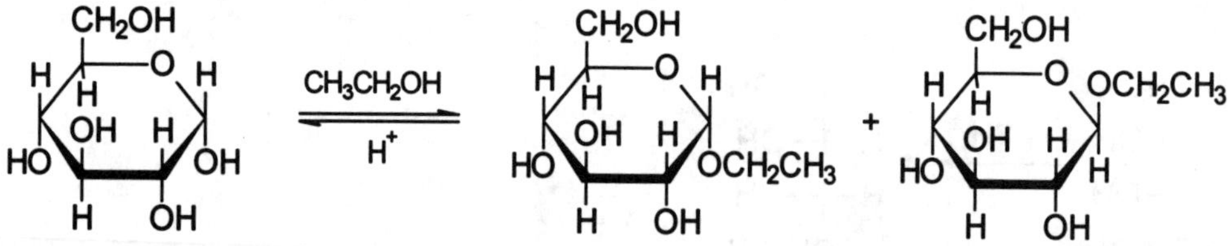

Solutions to Exercises

20.1 The classes of compounds are designated by aldo or keto if they contain an aldehyde or ketone group, respectively. The length of the carbon chain is indicated by terms *tetr*, *pent*, and *hex*. The ending of the class name is *ose*.

(a) aldopentose   (b) aldohexose   (c) ketohexose

20.2 The classes of compounds are designated by aldo or keto if they contain an aldehyde or ketone group, respectively. The length of the carbon chain is indicated by terms *tetr*, *pent*, and *hex*. The ending of the class name is *ose*.

(a) ketopentose   (b) aldohexose   (c) ketotetrose

20.3 The highest numbered stereogenic center is used to assign D and L configurations. If the —OH group at that point is on the right in the projection formula, it is D; if the —OH group is on the left, it is L. The chains are numbered to give the lower number to the most highly oxidized center. **(b)** and **(c)** are L carbohydrates; **(a)** is a D carbohydrate.

20.4 The highest numbered stereogenic center is used to assign D and L configurations. If the —OH group at that point is on the right in the projection formula, it is D; if the —OH group is on the left, it is L. The chains are numbered to give the lower number to the most highly oxidized center. **(a)** and **(c)** are L carbohydrates; **(b)** is a D carbohydrate.

20.5 There are two equivalent stereogenic centers in a 3-ketopentose. An enantiomeric pair of compounds and a meso compound are possible. The first structure has the D configuration. The second is its enantiomer and has the L configuration. The third compound is meso.

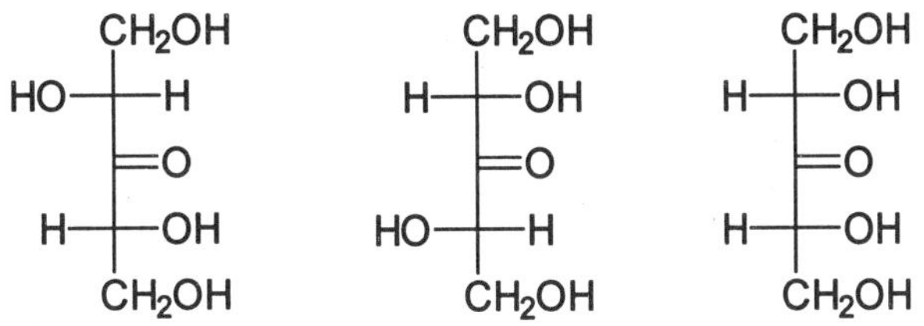

20.6 There are three nonequivalent stereogenic centers, so a total of 8 stereoisomers are possible. Four are of the D configuration and are shown below.

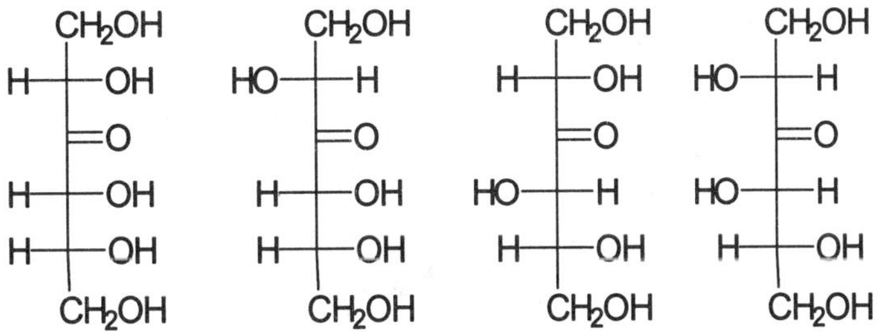

20.7

(a)
```
      CHO
HO ──┼── H
  H ──┼── OH
HO ──┼── H
      CH₂OH
```

(b)
```
      CHO
HO ──┼── H
HO ──┼── H
      CH₂OH
```

(c)
```
      CHO
HO ──┼── H
  H ──┼── OH
  H ──┼── OH
HO ──┼── H
      CH₂OH
```

(d)
```
      CHO
HO ──┼── H
HO ──┼── H
HO ──┼── H
      CH₂OH
```

(e)
```
      CH₂OH
       ═O
  H ──┼── OH
HO ──┼── H
HO ──┼── H
      CH₂OH
```

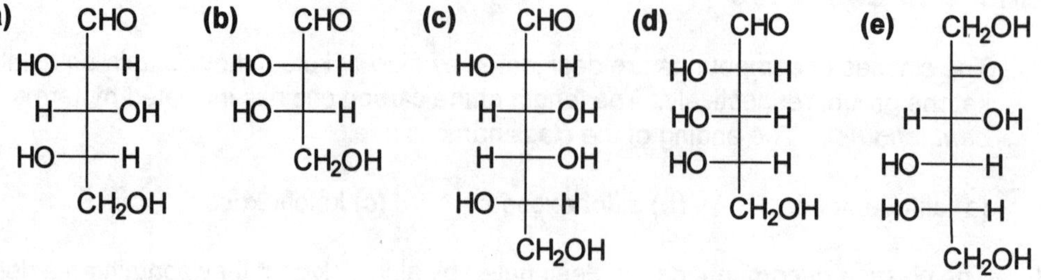

20.8

(a)
```
      CHO
HO ──┼── H
  H ──┼── OH
  H ──┼── OH
HO ──┼── H
      CH₃
```

(b)
```
      CHO
  H ──┼── OH
  H ──┼── H
  H ──┼── OH
      CH₂OH
```

(c)
```
      CHO
  H ──┼── H
  H ──┼── OH
  H ──┼── OH
  H ──┼── OH
      CH₃
```

(d)
```
      CHO
  H ──┼── OH
  H ──┼── OH
HO ──┼── H
HO ──┼── H
      CH₃
```

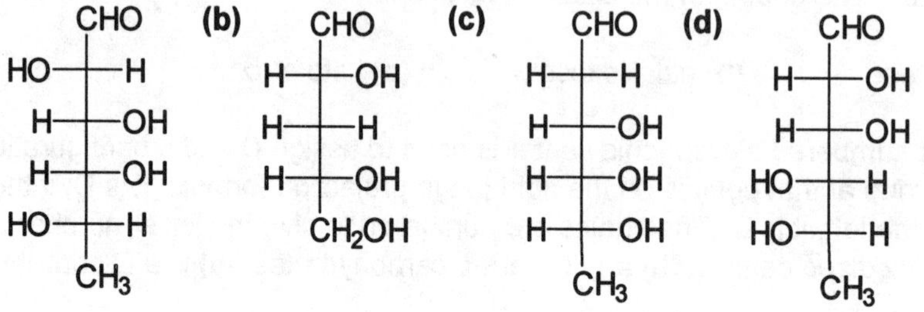

20.9   Two isomers are possible at the hemiacetal center.

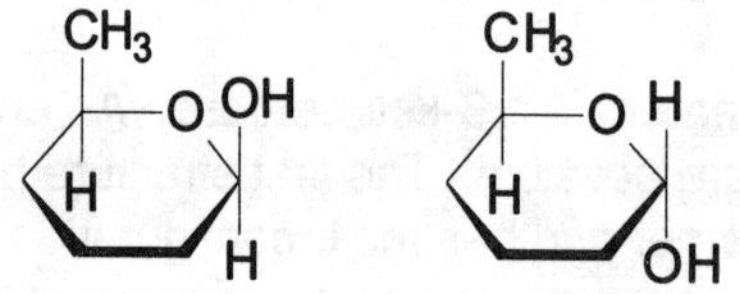

20.10   Two isomers are possible at the hemiketal center.

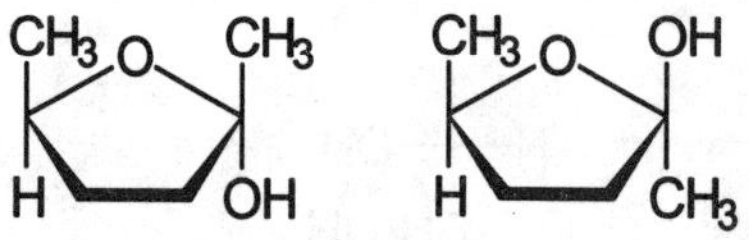

20.11   (a)

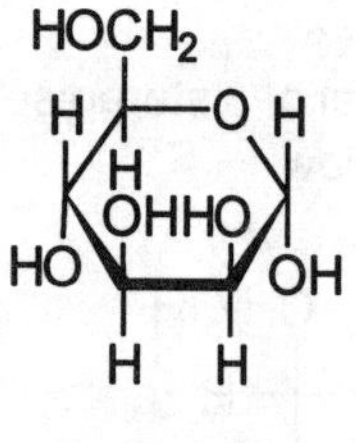

(b)

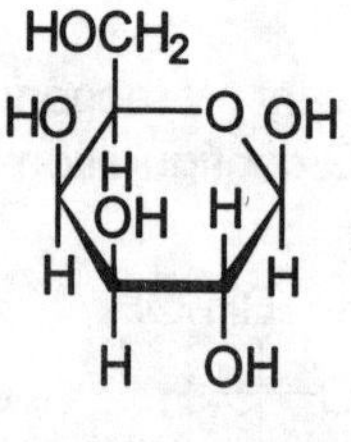

(c)

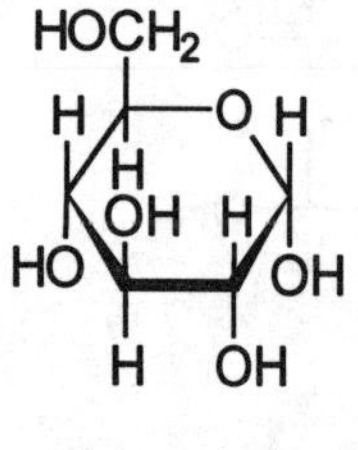

(d)

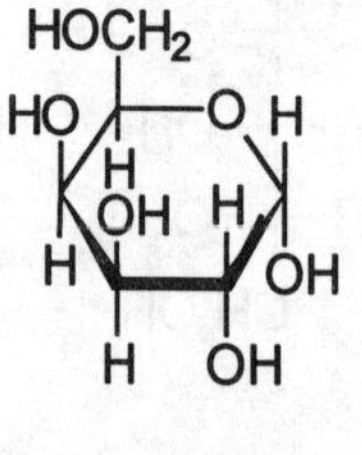

20.12 **(a)** 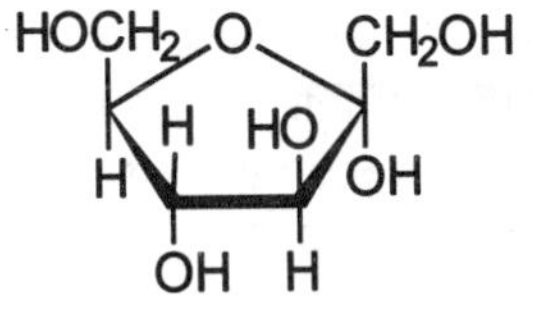   **(b)**

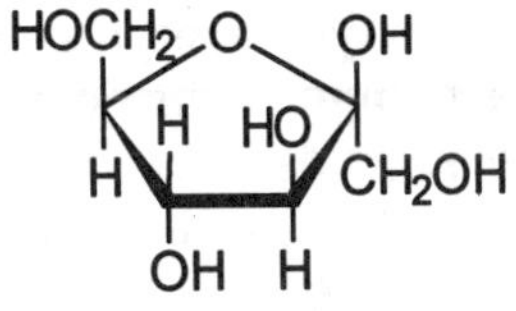

(c) 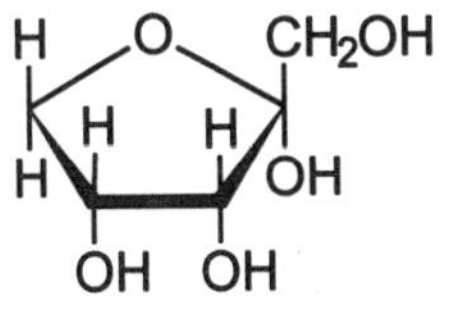   **(d)**

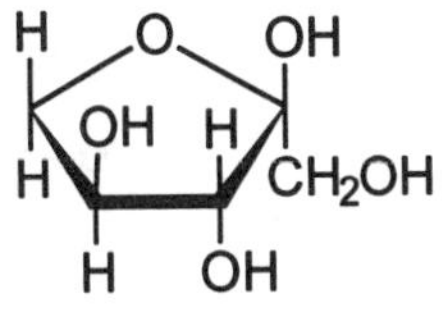

20.13 **(a)** α-D-galactopyranose **(b)** β-D-lyxopyranose **(c)** α-D-ribopyranose

20.14 **(a)** α-D-xylulofuranose **(b)** α-D-fructopyranose **(c)** β-D-ribulofuranose

20.15 The C-4 hydroxyl group of β-galactopyranose shown in the structure on the left is axial. The C-2 hydroxyl group of β-mannopyranose shown in the structure on the right is axial. They have the same number of axial hydroxyl groups.

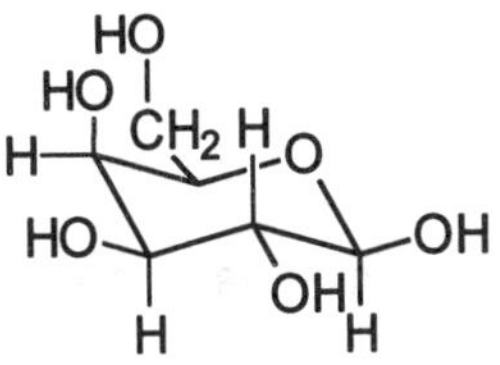

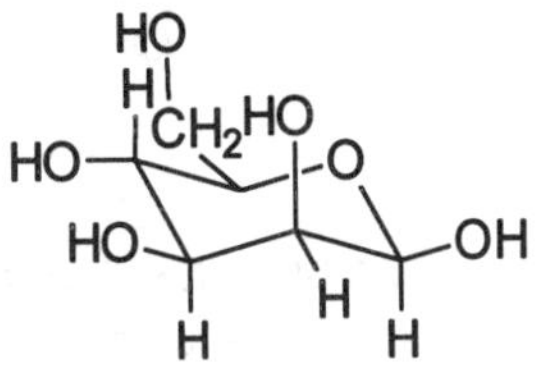

20.16 The C-2 and C-4 hydroxyl groups of β-talopyranose shown in the structure on the left are axial. The C-3 hydroxyl group of β-allopyranose shown on the right is axial. Thus, β-talopyranose has two axial hydroxyl groups and β-allopyranose has only one.

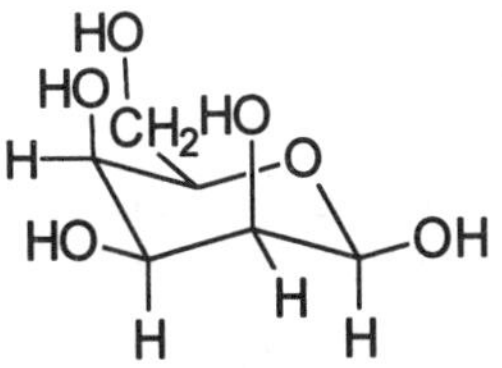

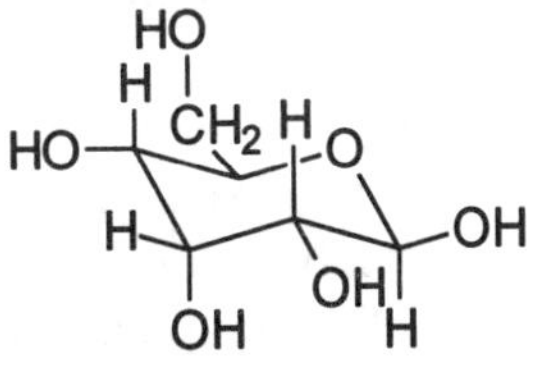

20.17 The rings of both (b) and (c) must be flipped into an alternate chair conformation to give the standard representation of the pyranose ring.

(a) β-D-6-deoxygalactose   **(b)** α-D-arabinose   **(c)** β-D-glucose

20.18 Flip ring (c) into an alternate chair conformation to give the standard representation of the pyranose ring. In either conformation, the hydroxyl groups on C-2, C-3, and C-4 are up, down, and up, respectively, which corresponds to left, right, left, respectively, in the Fischer projection. The compond is the mirror image of D-xylose, so it is L-xylose.

(a) β-D-lyxose   **(b)** β-D-6-deoxyallose   **(c)** α-L-xylose

20.19   All hydroxyl groups of α-idopyranose shown in the structure on the left are axial. The ring flip places all four hydroxyl groups in equatorial positions and the —CH₂OH group in an axial position in the structure on the right. The 1,3-diaxial interaction of two hydroxyl groups on one side of the ring as well as two hydroxyl groups on the other side of the ring is larger than the steric hindrance of the axial —CH₂OH group. The ring flip conformation is the more stable.

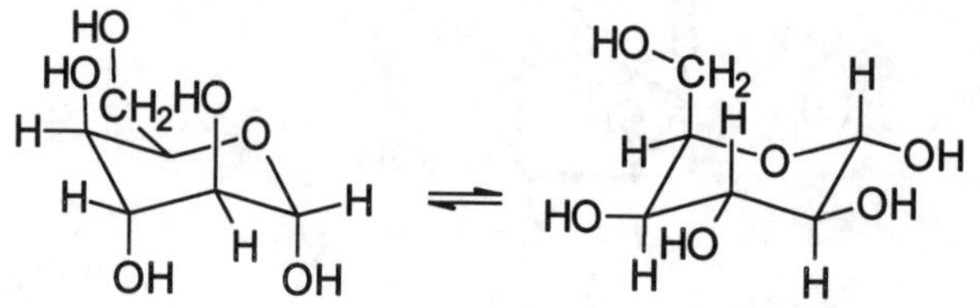

20.20   The ring must be flipped into an alternate chair conformation to give the standard representation of the pyranose ring. The compound is α-D-gulose.

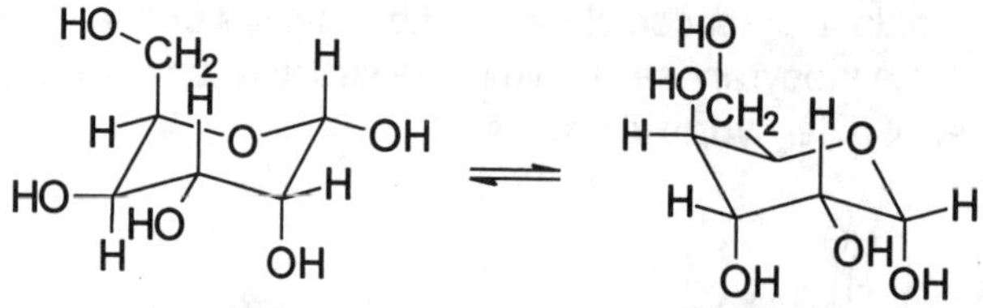

20.21   Both compounds I and II can mutarotate because they are a hemiacetal and a hemiketal, respectively. Compound III is a ketal and cannot mutarotate.

20.22   Compound I, which has a hemiacetal center on the right ring, can mutarotate. Compound II has acetal centers in both rings and cannot mutarotate.

20.23   The specific rotation, which is a weighted average of the rotation the two anomers, is closest to that of the β anomer. Thus, the β anomer predominates.

20.24   The specific rotation, which is a weighted average of the rotation the two anomers, is closest to that of the α anomer. Thus, the α anomer predominates. Let x equal the mole fraction of the α anomer. Then, the observed rotation is set equal to the specific rotations of each anomer multiplied by their respective mole fractions. The percent of the α anomer is 83.6%.

$(20.3) x + (-17.0) (1 - x) = 14.2$
$37.3 x = 31.2$
$x = 0.836$

20.25   Disregarding the substituents, the pyranose ring is more stable than a furanose ring. There is torsional strain in a five-membered ring and none in the six-membered pyranose. In ribose, the C-2 and C-4 hydroxyl groups are equatorial in the pyranose. In the β anomer, the C-1 hydroxyl group is also equatorial, but in the α anomer it would be axial. Thus, the β anomer, with three equatorial hydroxyl groups, predominates over the the α anomer, with two.

20.26 Disregarding the substituents, the pyranose ring is more stable than a furanose ring. There is torsional strain in a five-membered ring and none in the six-membered pyranose. In glucose, the C-2, C-3, and C-4 hydroxyl groups are equatorial, as is the C-6 atom. In mannose, galactose, and allose there are two equatorial hydroxyl groups and the C-6 atom is equatorial as well. The other diastereomers would have a majority of hydroxyl groups in axial positions in the standard representation. Thus, the higher energies of these conformations become comparable to that of the furanose form.

20.27 Conversion of the —CHO group into a —CH$_2$OH group in an aldotriose makes the C-1 and C-4 atoms equivalent. In the product from D-erythrose, the two hydroxyl groups are on the same side in the Fischer projection formula and on opposite sides of a place of symmetry placed perpendicular to the formula and bisecting the bond between C-2 and C-3 atoms, so the alditol shown on the left is a meso compound. In the product from D-threose, the hydroxyl groups on the C-2 and C-3 atoms are not symmetrical with respect to the described plane, so the alditol shown on the right is optically active.

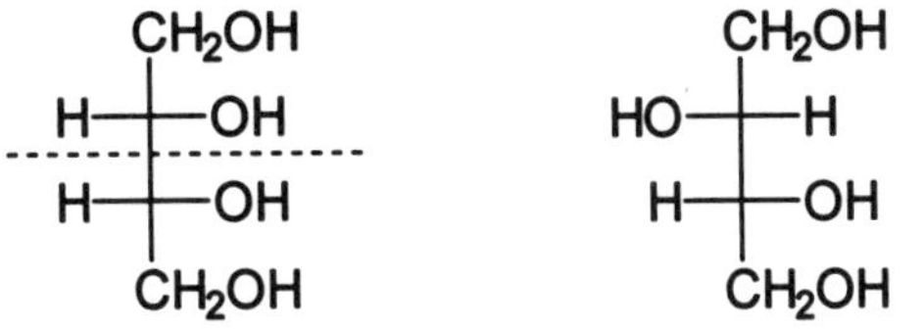

```
      CH2OH                    CH2OH
   H ─┼─ OH               HO ─┼─ H
   ----------             
   H ─┼─ OH                H ─┼─ OH
      CH2OH                    CH2OH
```

20.28 Conversion of the —CHO group into a —CH$_2$OH group in an aldopentose makes the C-1 and C-5 atoms equivalent. The potential plane of symmetry to be considered bisects the C-3 atom and its substituents. Only ribose and xylose have the C-2 and C-4 hydroxyl groups arranged in a mirror image relationship as required for a meso compound, so their alditols will be optically inactive.

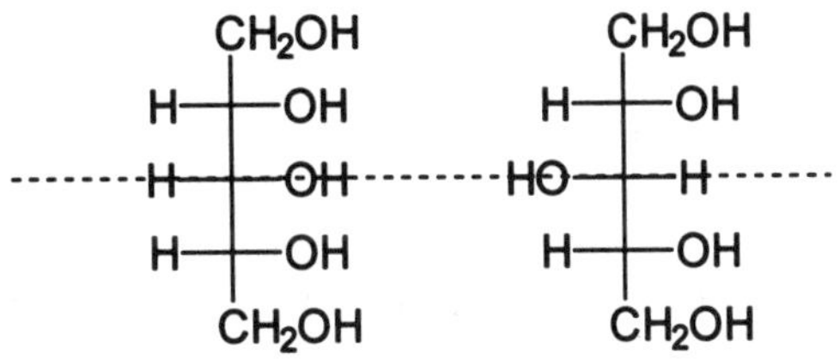

```
       CH2OH                    CH2OH
    H ─┼─ OH                 H ─┼─ OH
    ----H─┼─OH----       ----HO─┼─H----
    H ─┼─ OH                 H ─┼─ OH
       CH2OH                    CH2OH
```

20.29 Reduction of the carbonyl group gives a mixture of diastereomers that differ at the C-2 atom. One of these isomers has the same configuration as the alditol of glucose, glucitol. The other is the same as the alditol of mannose, mannitol.

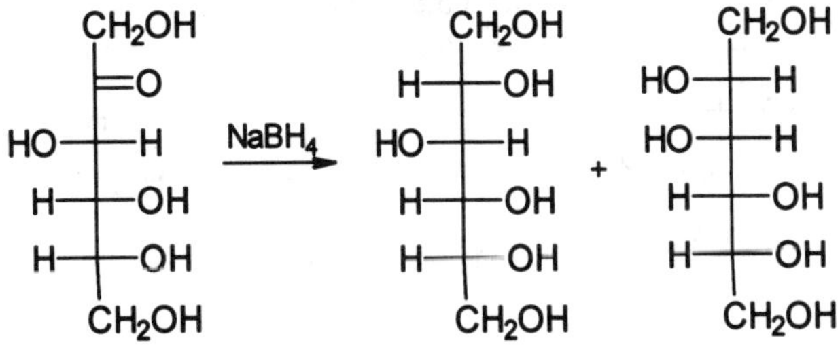

```
       CH2OH                      CH2OH                  CH2OH
       ┋O                      H ─┼─ OH              HO ─┼─ H
   HO ─┼─ H      NaBH4        HO ─┼─ H                HO ─┼─ H
    H ─┼─ OH    ────────►      H ─┼─ OH          +     H ─┼─ OH
    H ─┼─ OH                   H ─┼─ OH                 H ─┼─ OH
       CH2OH                      CH2OH                  CH2OH
```

20.30   Galactitol and talitol differ only in the configuration at the C-2 atom, so it must be the carbonyl carbon atom of tagatose. The configuration of the remaining stereogenic centers of tagatose are the same as in galactose and talose.

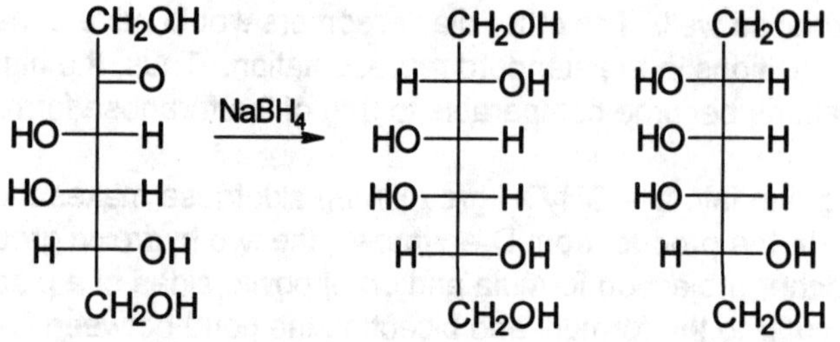

```
      CH2OH                    CH2OH              CH2OH
        ‖O                   H——OH            HO——H
  HO——H        NaBH4      HO——H            HO——H
  HO——H        ——>        HO——H            HO——H
   H——OH                   H——OH             H——OH
      CH2OH                    CH2OH              CH2OH
```

20.31   Both enantiomers give the same meso alditol.

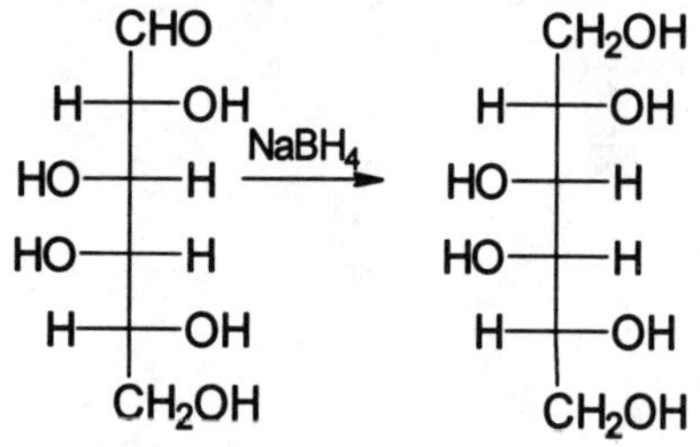

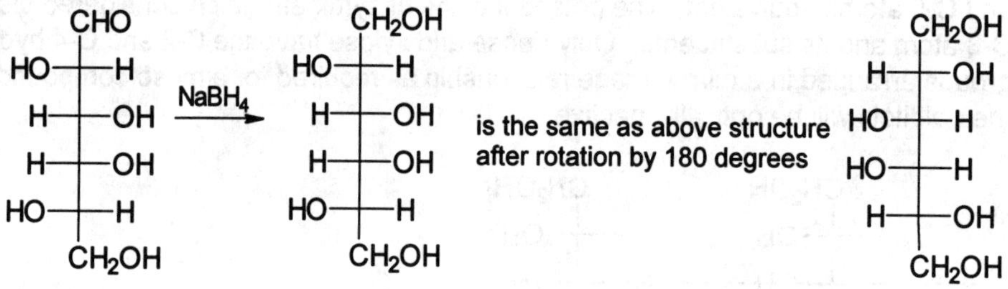

```
       CHO                     CH2OH
    H——OH                   H——OH
  HO——H       NaBH4      HO——H
  HO——H       ——>        HO——H
    H——OH                   H——OH
       CH2OH                   CH2OH
```

```
      CHO                   CH2OH                                       CH2OH
   HO——H                 HO——H                                      H——OH
    H——OH     NaBH4       H——OH       is the same as above structure   HO——H
    H——OH     ——>         H——OH       after rotation by 180 degrees    HO——H
   HO——H                 HO——H                                       H——OH
      CH2OH                 CH2OH                                        CH2OH
```

20.32   The same product is obtained from both compounds. To show this relationship, rotate the structure of the alditol of L-gulose by 180° and compare it to the structure of the alditol of D-glucose on the following page.

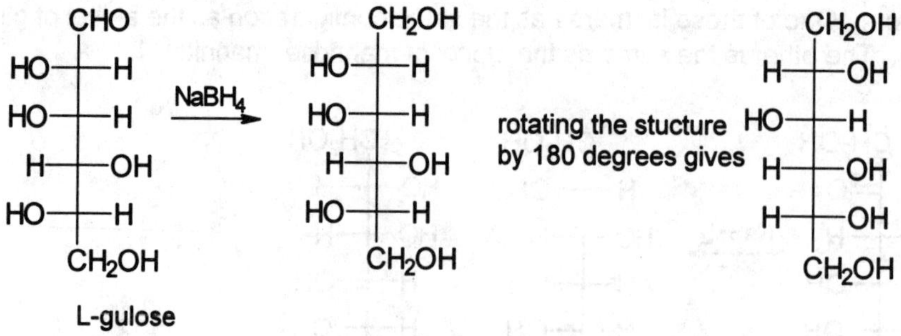

```
      CHO                   CH2OH                                   CH2OH
   HO——H                 HO——H                                   H——OH
   HO——H      NaBH4      HO——H      rotating the stucture      HO——H
    H——OH     ——>         H——OH     by 180 degrees gives        H——OH
   HO——H                 HO——H                                   H——OH
      CH2OH                 CH2OH                                    CH2OH
    L-gulose
```

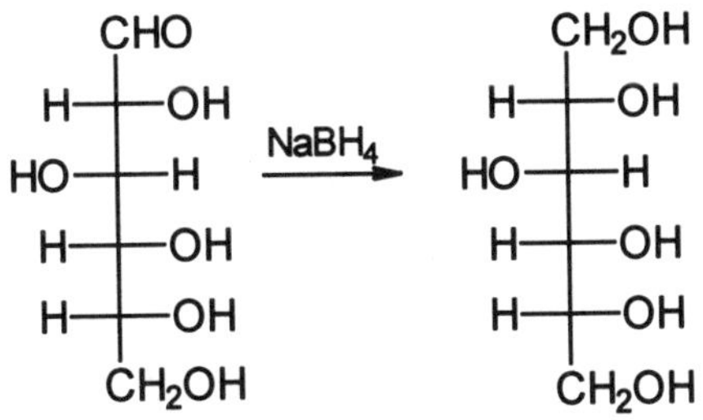

```
        CHO                      CH₂OH
   H──────OH               H──────OH
  HO──────H        NaBH₄  HO──────H
   H──────OH        ───▶   H──────OH
   H──────OH               H──────OH
       CH₂OH                   CH₂OH
     D-glucose
```

20.33 **(a)** 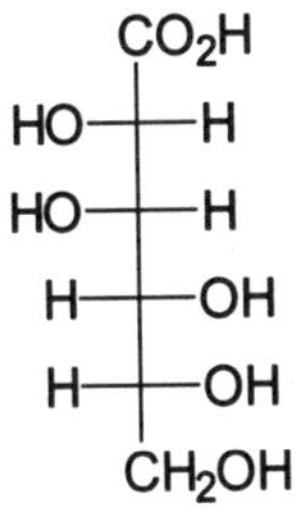
```
        CO₂H
  HO──────H
  HO──────H
   H──────OH
   H──────OH
      CH₂OH
```
(b) 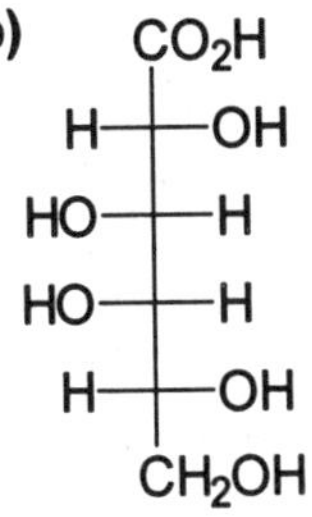
```
        CO₂H
   H──────OH
  HO──────H
  HO──────H
   H──────OH
      CH₂OH
```
(c) 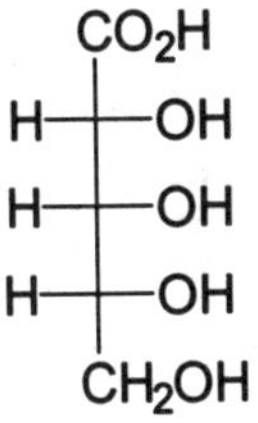
```
        CO₂H
   H──────OH
   H──────OH
   H──────OH
      CH₂OH
```
(d)
```
        CO₂H
  HO──────H
   H──────OH
   H──────OH
      CH₂OH
```

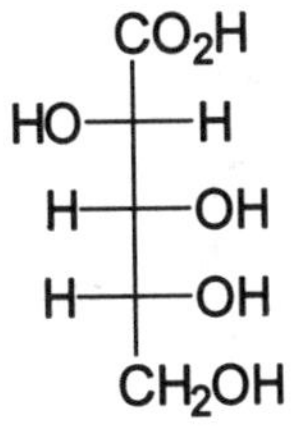

20.34 **(a)** 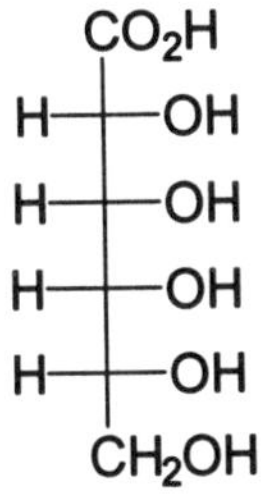
```
        CO₂H
   H──────OH
   H──────OH
   H──────OH
   H──────OH
      CH₂OH
```
(b) 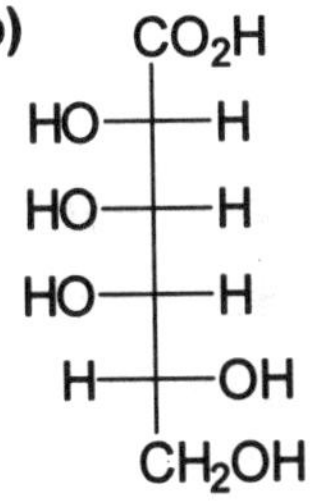
```
        CO₂H
  HO──────H
  HO──────H
  HO──────H
   H──────OH
      CH₂OH
```
(c) 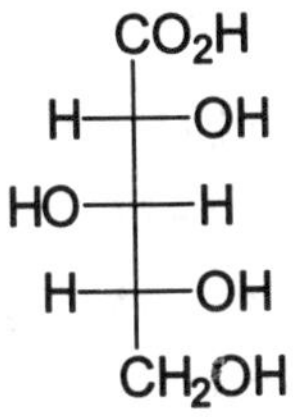
```
        CO₂H
   H──────OH
  HO──────H
   H──────OH
      CH₂OH
```
(d)
```
        CO₂H
  HO──────H
  HO──────H
   H──────OH
      CH₂OH
```

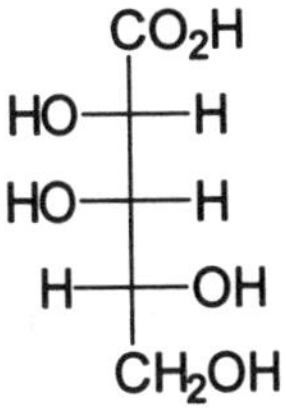

20.35 Conversion of the ─CHO group as well as the ─CH₂OH group into ─CO₂H groups makes the C-1 and C-4 atoms equivalent. In the product from D-erythrose (shown on the left) the two hydroxyl groups are on the same side in the Fischer projection formula and on opposite sides of a plane of symmetry placed perpendicular to the formula and bisecting the bond between the C-2 and C-3 atoms, so it is optically inactive. In the product from D-threose (shown on the right) the hydroxyl groups on the C-2 and C-3 atoms are not symmetrical with respect to the described plane, so it is optically active.

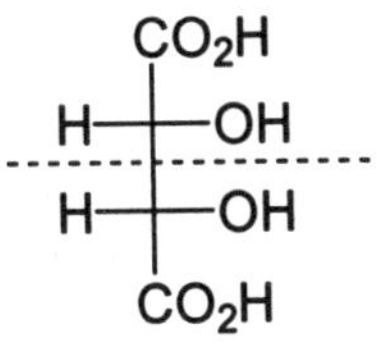

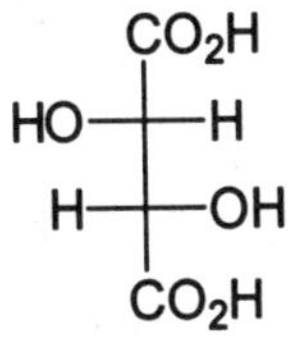

```
        CO₂H                     CO₂H
   H──────OH              HO──────H
  ----------------         H──────OH
   H──────OH
       CO₂H                    CO₂H
```

20.36 Conversion of the ─CHO group as well as the ─CH₂OH group into ─CO₂H groups makes the C-1 and C-5 atoms equivalent. The potential plane of symmetry to be considered bisects the C-3 atom and its substituents. Only ribose and xylose have the

C-2 and C-4 hydroxyl groups arranged in the mirror image relationship required for a meso compound.

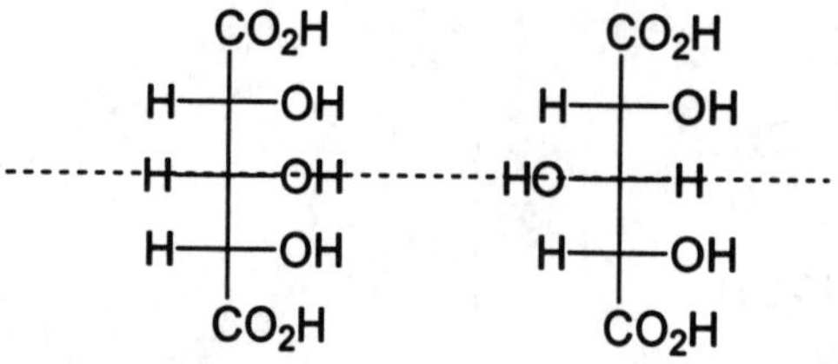

```
        CO₂H                    CO₂H
    H──┼──OH              H──┼──OH
----H──┼──OH----  ----HO──┼──H----
    H──┼──OH              H──┼──OH
        CO₂H                    CO₂H
```

20.37

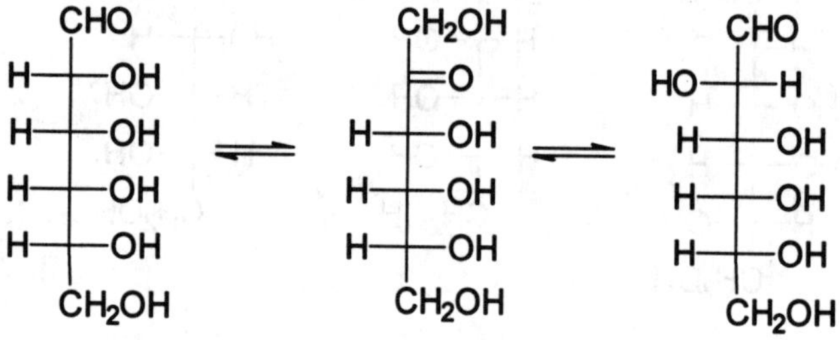

```
     CHO                 CH₂OH                CHO
 H──┼──OH                 ═O            HO──┼──H
 H──┼──OH    ⇌    H──┼──OH    ⇌    H──┼──OH
 H──┼──OH            H──┼──OH            H──┼──OH
 H──┼──OH            H──┼──OH            H──┼──OH
     CH₂OH               CH₂OH                CH₂OH
```

20.38

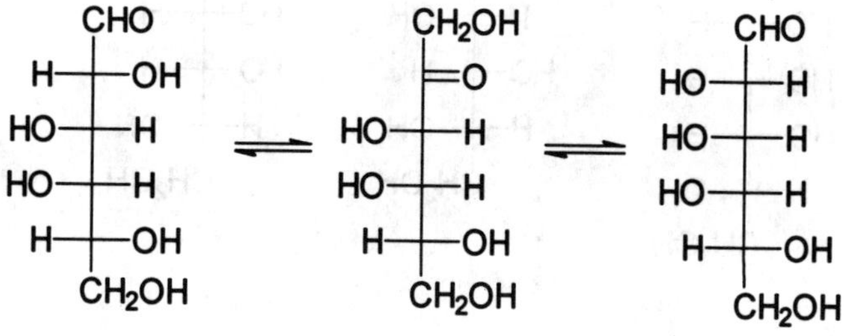

```
     CHO                 CH₂OH                CHO
 H──┼──OH                 ═O            HO──┼──H
HO──┼──H     ⇌    HO──┼──H    ⇌    HO──┼──H
HO──┼──H            HO──┼──H            HO──┼──H
 H──┼──OH            H──┼──OH            H──┼──OH
     CH₂OH               CH₂OH                CH₂OH
```

20.39

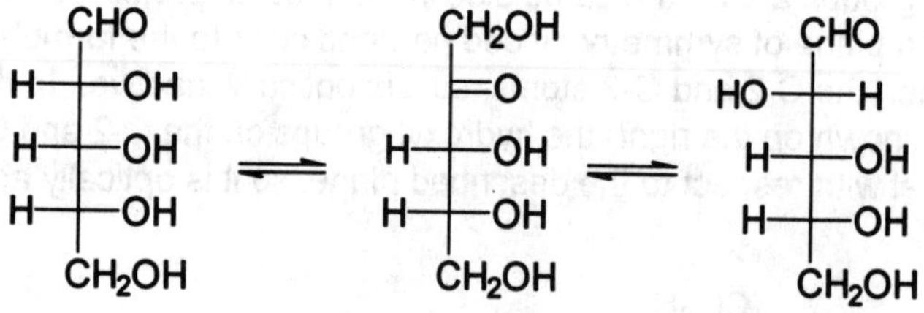

```
     CHO                 CH₂OH                CHO
 H──┼──OH                 ═O            HO──┼──H
 H──┼──OH    ⇌    H──┼──OH    ⇌    H──┼──OH
 H──┼──OH            H──┼──OH            H──┼──OH
     CH₂OH               CH₂OH                CH₂OH
```

20.40

```
     CHO                 CH₂OH                CHO
 H──┼──OH                 ═O            HO──┼──H
HO──┼──H     ⇌    HO──┼──H    ⇌    HO──┼──H
 H──┼──OH            H──┼──OH            H──┼──OH
     CH₂OH               CH₂OH                CH₂OH
```

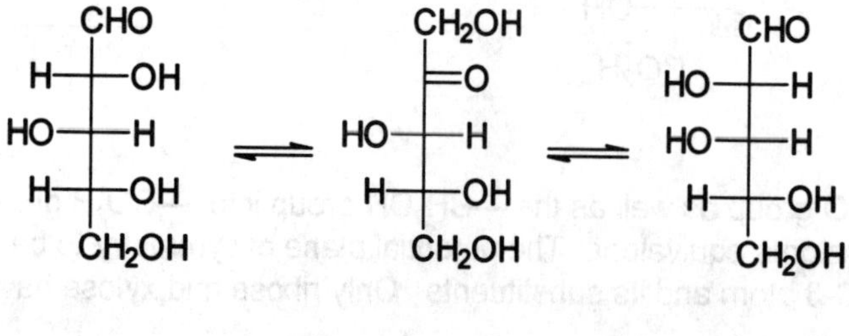

20.41 The carbonyl groups of ketones are generally more stable than those of aldehydes as the result of inductive effects of the two alkyl groups compared to a hydrogen atom and an alkyl group of an aldehyde. Thus, dihydroxyacetone phosphate (on the right) is more stable than D-glyceraldehyde-3-phosphate (on the left).

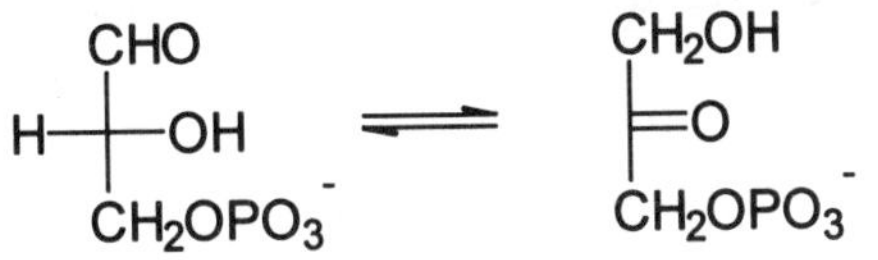

20.42 Glucose-6-phosphate and fructose-6-phosphate do not exist with their respective carbonyl groups but rather as cyclic hemiacetal and hemiketal derivatives, respectively. Thus, the relative stabilities reflect other differences beween the cyclic derivatives, such as structure and steric interactions.

20.43 (a)

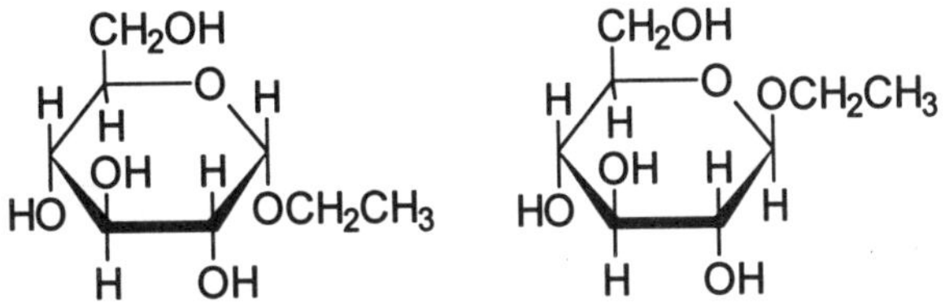

(b)

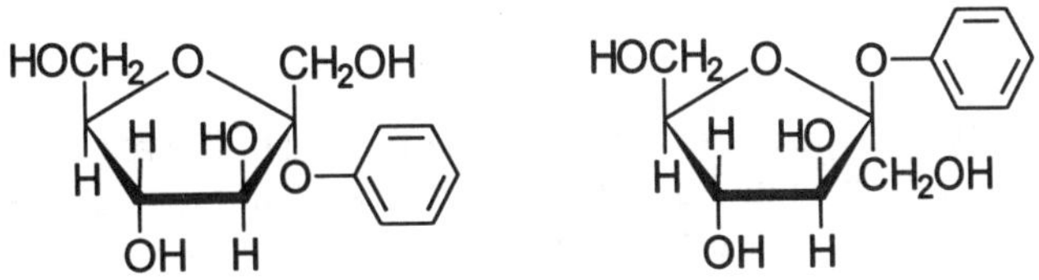

(c)

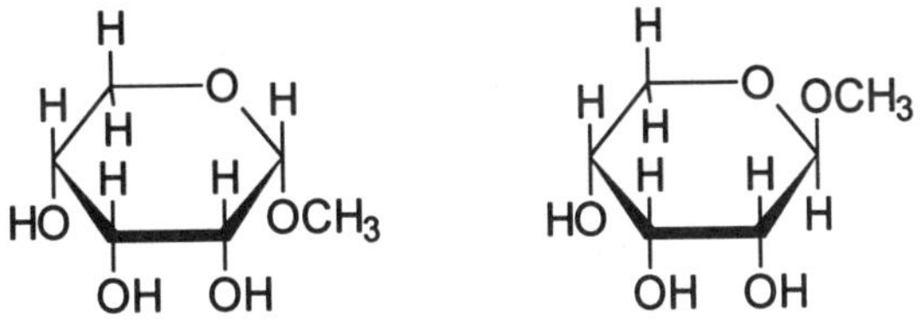

(d)

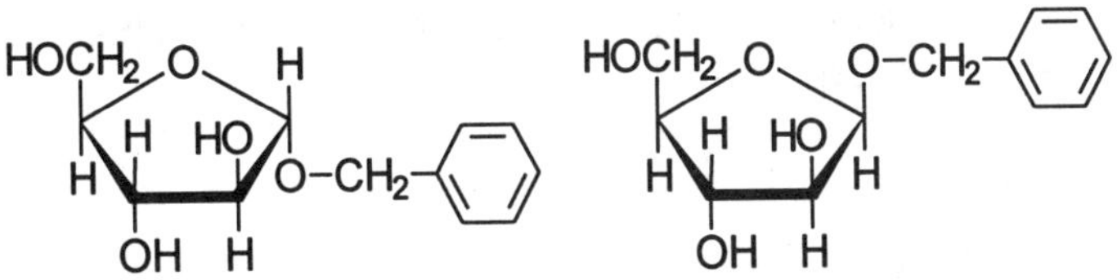

20.44 The reaction proceeds by a oxocarbocation in which the C-1 atom is achiral. Thus, a mixture of diastereomers must form when D-glucose reacts directly with methanol.

20.45 The aglycone of linamarin is the cyanohydrin of acetone. Because the cyanohydrin is unstable, its release by hydrolysis will lead to subsequent reversal of cyanohydrin formation, liberating HCN, which is poisonous.

20.46 Both monosaccharide units are D-glucose. The aglycone has three hydroxyl groups, so there are three isomeric diacetal derivatives that could give this compound.

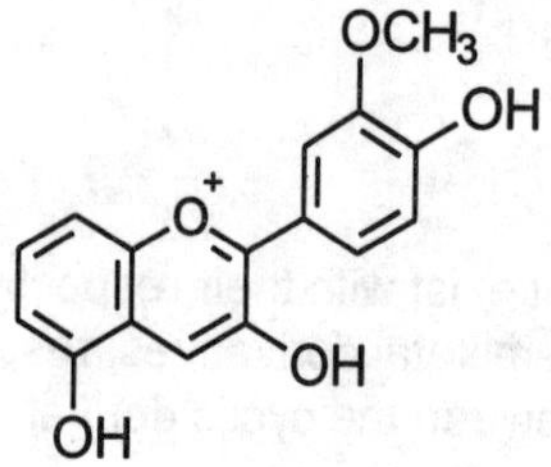

20.47

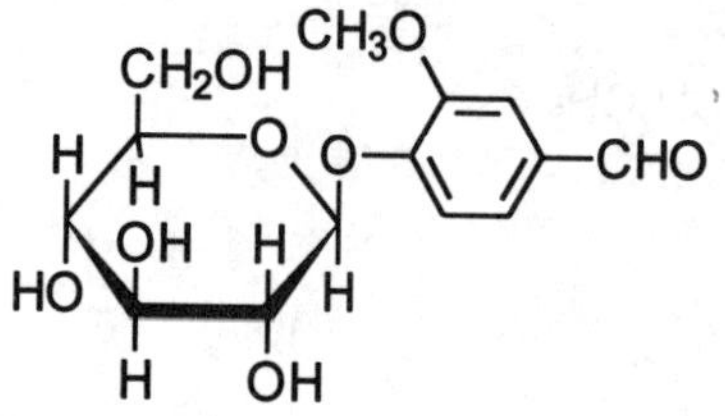

20.48 There are only four methyl groups in the carbohydrate portion of the antibiotic. The fifth methyl group had to be introduced in the aglycone portion of the molecule. Because hydrolysis gives p-methoxyphenol, the aglycone portion must be the following phenolic compound known as hydroquinone.

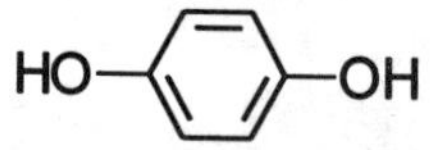

20.49 The hydrolysis products identify the carbohydrate part of salicin as glucose and the aglycone as 2-(hydroxymethyl)phenol. There are two hydroxyl groups in the aglycone that could be bonded to the acetal center of glucose in salicin. The oxidation reaction converts the —CH$_2$OH of the carbohydrate and the —CH$_2$OH of the aglycone. Because 2-hydroxybenzoic acid is obtained, the acetal must have had a bond between C-1 of glucose and the phenolic oxygen atom.

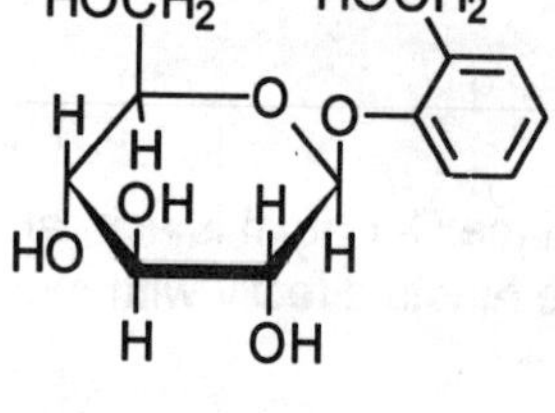

20.50 Two of the four hydroxyl groups in the aglycone are equivalent. Thus, the glycoside could be any of three possible isomers. Methylation of the glycoside followed by hydrolysis would give a trimethyl ether of the aglycone. Determination of its structure would indicate the position of the free hydroxyl group which must have formed the acetal linkage of the glycoside.

20.51 Both compounds give an oxocarbocation in the hydrolysis reaction. The anomeric carbon atom bearing the charge in compound I (an acetal) has a hydrogen atom bonded to it, whereas the related carbon atom in compound II (a ketal) has a carbon atom bonded. Thus, the carbocation of compound II is more stable and the transition state leading to its formation is lower in energy than that for compound I.

20.52 Both compounds give an oxocarbocation in the hydrolysis reaction. The anomeric carbon atoms bearing the charge in both compounds have the same degree of substitution. However, compound II has a hydroxyl group at the C-2 atom, which is inductively electron withdrawing and should destabilize the carbocation, thus raising the energy of the transition state leading to its formation.

20.53 The compound is an L carbohydrate, because the C-6 methyl group is "down". A ring flip to give the standard chair representation would place the C-6 methyl group in an axial position, which corresponds to the L configuration. Recall that D carbohydrates have a —CH$_2$OH group "up" in the Haworth projection and equatorial in the standard chair representation.

20.54 The anomers are diastereomers of one another, differing in configuration at one stereogenic center, so they are expected to have different physical properties. Methyl β-L-glucopyranoside is the enantiomer of methyl β-D-glucopyranoside and therefore has the same physical properties such as melting point, which is 105°C.

20.55 **(a)** a β linkage from C-1 of D-2-deoxyribose with a β linkage from C-1 of L-glucose
(b) a β linkage from C-1 of D-glucose to C-3 of D-galactose.
(c) a β linkage from C-1 of D-galactose to C-4 of D-mannose

20.56 **(a)** There is no glycosidic linkage. The compound is an ether formed between the C-2 atom of D-xylose and the C-6 atom of D-galactose.
(b) an α linkage from C-1 of D-glucose to C-1 of D-ribose
(c) a β linkage from C-1 of D-fructose to C-3 of D-glucose

20.57 **(a)** one mole of formaldehyde and four moles of formic acid
(b) two moles of formaldehyde, two moles of formic acid, and one mole of CO_2
(c) one mole of formaldehyde and five moles of formic acid
(d) two moles of formaldehyde, one mole of formic acid, and one mole of CO_2

20.58 **(a)** one mole of formaldehyde and four moles of formic acid
(b) two moles of formaldehyde, three moles of formic acid, and one mole of CO_2
(c) one mole of formaldehyde and three moles of formic acid
(d) one mole of formaldehyde and five moles of formic acid

20.59 **(a)** two moles of formaldehyde, two moles of formic acid, and one mole of CO_2

(b) one mole of formaldehyde, one mole of formic acid, and the following dialdehyde containing the original C-1, C-2, and C-3 atoms.

$$\begin{array}{c} CHO \\ H \!-\!\!\!\!-\!CH_2OH \\ CHO \end{array}$$

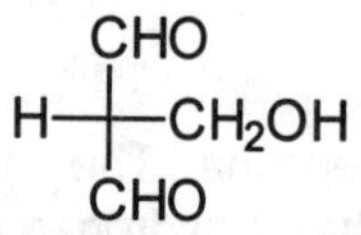

(c) acetaldehyde from the original C-4 and C-5 atoms, and the following dialdehyde containing the original C-1, C-2, and C-3 atoms

$$\begin{array}{c} CHO \\ H \!-\!\!\!\!-\!H \\ CHO \end{array}$$

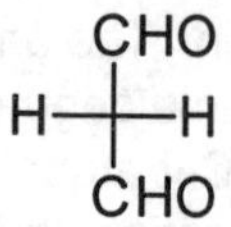

20.60 **(a)** one mole of formaldehyde, one mole of formic acid, one mole of carbon dioxide, and one mole of acetaldehyde from the original C-4 and C-5 atoms

(b) one mole of formaldehyde, three moles of formic acid, and one mole of acetic acid

(c) one mole of formaldehyde, three moles of formic acid, and one mole of acetic acid

20.61 Formation of the osazone decreases the number of stereogenic centers from four to three and cuts the number of possible stereoisomers in half. D-Allose and D-altrose form the same osazone; D-glucose and D-mannose form the same osazone; D-gulose and D-idose form the same osazone; D-galactose and D-talose form the same osazone.

20.62 A simple hydrazone derivative forms because there is no α-hydroxy group.

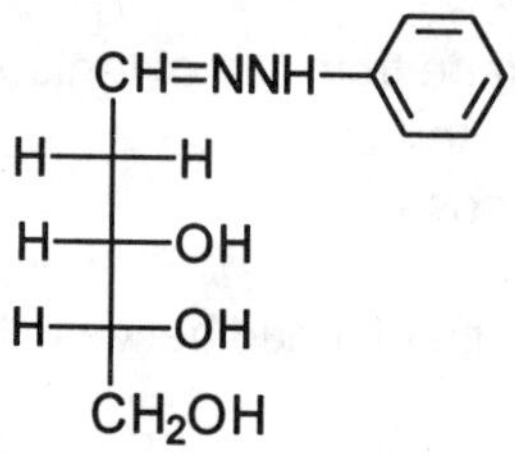

$$\begin{array}{c} CH\!=\!NNH\!-\!\bigcirc \\ H \!-\!\!\!\!-\!H \\ H \!-\!\!\!\!-\!OH \\ H \!-\!\!\!\!-\!OH \\ CH_2OH \end{array}$$

20.63 The Wohl degradation of D-idose gives D-xylose. Oxidation of D-xylose gives an optically inactive aldaric acid. The position of the methyl ether in the oxidation product described must be at C-3, which is where the plane of symmetry would be located. If the methyl ether were at C-2 or C-4, the oxidation product would be chiral. The methyl ether of the original idose was at C-4.

20.64 The position of the methyl ether in the chain-extended oxidation product described must be at C-4, which is where the plane of symmetry would be located. Any other position for the methyl ether would lead to an optically active product. The methyl ether of the original glucose was at C-3.

20.65 The aldohexose exists in a pyranose form, so methylation occurs at the hydroxyl groups of the C-1, C-2, C-3, C-4, and C-6 atoms. Treatment with acid hydrolyzes the acetal, restoring the —OH group at C-1. After methylation and hydrolysis, the "open chain" isomer in equilibrium with the methylated pyranose is oxidized at the C-1 atom giving a carboxylic acid and at the C-5 atom giving a ketone. Vigorous oxidation cleaves the C-5 to C-6 bond and the product is a dicarboxylic acid. In order to give an optically inactive product, the methylated sites at C-2, C-3, and C-4 have to be symmetrically arranged about a plane through the C-3 atom. The aldohexoses that have related symmetrical arrangements are allose, talose, glucose, and idose. The reaction for allose is given as an example.

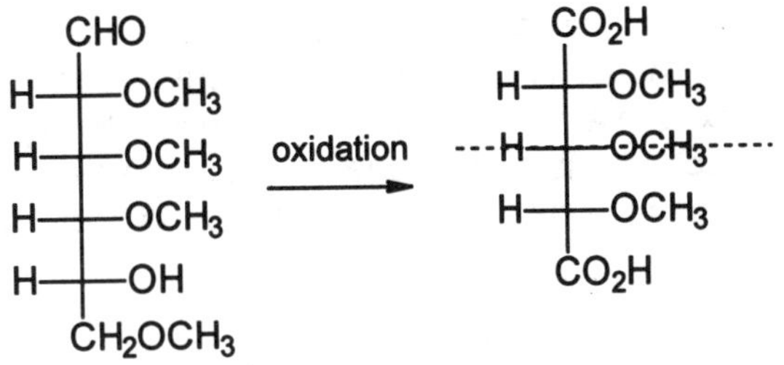

20.66 The aldohexose is in the furanose form as evidenced by the aldehyde group on the left side of the molecule. This C-5 atom was originally bonded to the C-6 atom in a —CH_2OH group. Oxidation by periodate released the C-6 atom as formaldehyde, leaving C-5 in an aldehyde group. The acetal center is located on the right side of the molecule as evidenced by the methoxy group, and it has the α configuration. The ring oxygen atom of the glycoside is derived from the hydroxyl group at C-4 of the aldohexose and is located on the right in the Fischer projection formula. If the aldohexose has the D configuration, then allose, altrose, glucose, and mannose are possible compounds.

20.67 The location of the three methoxy groups in the xylose derivative means that the C-5 oxygen atom was part of a pyranose ring and the compound was linked as a glycoside at its C-1 atom. The location of the three methoxy groups in the glucose derivative means that the C-5 oxygen atom was part of a pyranose ring. The compound must have been linked to the glycosidic center of the xylose by the C-6 oxygen atom.

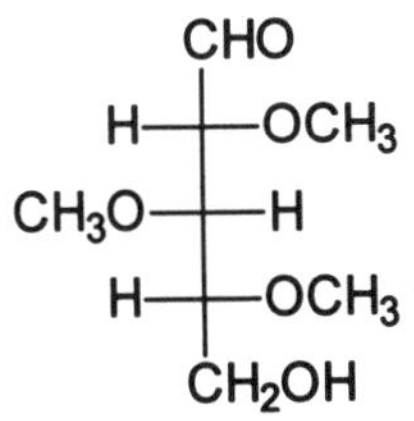

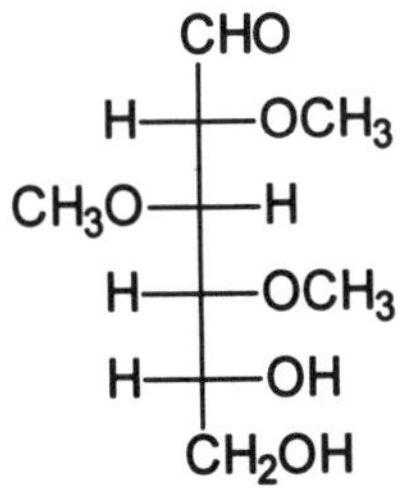

20.68 The location of the four methoxy groups in one glucose derivative means that the C-5 oxygen atom was part of a pyranose ring and the compound was linked as a glycoside at its C-1 atom. The location of the three methoxy groups in the other glucose derivative means that the C-5 oxygen atom was part of a pyranose ring. This compound must have been linked to the glycosidic center of the first glucose by the C-6 oxygen atom.

20.69 The location of the four methoxy groups in the glucose derivative means that the C-5 oxygen atom was part of a pyranose ring and the compound was linked as a glycoside at its C-1 atom. The location of the three methoxy groups in the fructose derivative means that the C-6 oxygen atom was part of a pyranose ring. The compound must have been linked to the glycosidic center of the glucose by the C-3 oxygen atom.

20.70 The location of the two methoxy groups in the glucose derivative could mean that either the C-4 or the C-5 oxygen atom was part of a furanose or pyranose ring, respectively. If this portion of the structure is a furanose, then the C-2 and C-5 oxygen atoms were part of glycosidic linkages to the other monosaccharide units. If this portion of the structure is a pyranose, then the C-2 and C-4 oxygen atoms were part of glycosidic linkages to the other monosaccharide units. For each possible case, one oxygen atom must be linked to the acetal center of another glucose molecule and the second to the acetal center of a galactose molecule.

21

Carboxylic Acids

In this chapter we consider a functional group that looks like two functional groups—the hydroxyl group and the carbonyl group—sharing the same carbon atom. Structurally this is correct but in terms of the chemistry the functional group is unique. It is not an alcohol, nor is it an aldehyde or ketone. The carboxyl functional group has some reactions in common with these "component" functional groups, but it behaves chemically as a unit. In this chapter you may have to refer to Chapter 19 to recall how the carbonyl carbon atom is affected by the carbon groups bonded to it, because the chemistry of the carboxyl group involves nucleophilic attack at the carbonyl carbon atom. In other words, there are inductive, resonance, and steric effects in the reactions of carboxylic acids and its derivatives (Chapter 22), just as there were for aldehydes and ketones.

21.1 The Carboxyl and Acyl Groups

Examine the structure of acetic acid and the electronic structure of the carbonyl group shown in Figure 21.1 in the text. The reactivity of the carboxyl group is interpreted based on its π electrons and the two sets of nonbonded electrons as well as the hydroxyl group. In addition, pay particular attention to the dipolar structures that are contributing resonance structures for the carboxyl group. There is a third resonance structure that occurs for carboxylic acids compared to only two contributing resonance structures for aldehydes and ketones. The lone pair electrons of the hydroxyl oxygen atom are donated to the electron deficient carbon atom in the second dipolar resonance structure. As a result, the electron density of the carbonyl carbon atom in a carboxylic acid is larger than for aldehydes and ketones.

Carboxylic acids are the "parents" of acid derivatives. The **acyl group** is only the RCO unit which is bonded to an electronegative atom in carboxylic acids and their derivatives. The acyl group of an **ester** is bonded to the oxygen atom of an alkoxy group. The acyl group of an **amide** is bonded to a nitrogen atom. Cyclic esters and amides are lactones and lactams, respectively. The acyl group is bonded to a chlorine atom, a carboxyl group, and a thiolate group in **acid chlorides**, **acid anhydrides**, and **thioesters**, respectively.

21.2 Nomenclature of Carboxylic Acids

The common names of the unbranched carboxylic acids do not provide information about the number of carbon atoms contained in the chain. Your instructor should indicate how many of the names must be memorized. See Table 21.1 in the text for the common names of carboxylic acids. Carbon atoms in the parent chain are designated α, β, γ, etc, starting with the carbon atom directly bonded to the carboxyl carbon atom. Branches on the chain are indicated using these Greek letters to identify the location of substituents or alkyl groups.

IUPAC names of carboxylic acids are based on the alkane names, with -oic acid replacing the final -e of the alkane. The carbon chain is numbered starting with the carboxyl carbon atom, but the number 1 is not included in the name. Examine the rules used to name carboxylic acids containing additional functional groups given in this section. There isn't much new if you have learned the rules for the nomenclature of other functional groups such as for aldehydes. The carboxyl group takes priority over halogen atoms, alkyl groups, double or triple bonds, and other functional groups containing a carbonyl group.

Salts of carboxylic acids are named using the name of the metal ion followed by the name of the acid, modified by changing -ic acid to -ate. A carboxylate anion is the conjugate base of a carboxylic acid.

21.3 Physical Properties of Carboxylic Acids

The carboxyl groups of neighboring carboxylic acid molecules form intermolecular hydrogen bonds. Thus carboxylic acids exist as dimers, and consequently they boil at higher temperatures than other hydrocarbon compounds of similar molecular weight. The boiling points of carboxylic acids are given in Table 21.2 of the text, but there is nothing exciting here and, as in the case of other physical constants, you don't have to memorize them. The melting points of the carboxylic acids containing an even number of carbon atoms are given in Table 21.3 in the text. The low molecular weight carboxylic acids are soluble in water, because the carboxyl group forms hydrogen bonds with water molecules.

21.4 Acidity of Carboxylic Acids

The structure of carboxylic acids and more importantly the structure of the carboxylate ion makes acids more acidic than alcohols. Loss of a proton from the carboxyl group leaves the carboxylate anion, which is stabilized by delocalization of two electron pairs over the carbon atom and two oxygen atoms. The formation of stable conjugate bases enhances the acidity of carboxylic acids.

Electronegative groups attached to the α-carbon atom tend to pull electron density away from the carboxyl group, making the proton easier to remove. This inductive effect further increases the acidity. Increased acidity increases the acid ionization constant and decreases the pK_a.

Note that both the methoxy group and the nitro group of substituted ethanoic acids increase the acidity. At first glance you might think that this result is contrary to what might be expected, based on the electronic properties of these two substituents in electrophilic aromatic substitution reactions. However, in those reactions it is the resonance donation of electrons by oxygen that affects the reactivity. In the case of these carboxylic acids, the only thing that counts is the inductive effect of the group, and oxygen is electronegative. Hence it withdraws electron density, as does the nitrogen atom of the nitro group with its formal positive charge.

The acidity of a substituted benzoic acid is affected by the change in the electron density of the aromatic ring. The substituents do not interact by resonance with the carboxylate group, nor does the carboxylate group interact by resonance with the aromatic ring.

21.5 Carboxylate Ions

In water, unsubstituted carboxylic acids are present largely as the nonionized form. If the pH of a solution is controlled by a buffer, then at pH 7 the carboxylic acid exists as its conjugate base. Work through the calculation to prove it to yourself, because this fact often comes as a surprise to students. In basic solution, the situation is a little easier to understand. A strong base reacts with the carboxylic acid to give the carboxylate ion. This method is used to dissolve carboxylic acids and separate them from other compounds that are not acidic.

21.6 Synthesis of Carboxylic Acids

Carboxylic acids are prepared from substrates with the proper hydrocarbon skeleton by oxidation of either an alcohol or an aldehyde. Because special methods are required to prepare aldehydes, the more common substrate for preparation of a carboxylic acid is the structurally related alcohol. The Jones reagent is used as the oxidizing agent.

Two other less general oxidative methods used to prepare carboxylic acids are the oxidation of the side chains of alkylbenzenes. Oxidation of an alkylbenzene by $KMnO_4$ produces a benzoic acid. The haloform reaction oxidizes a methyl ketone by removal of the methyl group. The reagent is a halogen in basic solution.

Preparation of carboxylic acids with one more carbon atom than the substrate is accomplished by two alternate procedures, both of which commence with an alkyl halide. Conversion of an alkyl halide into a Grignard reagent followed by reaction with carbon dioxide gives a carboxylic acid. Any class of haloalkane or even an aryl halide can be used. The second method is based on nucleophilic substitution of a halide ion by a cyanide ion. The resulting nitrile is hydrolyzed to form a carboxylic acid. The limitation on the reaction is the first step which is effective for primary haloalkanes and to a limited extent for secondary haloalkanes. The competing reaction is dehydrohalogenation of the haloalkane.

21.7 Reduction of Carboxylic Acids

Carboxylic acids are more difficult to reduce than aldehydes or ketones. Furthermore, the acidic proton reacts to destroy a portion of the metal hydride reagent commonly used for reduction of carbonyl groups. Only lithium aluminum hydride is sufficiently reactive to reduce a carboxylic acid, and the product is a primary alcohol. Sodium borohydride does not reduce carboxylic acids.

Diborane is a specialized reagent used to reduce carboxylic acids to give primary alcohols. The reagent does not readily reduce other unsaturated functional groups such as the nitrile group. In addition it does not reduce the nitro group. Note that diborane cannot be used if the substrate contains a double bond, because hydroboration would occur. The text doesn't mention this fact, but you should expect this competing reaction base on prior knowledge.

21.8 Decarboxylation of Carboxylic Acids

Loss of a carboxyl group and replacement by an atom such as hydrogen is **decarboxylation**. Carboxylic acids containing a β keto group undergo decarboxylation via a cyclic mechanism. Both β keto acids and malonic acids undergo this reaction.

The Hunsdiecker reaction replaces a carboxyl group by a halogen. The reagent is bromine, which reacts with the silver salt of the carboxylic acid. The reaction occurs via a free radical chain mechanism. Note that an unsaturated carboxylic acid cannot be used because bromine adds to double bonds.

21.9 Reactions of Carboxylic Acids and Derivatives—A Preview

The salts of carboxylic acids are weak nucleophiles but can be used to displace halide ion from a haloalkane to give esters. The most diverse reaction of carboxylic acids and their derivatives involves nucleophilic attack of the carbonyl carbon atom. However, in contrast to aldehydes and ketones, the tetrahedral intermediate is unstable and it ejects a leaving group. This overall process is **nucleophilic acyl substitution** or an **acyl transfer reaction**. This mechanism is another of the very important processes that are central to organic chemical reactions. Recall that others are nucleophilic substitution (at a saturated hydrocarbon), elimination, electrophilic aromatic substitution, and nucleophilic addition at carbonyl carbon atoms.

Although nucleophilic acyl substitution stoichiometrically resembles nucleophilic substitution at a saturated carbon atom, the mechanisms are very different. Nucleophilic acyl substitution occurs in two steps via a tetrahedral intermediate.

21.10 Conversion of Carboxylic Acids into Acyl Halides

The only method given for this conversion uses thionyl chloride. In many ways, the mechanism resembles that of the reaction of alcohols with thionyl chloride. That is because the reaction occurs via the electron pairs of the oxygen atom serving as a nucleophile to displace a chloride ion from thionyl chloride. However, the subsequent displacement of the sulfur-containing moiety occurs by a nucleophilic substitution in the case of alcohols and nucleophilic acyl substitution in the case of acids.

21.11 Conversion of Carboxylic Acids into Anhydrides

Anhydrides are prepared by removal of water from two moles of a carboxylic acid. Heating a dicarboxylic acid yields a cyclic anhydride. Intermolecular reaction of a carboxylic acid at a lower temperature requires a dehydrating agent such as phosphorus pentoxide. Mixed anhydrides can be made by displace of a halide from a haloalkane by a carboxylate ion.

21.12 Synthesis of Esters

Four reactions are described in this section of the text. These are:

1. Alkylation of a carboxylate ion using a haloalkane, which is akin to the Williamson ether synthesis. It occurs by a S_N2 mechanism and is largely limited to primary haloalkanes.

2. Methylation of a carboxylic acid using diazomethane. The reaction is largely limited to this simplest diazo compound but will occur with other diazo compounds.

3. Reaction of acyl chlorides with alcohols, which occurs by nucleophilic acyl substitution. The reaction is general for all alcohols and any acyl chloride.

4. The **Fischer esterification**, which uses a carboxylic acid and an alcohol in an acid-catalyzed equilibrium reaction that is driven to product by removal of the byproduct water.

21.13 Mechanism of Esterification

The mechanism of the Fischer esterification reaction occurs by nucleophilic attack of the alcohol on the carbonyl carbon atom of the carboxylic acid. The reaction requires protonation of the carbonyl oxygen atom to increase the partial positive charge of the carbonyl carbon atom. Isotope studies show that the oxygen atom of the alcohol is the bridging atom of the ether. Note that the configuration of a chiral alcohol would be retained in this esterification reaction, because the bond to the oxygen atom of the alcohol is not cleaved.

21.14 Spectroscopy of Carboxylic Acids

Carboxylic acids are characterized by the strong absorption due to the carbonyl group in the IR spectra of these compounds. The absorption occurs in the same region as for the carbonyl groups of aldehydes and ketones, but the absorption for carboxylic acids occurs at slightly higher wavenumber and tends to be somewhat broadened. The O—H bond of carboxylic acids absorbs in the same region as that for alcohols. However, the absorption is very much broader for carboxylic acids and it overlaps the C—H absorptions.

The proton NMR spectra of carboxylic acids show a characteristic absorption in the 9—12 δ region. The α hydrogen atoms of carboxylic acids occur in the 2.0—2.5 δ region, which is the same region for the α hydrogen atoms of aldehydes and ketones.

The α carbon atoms of carboxylic acids have ^{13}C NMR absorptions in the 20 δ region, which is at slightly higher field than for aldehydes and ketones. The carbonyl carbon atom is easily identified by its absorption in the 200 δ region.

Summary of Reactions

1. Synthesis of Carboxylic Acids by Oxidative Methods

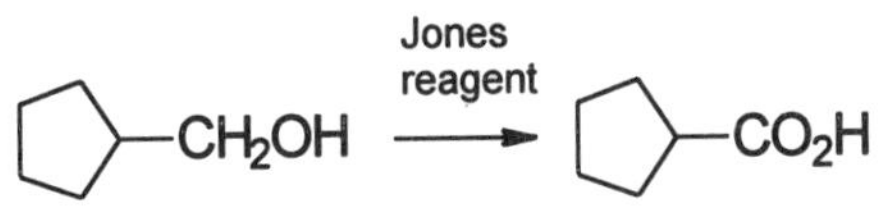

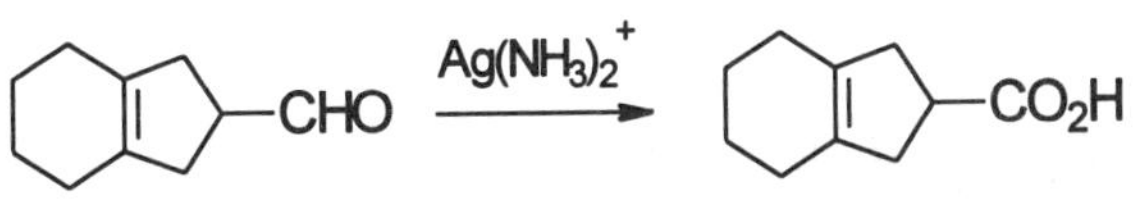

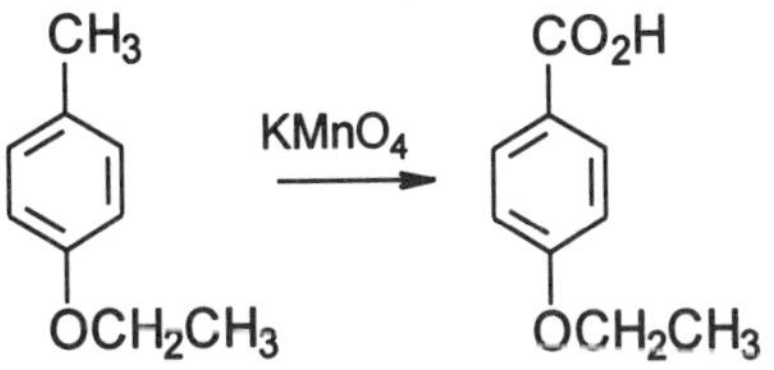

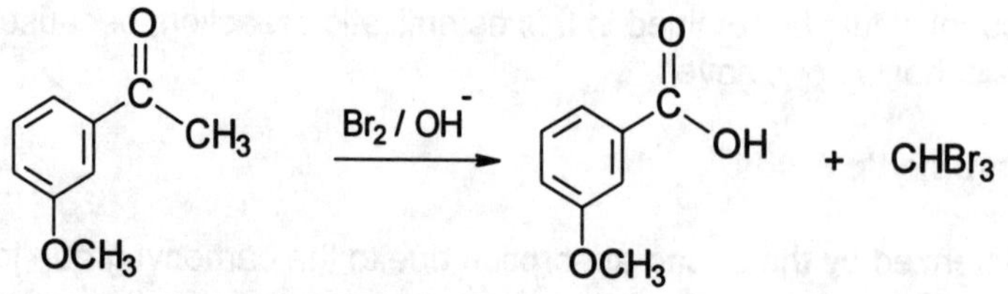

2. Synthesis of Carboxylic Acids from Haloalkanes

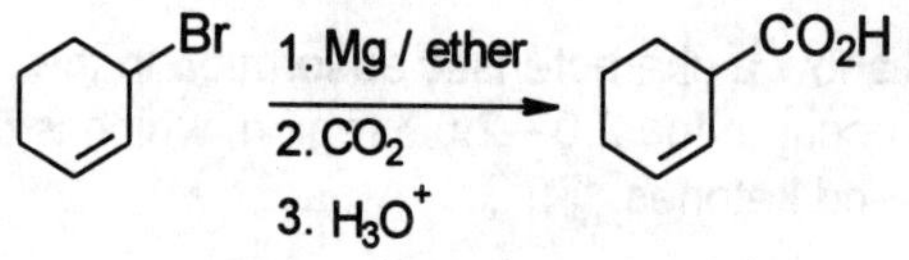

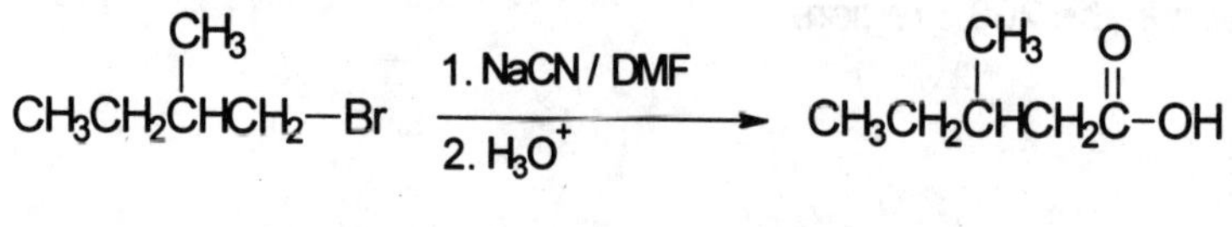

3. Reduction of Carboxylic Acids

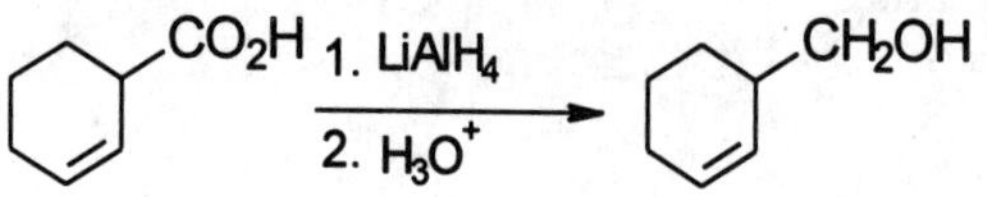

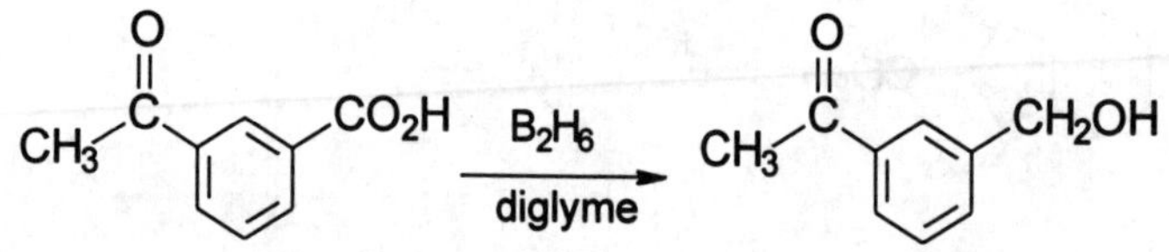

4. Decarboxylation of Carboxylic Acids

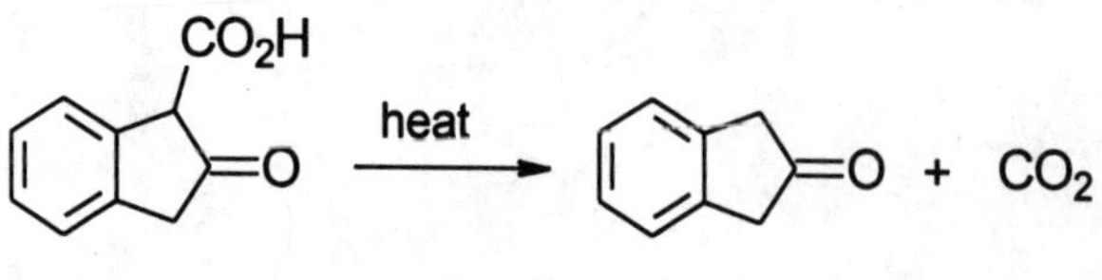

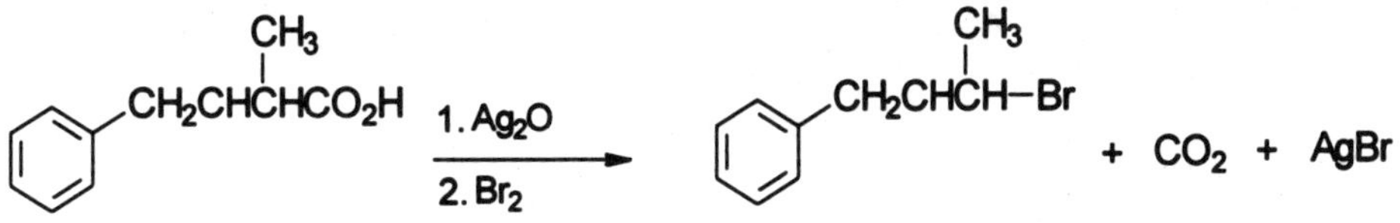

5. Synthesis of Acyl Halides

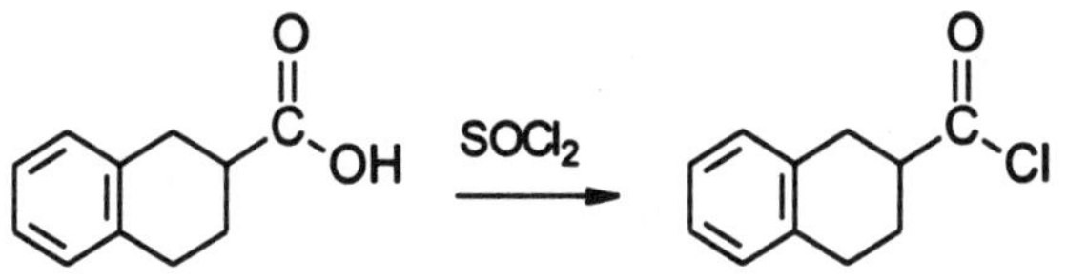

6. Synthesis of Acid Anhydrides

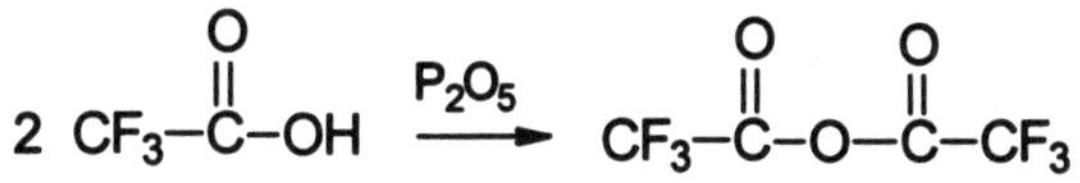

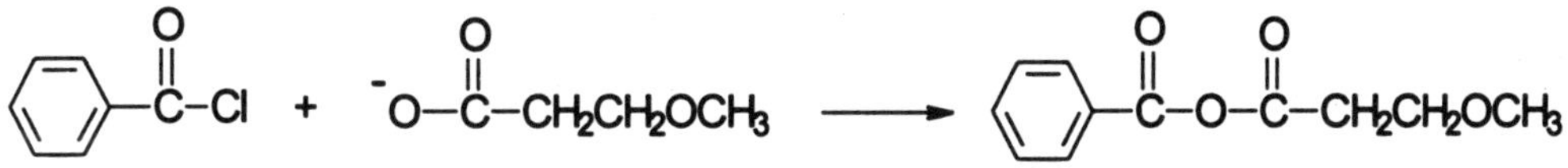

7. Synthesis of Esters

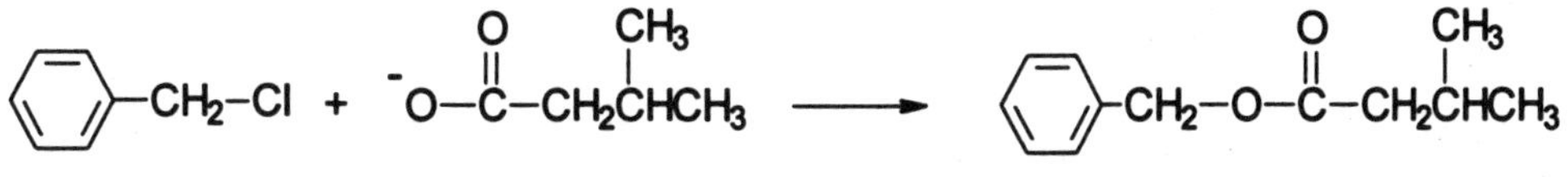

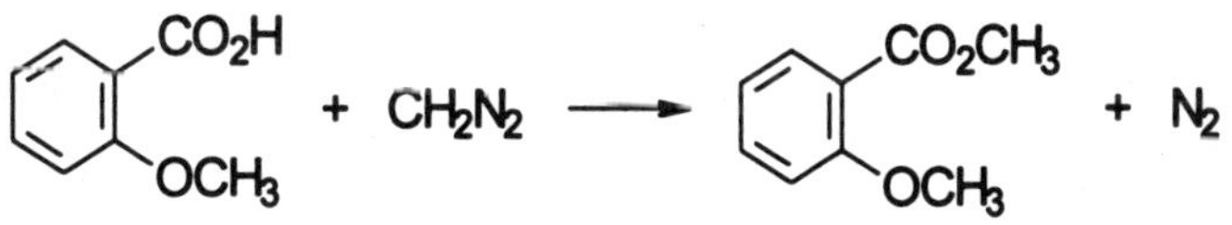

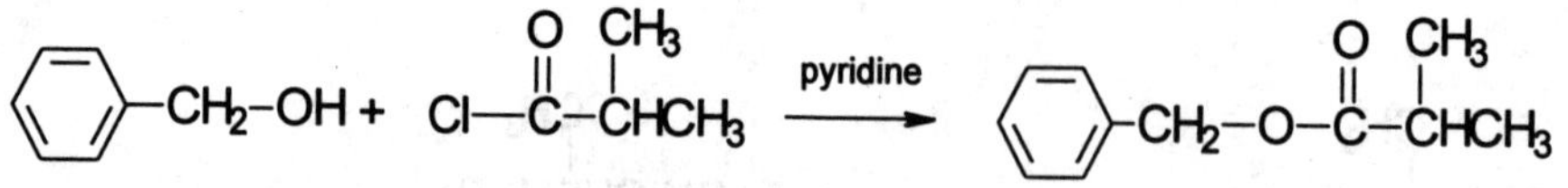

$$\text{C}_6\text{H}_5\text{-CH}_2\text{-OH} + \underset{\text{Cl}}{\overset{\displaystyle \text{O} \; \text{CH}_3}{\text{C}-\overset{|}{\text{CHCH}_3}}} \xrightarrow{\text{pyridine}} \text{C}_6\text{H}_5\text{-CH}_2\text{-O}-\underset{}{\overset{\displaystyle \text{O} \; \text{CH}_3}{\text{C}-\overset{|}{\text{CHCH}_3}}}$$

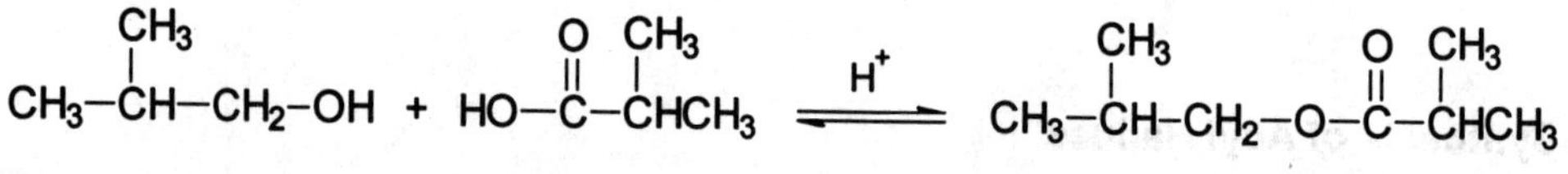

$$\underset{\text{CH}_3}{\overset{\displaystyle \text{CH}_3}{\text{CH}_3\text{-CH}-\text{CH}_2\text{-OH}}} + \text{HO}-\underset{}{\overset{\displaystyle \text{O} \; \text{CH}_3}{\text{C}-\text{CHCH}_3}} \underset{}{\overset{\text{H}^+}{\rightleftharpoons}} \text{CH}_3\text{-CH}-\text{CH}_2\text{-O}-\underset{}{\overset{\displaystyle \text{O} \; \text{CH}_3}{\text{C}-\text{CHCH}_3}}$$

Solutions to Exercises

21.1 **(a)** propionic acid **(b)** formic acid **(c)** caproic acid
(d) lauric acid **(e)** stearic acid **(f)** palmitic acid

21.2 **(a)** β-chlorobutyric acid **(b)** α-bromopropionic acid
(c) β-bromo-γ-methylvaleric acid **(d)** α-chloro-γ,γ-dimethylvaleric acid

21.3 **(a)** *trans*-2-hydroxycyclohexanecarboxylic acid
(b) (E)-3-chloro-3-phenyl-2-propenoic acid
(c) 5-hexynoic acid
(d) cyclodecanecarboxylic acid

21.4 **(a)** 3-methoxybenzoic acid
(b) 2-(3-methoxyphenyl)propanoic acid
(c) 3-ethyl-2-pentenoic acid
(d) cyclopentylethanoic acid

21.5 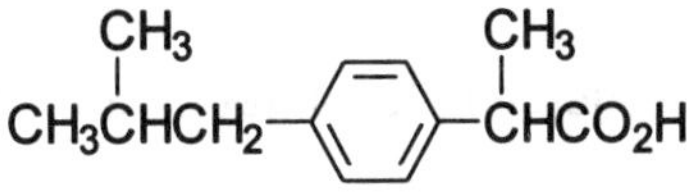

21.6 $CH_2{=}CH(CH_2)_8CO_2H$

21.7 **(a)** $C_nH_{2n}O_2$ **(b)** $C_nH_{2n-2}O_4$ **(c)** $C_nH_{2n-2}O_2$ **(d)** $C_nH_{2n-2}O_2$

21.8 **(a)** $HO_2CCH_2CH_2CO_2H$ $HO_2CCH(CH_3)CO_2H$

(b) $CH_3CH_2CH_2CO_2H$ $CH_3CH(CH_3)CO_2H$

(c) $CH_3CH_2CH_2CH_2CO_2H$ $CH_3CH_2CH(CH_3)CO_2H$ $CH_3CH(CH_3)CH_2CO_2H$ $CH_3C(CH_3)_2CO_2H$

(d) 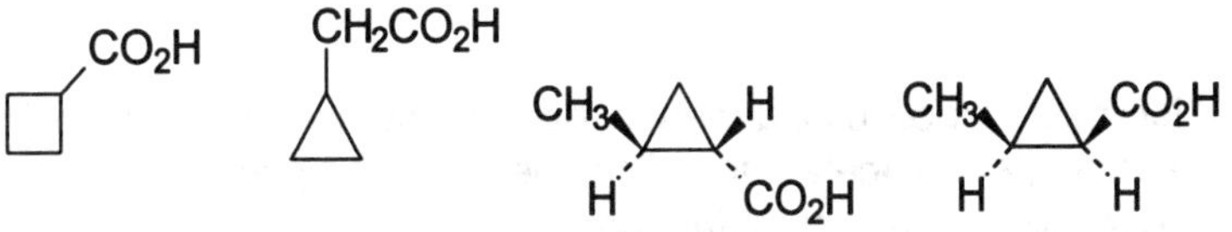

21.9 Because it has one more oxygen atom, butanoic acid has more unshared pairs of electrons that serve as hydrogen bond acceptors with water molecules.

21.10   Because it has two carboxylic acid groups, adipic acid has more unshared pairs of electrons that serve as hydrogen bond acceptors with water molecules, as well as another hydrogen atom bonded to oxygen that serves as a hydrogen bond donor to a water molecule.

21.11   The added methylene group of decanoic acid compared to nonanoic acid leads to increased London attractive forces between molecules.

21.12   The shape of 2,2-dimethylpropanoic acid is approximately spherical, so its points of contact with neighboring molecules are fewer than for the cylindrically shaped pentanoic acid. Thus the London forces are smaller for 2,2-dimethylpropanoic acid.

21.13   4-Methoxybenzoic acid has a high boiling point because it can form intermolecular hydrogen bonds between carboxylic acid groups, forming a dimer.

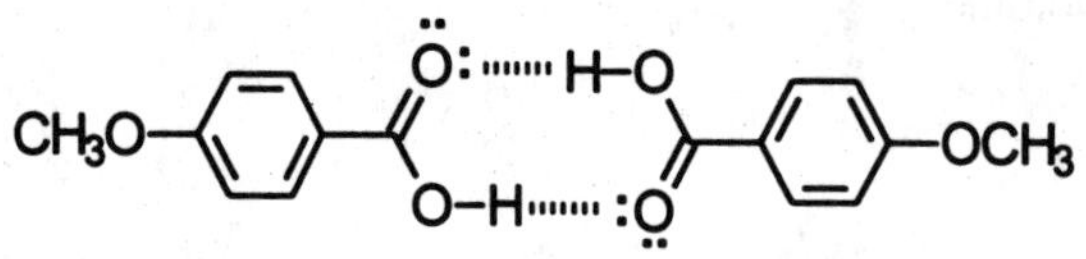

2-Methoxybenzoic acid can form intramolecular hydrogen bonds between the carboxylic acid group and the methoxy group.

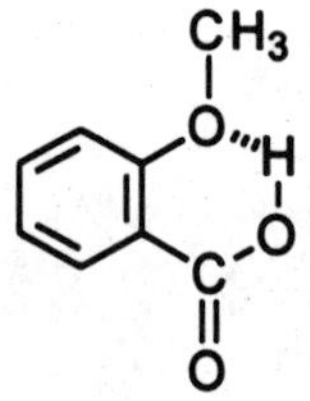

The intramolecularly bonded compound requires less energy to transfer it to the gas phase than the higher molecular weight intermolecularly bonded compound.

21.14   The trans isomer has a larger dipole moment due to the contribution of both the methyl and carboxyl groups.

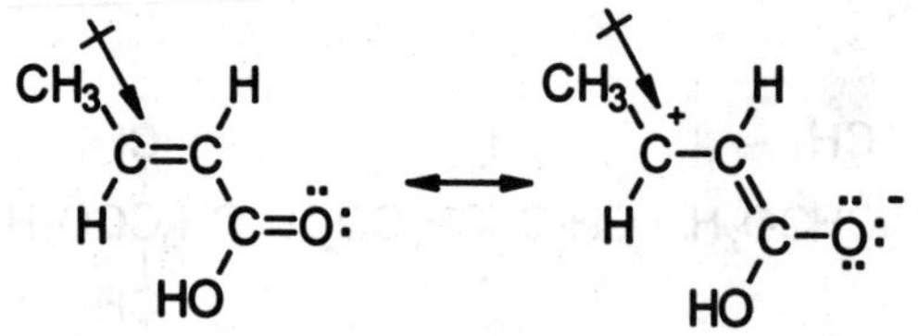

21.15   The higher K_a for methoxyacetic acid means than the methoxy group stabilizes its carboxylate ion compared to acetate ion. The methoxy group inductively attracts electron density, which leads to stabilization of the ion.

21.16   The higher K_a for p-nitrobenzoic acid means than the nitro group stabilizes its carboxylate ion compared to benzoate ion. The nitro group inductively attracts electron density, but there is a long distance between it and the carboxylate ion. The stabilization results from electron withdrawal from the aromatic ring by resonance. As a consequence of lower electron density at C-1 in the resonance stabilized structure, the electrons of the carboxylate ion are inductively pulled toward the aromatic ring.

21.17 The location of the chlorine atom with respect to the carboxylic acid of the C-1 atom corresponds to that in 3-chlorobutanoic acid, so the pK_a must be close to 4.06. The location of the chlorine atom with respect to the carboxylic acid of the C-6 atom corresponds to that in 4-chlorobutanoic acid, so the pK_a must be close to 4.52.

21.18 The pK_a increases from 4.06 to 2.84 for 3-chlorobutanoic acid and 2-chlorobutanoic acid, respectively. A similar increase would be expected for the cyano compounds. The estimated pK_a is 3.22.

21.19 They differ by 0.3 units on the log scale, which corresponds to a factor of 2 in K_a. The long chain dicarboxylic acids are twice as acidic as the long chain carboxylic acids because there are two hydrogen atoms per molecule in the dicarboxylic acids. The difference reflects a statistical factor, not an influence of structure.

21.20 The ionization constant for the transfer of a second proton from an acid to water is expected to be smaller than for the first proton, because it is more difficult to remove a proton from a negatively charged ion than from a neutral molecule. In the case of the oxalate ion and the malonate ion, the carboxylate group and carboxylic acid group are close enough to form a hydrogen bond. As a consequence, more energy is required to remove the proton.

21.21 The methoxy group cannot donate electrons by resonance in methoxyacetic acid as it does in an aromatic ring, because there is an intervening methylene group in the acetic acid. The smaller pK_a of methoxyacetic acid indicates that the methoxy group does stabilize the carboxylate ion, but this stabilization is the result of inductive electron withdrawal by the oxygen atom.

21.22 The groups cannot affect the acidity of the acetic acids by resonance, because there is an intervening methylene group. Therefore the difference indicates that the nitro group is inductively more electron withdrawing than the cyano group. Note that the nitro group has a formal positive charge on nitrogen. The carbon atom of the cyano group is partially positive as a result of the electronegativity difference between carbon and nitrogen atom. However, there is no formal charge.

21.23 The enhanced acidity of the o-hydroxy compound compared to the o-methoxy compound indicates that its carboxylate group is stabilized. The hydrogen atom of the hydroxyl group can form a hydrogen bond with the oxygen atom of the carboxylate group. This type of interaction does not exist in the o-methoxy compound.

21.24   The pK$_a$ value of the phosphorus compound is smaller than that of benzoic acid Thus, the phosphorus group withdraws electron density from the aromatic ring and stabilizes the carboxylate ion. The phosphorus group should be a deactivator in aromatic substitution reactions. The pK$_a$ value of the silicon compound is slightly larger than that of benzoic acid Thus, it donates electron density to the aromatic ring and destabilizes the carboxylate ion. The silicon group should be a weak activator in aromatic substitution reactions.

21.25   The smaller K$_a$ for p-methoxybenzoic acid means than the methoxy group destabilizes its carboxylate ion. The methoxy group inductively attracts electron density and could stabilize the ion, but there is a long distance between it and the carboxylate group. The destabilization results from electron donation to the aromatic ring by resonance. As a consequence of higher electron density at C-1 in the resonance stabilized structure, the carboxylate ion is destabilized. In the case of the methoxymethyl group, there is no resonance interaction between the oxygen atom and the aromatic ring. A methyl group is inductively electron donating, but the added methoxy group must make the methoxymethyl group somewhat inductively electron withdrawing. The net effect is a decrease in electron density at C-1, thus stabilizing the carboxylate ion.

21.26   The pK$_a$ values of o-, m-, and p-fluorobenzoic acid are 3.3, 3.9, and 4.1, respectively. All of the fluorine compounds are stronger acids than benzoic acid, suggesting that in each case the fluorine atom withdraws electron density from the ring and stabilizes the carboxylate ion. The effect is largest for the ortho compound and decreases with distance between the fluorine atom and the carboxylate ion, which is typical of an inductive effect. If the fluorine atom effectively donated electrons by resonance, the acidity of the para substituted compound would be greater than that of benzoic acid.

21.27   The two phenyl rings do not interact by resonance, because the ortho hydrogen atoms obstruct planarity between the two rings. As a result the observed effect is that of an sp^2 hybridized carbon atom of the phenyl group, which is inductively electron withdrawing. That effect stabilizes the carboxylate ion. The effect, although small, is greater in the case of the 3-carboxylic acid, because the phenyl ring is closer to the carboxyl group.

21.28   (a) The fluoro compound is the more acidic because the fluorine atoms inductively withdraw electron density from the aromatic ring and stabilize the carboxylate ion.
(b) Oxygen is more electronegative than nitrogen. As a result of inductive electron withdrawal, the carboxylate ion of the oxygen compound is more stable than the carboxylate ion of the nitrogen compound. The furan compound is the more acidic.
(c) The inductive electron withdrawal by oxygen decreases with distance from the carboxylate ion. Thus, the first compound is more acidic.

21.29   The nitrogen atom and the two oxygen atoms bonded to the sulfur atom make this group strongly electron withdrawing. As a result, the carboxylate ion is stabilized by a decrease in electron density at C-1.

21.30   There are no strongly electron withdrawing groups near the carboxyl group. However, there is an sp^2 hybridized carbon atom bonded to the sp^3 methylene group and it can withdraw electron density from that group, which is bonded to the carboxyl group. The effect on the acidity should be small, but the compound should be a stronger acid than acetic acid. Its pK_a should be less than 4.72.

21.31   Carboxylic acids exist to a larger degree as carboxylate salts in a more basic medium and hence are more soluble. Thus, the compound is more soluble in blood.

21.32   The pH of the foods is greater than the pK_a value of benzoic acid. Thus, the compound is present as the benzoate ion.

21.33   As the carboxylic acid, the two oxygen atoms are nonequivalent. However, in aqueous solution the carboxylate ion forms, and the two oxygen atoms become equivalent.

21.34   The orbitals should resemble those of the allyl system discussed in Chapter 12. The four electrons contributed by the two oxygen atoms are located in π_1 and π_2. The shortened carbon-oxygen bond is the result of the contribution of π_1. The charge distribution is accounted for by π_2, in which electron density occurs at the terminal oxygen atom.

π_1 π_2 π_3

21.35   **(a)** Prepare the Grignard reagent using magnesium and ether, and then add it to carbon dioxide, followed acidification in workup.
   (b) Convert the alcohol into bromocyclohexane using PBr_3, and then proceed as in part (a).
   (c) Add HBr to the double bond to form bromocyclohexane, and then proceed as in part (a).
   (d) Use ozone under oxidative workup conditions.
   (e) Oxidize the alcohol to the carboxylic acid using the Jones reagent.

21.36   **(a)** Prepare the Grignard reagent using magnesium and ether, and then add it to carbon dioxide, followed by acidification in workup.
   (b) Oxidize the alcohol to the carboxylic acid using the Jones reagent.
   (c) Oxidize the aldehyde to the carboxylic acid using the Jones reagent.
   (d) Prepare 1-hexanol using B_2H_6 followed by treatment with basic hydrogen peroxide (the hydroboration-oxidation procedure). Then oxidize the alcohol using the Jones reagent.
   (e) Use ozone under oxidative workup conditions.

21.37   **(a)** Prepare cyclohexylmethanol using B_2H_6 followed by treatment with basic hydrogen peroxide (the hydroboration-oxidation procedure). Then oxidize the alcohol to the carboxylic acid using the Jones reagent.
   (b) Add HBr in the presence of peroxide to give a primary halogen compound (anti-Markovnikov addition). Prepare the Grignard reagent using magnesium and ether, and then add it to carbon dioxide followed by acidification in workup. Alternatively, displace bromide ion by cyanide ion, followed by hydrolysis of the nitrile.

(c) Add HBr in the absence of peroxide to give a tertiary halogen compound. Prepare the Grignard reagent using magnesium and ether, and then add it to carbon dioxide, followed by acidification in workup.

21.38 (a) Oxidize the ethyl group to a carboxylic acid using a strong oxidizing agent such as potassium permanganate.
       (b) Use NBS to prepare the secondary bromo compound of the ethyl side chain as shown on page 451 of the text. Prepare the Grignard reagent using magnesium and ether, and then add it to carbon dioxide, followed by acidification in workup.
       (c) Use NBS to prepare the secondary bromo compound of the ethyl side chain as shown on page 451 of the text. Prepare the Grignard reagent using magnesium and ether, and then add it to ethylene oxide to give 3-(p-methoxyphenyl)butanol. Oxidize the alcohol using the Jones reagent.

21.39 Reduce the carboxylic acid to dodecanol. Use PBr_3 to give 1-bromododecane. Prepare the Grignard reagent using magnesium and ether, and then add it to carbon dioxide, followed by acidification in workup.

21.40 Prepare the Grignard reagent using magnesium and ether, and then add it to carbon dioxide, followed by acidification in workup. The displacement of bromide ion by cyanide ion to give a nitrile cannot be accomplished with a tertiary halide.

21.41 (a) 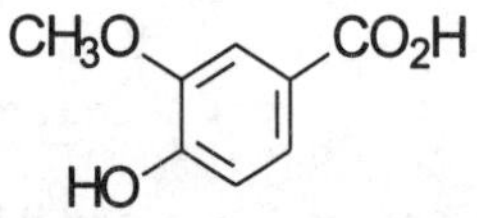   (b) 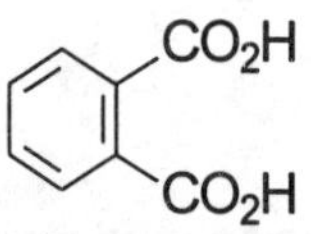   (c)

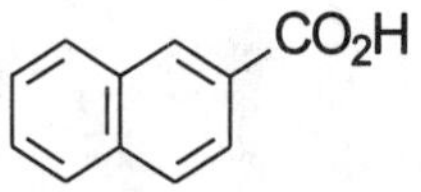

       (d) 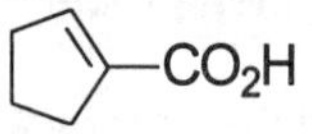   (e)

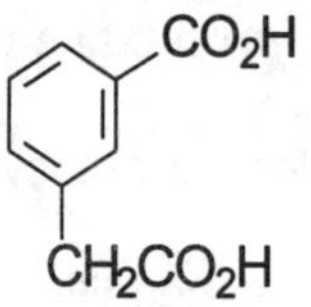

21.42 (a) 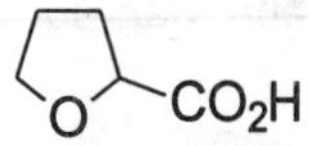   (b) 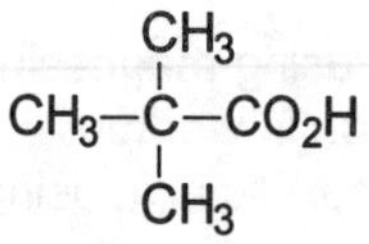   (c)

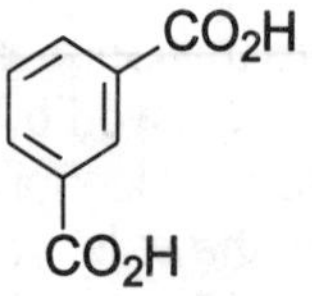

       (d) 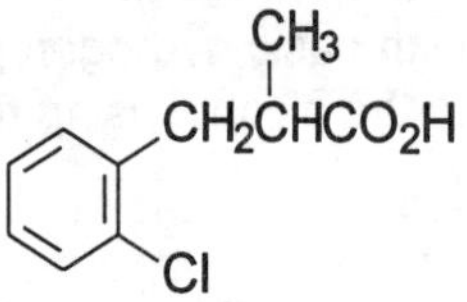   (e)

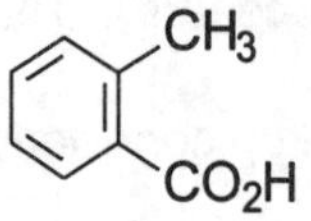

21.43 Boron is electron deficient in BH_3 and can form a coordinate covalent bond to the oxygen atom of the carbonyl group. As a result, the hydrogen atom can be transferred to the carbonyl carbon atom. That cyclic intermediate would resemble that of the reaction of BH_3 with an alkene in the hydroboration reaction.

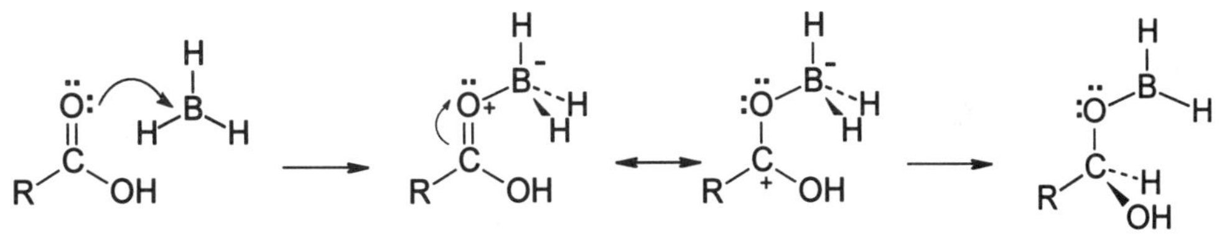

21.44 The unshared pair of electrons of the nitrile can coordinate to the boron atom, but the linear geometry of the nitrile group does not allow close approach of the hydrogen atom to the carbon atom.

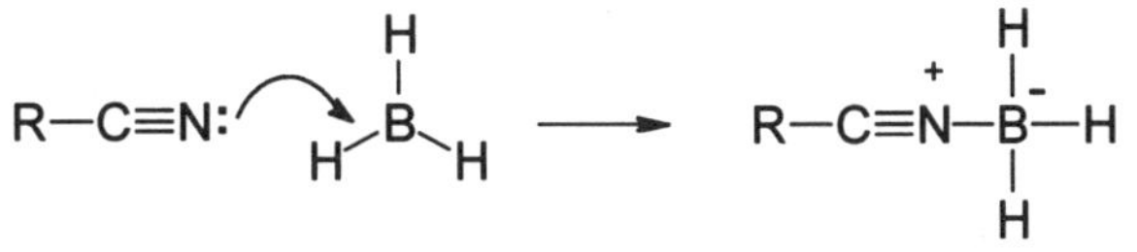

21.45 Aluminum is more electropositive than boron. As a result, the electron pair in an Al—H bond is more available to the hydrogen atom, which can depart as a hydride ion in a reaction. Lithium borohydride is more active than sodium borohydride because the coordination of the smaller lithium ion is stronger than for the larger sodium ion. Coordination of the metal ion polarizes the carbonyl bond and increases the partial positive charge on the carbonyl carbon atom. As a result, the center is more reactive toward hydride ion derived from the borohydride ion.

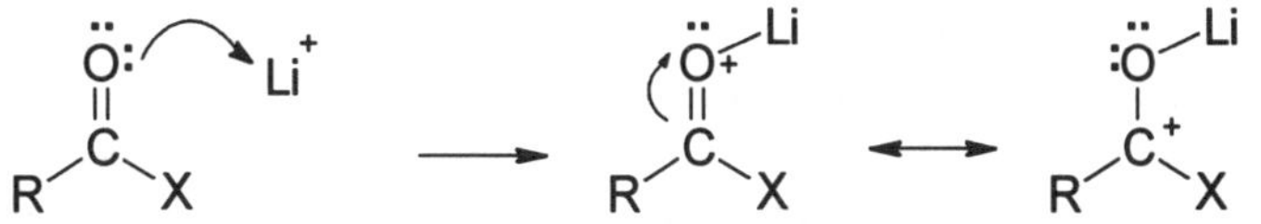

21.46 The BH_3 reduces the carboxylic acid group but the ester group is unchanged. This compound is the methyl ester of the enantiomer of the product in Exercise 21.45.

21.47 No, because the bromine would react with the unsaturated carbon-carbon bond and give an addition product.

21.48 Cyclohexanecarboxylic acid decarboxylates more readily because the carboxyl carbon atom is bonded to an sp^3 hybridized carbon atom, which is a weaker bond than the bond to the sp^2 hybridized carbon atom of benzoic acid.

21.49 The cyclic mechanism for decarboxylation of β-ketoacids forms a enol which isomerizes to a carbonyl compound. The double bond at the bridgehead atom for the enol in the bicyclic compound would be highly strained.

21.50 A cyclic concerted mechanism gives an alkene with a double bond between the original α and β carbon atoms of 3-butenoic acid.

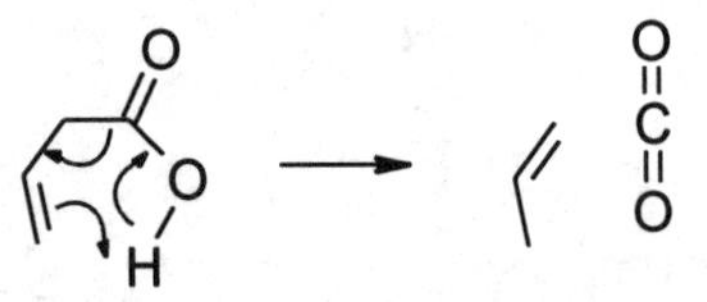

The product of the decarboxylation of (E)-4-methyl-3-pentenoic acid is 3-methyl-1-butene.

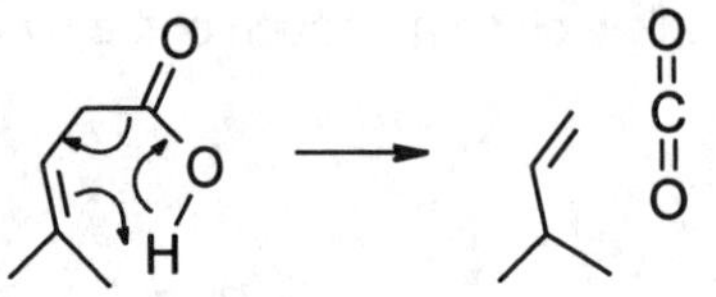

21.51 The carboxylic acid first reacts with oxalyl chloride to form HCl and an anhydride, which then decomposes by a cyclic mechanism in which the chlorine atom of the intermediate attacks the carbonyl carbon atom of the original carboxylic acid. A concerted process in the third step releases the acyl chloride, CO_2, and CO.

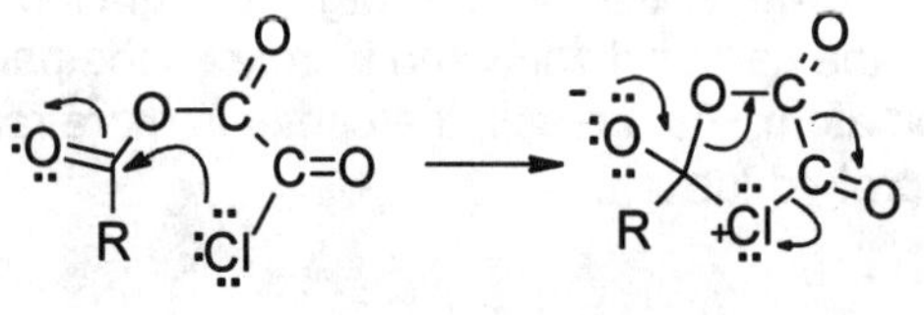

21.52 Thionyl chloride will react with the hydroxyl group as well.

21.53 The probability of forming a nine-membered ring is very small. That is, the entropy of that cyclization process is strongly negative. Under these circumstances, even a bimolecular reaction such as polymerization competes favorably.

21.54 The cis isomer has two carboxylic acid groups sufficiently close to form an anhydride when heated. The trans isomer cannot form an anhydride.

21.55 **(a)** The nucleophilic carboxylate group displaces a halide ion to form an ester. The iodide ion is a better leaving group than the chloride ion. The resulting ester product retains the chlorine atom, as shown on the next page.
(b) Diazomethane reacts with the acid to give a carboxylate ion and the methyldiazonium ion. Subsequent reaction of these two ions gives a methyl ester.
(c) The reaction of an acid chloride with an alcohol occurs via attack of the nucleophilic oxygen atom of the alcohol on the carbonyl group. The diol has a tertiary and a primary alcohol. The primary alcohol is less sterically hindered and reacts at a faster rate to produce an ester.
(d) A Fischer esterification reaction requires that the equilibrium be shifted toward the ester product. The resonance stabilization of the phenol favors the reactant side of the reaction, so esterification of phenols is not favorable. Esterification of the benzyl alcohol occurs.

(a)

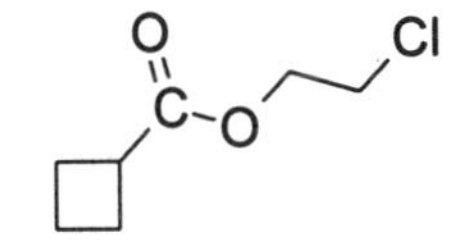

(b) 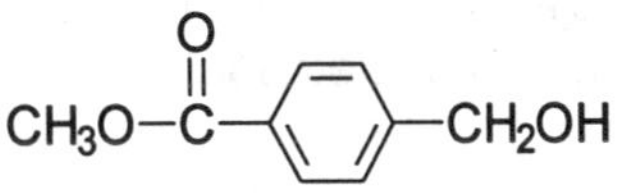 $CH_3O-\overset{\overset{\displaystyle O}{\|}}{C}-\underset{}{\text{[benzene ring]}}-CH_2OH$

(c) $CH_3-\overset{\overset{\displaystyle OH}{|}}{\underset{\underset{\displaystyle CH_3}{|}}{C}}-CH_2CH_2CH_2-O-\overset{\overset{\displaystyle O}{\|}}{C}-CH_3$

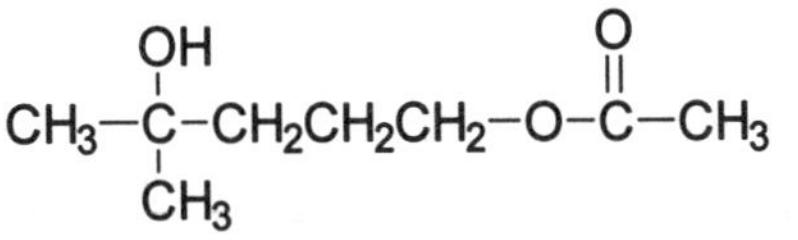

(d)

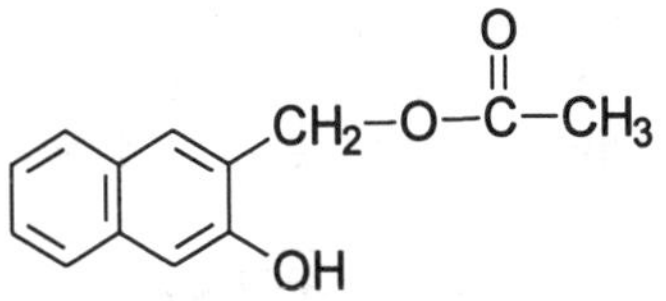

21.56 **(a)** The reagent resembles diazomethane and has phenyl groups in place of hydrogen atoms. It reacts to give a carboxylate ion and a diphenylmethyldiazonium ion. Subsequent reaction of these two ions give a diphenylmethyl ester, as shown below.

(b) The nucleophilic carboxylate group displaces a halide ion to form an ester. Although the bromide ion is a better leaving group than the chloride ion, the reaction does not occur at that site because the carbon atom is sp^2 hybridized. The ester product results from nucleophilic attack at the sp^3 carbon atom to give a product that retains the bromine atom.

(c) A Fischer esterification reaction occurs via an equilibrium process that is easily shifted toward the ester product by adding excess methanol.

(d) The reaction of an acid bromide with an alcohol occurs via nucleophilic attack of the nucleophilic oxygen atom of the alcohol on the carbonyl group. The diol has a secondary and a primary alcohol. The primary alcohol is less sterically hindered and reacts at a faster rate to produce an ester.

(a) 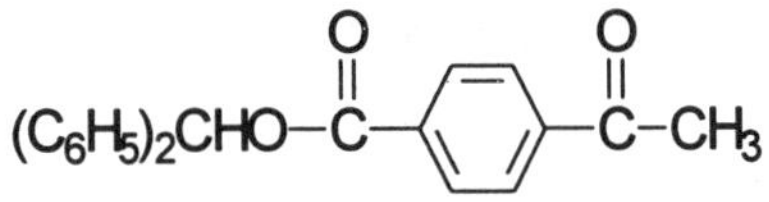 $(C_6H_5)_2CHO-\overset{\overset{\displaystyle O}{\|}}{C}-\text{[benzene ring]}-\overset{\overset{\displaystyle O}{\|}}{C}-CH_3$

(b)

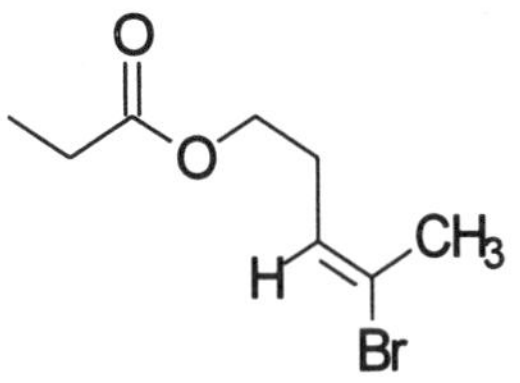

(c) 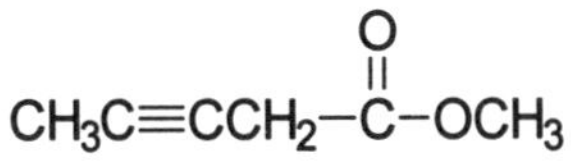 $CH_3C\equiv CCH_2-\overset{\overset{\displaystyle O}{\|}}{C}-OCH_3$

(d) 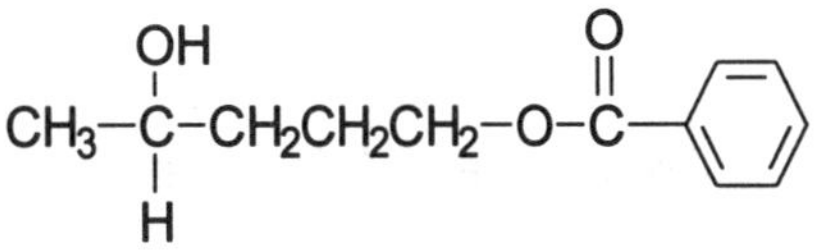 $CH_3-\overset{\overset{\displaystyle OH}{|}}{\underset{\underset{\displaystyle H}{|}}{C}}-CH_2CH_2CH_2-O-\overset{\overset{\displaystyle O}{\|}}{C}-\text{[benzene ring]}$

21.57 Reduce one portion of the carboxylic acid using lithium aluminum hydride to obtain cyclohexylmethanol. Convert another portion of the carboxylic acid to an acid chloride using thionyl chloride. React the acid chloride with the alcohol in the presence of pyridine to neutralize the HCl formed.

21.58 Prepare 2-phenylethanol from one portion of benzoic acid. One possible method is the Hunsdiecker reaction to give bromobenzene. Formation of the Grignard reagent followed by addition to ethylene oxide gives the alcohol. Then react the alcohol with another portion of benzoic acid in a Fischer esterification reaction.

21.59 **(a)** butanoic acid and 1-butanol
(b) butanoic acid and methanol
(c) butanoic acid and ethanol

21.60 **(a)** 3-phenylpropanoic acid and 2-propanol
(b) cycloheptanecarboxylic acid and 2-propanol
(c) m-bromobenzoic acid and cyclobutanol
(d) p-methoxybenzoic acid and cyclopentanol

21.61 **(a)** Oxidation gives a carboxylic acid, which is converted into an acid chloride by $SOCl_2$. Subsequent reaction with methanol give a methyl ester, as shown below.
(b) Reduction by lithium aluminum hydride gives a primary alcohol, which is then converted to a chloro compound by $SOCl_2$. Formation of a Grignard reagent followed by addition to carbon dioxide gives a carboxylic acid with one more carbon atom than the starting material.
(c) Oxidation by PCC gives an aldehyde. Subsequent reaction with methanol give an acetal.

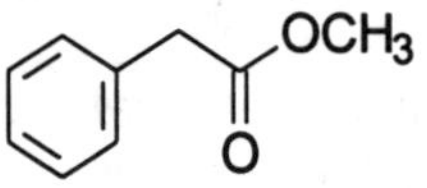

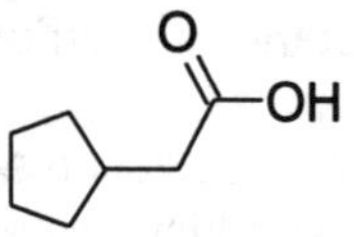

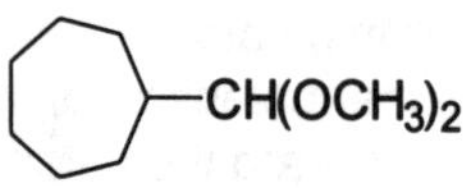

21.62 **(a)** Oxidation gives a carboxylic acid. Subsequent reaction with methanol in a Fischer esterification reaction gives a methyl ester as shown below.
(b) Reduction by lithium aluminum hydride give a secondary alcohol, which is then converted to an acetate ester by a Fischer esterification process.
(c) Reduction by the hydride reagent gives an aldehyde. Subsequent reaction with ethylene glycol gives a cyclic acetal.

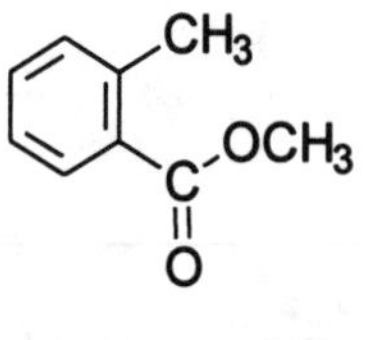

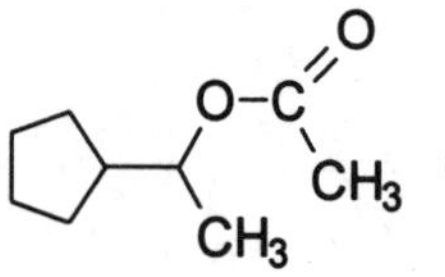

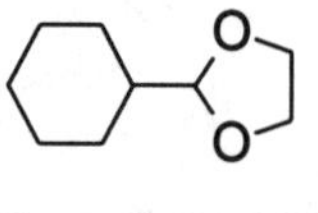

21.63 **(a)** glutaric acid and 1,4-butanediol

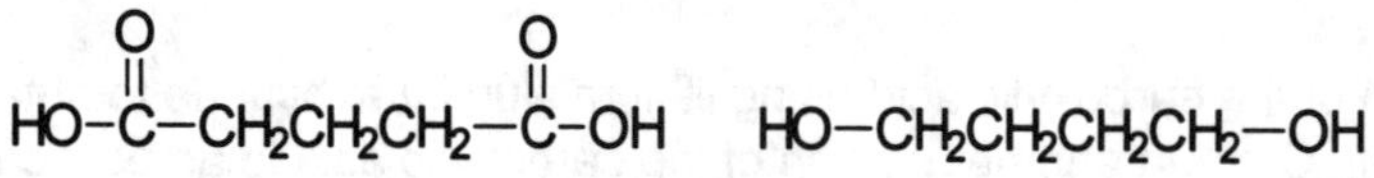

(b) adipic acid and 1,3-propanediol

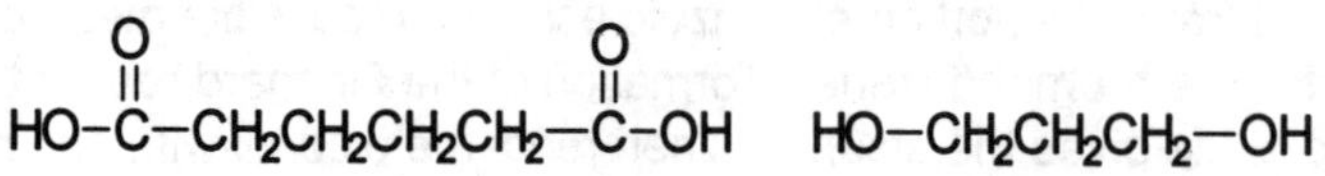

21.64 Oxidize xylene to give terephthalic acid using a strong oxidizing agent. Reduce both the benzene ring and the two carboxyl groups to give a diol. Then react the diol with the dicarboxylic acid or a derivative of the dicarboxylic acid.

21.65 Each molecular formula given has at least two oxygen atoms, so the resonance lower than 10 δ is most likely due to a carboxyl group proton in each compound.

(a) The singlet of intensity 9 is due to nine equivalent protons. The compound is 2,2-dimethylpropanoic acid.

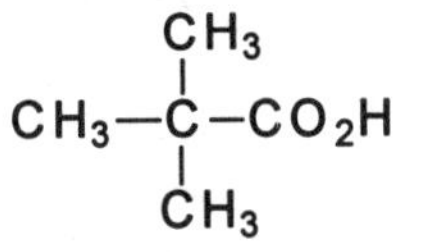

$$CH_3-\underset{\underset{CH_3}{|}}{\overset{\overset{CH_3}{|}}{C}}-CO_2H$$

(b) The doublet of intensity 3 and the quartet of intensity 1 indicate three protons and one proton, respectively, splitting each other on neighboring carbon atoms. The resonance at 1.75 ppm is due to a methyl group, and the resonance at 4.45 ppm is due to a proton bonded to a carbon atom which also has a chlorine atom bonded to it. The compound is 2-chloropropanoic acid.

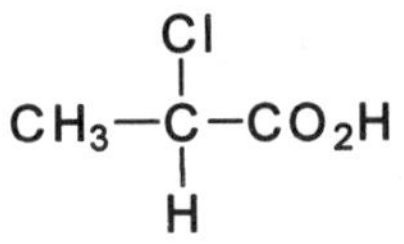

$$CH_3-\underset{\underset{H}{|}}{\overset{\overset{Cl}{|}}{C}}-CO_2H$$

(c) The molecular formula indicates that the compound has a total of four double bonds and/or rings, so a benzene ring is possible. The singlet of intensity 3 at 1.4 ppm indicates a methyl group with no neighboring protons. The doublets of intensity 2 at 7.25 and 8.0 ppm, respectively, indicate two sets of equivalent protons on a benzene ring. The compound is p-methylbenzoic acid.

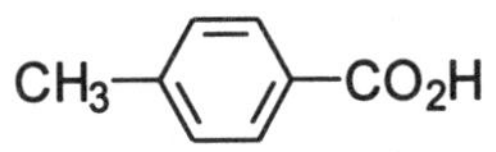

$$CH_3-\langle\ \rangle-CO_2H$$

(d) The extra oxygen atom in the molecular formula could indicate an alcohol or an ether group, but there is no resonance of intensity 1, so it cannot be an alcohol. The singlet of intensity 2 at 3.4 ppm and the singlet of intensity 3 at 4.0 ppm are due to a methylene and a methyl group, respectively, bonded to an oxygen atom. The compound is methoxyethanoic acid.

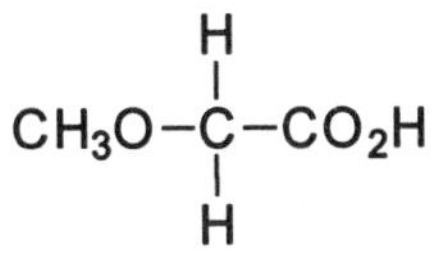

$$CH_3O-\underset{\underset{H}{|}}{\overset{\overset{H}{|}}{C}}-CO_2H$$

(e) The molecular formula indicates that the compound has a total of four double bonds and/or rings, so a benzene ring is possible. The extra oxygen atom in the molecular formula could indicate an alcohol or an ether group, but there is no resonance of intensity 1, so it cannot be an alcohol. The triplets of intensity 2 at 2.7 and 4.2 ppm are due to neighboring methylene groups that are also bonded to a carbonyl carbon atom and an oxygen atom, respectively. The complex multiplet of intensity 5 at 7.4 ppm is due to a phenyl group. The compound is 3-phenoxypropanoic acid.

$\langle\!\bigcirc\!\rangle-OCH_2CH_2CO_2H$

21.66 Each molecular formula given has at least two oxygen atoms, so the resonance lower than 10 δ is most likely due to a carboxyl group proton in each compound.

(a) The singlet of intensity 9 is due to nine equivalent protons. The singlet of intensity 2 at 2.21 ppm is due to a methylene group that is bonded to a carbonyl carbon atom and has no neighboring protons. The compound is 3,3-dimethylbutanoic acid.

$$CH_3-\underset{\underset{CH_3}{|}}{\overset{\overset{CH_3}{|}}{C}}-CH_2CO_2H$$

(b) The triplets of intensity 2 at 2.85 and 3.80 ppm are due to neighboring methylene groups that are also bonded to a carbonyl carbon atom and a chlorine atom, respectively. The compound is 3-chloropropanoic acid.

$ClCH_2CH_2CO_2H$

(c) The molecular formula indicates that the compound has a total of four double bonds and/or rings, so a benzene ring is possible. The singlet of intensity 2 at 3.6 ppm indicates a methylene group with no neighboring protons. The singlet of intensity 5 at 7.25 ppm is due to a phenyl group. The compound is phenylethanoic acid.

$\langle\!\bigcirc\!\rangle-CH_2CO_2H$

(d) The extra oxygen atom in the molecular formula could indicate an alcohol or an ether group, but there is no resonance of intensity 1, so it cannot be an alcohol. The triplet of intensity 3 at 1.27 ppm and the quartet of intensity 2 at 3.55 ppm indicate three protons and two protons, respectively, splitting each other on neighboring carbon atoms, with the methylene group also bonded to an oxygen atom. The singlet of intensity 2 at 4.13 ppm is due to a methylene group that is bonded to a carbonyl carbon atom and an oxygen atom. The compound is ethoxyethanoic acid.

$CH_3CH_2-O-CH_2CO_2H$

(e) The molecular formula indicates that the compound has a total of four double bonds and/or rings, so a benzene ring is possible. The extra oxygen atom in the molecular formula could indicate an alcohol or an ether group, but there is no singlet of intensity 1, so it cannot be an alcohol. The doublet of intensity 3 at 1.72 ppm and the quartet of intensity 1 at 4.95 ppm indicate three protons and one proton, respectively, splitting each other on neighboring carbon atoms. The resonance of intensity 1 at 4.95 ppm also indicates that the proton is bonded to a carbon atom that is bonded to an oxygen atom and one other atom. The complex multiplet of intensity 5 at 7.4 ppm is due to a phenyl group. The compound is 2-phenoxypropanoic acid.

21.67 **(a)** The compound has six carbon atoms, but five signals, so two of the carbon atoms are equivalent. The quartets at 9.3 and 24.6 ppm are due to methyl groups isolated from and close to the carbonyl carbon atom, respectively. The triplet at 33.5 ppm is due to a methylene group. The singlet at 42.7 ppm must be due to a quaternary carbon atom, because it is not split by any bonded hydrogen atoms. The singlet at 185.5 ppm must be due to a carboxyl carbon atom. The compound is 2,2-dimethylbutanoic acid.

(b) The molecular formula, $C_7H_6O_2$, indicates that the compound has a total of four double bonds and/or rings, so a benzene ring is possible. The compound has seven carbon atoms, but five signals, so there must be two sets of two equivalent carbon atoms or one set of three equivalent carbon atoms. The doublets at 128.7, 129.6, and 131.2 ppm are due to carbon atoms with one hydrogen atom and a double bond to another carbon atom, so they indicate carbon atoms in a benzene ring. The singlet at 133.0 ppm is due to a carbon atom with a double bond to another carbon atom, but no hydrogen atom, so it indicates a connecting carbon atom in a benzene ring. The singlet at 167.7 ppm must be due to a carboxyl carbon atom. The compound is benzoic acid. Note that the compound contains two sets of two equivalent carbon atoms in the benzene ring, C-2 and C-6 in one set and C-3 and C-5 in the other.

(c) The compound has four nonequivalent carbon atoms with signals corresponding to methyl (quartet at 13.4 ppm), methylene (triplets at 18.5 and 36.3 ppm), and carbonyl (singlet at 179.6) carbon atoms. The compound is butanoic acid.

$CH_3CH_2CH_2CO_2H$

21.68 **(a)** The compound has five nonequivalent carbon atoms with signals corresponding to methyl (quartet at 13.5 ppm), methylene (triplets at 22.0, 27.0, and 34.1 ppm), and carbonyl (singlet at 179.7) carbon atoms. The compound is pentanoic acid.

$CH_3CH_2CH_2CH_2CO_2H$

(b) The molecular formula, $C_7H_6O_3$, indicates that the compound has a total of four double bonds and/or rings, so a benzene ring is possible. The extra oxygen atom in the molecular formula could indicate an alcohol or an ether group. The compound has seven carbon atoms, but five signals, so there must be two sets of two equivalent carbon atoms or one set of three equivalent carbon atoms. The doublets at 115.8 and 121.9 ppm are due to carbon atoms with one hydrogen atom and a double bond to another carbon atom, so they indicate carbon atoms in a benzene ring. The singlets at 132.7 and 162.5 ppm are due to carbon atoms with a double bond to another carbon atom, but no hydrogen atom, so they indicate connecting carbon atoms in a benzene ring, one of which is shifted further downfield because it is deshielded by an oxygen atom. The singlet at 169.0 ppm must be due to a carboxyl carbon atom. The compound is p-hydroxybenzoic acid. Note that the compound contains two sets of two equivalent carbon atoms in the benzene ring, C-2 and C-6 in one set and C-3 and C-5 in the other.

$$HO-\langle benzene \rangle-CO_2H$$

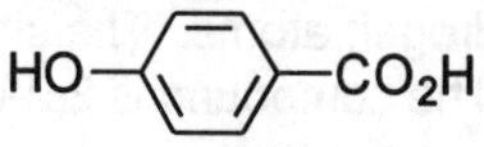

(c) The molecular formula, $C_7H_{12}O_2$, indicates that the compound has one double bond or a ring. The compound has seven carbon atoms, but five signals, so there must be two sets of two equivalent carbon atoms or one set of three equivalent carbon atoms. The triplets at 26.0, 26.2, and 29.6 ppm are due to carbon atoms with two hydrogen atoms each. The doublet at 43.7 ppm is due to a carbon atom bonded to a hydrogen atom and a carbonyl carbon atom. The singlet at 182.1 ppm must be due to a carboxyl carbon atom. The compound is cyclohexanecarboxylic acid. Note that the compound contains two sets of two equivalent carbon atoms in the cyclohexane ring, C-2 and C-6 in one set and C-3 and C-5 in the other.

$$\langle cyclohexane \rangle-CO_2H$$

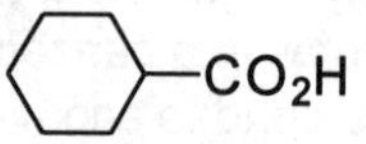

22

Carboxylic Acid Derivatives

No other chapter contains as many functional groups nor as many reagents for the conversion of functional groups. However, there is a common theme for the mechanisms of the reactions of the various acid derivatives. Each compound has a potential leaving group attached to the carbonyl carbon atoms, and each carbonyl group can react with a nucleophile in addition reactions. The properties of leaving groups and nucleophiles are already well understood based on other reactions studied in this text. Furthermore, the reactivity of the carbonyl group toward nucleophilic addition and the effect of structure on that reactivity has also been studied. All that remains is to put together a common picture for all of the various acid derivatives with the several reagents that attack the carbonyl carbon atom and account for the relatively minor differences based on the theory of nucleophilic acyl substitution. However, you can't avoid learning the reagents and the net result of their reaction with each acid derivative.

Keys to the Chapter

22.1 Nomenclature of Carboxylic Acid Derivatives

The names of the acid derivatives with the exception of nitriles resemble those of the structurally related carboxylic acid. The acid halides are named using **oyl halide** in place of oic acid of carboxylic acids. Compounds with an acid halide bonded to a ring carbon atom are named by appending **carbonyl halide** to the name of the alkane. The acid anhydrides are named using **oic anhydride** in place of oic acid of carboxylic acids.

Esters are named by the name of the alkyl group bonded to the bridging oxygen atom followed by the name of the acid, with **oate** replacing oic acid. Compounds with the acid portion bonded to a ring carbon atom are named by appending **carboxylate** to the name of the alkane. Lactones are named by adding **lactone** as a separate word to the name of the hydroxyacid.

Amides are named using **amide** in place of oic acid of carboxylic acids. Compounds with an amide bonded to a ring carbon atom are named by appending **carboxamide** to the name of the alkane. The prefix N- is used to identify alkyl or aryl groups bonded to the nitrogen atom of the amide. Lactams are named by adding **lactam** as a separate word to the name of the amino acid.

Nitriles are named by adding **nitrile** to the name of the related alkane that included the carbon atom of the nitrile group.

22.2 Physical Properties of Acyl Derivatives

Acid halides and acid anhydrides are polar compounds and have physical properties that resemble those of structurally similar carbonyl derivatives of similar molecular weight.

Esters are polar molecules but cannot form intermolecular hydrogen bonds like carboxylic acids do. As a consequence, esters have lower solubility in water and have lower boiling points compared to carboxylic acids.

The lower molecular weight amides are soluble in water as a result of hydrogen bonding to water molecules. The melting points of primary and secondary amides are higher than those for alkanes of similar molecular weight because of intermolecular hydrogen bonding. Tertiary amides have lower melting points than isomeric primary and secondary amides, because they do not have an N-H bond to form intermolecular hydrogen bonds.

Nitriles are very polar acid derivatives due to the large bond moment of the triple bond between carbon and nitrogen. Lower molecular weight nitriles are soluble in water.

22.3 Basicity of Acyl Derivatives

Protonation of the carbonyl oxygen atom is a reaction common to all acid-catalyzed reactions of acid derivatives. The lone pair electrons of oxygen are in sp^2 hybridized orbitals and the oxygen atom is less basic than the oxygen atom of alcohols or ethers, whose lone pair electrons are in sp^3 hybridized orbitals.

The basicity of acid derivatives is related to the ability of the electronegative atom bonded to the carbonyl carbon atom to stabilize the conjugate acid by resonance donation of electron density from that atom. Amides are the most basic, because nitrogen is an effective donor of electrons. Esters are less basic than amides, because oxygen is more electronegative and is a less effective donor of electrons. Acid chlorides are very much less basic, because chlorine is not effective at stabilizing positive charge.

Nitriles are very weak bases, because the lone pair electrons of the nitrogen atom are in an sp hybridized orbital. In addition, there is no alternative stabilized resonance form for the conjugate acid.

22.4 Nucleophilic Acyl Substitution

Nucleophilic acyl substitution (acyl transfer reaction) occurs by a two-step mechanism. Attack of the carbonyl carbon atom of an acyl derivative by a nucleophile yields a tetrahedral intermediate which can then lose a leaving group. The net result is a substitution reaction. The first step is rate determining. Thus, the rate of the reaction depends on the effect of the hydrocarbon group and the attached electronegative atom on the stability of the acid derivative and the transition state, not on the leaving group characteristics of the group displaced by the nucleophile.

The acid-catalyzed reaction of acid derivatives occurs by protonation of the carbonyl oxygen atom, which increases the electrophilicity of the carbonyl carbon atom. Attack of a nucleophile gives a tetrahedral intermediate that subsequent ejects the leaving group. Finally, loss of a proton from the carbonyl oxygen atom completes the mechanism.

The base-catalyzed reaction of acid derivatives occurs by abstraction of a proton from the nucleophile H—Nu, which prepares the nucleophile for attack of the carbonyl oxygen atom. The attack of the nucleophile gives a tetrahedral intermediate that has a negative charge on the oxygen atom. Return of an electron pair from that oxygen atom to reform the carbon-oxygen double bond results in ejection of the leaving group.

The different effects of the electronegative atoms on the reactivity of acid derivatives are explained using resonance and inductive effects on the stability of the reactant. The transition state resembles the tetrahedral intermediate, which cannot be stabilized by resonance. The stability of the transition states of all acyl derivatives for reaction with the same nucleophile is approximately the same. Thus, the stability of the reactant controls the rate of its reaction. The most stable acyl derivatives are the least reactive. Amides are the most stable and the least reactive, because nitrogen is an effective donor of electrons to the carbonyl group. Anhydrides and esters are somewhat less stable, because oxygen is more electronegative than nitrogen and is a less effective donor of electrons. Anhydrides are less stable because the donation of electrons to one carbonyl group is in competition with the donation of electrons to the second carbonyl group. Thus, in comparison to esters, where the oxygen atom need only stabilize one carbonyl group, anhydrides are more reactive than esters. Acid chlorides are very much less stable, because chlorine is not effective at stabilizing positive charge by donation of electron density by resonance.

22.5 Hydrolysis of Acyl Derivatives

Acyl halides and acid anhydrides react readily with water to give carboxylic acids. Esters react with water in an equilibrium reaction to give an alcohol and a carboxylic acid. Amides are stable to water under neutral conditions.

The degree of completion of the reaction of an ester with water is increased by use of an equivalent amount of hydroxide ion. The **saponification** reaction is spontaneous, because the product is a carboxylate ion that is a weaker base than hydroxide ion.

Amides are difficult to hydrolyze and an equivalent amount of acid or base must be used. In the case of acid, the products are the carboxylic acid and the conjugate acid of the amine. In the case of base, the product are the amine and the conjugate base of the carboxylic acid.

Nitriles hydrolyze with difficulty using either concentrated acid or base. In the case of acid, the product is the ammonium ion and the carboxylic acid. In the case of base, the product is the conjugate base of the carboxylic acid and ammonia.

22.6 Reaction of Acyl Derivatives with Alcohols

The reactivity of acyl derivatives with alcohols parallels their reactivity with water. The reaction of either acid chlorides or acid anhydrides with alcohols is an excellent way to prepare esters. Esters react with alcohols to interchange alkoxy groups in a **transesterification** reaction. The reaction is one to be avoided and, as a result, reactions of esters in an alcohol solvent are selected so that the alcohol and the ester contain the same alkoxy group.

22.7 Reaction of Acyl Derivatives with Amines

Amines (or ammonia) are better nucleophiles than alcohols (or water), so the reactions of amines with acyl derivatives are faster than the corresponding reaction with alcohols. Acid chlorides react with ammonia, primary amines, and secondary amines to produce primary, secondary, and tertiary amides, respectively. A second mole of the nitrogen compound is required to neutralize the HCl formed. Usually pyridine is added to conserve the amine and allow the use of only one mole of the amine.

Acid anhydrides react with ammonia, primary amines, and secondary amines to produce primary, secondary, and tertiary amides, respectively. The byproduct of the reaction is one mole of the carboxylic acid related to the acid anhydride. Esters react similarly with ammonia, primary amines, and secondary amines to give an alcohol as the byproduct.

22.8 Reduction of Acyl Derivatives

All acyl derivatives can react with an appropriate metal hydride reducing agent by attack of a hydride ion at the carbonyl carbon atom. Acid chlorides are extremely reactive and readily lose a chloride ion from the tetrahedral intermediate formed by transfer of a hydride ion from lithium aluminum hydride. The product aldehyde then rapidly reacts with additional hydride to give a primary alcohol. This second step is avoided by using lithium aluminum tri(tert-butoxy) hydride and the aldehyde is the final product.

Esters are reduced only by lithium aluminum hydride to give a primary alcohol related to the acid portion of the ester, with the alcohol of the original ester as a byproduct. The process occurs by nucleophilic attack of hydride at the carbonyl carbon atom to give a tetrahedral intermediate that subsequently ejects an alkoxide ion. An aldehyde results, which is then rapidly reduced to the primary alcohol. Diisobutylaluminum hydride (DIBAL) reacts with esters to give the aldehyde.

Amides are reduced by lithium aluminum hydride to give amines containing the groups originally bonded to the nitrogen atom and an alkyl group derived from the acid minus its oxygen atom.

Nitriles are reduced by a variety of reagents. Raney nickel and hydrogen at high pressures give primary amines. Lithium aluminum hydride also reduces nitriles to amines.

22.9 Reaction of Acyl Derivatives with Organometallic Reagents

The carbanion derived from an organometallic compound is a nucleophile that can attack the carbonyl carbon atom of acyl derivatives. Ejection of a leaving group from the tetrahedral intermediate gives the final product, which may or may not react with the organometallic compound.

Acid chlorides react with Grignard reagents in a manner similar to that for esters to give an intermediate aldehyde that subsequently reacts further to give an alcohol. However, this reaction is not used because esters are more convenient to use. Reaction of acid chlorides with the Gilman reagent gives ketones which do not react further with the reagent.

Esters react with Grignard reagents to add an alkyl (or aryl) group to the carbonyl carbon atom and eject an alkoxide ion. The resulting aldehyde reacts further with the reagent to give a tertiary alcohol containing two equivalents of the alkyl (or aryl) group of the Grignard reagent.

22.10 Spectroscopy of Acid Derivatives

The IR spectra of nitriles contain an absorption for the carbon-nitrogen triple bond in the 2200-2250 cm^{-1} region. The absorption is more intense than that for carbon-carbon triple bond in the 2100-2200 cm^{-1} region.

Acyl derivatives are characterized by a strong absorption due to the carbonyl group in the IR spectra of these compounds. The absorption occurs in the same region as for the carbonyl groups of aldehydes and ketones. However, the position of the absorption is strongly affected by the electronegative atom and its contribution to the stability of the dipolar resonance form by a combination of inductive and resonance affects. Esters have absorptions at 1735 cm^{-1}, but that position is affected by conjugation of double bonds with the acyl group. The absorptions of lactones show the same changes with ring size as for cycloalkanones.

Acid chlorides absorb at 1800 cm^{-1}, because the chlorine atom cannot stabilize the dipolar resonance form by donation of electrons. In fact, the inductive electron withdrawal of electrons destabilizes that resonance form and increases the double bond character of the carbonyl group.

Amides very effectively donate electrons to the carbonyl carbon atom by resonance. As a consequence, the double bond character of the carbonyl group is reduced and the resulting absorption of the carbonyl group occurs in the 1650-1655 cm^{-1} region.

The proton NMR of the α hydrogen atoms of acyl derivatives occurs in the 2 δ region, which is the same region for the α hydrogen atoms of aldehydes and ketones. The chemical shift of the hydrogen atoms on the carbon atom bonded to the oxygen atom of the alkoxyl part of esters are at somewhat lower field (4 δ) than alcohols. The chemical shift of the hydrogen atoms bonded to the alkyl carbon atom bearing the nitrogen atom in amides occur in the 2.6—3.0 δ region.

The absorption of α carbon atoms of acid derivatives have ^{13}C NMR absorptions in the 20 δ region. The carbonyl carbon atom is easily identified by its absorption in the 165-180 δ region. The absorptions of the carbon atom bonded to the oxygen atom of the alkoxyl part of an ester are in the 60 δ region.

Summary of Reactions

1. Hydrolyis of Acid Derivatives

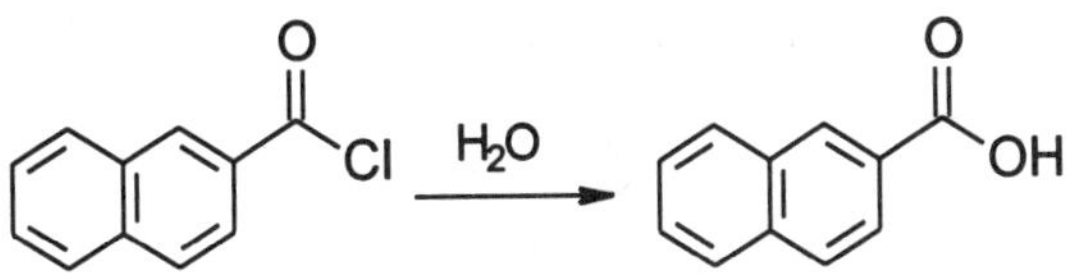

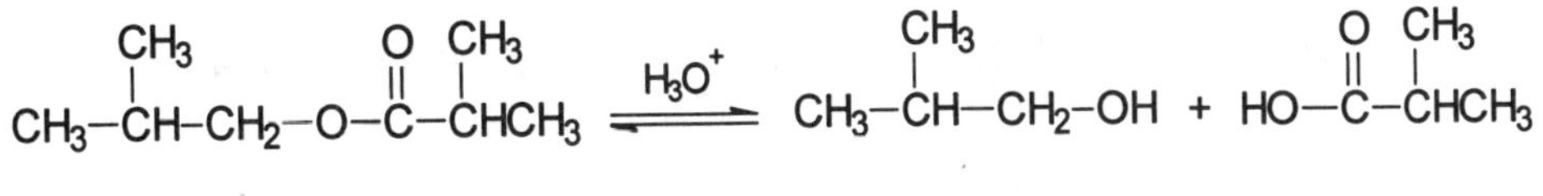

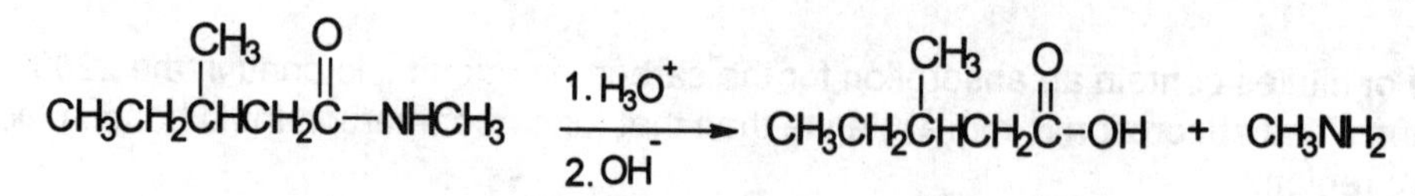

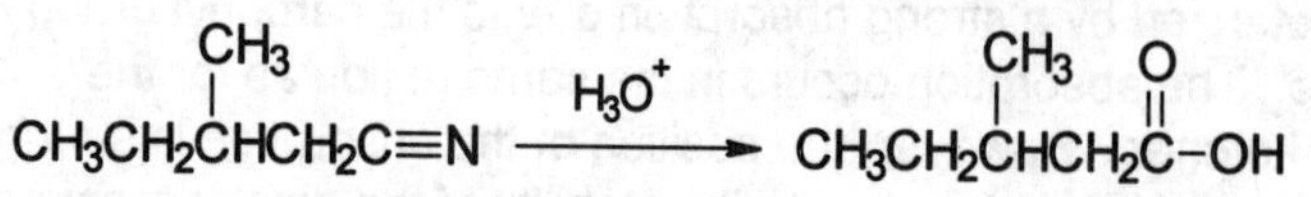

$$CH_3CH_2CHCH_2\overset{O}{\underset{\underset{CH_3}{|}}{C}}-NHCH_3 \xrightarrow[\text{2. OH}^-]{\text{1. H}_3O^+} CH_3CH_2CHCH_2\overset{O}{\underset{\underset{CH_3}{|}}{C}}-OH + CH_3NH_2$$

$$CH_3CH_2\underset{\underset{CH_3}{|}}{CH}CH_2C\equiv N \xrightarrow{H_3O^+} CH_3CH_2\underset{\underset{CH_3}{|}}{CH}CH_2\overset{O}{C}-OH$$

2. Reactions of Acid Derivatives with Alcohols and Phenols

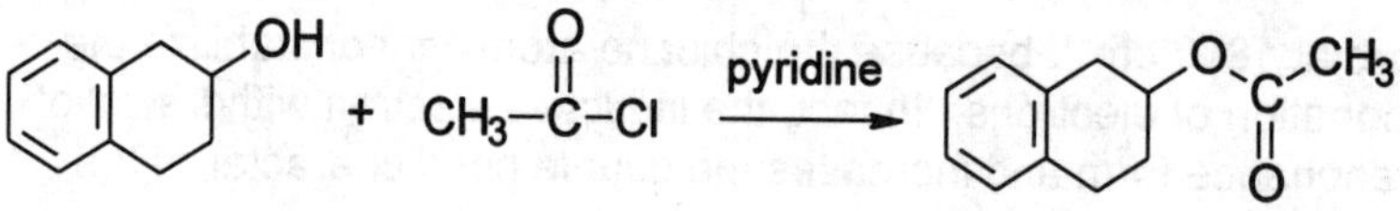

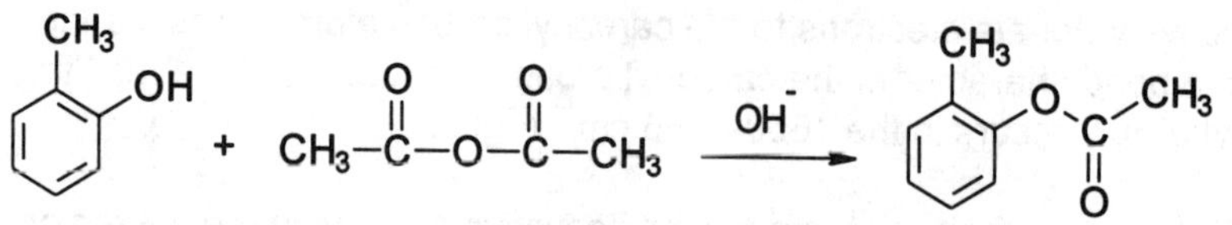

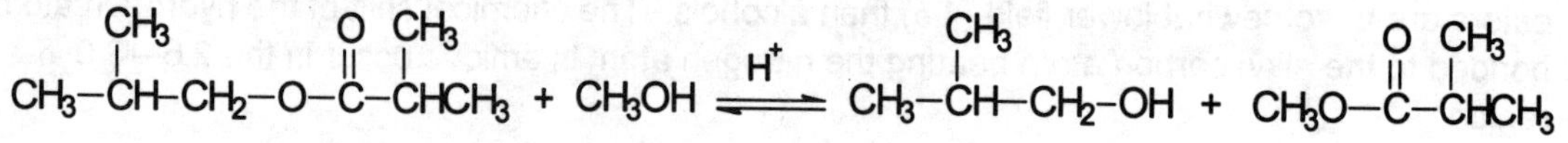

$$\text{(naphthalenol)} + CH_3-\overset{O}{C}-Cl \xrightarrow{\text{pyridine}} \text{(ester)}$$

$$\text{(o-cresol)} + CH_3-\overset{O}{C}-O-\overset{O}{C}-CH_3 \xrightarrow{OH^-} \text{(o-cresyl acetate)}$$

$$CH_3-\overset{CH_3}{\underset{|}{CH}}-CH_2-O-\overset{O}{C}-\overset{CH_3}{\underset{|}{CH}}CH_3 + CH_3OH \rightleftharpoons CH_3-\overset{CH_3}{\underset{|}{CH}}-CH_2-OH + CH_3O-\overset{O}{C}-\overset{CH_3}{\underset{|}{CH}}CH_3$$

3. Reactions of Acid Derivatives with Amines

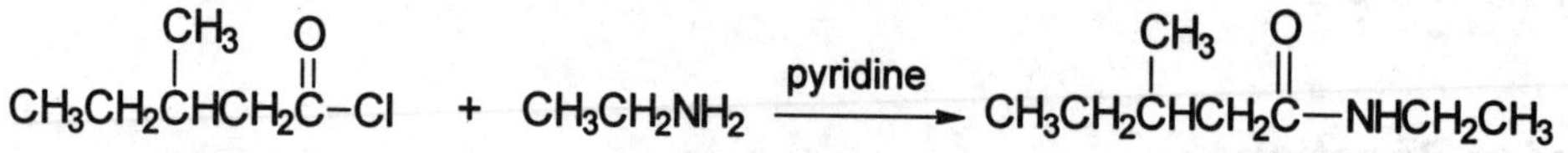

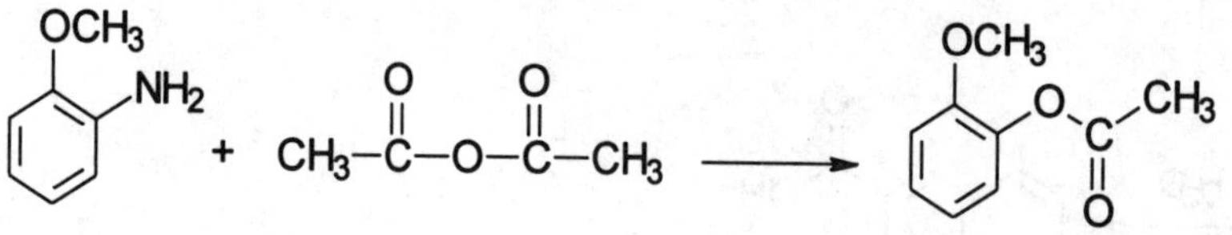

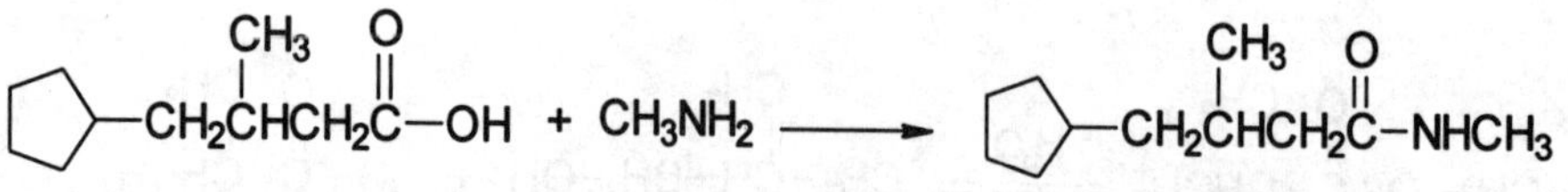

$$CH_3CH_2\overset{CH_3}{\underset{|}{CH}}CH_2\overset{O}{C}-Cl + CH_3CH_2NH_2 \xrightarrow{\text{pyridine}} CH_3CH_2\overset{CH_3}{\underset{|}{CH}}CH_2\overset{O}{C}-NHCH_2CH_3$$

$$\text{(o-anisidine)} + CH_3-\overset{O}{C}-O-\overset{O}{C}-CH_3 \longrightarrow \text{(product)}$$

$$\text{(cyclopentyl)}-CH_2\overset{CH_3}{\underset{|}{CH}}CH_2\overset{O}{C}-OH + CH_3NH_2 \longrightarrow \text{(cyclopentyl)}-CH_2\overset{CH_3}{\underset{|}{CH}}CH_2\overset{O}{C}-NHCH_3$$

4. Reduction of Acid Chlorides

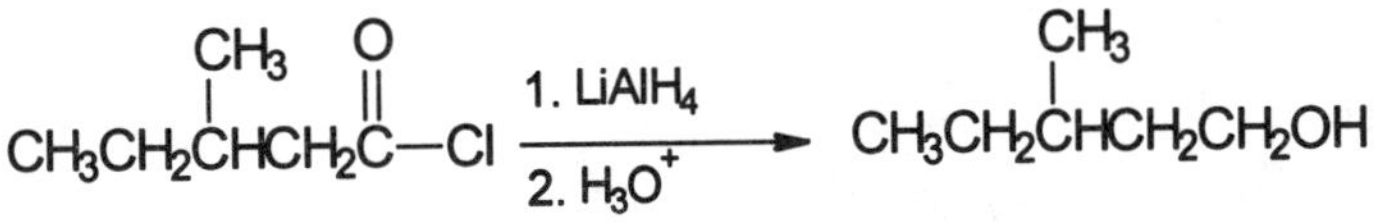

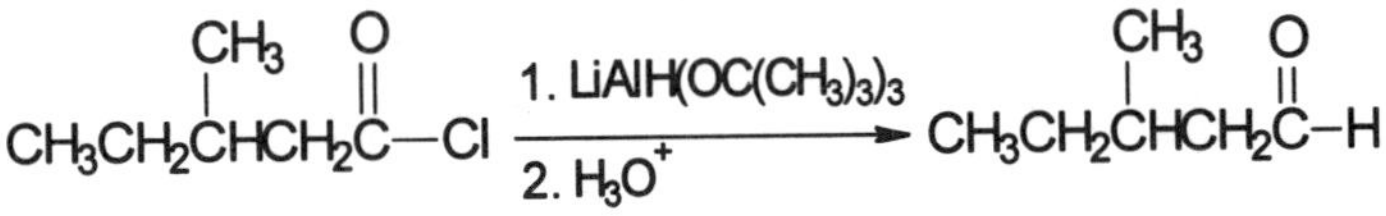

CH3 O
CH3CH2CHCH2C—Cl
1. LiAlH4
2. H3O+
CH3
CH3CH2CHCH2CH2OH

CH3 O
CH3CH2CHCH2C—Cl
1. LiAlH(OC(CH3)3)3
2. H3O+
CH3 O
CH3CH2CHCH2C–H

5. Reduction of Esters

O
C
OCH3
1. LiAlH4
2. H3O+
CH2OH
+ CH3OH

O
C
OCH3
1. DIBAL
2. H3O+
O
C
H
+ CH3OH

5. Reduction of Amides

O
NHCH3
1. LiAlH4
2. H3O+
NHCH3

6. Reduction of Nitriles

CH3
CH3CH2CHCH2C≡N
1. LiAlH4
2. H3O+
CH3
CH3CH2CHCH2CH2—NH2

CH3
CHCH2C≡N
Ni
H2
CH3
CHCH2CH2NH2

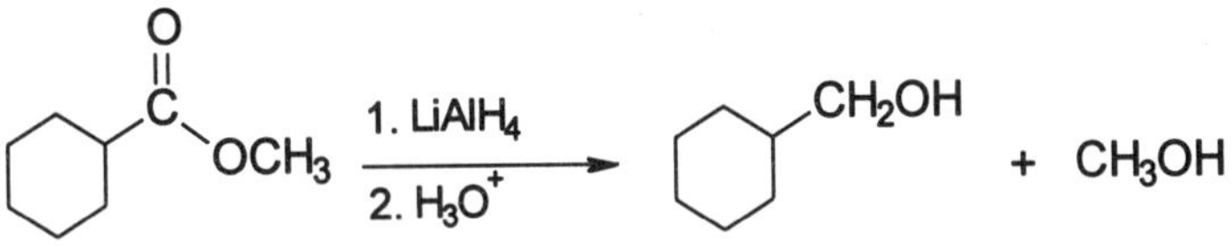

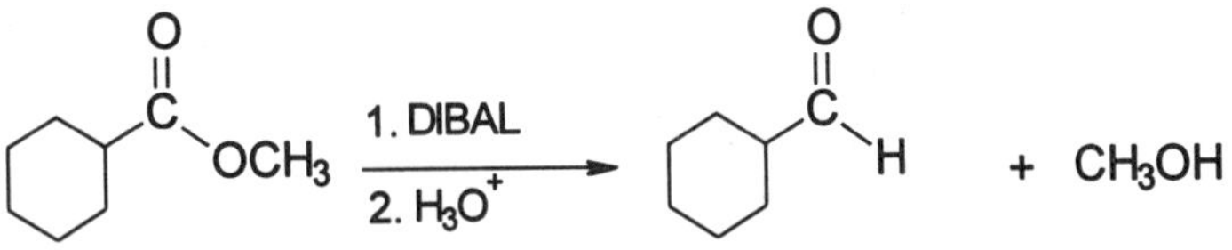

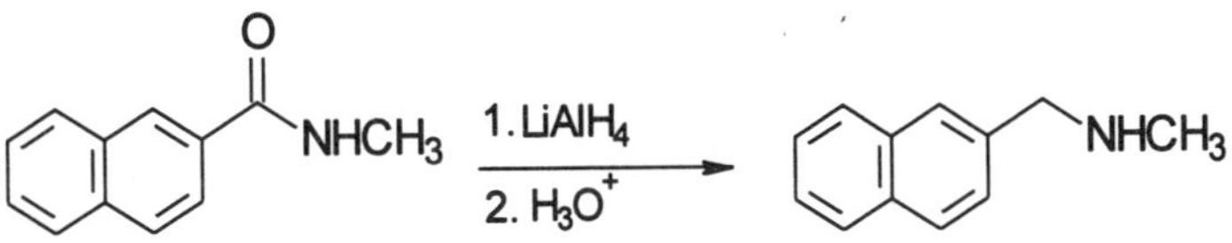

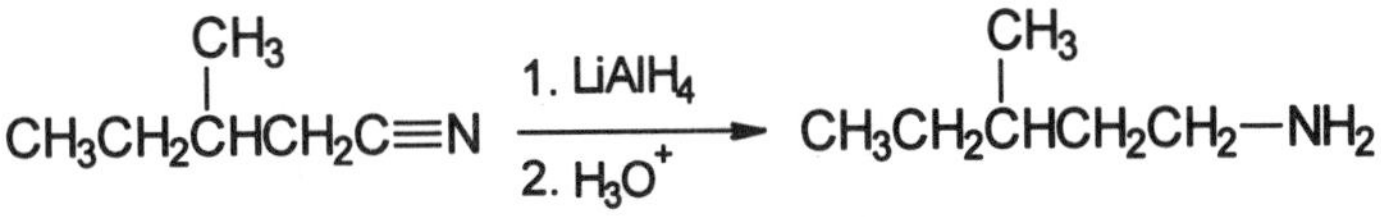

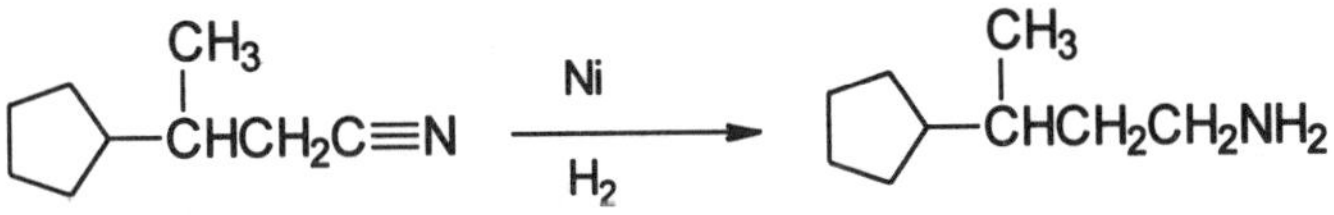

7. Reaction of Acid Derivatives with Organometallic Reagents

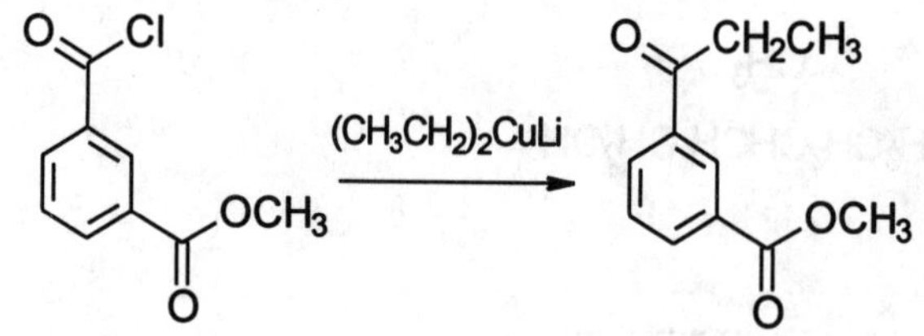

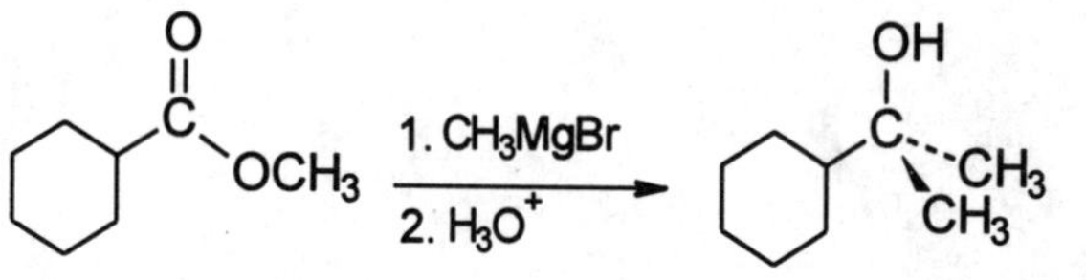

Solutions to Exercises

22.1 **(a)** ethyl phenylethanoate
(c) N,N-diethylcyclobutanecarboxamide
(e) 3,4-dimethoxybenzoyl chloride

(b) 3-cyclohexylbutanenitrile
(d) 2-bromoethyl 3-bromobenzoate

22.2 **(a)** cyclohexyl benzoate
(c) N-cyclohexyl-2-fluoroethanamide
(e) 2-methylbutyl 5-cyclopentylpentanoate

(b) 4-methylbenzonitrile
(d) 3,4-dimethylbenzoyl chloride

22.3 **(a)**

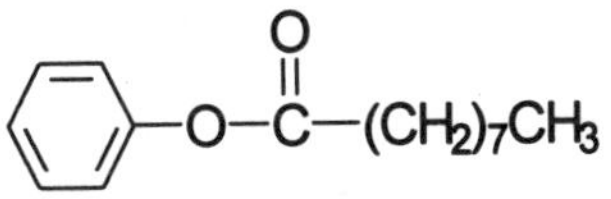

(b) 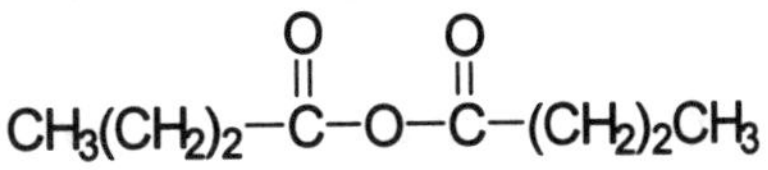 $CH_3(CH_2)_2-\overset{O}{\overset{\|}{C}}-O-\overset{O}{\overset{\|}{C}}-(CH_2)_2CH_3$

(c)

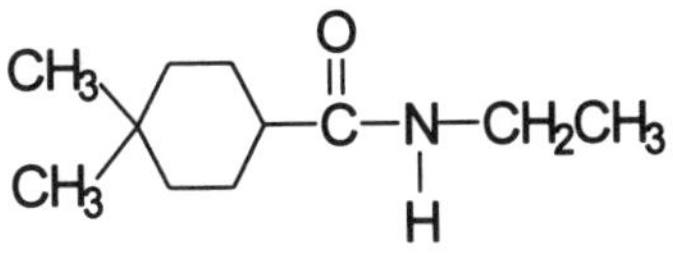

(d)

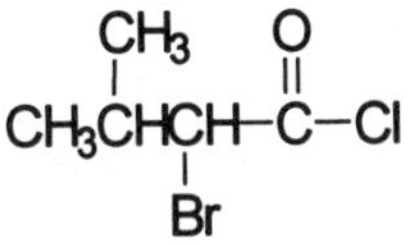

(e)

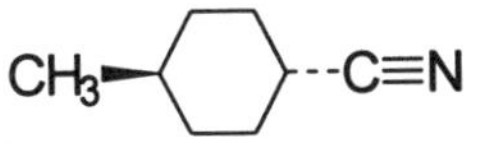

22.4 **(a)**

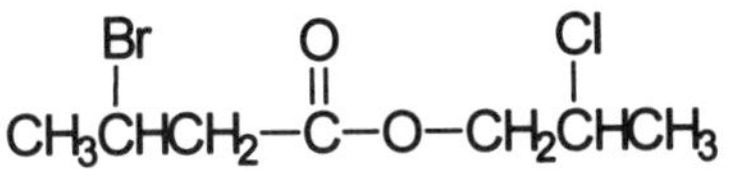

(b)

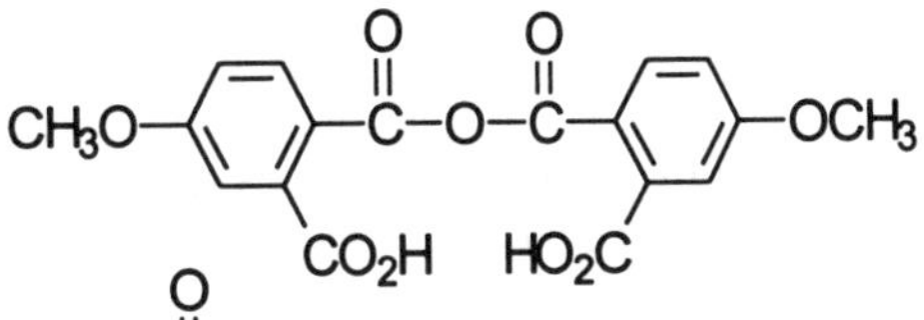

(c)

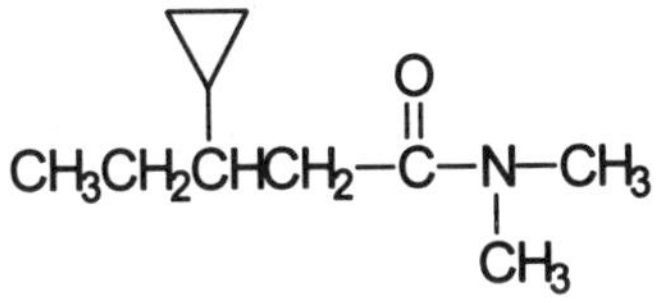

(d)

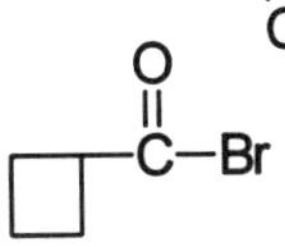

(e)

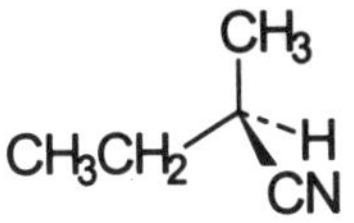

22.5 The carboxylic acid is derived from that portion of the compound that contains the carbonyl group. The name of the acid is mandelic acid, which is obtained by replacing the -ate with -ic acid.

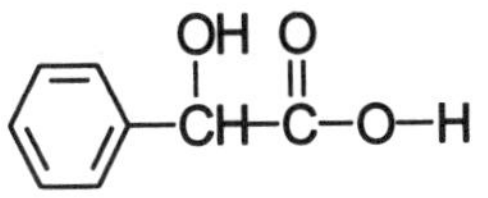

22.6   The carboxylic acid is derived from that portion of the compound that contains the carbonyl group. The name of the acid is furanoic acid, which is obtained by replacing the -ate with -ic acid.

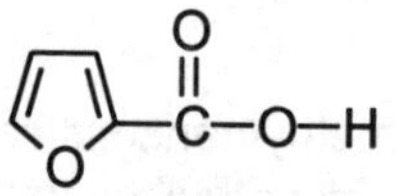

22.7   The functional group is an ester contained within a ring and is thus called a lactone. The configuration about the double bond is (Z).

22.8   The functional group is an amide contained within the four-membered ring and is thus called a lactam.

22.9   **(a)** 5-hydroxypentanoic acid lactone
          (b) 9-hydroxynonanoic acid lactone
          (c) 4-hydroxypentanoic acid lactone

22.10 (a) 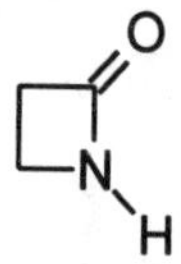   **(b)** 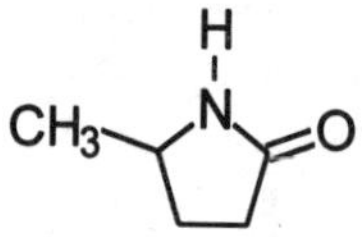 CH₃   **(c)**

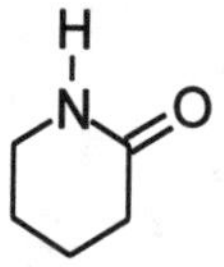

22.11   Compound I has two carbonyl groups joined by a common oxygen atom and is a cyclic anhydride. Only II and III are lactones.

22.12   Compound II has a methylene group separating the nitrogen atom and the carbonyl group. It is not a lactam. Only I and III are lactams.

22.13   The compounds are isomeric esters and thus have similar polarities. In addition they have similar molecular shapes. Thus, the dipole-dipole as well as London forces are similar, and as a result the boiling points are close.

22.14   The compounds are isomeric methyl esters and thus have similar polarities. However, they have different molecular shapes. The acid portion of methyl 2,2-dimethylpropanoate has a nearly spherical shape as compared to the cylindrical shape of the acid portion of methyl pentanoate. Thus, the London forces of methyl 2,2-dimethylpropanoate are smaller and the boiling point is lower.

22.15   The molecular weights differ only by 1 amu and both compounds have similar linear shapes. However, the nitrile has a polar carbon-nitrogen triple bond and as a result has larger dipole-dipole forces than propyne, which has a much less polar carbon-carbon triple bond.

22.16   Both compounds can form intermolecular hydrogen bonds between a hydrogen atom bonded to an electronegative atom on one molecule and the carbonyl group on a neighboring molecule. The N—H group is a better hydrogen bond donor and forms stronger hydrogen bonds, so the amide has a higher boiling point.

22.17 The amide functional group is resonance stabilized. As a result the oxygen atom has increased electron density and the nitrogen atom has decreased electron density.

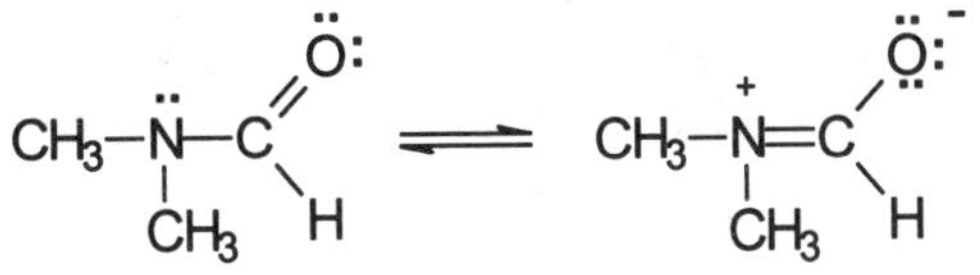

Protonation at the electron pair of nitrogen would give a conjugate acid in which there is no resonance stabilization of the carbonyl group. Resonance stabilization of the carbonyl group is still possible in the conjugate acid obtained by protonating oxygen.

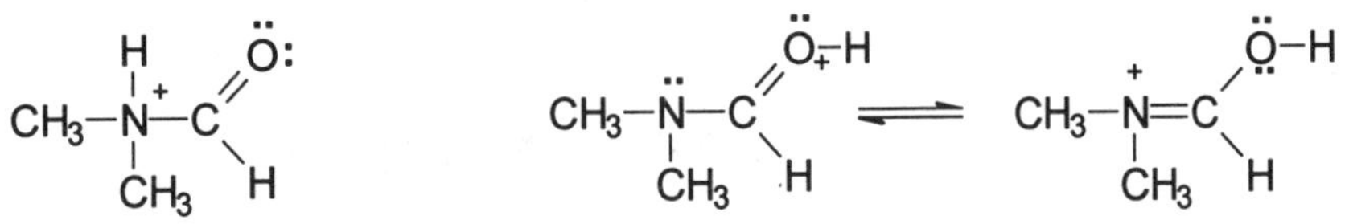

22.18 The carbon-nitrogen bond has some double bond character as a result of resonance stabilization of the carbonyl group by donation of electrons from the nitrogen atom, as depicted in the answer to Exercise 22.17.

22.19 (a) Esters are more stable than acid chlorides, because the carbonyl group is more stabilized by donation of electrons by resonance from oxygen than from chlorine. The reaction occurs to give the more stable ester.

(b) Amides are more stable than esters, because the carbonyl group is more stabilized by donation of electrons by resonance from nitrogen than from oxygen. The reaction will not occur.

(c) Amides are more stable than esters, because the carbonyl group is more stabilized by donation of electrons by resonance from nitrogen than from oxygen. The reaction will occur.

22.20 (a) Amides are more stable than acid anhydrides, because the carbonyl group is more stabilized by donation of electrons by resonance from nitrogen than from oxygen. The reaction will occur.

(b) Acid anhydrides are more stable than acid chlorides because the carbonyl group is more stabilized by donation of electrons by resonance from oxygen than from chlorine. The reaction will not occur.

(c) Esters are more stable than acid anhydrides because the carbonyl group is more stabilized by donation of electrons by resonance from oxygen bonded to an alkyl group than from oxygen bonded to another carbonyl group. The reaction will occur.

22.21 Esters are more stable than thioesters, because the carbonyl group is more stabilized by donation of electrons by resonance from oxygen than from sulfur. The electrons of sulfur are in the third energy level, and overlap of 3p and 2p orbitals in thioesters is less effective than overlap of 2p and 2p orbitals in esters.

22.22 Both sulfur and chlorine are third row elements and neither contributes electrons very effectively by resonance. However, to the extent that they do, sulfur is a better donor of electrons by resonance because it is less electronegative than chlorine. Thus, the resonance effect makes the thioester more stable than the acid chloride. In addition, there is a stronger inductive electron withdrawal of electrons by the more electronegative chlorine, which destabilizes the carbonyl group leading to increased reactivity of the acid chloride.

22.23 (a) The *tert*-butyl group of the alcohol portion of the ester on the right sterically hinders approach of a nucleophile to the carbonyl group. Thus, the compound on the left is the more reactive.

(b) The resonance interaction of the nonbonded electrons of the oxygen atom with the aromatic ring of the phenolic portion of the ester decreases the electron density at oxygen. Thus, there is a decreased availability of electrons to the carbonyl group; the carbonyl group is not as stable for the compound on the left and it is more reactive.

(c) The axial carbomethoxy group of the compound on the right is sterically hindered and approach of a nucleophile is also sterically hindered. Thus, the compound on the left is the more reactive.

(d) The carbonyl group of the six-membered lactone can be approached in an equatorial direction. The five-membered lactone has its carbonyl group in the plane of the ring, so approach of a nucleophile must occur perpendicular to that plane and is sterically hindered.

22.24 (a) The *tert*-butyl group bonded to the carbonyl carbon atom of the compound on the right sterically hinders approach of a nucleophile. Thus, the compound on the left is the more reactive.

(b) The resonance interaction of the nonbonded electrons of the oxygen atom is decreased by the resonance electron withdrawal of those electrons by the p-nitro group. As a result, the carbonyl group is not as stable for the compound on the left and it is more reactive.

(c) The trifluoromethyl group bonded to the carbonyl carbon atom inductively withdraws electron density. As a result, the carbonyl group is not as stable in the compound on the left and it is more reactive toward nucleophiles.

(d) Ring strain of the compound is increased by the incorporation of an sp^2 hybridized carbon atom. In the transition state that carbon atom becomes sp^3 hybridized and the ring strain is decreased. Thus, the activation energy is smaller for the compound on the left.

22.25 Esters are more stable than thioesters, because the carbonyl group is more stabilized by donation of electrons by resonance from oxygen than from sulfur. The electrons of sulfur are in the third energy level, and overlap of 3p and 2p orbitals in thioesters is less effective than overlap of 2p and 2p orbitals in esters. As a consequence, the equilibrium favors the more resonance stabilized ester.

22.26 In both reactions a more stable amide is formed from an ester, which is responsible for the negative $\Delta G°_{rxn}$. The resonance interaction of the nonbonded electrons of the oxygen atom with the aromatic ring of the phenolic portion of the aromatic ester decreases the electron density at oxygen. Thus, there is a decreased availability of electrons to the carbonyl group. As a result, the carbonyl group is not as stable for the first compound. Thus, the $\Delta G°_{rxn}$ is more negative for the phenyl ester than for the cyclohexyl ester.

22.27 The more polar carbon-oxygen double bond is more stable than the carbon-nitrogen double bond. In addition, the nitrogen atom of the amide stabilizes the carbonyl group by resonance donation of electrons. Because the oxygen atom of the imidic acid is more electronegative than a nitrogen atom, it is less effective in donation of electrons by resonance. Thus, the amide is more stable than the imidic acid.

22.28 The carbonyl group of the hydroxamic acid is stabilized by donation of electrons of nitrogen by resonance. Because the oxygen atom of the O-acyl hydroxylamine is more electronegative than a nitrogen atom, it is less effective in donation of electrons by resonance.

22.29 The other possible product results from displacement of the ethoxy group as ethanol. The activation energy of the second step of the two step nucleophilic acyl substitution reaction is affected by the identity of the leaving group. Thiols are stronger acids than alcohols. Hence, the thiolate ion is a weaker base than the alkoxide ion. Weaker bases are generally better leaving groups, so thiolate is a better leaving group than methoxide ion.

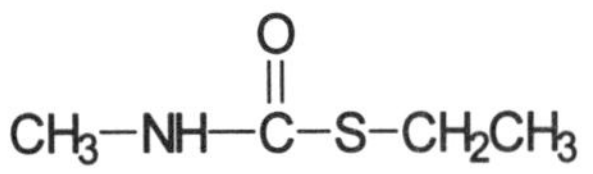

$$CH_3-NH-\overset{\overset{\displaystyle O}{\|}}{C}-S-CH_2CH_3$$

22.30 Carboxylic acids are more stable than acid anhydrides, because the carbonyl group is more stabilized by donation of electrons by resonance from oxygen bonded to hydrogen than from oxygen bonded to another carbonyl group. Thus, anhydrides are more reactive toward nucleophilic attack by methanol than is the carboxylic acid group of the product, shown below.

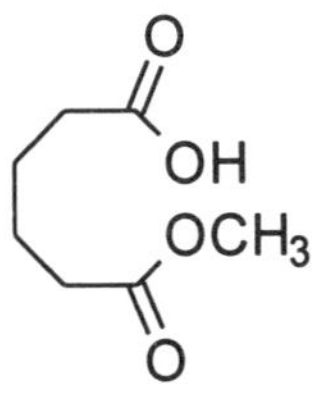

22.31 The trifluoroacetate portion of the compound withdraws electrons from the carbonyl carbon atom on the left to a greater extent than the acetate portion of the compound withdraw electrons from the carbonyl carbon atom on the right. Thus, the nucleophile attacks the carbonyl group on the left to release the trifluoroacetate group.

22.32 A methyl group is sterically smaller than an ethyl group. Thus attack of a nucleophile should occur more favorably at the carbonyl group on the left to release a propanoate ion.

The major ester should be the ethyl ethanoate, the minor ester ethyl propanoate.

22.33 The nitrogen atom of an amino group is more basic than the oxygen atom of a hydroxyl group. For atoms in the same period, the more basic atom is the better nucleophile.

22.34 The amide cannot be resonance stabilized by donation of electrons from nitrogen to the carbonyl group. Such a resonance contributor would have a carbon-nitrogen double bond at the bridgehead of the bicyclic structure, which would be highly strained. Resonance stabilization is possible for 5-aminopentanoic acid lactam, so it is more stable and is hydrolyzed more slowly.

22.35 The lactone is formed from the primary alcohol to give a six-membered ring. Reaction with the tertiary alcohol would give a more strained four-membered lactone.

22.36 The approach of the nucleophile to the carboxyl carbon atom is sterically hindered by the quaternary center at the C-2 atom.

22.37 (a) $CH_3-\overset{O}{\overset{\|}{C}}-OH$ $HO-CH_2CH_2CH_3$

(b) $CH_3CH_2CH_2CH_2-\overset{O}{\overset{\|}{C}}-OH$ $HO-CH_2CH_2CH_2CH_3$

(c) CH_3CH_2-OH $HO-\overset{O}{\overset{\|}{C}}-CH_2CH_3$

(d) $CH_3CH_2CH_2CH_2-OH$ $HO-\overset{O}{\overset{\|}{C}}-CH_2CH_2CH_3$

22.38 (a) $HO-CH_2CH_3$ (b) $(CH_3)_2CH-OH$

(c)   (d) CH_3-OH

(e) 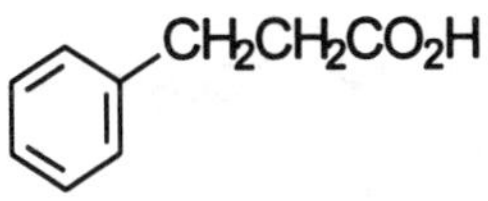 CH₂CH₂CO₂H   (CH₃)₂CH—OH **(f)** 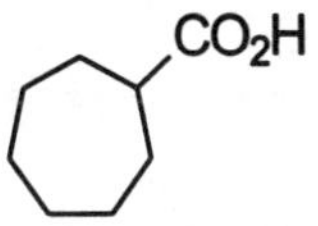 CO₂H   (CH₃)₃C—OH

22.39 Because there is only one product, the ester must be a lactone.

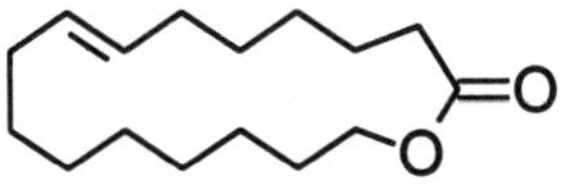

22.40 $CH_3(CH_2)_{24}$—$\overset{\overset{\displaystyle O}{\|}}{C}$—O—$(CH_2)_{29}CH_3$ $CH_3(CH_2)_{26}$—$\overset{\overset{\displaystyle O}{\|}}{C}$—O—$(CH_2)_{29}CH_3$

$CH_3(CH_2)_{24}$—$\overset{\overset{\displaystyle O}{\|}}{C}$—O—$(CH_2)_{31}CH_3$ $CH_3(CH_2)_{26}$—$\overset{\overset{\displaystyle O}{\|}}{C}$—O—$(CH_2)_{31}CH_3$

22.41 **(a)** 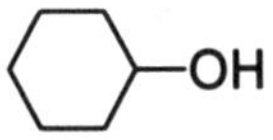 —OH CO_2 **(b)** 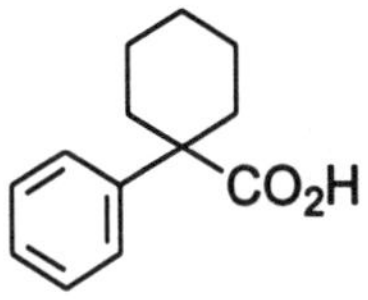 CO_2H

(c) 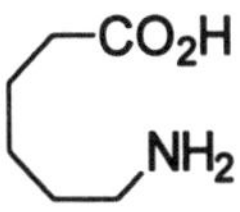 —CO₂H / NH₂   **(d)** 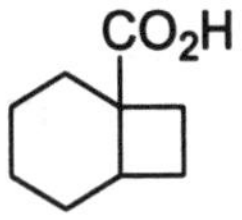 CO₂H

22.42 **(a)** 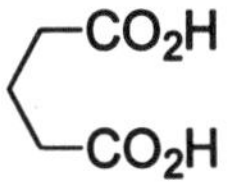 —CO₂H / —CO₂H   **(b)** 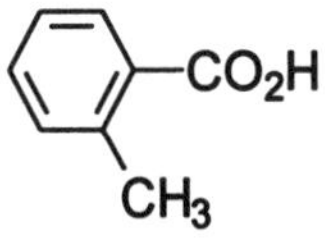 —CO₂H / CH₃

(c) 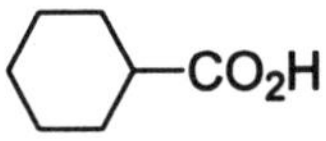 —CO₂H $NH_2CH(CH_3)_2$ **(d)** 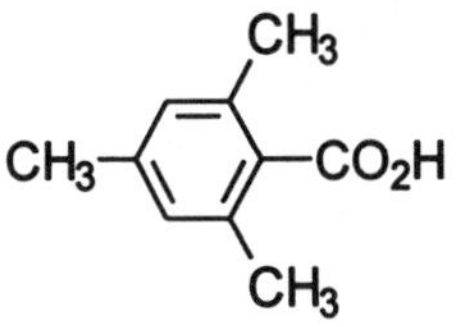 CH_3——CO₂H CH_3OH

22.43 **(a)** 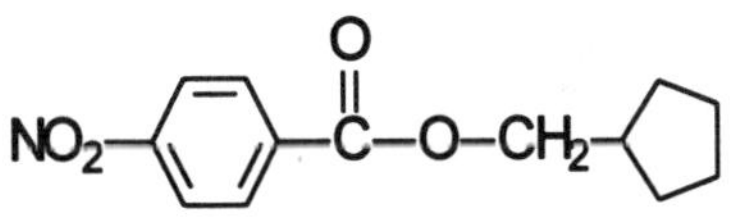 NO_2——$\overset{\overset{\displaystyle O}{\|}}{C}$—O—CH₂—   **(b)** 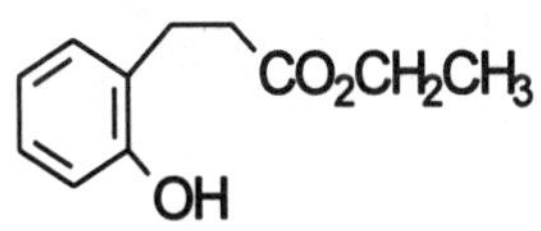 CO₂CH₂CH₃ / OH

(c) 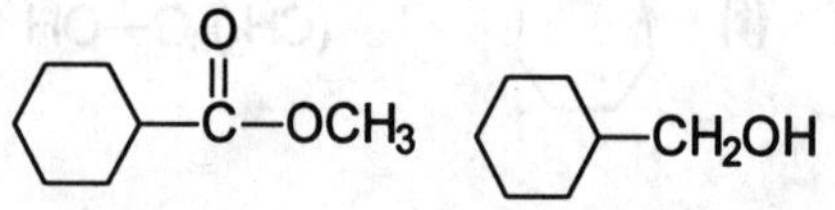

(d) 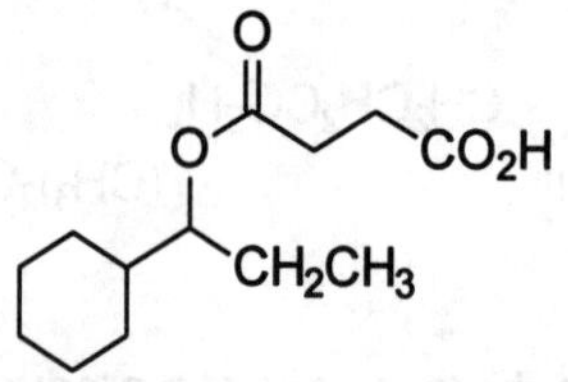

22.44 (a) 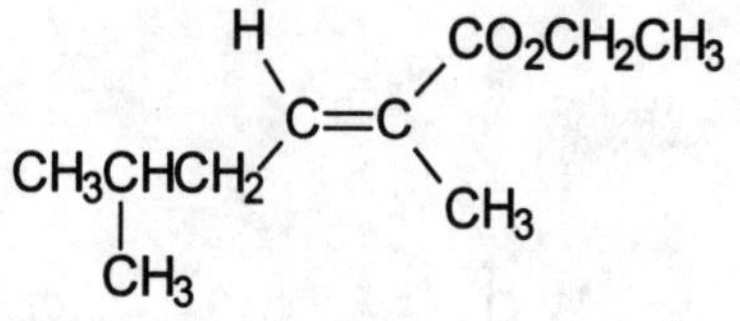

(b) 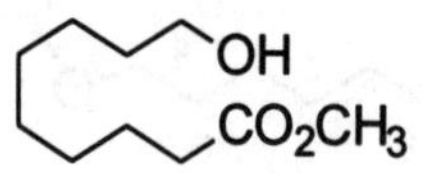

(c) 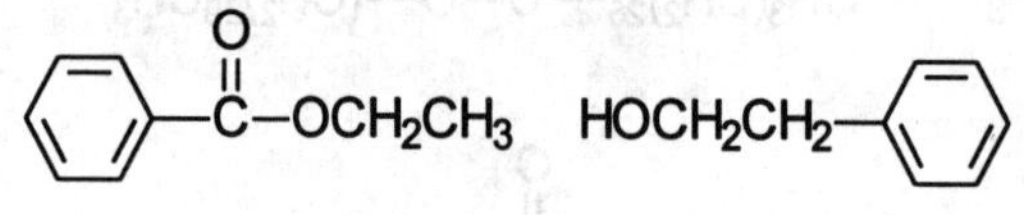

(d) 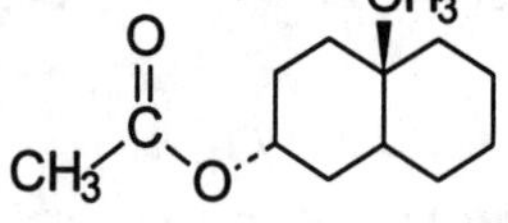

22.45 (a) 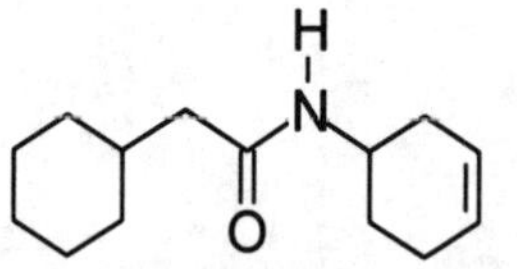

(b) 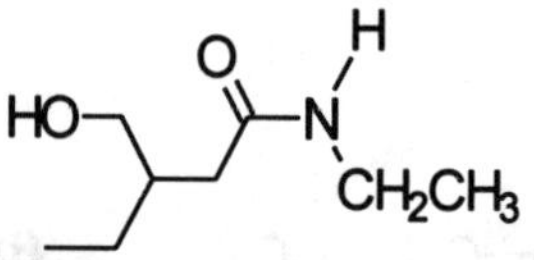

(c) 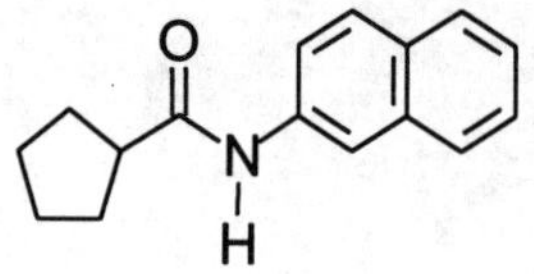

(d) 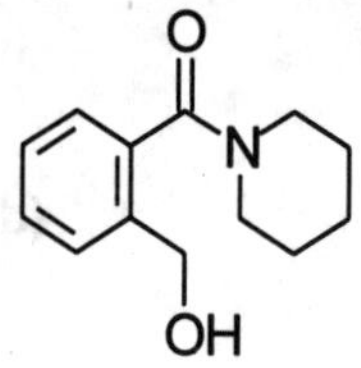

22.46 (a) 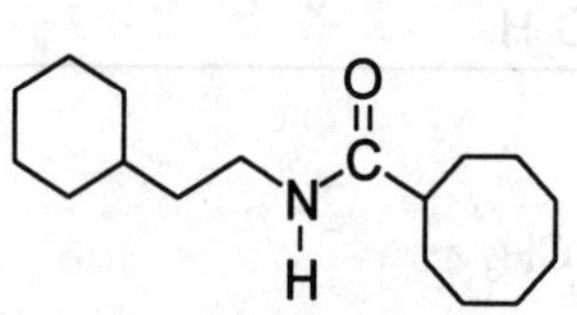

(b) 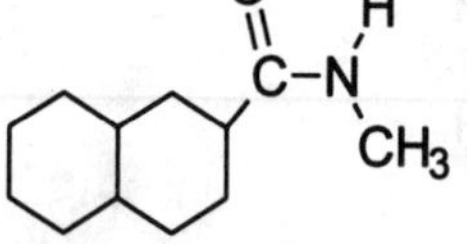

(c) 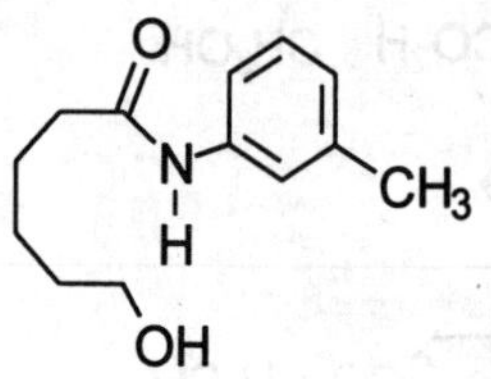

(d) 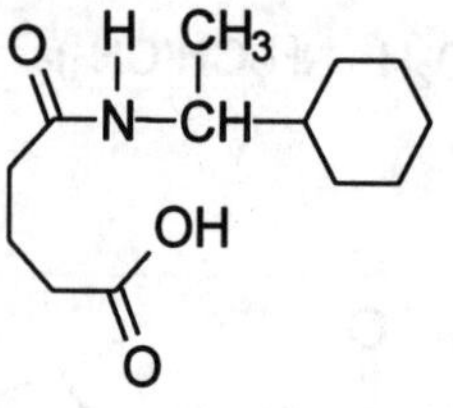

22.47 (a) 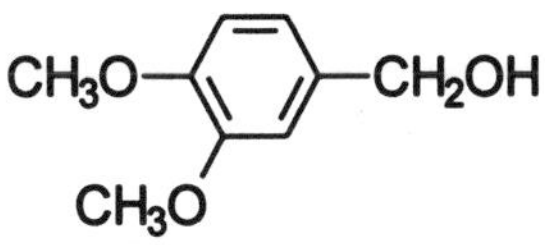 [structure: CH₃O-disubstituted benzene with CH₂OH]

CH₃O—⟨benzene⟩—CH₂OH, with CH₃O substituent

(b) 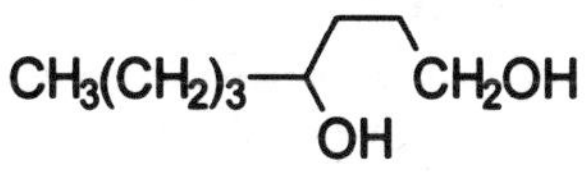 CH₃(CH₂)₃—CH(OH)—CH₂—CH₂OH

(c) 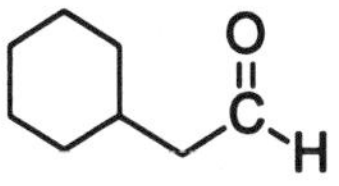 [cyclohexane with CH₂—C(=O)H side chain]

(d) 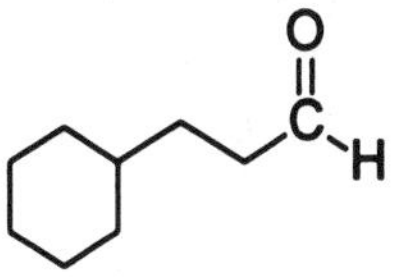 [cyclohexane with CH₂CH₂—C(=O)H side chain]

(e)  [cyclobutane with CH₂—N(CH₂CH₃)₂]

(f) 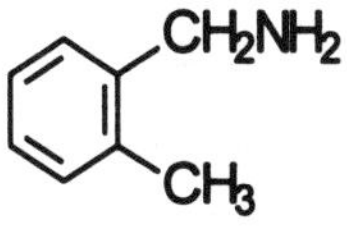 [benzene with CH₂NH₂ and CH₃ substituents]

22.48 (a) 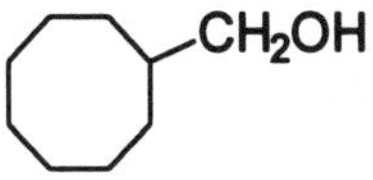 [cyclooctane with CH₂OH]

(b) 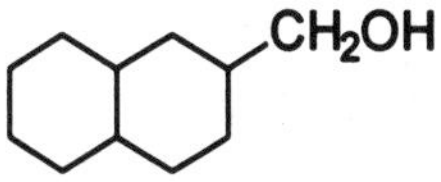 [decalin with CH₂OH]

(c) 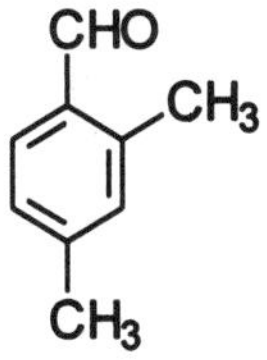 [benzene with CHO and two CH₃ groups]

(d) 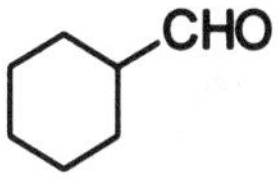 [cyclohexane with CHO]

(e) 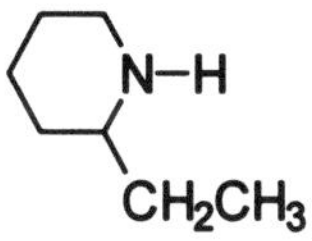 [piperidine, N—H, with CH₂CH₃]

(f) 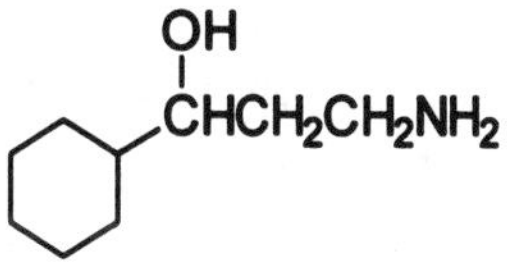 [cyclohexane with CH(OH)CH₂CH₂NH₂]

22.49 Since only one 16-carbon product is obtained, the compound must be an ester with 16 carbon atoms in the acyl group as well as in the alkyl group.

$$CH_3(CH_2)_{14}-\overset{O}{\overset{\|}{C}}-O-(CH_2)_{15}CH_3$$

22.50 [two macrocyclic lactone structures]

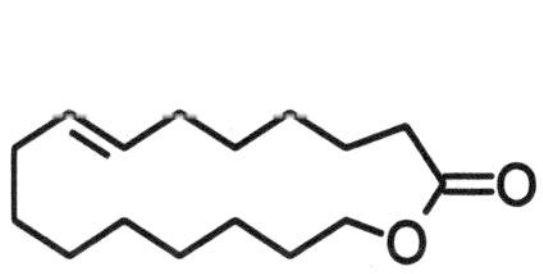

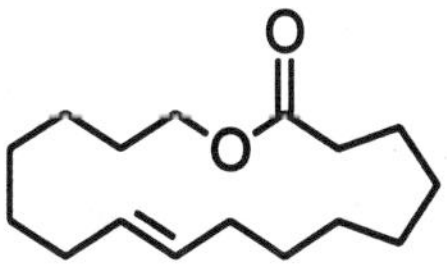

22.51 The two R groups are derived from the Grignard reagent. The third group on the carbon atom bearing oxygen is hydrogen and was originally bonded to the carbonyl carbon atom of the ester. Thus, the ester must be a derivative of formic acid, such as ethyl formate. Note that the alkyl group in the ester is not part of the alcohol product.

22.52 The first mole reacts to give a tetrahedral intermediate which loses methoxide ion and gives the methyl ester shown on the following page. After this point, the subsequent reactions are those of an ester with a Grignard reagent. The second intermediate is a ketone.

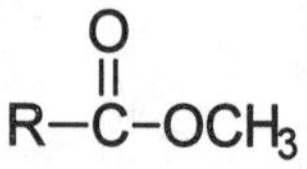

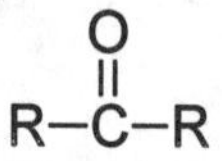

$$R-\overset{\overset{\displaystyle O}{\|}}{C}-OCH_3 \qquad R-\overset{\overset{\displaystyle O}{\|}}{C}-R$$

22.53 The acid chloride is much more reactive toward the Grignard reagent than the product ketone. At the low temperature of the reaction, the ketone does not react. At a higher temperature it would add.

22.54 React an acid chloride with a lithium dialkylcuprate. The acid chloride consists of the carbonyl carbon atom and one of the two alkyl groups bonded to it in the product. The lithium dialkylcuprate contains the other alkyl group. One possible combination is cyclopentylethanoyl chloride and lithium di(2-methyl propyl)cuprate. The second combination that give the same compound is 3-methylbutanoyl chloride and lithium di(cyclopentylmethyl)cuprate.

22.55 **(a)** Oxidation gives a carboxylic acid, which is then converted into an acid chloride by PCl_3. Reaction with the lithium dialkylcuprate gives a ketone, as shown below.
(b) Reduction by lithium aluminum hydride gives an alcohol. Reaction with $SOCl_2$ gives a chloro compound which when converted to a Grignard adds to the methyl acetate to give a tertiary alcohol.
(c) Reduction by the complex aluminum hydride gives an aldehyde. Reaction with methanol gives an acetal.

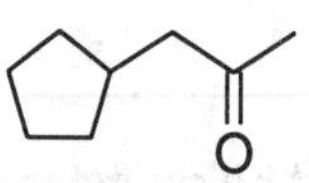

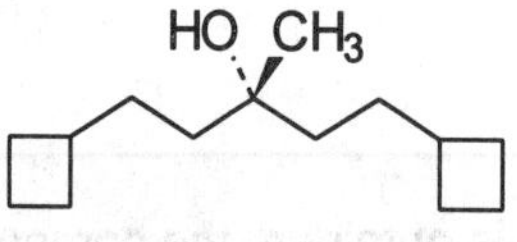

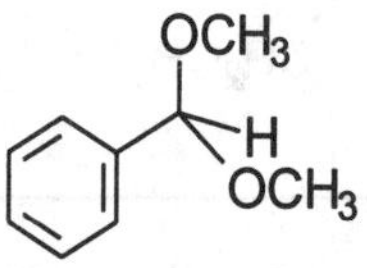

22.56 **(a)** Oxidation gives a carboxylic acid, which is then converted into an acid chloride. Reaction with the methanol gives an ester.
(b) Reaction with the amine gives an amide. Reduction by lithium aluminum hydride gives an amine.
(c) Reduction by the complex aluminum hydride gives an aldehyde. A Clemmensen reduction then gives an alkane.

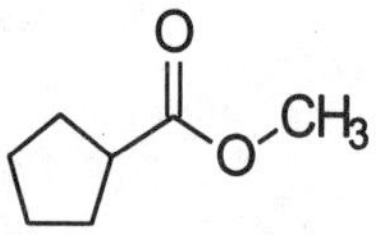

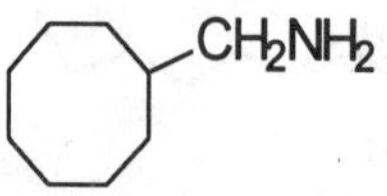

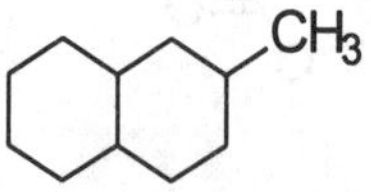

22.57 The bromine atom cannot effectively contribute electrons by resonance to the carbonyl group. It is less electronegative than chlorine and should not destabilize the dipolar resonance form as much as chlorine does. Thus the carbonyl group will have more single bond character in the case of the bromine atom and the bond will require less energy to stretch. This effect will be seen as an absorption at a lower wavenumber for the acyl bromide.

22.58 Neither the sulfur nor the chlorine atom can effectively contribute electrons by resonance to the carbonyl group. Sulfur is less electronegative than chlorine and should not destabilize the dipolar resonance form as much as chlorine does. Thus, the carbonyl group will have more single bond character in the case of the thioester and requires less energy to stretch. This effect results in an absorption at a lower wavenumber for the thioester.

22.59 The absorption is due to a nitrile group. There is an additional deficiency of two hydrogen atoms which could be due to a double bond or a ring. The value is at the upper value of the range for nitriles (2200-2250 cm^{-1}). Thus, the double bond cannot be in conjugation with the nitrile group because such a resonance stabilized functional group would absorb at lower wavenumber. Only one unsaturated nitrile is possible. In addition, only one cycloalkylnitrile is possible.

$$CH_2=CHCH_2-C\equiv N \qquad \triangleright-C\equiv N$$

22.60 The absorption of II occurs at lower wavenumber than I, because conjugation of the carbon-carbon double bond with the carbonyl group increases the single bond character of the carbonyl group. The absorption of III occurs at higher wavenumber that I, because the electron withdrawal of the carbon-carbon double bond decreases the electron density of the oxygen atom and hence the availability of electrons to stabilize the dipolar resonance form of the carbonyl group. Thus, the order for compounds I, II, and III is 1775, 1741, and 1806 cm^{-1}, respectively.

22.61 **(a)** The singlet of intensity 9 at 1.5 ppm is due to nine equivalent protons, a good indicator of a *tert*-butyl group. The singlet of intensity 2 at 3.9 ppm is due to two equivalent protons on a carbon atom bonded to an oxygen atom or a chlorine atom. The compound is *tert*-butyl chloroethanoate, as shown on the next page. Note that, although chloromethyl 2,2-dimethylpropanoate might seem to fit the spectrum also, the singlet due to the protons in the —CH_2Cl group would be shifted further downfield, because they would be deshielded by both the ester oxygen atom and the chlorine atom.

(b) The extra oxygen atom in the molecular formula could indicate an alcohol or an ether group in addition to the ester. The triplet of intensity 6 indicates six equivalent protons (two methyl groups) with a neighboring methylene group. The quartet of intensity 4 indicates four equivalent protons (two methylene groups) with a neighboring methyl group; the signal's position at 4.2 ppm indicates that the protons are bonded to a carbon atom that is also bonded to an oxygen atom. Thus there are two equivalent —CH_2CH_3 groups in the compound. The compound is diethyl carbonate.

(c) The molecular formula indicates that the compound has a total of four double bonds and/or rings, so a benzene ring is possible. The triplet of intensity 3 at 1.2 ppm and the quartet of intensity 2 at 4.1 ppm indicate three protons and two protons, respectively, splitting each other on neighboring carbon atoms, with the methylene group also bonded to an oxygen atom. The singlet of intensity 2 indicates a methylene group with no neighboring protons; its position at 3.6 ppm indicates that the protons are bonded to a carbon atom that is also bonded to an oxygen atom or a benzene ring. The singlet of intensity 5 at 7.1 ppm is due to a phenyl group. The compound is ethyl phenylethanoate.

(a) $ClCH_2-\overset{\overset{\displaystyle O}{\|}}{C}-OC(CH_3)_3$ (b) $CH_3CH_2-\overset{\overset{\displaystyle O}{\|}}{C}-OCH_2CH_3$ (c) [structure of phenyl group with OCH_2CH_3 ester]

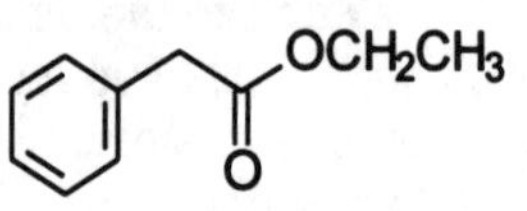

22.62 **(a)** The correct formula is $C_6H_{11}ClO_2$. The singlet of intensity 9 at 1.3 ppm is due to nine equivalent protons, a good indicator of a *tert*-butyl group. The singlet of intensity 2 is due to two equivalent protons; its position at 5.7 ppm indicates that the protons are bonded to a carbon atom bonded to an oxygen atom and a chlorine atom. The compound is chloromethyl 2,2-dimethylpropanoate, as shown below.

(b) The singlet of intensity 3 at 2.1 ppm is due to a methyl group bonded to the carbonyl carbon atom. The triplets of intensity 2 at 3.5 and 4.4 ppm are due to neighboring methylene groups that are also bonded to an oxygen atom and a bromine atom, respectively. The compound is bromoethyl ethanoate.

(c) The extra oxygen atom in the molecular formula could indicate an alcohol or an ether group in addition to the ester. The molecular formula indicates that the compound has a total of four double bonds and/or rings, so a benzene ring is possible. The triplet of intensity 3 at 1.15 ppm indicates a methyl group split by a neighboring methylene group. The triplet of intensity 2 at 4.25 ppm indicates a methylene group split by a neighboring methylene group and bonded to an oxygen atom. The complex multiplet of intensity 2 at 1.7 ppm is due to a methylene group split by neighboring methylene and methyl groups. The doublets of intensity 2 at 6.9 and 7.9 ppm are due to two sets of equivalent protons on a benzene ring. The singlet of intensity 1 at 9.5 ppm is due to a proton bonded to an oxygen atom on the benzene ring. The compound is propyl *p*-hydroxybenzoate.

(a) $CH_3-\overset{\overset{\displaystyle CH_3}{|}}{\underset{\underset{\displaystyle CH_3}{|}}{C}}-CO_2CH_2Cl$ (b) $CH_3-\overset{\overset{\displaystyle O}{\|}}{C}-OCH_2CH_2Br$ (c) [structure of *p*-hydroxybenzoate ester with HO and $O-CH_2CH_2CH_3$]

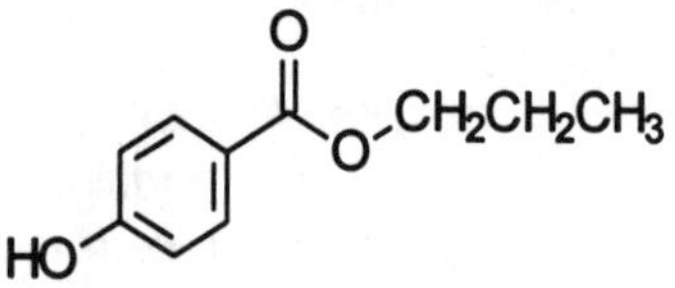

22.63 **(a)** The compound has six carbon atoms, but four signals, so there must be two sets of two equivalent carbon atoms or one set of three equivalent carbon atoms. The quartets at 27.3 and 51.5 ppm are due to methyl groups, the former fairly close to the carbonyl carbon atom and the latter bonded to an ester oxygen atom. The singlet at 38.7 ppm must be due to a quaternary carbon atom, because it is not split by any bonded hydrogen atoms. The singlet at 178.8 ppm is due to a carbonyl carbon atom. The compound is methyl 2,2-dimethylpropanoate, as shown below.

(b) The molecular formula, $C_5H_8O_2$, indicates that the compound has one double bond or a ring. The compound has five nonequivalent carbon atoms. The triplets at 19.1, 22.7, 29.9, and 69.4 ppm are due to a series of methylene groups, the first bonded to a carbonyl carbon atom and the last bonded to an ester oxygen atom. The singlet at 171.2 ppm is due to a carbonyl carbon atom. The compound is 5-hydroxypentanoic acid lactone.

(c) The molecular formula, $C_9H_{10}O_2$, indicates that the compound has a total of four double bonds and/or rings, so a benzene ring is possible. The compound has nine carbon atoms, but seven signals, so there must be two sets of two equivalent carbon atoms or one set of three equivalent carbon atoms. The triplet at 20.7 ppm is due to a methylene group bonded to the carbonyl carbon atom. The quartet at 66.1 ppm is due to a methyl group bonded to an ester oxygen atom. The singlet at 128.1 ppm and the doublets at 128.4, 128.8, and 136.1 are due to carbon atoms in a benzene ring. The singlet at 170.6 ppm is due to a carbonyl carbon atom. The compound is methyl phenylethanoate.

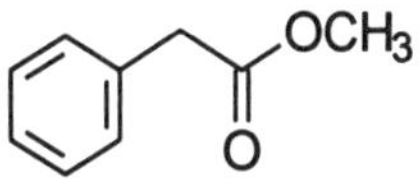

22.64 **(a)** The compound has six nonequivalent carbon atoms. The quartets at 11.4 and 16.9 ppm are due to methyl groups isolated from the carbonyl carbon atom. The triplet at 27.6 ppm is due to a methylene group close to the carbonyl carbon atom. The doublet at 41.4 ppm is due to a carbon atom bonded to a hydrogen atom and a carbonyl carbon atom. The quartet at 51.1 ppm is due to a methyl group bonded to an ester oxygen atom. The singlet at 176.2 ppm is due to a carbonyl carbon atom. The compound is methyl 2-methylbutanoate, as shown on the following page.

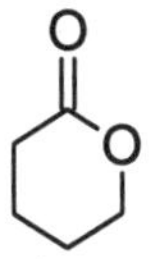

(b) The molecular formula, $C_6H_{10}O_2$, indicates that the compound has one double bond or a ring. The compound has six carbon atoms, but five signals, so there must be a set of two equivalent carbon atoms. The triplets at 18.4 and 25.2 ppm are due to methylene groups. The doublet at 37.9 ppm is due to a tertiary carbon atom. The quartet at 51.4 ppm is due to a methyl group bonded to an ester oxygen atom. The singlet at 175.7 ppm is due to a carbonyl carbon atom. The compound is methyl cyclobutylethanoate.

(c) The compound has seven carbon atoms, but five signals, so there must be two sets of two equivalent carbon atoms or one set of three equivalent carbon atoms. The quartets at 14.2 and 27.2 ppm are due to methyl groups isolated from and close to the carbonyl carbon atom, respectively. The singlet at 38.7 ppm is due to a quaternary carbon atom. The triplet at 60.2 ppm is due to a methylene group bonded to an ester oxygen atom. The singlet at 178.4 ppm is due to a carbonyl carbon atom. The compound is ethyl 2,2-dimethylpropanoate.

(a) 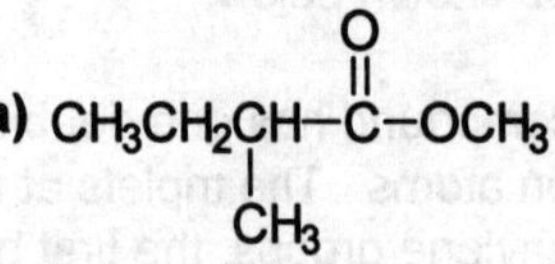 $CH_3CH_2CH-\overset{\displaystyle O}{\overset{\|}{C}}-OCH_3$ (b) ⬜$-\overset{\displaystyle O}{\overset{\|}{C}}OCH_3$ (c) $(CH_3)_3C-\overset{\displaystyle O}{\overset{\|}{C}}-OCH_2CH_3$

$\underset{CH_3}{|}$

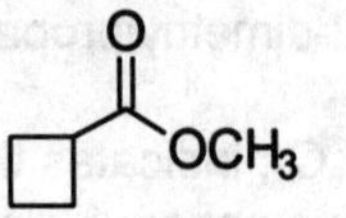

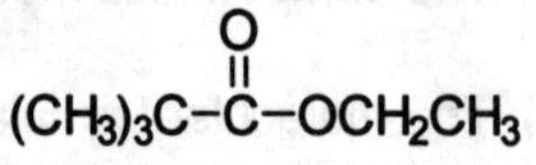

23

Enols and Enolates Condensation Reactions

The synthetic concepts presented in this chapter are the most sophisticated in the text. The common feature of the many reactions considered is that they occur at the α carbon atom. However, these reactions often join molecules with several functional groups to one another to form even more complex structures. Moreover, the results of those reactions are not always readily evident to students, because they are accompanied by several subsequent reactions that change one or more of the functional groups or even result in the elimination of functional groups.

Keys to the Chapter

23.1 Synthesis and Retrosynthesis

Reactions to form carbon-carbon bonds are required to prepare molecules with complex structures from low-molecular weight starting materials. A small number of such bond formation reactions have been presented in earlier chapters. In this chapter, many reactions are presented that involve joining atoms at the carbon atom adjacent to a carbonyl group.

Recognizing what starting materials to use to prepare a desired compound requires working backward in a mental process called **retrosynthesis**. The various methods to prepare the compound are considered in terms of potential last steps. Then the immediately preceding step is considered and so on. The process of proposing a synthesis by retrosynthesis requires an excellent command of the types of carbon-carbon bonds that may be formed in all of the reactions that form such bonds. In other words, you need to know what are the special characteristics of each type of reaction that lends itself to the synthesis of certain assemblies of atoms.

23.2 The α Carbon Atom of Carbonyl Compounds

The carbon atom bonded to the carbonyl carbon atom is known as the α carbon atom. It is a reactive site because its hydrogen atom is acidic and results in formation of a carbanion which serves as a nucleophile. The pK_a of the α hydrogen atom is approximately 18, which means that the K_a is approximately 30 powers of ten large than the K_a for hydrocarbons. The increased acidity is the result of resonance stabilization of the **enolate ion**. One of the two resonance forms of the enolate ion has the negative charge on the α carbon atom; the other resonance form has the negative charge on the oxygen atom.

Enolates are formed by reaction of a carbonyl compound with a base. The concentration of enolate formed depends on the K_b of the base, which in turn is related to the K_a of the conjugate acid of the base. Sodium hydroxide is a weaker base than the enolate ion, and it is not sufficiently basic to give a high concentration of enolate. Sodium amide is a much stronger base, and it quantitatively converts carbonyl compounds to their enolates.

Enolates can react as nucleophiles by combination with an electrophilic reagent at the oxygen atom or the carbon atom of the enolate. Although the electron density of the enolate is highest at the oxygen atom, the most common reaction site of enolates is at the carbon atom. This selectivity is related to the bonds formed in the transition state. Reaction at the oxygen atom forms an **enol product** that contains a carbon-carbon double bond. Reaction at the carbon atom forms a **keto product** that contains a carbon-oxygen double bond. The greater stability of the carbonyl group favors formation of the keto product.

23.3 Keto-Enol Equilibria of Aldehydes and Ketones

Aldehydes and ketones both exist in two isomeric forms known as **keto** and **enol** tautomers. They differ in the location of a hydrogen atom and the type of double bond. The keto form is more stable than the enol form. **Tautomerization** is a net process by which protons are transferred from one site to another by a series of steps. In acidic solution, the steps are protonation of the carbonyl oxygen atom by an acid to give a conjugate acid, followed by deprotonation of the α carbon atom by the conjugate base of the acid. The acid and base are the hydronium ion and water, respectively. In basic solution, the steps are deprotonation of the α carbon atom to give an enolate ion, followed by protonation of the oxygen atom by the conjugate acid of the base. The base and acid are the hydroxide ion and water, respectively.

The stability of an enol is reflected in its concentration in equilibrium with the keto form. Ketones have a smaller concentration of enol than do aldehydes. This fact reflects the greater stability of ketones compared to aldehydes as a result of the electron donation of alkyl groups to the carbonyl carbon atom. The stability of isomeric enols from a ketone reflects the stability due to the degree of substitution of the double bond. Conjugation of the double bond of the enol increases its stability.

23.4 Consequences of Enolization

The hydrogen atom of the α carbon atom of a carbonyl compound is called an **enolizable hydrogen atom**. If the α carbon atom is chiral, the formation of the isomeric enol results in loss of optical activity, because a racemic mixture is formed when the keto form is regenerated. If the keto-enol equilibrium occurs in a protic solvent containing deuterium in place of hydrogen, then the α hydrogen atoms are exchanged by deuterium.

23.5 α-Halogenation Reactions of Aldehydes and Ketones

The α hydrogen atoms of carbonyl compounds can be replaced by halogen atoms. The regioselectivity and the number of hydrogen atoms substituted depend on whether acidic or basic conditions are used. Under acidic conditions, one hydrogen atom is substituted without the complication of multiple substitution. Under basic conditions, multiple substitution occurs.

Acid-catalyzed halogenation occurs by halogenation of the enol, whose formation is the rate determining step. The double bond of the enol is attacked by the halogen in a reaction similar to the electrophilic attack of a simple alkene. However, the resulting intermediate is not the bromonium ion but an oxocarbocation. Loss of a proton from the hydroxyl group gives the halogenated product. Multiple substitution is disfavored, because the halogen atom makes the carbonyl oxygen atom less basic and decreases the rate of formation of the enol. The halogen also destabilizes the conjugate acid, because it tends to withdraw electron density from the oxocarbocation. In ketones with two nonequivalent α carbon atoms, the more substituted carbon atom is halogenated. This regioselectivity reflects the greater stability of the enol with the more substituted carbon-carbon double bond.

Halogenation under basic conditions occurs via nucleophilic attack of the enolate at the halogen molecule. Because the halogen atom is inductively electron withdrawing, the α hydrogen atom of the halogenated ketone is more acidic. Therefore, not only does multiple substitution occur, but it continues at the α carbon atom originally substituted in preference to a second α carbon atom. Continued halogenation of a methyl ketone forms a trihalomethyl derivative that is cleaved into a carboxylate and a haloform as the result of nucleophilic attack of the carbonyl carbon atom.

23.6 Alkylation of Enolate Ions

Enolates, formed by the abstraction of the α hydrogen atom by a strong base, are nucleophiles. Lithium diisopropylamide (LDA) or sodium hydride are required as bases. The site of proton abstraction is related to the acidity of the two possible α hydrogen atoms, which is in the order primary > secondary > tertiary. Reaction of the enolate with an alkyl halide forms α alklylated ketones. Multiple alkylation can occur as the result of proton exchange between the original enolate and the alkylated ketone, followed by alkylation of that enolate ion.

23.7 The Aldol Condensation of Aldehydes

The **aldol condensation** is the reaction of two moles of an aldehyde to form a β-hydroxyaldehyde, or aldol, in the presence of a base. The product is formed by addition of the enolate, formed by abstraction of an α hydrogen atom of one aldehyde by hydroxide ion, to the carbonyl carbon atom of the second aldehyde. Protonation of the alkoxide by exchange of a proton from water gives the aldol. The first step is an addition reaction to form a tetrahedral product. Subsequent dehydration in the reaction mixture often occurs to give an α, β unsaturated aldehyde. The combination of the two steps constitutes a condensation reaction.

Based on an analysis of the bond energies of the reactants and products, the addition step is slightly exothermic. However, ΔS°_{rxn} is negative for the reaction, because two molecules combine to give one. As a result, the ΔG°_{rxn} for the reaction is approximately zero and the equilibrium constant is close to 1. (Because ketones are more stable than aldehydes, the aldol reaction is unfavorable for ketones.) Under the basic conditions of the aldol condensation reaction, a dehydrated product forms. This reaction has $\Delta H^\circ_{rxn} < 0$ and $\Delta S^\circ_{rxn} > 0$ and the reaction is spontaneous. As a result of this step, the overall ΔG°_{rxn} for the formation of an α, β unsaturated aldehyde is favorable.

23.8 Mixed Aldol Condensation

A **mixed aldol condensation** is the formation of an aldol incorporating two different aldehydes. The reaction gives a mixture of four possible products if both aldehydes have α hydrogen atoms. If only one has α hydrogen atoms, then two products can result. If the carbonyl group of the aldehyde without α hydrogen atoms is more reactive toward nucleophiles, then one product results.

23.9 Intramolecular Aldol Condensation

Intramolecular aldol condensations are more favorable than intermolecular aldol condensations because $\Delta S°_{rxn}$ = 0. Cyclization occurs if the α carbon atom and the second carbonyl carbon atom can bond to form a five- or six-membered ring. If two or more processes can yield these types of rings, it is necessary to consider which process is favored. Remember that the various possible enolates are formed in low concentration under equilibrium conditions. Thus, the enolate that is the better nucleophile attacks the more reactive carbonyl carbon atom and should dominate the product formed. In general, for example, intramolecular aldol condensations where the enolate attacks the carbonyl carbon atom of an aldehyde are favored over addition to the carbonyl carbon atom of a ketone.

23.10 Conjugation in α,β Unsaturated Aldehydes and Ketones

The carbon-carbon double bond in an α,β-unsaturated aldehyde or ketone affects the stability of the carbonyl group and hence its reactivity. The positive charge on the carbonyl carbon atom in the dipolar resonance form can be stabilized by donation of π electrons from the carbon-carbon double bond. As a consequence, there is some partial positive charge at the β carbon atom. This contributing resonance structure decreases the reactivity of the carbonyl carbon atom toward nucleophiles and offers an alternate site for reactivity at the β carbon atom.

23.11 Conjugate Addition Reactions

Addition of compounds represented by H—Nu to the carbon-oxygen double bond of an α,β-unsaturated aldehyde or ketone gives a 1,2-addition product. Addition of the nucleophilic portion of the reagent at the β carbon atom and a hydrogen atom at the carbonyl oxygen atom is a 1,4-conjugate addition. Tautomerization gives a product that appears to be the result of 1,2-addition across the carbon-carbon double bond.

Strong nucleophiles such as the hydride ion of metal hydrides and the carbanion of a Grignard reagent react to give 1,2-addition products with α,β-unsaturated aldehydes and ketones. Weak electrophiles such as cyanide ion, amines, alcohols, and thiols give 1,4-addition products. In contrast to the Grignard reagent, the Gilman reagent also adds to give 1,4-addition products.

23.12 The Michael Reaction and Robinson Annulation

The reaction of an enolate with α,β- unsaturated aldehydes and ketones is termed the **Michael reaction**. The enolate is called a **Michael donor** and the α,β-unsaturated carbonyl compound is called a **Michael acceptor**. The carbonyl group of the Michael donor remains in the condensation product and may undergo an intramolecular aldol condensation reaction with the α carbon atom of the original Michael acceptor. This process, which forms a ring, is termed the **Robinson annulation**.

23.13 α Hydrogen Atoms of Acid Derivatives

The acidity of α hydrogen atoms of acid derivatives is affected by the amount of positive charge on the carbonyl carbon atom, which in turn is affected by the stabilization of that charge by the electronegative atom bonded to the carbonyl carbon atom. The acidity of α hydrogen atoms of esters (pK_a = 25) is less than that of aldehydes and ketones (pK_a = 20), because the oxygen atom of the alkoxy group supplies electrons by resonance to the partially positive carbonyl carbon atom. The resulting delocalization places some positive charge on oxygen and decreases the positive charge on the carbonyl carbon atom. As a consequence, the acidity of α hydrogen atoms of esters is decreased.

Enolates of esters can be prepared in low concentration at equilibrium by using the alkoxide ion in an alcohol corresponding to the alkoxy group contained in the ester. The equilibrium constant for the reaction is approximately 10^{-9}, because the pK_a values of the ester and the alcohol differ by 9. High concentrations of the ester enolate are prepared by using LDA, which is a poor nucleophile and a strong base. The pK_a of diisopropyl amine is 40. Thus, the equilibrium constant for the reaction is approximately 10^{15}.

Dicarbonyl compounds such as ethyl acetoacetate and dimethyl malonate are significantly stronger acids (the pK_a values are 11 and 13, respectively). The increased acidity is the result of delocalization of negative charge in the conjugate base by the additional carbonyl group.

23.14 Reactions at the α Carbon Atom of Acid Derivatives

The hydrogen atom of the α carbon atom of an ester is an **enolizable hydrogen atom**. If the α carbon atom is chiral, the formation of the ester enolate results in loss of optical activity because a racemic mixture is formed when the ester is regenerated. If the equilibrium occurs in a protic solvent containing deuterium in place of hydrogen, then the α hydrogen atoms are exchanged by deuterium.

Ester enolates, formed by the abstraction of the α hydrogen atom by a strong base, are nucleophiles. Lithium diisopropylamide (LDA) is required as the base Reaction of the ester enolate with an alkyl halide forms α alkylated esters. Only primary haloalkanes can be used.

The α hydrogen atoms of carboxylic acids can be replaced by a single halogen atom using bromine and a small amount of phosphorus in the **Hell-Volhard-Zelinsky reaction**. The reaction occurs by bromination of a small amount of the acyl bromide. Under the reaction conditions the bromoacyl bromide is converted into the bromocarboxylic acid. If one equivalent of PBr_3 is used with the Br_2, an α bromoacyl bromide is formed. Reaction of this compound with an alcohol gives an α bromo ester.

23.15 The Claisen Condensation

The reaction of two moles of an ester in the presence of the alkoxide base corresponding to the alkoxyl group of the ester produces a β keto ester. The ester must have two α hydrogen atoms and one equivalent of base is required.

The $\Delta H°_{rxn}$ for the Claisen condensation as given by the balanced chemical equation is slightly positive and the $\Delta S°_{rxn}$ is approximately zero. Although the $\Delta G°_{rxn} > 0$, the reaction as done experimentally produces the conjugate base of the β keto ester. The $\Delta H°_{rxn}$ for the exchange of a proton between the β-keto ester and the alkoxide is sufficiently negative to give a net negative $\Delta H°_{rxn}$ for the combined condensation reaction and proton exchange reaction.

The Claisen reaction occurs in four steps which are:

1. abstraction the α hydrogen atom by the alkoxide ion
2. attack of the carbonyl carbon atom of the ester by an ester enolate
3. ejection of an alkoxide ion from the conjugate base of a hemiketal
4. abstraction of an α hydrogen atom of the keto ester by the alkoxide ion

The addition of dilute acid at the end of the reaction protonates the conjugate base of the keto ester.

An intramolecular variation of the Claisen condensation is the **Dieckmann condensation.** The ester must have two α hydrogen atoms and one equivalent of base is required. The intramolecular reaction has $\Delta S°_{rxn} > 0$, because one mole the diester gives one mole of keto ester and one mole of alcohol.

A mixed Claisen condensation is the formation of a keto ester incorporating two different esters. The reaction gives a mixture of four possible products if both esters have α hydrogen atoms. If only one has α hydrogen atoms, then two products can result. If the carbonyl group of the ester without α hydrogen atoms is more reactive toward nucleophiles, then one product results.

Nonenolizable esters react with the enolate of ketones to give β diketones. The enolate of the ketone attacks the carbonyl carbon atom of the ester in a reaction similar to that of the Claisen condensation.

23.16 Aldol-Type Condensations of Acid Derivatives

Aldol-type condensations occur between a carbonyl compound and the enolate of an ester. The product is either a β-hydroxy ester or its related α,β-unsaturated ester. The **Knoevenagel condensation** occurs between a malonate ester and an aldehyde or ketone. This aldol-type reaction gives an α,β-unsaturated ester. The **Reformatskii reaction** occurs between a zinc enolate of an ester and an aldehyde or ketone to produce a β-hydroxy ester.

23.17 β Dicarbonyl Compounds in Synthesis

The alkylation of acetoacetate or malonate esters is a useful synthetic process that is synthetically equivalent to the direct alkylation of a ketone or an ester, respectively. The acidity of both compounds is higher than that of ketones and esters and allows the abstraction of the α proton by an alkoxide ion to quantitatively form their respective conjugate bases. Alkylation of the conjugate base occurs by its nucleophilic attack of a haloalkane. Hydrolysis of the alkylated acetoacetate leads to decarboxylation of the keto acid and formation of a ketone. Hydrolysis of the alkylated malonate ester leads to decarboxylation of the diacid and formation of an acid.

23.18 Michael Condensation of Acid Derivatives

The reaction of an enolate with α,β-unsaturated aldehydes and ketones is termed the **Michael reaction.** The enolate is called a **Michael donor** and the α,β-unsaturated carbonyl compound, such as 3-buten-2-one, is called a **Michael acceptor.** 1,3-Dicarbonyl derivatives, such as dimethyl malonate, easily form enolates that act as Michael donors. Subsequent hydrolysis of the addition product and decarboxylation yields 1,5-dicarbonyl compounds.

Summary of Reactions

1. Exchange of α Hydrogen Atoms

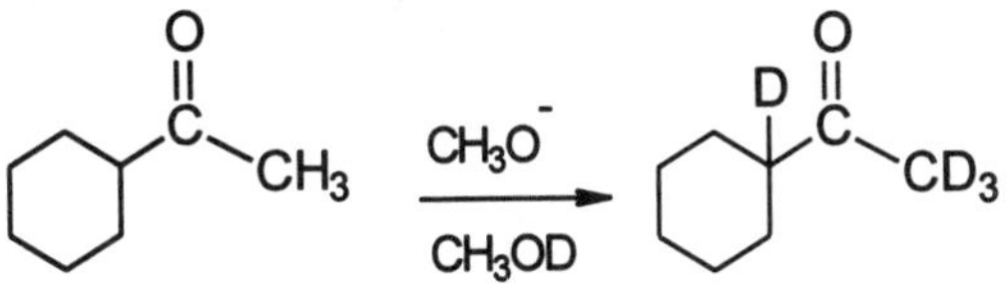

2. Isomerization of Carbonyl Compounds

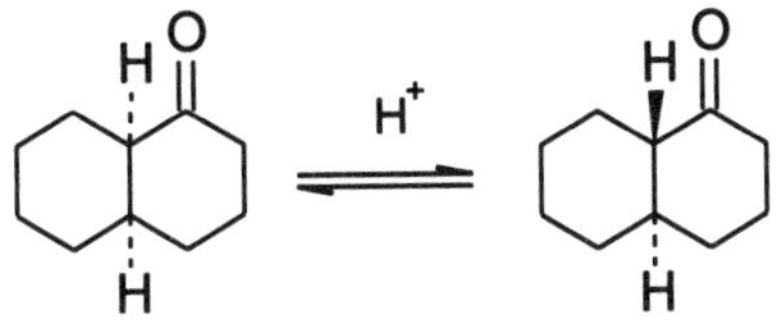

3. α Halogenation of Carbonyl Compounds

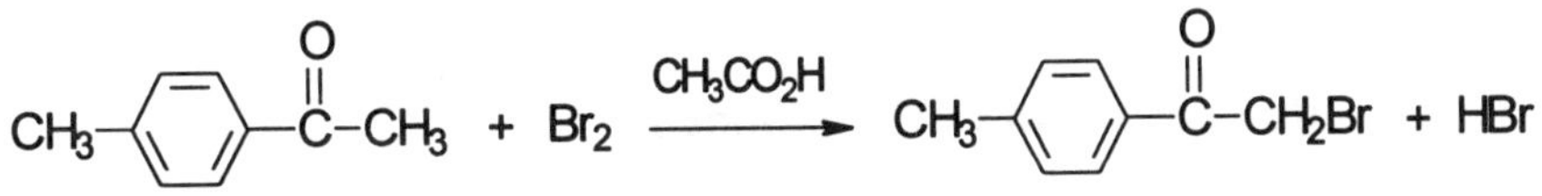

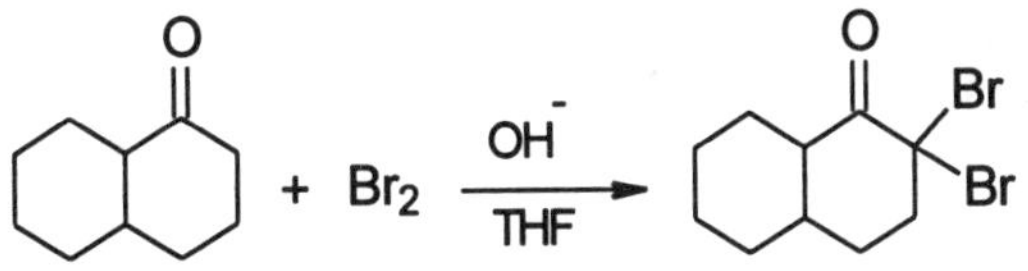

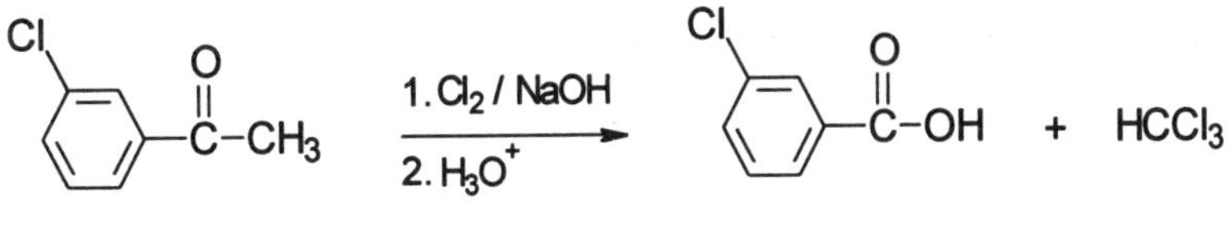

3. α Alkylation of Carbonyl Compounds

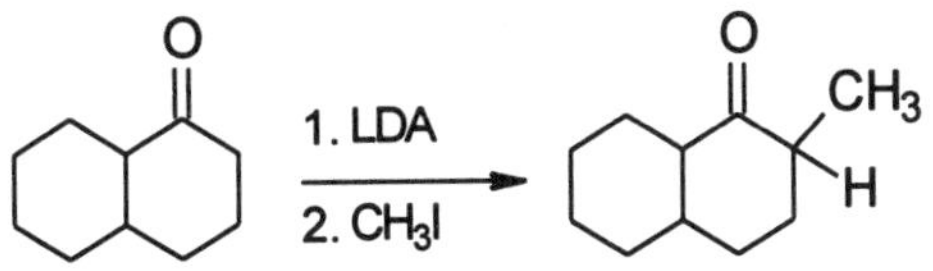

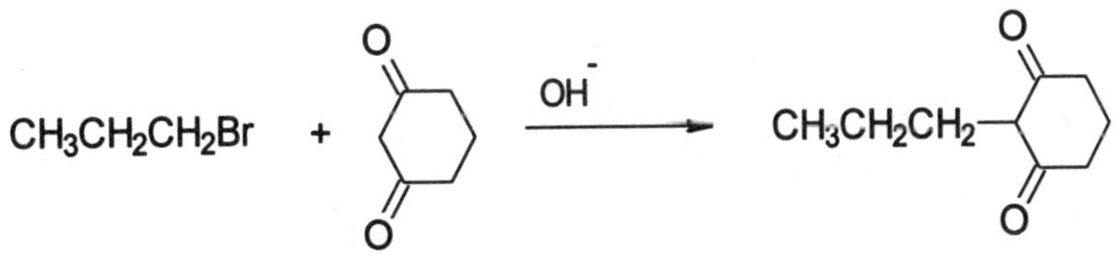

4. Aldol Condensation

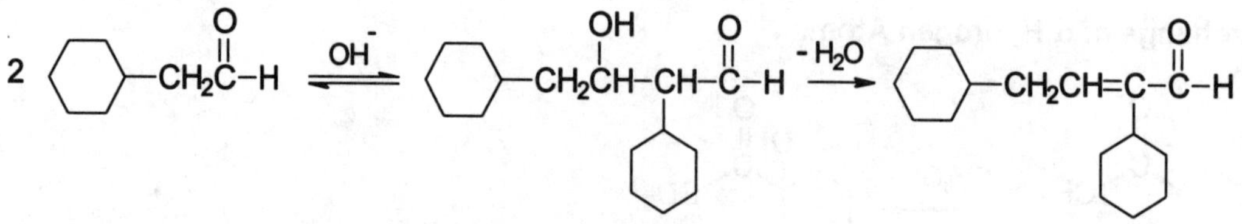

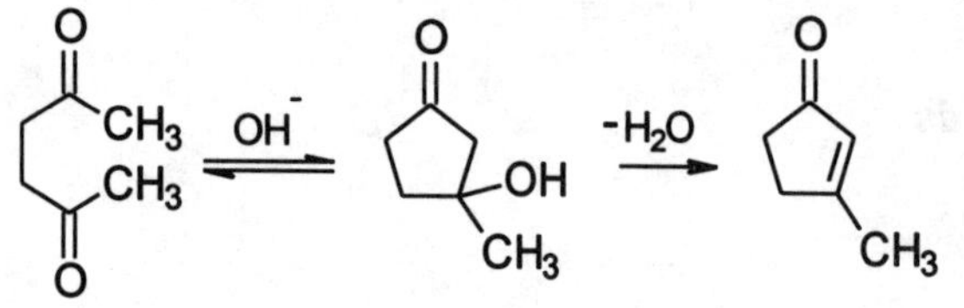

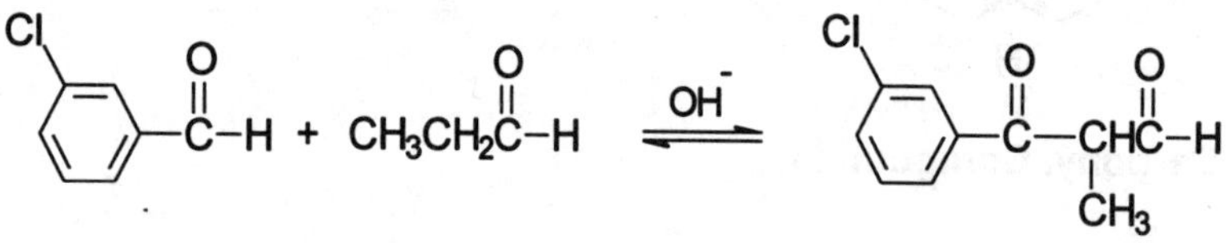

5. Conjugate Addition Reactions of Carbonyl Compounds

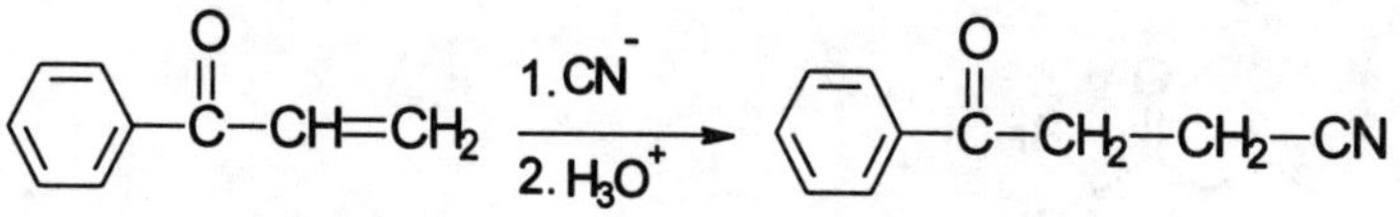

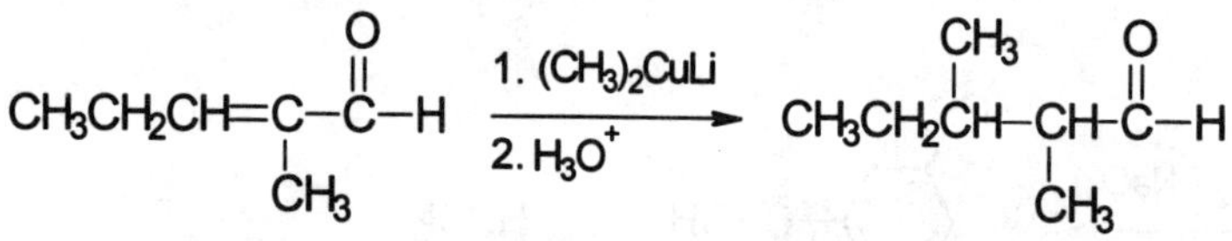

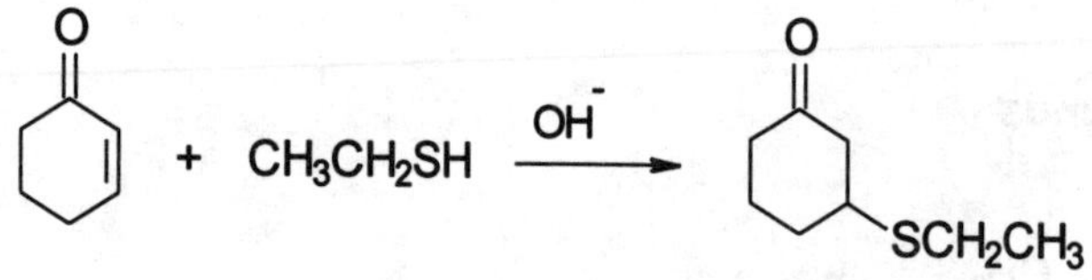

6. Michael Condensation of Carbonyl Compounds

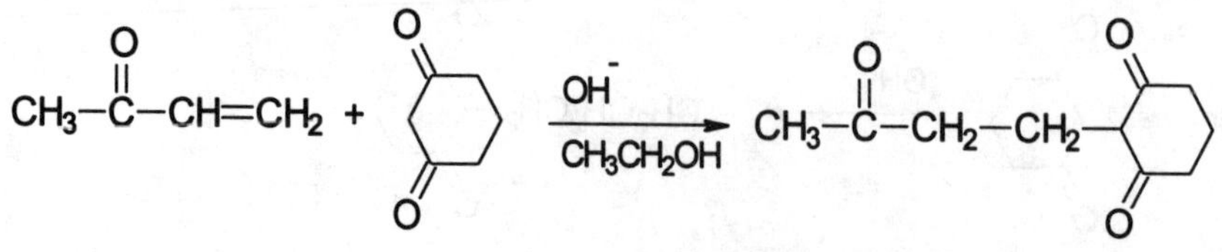

7. Exchange of α Hydrogen Atoms of Esters

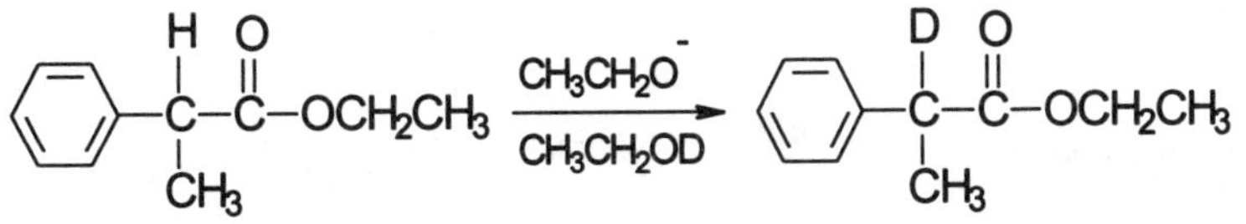

8. α Alkylation of Esters

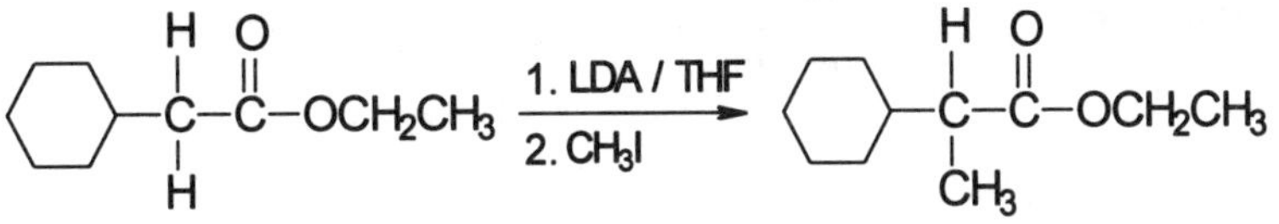

9. α Halogenation of Carboxylic Acids (Hell-Volhard-Zelinsky Reaction)

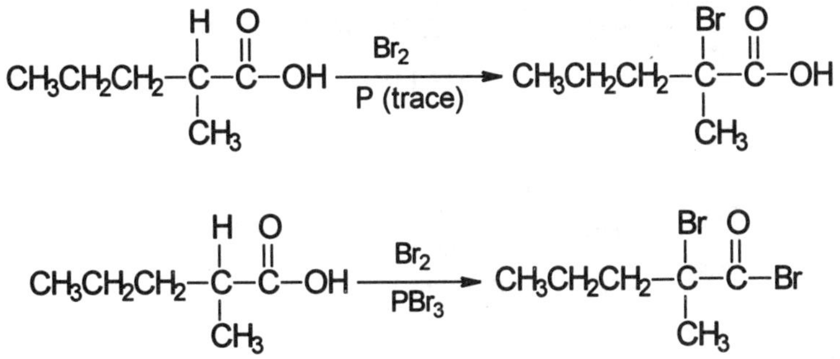

10. Claisen Condensation

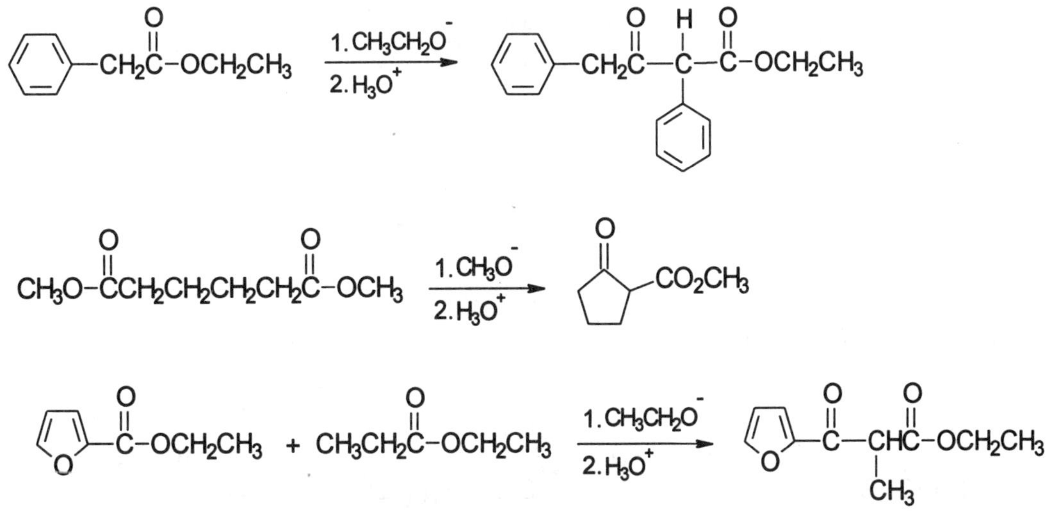

11. Aldol-Type Condensations of Acid Derivatives

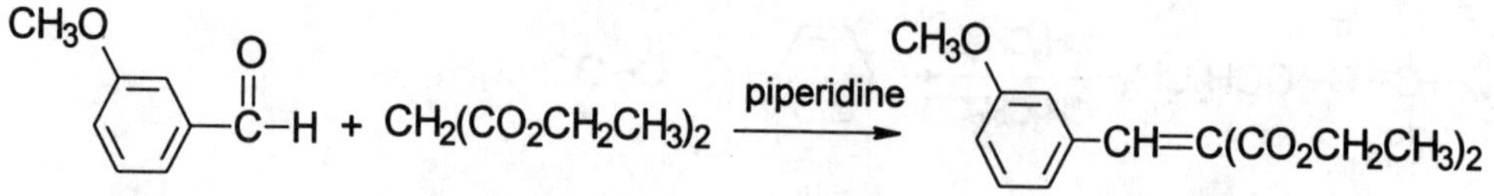

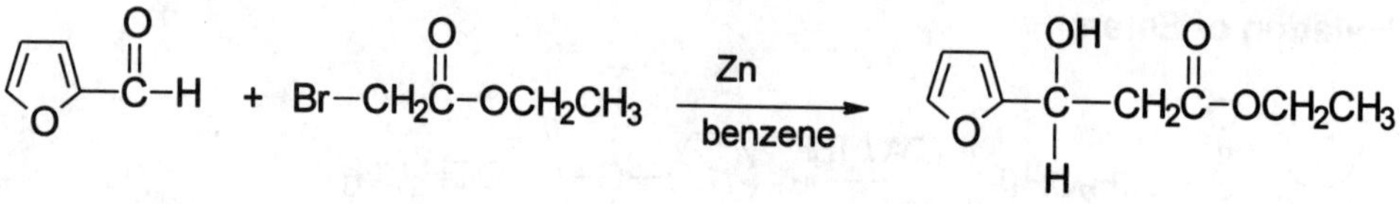

$$CH_3O-C_6H_4-\overset{O}{\overset{\|}{C}}-H + CH_2(CO_2CH_2CH_3)_2 \xrightarrow{\text{piperidine}} CH_3O-C_6H_4-CH=C(CO_2CH_2CH_3)_2$$

$$\text{furan}-\overset{O}{\overset{\|}{C}}-H + Br-CH_2\overset{O}{\overset{\|}{C}}-OCH_2CH_3 \xrightarrow[\text{benzene}]{Zn} \text{furan}-\overset{OH}{\underset{H}{\overset{|}{C}}}-CH_2\overset{O}{\overset{\|}{C}}-OCH_2CH_3$$

12. Acetoacetate Ester Synthesis

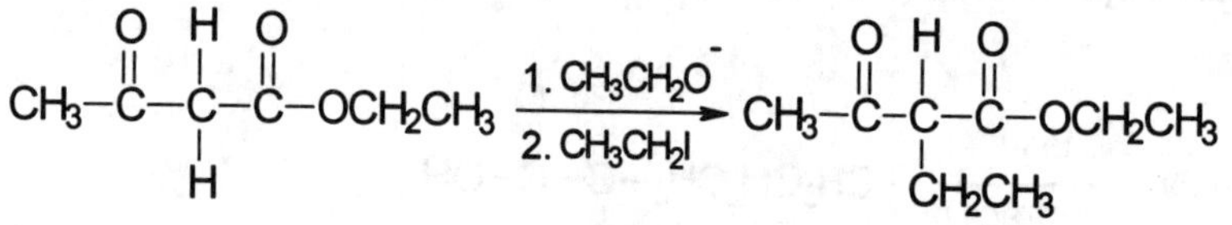

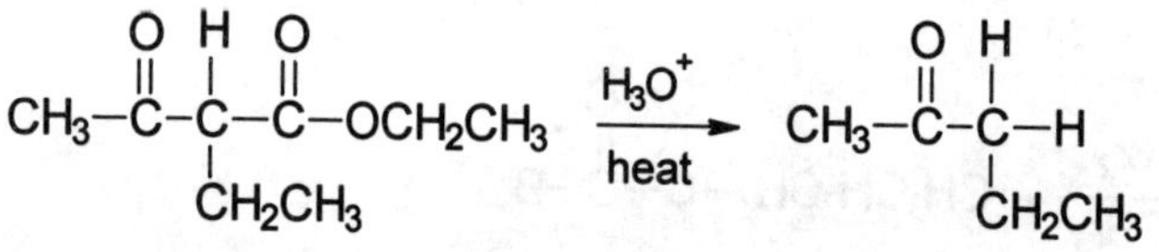

$$CH_3-\overset{O}{\overset{\|}{C}}-\overset{H}{\underset{H}{\overset{|}{C}}}-\overset{O}{\overset{\|}{C}}-OCH_2CH_3 \xrightarrow[\text{2. CH}_3\text{CH}_2\text{I}]{\text{1. CH}_3\text{CH}_2\text{O}^-} CH_3-\overset{O}{\overset{\|}{C}}-\overset{H}{\underset{CH_2CH_3}{\overset{|}{C}}}-\overset{O}{\overset{\|}{C}}-OCH_2CH_3$$

$$CH_3-\overset{O}{\overset{\|}{C}}-\overset{H}{\underset{CH_2CH_3}{\overset{|}{C}}}-\overset{O}{\overset{\|}{C}}-OCH_2CH_3 \xrightarrow[\text{heat}]{H_3O^+} CH_3-\overset{O}{\overset{\|}{C}}-\overset{H}{\underset{CH_2CH_3}{\overset{|}{C}}}-H$$

13. Malonate Ester Synthesis

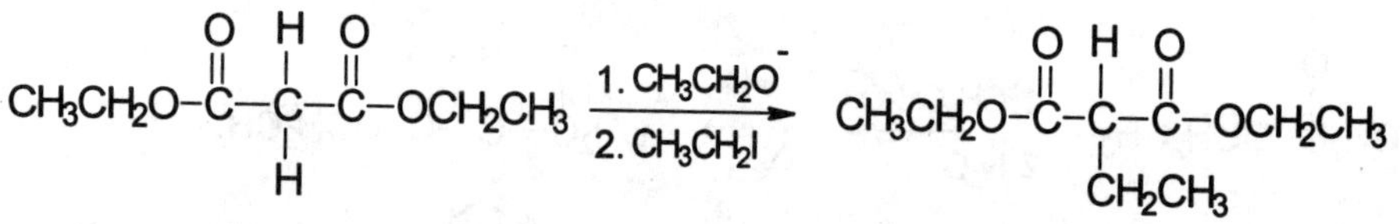

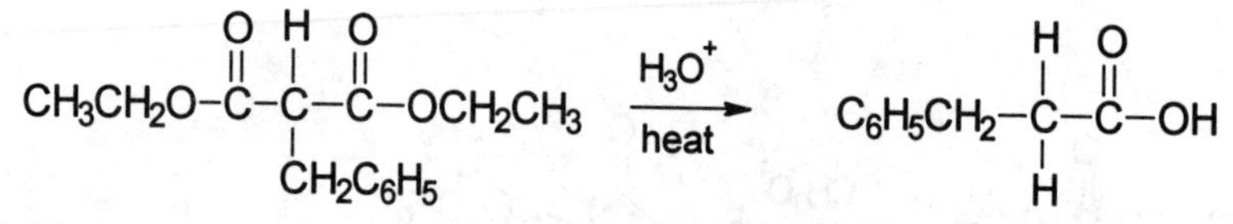

$$CH_3CH_2O-\overset{O}{\overset{\|}{C}}-\overset{H}{\underset{H}{\overset{|}{C}}}-\overset{O}{\overset{\|}{C}}-OCH_2CH_3 \xrightarrow[\text{2. CH}_3\text{CH}_2\text{I}]{\text{1. CH}_3\text{CH}_2\text{O}^-} CH_3CH_2O-\overset{O}{\overset{\|}{C}}-\overset{H}{\underset{CH_2CH_3}{\overset{|}{C}}}-\overset{O}{\overset{\|}{C}}-OCH_2CH_3$$

$$CH_3CH_2O-\overset{O}{\overset{\|}{C}}-\overset{H}{\underset{CH_2C_6H_5}{\overset{|}{C}}}-\overset{O}{\overset{\|}{C}}-OCH_2CH_3 \xrightarrow[\text{heat}]{H_3O^+} C_6H_5CH_2-\overset{H}{\underset{H}{\overset{|}{C}}}-\overset{O}{\overset{\|}{C}}-OH$$

14. Michael Condensation of Acid Derivatives

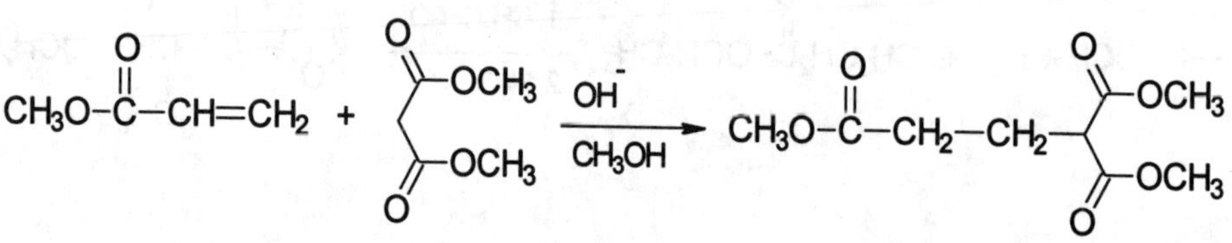

$$CH_3O-\overset{O}{\overset{\|}{C}}-CH=CH_2 + \overset{O}{\overset{\|}{CH_2}}\overset{-OCH_3}{\underset{-OCH_3}{}} \xrightarrow[\text{CH}_3\text{OH}]{OH^-} CH_3O-\overset{O}{\overset{\|}{C}}-CH_2-CH_2-\overset{-OCH_3}{\underset{-OCH_3}{}}$$

Solutions to Exercises

23.1   The K_a of 2,4-pentanedione is 1×10^{-9}, so the compound is a stronger acid than ethanol, whose K_a is approximately 1.3×10^{-16}. Thus the reaction of 2,4-pentanedione with sodium ethoxide has $K = 7.7 \times 10^6$.

23.2   The K_a values of acetonitrile and diisopropylamine are 1×10^{-25} and 1×10^{-40}, respectively. Thus acetonitrile is a stronger acid than diisopropylamine, and the reaction of acetonitrile with lithium diisopropylamide has $K = 1 \times 10^{15}$.

23.3   The K_a values of acetophenone and diisopropylamine are 1×10^{-16} and 1×10^{-40}, respectively. Thus acetophenone is a stronger acid than diisopropylamine, and the reaction of acetophenone with lithium diisopropylamide has $K = 1 \times 10^{24}$.

23.4   The K_a of nitromethane is 6.3×10^{-11}, so the compound is a stronger acid than ethanol, whose K_a is approximately 1.3×10^{-16}. Thus the reaction of nitromethane with sodium ethoxide has $K = 4.8 \times 10^5$.

23.5   Acetone is the stronger acid, which suggests that its conjugate base is more stable. The conjugate base of acetone has a contributing resonance form with a negative charge on a primary carbon atom. For 3-pentanone the charge of the carbanion is on a secondary carbon atom. Because primary carbanions are more stable than secondary carbanions, acetone gives a more stable conjugate base than does 3-pentanone.

23.6   1-Phenyl-2-propanone is the stronger acid, which suggests that its conjugate base is more stable. The conjugate base of 1-phenyl-2-propanone has a contributing resonance form with a negative charge on a secondary carbon atom that is also benzylic. Thus, the charge is delocalized and this conjugate base is resonance stabilized. For the conjugate base of acetone, the charge is localized on a primary carbon atom, so it is less stable.

23.7   Both enols have the double bond within their respective rings in the enol form, and both have the same degree of substitution. However, the double bond of the enol of cyclobutanone increases the strain energy of the small ring, so this enol is less stable compared to the enol of cyclohexanone. Thus, cyclohexanone has a larger percent of enol at equilibrium.

23.8   Both enols have the double bond within a six membered ring in the enol form. However, the double bond of the enol of 1,3-cyclohexanedione is conjugated with the second carbonyl group and is resonance stabilized. The enol of 1,4-cyclohexanedione is not resonance stabilized. Thus, 1,3-cyclohexanedione has a larger percent of enol at equilibrium.

23.9   There is only one α-hydrogen atom and it is located at the C-4 atom. However, two enols that are geometric isomers can form. The most stable enol has the large *tert*-butyl trans to the methyl group.

23.10   There are two α-carbon atoms, the C-2 and C-5 atoms. The double bond of the enol at C-2 has a higher degree of substitution and is the more stable.

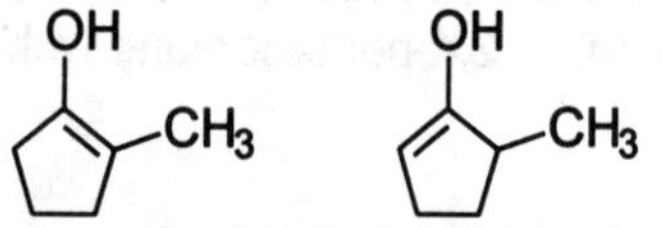

23.11   The enol of 1,2-diphenylethanone has extended conjugation between the two phenyl rings via the carbon-carbon double bond. The enol of 1,3-diphenyl-3-propanone has only one ring conjugated with a carbon-carbon double bond. 1,2-Diphenylethanone has a larger percent of enol at equilibrium.

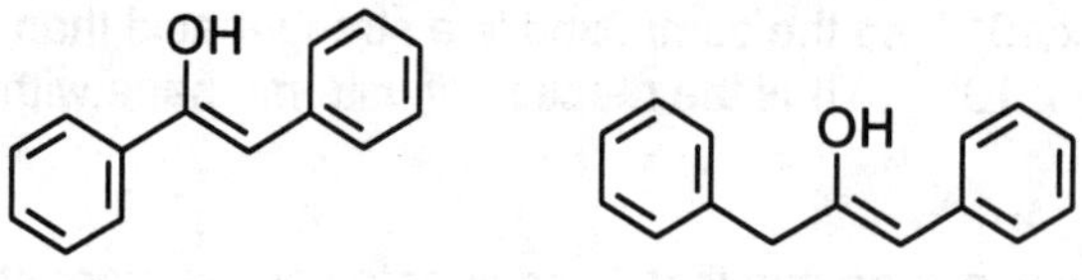

23.12   The keto form has its carbonyl group conjugated with a benzene ring. However, the enol tautomer has a double bond that is part of the aromatic ring system of phenanthrene. The resonance stabilization of the fused ring system favors formation of the enol, which is a phenol.

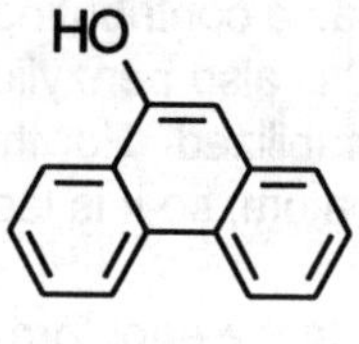

23.13   In each case there is only one α-carbon atom with enolizable hydrogen atoms, so only one enolate ion forms.

(a) $(CH_3)_3C$ (b) 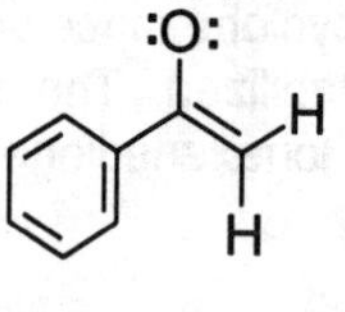   (c)

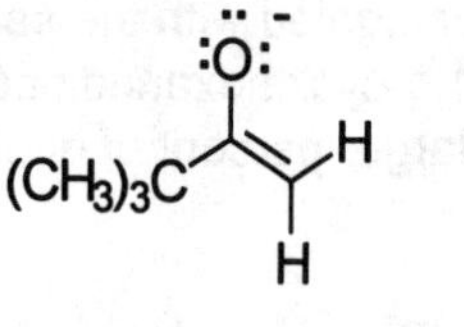

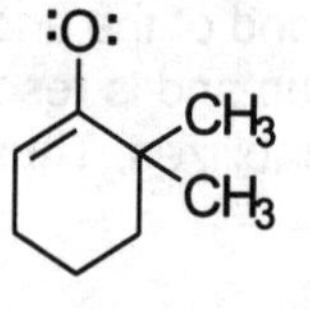

23.14   (a) There are two α-carbon atoms with enolizable hydrogen atoms. The C-1 atom gives a single enolate. The C-3 atom gives a pair of enolates that are geometric isomers. The more substituted double bond of the enolates at the C-3 atom is more stable than the enolate with a double bond at the C-1 atom. Of the two enolates involving the C-3 atom, the isomer with the alkyl groups trans to one another is the more stable. The order of most stable to least stable is shown on the following page.

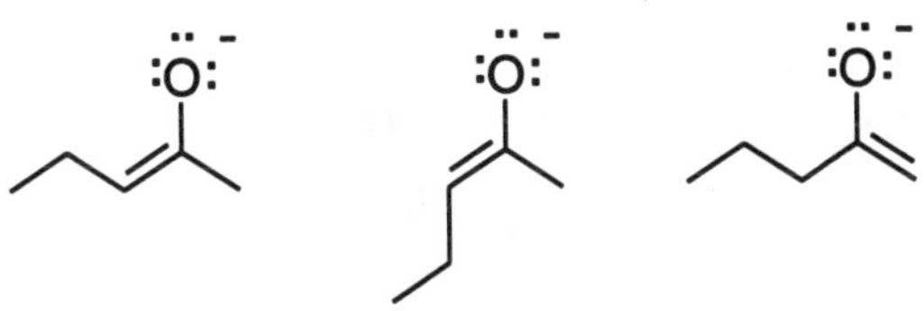

(b) There are two α-carbon atoms with enolizable hydrogen atoms. The C-3 atom gives a single enolate. The C-1 atom gives a pair of enolates that are geometric isomers. The more substituted double bond of the enolates involving the C-1 atom is more stable than the enolate with a double bond at the C-3 atom, because the double bond is conjugated with the benzene ring. Of the two enolates involving the C-3 atom, the isomer with the alkyl and aryl groups trans to one another is the more stable. The order of most stable to least stable is shown below.

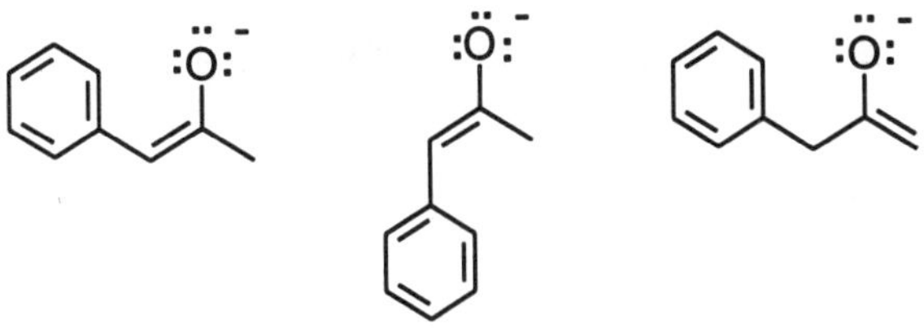

(c) There are two enolates. The enolate involving the C-6 atom has a localized double bond. The enolate involving the C-2 atom is conjugated with the second carbonyl group and is resonance stabilized. The order of more stable to less stable is shown below.

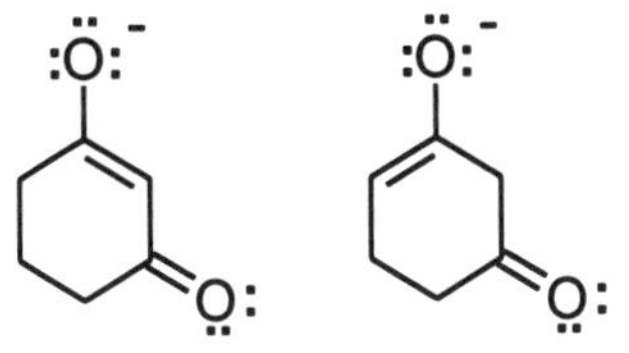

23.15

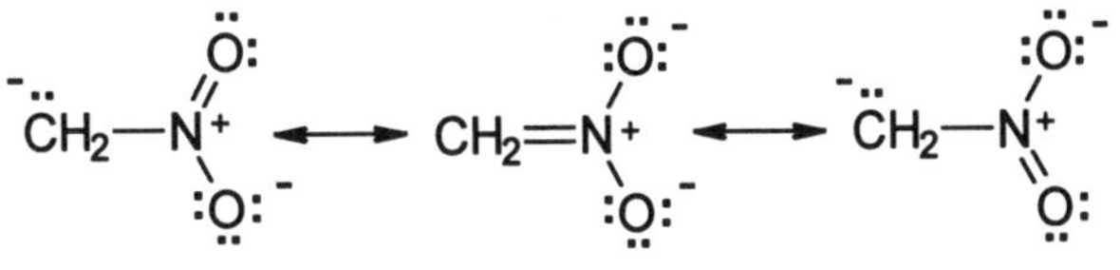

23.16

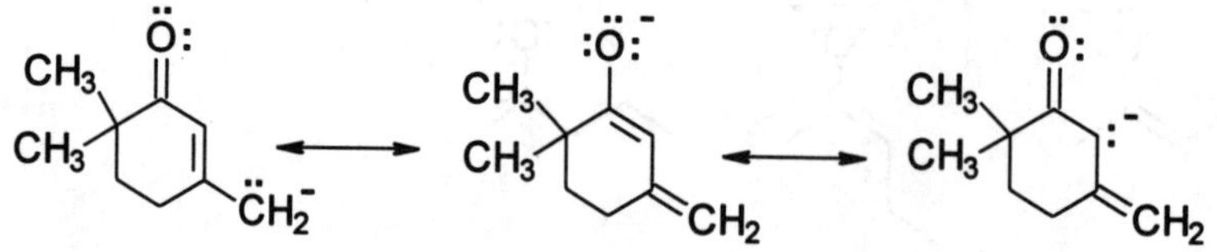

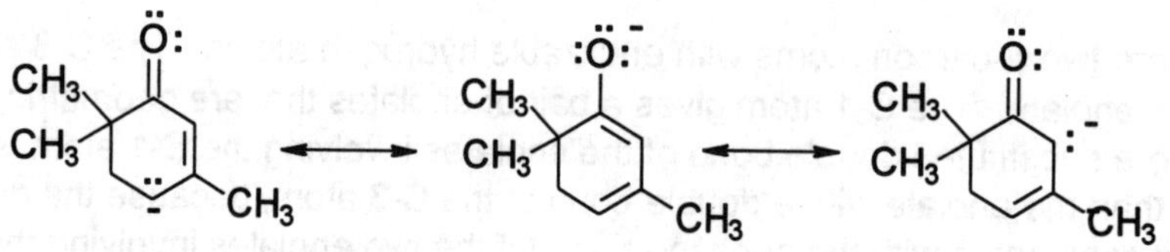

23.17 The double bond of each enolate has the same degree of substitution.

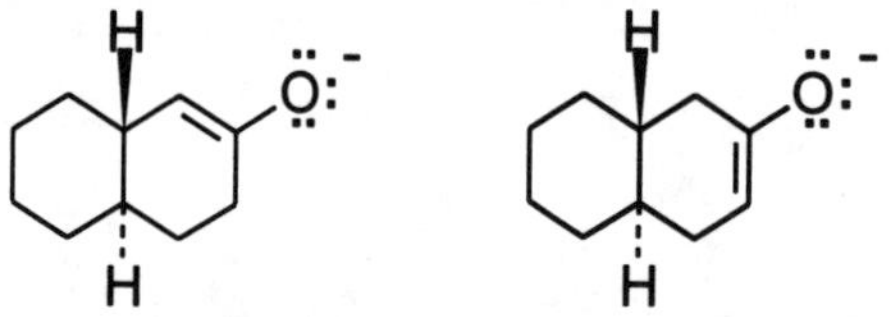

23.18 There are two α-carbon atoms with enolizable hydrogen atoms. They are located at the C-2 and C-5 atoms. The double bond of the enolate to the C-2 atom has a higher degree of substitution and is the more stable. It constitutes 94% of the mixture of enolates.

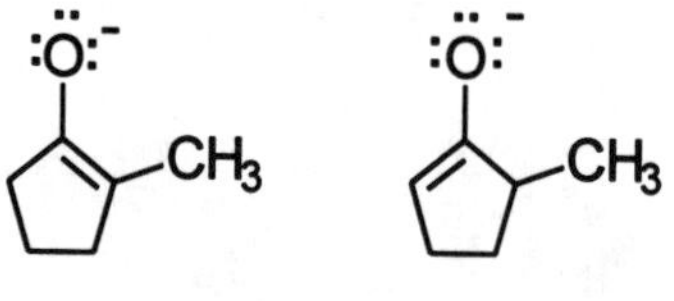

23.19 There are two α-carbon atoms with enolizable hydrogen atoms but they are equivalent. The C-2 atom gives a pair of enolates that are geometric isomers. Of the two enolates, the isomer with the alkyl groups trans to one another is the more stable and constitutes 84% of the enolate mixture.

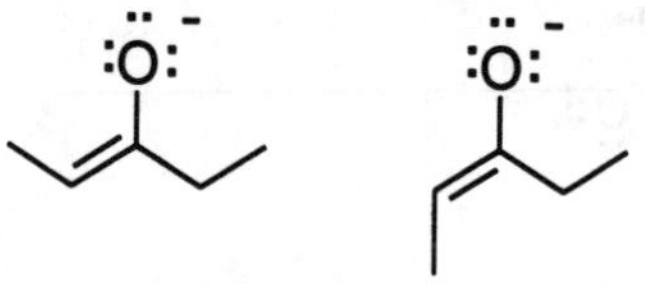

23.20 There is only one α-carbon atom with enolizable hydrogen atoms. The C-3 atom gives a pair of enolates that are geometric isomers. Of the two enolates, the isomer with the alkyl groups trans to one another is the more stable. However, in this compound the alkyl groups are a *tert*-butyl and a methyl group, so the steric hindrance is larger than for the ethyl and methyl groups of the enolate of 3-pentanone. Thus, the ratio of the two isomers is much larger and the trans isomer will constitute more than 84% of the enolate mixture.

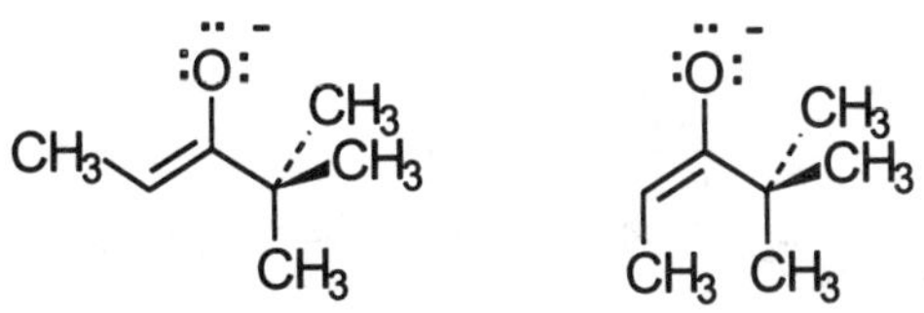

23.21 Formation of an enolate occurs by abstraction of a proton from the bridgehead position by ethoxide ion. Protonation in a reverse step by ethanol can occur to give either the original ketone or its isomer. The isomer is a ketone derived from *trans*-decalin, which has its rings fused by diequatorial bonds and is more stable than *cis*-decalin.

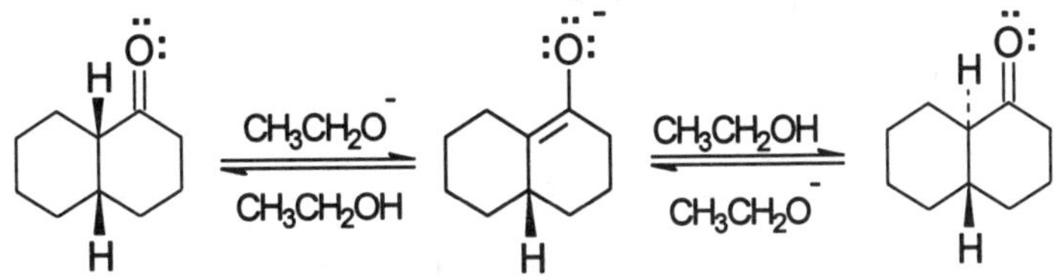

23.22 The enolate formed is resonance stabilized, with charge distributed between a secondary and a tertiary carbon atom. Protonation can occur to give the original ketone or the isomer. This isomeric ketone is resonance stabilized by the extended conjugation of two double bonds with the carbonyl group, so it is more stable than the original ketone, whose carbonyl group is not conjugated with a double bond.

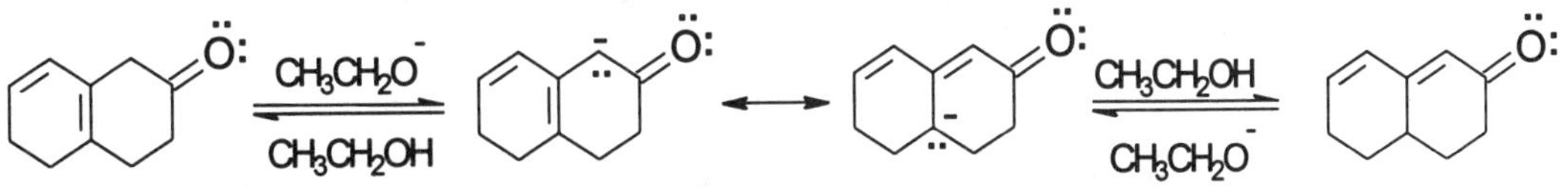

23.23 The enolate formed is resonance stabilized, with charge distributed between two secondary carbon atoms. Protonation can occur to give the original ketone or the isomer. This isomeric ketone is resonance stabilized by conjugation of the ketone with the double bond, so it is more stable than the original nonconjugated ketone.

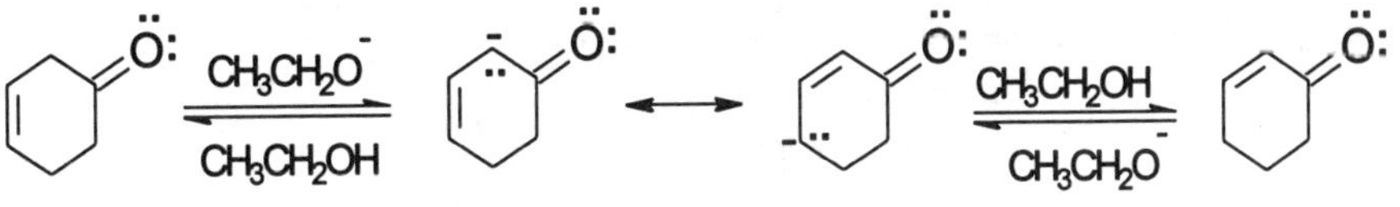

23.24   The enolate formed is resonance stabilized, with charge distributed between two secondary carbon atoms. Protonation can occur to give the original ketone or the isomer. This isomeric ketone is not resonance stabilized by conjugation of the ketone with the double bond, so it is less stable than the original conjugated ketone. Nevertheless, at equilibrium there is a low concentration of the less stable 2-methyl-3-cyclopentenone.

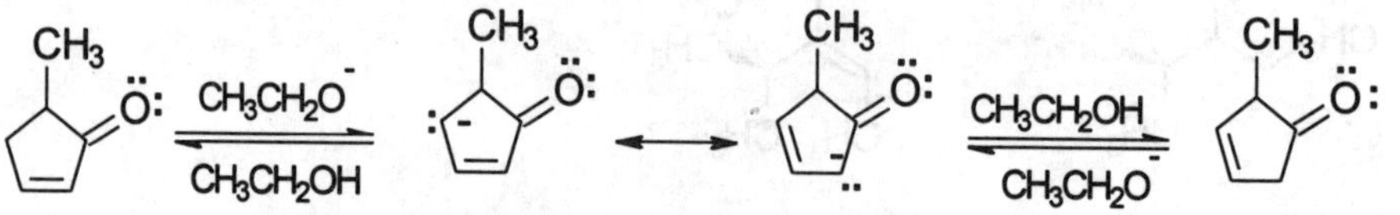

A proton can be abstracted from either the C-2- or C-5 atom of the nonconjugated product 2-methyl-3-cyclopentenone. Loss of a proton from the C-5 atom generates the resonance stabilized enolate of the above reaction and leads to formation of the original 5-methyl-2-cyclopentenone. However, abstraction of a proton from C-2 gives a different resonance stabilized enolate with negative charge distributed between the C-2 and C-4 atoms.

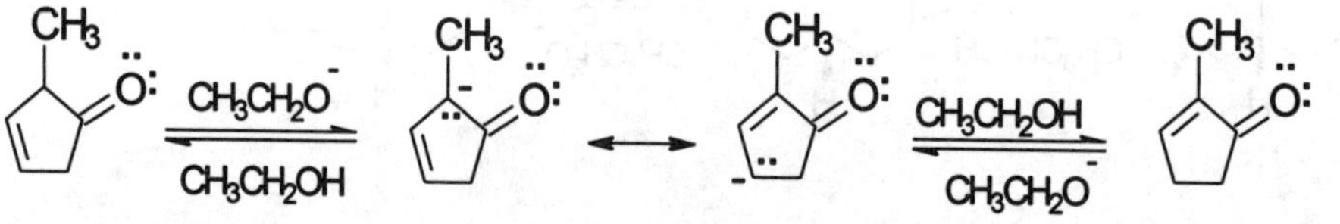

Protonation can occur to give the original 2-methyl-3-cyclopentenone or an isomeric conjugated ketone. 2-Methyl-2-cyclopentenone is the most stable of the three isomers in equilibrium in this series of reactions, because it is both conjugated and the more highly substituted ketone. Thus, 5-methyl-2-cyclopentenone is isomerized by the base into 5-methyl-3-cyclopentenone, which in turn is isomerized into 2-methyl-2-cyclopentenone. Note that although the number of the methyl group changes, it does not rearrange. The numbering is controlled by the location of the double bond relative to the ketone.

23.25   The chiral center of (R)-3-methyl-1-phenyl-1-pentanone is at the β carbon atom relative to the ketone, so its hydrogen atom is not sufficiently acidic to be abstracted by ethoxide ion. Thus, its stereochemistry is not affected in the basic solution. The chiral center of the isomeric (R)-2-methyl-1-phenyl-1-pentanone is at the α carbon atom relative to the ketone, so its hydrogen atom is abstracted to form an enolate. Reprotonation can give either the (R) or (S) isomer.

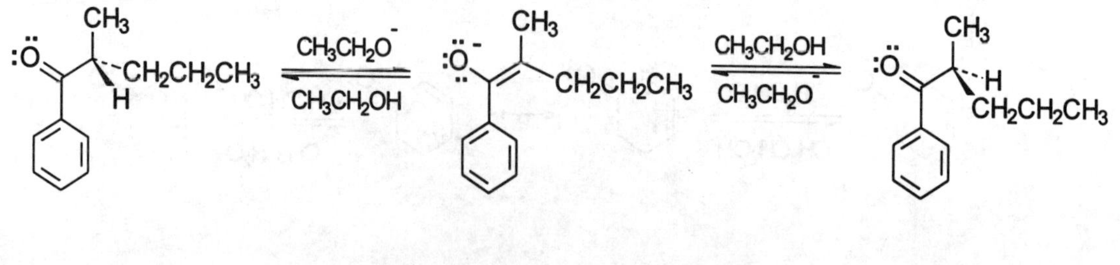

23.26 **(a)** The chiral center of (R)-2-methylcyclohexanone is at the α carbon atom relative to the ketone, and it is abstracted to form an enolate. Reprotonation can give either the (R) or (S) isomer, so the solution gradually loses its optical activity.

(b) The chiral center of (R)-3-methylcyclohexanone is at the β carbon atom relative to the ketone, and its hydrogen atom is not sufficiently acidic to be abstracted by ethoxide ion. Thus, its stereochemistry is not affected in the basic solution.

(c) There is no hydrogen atom at the chiral center of (R)-2-methyl-2-ethylcyclohexanone, so an enolate can form only at the C-6 atom, which does not affect the chirality of the molecule.

23.27 The α carbon atom of 7-bicyclo[2.2.1]heptanone is at a bridgehead position. The carbanion if formed could not form the carbon-carbon double bond of the enolate because the orbital of the carbanion and those of the carbonyl group are perpendicular to one another.

However, the proton at the C-3 position of 2-bicyclo[2.2.1]heptanone is acidic, because an enolate ion can form and is stabilized by resonance. Note that the proton at the bridgehead position is not acidic in this compound either. Deuterium exchange will occur at the C-3 position.

23.28 The α positions in this molecule are C-3 and C-1. There is no proton at the C-3 position of 3,3-bicyclo[2.2.1]heptanone. The proton at the C-1 atom is at a bridgehead position and is not acidic because a resonance stabilized enolate would require a double bond at the bridgehead position, which is geometrically impossible.

23.29 Compound I has two methylene carbon atoms at the α positions, so four hydrogen atoms can be exchanged by four deuterium atoms. Compound II has two C—H bonds at a methylene carbon atom and a C—H bond at a tertiary center. Thus, only three deuterium atoms can be incorporated into compound II.

23.30   2-Pentanone has a methyl carbon atom and a methylene carbon atom at the α positions, so five hydrogen atoms can be exchanged by five deuterium atoms. 3-Pentanone has two methylene carbon atoms at the α positions, so only four hydrogen atoms can be exchanged by deuterium atoms.

23.31   The most acidic proton of 3-methyl-2,4-pentanedione is at the C-3 atom because the enolate formed is resonance stabilized by conjugation with the second carbonyl group. Thus, this proton is exchanged rapidly.

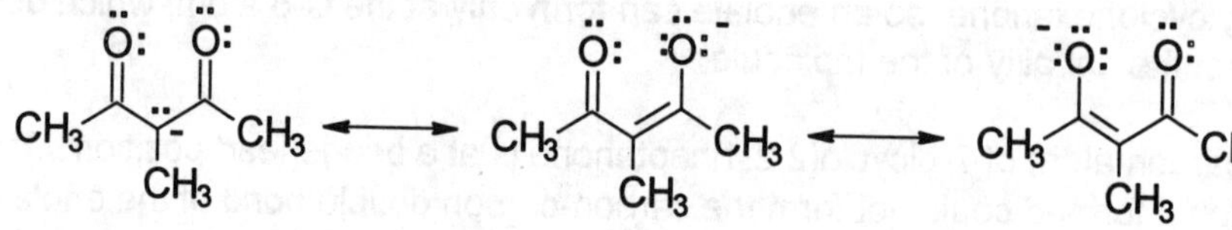

The protons of the C-1 and C-5 atoms can also be exchanged, but they are not as acidic as the proton at the C-3 atom. The enolate is not resonance stabilized. Eventually all six hydrogen atoms of the two equivalent methyl groups are exchanged by deuterium atoms.

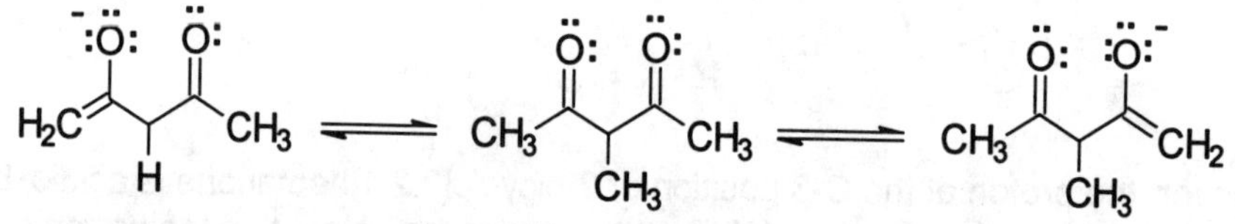

23.32   There are four possible sites where protonation can occur based on the resonance forms that can be written for the three isomeric enolates.

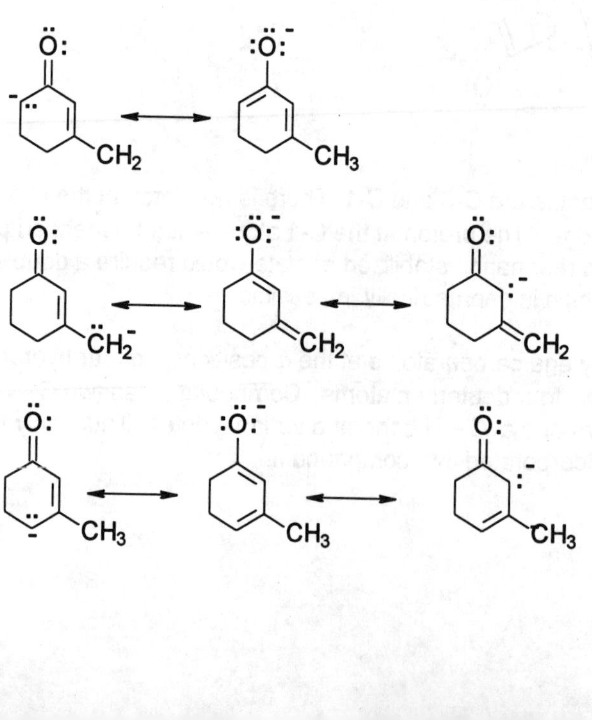

The two hydrogen atoms are C-6 are exchanged via the first enolate shown. The hydrogen atom at C-2 as well as those of the methyl group are exchanged via the second enolate. The two hydrogen atoms at C-4 are exchanged via the third enolate.

23.33 The most stable enol is produced by protonating one of the two equivalent carbonyl oxygen atoms and forming a double bond between the C-2 and C-3 atoms. Bromination of the enol thus occurs at the C-3 atom.

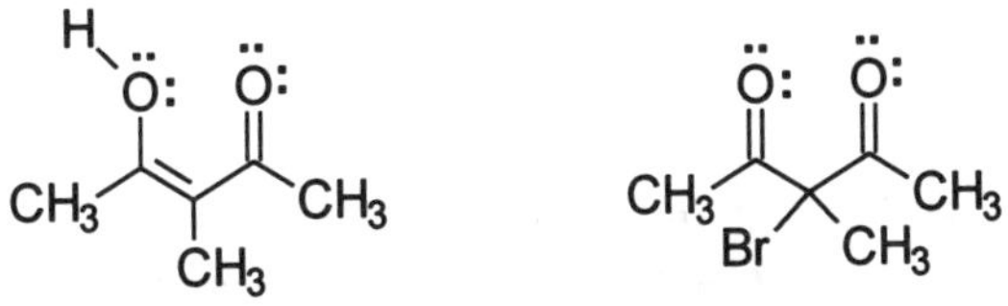

23.34 There are two possible enols. The one with a double bond between the C-2 and C-3 atoms is the more stable because it is more highly substituted. This enol give the major product. The enol with a double bond between the C-1 and C-2 atoms accounts for the minor product.

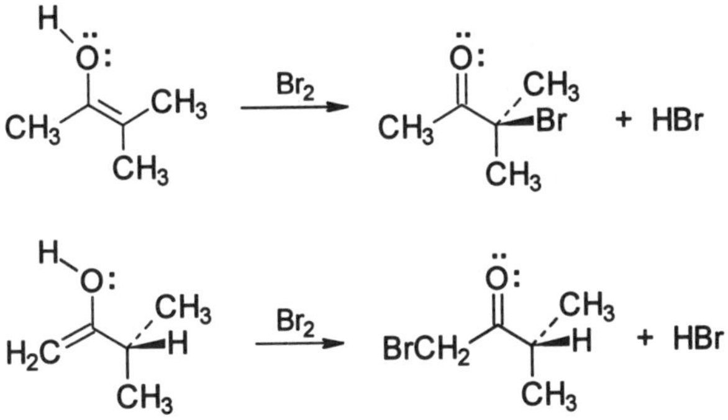

23.35 The formation of CHI_3 requires a methyl group bonded to the carbonyl carbon atom. Only (c) and (d) have such structures.

23.36 Adipic acid is hexanedioic acid, a six-carbon acid. The loss of two carbon atoms in the molecular formula of the product relative to the reactant, as well as the formation of a dicarboxylic acid, means that the reactant was a diketone with methyl groups bonded to each carbonyl carbon atom before the bromination reaction.

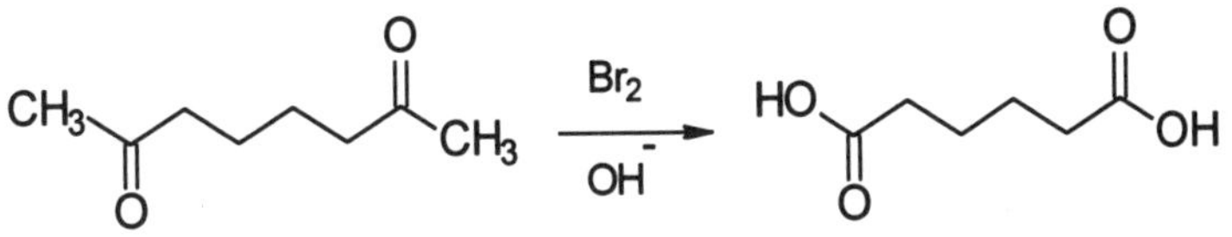

23.37 An enolate cannot form because a double bond cannot be located at the bridgehead carbon atom. The orbital containing the electron pair of a possible bridgehead carbanion is approximately perpendicular to the plane of the π bond of the carbonyl groups, which is not suitable for π bond formation. Either of the two methylene groups that are α to the two carbonyl groups could be brominated.

23.38 Bromination occurs at the site of the most acidic hydrogen atom, which in this case is the methylene group which is secondary rather than the bridgehead carbon which is tertiary. The subsequent bromination occurs at a faster rate than the first and occurs at the site of the first bromination.

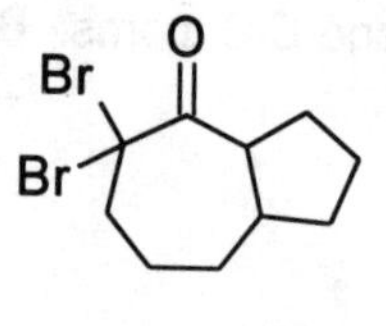

23.39 The methylene groups at the C-2 and C-6 atoms are equivalent. The enol formed under acid conditions has a double bond that may be attacked by the electrophilic bromine from either side of the ring. The isomeric *cis*- and *trans*-2-bromo-4-*tert*-butylcyclohexanone compounds are formed.

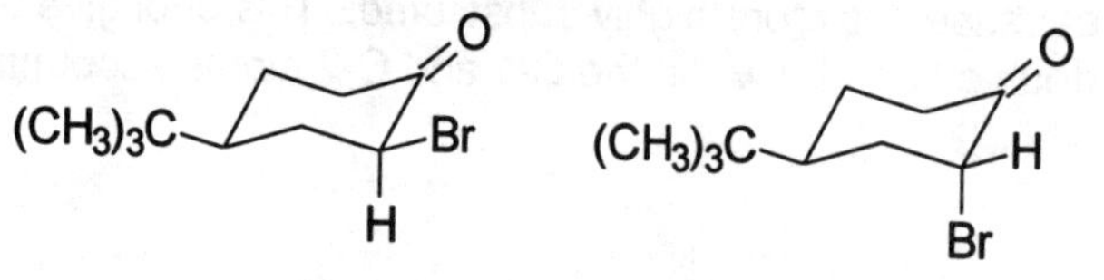

23.40 There are two possible enols and each can react with the electrophilic bromine from either side of the ring.

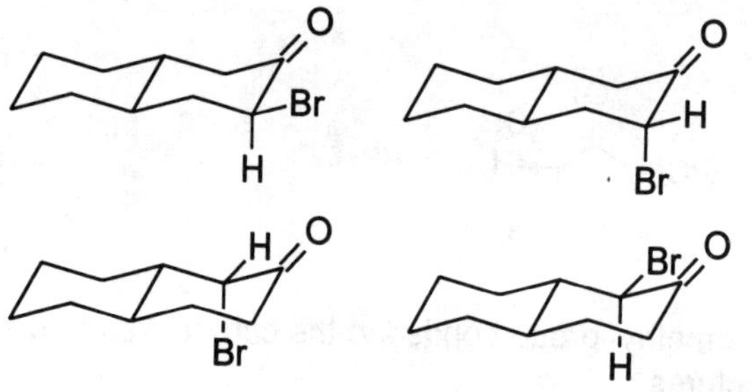

23.41 Alkylation occurs at the C-4 atom, whose hydrogen atom is abstracted by base to form an enolate ion. The alkyl iodide is primary and undergoes substitution without significant competing elimination reaction.

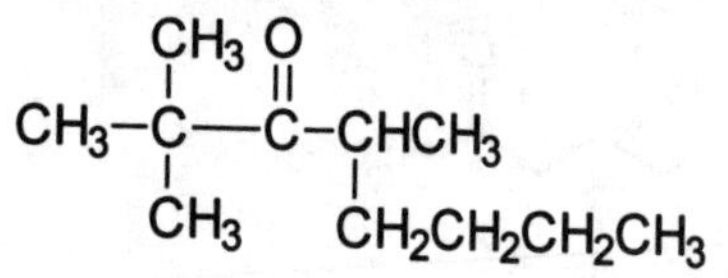

23.42 The alkyl bromide is secondary, so it undergoes an elimination reaction in the presence of the strong conjugate base formed abstraction of the proton at C-2 of cyclohexanone. Protonation of the conjugate base gives the original ketone.

23.43 The methylene groups at the C-2 and C-6 atoms are equivalent. The enolate formed under basic conditions has electron density at the C-2 atoms, which serves as a nucleophile and displaces iodide ion from iodoethane. The trigonal pyramidal carbanion can invert, so displacement can occur with the newly formed bond on either side of the ring. Isomeric cis- and trans compounds result.

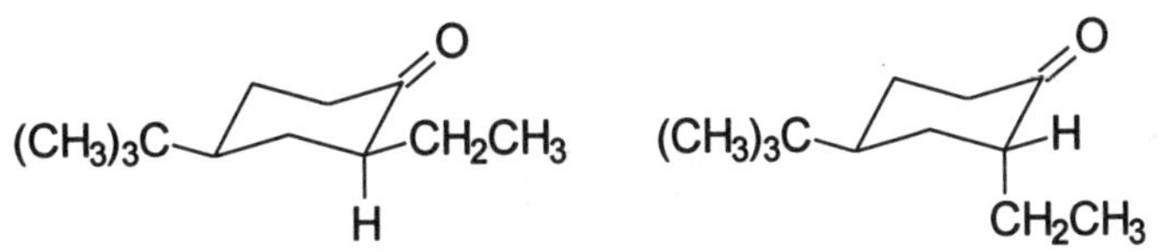

23.44 There are two possible enolates, because two nonequivalent methylene groups are located α to the carbonyl group. Each can react with methyl iodide, and the methyl group may bond to either side of the ring.

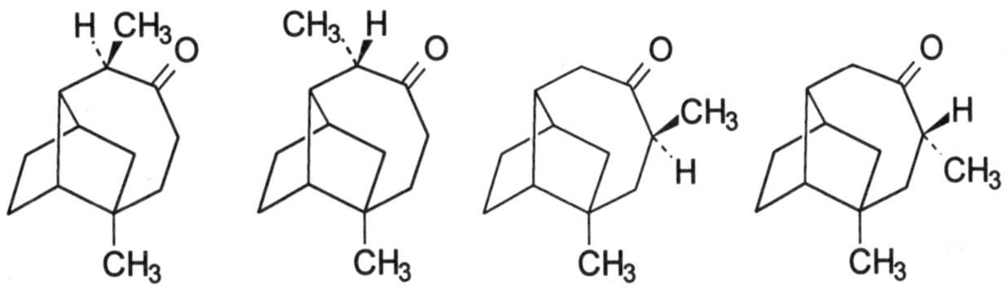

23.45 The enolate has some negative charge at the C-1 atom and that atom intramolecularly displaces bromide ion from the C-6 atom.

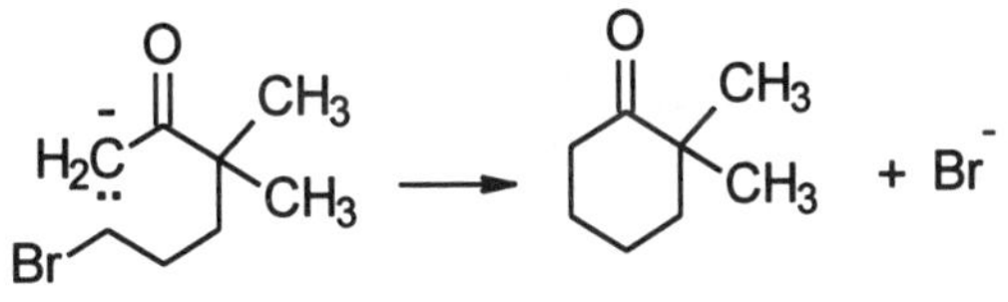

23.46 The enolate formed by abstraction of a hydrogen atom of the methylene group of the α carbon atom intramolecularly displaces bromide ion from the primary carbon atom.

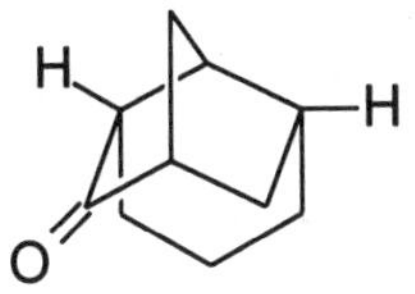

23.47 Compound II is the more stable because the double bond is more highly substituted. This compound is the major product because the enolates formed are in equilibrium with the weak base. The major product thus results from reaction of the major enolate in solution. The products are shown on the following page.

23.48   There are two isomeric enolates. The enolate derived from abstraction of a proton at the C-3 atom is more highly substituted than the enolate derived from abstraction of a proton at the C-1 atom. The major product is the isomer with the double bond at the C-2 atom.

23.49   The strong base removes a proton from the more acidic secondary carbon atom at a faster rate than from the tertiary carbon atom. Because the base is very strong, neither enolate reverts to the carbonyl compound and the two are not in equilibrium. The product is the result of kinetic control.

23.50   The strong base removes a proton from the more acidic primary carbon atom at a faster rate than from the secondary carbon atom. Because the base is very strong, neither enolate reverts to the carbonyl compound and the two are not in equilibrium. The major product, which has a double bond at C-1, is the result of kinetic control.

23.51 The strong base removes a proton from the more acidic secondary carbon atom at a faster rate than from the tertiary carbon atom. Because the base is very strong, neither enolate reverts to the carbonyl compound and the two are not in equilibrium. The major alkylation product is derived from the substitution at the original C-6 atom. The compound is named 2-benzyl-6-methylcyclohexanone because the benzyl group takes alphabetic preference over the methyl group.

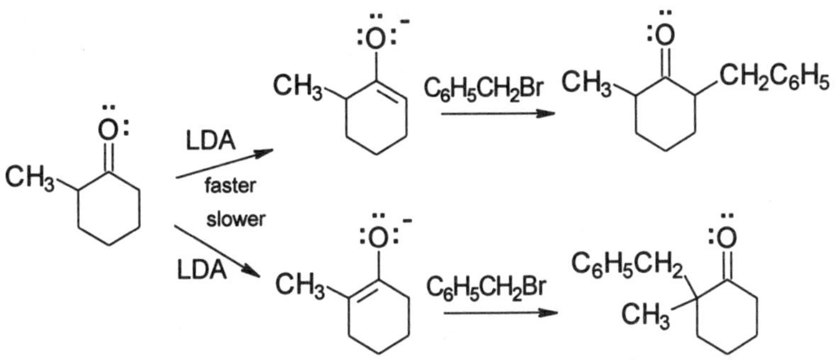

23.52 Reaction of 2-methylcyclohexanone with benzyl bromide using triethylamine as a base would generate two isomeric enolates, which exist in equilibrium with the ketone in the weak base. The major enolate is the more highly substituted compound with a double bond at the C-2 atom. Reaction of this enolate with benzyl bromide gives 2-benzyl-2-methylcyclohexanone as the major product.

23.53 The strong base removes a proton from the more acidic primary carbon atom at a faster rate than from the tertiary carbon atom. Intramolecular attack of the primary carbanion at the primary bromoalkane center forms a seven-membered ring. Although formation of seven-membered rings is generally not favored, this process occurs because four of the bonds are restricted by the five-membered ring and there is a larger probability of ring closure.

23.54 The weaker base used can give two enolates that are in equilibrium with the ketone, so two products can result. The enolate with a double bond to the atom of the five-membered ring can intramolecularly react to produce a second five-membered ring.

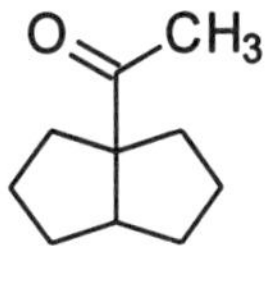

23.55 **(a)** 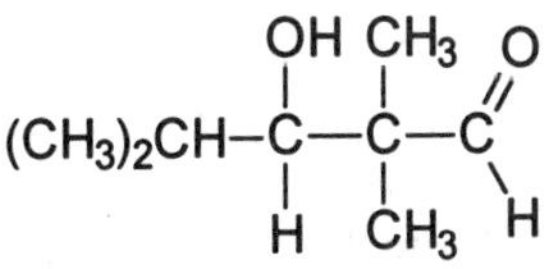   **(b)**

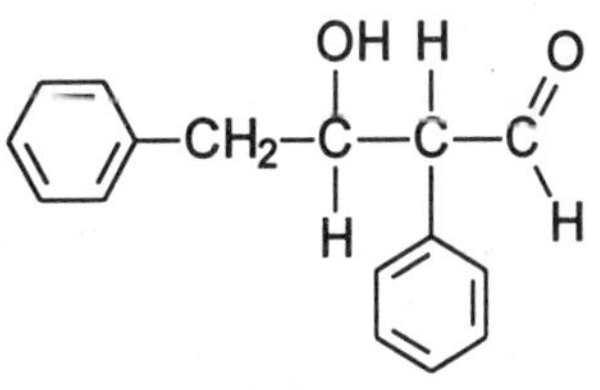

(c) 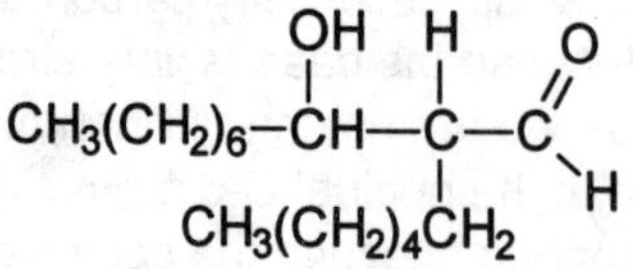 $CH_3(CH_2)_6-\underset{\underset{CH_3(CH_2)_4CH_2}{|}}{\overset{\overset{OH}{|}}{CH}}-\underset{\underset{H}{|}}{\overset{\overset{H}{|}}{C}}-\overset{\overset{O}{\|}}{C}\diagdown_H$

23.56 (a) cyclohexanone and ethanal   (b) cyclohexanone and 2,4-pentanedione
    (c) acetophenone and 3-pentanone   (d) benzaldehyde and acetophenone

23.57

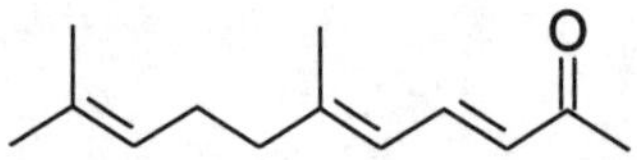

23.58 Either the C-1 or C-3 atoms of 2-butanone can form an enolate. Subsequent reaction of each enolate at the carbonyl carbon of citral gives an unsaturated product. The two possible products are:

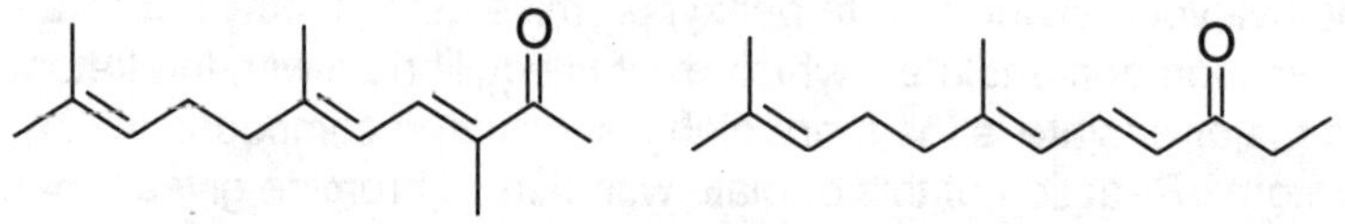

23.59 The compound is 3-hydroxy-2,2-dimethylpropanal, which is the product of a mixed aldol condensation of 2-methylpropanal and methanal (formaldehyde).

$H-\underset{\underset{H}{|}}{\overset{\overset{OH}{|}}{C}}-\underset{\underset{CH_3}{|}}{\overset{\overset{CH_3}{|}}{C}}-\overset{\overset{O}{\|}}{C}\diagdown_H$

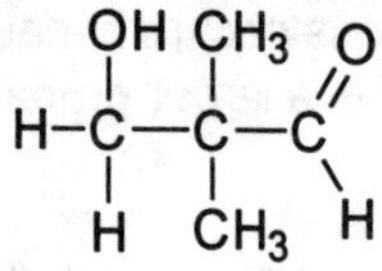

23.60 A mixed aldol condensation of acetophenone and methanal (formaldehyde) gives a product that can undergo a mixed aldol condensation two more times.

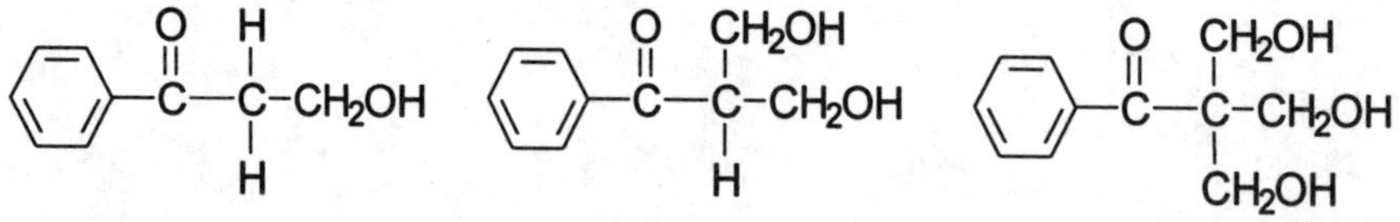

23.61 Formation of an enolate by abstraction of a proton at C-3 of 2,5-hexanedione followed by ring closure would give a cyclopropane ring in the product shown on the next page. This process doesn't occur due to strain in the ring formed. In addition, subsequent dehydration is disfavored because a double bond in the ring would further increase the ring strain. Formation of an enolate by abstraction of a proton at C-3 of 2,6-hexanedione followed by ring closure would give a cyclobutane in the product shown on the next page. This process doesn't occur for the same reasons cited for the intramolecular reaction of 2,5-hexanedione.

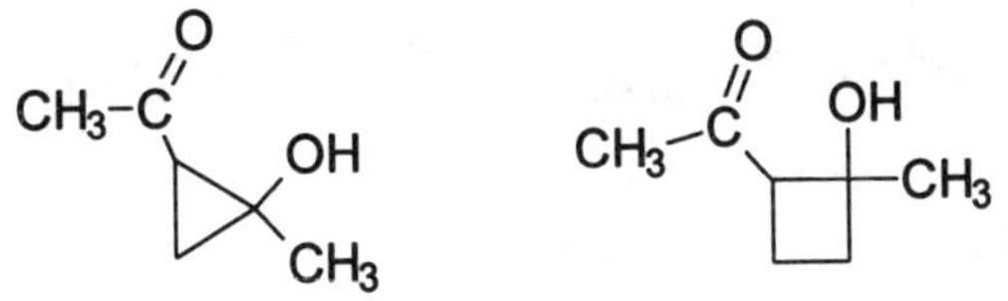

23.62   Enolates can form as the result of proton abstraction from any of four nonequivalent α carbon atoms. However, six-membered rings result only from the enolates derived from abstraction of protons from the C-1 and C-7 atoms. All of the enolates are in equilibrium with the diketone. Thus, the products formed are the result of favorable rates of ring closure and their individual stability. Reaction of the enolate derived from C-1 with the C-6 carbonyl group gives the first of the two products listed below. Reaction of the enolate derived from C-7 with the C-2 carbonyl group gives the second product.

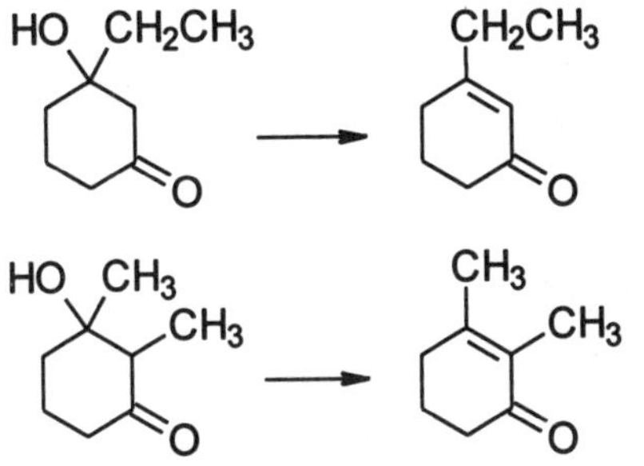

Formation of the first product requires nucleophilic attack at the more hindered carbonyl carbon atom. In addition, the dehydration product has the less substituted double bond. Thus, the second product should be the major product.

23.63

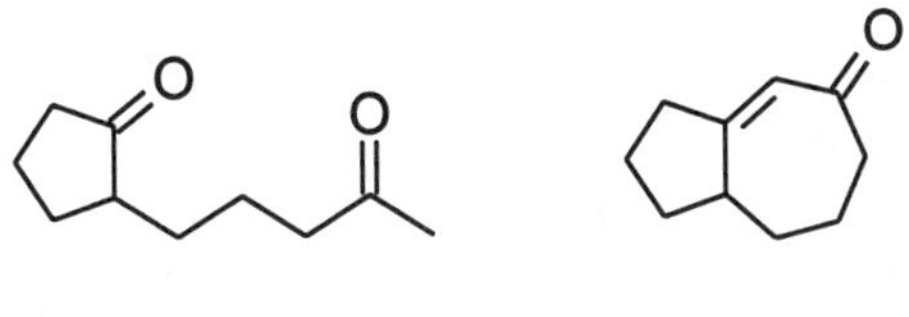

23.64

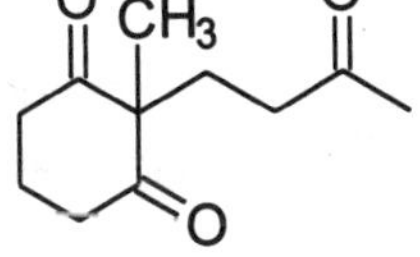

23.65 **(a)** 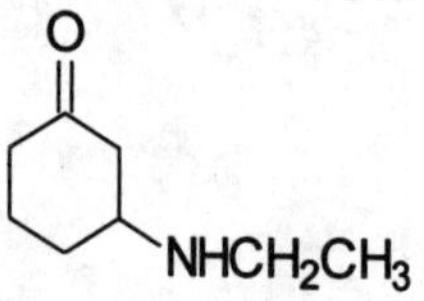   **(b)** 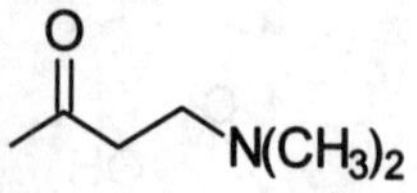   **(c)**

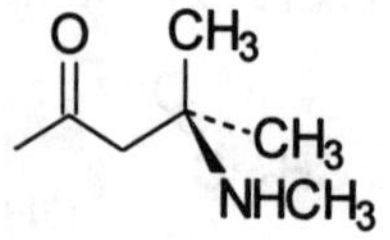

23.66 **(a)** 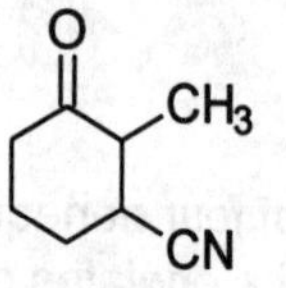   **(b)** 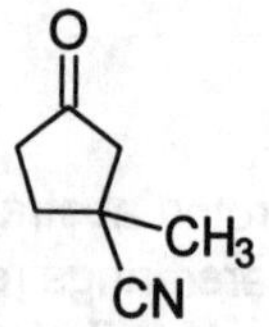   **(c)**

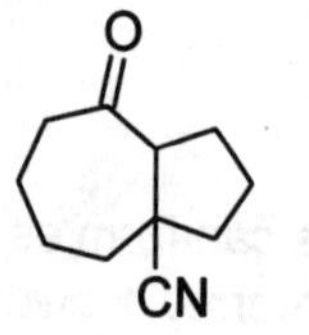

23.67   Addition of an ethyl Grignard reagent occurs 1,2 to give a tertiary alcohol. Addition of lithium diethylcuprate gives a conjugate addition product.

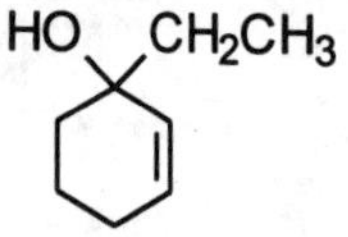

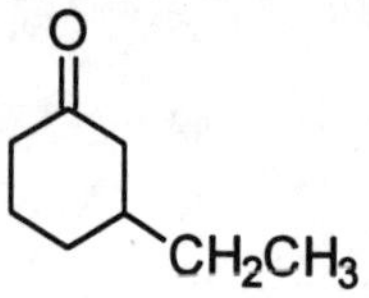

23.68   **(a)** 2-cycloheptenone and lithium diphenylcuprate
(b) 3-buten-2-one and lithium diethylcuprate
(c) 2-cyclohexenone and lithium divinylcuprate

23.69

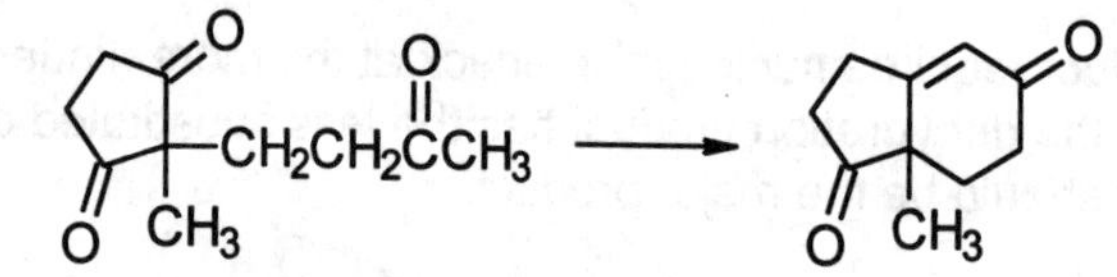

23.70   **(a)** 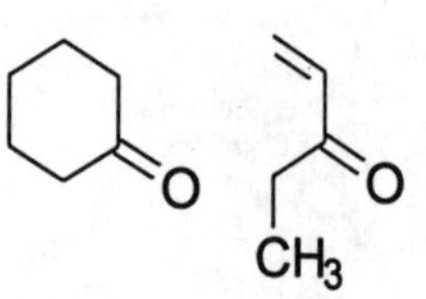   **(b)** 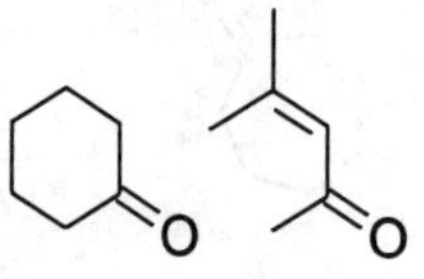   **(c)**

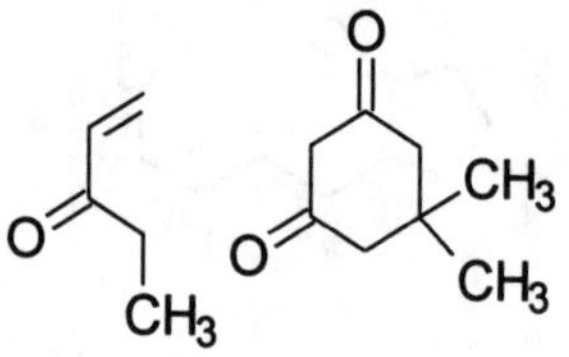

23.71   Ethyl β-cyanoacetate is the more acidic, because the cyano group is more electron withdrawing than the carboethoxy group and hence stabilizes the conjugate base.

23.72   Nitromethane is more acidic than methyl acetate ($pK_a = 25$). Thus, the nitro group is more electron withdrawing than a carbomethoxy group as indicated by its stabilization of the conjugate base. The same effect should be observed in nitroacetone, which should be a stronger acid than ethyl acetoacetate.

23.73   The K_a of malonitrile is 1×10^{-11} and the compound is a stronger acid than ethanol, whose K_a is approximately 1×10^{-16}. Thus the equilibrium constant for the reaction of nitromethane with sodium ethoxide is 1×10^5.

23.74   Because $K > 1$, the K_a of ethyl 2-cyanoacetate should be larger than the K_a of ethanol, which is approximately 1×10^{-16}. The equilibrium constant indicates the factor by which the K_a values of the two acids differ, which is 10^7. Thus, the K_a pf ethyl 2-cyanoacetate is 1×10^{-9}, so the pK_a is 9.

23.75

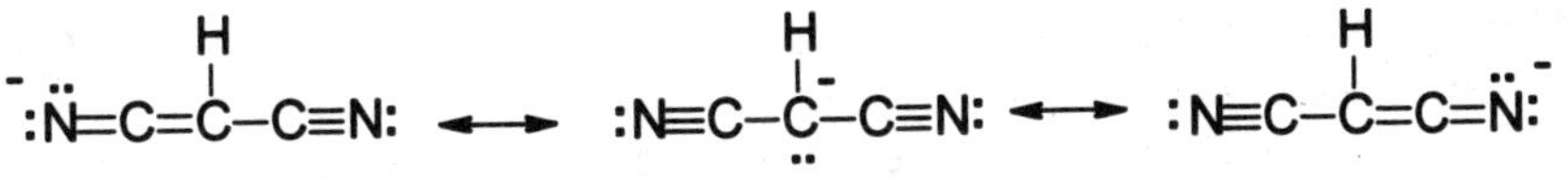

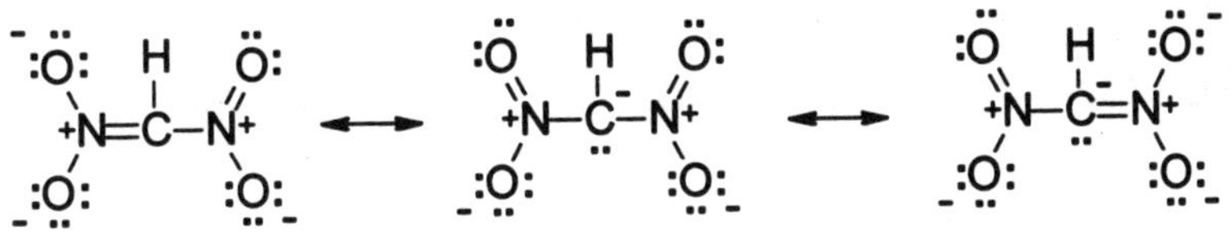

23.76

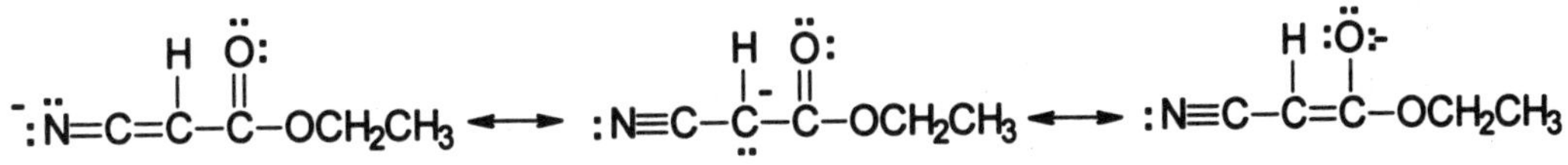

23.77

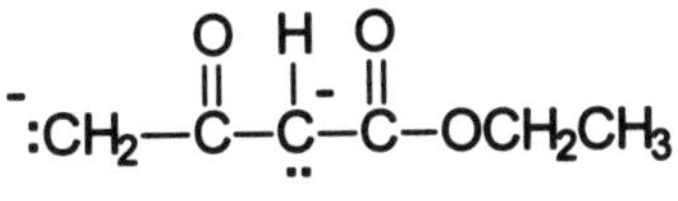

23.78   **(a)**

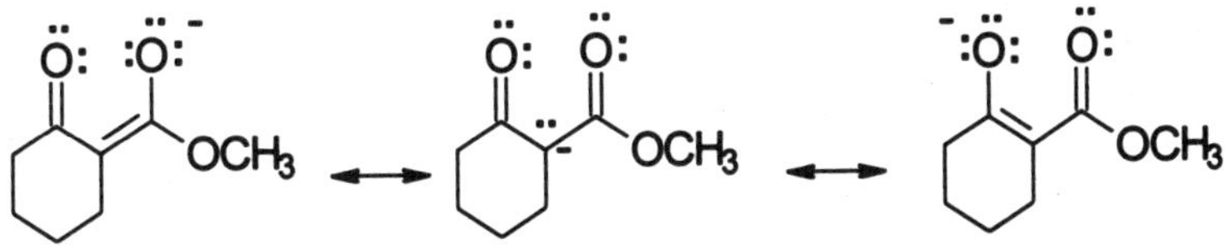

   (b)

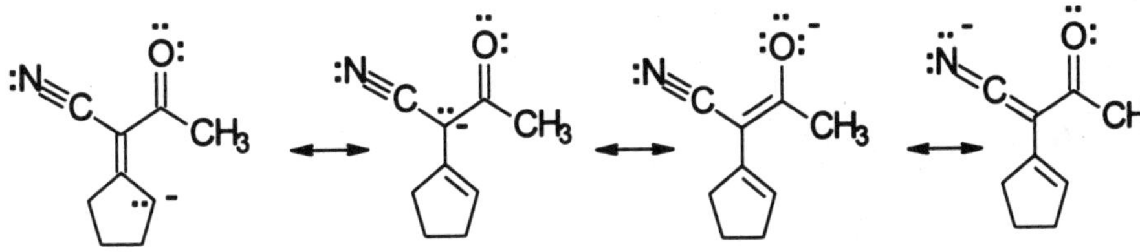

   (c)

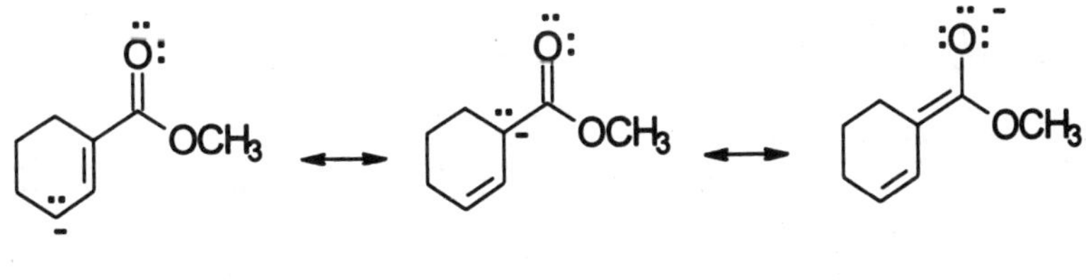

23.79   **(a)** $CH_3CH_2CH_2CD_2CO_2CH_2CH_3$

(b) $CH_3CH_2\overset{\overset{\displaystyle CH_3}{|}}{C}DCO_2CH_2CH_3$

(c) $CH_3\overset{\overset{\displaystyle CH_3}{|}}{C}HCD_2CO_2CH_2CH_3$

23.80   The first two hydrogen atoms that are exchanged are the most acidic ones ($pK_a = 11$), at the C-2 atom between the two carbonyl groups. The hydrogen atoms of the methyl group are less acidic ($pK_a = 19$) and are exchanged at a slower rate. The products are

$$CH_3-\overset{\overset{\displaystyle O}{||}}{C}-CD_2-\overset{\overset{\displaystyle O}{||}}{C}-OCH_2CH_3 \qquad CD_3-\overset{\overset{\displaystyle O}{||}}{C}-CD_2-\overset{\overset{\displaystyle O}{||}}{C}-OCH_2CH_3$$

23.81   **(a)** $CH_3CH_2CH_2CO_2H \; + \; Br_2 \; \xrightarrow{\;PBr_3\;} \; CH_3CH_2\overset{\overset{\displaystyle Br}{|}}{C}HCO_2H$

(b) $CH_3CH_2CH_2CO_2H \; + \; Br_2 \; + \; PBr_3 \; \longrightarrow \; CH_3CH_2\overset{\overset{\displaystyle Br}{|}}{C}H-\overset{\overset{\displaystyle O}{||}}{C}-Br$

(c) $CH_3CH_2\overset{\overset{\displaystyle Br}{|}}{C}HCO_2H \; + \; NH_3 \; \longrightarrow \; CH_3CH_2\overset{\overset{\displaystyle NH_2}{|}}{C}HCO_2H$

(d) $CH_3CH_2\overset{\overset{\displaystyle Br}{|}}{C}HCO_2H \; + \; NaCN \; \longrightarrow \; CH_3CH_2\overset{\overset{\displaystyle CN}{|}}{C}HCO_2H$

23.82   **(a)** $CH_3\overset{\overset{\displaystyle Br}{|}}{C}HCO_2H$ then $CH_3\overset{\overset{\displaystyle NHCH_3}{|}}{C}HCO_2H$

(b) $CH_3\overset{\overset{\displaystyle Br}{|}}{C}H-\overset{\overset{\displaystyle O}{||}}{C}-Br$ then $CH_3\overset{\overset{\displaystyle CN}{|}}{C}H-\overset{\overset{\displaystyle O}{||}}{C}-Br$

(c) 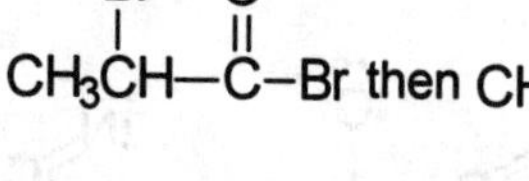 ⟨benzene ring⟩$-O-CH_2CO_2H$

23.83   **(a)** 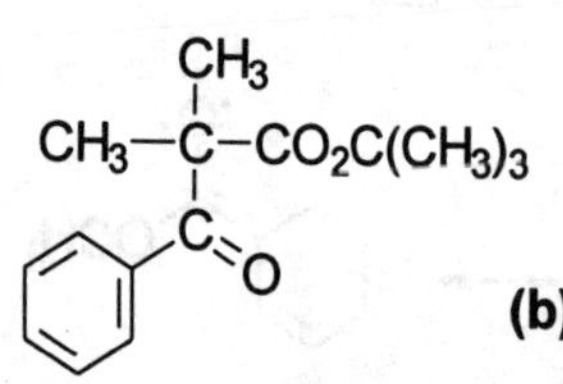 ⟨benzene ring with⟩ $\overset{\displaystyle C=O}{}$ bonded to $CH_3-\overset{\overset{\displaystyle CH_3}{|}}{\underset{}{C}}-CO_2C(CH_3)_3$

(b) $CH_3-\overset{\overset{\displaystyle CH_3}{|}}{\underset{\underset{\displaystyle CH_2CH_3}{|}}{C}}-CO_2CH_2CH_3$

(c) $CH_3-\overset{\overset{\displaystyle CH_3}{|}}{\underset{\underset{\displaystyle CH_2CH_2CH_2Cl}{|}}{C}}-CO_2C(CH_3)_3$

23.84　An S_N2 displacement reaction cannot occur at the sp^2 hybridized carbon atom of an aryl halide such as bromobenzene. The compound can be made by reaction of the enolate of ethyl 2-phenylacetate with diethyl carbonate.

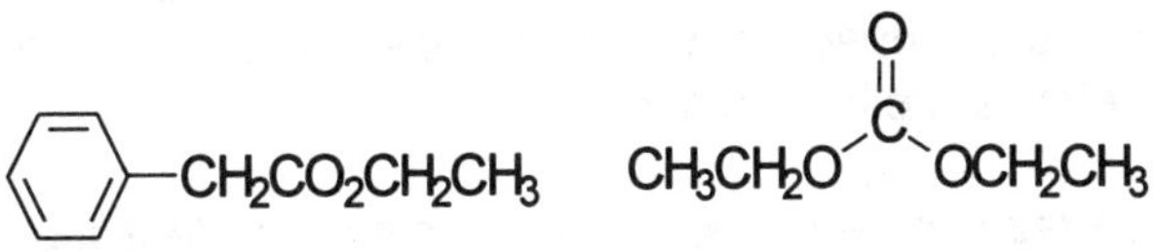

23.85　Prepare ethyl 2-bromoacetate from acetic acid. Then react it with 2,4-dichlorophenol using ethoxide ion as a base to convert the phenol to a phenoxide which then displaces bromide ion to give the ether bond. Subsequent hydrolysis of the ethyl ester gives the product.

23.86　Prepare the enolate of ethyl acetate using LDA and then alkylate 1-bromopropane. Isolate the product and prepare its enolate using LDA again. Then alkylate a second time using 1-bromopropane. Hydrolyze the dialkylated product to give valproic acid.

23.87　**(a)**

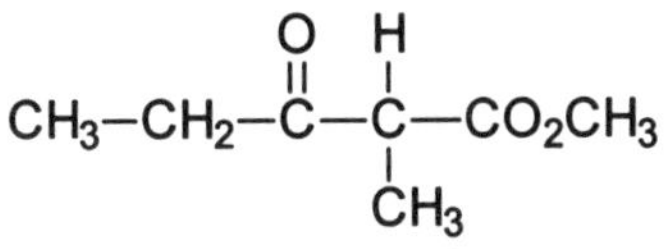

(b)

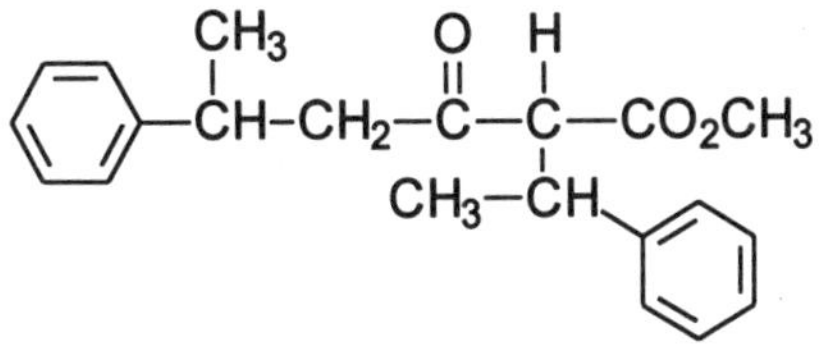

(c)

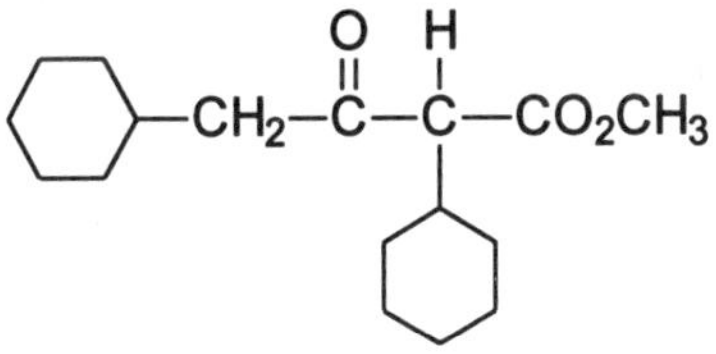

23.88　**(a)**

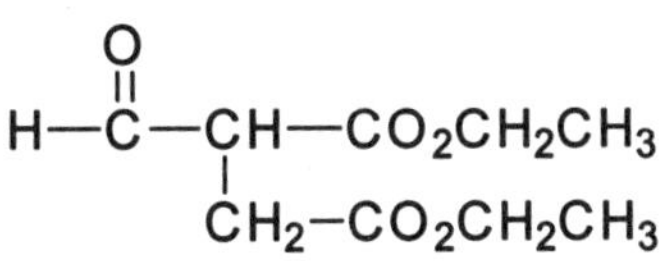

(b)

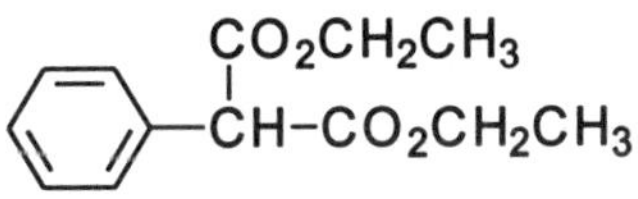

(c) 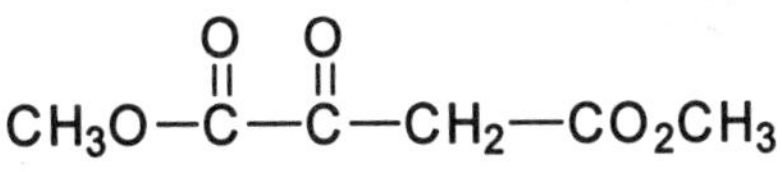　$CH_3O-\overset{\overset{O}{\|}}{C}-\overset{\overset{O}{\|}}{C}-CH_2-CO_2CH_3$

23.89 This compound is a product that cannot be obtained by a Claisen condensation of ethyl 2-methylpropanoate, because it does not have an α hydrogen atom to be abstracted by base in the final step that drives the reaction to completion. As a result, in a solution containing ethoxide ion, the reverse reaction occurs. The ethoxide ion attacks the ketone to form a tetrahedral intermediate which then releases the enolate of ethyl 2-methylpropanoate and ethyl 2-methylpropanoate.

23.90 The reverse of the given reaction would be a mixed Claisen condensation of methyl 2-methylpropanoate and methyl ethanoate to give the ketoester. This reaction is unfavorable because the ketoester does not have an α hydrogen atom to be abstracted by base in the final step that drives the reaction to completion. As a result, in a solution containing methoxide ion, the forward reaction occurs. The methoxide ion attacks the ketone to form a tetrahedral intermediate which then releases the enolate of methyl 2-methylpropanate and methyl ethanoate.

23.91 (a)

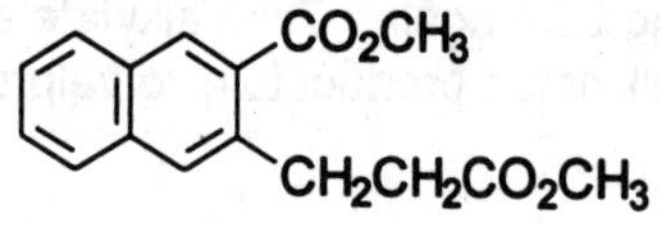

(b)

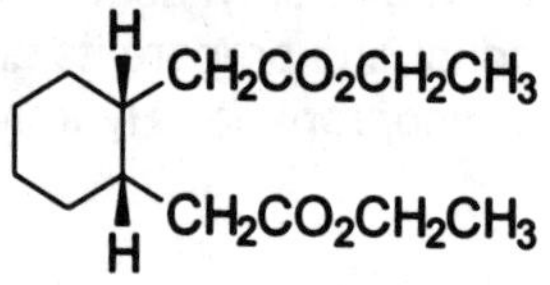

(c)

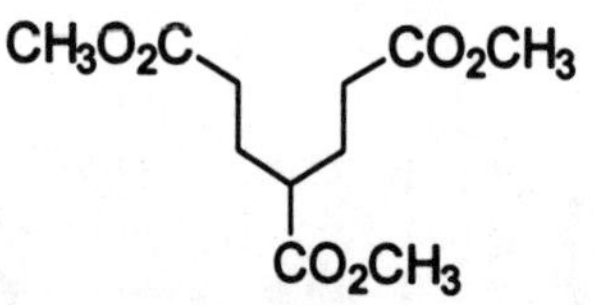

23.92 (a)

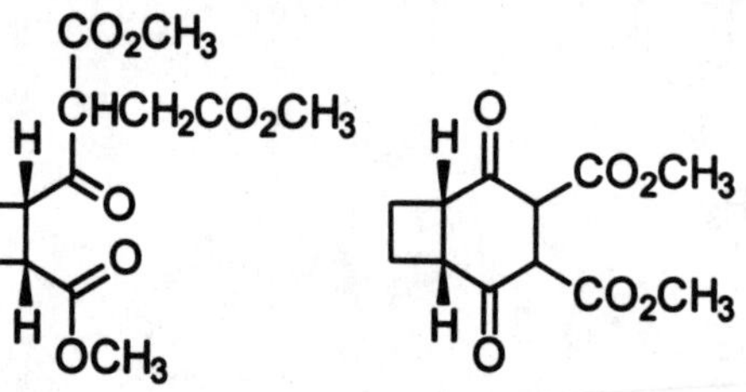

(b)

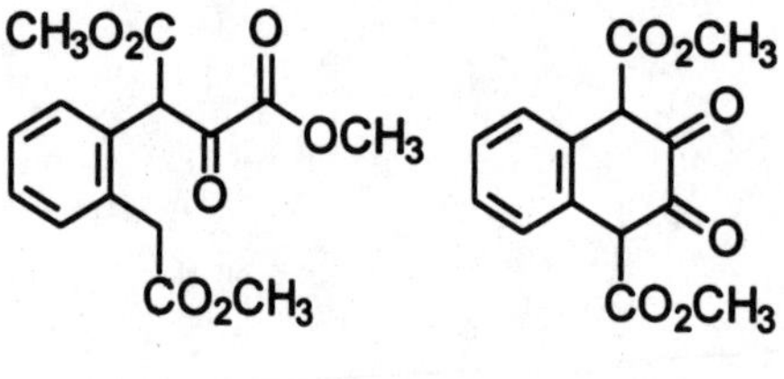

(c)

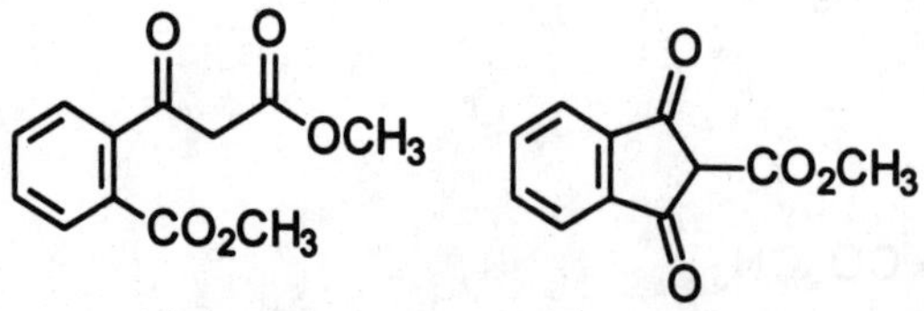

23.93 **(a)** ethyl 2,2-dimethylpropanoate and ethyl butanoate
(b) methyl benzoate and methyl propanoate
(c) dimethyl carbonate and methyl 2-phenylethanoate

23.94 **(a)** A six membered ring results from either of the two possible Dieckmann condensation reactions. The acidity of the protons of the methyene groups are quite different. The protons of the methylene group of the side chain at the upper position are more acidic because the conjugate base is resonance stabilized by the aromatic ring. The larger quantity of this enolate at equilibrium should favor formation of the first product listed below, which results from a cyclization reaction with the carbonyl group of the side chain at the lower position. The second product results from abstraction of the α hydrogen atom of the side chain in the lower position, followed by a cyclization reaction with the carbonyl group of the side chain at the upper position.

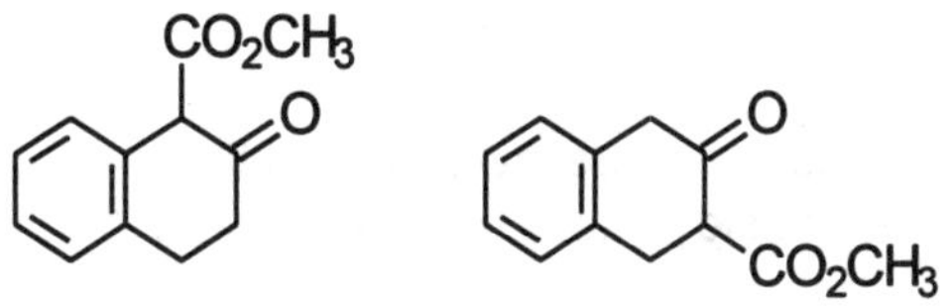

(b) A five membered ring results from either of the two possible Dieckmann condensations. Abstraction of a proton from the methylene group α to the carbonyl group on the left, followed by cyclization with the carbonyl group on the right gives the first compound listed. Abstraction of a proton from the carbon atom α to the carbonyl group on the right, followed by cyclization with the carbonyl group on the left could give the second compound listed. However, this product is unlikely because it does not have an acidic site to react with methoxide ion to stabilize the product.

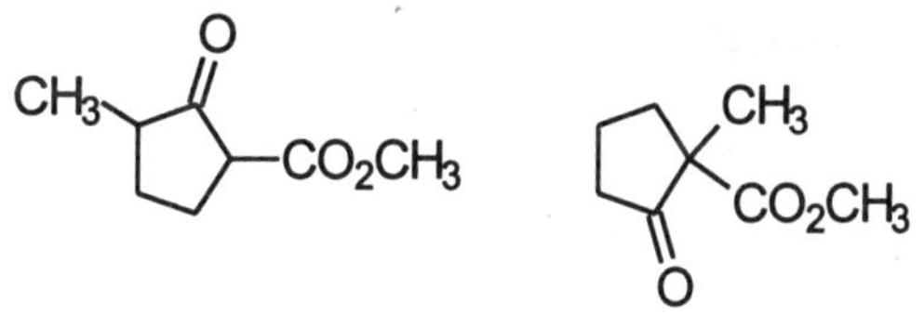

23.95 The major product results from abstraction of the more acidic hydrogen atom, which is located at the C-6 atom, followed by attack at one of the two equivalent carbonyl carbon atoms of diethyl oxalate. Loss of an ethoxide ion gives the following product.

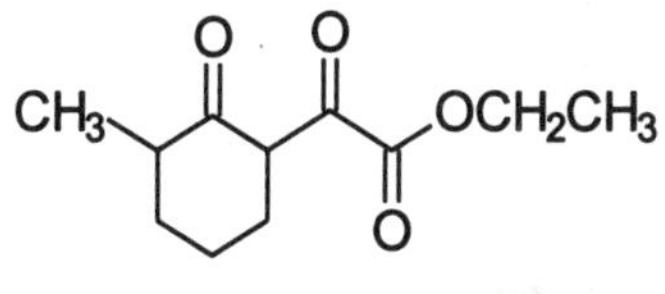

23.96 Abstraction of the acidic hydrogen atom located at the C-2 atom gives an enolate, which then attacks the carbonyl carbon atom of diethyl carbonate. Loss of an ethoxide ion gives the following product.

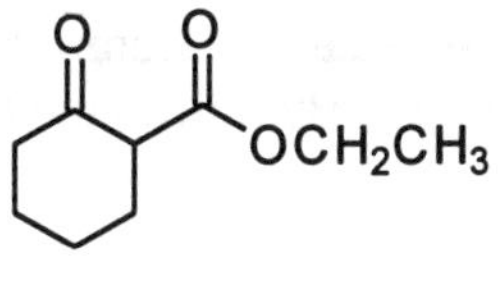

23.97   (a)

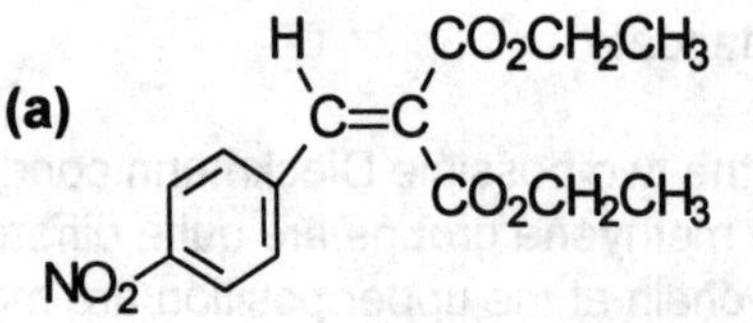

(b)

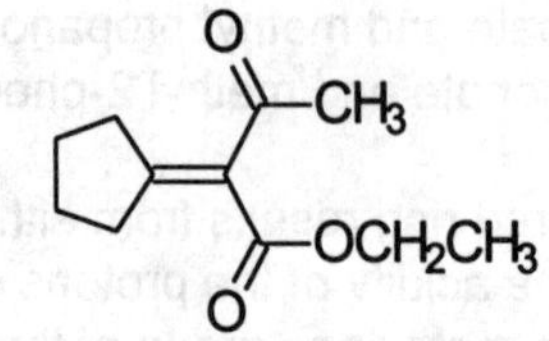

(c)

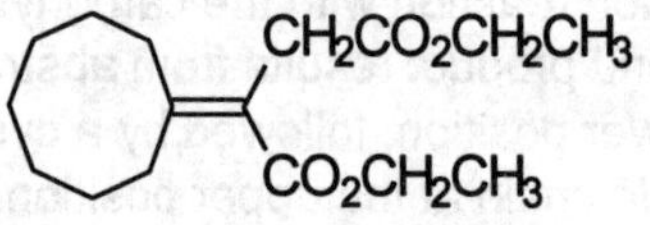

23.98   The Reformatskii reaction gives a secondary benzylic alcohol that dehydrates to give an unsaturated carboxylic acid whose double bond is conjugated with the aromatic ring. Two geometric isomers result.

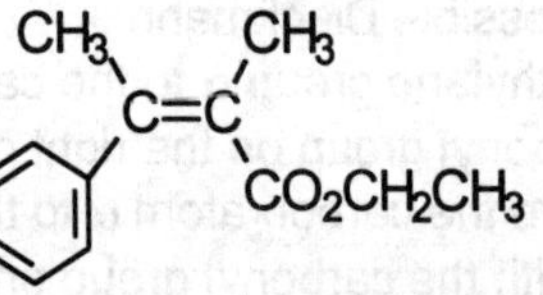

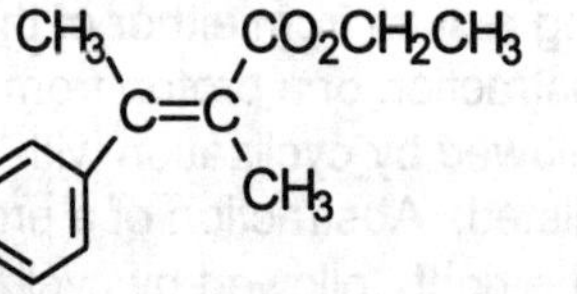

23.99   (a) butanal and ethyl acetoacetate
(b) benzaldehyde and diethyl malonate
(c) cyclohexanone and ethyl 2-cyanoacetate

23.100   (a)

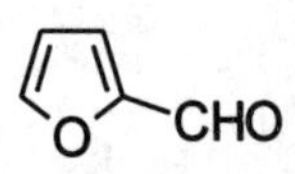

$BrCH_2CO_2CH_3$

(b)

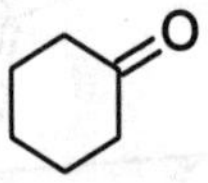

$BrCH_2CO_2CH_2CH_3$

(c)

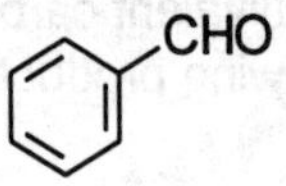

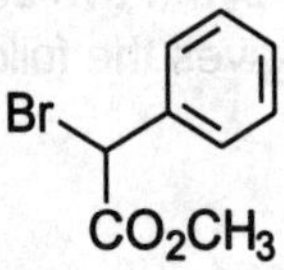

23.101   (a) Dialkylate diethyl malonate by first reacting the enolate with 3-bromopropene and then reacting the enolate of this product with iodomethane. Hydrolyze the diester and decarboxylate.
(b) Alkylate the enolate of diethyl malonate with 3-bromohexane. Hydrolyze the diester and decarboxylate.
(c) Dialkylate diethyl malonate by first reacting the enolate with benzyl bromide and then reacting the enolate of this product with iodoethane. Hydrolyze the diester and decarboxylate.

(d) Alkylate the enolate of diethyl malonate with benzyl bromide. Hydrolyze the diester and decarboxylate.

(e) Dialkylate diethyl malonate by first reacting the enolate with 3-bromopropyne and then reacting the enolate of this product with iodoethane. Hydrolyze the diester and decarboxylate.

23.102 (a) Alkylate the enolate of ethyl acetoacetate with benzyl chloride. Hydrolyze the ketoester and decarboxylate.

(b) Alkylate the enolate of ethyl acetoacetate with 3-bromopropene. Hydrolyze the ketoester and decarboxylate.

(c) Alkylate the enolate of ethyl acetoacetate with 1-bromo-2-methylpropane. Hydrolyze the ketoester and decarboxylate.

(d) Dialkylate ethyl acetoacetate by first reacting the enolate with 3-bromopropene and then reacting the enolate of this product with iodopropane. Hydrolyze the ketoester and decarboxylate.

(e) Dialkylate ethyl acetoacetate by first reacting the enolate with bromomethyl-cyclopentane and then reacting the enolate of this product with iodomethane. Hydrolyze the ketoester and decarboxylate.

23.103 The enolate of dimethyl malonate displaces a bromide ion from the primary carbon atom. Addition of a second mole of base forms an enolate of the product which displaces a chloride ion from the secondary carbon atom and forms a four-membered ring. Hydrolysis and decarboxylation occurs in the third step to give a carboxylic acid.

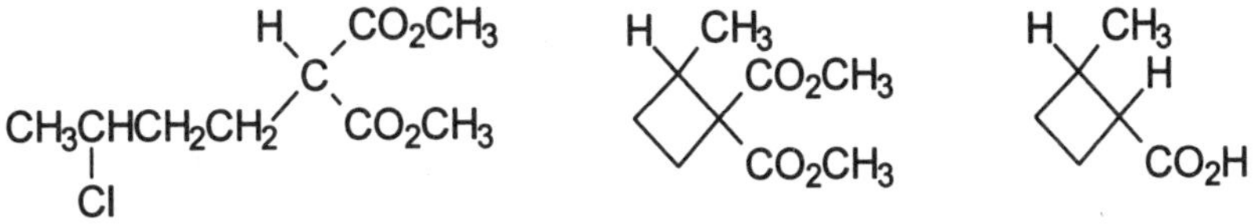

23.104 Saponification yields a dicarboxylate salt which cannot decarboxylate. Careful hydrolysis yields an alkylated malonic acid.

23.105 The enolate of ethyl acetoacetate attacks the primary carbon atom of 2,2-dimethyloxirane to give a ring opened primary alkoxide. The alkoxide ion intramolecularly attacks the carbonyl carbon atom of the ester to give a tetrahedral intermediate which subsequently loses an ethoxide ion. The structures of the primary alkoxide and the cyclic ester product are shown.

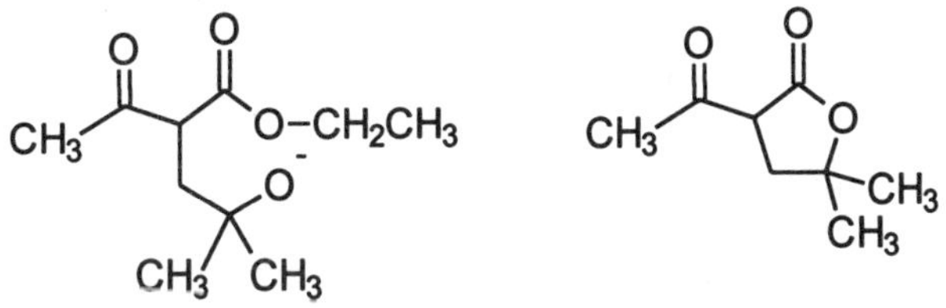

23.106 (a)

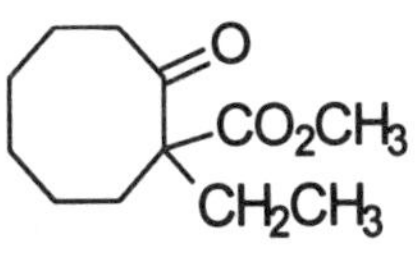

(b)

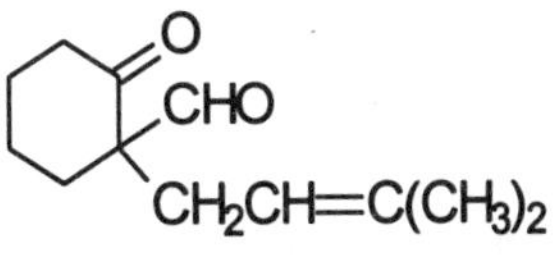

(c) $CH_3C\equiv CCH_2CH_2-\overset{\overset{\displaystyle CN}{|}}{C}H-CO_2CH_2CH_3$

23.107 (a) The ketone is the more reactive, because the intermediate enolate is more stable. Recall that the α carbon atoms of ketones are more acidic than those of esters, because the ketone carbonyl group stabilizes the negative charge better than does the carbonyl group of an ester.

(b) The second compound is the more reactive because its β carbon atom is less sterically hindered for attack by a nucleophile. Note that the charge of the intermediate cannot be delocalized by the aromatic ring of the first compound.

(c) The second compound is the more reactive because the intermediate enolate is

(a)

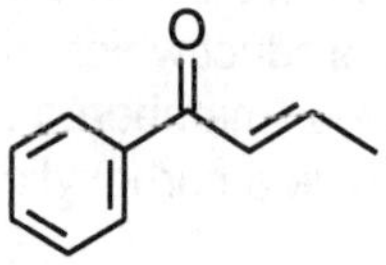

CO_2CH_3
$CH_2CO_2CH_3$

(b) 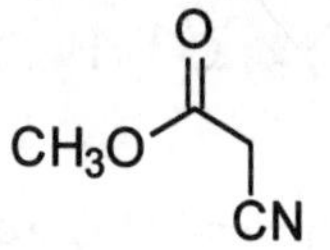 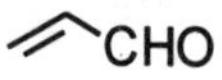 CH_3O

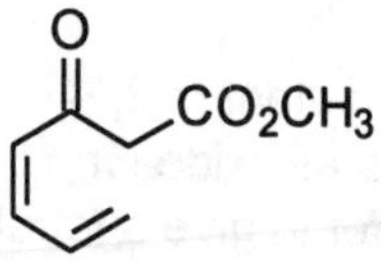

CN CHO

(c)

CO_2CH_3

(a) CN $\overset{\overset{\displaystyle CN}{|}}{C}H_2CN$ **(b)** $CH_3-\overset{\overset{\displaystyle O}{||}}{C}-\overset{\underset{\displaystyle CH_3}{|}}{C}H-CO_2CH_3$

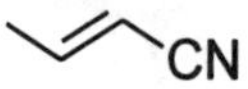

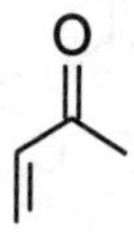

(c) 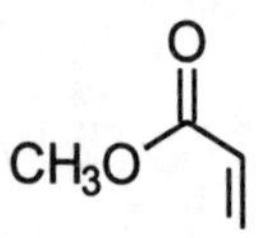 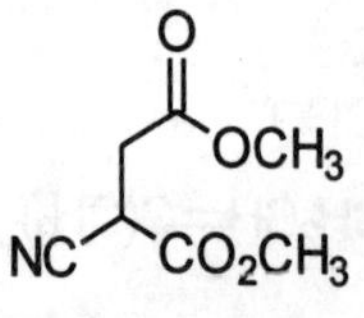 CH_3O OCH$_3$

NC   CO$_2$CH$_3$

24

Lipids

This chapter is arranged to present the structures of diverse classes of compounds that are predominately ester derivatives found in virtually all cells. No synthetic methods are given, but a limited number of naturally occurring reactions of lipids are considered. Although the molecules are large, the functional groups are easily identified and should look familiar to you.

Keys to the Chapter

24.1 Classification of Lipids

Simple lipids are not hydrolyzed by aqueous base, whereas complex lipids are hydrolyzed to form several components. The complex lipids consist of esters of an alcohol and long-chain fatty acids. Some contain phosphoric acid as a phosphate, as well as additional units unique to each class of compounds. The list given on page 929 of the text gives this information, and the relationships between each class of lipids are shown in Figure 24.1 of the text.

24.2 Fatty Acids

The saturated **fatty acids** are recognized by the even number of carbon atoms and the absence of double bonds. They are usually written with a terminal methyl group, an even number of methylene groups enclosed within parentheses, and a carboxylic acid group. The compounds lauric, myristic, palmitic, stearic, and arachidic acid contain 12, 14, 16, 18, and 20 carbon atoms, respectively.

The unsaturated fatty acids also contain an even number of carbon atoms; the more common ones have 18 carbon atoms. Oleic, linoleic, and linolenic acid have 18 carbon atoms and contain 1, 2, and 3 double bonds, respectively. Arachidonic acid contains 20 carbon atoms and has 4 double bonds. The structures of common saturated and unsaturated fatty acids are given in Table 24.1 in the text.

The melting points of the long chain fatty acids found in nature increase with chain length as a result of an increase in London forces. The presence of a double bond with the *cis* configuration causes a "bend" in the molecule which prevents efficient packing in the solid state. As a consequence, *cis* fatty acids have lower melting points than the saturated carboxylic acids with the same number of carbon atoms. The melting points of common saturated and unsaturated fatty acids are given in Table 24.2 in the text.

Essential fatty acids are those component fatty acids in cells that cannot be synthesized by the organism. As a result the organism must obtain these essential fatty acids from food sources. Humans cannot synthesize linoleic and linolenic acids. Higher unsaturated acids such as arachidonic acid can be produced from these essential fatty acids.

24.4 Waxes

Waxes are esters of long-chain alcohols and fatty acids, both of which contain an even number of carbon atoms. The properties of different waxes vary depending on the alcohol and acid components. Waxes are low-melting solids commonly found on the surface of plant leaves and fruits. The uses of waxes in our daily lives are based in part on the melting point of each wax.

24.4 Triacylglycerols

Triacylglycerols or triglycerides are triesters of glycerol and fatty acids. Triacylglycerols that contain a large percentage of saturated fatty acids are called **fats**. **Oils** are triacylglycerols with a high degree of unsaturation in the fatty acids. Fats are obtained from animal sources, whereas oils are obtained from plant sources. The unsaturated triacylglycerols can be hydrogenated to produce saturated compounds.

Triacylglycerols are nonpolar and accumulate in nonpolar sites in the body. They are found in adipocytes and in adipose tissue. Their function is two-fold. They physically insulate and protect some organs against shock. In addition they serve as sources of long-term energy.

24.5 Glycerophospholipids

Glycerophospholipids contain a phosphate group bonded to the C-3 atom of glycerol, and this functional group is in turn bonded to an alcohol unit such as choline, serine, or ethanolamine. The other two alcohol functional groups of glycerol in glycerophospholipids are part of esters formed with fatty acids. These compounds have polar sites as a result of ionization of the one remaining —OH group of the phosphate unit. In addition there are site of either positive or negative charge or both in the alcohol unit that is bonded as an ester to the phosphate group.

24.6 Sphingolipids

Sphingolipids have sphingosine, an amino diol, as the backbone of the structure. In a **ceramide**, the amino group exist as an amide formed by reaction with a fatty acid. This unit and the hydrocarbon chain of sphingosine itself constitute two long chains that make up the nonpolar portion of the sphingolipid. In a **sphingophospholipid**, the hydroxyl group at the C-1 atom is bonded to a phosphate unit which in turn forms an ester with an alcohol unit such as choline, serine, or ethanolamine.

Glycosphingolipids are similar to sphingophospholipids in the sphingosine backbone and the fatty acid bonded to the amino group to form an amide; however, glycosphingolipids have a glycosidic bond to a carbohydrate at the C-1 atom of sphingosine rather than a phosphate unit. **Gangliosides** contain an oligosaccharide; **cerebrosides** contain only galactose.

24.7 Biological Membranes

Glycerophospholipids, sphingophospholipids, and glycosphingolipids all contain two nonpolar chains that can associate and form lipid bilayers that are stabilized by London forces. Membrane fluidity depends on the composition of the bilayer. Unsaturation in the component fatty acids of the lipids increases the fluidity of the bilayer. Proteins are bonded to the surface of the bilayer and incorporated within the bilayer. The proteins extending across the bilayer are involved in transport of nutrients and waste products across the bilayer. Carbohydrates associated in the form of glycolipids are involved in intercellular recognition and are part of the immune system.

Transport across cell membranes in a direction dictated by concentration differences does not require cellular energy in **facilitated diffusion**. **Active transport** requires cellular energy and occurs in a direction from low concentration to high concentration.

24.8 Catabolic Reactions of Fatty Acids

Catabolic reactions are degradative sequences that release stored chemical energy. Fatty acids are degraded two carbon atoms at a time to give acetyl units. The process is initiated by conversion of the fatty acid into a thioester of coenzyme A. The subsequent degradation process occurs in four steps listed below. The **fatty acid cycle** occurs repeatedly to eventually cleave a four carbon derivative into two acetyl units.

1. dehydrogenation of the carboxyl derivative to give a conjugated unsaturated carboxyl derivative
2. hydration of the trans carbon-carbon double bond to give a 3-hydroxy derivative
3. oxidation of the L-hydroxylacyl group to give a keto acid derivative
4. a reverse Claisen condensation to give an acetyl derivative and an acyl derivative containing two fewer carbon atoms.

24.9 Catabolism of Unsaturated Fatty Acids

Catabolic reactions of unsaturated fatty acids proceed by the same process as for saturated fatty acids until one of the "normal" reactions would fail because of the placement or stereochemistry of the double bond. Enzymes are available to get around these difficulties. One such enzyme isomerizes the cis double bond, commonly found in unsaturated acids, into the trans isomer. Epimerases can isomerize hydroxy compounds of the improper configuration into the isomer required for step 3 listed above.

24.10 Biosynthesis of Fatty Acids

Fatty acids are biosynthesized from acetyl groups by a series of reactions that is not the microscopic reverse of the catabolic reactions. The process occurs while the growing fatty acid is associated with an **acyl carrier protein**. The enzymes required are different from those used in the catabolic reaction. A different dinucleotide is involved—NADPH is required for biosynthesis, whereas NAD^+ is used in catabolism of fatty acids.

Initiation of the biosynthesis occurs by a condensation reaction of acetyl CoA with bicarbonate to give a malonyl derivative and which requires energy supplied by ATP. The chain elongation steps are:

1. condensation of the growing acyl derivative with acetyl ACP to give an ACP derivative with two additional carbon atoms
2. reduction of the ketone group of the keto ester to give a hydroxyl group with the D configuration
3. dehydration of the alcohol to give a trans unsaturated conjugated acid derivative
4. reduction of the double bond to give a saturated acid derivative.

Solutions to Exercises

24.1   The molecular formulas are $C_{26}H_{52}O_2$ and $C_{30}H_{62}O$ for the acid and alcohol components, respectively. The structures for the acid and the alcohol components are represented by $CH_3(CH_2)_{24}CO_2H$ and $CH_3(CH_2)_{28}CH_2OH$, respectively.

24.2   Whale oil can be saponified to give a salt of palmitic acid which can function as a soap. The other product is a 16 carbon alcohol which could be oxidized to palmitic acid.

24.3   **(a)** There is a fatty acid with an even number of carbon atoms (32) in the ester. However, the alcohol component is propanol, which contains an odd number of carbon atoms. The compound should not be a naturally occurring wax.
(b) There is a fatty acid with an even number of carbon atoms (30) in the ester. However, the alcohol component contains an odd number of carbon atoms (21). The compound should not be a naturally occurring wax.
(c) There is a fatty acid with an odd number of carbon atoms (29) in the ester. The alcohol component contains an even number of carbon atoms. The compound should not be a naturally occurring wax.

24.4   At the low temperature of its environment, the wax remains a liquid because unsaturated fatty acids contribute to lower melting points for their derivative compounds due to inefficient packing resulting from "bends" in the unsaturated acids.

24.5   Based on the name, the compound must contain sixteen carbon atoms. The point of unsaturation is at the C-9 atom, as is the case for oleic acid and the double bond most likely has a cis configuration, as indicated in Table 24.1 of the text.

$$CH_3(CH_2)_4CH_2 \quad CH_2(CH_2)_6CO_2H$$
$$C=C$$
$$H \qquad H$$

24.6   The compound has a triple bond between the C-9 and C-10 atoms.

$$CH_3(CH_2)_6CH_2-C\equiv C-CH_2(CH_2)_6CO_2H$$

24.7   The compound has a double bond between the C-7 and C-8 atoms of the sixteen carbon atom chain. It is isomeric with palmitoleic acid, whose melting point is -0.5°C. If the double bond were cis, the melting point of hypogeic acid should be similar to that of palmitoleic acid. The higher melting point suggests that the double bond in hypogeic acid is trans.

$$H \qquad CH_2(CH_2)_4CO_2H$$
$$C=C$$
$$CH_3(CH_2)_6CH_2 \qquad H$$

24.8   The compound has a double bond between the C-9 and C-10 atoms of the eighteen carbon atom chain as does oleic acid, but the double bond is trans in elaidic acid and cis in oleic acid. The melting point of elaidic acid (45°C) is higher than that of oleic acid (13°C), but lower than that of stearic acid (70°C), which is saturated. The melting point suggests that the trans double bond allows stronger London forces between the chains of elaidic acid because it exists in a more extended shape similar to that of stearic acid.

24.9

$$
\begin{array}{l}
CH_2-O-\overset{\displaystyle O}{\overset{\|}{C}}-(CH_2)_{12}CH_3 \\
CH-O-\overset{\displaystyle O}{\overset{\|}{C}}-(CH_2)_{14}CH_3 \quad +\ 3\ OH^- \longrightarrow \\
CH_2-O-\overset{\displaystyle O}{\overset{\|}{C}}-(CH_2)_{16}CH_3
\end{array}
\qquad
\begin{array}{l}
CH_2-OH \\
CH-OH \quad + \\
CH_2-OH
\end{array}
\qquad
\begin{array}{l}
\overset{-}{O}-\overset{\displaystyle O}{\overset{\|}{C}}-(CH_2)_{12}CH_3 \\
\overset{-}{O}-\overset{\displaystyle O}{\overset{\|}{C}}-(CH_2)_{14}CH_3 \\
\overset{-}{O}-\overset{\displaystyle O}{\overset{\|}{C}}-(CH_2)_{16}CH_3
\end{array}
$$

24.10   The second oil is more unsaturated, because it contains a higher percentage of linoleic acid, which has two double bonds.

24.11   The compound is highly unsaturated, so it is classified as an oil. From top to bottom in the structure, the fatty acid components are linoleic, oleic, and linolenic acids, respectively.

24.12   The compound is largely saturated, so it is classified as a fat. From top to bottom, the fatty acid components are palmitic, lauric, and oleic acids, respectively.

24.13   The two ester groups bonded via methylene groups to the C-2 atom are equivalent. Thus, the C-2 atom is not chiral, and optically active compounds are not possible.

$$
\begin{array}{l}
CH_2-O-\overset{\displaystyle O}{\overset{\|}{C}}-(CH_2)_{16}CH_3 \\
CH-O-\overset{\displaystyle O}{\overset{\|}{C}}-(CH_2)_{14}CH_3 \\
CH_2-O-\overset{\displaystyle O}{\overset{\|}{C}}-(CH_2)_{16}CH_3
\end{array}
$$

24.14   The C-1 and C-3 atoms must contain esters of different carboxylic acids. One is oleic acid and the other is stearic acid. The C-2 atom must be an ester of stearic acid.

$$
\begin{array}{l}
CH_2-O-\overset{\displaystyle O}{\overset{\|}{C}}-(CH_2)_7CH=CH(CH_2)_7CH_3 \\
CH-O-\overset{\displaystyle O}{\overset{\|}{C}}-(CH_2)_{16}CH_3 \\
CH_2-O-\overset{\displaystyle O}{\overset{\|}{C}}-(CH_2)_{16}CH_3
\end{array}
$$

24.15   The carboxylic acid group at the top part of the molecule is palmitic acid. The carboxylic acid group at the C-2 atom is oleic acid. The grouping at the bottom part of the molecule is a phosphate ester and contains ethanolamine.

24.16   The carboxylic acid group at the top part of the molecule is palmitic acid. The carboxylic acid group at the C-2 atom is oleic acid. The grouping at the bottom part of the molecule is a phosphate ester and contains serine. Hydrolysis of this glycerophospholipid will produce one mole each of glycerol, palmitic acid, oleic acid, serine, and phosphate.

24.17   The amide group of sphingophospholipids is more stable toward hydrolysis that is the ester group of glycerophospholipids.

24.18   The other "chain" is the C-4 through C-18 atoms of sphingosine.

24.19   Both contain an amide group formed from the amine of sphingosine and a fatty acid. However, they differ because sphingophospholipids contain a phosphate unit bonded to an alcohol, whereas glycosphingolipids are bonded directly to a carbohydrate.

24.20   Cerebrosides are glycosides of galactose. Gangliosides are glycosides of oligosaccharides containing glucose and galactose..

24.21   The bond from the alcohol of sphingosine to the acetal center of the carbohydrate unit is classifed as a glycosidic bond.

24.22   Glycosides are stable in basic solution but are hydrolyzed in acidic solution.

24.23   Fatty acids that do not have "bends", such as at the site of unsaturation, can pack efficiently in a cell membrane and have strong London attractive forces. Such membranes are more rigid. Membranes containing a high percentage of unsaturated fatty acids are more flexible.

24.24   The primary forces are London forces between neighboring alkyl group chains of the fatty acids forming the interior of the membrane.

24.25   The proteins are attached to the membrane only by electrostatic forces or hydrogen bonds, so they can easily dissolve in the micelles of the detergent.

24.26   The sugar portion of glycolipids is located on the cell surface. The hydrocarbon chains of the carboxylic acid residues are imbedded in the membrane.

24.27   The dehydrogenation occurs to give a trans configuration in the fatty acid.

24.28   Hydration of the unsaturated fatty acid occurs to give an alcohol with the L configuration.

24.29   For a carboxylic acid with 18 carbon atoms, a total of 9 moles of acetyl CoA are produced. However, the fatty acid cycle occurs only 8 times because the last cycle produces two moles of acetyl CoA.

24.30   For a carboxylic acid with 16 carbon atoms, a total of 8 moles of acetyl CoA are produced. The fatty acid cycle occurs 7 times.

24.31   Five moles of acetyl CoA are produced from decanoic acid. Myristic acid contains 14 carbon atoms, so it yields 7 moles of acetyl CoA.

24.32   10-Phenyldecanoic acid yields phenylethanoic acid in the last turn of the fatty acid cycle. If 9-phenylnonanoic acid were used, benzoic acid would be the final oxidation product.

24.33   The steps are condensation, reduction, dehydration, and reduction.

24.34   Myristic acid contains 14 carbon atoms, so its biosynthesis requires 7 moles of acetyl CoA.

25

Amines and Amides

Amines and amides are the last of the functional groups presented in this text, but that isn't because nitrogen-containing compounds are less important than oxygen-containing compounds. Indeed amides contained in proteins are important for the maintenance of life. In this chapter the behavior of nitrogen-containing compounds are contrasted to oxygen-containing compounds because there are substantial differences even though the two elements are neighbors in the periodic table.

Keys to the Chapter

25.1 Organic Nitrogen Compounds

Because nitrogen has one less electron than oxygen, it can form one additional bond. Three bonds characterize neutral nitrogen compounds. These bonds may be three single bonds as in amines and amides, one double bond and a single bond in imines, or a triple bond in nitriles.

Nitrogen is incorporated in a large number of biologically important compounds that have a wide range of physiological properties. However, once the functional group is identified, the chemistry is reasonably predictable based on the ideas that functional groups have certain characteristic reactivities.

25.2 Bonding and Structure of Amines

Amines are pyramidal at the nitrogen atom, with approximately tetrahedral bond angles to all bonded atoms. However, the configuration of an amine is not static. Amines undergo **nitrogen inversion** to give mixtures of mirror images. The process occurs via a planar transition state. Thus, you should be able to answer questions on how the structure of the amine affects the rate of the inversion process.

25.3 Classification and Nomenclature of Amines

Amines are classified according to the number of alkyl or aryl groups bonded to the nitrogen atom. Primary, secondary, and tertiary amines have 1, 2, and 3 groups bonded, respectively. (Amides are classified the same way, with the acyl group counting as one of the carbon groups bonded to the nitrogen atom.) Abbreviations for the classes of amines (or amides) are 1°, 2°, and 3°.

The common names of simple amines are based on the identity of the alkyl or aryl groups bonded to the nitrogen atom. The names of the alkyl groups are written in alphabetical sequence as one word, followed by the word amine. You should know the common names and the structures of simple heterocyclic amines such as pyrrolidine, pyrrole, piperidine, and pyridine.

The common name of a complex amine is based first on identifying the longest continuous chain containing an attached nitrogen atom. The chain is numbered to assign the lowest number to the carbon atom bonded to the nitrogen atom. The nitrogen atom may be contained in an amino group ($-NH_2$), an N-alkylamino group (-NHR), or an N,N-dialkylamino group ($-NR_2$). Aryl groups may be present in place of alkyl groups, and the same procedure is followed. In naming alkylamino or arylamino groups, the prefix N- indicates that the alkyl or aryl group is attached to the nitrogen atom and not to the parent chain.

The IUPAC name of an amine is also based on the longest continuous chain containing an attached nitrogen atom. The -e ending of the parent alkane is changed to -amine. The chain is numbered to give the lowest number to the carbon atom bearing the nitrogen atom. Alkyl groups attached to nitrogen are designated with N-, but they are named along with other substituents on the parent chain.

Heterocyclic aromatic amines have rings that are numbered using a selected nitrogen atom as the number one atom. However, that selection must be memorized, as well as the method of continuing the numbering of the ring.

25.4 Physical Properties of Amines

Amines may be gases, liquids, or solids depending on the molecular weight and structure. The boiling points of primary and secondary amines are higher than those for alkanes of similar molecular weight because of intermolecular hydrogen bonding. Tertiary amines have lower boiling points than isomeric primary and secondary amines, because tertiary amines do not have an N-H bond to form intermolecular hydrogen bonds. The lower molecular weight amines are soluble in water as a result of hydrogen bonding to water molecules.

25.5 Basicity of Amines

The basicity of amines of different classes is difficult to "explain", because the number of groups affects the electron density at the nitrogen atom, but also the stabilization of the conjugate acid by solvent. Thus, the basicity of amines can be "explained" only for amines with similar structures at the nitrogen atoms.

The basicity of an amine is increased by electron-donating groups and decreased by electron-withdrawing groups. Aryl amines are less basic than alkyl-substituted amines because some electron density provided by the nitrogen atom is distributed throughout the aromatic ring. Basicity is expressed using K_b values measured from the reaction of the amine with water. An alternate indicator of basicity is pK_b, which is -log K_b. A strong base has a large K_b and a small pK_b. The basicity of amines is also expressed by the acidity of their conjugate acids. A strong base has a weak conjugate acid as given by a small value of K_a and a large pK_a.

The basicity of heterocyclic amines depends on the location of the electron pair of the nitrogen atom, its hybridization, and whether or not resonance stabilization is possible. In pyrrole the electron pair is part of the aromatic system, and as a result pyrrole is a very weak base. Pyridine is a weaker base than saturated amines of similar structure, because its electron pair is in an sp^2 hybridized orbital and the pair is more tightly held by the atom. Protonation of a similar nitrogen atom in pyrimidine is more favorable, because the charge is delocalized to the second nitrogen atom.

25.6 Solubility of Ammonium Salts

Formation of the conjugate acid of an amine gives an ionic substance which is more soluble in water than the original amine. Amines may be separated from other nonbasic organic compounds by the addition of acid to dissolve the amine. After physical separation and subsequent neutralization with base, the free amine separates from water.

25.7 Synthesis of Amines by Displacement Reactions

Ammonia and amines are nucleophiles, and they displace halide ion from haloalkanes to give more highly substituted amines. However, due to the multiple displacements possible as each product successively acts as a nucleophile, the reaction is not synthetically useful to obtain a single product.

      The Gabriel synthesis is used to convert a haloalkane into an amine in which the —NH_2 group replaces the halogen. However only primary amines can be prepared because of competing elimination reactions in one of the steps. The several steps required are:

1. Convert phthalimide into its conjugate base
2. Add a primary haloalkane to form an alkylated phthalimide
3. Release the amine using either strong base or hydrazine

25.8 Synthesis of Amines by Reduction

Any functional group containing nitrogen in a higher oxidation state can be reduced to give an amine, which contains nitrogen in its lowest oxidation state. These include azides, imines, nitriles, amides, and nitro compounds.

      Azides are prepared by the S_N2 displacement of a halide from a haloalkane by the azide ion. Reduction of the azide by either hydrogen and platinum as catalyst or by lithium aluminum hydride in ether gives a primary amine.

      The double bond of an imine can be reduced by hydrogen and Raney nickel or by a metal hydride. Sodium borohydride is sufficiently reactive as a reducing agent, so lithium aluminum hydride is seldom used. The **reductive amination** reaction forms an imine, by reaction of a primary amine with either an aldehyde or ketone, which is reduced immediately in the reaction by hydrogen in the presence of a nickel catalyst. Secondary amines also react by formation of an iminium ion followed by reduction.

      Nitriles are reduced to primary amines using lithium aluminum hydride. The nitrile can be made by displacement of a halide ion from a haloalkane by cyanide ion. This method allows the formation of primary amines having one additional carbon atom.

      Reduction of amides using lithium aluminum hydride is the most versatile way of producing amines. Amides are easily prepared by reaction of an acyl chloride and an amine. However, the reaction is most versatile because primary, secondary, and tertiary amines can be synthesized using primary, secondary, and tertiary amides, respectively.

      Reduction of nitroaromatic compounds is used to produce anilines. Tin and HCl is the usual reducing agent.

25.9 Hofmann Rearrangement

The Hofmann rearrangement occurs when a primary amide reacts with a basic solution of a halogen such as chlorine or bromine. In this process the carboxyl carbon atom is lost as carbonate ion and a primary amine results. The rearrangement occurs when an alkyl group is transferred from the carboxyl carbon atom to the nitrogen atom in one of the several intermediates involved in the reaction mechanism. The rearrangement occurs with retention of configuration of the alkyl group.

25.10 Overview of Reactions

The reactions of amines are distinctly different than the reactions of alcohols. Amines are substantially stronger bases than alcohols. Amines are sufficiently basic to exist to some degree as the conjugate acid in water. Alcohols require strong acids to form the conjugate acid. Amines are much less acidic than alcohols—the pK_a values of amines and alcohols are 35 and 16, respectively.

Within a period of the periodic table, the nucleophilicity decreases from left to right for the elements in compounds of similar structure. Ammonia is a distinctly better nucleophile than water. because the electronegativity of nitrogen is smaller than that of oxygen. Likewise, amines are better nucleophiles than alcohols. Usually it is necessary to convert an alcohol to its alkoxide ion to make it sufficiently nucleophilic to displace a leaving group such as a halide ion from a haloalkane. The neutral amine is sufficiently nucleophilic for this type of displacement reaction.

Substitution reactions to replace oxygen as a leaving group occur if the oxygen is protonated to provide for water as a leaving group. The NH_2^- is a much stronger base than the hydroxide ion and as such is a much poorer leaving group. Even protonation of the amine to give an ammonium ion doesn't allow for the loss of ammonia as a leaving group.

Elimination reactions of amine compounds are similarly less likely than elimination reactions of alcohols. The reason is similar to that just described for substitution reactions. Nitrogen moieties are more basic than the oxygen analogs and are poorer leaving groups.

Amines can be oxidized, but the resulting imines, which correspond to the carbonyl groups obtained by the oxidation of alcohols, are much more sensitive to further reaction.

25.11 Enamines

Enamines correspond in structure to enols. They are formed by the reaction of secondary amines with carbonyl compounds. Common secondary amines used to form enamines include pyrrolidine, piperidine, and morpholine.

Enamines react as nucleophiles, resulting in alkylation at the position equivalent to the α carbon atom of the original carbonyl compound. The reaction is restricted to displacement of halide ion from primary compounds. The product is an alkylated imine which upon hydrolysis gives an alkylated carbonyl compound.

One advantage of using enamines to alkylate carbonyl compounds is avoidance of strong base. The second advantage is the formation of a singly alkylated product. In contrast, multiple alkylation of ketones occurs because proton exchange generates the enolate of the alkylated product.

25.12 Formation of Amides

Amides are synthesized by reaction of an acyl halide with ammonia, a primary amine, or a secondary amine to give a primary, secondary, or tertiary amide, respectively. Amides are used to modify the reactivity of amines and as such also serve as protecting groups. The reactivity of anilines toward aromatic substitution is decreased by formation of the amide. Subsequent hydrolysis using acidic or basic conditions frees the amine.

25.13 Sulfonamides

Sulfonyl chlorides react with amines much like acyl halides do. The resulting sulfonamides are more acidic than amides, because the sulfonyl group is more electron withdrawing than an acyl group. (Note also that the corresponding oxygen compounds would be a sulfonic acid and a carboxylic acid, respectively, and we see the same effect on the acidity of the O—H bond.)

The **Hinsberg test** can be used to classify amines. Tertiary amines do not react with benzenesulfonyl chloride in the presence of base. There are no signs of a reaction. Secondary amines give a water insoluble sulfonamide because there are no acidic N—H bonds. Primary amines give a soluble conjugate base of the sulfonamide. Addition of acid results in protonation of the conjugate base, and the insoluble sulfonamide forms.

25.14 Quaternary Ammonium Salts

Continued reaction of an amine with an haloalkane eventually gives a quaternary ammonium ion. Reaction of an amine with methyl iodide to give a quaternary ammonium ion is termed **exhaustive methylation**.

Quaternary ammonium salts containing one long carbon chain are **invert soaps**, because their polar end is positively charged in contrast to a negative charge on soaps. These compounds are effective against bacteria and are used in hospitals.

Quaternary ammonium hydroxide salts formed by exhaustive methylation followed by exchange of the halide ion by hydroxide undergo an elimination reaction called the **Hofmann elimination**. The elimination occurs by an anti periplanar transition state and gives the least substituted alkene. In cases where both E and Z isomers are possible, the E isomer predominates.

25.15 Reactions of Amines with Nitrous Acid

Recall that aromatic amines react with nitrous acid (HNO_2) to give diazonium ions that are sufficiently stable to be substituted by other groups. Saturated amines have characteristic reactivities toward nitrous acid. Primary amines produce nitrogen gas and products of the reaction of the carbocation intermediate . Secondary amines form N-nitrosoamines (R_2N-N=O), which separate from the reaction mixture as a yellow oil. Tertiary amines do not give any visible reaction with nitrous acid.

25.16 Spectroscopy of Amines

Infrared spectroscopy is usually not used to confirm the presence of the C—N bond, because the stretching vibrations occur in a region complicated by other absorptions. The N—H stretching vibration of amines is easily seen as a broad absorption, similar to that found for the O—H vibration of alcohols, on the "left" of the spectrum in the 3200-3375 cm^{-1} region. Primary amines give two absorptions; secondary amines give one absorption.

In NMR spectra, the chemical shift of hydrogen atoms bonded to the carbon atom bearing the nitrogen atom of amines occurs in the 2—3 δ region. The N—H group has a variable chemical shift due to rapid exchange among various hydrogen bonding species whose identities are concentration dependent.

The α carbon atom of an amine has a ^{13}C chemical shift that reflects the smaller deshielding effect of the nitrogen atom relative to the more electronegative oxygen atom. The carbon absorptions are in the 30—50 δ region.

Summary of Reactions

1. Synthesis of Amines by Displacement Reactions of Haloalkanes

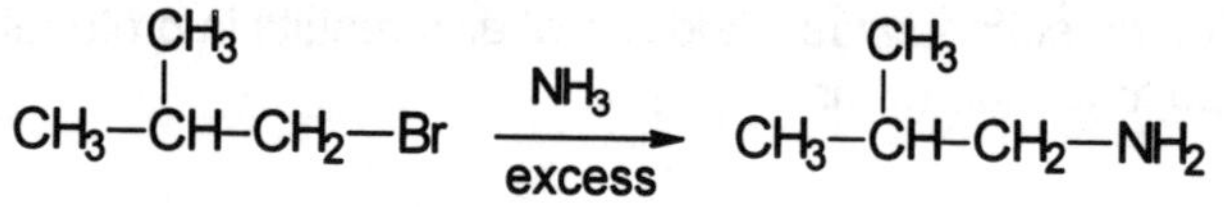

2. Gabriel Synthesis

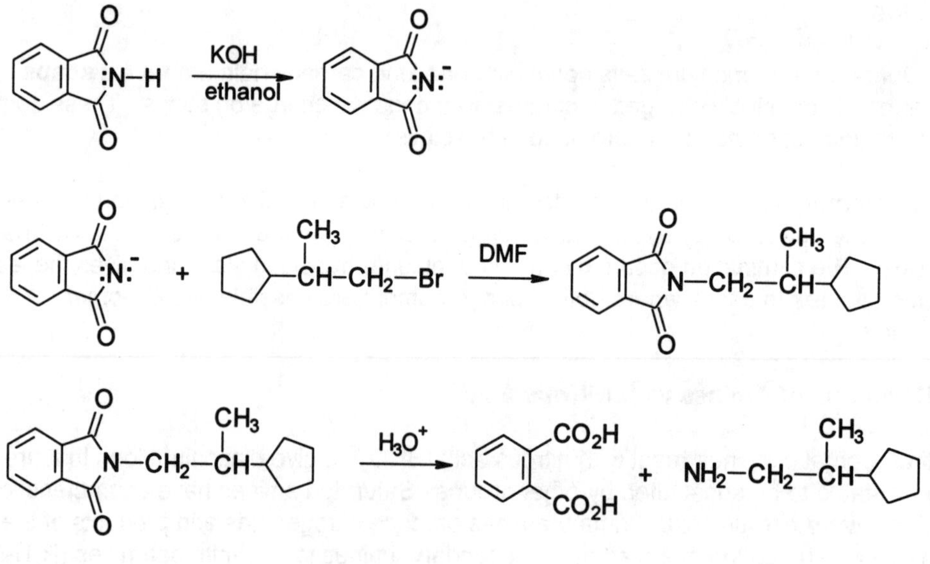

3. Synthesis of Amines by Reductive Methods

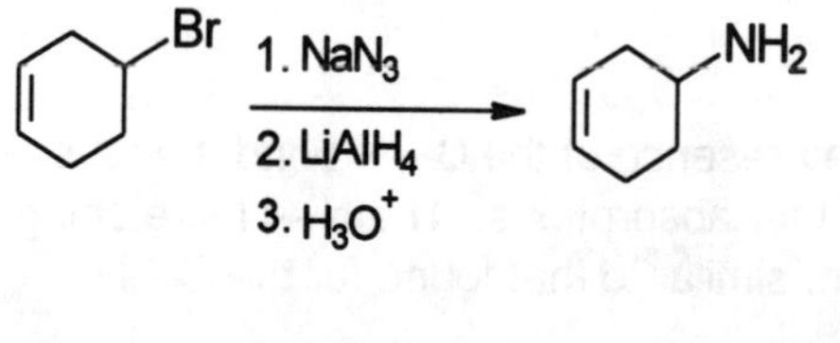

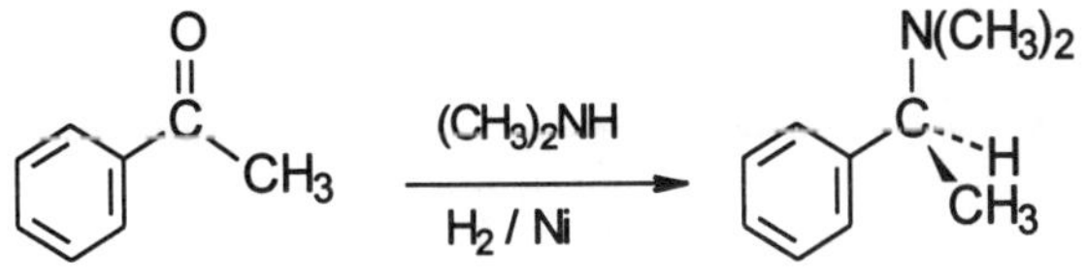

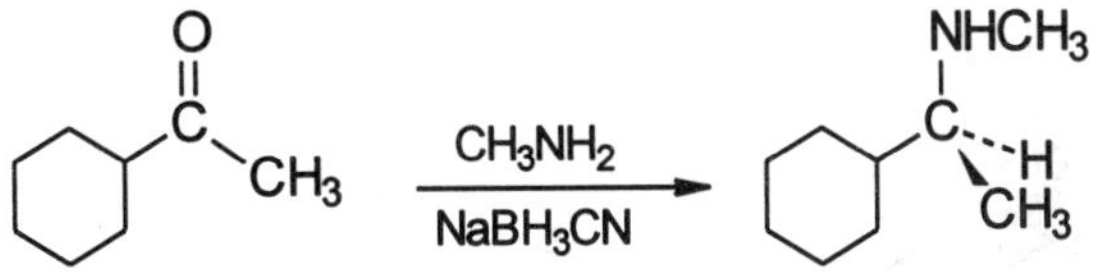

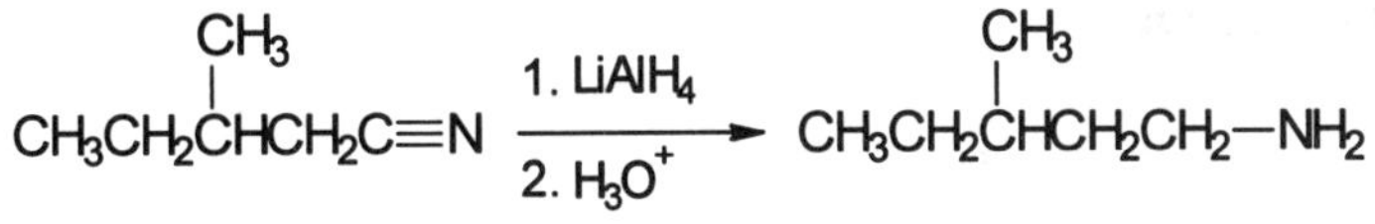

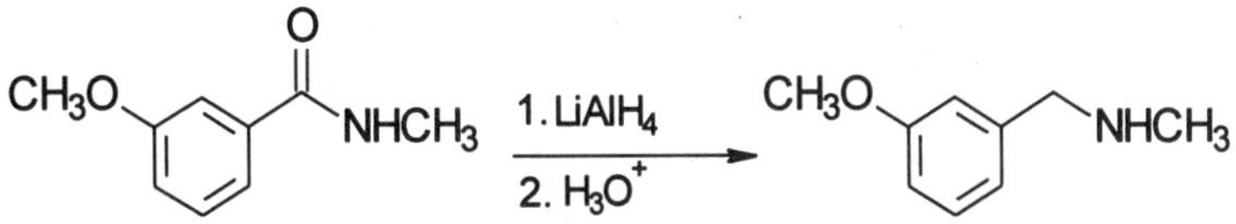

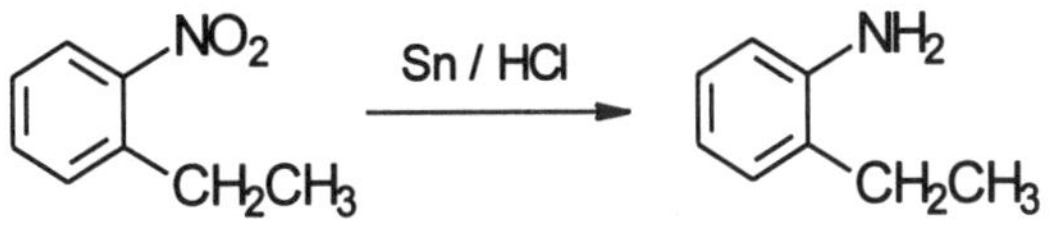

O
C
CH3
(CH3)2NH
H2 / Ni
N(CH3)2
C
H
CH3

O
C
CH3
CH3NH2
NaBH3CN
NHCH3
C
H
CH3

CH3
CH3CH2CHCH2C≡N
1. LiAlH4
2. H3O+
CH3
CH3CH2CHCH2CH2—NH2

O
CH3O
NHCH3
1. LiAlH4
2. H3O+
CH3O
NHCH3

NO2
CH2CH3
Sn / HCl
NH2
CH2CH3

4. Hofmann Rearrangement

H
O
C
NH2
CH3
H
NaOH
Br2
H
NH2
CH3
H

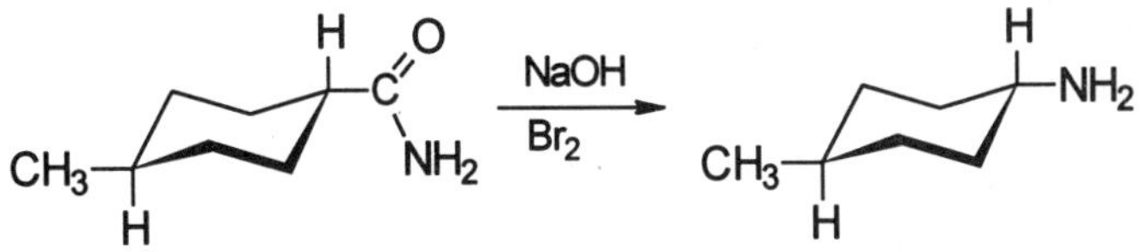

5. Alkylation of Enamines

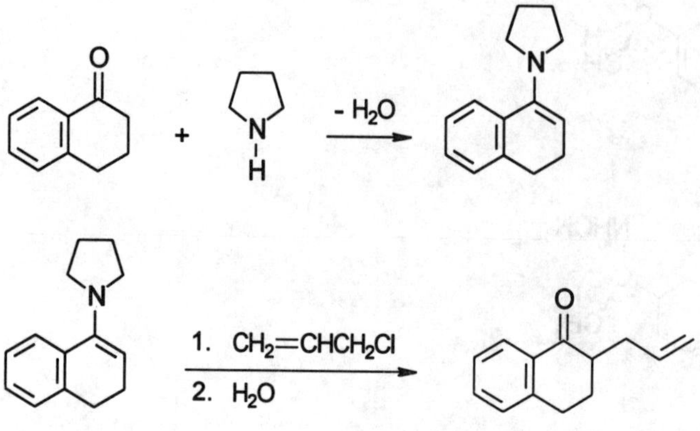

6. Formation of Amides as Protecting Groups

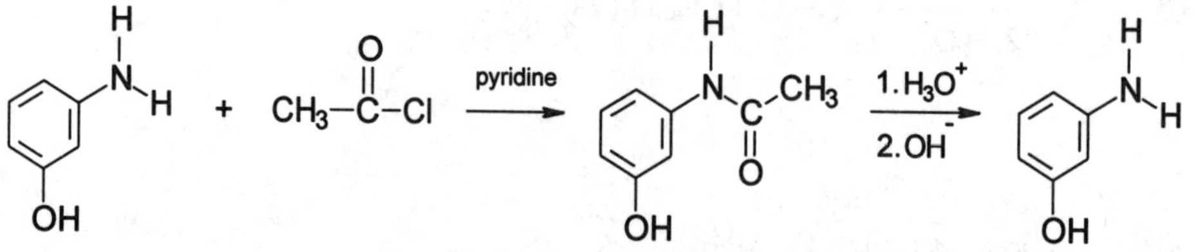

7. Hofmann Elimination

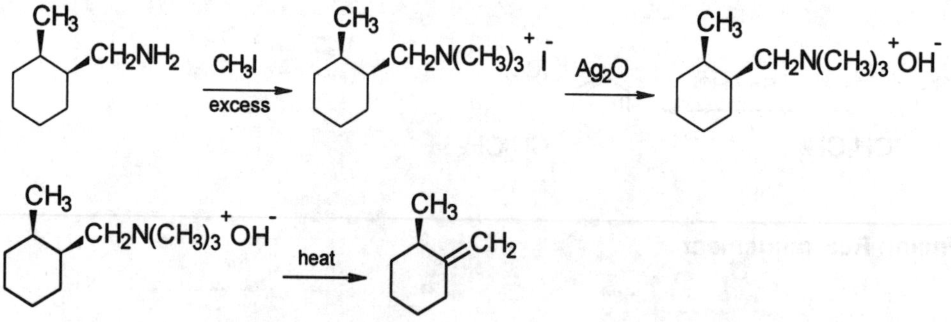

Solutions to Exercises

25.1 The N—H bond of pyrrolidine is longer because the nitrogen atom is sp^3 hybridized. The nitrogen atom of pyrrole is sp^2 hybridized. Bond length decreases with increasing percent s character.

25.2 The hybridization of the nitrogen atom changes from sp^3 to sp^2 in forming the transition state for inversion. Thus, groups bonded to the nitrogen atom are less sterically crowded in the transition state where the bond angles between groups is 120°, compared to 109° for the compound. The compound with the *tert*-butyl group has a greater degree of steric hindrance in the initial state and that steric hindrance is decreased in the transition state. Thus, *tert*-butyldimethylamine inverts at a faster rate.

25.3 (a) secondary (b) tertiary (c) secondary (d) primary

25.4 (a) tertiary (b) secondary (c) tertiary (d) secondary

25.5 (a) The nitrogen atom located on the right in the structure has two methyl groups and a complex alkyl group bonded to it. The amine is tertiary.
(b) The nitrogen atom located in the ring of the structure has two carbon atoms as part of the ring bonded to it. The amine is secondary.
(c) The nitrogen atom is bonded to a carbonyl group and is classified as an amide.

25.6 (a) The nitrogen atom located in the ring on the right in the structure has two methylene groups that are part of the ring and a quaternary carbon atom bonded to it. The amine is tertiary.
(b) The nitrogen atom bonded to the aromatic ring is also bonded to a carbonyl group and is classified as an amide. The nitrogen atom located in the ring on the right in the structure has two methylene groups that are part of the ring and a methyl group bonded to it. The amine is tertiary.
(c) The nitrogen atom located on the right of the structure has two alkyl groups and a hydrogen atom bonded to it. The amine is secondary. The nitrogen atom bonded to the aromatic ring is an amide.

25.7 (a) 3-hexanamine   (b) N,N-dimethyl-1-butanamine (c) 4-methyl-2-pentanamine

25.8 (a) 2-cyclohexyl-1-ethanamine (b) 3-cyclohexenamine
(c) 1-cyclohexyl-1-ethanamine (d) N.N-dimethylcycloheptanamine

25.9

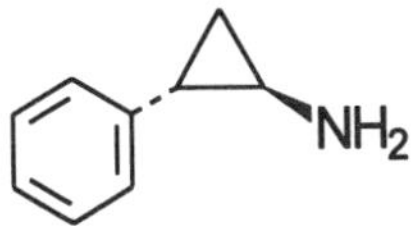

25.10

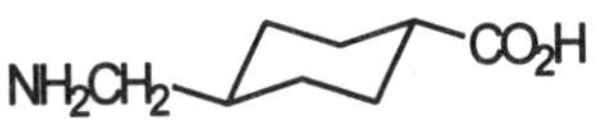

25.11 **(a)** 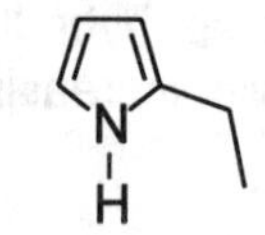 **(b)** 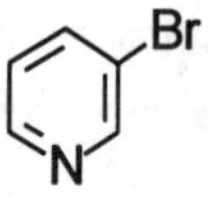 **(c)** 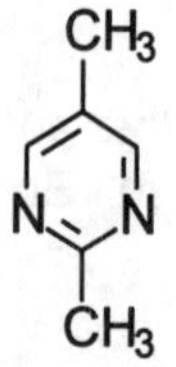 **(d)**

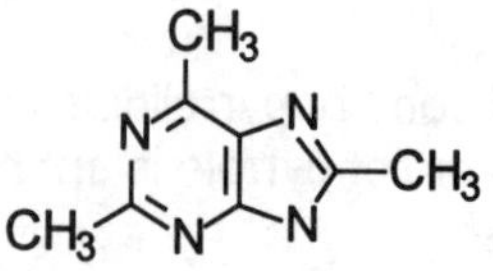

25.12 **(a)** 3-chloropyrrole **(b)** 4-ethylpyrimidine
(c) 3,5-dimethylpyridine **(d)** 1,2-dimethylindole

25.13 A saturated amine has one more hydrogen atom than an alkane with the same number of carbon atoms. The general formula is C_nH_{2n+3}. A cyclic amine has two fewer hydrogen atoms, so the general formula is C_nH_{2n+1}.

25.14 **(a)** two; they are ethylamine and dimethylamine
(b) four; they are propylamine, isopropylamine, ethylmethylamine and trimethylamine
(c) eight; however, three are enamines

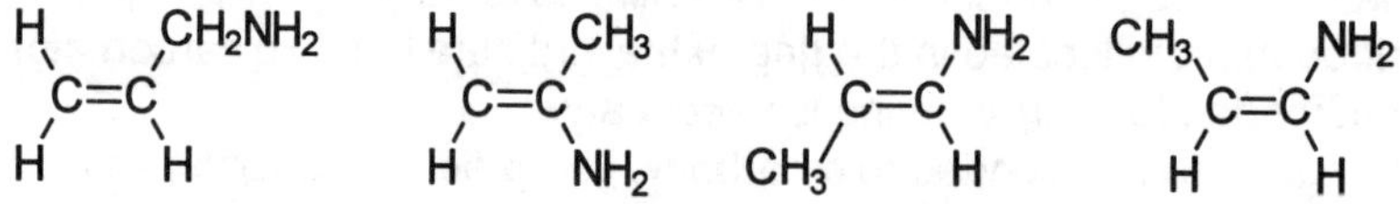

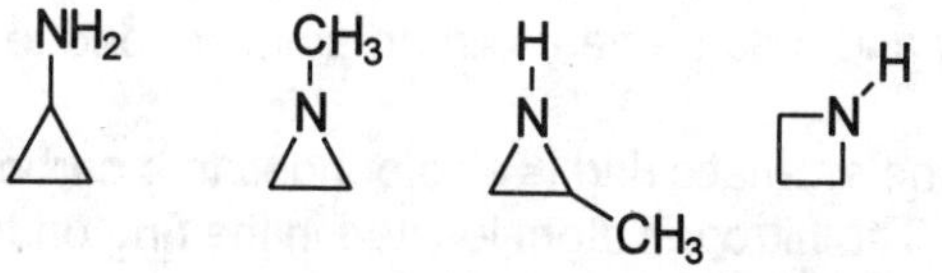

25.15 There are four possible isomeric alkyl groups that can be part of the isomeric primary amines. The alkyl groups are butyl, *sec*-butyl, isobutyl, and *tert*-butyl.

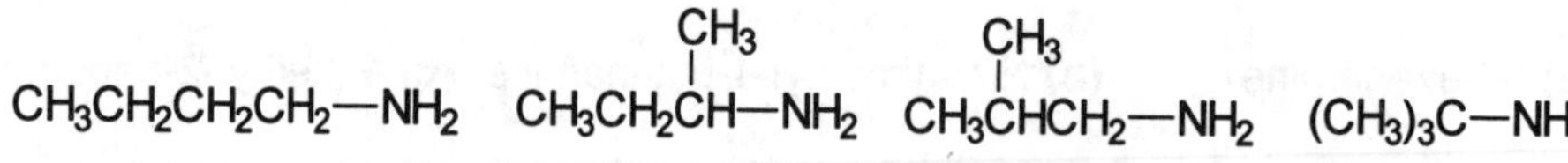

25.16 The five carbon atoms must be distributed among three alkyl groups. If two of the groups are methyl then the remaining three carbon atoms may be in either a propyl or an isopropyl group. The only other possible distribution of atoms is among two ethyl groups and a methyl group. Thus, there are three isomeric tertiary amines.

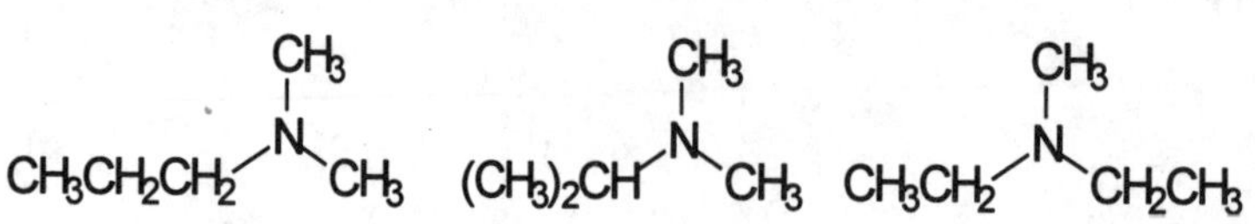

25.17 Propylamine can form hydrogen bonds between an N—H bond and an electron pair on a neighboring molecule. Trimethylamine has no N—H bond and cannot form hydrogen bonds. Hydrogen bonding increases the boiling point of propylamine.

25.18   1,2-Diaminoethane has two sites within the molecule that can form hydrogen bonds. Both nitrogen atoms provide hydrogen bond donor sites and hydrogen bond acceptor sites. The extent of hydrogen bonding is greater than for propylamine which has only one site within the molecule where hydrogen bonds may form.

25.19   Hydrogen bonds form more readily at the nitrogen atom of 1-butanamine because the site is sterically unhindered. In dibutylamine, the nitrogen atom is secondary and the site is more crowded. Hydrogen bonding does not occur as easily.

25.20   Tertiary amines have no N—H bond and cannot form hydrogen bonds. There is some small polarity in amines compared to alkanes but the dipole-dipole forces are small.

25.21   The stronger base will have a smaller pK_b value. Thus, triethylamine is the stronger base.

25.22   The stronger base will have a larger K_b value. Thus, dimethylamine is the stronger base.

25.23   The smaller pK_a of the conjugate acid of the second compound means that it is the stronger acid. Thus, the second amine is the weaker base. The cyano group is inductively electron withdrawing and it decreases the electron density at the nitrogen atom of the amine. In the second compound the distance between the nitrogen atom and the cyano group is smaller, so the inductive effect is more pronounced.

25.24   The smaller pK_a of the conjugate acid of the second compound means that it is the stronger acid. Thus, the second amine is the weaker base. The methoxy group is inductively electron withdrawing and it decreases the electron density at the nitrogen atom of the amine. Note however that the effect is relatively small due to the distance between the methoxy group and the nitrogen atom.

25.25   The acidity of the N—H bond depends on the hybridization of the nitrogen atom. The tendency to hold electrons closer to the nitrogen atom increases with increasing %s character of the bond, which is in the order $sp^3 < sp^2 < sp$. The %s character changes from 25% to 33% to 50% in this series. The difference in % s character is larger between sp^2 and sp than between sp^3 and sp^2.

25.26   Based on information of Exercise 25.25, we would expect that the sp^3 hybridized nitrogen atoms would be more basic than the imine nitrogen atom. However, the amine groups are part of an electron delocalized system with the multiple bond of the imine.

As a consequence, protonation at either amino group destabilizes the conjugate acid because the number of resonance contributors is reduced.

Protonation at the imine nitrogen atom gives a conjugate acid that is resonance stabilized. Note that all three contributors are equivalent.

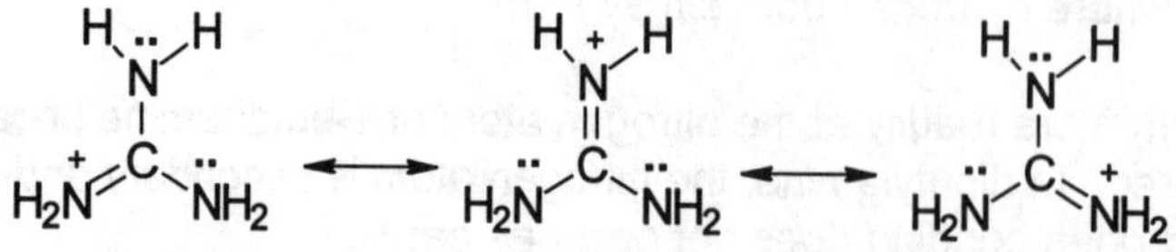

25.27 The site of protonation is at the sp^2 orbital of nitrogen in oxazole, because nitrogen is more basic than oxygen. The oxygen atom of oxazole decreases the electron density at the nitrogen atom by an inductive effect. Thus, the compound is a weaker base than the sulfur analog. Sulfur is less electronegative than oxygen and inductively is a weaker electron withdrawing atom.

25.28 The site of protonation is at the sp^2 orbital of nitrogen in both pyrazole and imidazole. The nitrogen atom bonded to hydrogen is not basic, because its lone pair of electrons is part of the aromatic ring system. However, that nitrogen atom withdraws electrons inductively from the atom where protonation occurs. The inductive effect is larger in pyrazole because the two atoms are closer. Inductive effects decrease with distance separating an electron withdrawing group and a reaction center.

25.29 **(a)** $CH_3CH_2CH_2CH_2CH_2OH$ $\xrightarrow[H_2SO_4]{CrO_3}$ $CH_3CH_2CH_2CH_2CO_2H$ $\xrightarrow[pyridine]{SOCl_2}$ $CH_3CH_2CH_2CH_2\overset{O}{\overset{||}{C}}Cl$

$\xrightarrow{NH_3}$ $CH_3CH_2CH_2CH_2\overset{O}{\overset{||}{C}}NH_2$ $\xrightarrow[NaOH]{Br_2}$ $CH_3CH_2CH_2CH_2NH_2$

(b) $CH_3CH_2CH_2CH_2CH_2OH$ $\xrightarrow{PBr_3}$ $CH_3CH_2CH_2CH_2CH_2Br$

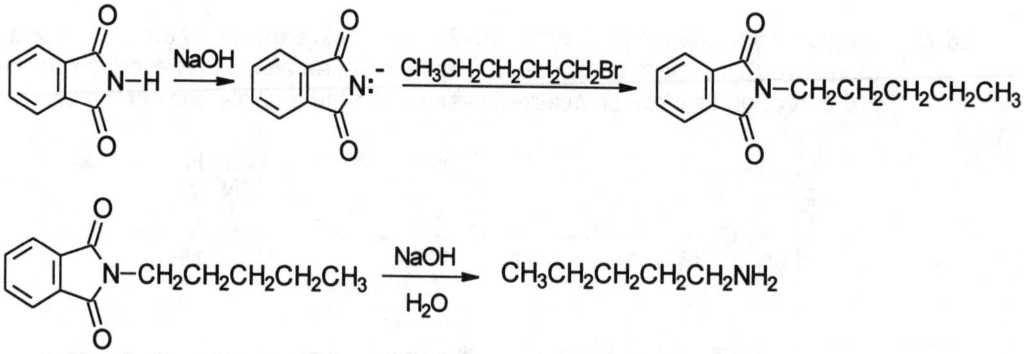

(c) $CH_3CH_2CH_2CH_2CH_2OH$ $\xrightarrow{PBr_3}$ $CH_3CH_2CH_2CH_2CH_2Br$ $\xrightarrow[DMF]{NaCN}$ $CH_3CH_2CH_2CH_2CH_2CN$

$\xrightarrow[\text{2. } H_3O^+]{\text{1. LiAlH}_4\text{/ether}}$ $CH_3CH_2CH_2CH_2CH_2CH_2NH_2$

25.30 **(a)**

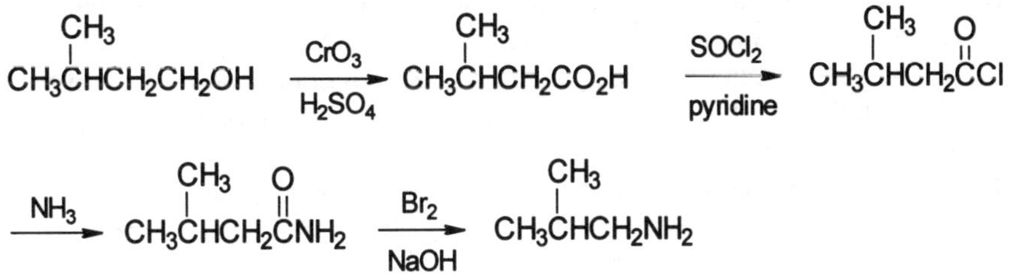

$$CH_3\underset{\underset{CH_3}{|}}{C}HCH_2CH_2OH \xrightarrow[H_2SO_4]{CrO_3} CH_3\underset{\underset{CH_3}{|}}{C}HCH_2CO_2H \xrightarrow[pyridine]{SOCl_2} CH_3\underset{\underset{CH_3}{|}}{C}HCH_2\overset{\overset{O}{||}}{C}Cl$$

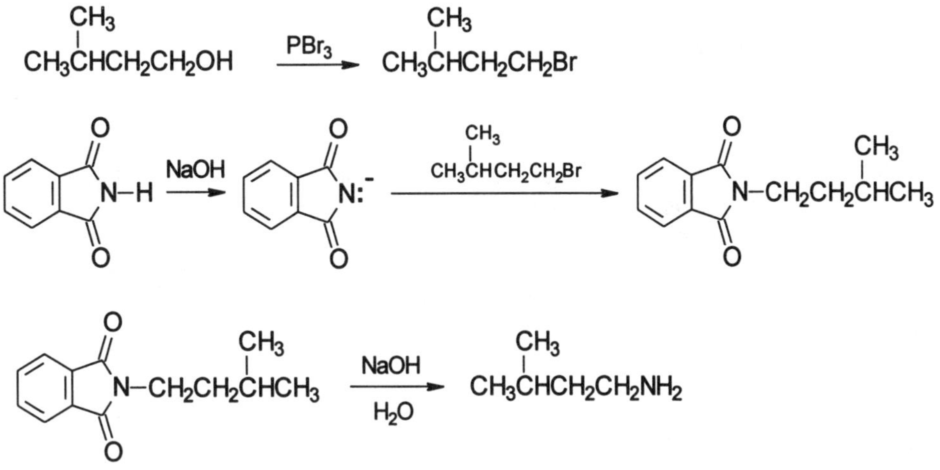

$$\xrightarrow{NH_3} CH_3\underset{\underset{CH_3}{|}}{C}HCH_2\overset{\overset{O}{||}}{C}NH_2 \xrightarrow[NaOH]{Br_2} CH_3\underset{\underset{CH_3}{|}}{C}HCH_2NH_2$$

(b)

$$CH_3\underset{\underset{CH_3}{|}}{C}HCH_2CH_2OH \xrightarrow{PBr_3} CH_3\underset{\underset{CH_3}{|}}{C}HCH_2CH_2Br$$

Phthalimide $\xrightarrow{NaOH}$ phthalimide anion $\xrightarrow{CH_3CHCH_2CH_2Br}$ N-substituted phthalimide

N-CH_2CH_2CHCH_3 $\xrightarrow[H_2O]{NaOH}$ $CH_3\underset{\underset{CH_3}{|}}{C}HCH_2CH_2NH_2$

(c)

$$CH_3\underset{\underset{CH_3}{|}}{C}HCH_2CH_2OH \xrightarrow{PBr_3} CH_3\underset{\underset{CH_3}{|}}{C}HCH_2CH_2Br \xrightarrow[DMF]{NaCN} CH_3\underset{\underset{CH_3}{|}}{C}HCH_2CH_2CN$$

$$\xrightarrow[2.\ H_3O^+]{1.\ LiAlH_4/ether} CH_3\underset{\underset{CH_3}{|}}{C}HCH_2CH_2CH_2NH_2$$

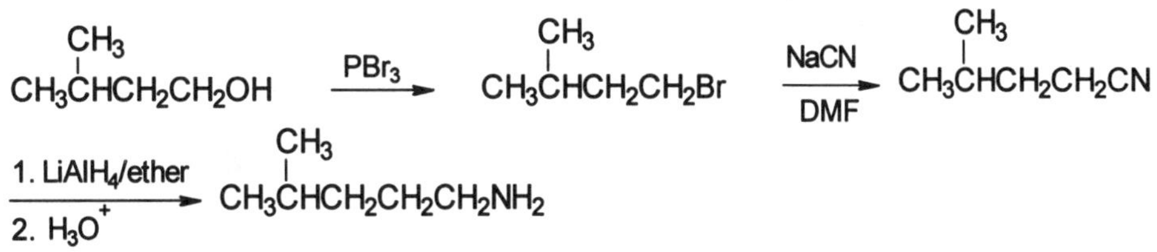

25.31 **(a)** Convert the alcohol to the trans chloro compound using thionyl chloride in dioxane, which will retain the configuration. Prepare the conjugate base of phthalimide and react it with the halogen compound. The reaction occurs with inversion of configuration. Hydrolyze the product.

(b) Convert the carboxylic acid to an amide by reacting it with $SOCl_2$ /pyridine followed by ammonia. Treat the amide with bromine and sodium hydroxide in the Hofmann rearrangement, which occurs with retention of configuration.

(c) Prepare the conjugate base of phthalimide and react it with the halogen compound. The reaction occurs with inversion of configuration. Hydrolyze the product.

(d) Convert the alcohol to the trans chloro compound using thionyl chloride in pyridine which will invert the configuration. Prepare the conjugate base of phthalimide and react it with the halogen compound. This reaction also occurs with inversion of configuration, so the methyl group and the phthalimide group are cis to one another. Hydrolyze the product.

25.32 **(a)** Convert the acid into an acid chloride using thionyl chloride and then react it with ethylamine to give N-ethylbenzamide. Reduce the product with $LiAlH_4$

(b) React the chloro compound with sodium cyanide in dimethylformamide. Reduce the product with lithium aluminum hydride.

(c) React the dibromo compound with two equivalents of sodium cyanide in dimethylformamide. Reduce the product with lithium aluminum hydride.

25.33 The displacement by azide ion occurs with inversion of configuration. Subsequent reduction of the azide does not affect the configuration of the stereogenic center. The product has the (S) configuration.

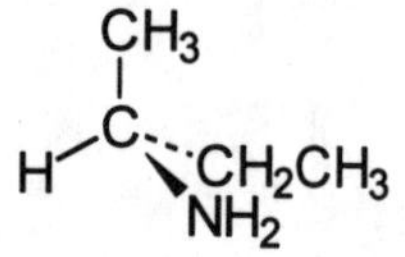

25.34 The ring opening of the epoxide by azide ion occurs with inversion of configuration at the site of the attack by the azide ion. The configuration of the carbon atom of the alkoxide ion and subsequent alcohol is retained. Thus, the azido alcohol is the trans isomer. Subsequent reduction of the azide does not affect the configuration of the stereogenic center. The product is the trans isomer.

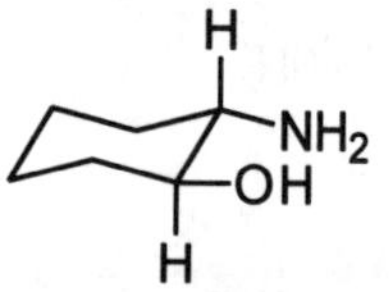

25.35 (a) 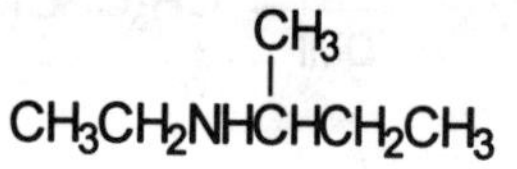 $CH_3CH_2NHCHCH_2CH_3$ (with CH_3 substituent) (b) 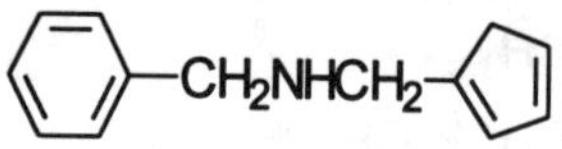 $\bigcirc-CH_2NHCH_2-\bigcirc$ (c) 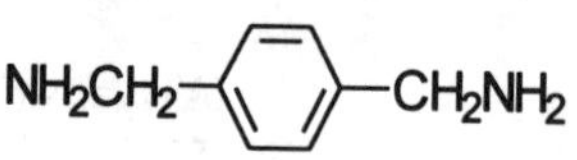 $NH_2CH_2-\bigcirc-CH_2NH_2$

25.36 (a) 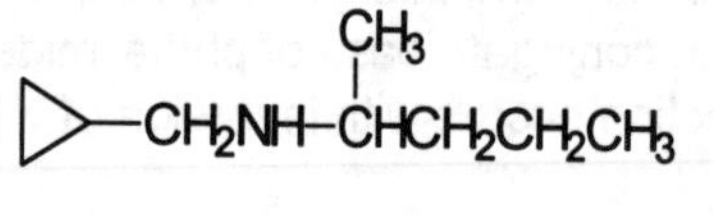 $\triangleright-CH_2NH-CHCH_2CH_2CH_3$ (with CH_3 substituent) (b)  $NH_2CH_2-\bigcirc-CH_2NH_2$ (c) 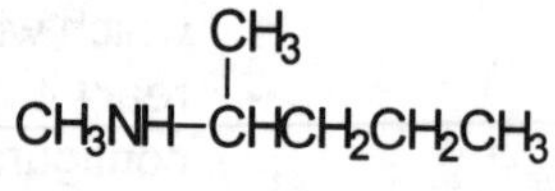 $CH_3NH-CHCH_2CH_2CH_3$ (with CH_3 substituent)

25.37 (a) 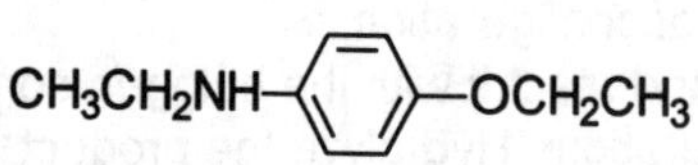 $CH_3CH_2NH-\bigcirc-OCH_2CH_3$ (b) 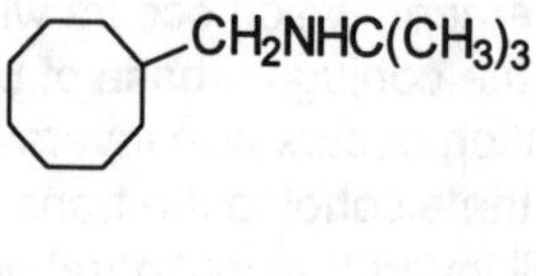 $\bigcirc-CH_2NHC(CH_3)_3$

(c) 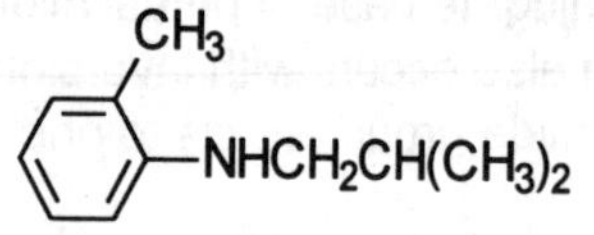 $\bigcirc-NHCH_2CH(CH_3)_2$ (with CH_3 substituent) (d) 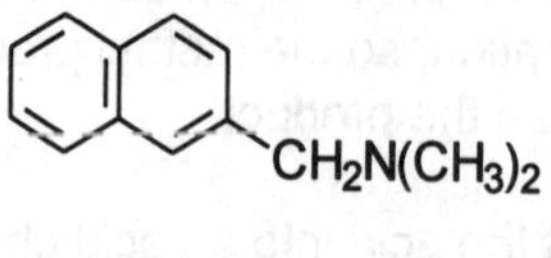 $\bigcirc\bigcirc-CH_2N(CH_3)_2$

25.38 **(a)** 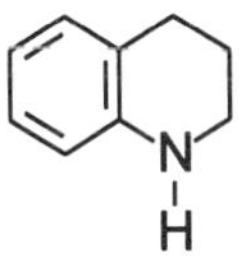 **(b)** 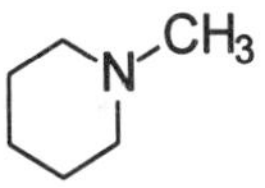 **(c)**

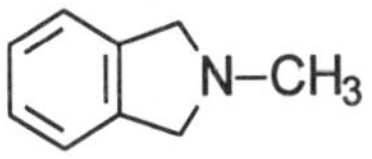

25.39 **(a)** Reaction with thionyl chloride yields an acid chloride, which is converted into an amide by the reaction with ammonia. Reduction of the amide gives a primary amine.
(b) Reaction with cyanide displaces the bromide ion to give a nitrile. Reduction of the nitrile gives a primary amine.
(c) Reaction with lithium aluminum hydride gives an alcohol. The subsequent reaction with phosphorus tribromide gives a bromoalkane. Excess ammonia favors formation of a primary amine.

(a) $CH_3(CH_2)_4CH_2NH_2$ **(b)** 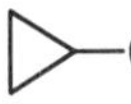 $\triangleright\!\!-CH_2CH_2CH_2NH_2$ **(c)** $CH_3(CH_2)_3CH_2NH_2$

25.40 **(a)** Reaction with PCC gives a ketone. Subsequent reaction with methylamine under reductive amination conditions gives a secondary amine.
(b) Reaction with thionyl chloride yields an acid chloride, which is converted into an amide by the reaction with ammonia. Reduction of the amide gives a primary amine.
(c) Reaction with excess HBr gives a dibromo compound. Reaction with excess cyanide ion gives a dinitrile, which when reduced gives a diamine.

(a) 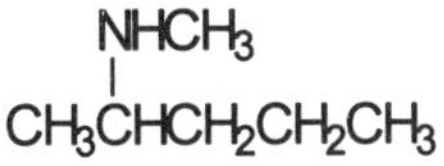 $\overset{\displaystyle NHCH_3}{\underset{\displaystyle |}{CH_3CHCH_2CH_2CH_3}}$ **(b)** 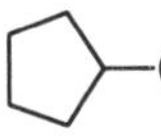 $\bigcirc\!\!-CH_2CH_2NH_2$ **(c)** $NH_2CH_2CH_2(CH_2)_4CH_2CH_2NH_2$

25.41 **(a)** React the 1-phenyl-2-bromopropane with excess aminoethane. Multiple alkylation is less likely for this process because the secondary amine product is moderately hindered.
(b) React the 1-phenyl-2-propanone with aminoethane under conditions of reductive amination. Use sodium cyanoborohydride in methanol as the solvent.
(c) React the 1-phenyl-2-propanamine with acetyl chloride to produce an amide. Reduce the amide with lithium aluminum hydride in ether as the solvent.
(d) Prepare the amide of 2-methyl-3-phenylpropanoic acid by reacting it with $SOCl_2$ and then reacting the acid chloride with ammonia. React the amide with bromine and base under the conditions of the Hofmann rearrangement to give 1-phenyl-2-propanamine. Then proceed as in (c).

25.42 Form the ester by reaction with 2-(N,N-dimethyl)ethanol. Reduce the nitro group using tin and HCl. React the amine with butanal in the presence of hydrogen and Raney nickel.

25.43   (a) 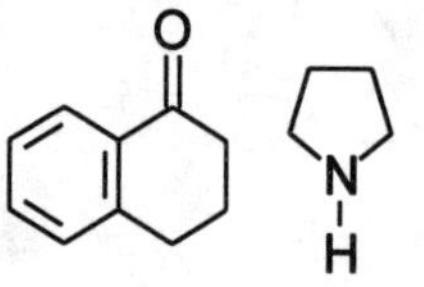   (b) 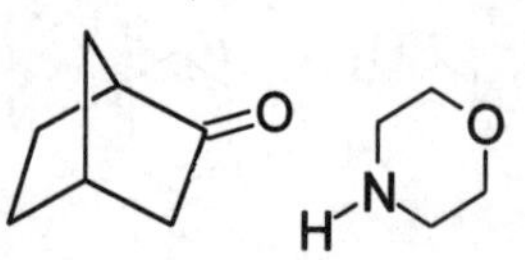   (c)

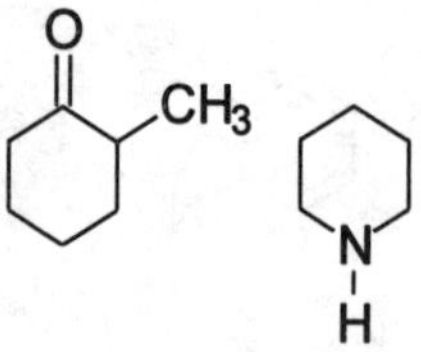

25.44   (a) 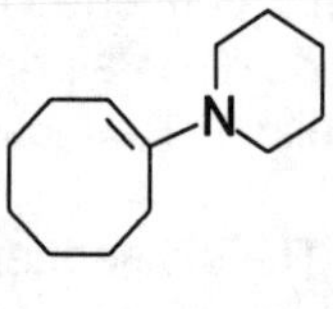   (b) 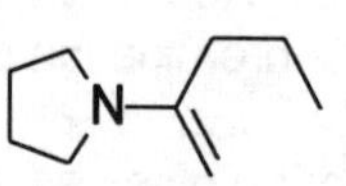   (c)

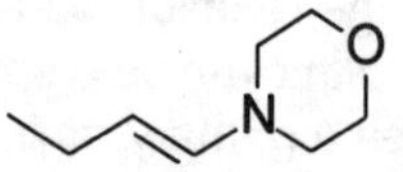

25.45   In each reaction, the double bond of the enamine acts as a nucleophile. Subsequent hydrolysis of the iminium ion obtained gives the alkylated product.

(a) 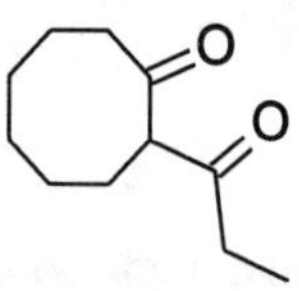   (b) 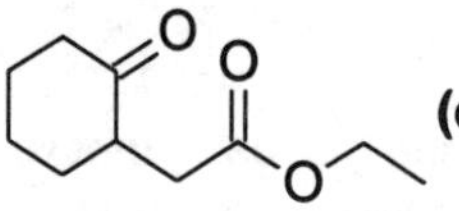   (c)

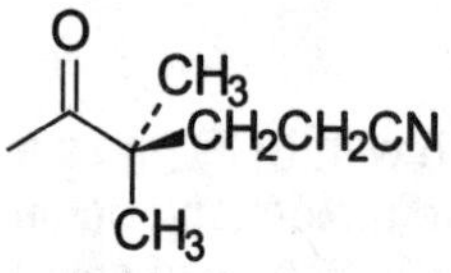

25.46   The enamine with a double bond in conjugation with the aromatic ring is formed. The allyl group alkylates the enamine.

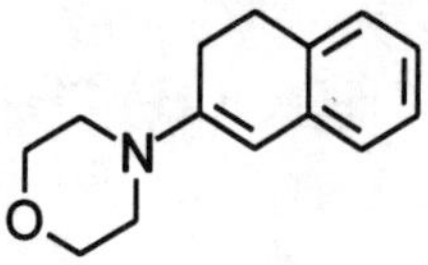

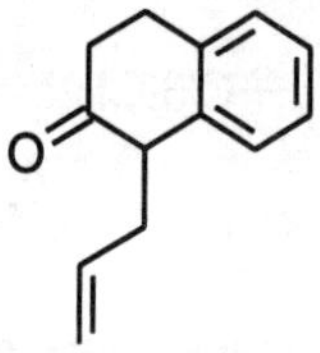

25.47   (a)

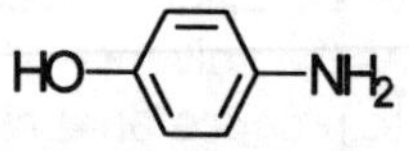

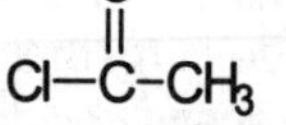

(b)

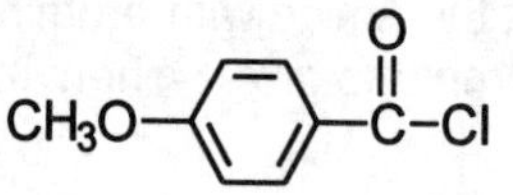

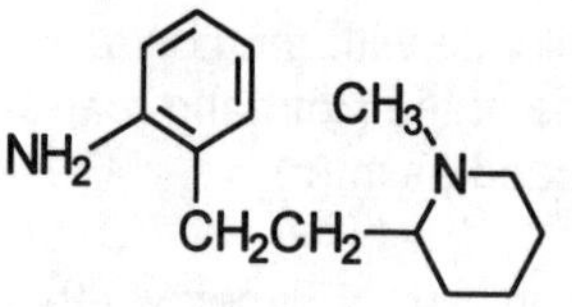

(c)

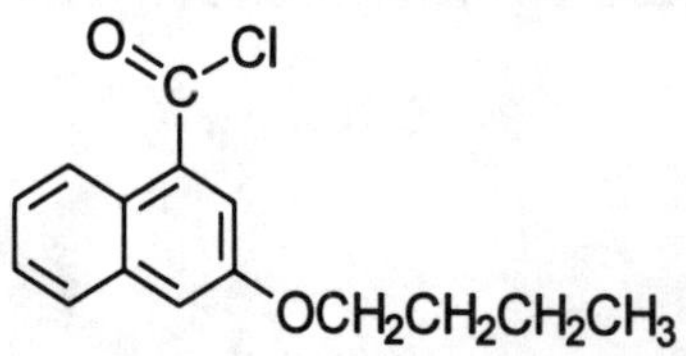

$NH_2CH_2CH_2N(CH_2CH_3)_2$

25.48 **(a)**

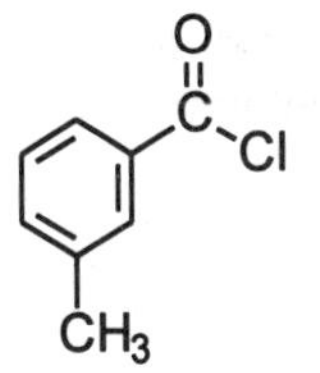

(CH$_3$CH$_2$)$_2$NH

(b)

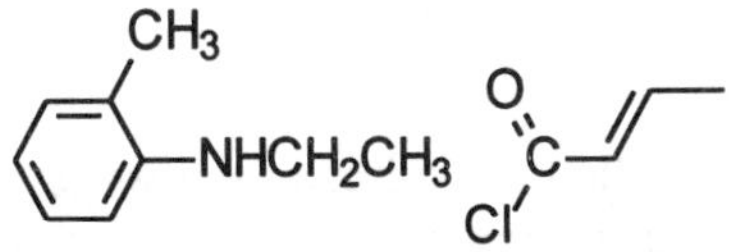

(c)

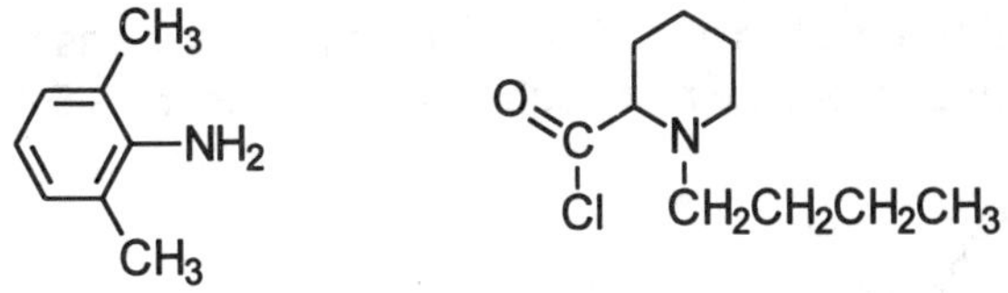

25.49 The amide bond is more difficult to hydrolyze than an ester bond.

25.50

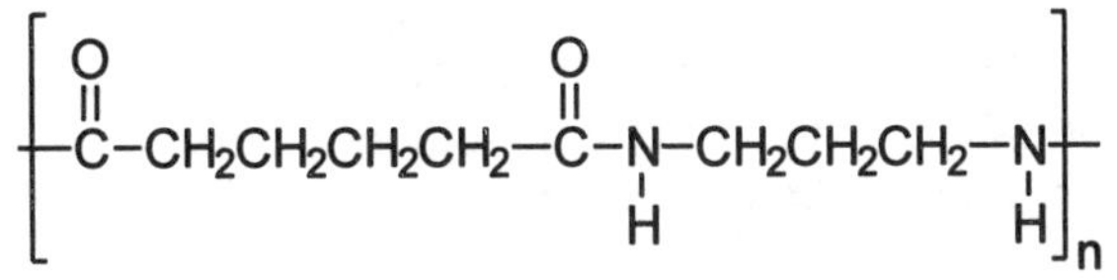

25.51 Neither I nor III give evidence of a reaction. Compound II, which is a secondary amine, gives an insoluble sulfonamide. Addition of acid to the solution of III will yield a precipitate of a water insoluble sulfonamide. Compound I remains in solution even after acidification because it did not react with the reagent.

25.52 Compound I reacts with the benzenesulfonyl chloride to give a base soluble compound. Addition of acid gives a precipitate. Compound II, which is a secondary amine, gives an insoluble sulfonamide. Compound III gives no evidence of a reaction and no precipitate forms when acid is added.

25.53 The acidity of an acid is increased if its conjugate base is stabilized by electron withdrawing substituents. In the case of sulfamethazine, the methyl groups donate electron density and their effect is opposite that required to stabilize the conjugate base.

25.54 The acidity of an acid is increased if its conjugate base is stabilized by electron withdrawing substituents. In the case of sulfadiazine, the "extra" nitrogen atom compared to the structure of sulfapyridine causes greater withdrawal of electron density from the nitrogen atom of the sulfonamide.

25.55 React 1,6-dibromohexane with two moles of the following amine. Then exhaustively methylate with methylamine to place two methyl groups on each nitrogen atom..

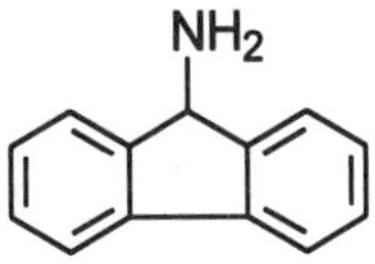

25.56   There are two possible primary amines. One is benzylamine. The structure of the second is shown below. This is compound is the more likely candidate, because it is relatively easy to alkylate amines with the very reactive benzylchloride.

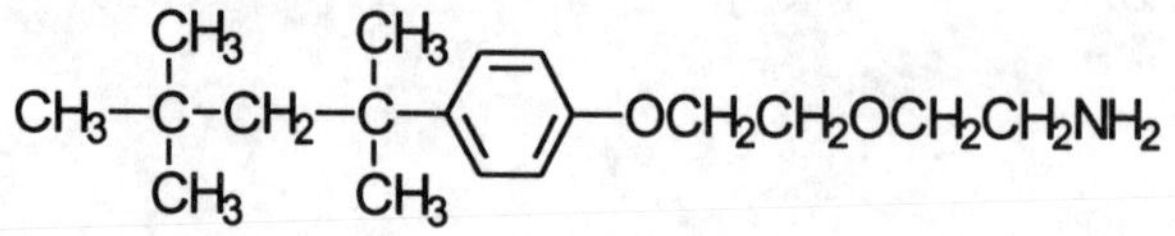

$$CH_3-\underset{\underset{CH_3}{|}}{\overset{\overset{CH_3}{|}}{C}}-CH_2-\underset{\underset{CH_3}{|}}{\overset{\overset{CH_3}{|}}{C}}-\bigcirc-OCH_2CH_2OCH_2CH_2NH_2$$

25.57   The major product of the Hofmann elimination is the less substituted alkene, derived from elimination of a hydrogen atom from either of the two methyl groups bonded to nitrogen. The Hofmann is regioselective, resulting in a 12:1 product ratio in this case. The β elimination of the bromoalkane favors the more substituted product by a 6:4 ratio.

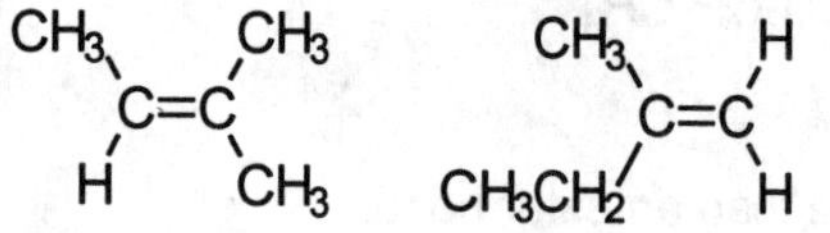

$$\underset{H}{\overset{CH_3}{\diagdown}}C=C\underset{CH_3}{\overset{CH_3}{\diagup}} \qquad \underset{CH_3CH_2}{\overset{CH_3}{\diagdown}}C=C\underset{H}{\overset{H}{\diagup}}$$

25.58   The conformation required to form propene has a gauche arrangement of the methyl group of the propyl group and the dibutylpropylamino group. This conformation is less sterically hindered than the conformation required to form 1-butene. It has a gauche arrangement of an ethyl group and the butyldipropylamino group.

25.59   Ethylene forms in the largest amount because it is the least substituted alkene and there is no steric hindrance in the conformation required for the elimination reaction. There are also two alkenes containing a phenyl ring, which form in lesser amounts. Although the alkene derived from loss of hydrogen from the methyl group is less substituted, the alkene derived from loss of hydrogen at the methylene group is conjugated with the phenyl ring, so it would form in a greater amount.

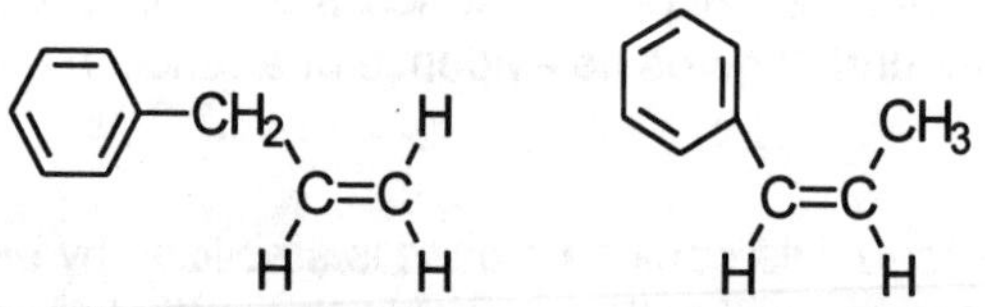

$$\bigcirc-CH_2\underset{H}{\overset{H}{\diagdown}}C=C\underset{H}{\overset{H}{\diagup}} \qquad \bigcirc\underset{H}{\overset{}{\diagdown}}C=C\underset{H}{\overset{CH_3}{\diagup}}$$

25.60   To visualize the product, write the structure in the eclipsed conformation derived by "removing" the structure from its two dimensional Fischer projection. Then rotate about the carbon-carbon bond to form a staggered conformation with a hydrogen atom and a nitrogen atom in an anti periplanar arrangement. The quaternary ammonium ion derived from this amine is used to predict the stereochemistry of the product alkene, in which the phenyl groups are cis to one another.

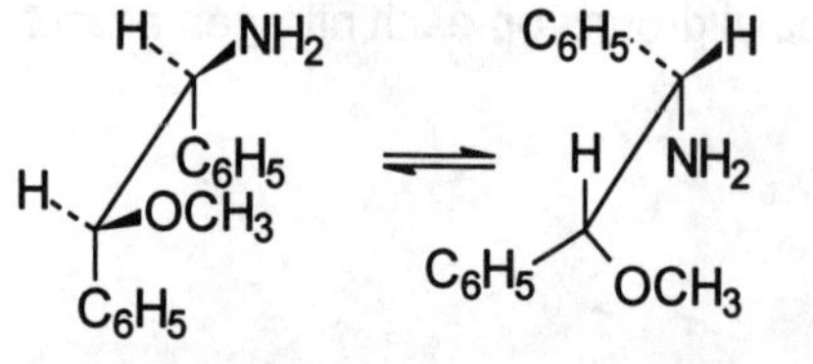

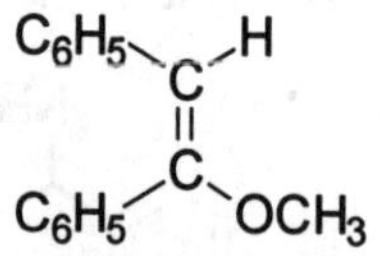

25.61 **(a)** $CH_2\!=\!CH_2$ **(b)**  [cyclohexene structure] **(c)** 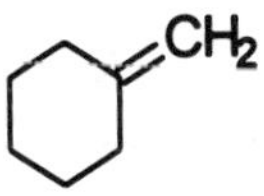 [methylenecyclohexane structure with CH_2]

25.62 [1,2-divinyl / 3-methylene-1,4-pentadiene type structure]

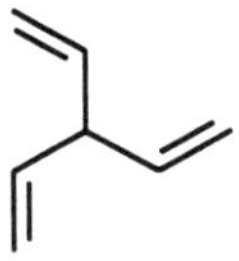

25.63 (a) 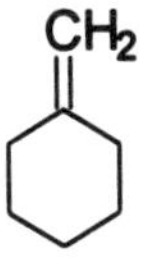 [methylenecyclohexane structure with CH_2] **(b)** 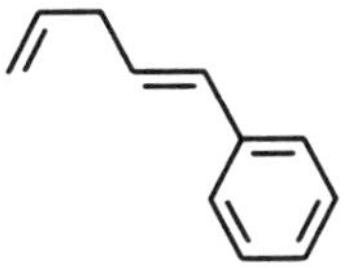 [1-phenyl-1,3-butadiene structure] **(c)** 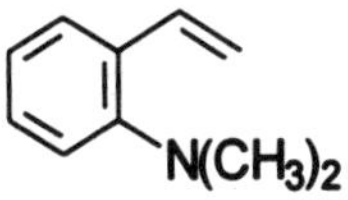 [benzene ring with vinyl group and $N(CH_3)_2$ substituent]

25.64 $CH_3\!-\!\underset{\underset{\text{(piperidine ring)}}{}}{\overset{\overset{H}{|}}{N}}\!-\!CH_2(CH_2)_9CH_3$

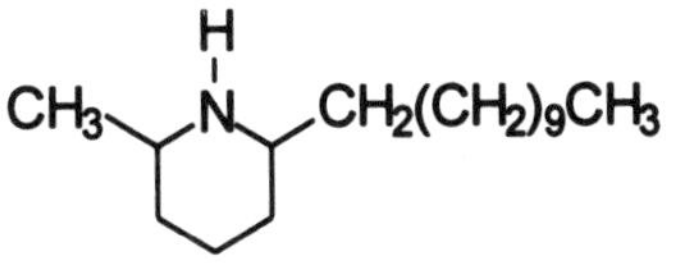

25.65 The compound must be a secondary amine. There are three possible reactants.

$CH_3CH_2NHCH_2CH_3$ $CH_3CH_2CH_2NHCH_3$ $(CH_3)_2CHNHCH_3$

25.66 The compound must be a primary amine. There are four possible reactants.

$CH_3CH_2CH_2CH_2NH_2$ $CH_3\overset{\overset{CH_3}{|}}{CH}CH_2NH_2$ $(CH_3)_3CNH_2$ $CH_3CH_2\overset{\overset{NH_2}{|}}{CH}CH_3$

25.67 The carbocation formed by loss of nitrogen from the diazonium ion reacts with water, which is a nucleophile, to give the substitution product cyclohexanol. It can also lose a proton to a base to give the elimination product cyclohexene.

25.68 The cyclopentylmethyl carbocation is primary. It rearranges by migration of a methylene group to give the cyclohexyl carbocation, which is secondary and which reacts with water to give the substitution product cyclohexanol.

25.69 After the formation of the diazonium ion, the product is formed by a mechanism similar to that of the pinacol rearrangement. Nitrogen is the leaving group in this reaction whereas water is the leaving group in the pinacol rearrangement (Section 16.5). A methyl group migrates to the tertiary carbocation center to give a secondary carbocation stabilized by an electron pair of the oxygen atom. Loss of a proton from the hydroxyl group gives a ketone.

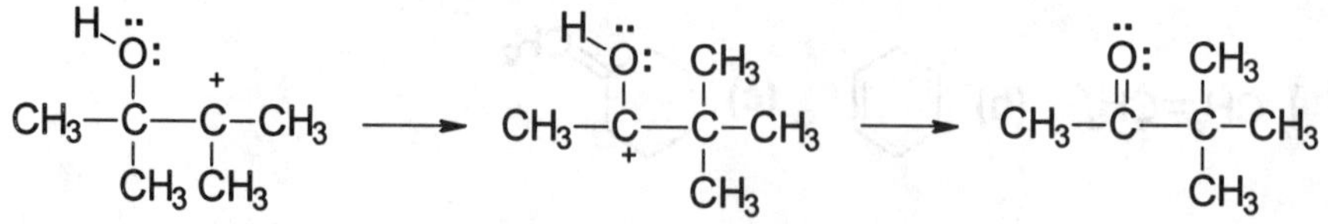

25.70   After formation of the diazonium ion, the product is formed by a mechanism similar to that of the pinacol rearrangement. Nitrogen is the leaving group in this reaction. A methylene group migrates to the primary carbocation center to give a secondary carbocation stabilized by an electron pair of the oxygen atom. Loss of a proton from the hydroxyl group gives a ketone product.

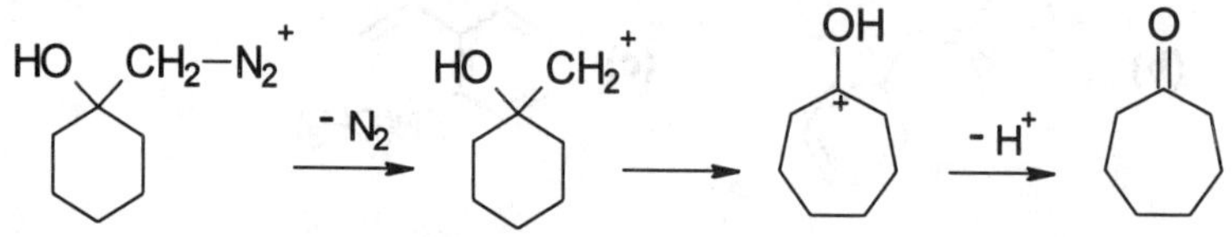

25.71   **(a)** There are two absorptions in the 3200-3375 cm^{-1} region due to the N—H bonds in the NH_2 group.

   (b) There is one absorption in the 3200-3375 cm^{-1} region due to the single N—H bond.

   (c) There are no absorptions in the 3200-3375 cm^{-1} region because there are no N—H bonds in the molecule.

25.72   **(a)** Only ethyldimethylamine has no N—H bonds in the molecule and therefore no absorption in the 3200-3375 cm^{-1} region.

   (b) Secondary amines have a single N—H bond in the molecule and give a single absorption in the 3200-3375 cm^{-1} region. There are three possible isomers whose structures are given below.

$$CH_3CH_2-NH-CH_2CH_3 \quad CH_3CH_2CH_2-NH-CH_3 \quad \underset{\underset{CH_3}{|}}{CH_3CH}-NH-CH_3$$

   (c) Primary amines have two N—H bonds in the molecule and give two absorptions in the 3200-3375 cm^{-1} region.   There are four possible isomers whose structures are given below.

$$CH_3CH_2CH_2CH_2-NH_2 \quad \underset{\underset{CH_3}{|}}{CH_3CH_2CH}-NH_2 \quad \underset{\underset{CH_3}{|}}{CH_3CHCH_2}-NH_2 \quad (CH_3)_3C-NH_2$$

25.73   **(a)** The quartet and triplet pattern is typical of an ethyl group. The six carbon atoms are distributed in three equivalent ethyl groups in triethylamine. Note that the peak for the methylene groups appears at 2.5 δ as expected for groups attached directly to the nitrogen atom.

(b) The pattern is typical of an isopropyl group, although the expected heptet is such a weak signal that it appears as a quintet. The six carbon atoms are distributed in two equivalent isopropyl groups in diisopropylamine.

25.74 **(a)** The singlet of intensity 12 is due to twelve equivalent protons, a good indicator of four methyl groups; its position at 2.25 δ indicates that the protons are bonded to carbon atoms bonded to nitrogen. The singlet of intensity 2 is due to two equivalent protons; its position at 2.7 δ indicates that these protons are also bonded to a carbon atom bonded to nitrogen. The compound is shown below.

(b) The triplet of intensity 4 at 2.7 δ is due to four equivalent protons on carbon atoms bonded to nitrogen and split by a neighboring methylene group. The singlet of intensity 4 at 1.1 δ is due to four equivalent protons bonded to nitrogen. The quintet of intensity 2 at 1.6 δ is due to two equivalent protons split by a two neighboring methylene groups. The compound is shown below.

(c) The singlet of intensity 2 at 2.5 δ is due to two equivalent protons on a carbon atom that has no neighboring protons and is bonded to nitrogen. The singlet of intensity 6 at 1.1 δ is due to six equivalent protons on carbon atoms that have no neighboring protons and are not bonded to nitrogen. The singlet of intensity 4 at 1.2 δ is due to four protons bonded to nitrogen atoms. The compound is shown below. Note that a single peak is observed for the protons bonded to nitrogen, although the nitrogen atoms are not equivalent. This is because these protons undergo such rapid exchange that a single "time-averaged" signal is obtained.

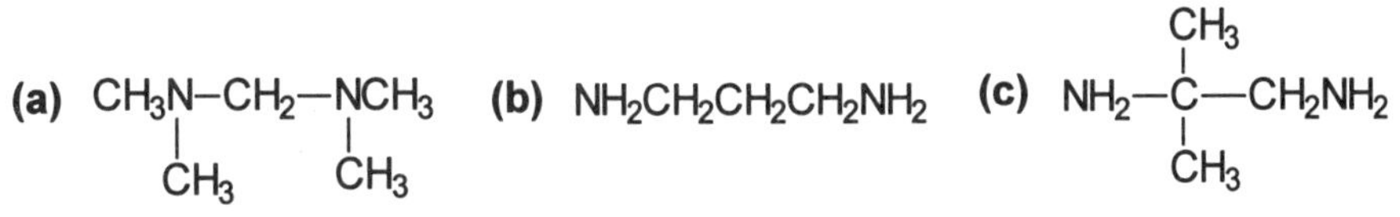

(a) $CH_3N-CH_2-NCH_3$
　　　　$|$　　　$|$
　　　CH_3　　CH_3

(b) $NH_2CH_2CH_2CH_2NH_2$

(c)
　　　　　　CH_3
　　　　　　$|$
　$NH_2-C-CH_2NH_2$
　　　　　　$|$
　　　　　CH_3

25.75 The absence of an absorbance in the 3200-3600 cm^{-1} region of the IR spectrum indicates that the amines are tertiary.

(a) The compound has six carbon atoms but only two signals, so in the tertiary amine there must be two sets of three equivalent carbon atoms. The quartet at 12.6 ppm is due to methyl groups, but they are not bonded to a nitrogen atom. The triplet at 46.9 ppm is due to methylene groups bonded to a nitrogen atom. The compound is triethylamine, as shown below.

(b) The compound has six carbon atoms but only three signals, so there must be three sets of equivalent carbon atoms. The quartet at 25.6 ppm is due to one or more methyl groups, but they are not bonded to a nitrogen atom. The quartet at 38.7 ppm is due to one or more methyl groups which are bonded to a nitrogen atom. The singlet at 53.2 ppm is due to a carbon atom bonded to a nitrogen atom but no hydrogen atoms. The compound is *tert*-butyldimethylamine.

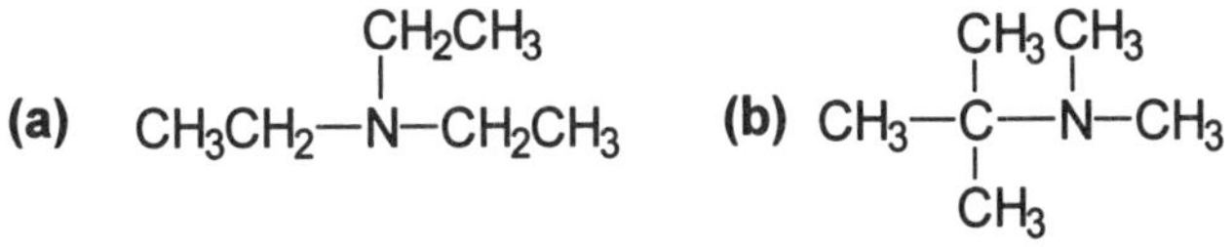

(a)
　　　　　CH_2CH_3
　　　　　　$|$
　$CH_3CH_2-N-CH_2CH_3$

(b)
　　　　　CH_3　CH_3
　　　　　　$|$　　$|$
　$CH_3-C-N-CH_3$
　　　　　　$|$
　　　　　CH_3

25.76   The presence of a single absorbance in the 3200-3600 cm^{-1} region of the IR spectrum indicates that the amines are secondary.

(a) The compound has six carbon atoms but only two signals, so in the secondary amine there must be two sets of equivalent carbon atoms. The quartet at 23.7 ppm is due to methyl groups, but they are not bonded to a nitrogen atom. The doublet at 45.3 ppm is due to a carbon atom bonded to a nitrogen atom and one hydrogen atom. The compound is diisopropylamine, as shown below.

(b) The compound has six carbon atoms but only three signals, so in the secondary amine there must be three sets of two equivalent carbon atoms. The quartet at 12.0 ppm is due to methyl groups, but they are not bonded to a nitrogen atom. The triplet at 23.9 ppm is due to methylene groups, but they are not bonded to a nitrogen atom either. The triplet at 52.3 ppm is due to methylene groups bonded to a nitrogen atom. The compound is dipropylamine.

(a)
$$CH_3CH-\underset{\underset{}{\overset{\overset{H}{|}}{N}}}{}-CHCH_3$$
$$\underset{CH_3}{|} \qquad \underset{CH_3}{|}$$

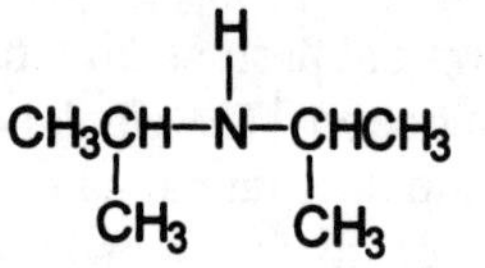

(b) $CH_3CH_2CH_2-NH-CH_2CH_2CH_3$

26

Amino Acids and Proteins

This chapter illustrates the bridge between organic chemistry and biochemistry, which is a science that deals with the chemistry of life processes. The reactions of amino acids and the related polymeric proteins are understandable in an organic chemistry sense based on the chemistry of the amide functional groups and the component amines and carboxylic acids. Although the "reagents" of these organic systems are different than those used in the laboratory, the mechanisms of the reactions are comparable. Finally, chemists have learned from nature, and they now use enzymes in the laboratory to carry out synthetic reactions.

Keys to the Chapter

26.1 Proteins and Polypeptides are Polymers

Proteins and smaller polymers called **polypeptides** are polyamides formed from relatively few types of amino acids. The amide bonds in these compounds are called **peptide bonds**. Proteins are centrally important biomolecules with a range of functions from structural components such as muscles to the catalytic activity of enzymes to the protective function of antibodies.

26.2 Amino Acids

Amino acids contain both an amino group and a carboxylic acid group. A group of 20 important amino acids with the amino group at the α carbon atom is the subject of this chapter. With the exception of glycine these amino acids have the S configuration and are designated as L using the Fischer configurational system.

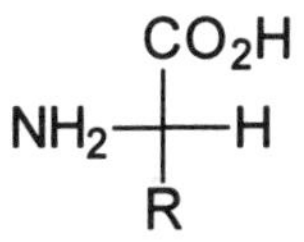

$$
\begin{array}{c}
CO_2H \\
NH_2 \!-\!\!\!\!\!-\!\!\!\!\!- H \\
R
\end{array}
$$

Amino acids are classified according to the relative numbers of amino and carboxylic acid groups in the molecules. If there is one amino group and one carboxylic acid group, the substance is a **neutral amino acid**, although it may contain polar or nonpolar side chains. Compounds that have an additional amino group are **basic amino acids**; compounds with a second carboxylic acid group are **acidic amino acids**.

The side chains of amino acids may interact favorably with water or repel water molecules. These amino acids are said to be **hydrophilic** and **hydrophobic**.

26.3 Acid-Base Properties of Amino Acids

Near a neutral pH, a neutral amino acid exists in an electrically neutral form. However the species does have charges located at the amino and carboxylic acid groups. There is both an ammonium group and a carboxylate group, so the amino acid exists as a **zwitterion**.

$$CH_3-\underset{\underset{NH_3^+}{|}}{\overset{\overset{H}{|}}{C}}-CO_2^-$$

As base is added to a neutral solution, a proton is lost from the ammonium group and a free amino group results. There is still a charge at the carboxylate group, so the resultant species has a negative charge.

$$CH_3-\underset{\underset{NH_2}{|}}{\overset{\overset{H}{|}}{C}}-CO_2^-$$

As acid is added to a neutral solution, a proton is gained by the carboxylate group and a carboxylic acid group results. There is still a charge at the ammonium group, so the resultant species has a positive charge.

$$CH_3-\underset{\underset{NH_3^+}{|}}{\overset{\overset{H}{|}}{C}}-CO_2H$$

A zwitterion is a double ion that has sites of both negative and positive charge in the species. In the case of amino acids, the negative site is a carboxylate ion and the positive site is an ammonium ion. The pH at which the concentration of the zwitterion predominates is the **isoionic point**. For neutral amino acids the isoionic point lies halfway between the pK_a values of the carboxylic acid group and the ammonium group. The pK_a values of amino acids appear in Table 26.4 in the text.

For neutral amino acids, the isoionic point is near 7. Acidic amino acids have isoionic points smaller than 7. For acidic amino acids, the second carboxylic acid group tends to dissociate and exist as a carboxylate ion at a pH near 7. In order to protonate that carboxylate ion, it is necessary to have a low pH (acidic solution). Basic amino acids have isoionic points larger than 7. For basic amino acids, the second amino group tends to be protonated at a pH near neutrality. Thus there are two ammonium ions and only one carboxylate ion. In order to remove a proton from the ammonium ion, it is necessary to have a high pH (basic solution). The isoionic points of selected acidic, neutral, and basic amino acids appear in Table 26.5 in the text.

26.5 Synthesis of Amino Acids

Amino acids can be prepared by the nucleophilic substitution of a halide ion of α-halocarboxylic acids by ammonia. The α-halocarboxylic acids are prepared by the Hell-Volhard-Zelinsky reaction.

The Strecker synthesis occurs by the addition of ammonia to an aldehyde to give an imine which then adds HCN across the carbon-nitrogen double bond to give an α-amino nitrile. The nitrile is then hydrolyzed in a second step to give the α-amino carboxylic acid.

Reductive amination of an α-keto acid using ammonia yields an α-imino acid which is then reduced by hydrogen and a palladium catalyst. The entire reductive amination is carried out in a single step with all reagents present.

The acetamidomalonate synthesis uses a malonate ester-type synthesis. Diethyl acetamidomalonate has an α hydrogen atom that is acidic, and the compound can be converted to a carbanion. The carbanion displaces a halide ion from a haloalkane to form an alkylated product. This alkyl group corresponds to the side chain of the amino acid. Acid-catalyzed hydrolysis cleaves the acetyl group of the amide and also hydrolyzes the ester of the malonate. Under the reaction conditions, the malonic acid undergoes decarboxylation, yielding an amino acid..

26.6 Reactions of Amino Acids

The individual functional groups of amino acids have their own characteristic reactivity. Carboxylic acids are esterified by reaction of an alcohol under acid-catalyzed conditions. Ethyl and benzyl esters are commonly used to protect the carboxyl group. Acid-catalyzed hydrolysis readily reverses the process to unprotect the carboxylic acid. Benzyl esters are cleaved under neutral conditions using hydrogen and palladium. The process, called **hydrogenolysis**, gives toluene as the byproduct.

The α-amino group of an amino acid is protected by conversion to an amide. Benzyl chloroformate is similar to an acyl chloride in its reactivity. It converts an α-amino group into a benzyloxycarbonyl (**Cbz**) derivative. Di-tert-butyldicarbonate is an anhydride that is also used to protect α-amino groups by converting them into a *tert*-butoxycarbonyl (**Boc**) derivative. Either derivative is easily hydrolyzed using trifluoroacetic acid to give a carbamic acid which decarboxylates to give the free α-amino group. The Cbz derivative also reacts with hydrogen and palladium in a hydrogenolysis reactions that gives a carbamic acid which decarboxylates.

26.7 Peptides

The combination of two amino acids joined by an amide bond is a **dipeptide**. Larger numbers of amino acids are **tripeptides, tetrapeptides**, etc. Peptides have two ends. The ends containing the free α-amino acid and the free carboxylic acid are the **N-terminal** and **C-terminal** amino acids, respectively. Peptides are named from the N-terminal amino acid in sequence toward the C-terminal amino acid. The shorthand three-letter representations of the amino acids are used in the name.

The number of possible isomeric peptides with n different amino acids is equal to $n!$ where n! is equal to the product of all numbers from 1 to n.

The function of hormonal peptides depends on the amino acid composition and their sequence in the structure. The differences in function may be the result of very few differences in the structure. Different amino acids that are in different classes such as neutral versus acidic amino acids can cause substantial differences in the function of the peptide.

26.8 Synthesis of Peptides

Peptide synthesis requires precise amide bond formation between specific amino acids. Because each amino acid is both an amine and a carboxylic acid, direct reaction between two amino acids can form the four dipeptides A-A, B-B, A-B, and B-A. In order to obtained a desired product such as B-A, it is necessary to protect the amino group of B and the carboxylic acid group of A. In this way, only the carboxylic acid group of B and the amino group of A can react with each other.

Synthesis of higher peptides such as C-B-A is accomplished by removing the protecting group of B in the dipeptide B-A and reacting it with the amino acid C which is protected at the amino group. Then the free carboxylic acid group of C can react with the free amino group of B-A which is protected at the carboxylic acid group of A.

The carboxyl group of an amino acid is protected by forming a benzyl ester. The amino group is protected by forming a Boc derivative. These two protected amino acids are joined in an amide bond using DCCI. Continued extension of the peptide is accomplished by hydrolysis of the Boc group using trifluoroacetic acid. The free amino group is now available for reaction with another amino acid protected at the amino group. Finally, the polypeptide is unprotected by hydrolysis of both protecting groups using basic hydrolysis.

26.9 Solid Phase Synthesis

The efficiency of the synthesis of a polypeptide is improved by attaching the C-terminal amino acid to a solid polymer. As a consequence, the products of either the joining of amino acid residues or the hydrolysis step stay attached to the polymer. The polymer with the growing polypeptide chain is easily handled by physical methods without mechanical loss. The result is a substantial net yield. Finally, the peptide chain is liberated from the polymer by hydrolysis with anhydrous hydrogen fluoride.

26.10 Determination of Amino Acid Composition in Proteins

The composition of peptides or proteins is determined by complete hydrolysis using 6 M HCl. The identity and number of each amino acid is determined by chromatographic methods. The composition is written using subscripts located on the three letter abbreviation for the amino acid separated by commas between each of the component amino acids.

26.11 Determination of Amino Acid Sequences in Proteins

Partial hydrolysis of a peptide results in the formation of lower molecular weight peptide fragments of varying lengths. A mixture of di-, tri-, and tetrapeptides might be formed. After the structures of the peptide fragments are determined, the fragments are examined for common sequences of amino acids. These common sequences are then aligned and the entire amino acid sequence can be deduced.

The products of an enzyme-catalyzed hydrolysis depend on the enzyme chosen and the amino acid sequence in the peptide. Chymotrypsin catalyzes the hydrolysis of peptide bonds only at the carboxyl end of the aromatic amino acids. Trypsin catalyzes the hydrolysis of peptide bonds only at the carboxyl end of the basic amino acids lysine and arginine. After the structures of the peptide fragments are determined, the manner in which they were originally joined may be determined based on the identity of the terminal amino acids and knowledge of the catalytic activity of the enzyme used.

Carboxypeptidases and aminopeptidases sequentially hydrolyze entire peptides, starting at the free carboxyl end and the free amino end, respectively. The "products" of both carboxypeptidase- and aminopeptidase-catalyzed hydrolysis reactions are complex, because the enzyme continues to hydrolyze the protein chain.

The N-terminal amino acid of a peptide is identified by reaction with phenyl isothiocyanate (the **Edman reagent**) followed by hydrolytic release of a heterocyclic compound called a phenylthiohydantoin. In these compounds, the "R" group of the N-terminal amino acid is attached to a ring carbon atom located between the carbonyl carbon atom and the nitrogen atom of the original amino acid. The amino acid is identified by comparing the phenylthiohydantoin obtained to those of known amino acids. The Edman degradation may be repeated sequentially to identify the amino acids one by one from the N-terminal end.

26.12 Bonding in Proteins

The covalent bonds that "hold" proteins together are the peptide bond and the disulfide bond between cysteine residues. Additional bonds include ionic bonds between charged side chains, as in the case of acidic and basic amino acids. Intramolecular hydrogen bonds occur between amide groups as well as between some side chains with groups such as the hydroxyl group. Hydrophobic interactions between nonpolar side chains also provide some stabilization to the structure of the protein.

26.13 Protein Structure

The biological function of a protein is observed only in its **native conformation**. The entire structure is viewed at various levels called primary through quaternary. The **primary structure** of a protein is the sequence of amino acids and the location of the disulfide bonds. The arrangement of proximate amino acids in a protein chain, which is frequently the result of hydrogen bonds, constitutes its **secondary structure**. The spatial arrangement of amino acids that are far apart in the protein chain is called **tertiary structure**. **Quaternary structure** is the association of several protein chains or subunits into a closely packed arrangement.

Solutions to Exercises

26.1   In the Fischer projection formula, the amino group is on the right for an amino acid with the D configuration.

$$CO_2H$$
$$H-\!\!-\!\!-NH_2$$
$$CH_2CH_2CO_2H$$

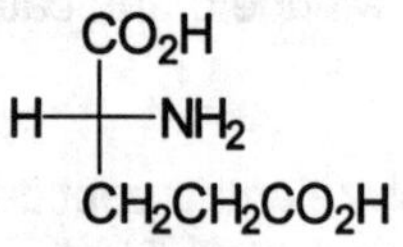

26.2   In the Fischer projection formula, the amino group is on the right for an amino acid with the D configuration.

$$CO_2H$$
$$H-\!\!-\!\!-NH_2$$
$$CH_2-\!\!\bigcirc$$

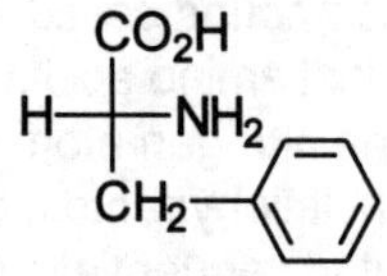

26.3   The amino acid resembles lysine. It has a hydroxyl group bonded to the C-5 atom of the naturally occurring amino acid.

26.4   The amino acid resembles cysteine. It has an allyl group in place of the hydrogen of the S—H group. In addition, the sulfur is in an oxidized state known as a sulfoxide.

26.5   The compound has one carboxylic acid group and one amino group. Thus, it is a neutral amino acid. Its IUPAC name is 4-aminobutanoic acid.

26.6   The compound has two carboxylic acid groups and one amino group. Thus, it is an acidic amino acid. The common name of the dicarboxylic acid is adipic acid. The amino group is on an α carbon atom. The common name is α-aminoadipic acid.

26.7   At pH 1 both compounds exist as their respective conjugate acids.

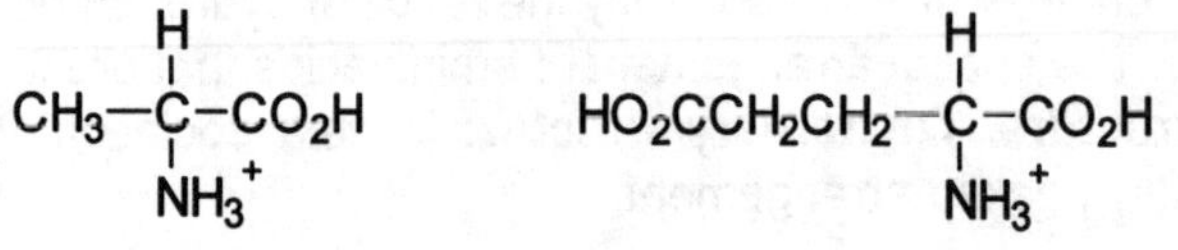

$$CH_3-\overset{H}{\underset{NH_3^+}{C}}-CO_2H \qquad HO_2CCH_2CH_2-\overset{H}{\underset{NH_3^+}{C}}-CO_2H$$

At pH 12 both compounds exist as their respective conjugate bases. Both carboxyl groups of glutamic acid are in the form of carboxylate ions.

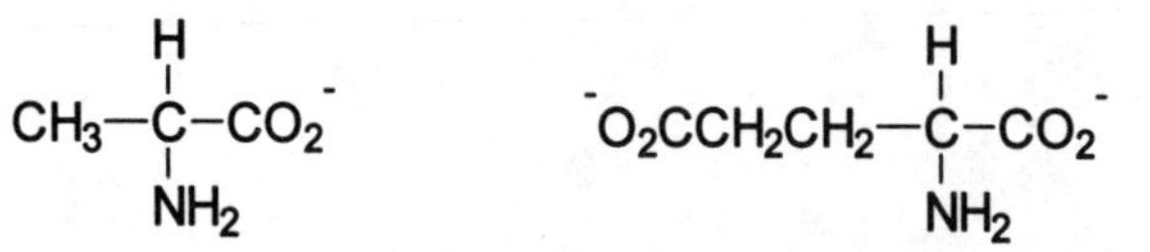

$$CH_3-\overset{H}{\underset{NH_2}{C}}-CO_2^- \qquad {}^-O_2CCH_2CH_2-\overset{H}{\underset{NH_2}{C}}-CO_2^-$$

26.8 The zwitterions have a negative charge on the carboxylate ion and a positive charge on the nitrogen atom of the ammonium ion.

$$\underset{\underset{NH_3^+}{|}}{HOCH_2-\overset{\overset{H}{|}}{C}-CO_2^-} \qquad\qquad \underset{\underset{NH_3^+}{|}}{(CH_3)_2CH-\overset{\overset{H}{|}}{C}-CO_2^-}$$

26.9 Aspartic acid is an acidic amino acid, so its solution has a lower pH than a solution of asparagine, which is a neutral amino acid.

26.10 Lysine is a basic amino acid, so its solutions will be basic.

26.11 The pK_a value is for the $-NH_3^+$ group. The pK_a of NH_4^+ is 9.25. The phenolic hydroxyl group of tyrosine has a pK_a of 10.07. The pK_a of phenol is 10.

26.12 The pK_a value is for the $-SH$ group. The pK_a values of thiols are approximately 8.

26.13 The decrease in pK_a means that the "group" bonded to the $-NH_3^+$ group withdraws electron density inductively. This effect increases the acidity of the $-NH_3^+$ group. The hydroxyphenyl group is inductively more electron withdrawing than a phenyl group.

26.14 The decrease in pK_a means that the "group" bonded to the carboxyl group withdraws electron density inductively. This effect increases the acidity of carboxylic acids. In the case of aspartic acid, the electron withdrawing group is the ammonium group, which is closer to the acidic site than is the ammonium group in glutamic acid.

26.15 The difference in the pK_a values for aspartic acid and the structurally related asparagine reflects the difference in the inductive effect of a carboxylate group and an amide group on the ionization sites. Whatever that difference, the effect will be smaller for glutamic acid compared to glutamine because the groups are further removed from the ionization sites. As the chain between the groups and the reaction site increases, the difference between the effects of the groups decreases until eventually there is no effect.

26.16 Based on the hybridization of the atoms, we would expect the site of protonation to be the amino group because its hybridization is sp^3. The lone-pair electrons of the imino group are sp^2 hybridized, so they are more tightly held to the nitrogen atom because the %s character of the orbital is greater. However, the effect of resonance stabilization of the base and its conjugate acid is very important. The basic form of the side chain group is resonance stabilized.

$$\underset{H_2N-\overset{\overset{NH}{\|}}{C}-NH-R}{} \longleftrightarrow \underset{H_2N=\overset{\overset{:NH^-}{|}}{C}-NH-R}{}$$

Protonation at the amino group eliminates the possibility of delocalization of electrons.

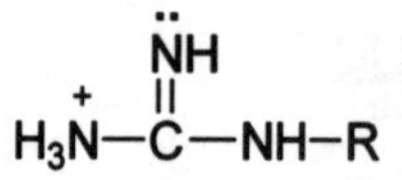

$$
\overset{\displaystyle \overset{..}{N}H}{\underset{\displaystyle H_3\overset{+}{N}-C-NH-R}{\parallel}}
$$

However, protonation of the nitrogen of the imino group gives a resonance stabilized conjugate acid with two equivalent groups bonded to carbon.

$$
\overset{\overset{+}{N}H_2}{\underset{H_2\ddot{N}-C-NH-R}{\parallel}}
\quad\longleftrightarrow\quad
\overset{\ddot{N}H_2}{\underset{H_2\overset{+}{N}=C-NH-R}{|}}
$$

Resonance stabilization of this conjugate acid favors its formation compared to the alternate conjugate acid resulting from protonation of the amino group.

26.17 **(a)** It should be close to 7, because the component amino acid residues are all neutral amino acids.
(b) It should be less than 7, because there is an acidic amino acid (Asp) in the tripeptide.
(c) It should be greater than 7, because there is a basic amino acid (Lys) in the tripeptide.

26.18 **(a)** It should be less than 7, because there is an acidic amino acid (Glu) in the tripeptide..
(b) It should be greater than 7, because there is a basic amino acid (Arg) in the tripeptide.
(c) It should be close to 7, because two of the amino acid residues are neutral amino acids, and although histidine is a basic amino acid, its isoionic point is only 7.59.

26.19 Vasopressin has an arginine residue, so its isoionic point should be higher. Oxytocin contains only neutral amino acids.

26.20 It should be close to 7 because the component amino acid residues—Ala, Gly, Phe, and Leu—are all neutral amino acids.

26.21 Chymotrypsin must contain one or more basic amino acids that contribute to an isoionic point that is larger than 7.

26.22 Pepsin must contain one or more acidic amino acids that contribute to an isoionic point that is smaller than 7.

26.23 2-Bromobutane is required to supply a *sec*-butyl group for the synthesis. Secondary halides competitively undergo dehydrohalogenation in syntheses designed to substitute the halide by a nucleophile.

26.24 The reaction requires the structurally related α-keto acid and ammonia. The structure of the keto acid is

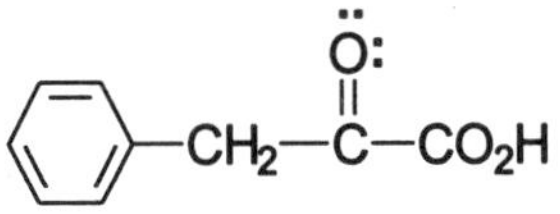

$$\langle \bigcirc \rangle - CH_2 - \overset{\overset{\displaystyle \ddot{O}:}{\|}}{C} - CO_2H$$

26.25 Conjugate addition of ammonia to acrylonitrile gives 3-aminopropanenitrile. Subsequent hydrolysis of the nitrile gives 3-aminopropanoic acid. Competitive 1,2-addition would give an iminoamide.

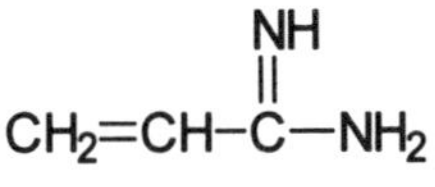

$$CH_2 = CH - \overset{\overset{\displaystyle NH}{\|}}{C} - NH_2$$

The triple bond of the nitrile is quite stable and reactions with nucleophiles such as in the hydrolysis reaction require somewhat drastic conditions. Thus, the related nucleophilic attack of ammonia at the nitrile carbon atom is also not expected to be as favorable as the conjugate addition reaction.

26.26 A conjugate addition of methylthiol gives a compound that introduces the thiomethyl group.

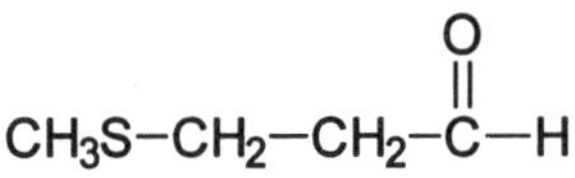

$$CH_3S - CH_2 - CH_2 - \overset{\overset{\displaystyle O}{\|}}{C} - H$$

Subsequent reaction with sodium cyanide and ammonium chloride (Strecker synthesis) adds a carbon atom to the chain and gives methionine.

26.27 Reaction of cysteine with allyl bromide introduces the allyl group via an S_N2 displacement reaction of bromide by the thiol group. Subsequent oxidation by hydrogen peroxide gives the sulfoxide.

26.28 The two carboxylic acid groups on the left of the structure constitute a malonic acid. Decarboxylation can occur to give glutamic acid.

26.29

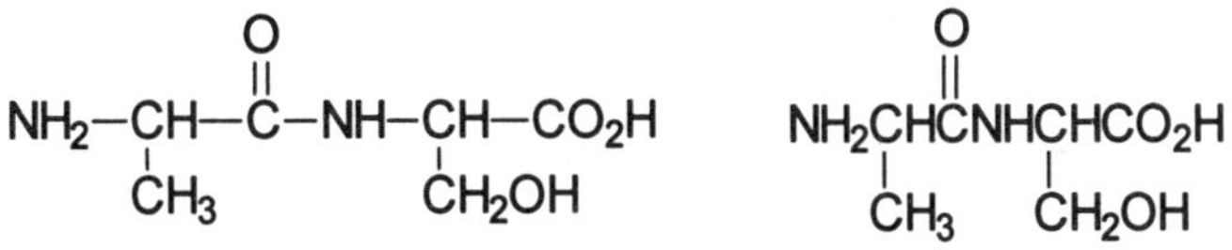

$$NH_2 - \underset{\underset{\displaystyle CH_3}{|}}{CH} - \overset{\overset{\displaystyle O}{\|}}{C} - NH - \underset{\underset{\displaystyle CH_2OH}{|}}{CH} - CO_2H \qquad NH_2 \underset{\underset{\displaystyle CH_3}{|}}{CH} \overset{\overset{\displaystyle O}{\|}}{C} NH \underset{\underset{\displaystyle CH_2OH}{|}}{CH} CO_2H$$

26.30 The C-terminal amino acid of glycylserine is serine, whereas the C-terminal amino acid of serylglycine is glycine. Peptides are always named starting with the N-terminal amino acid.

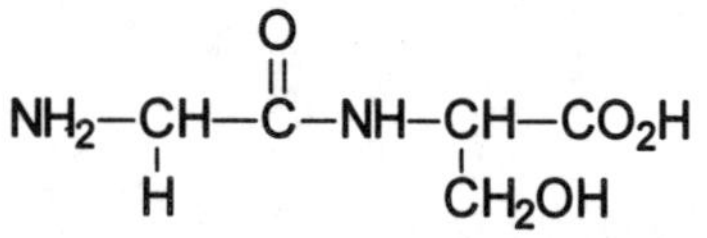

$$\text{NH}_2\text{-CH-}\overset{\overset{\displaystyle O}{\|}}{\text{C}}\text{-NH-CH-CO}_2\text{H}$$

NH₂—CH—C—NH—CH—CO₂H
| |
H   CH₂OH

Gly-Ser

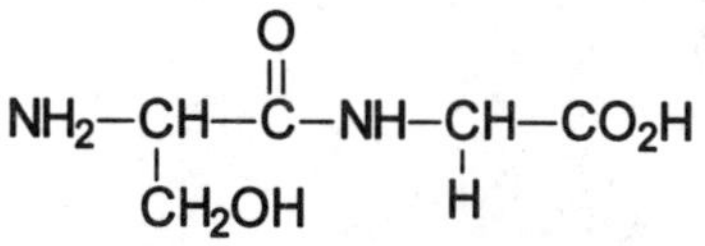

NH₂—CH—C—NH—CH—CO₂H
| |
CH₂OH   H

Ser-Gly

26.31   The acidic amino acids have carboxylic acid groups as part of the side chain that exist as the acid or its conjugate base in peptides.

26.32   The basic amino acids have nitrogen atoms as part of the side chain that exist as amino groups or ammonium groups in peptides.

26.33   From left to right, the amino acid residues are serine, glycine, and alanine. The name of the tripeptide is serylglycylalanine.

26.34   From left to right, the amino acid residues are glycine, cysteine, and valine. The name of the tripeptide is glycylcysteylvaline.

26.35   The compound is a tripeptide containing glutamic acid, histidine, and proline. Proline is the C-terminal amino acid and it exists as an amide. The N-terminal amino acid is glutamic acid, but the carboxyl group in its side chain exists as a cyclic amide (lactam) by reaction with the α amino group.

26.36   The compound is a tripeptide containing glutamic acid, cysteine, and glycine. Glutamic acid forms an amide with the amino group of cysteine using the carboxyl group of its side chain rather than the carboxyl group of the C-1 atom.

26.37   There are six isomeric tetrapeptides. They are

Gly—Gly—Ala—Ala   Gly—Ala—Gly—Ala   Gly—Ala—Ala—Gly

Ala—Ala—Gly—Gly   Ala—Gly—Ala—Gly   Ala—Gly—Gly—Ala

26.38   There are twelve isomeric tetrapeptides. They are

Gly—Gly—Ala—Leu   Gly—Ala—Gly—Leu   Gly—Ala—Leu—Gly

Ala—Leu—Gly—Gly   Ala—Gly—Leu—Gly   Ala—Gly—Gly—Leu

Gly—Gly—Leu—Ala   Gly—Leu—Gly—Ala   Gly—Leu—Ala—Gly

Leu—Ala—Gly—Gly   Leu—Gly—Ala—Gly   Leu—Gly—Gly—Ala

26.39   Four dipeptide must be identified. For the general formula A—B—C—D—E, the necessary dipeptide are A—B, B—C, C—D, and D—E.

26.40   For the general formula A—B—C—D—E—F—G—H, one combination of tripeptide is A—B—C, C—D—E, E—F—G, and F—G—H. Four tripeptide must be identified.

26.41 No, because there are two ways that the two dipeptide may have been combined in the tetrapeptide. They are Pro—Arg—The—Lys and The—L.ys—Pro—Arg.

26.42 Line up the hydrolysis products to list common amino acids In a vertical column.

    Asp—Arg—Val
           Val—Try—Ile
                 Ile—His—Pro
                       Pro—He

The octapeptide is Asp—Arg—Val—Try—Ile—His—Pro—He.

26.43 Treatment with the Edman reagent identifies alanine as the N-terminal amino acid. Line up the hydrolysis products, starting with alanine, to list common amino acids in a vertical column.

    Ala—Gly—Cys—Lys—Asn—Phe
               Asn—Phe—Phe—Trp—Lys
                     Phe—Trp
                          Lys—Thr
                               Thr—Phe—Thr—Ser—Cys
                               Thr—Ser—Cys

These correspond to the following peptide sequence.

    Ala—Gly—Cys—Lys—Asn—Phe—Phe—Trp—Lys—Thr—Phe—Thr—Ser—Cys

26.44 Treatment with an aminopeptidase identifies arginine as the N-terminal amino acid. Line up the hydrolysis products, starting with arginine, to list common amino acids in a vertical column.

               Gly—Phe—Ser
    Arg—Pro—Pro—Gly
                    Ser—Pro—Phe—Arg

The structure of the polypeptide is Arg—Pro—Pro—Gly—Phe—Ser—Pro—Phe—Arg.

26.45 Trypsin cleaves peptides at the C-terminal side of the basic amino acids lysine and arginine.

(a) Arg and Gly—Tyr   **(b)** no reaction occurs
(c) no reaction occurs   **(d)** no reaction occurs

26.46 Trypsin cleaves peptides at the C-terminal side of the basic amino acids lysine and arginine.

(a) Ser and Asp—Lys   **(b)** Lys and Tyr—Cys
(c) no reacton occurs   **(d)** Arg and Glu—Ser

26.47 Chymotrypsin cleaves peptides at the C-terminal side of the aromatic amino acids phenylalanine, tyrosine, and tryptophan.

       (a) no reaction occurs   **(b)** no reaction occurs
       (c) Ser, Trp, and Ser   **(d)** Asp and Ser—Phe

26.48   Chymotrypsin cleaves peptides at the C-terminal side of the aromatic amino acids phenylalanine, tyrosine, and tryptophan.

       (a) no reaction occurs   **(b)** Cys and Lys—Tyr
       (c) no reaction occurs   **(d)** no reaction occurs

26.49   No, it does not. Trypsin cleaves peptides at the C-terminal side of the basic amino acids lysine and arginine. Both dipeptides have a C-terminal basic amino acid. The tetrapeptide could be either of the following two structures. The site of cleavage is indicated by bold face.

       Pro—**Arg**—Thr—Lys   Thr—**Lys**—Pro—Arg

26.50   No, it does not. Chymotrypsin cleaves peptides at the C-terminal side of the aromatic amino acids phenylalanine, tyrosine, and tryptophan. The hydrolysis products have two aromatic amino acids and they can results from either of the following two structures. The sites of cleavage are indicated by bold face.

       Tyr—Gly—Gly—**Phe**—Met   Gly—Gly—**Phe**—**Tyr**—Met

26.51   Chymotrypsin cleaves peptides at the C-terminal side of the aromatic amino acids phenylalanine, tyrosine, and tryptophan. The hydrolysis products have two aromatic amino acids, and they can result from either of the following two structures.

       Ala—Arg—Gly—**Tyr**—**Trp**—Ala—Ser—Gly—Glu

       Trp—Ala—Arg—Gly—**Tyr**—Ala—Ser—Gly—Glu

26.52   Trypsin cleaves peptides at the C-terminal side of the basic amino acids lysine and arginine. Only one of the peptides has a C-terminal basic amino acid. The structure of the sleep peptide is the second of the possible structures given in Exercise 26.51.

       Trp—Ala—**Arg**—Gly—Tyr—Ala—Ser—Gly—Glu

26.53   Chymotrypsin cleaves peptides at the C-terminal side of the aromatic amino acids phenylalanine, tyrosine, and tryptophan. Four of the five glutamic acid residues are in one octapeptide. Thus, based on the end group analysis and fragment III, the sequence of amino acids from the N-terminal position is Glu—Gly—Pro—Trp. Based on the end group analysis and fragment II, the sequence of amino acids at the C-terminal position is Met—Asp—Phe, because there is only one phenylalanine in the peptide. These two peptides can combine with the octapeptide and dipeptide I in two ways. The two possible locations of the dipeptide fragment I are shown in bold face.

       Glu–Gly–Pro–Trp–**Gly–Trp**–Leu–Glu–Glu–Glu–Glu–Ala–Ala–Tyr–Met–Asp–Phe

       Glu–Gly–Pro–Trp–Leu–Glu–Glu–Glu–Glu–Ala–Ala–Tyr–**Gly–Trp**–Met–Asp–Phe

26.54 Both peptide IV and peptide V from hydrolysis by chymotrypsin are contained with peptide V from hydrolysis by trypsin. Both peptide II and peptide III from hydrolysis by trypsin are contained in peptide VI from hydrolysis by chymotrypsin. Both peptide III and peptide II from hydrolysis by chymotrypsin are contained in peptide IV from hydrolysis by trypsin. Therefore the following three peptides contain the majority of the amino acids.

A Val–Tyr–Pro–Asp–Ala–Gly–Glu–Asp–Gln–Ser–Ala–Glu–Ala–Phe–Pro–Leu–Glu–Phe

B Gly–Lys–Pro–Val–Gly–Lys–Lys–Arg–Pro–Lys–Val–Tyr

C Ser–Tyr–Ser–Met–Glu–His–Phe–Arg

Tryptophan occurs in only two fragments. Thus, the dipeptide I from hydrolysis by chymotrypsin and peptide I from hydrolysis by trypsin can be conbined to give an amino acid sequence Arg–Trp–Gly–Lys. Part of this sequence is duplicated in the two amino acids at the N-terminal positions of peptide B. The appropriate overlap gives

Arg–Trp–Gly–Lys–Pro–Val–Gly–Lys–Lys–Arg–Pro–Lys–Val–Tyr

The N-terminal amino acid of this fragment is common with the arginine of peptide C, which does not match the arginine of peptide B. The resulting peptide with appropriate overlap is

Ser–Tyr–Ser–Met–Glu–His–Phe–Arg–Trp–Gly–Lys–Pro–Val–Gly–Lys–Lys–Arg–Pro–Lys–Val–Tyr

The two amino acids at the C-terminal end of this peptide match the two amino acids at the N-terminal positions of peptide A. The resulting joined peptide with appropriate overlap gives the structure of corticotropin.

Ser–Tyr–Ser–Met–Glu–His–Phe–Arg–Trp–Gly–Lys–Pro–Val–Gly–Lys–Lys–Arg–Pro–Lys–Val–Tyr–Pro–Asp–Ala–Gly–Glu–Asp–Gln–Ser–Ala–Glu–Ala–Phe–Pro–Leu–Glu–Phe

26.55 The two possible tetrapeptides based on information given in Exercise 26.49 are

Pro—Arg—Thr—Lys   Thr—Lys—Pro—Arg

Hydrolysis by an aminopeptidase frees the N-terminal amino acid. Because threonine forms, the second tetrapeptide is the correct structure.

26.56 The two possible pentapeptides based on information given in Exercise 26.50 are

Tyr—Gly—Gly—Phe—Met   Gly—Gly—Phe—Tyr—Met

Hydrolysis by a carboxypeptidase frees the C-terminal amino acid. Because methionine is the C-terminal amino acid in both cases, the structure is still not established.

26.57 There are two peptide chains in insulin joined by disulfide linkages. One chain has glycine as the N-terminal amino acid. The other chain has phenylalanine as the N-terminal amino acid.

26.58

26.59   The N-terminal amino acid is aspartic acid.

26.60   The N-terminal amino acid is serine.

26.61   Amino acids without polar side chains exist in the interior of the protein. Glycine and phenylalanine do not have polar side chains.

26.62   Amino acids without polar side chains exist in the interior of the protein. Proline and cysteine do not have polar side chains.

26.63   Amino acids without polar side chains exist in the interior of the bilayer. Glycine and phenylalanine do not have polar side chains.

26.64   Amino acids without polar side chains exist in the interior of the bilayer. Proline and cysteine do not have polar side chains.

26.65   Once incorporated in a protein, proline does not have an N—H bond. Thus, it cannot form intramolecular hydrogen bonds with other amino acids within the helix.

26.66   Valine is a neutral amino acid, whereas glutamic acid is an acidic amino acid. The structure and properties of the side chains of these two amino acids differ significantly and should affect the tertiary structure of hemoglobin.

27

Aryl Halides, Phenols, and Anilines

This chapter is the final one that is organized around the chemistry of functional groups. The unique aspects of this chapter are that it considers not one functional group but three, and it is limited to the chemistry of these groups when bonded to an aromatic ring. Moreover, the majority of the chapter focuses on the chemistry unique to those functional groups precisely because they are bonded to an aromatic ring. A comparison of some of the chemical reactions which you have encountered before illustrates the attenuation of reactivity that is a consequence of the aromatic ring.

Keys to the Chapter

27.1 Properties of Aromatic Compounds

Based on Chapter 14 we know that electronegative atoms with lone pair electrons affect the electron density of the aromatic ring and therefore the reactivity of the ring toward electrophilic aromatic substitution. The electron withdrawing inductive effect of the electronegative atom and its ability to donate its lone pair electrons by resonance are both important. Recall that numerous times in succeeding chapters we found that chlorine inductively withdraws electron density and is a poor donor of electrons by resonance. Both oxygen and nitrogen inductively withdraw electron density, with oxygen being the strongest because it is more electronegative. Both oxygen and nitrogen are electron donors by resonance, but nitrogen is the more effective because it is the least electronegative. The net effect is that both oxygen and nitrogen are net donors of electron density with nitrogen being the best. Reversing the focus of the discussion, it follows that the properties of the chlorine, oxygen, and nitrogen atoms bonded to the aromatic ring are themselves changed.

The C—Cl bond in aryl chlorides is shorter than in alkyl chlorides, because the sp^2 hybridized bond of the aromatic ring carbon atom has a larger % s character. In addition the bond energy is larger. The C—O and C—N bonds are also shorter in aromatic compounds than in saturated compounds for the same reason. In addition, the bonds are shorter because they both have some double bond character as a result of contributing resonance forms based on the donation of their lone pair electrons to the aromatic ring.

The bond polarity of the carbon bond to chlorine, oxygen, or nitrogen is affected by both the hybridization of the bond and the donation of electron density to the ring by resonance. The higher % s character to some degree withdraws electrons away from the heteroatom, resulting in some decrease in the polarity of the bond. This effect is seen in the chlorine compound. In addition, there is a further effect in the same direction as a result of donation of electrons by resonance. In both oxygen and nitrogen, there is a net reversal of polarity of their respective bonds.

27.2 Acid-Base Properties

The pK_a values of phenols are smaller than the pK_a values of alcohols. This difference is the result of delocalization of the negative charge of the phenoxide ion by the benzene ring. Those groups that are electron-withdrawing, such as the nitro group, more effectively stabilize the negative charge in the conjugate base and thus increase the acidity of the phenol. The pK_a values thus decrease.

The pK_b values of anilines are larger than the pK_b values of saturated amines. In this case the difference is the result of delocalization of the lone pair electron of the nitrogen atom in the aniline itself. Stabilization of the compound as a result of resonance diminishes the availability of the lone pair electrons for protonation by an acid. Those groups that are electron-withdrawing, such as the nitro group, more effectively delocalize the lone pair electrons in the aniline itself and thus further decrease the basicity of the aniline. The pK_b values increase.

27.3 Formation of Organometallic Reagents

The reaction to form aryl Grignard reagents is similar to the formation of alkyl Grignard reagents. However, in the case of the chloro compounds, it is necessary to use THF as the solvent rather than diethyl ether. Bromo compounds form Grignard reagents in ether. As a result, it is possible to prepare the Grignard reagent from a C—Br bond without affecting a C—Cl bond. Once formed, both types of Grignard reagents react in the same way and add to the typical carbon-oxygen double bonds of carbonyl compounds, esters, and carbon dioxide.

Aryllithium compounds are prepared by direct reaction of aryl halides with lithium metal or by a transmetallation reaction between an aryl halide and an alkyllithium compound.

27.4 Nucleophilic Substitution

The typical S_N2 and S_N1 mechanisms of alkyl halides do not apply to aryl halides. Nucleophilic substitution does occur, but by two different mechanisms termed **addition-elimination** and **elimination-addition** reactions.

The addition-elimination reaction results from attack of a nucleophile at the carbon atom bearing a leaving group, forming a tetrahedral intermediate. Remember that an intermediate is not a transition state. Although the intermediate in this addition-elimination reaction may resemble the transition state structure in an S_N2 mechanism, remember that an intermediate has a lifetime that in some cases may allow for its isolation. In addition, note that the intermediate is not formed by attack of the nucleophile from the back of the carbon-halogen bond but rather from "front" side.

The addition-elimination reaction, also known as **nucleophilic aromatic substitution**, occurs only if electron withdrawing groups are bonded to the ring to stabilize the negative charge of the cyclohexadienyl anion. The strongly electron withdrawing nitro group is most effective, but it must be ortho or para to the carbon-halogen bond for resonance stabilization to occur.

The second step in the addition-elimination reaction is the ejection of the halide ion as the leaving group. This step is not rate determining. Thus, the normal order of leaving group tendencies of the halide ions is not observed. Rather, the effect of the halogen atom in stabilizing the cyclohexadienyl anion is observed. As a result of inductive electron withdrawal, the fluorine atom is the most effective in stabilizing the anion and hence the transition state leading to that anion is lowest in energy for the fluoro compound.

The elimination-addition mechanism of aryl halides involves the reaction of an amide ion, a very strong base, to abstract a proton from the position ortho to the halogen atom. An elimination reaction occurs as the halide ion serves as the leaving group. A very reactive intermediate called **benzyne** results, in which the two carbon atoms become equivalent. Subsequent attack of the amide ion at either of the carbon atoms gives an aryl anion which is subsequently protonated by transfer of a proton from ammonia and the regeneration of an amide ion.

When substituents such as a methyl group are also present, the two carbon atoms of the benzyne intermediate are no longer structurally equivalent and mixtures of anilines result. There is little regioselectivity, so mixtures of isomers usually result. Some regioselectivity may result, but only because the ring substituent inductively stabilizes charge. Note that because the charge is located in an sp^2 hybridized orbital, resonance stabilization of charge cannot occur.

27.5 Reactions of Phenols—A Review

Phenols undergo some of the same reactions as alcohols. They form ethers and esters. Ethers are synthesized by the Williamson synthesis. Unlike the formation of alkoxides, which require a strong base such as sodium hydride, phenoxides are easily produced using hydroxide ion. The phenoxide ion then acts as a nucleophile and displaces a halide ion from an alkyl halide. Aryl methyl ethers can be obtained using dimethyl sulfate.

Esters of phenols cannot be obtained by the Fischer ester synthesis, because the equilibrium constant for esters of phenols is even more unfavorable than for esters of alcohols. Either an acyl halide or an acid anhydride must be used.

Phenols undergo electrophilic aromatic substitution. However, the activation of the aromatic ring by the hydroxyl group makes monosubstitution difficult unless the electrophilic group introduced strongly deactivates the ring.

27.6 Reactions of Phenoxide Ions

Based on the resonance forms of the phenoxide ion, we can write structures with some negative charge on the carbon atoms ortho and para to the oxygen atom. This carbanion structure reacts with certain carbon-oxygen double bonds in a manner similar to the reaction of the carbanion of Grignard reagents. Reaction with formaldehyde gives a hydroxymethyl derivative which can dehydrate to give a conjugate ketone, which can undergo conjugate addition reactions that resemble the aldol reaction. As a consequence, a polymeric product known as Bakelite results.

Reaction of the phenolate ion at the ortho position with carbon dioxide resembles the reaction of a Grignard reagent with carbon dioxide in that a carboxylic acid is formed. The **Kolbe synthesis** is used to produce salicylic acid.

27.7 Quinones

Quinones are cyclohexadienediones that can be prepared by the reaction of diphenols. Quinones are oxidizing agents as they are reduced to regenerate the aromatic ring. The reduction potential of quinones depends on the stability of the aromatic ring formed and the electron-attracting characteristics of substituents. Quinones with several electron-withdrawing groups are sufficiently strong oxidizing agents to dehydrogenate some suitably substituted hydrocarbons.

27.8 Substitution Reactions of Aryldiazonium Salts

Primary amines form diazonium salts when reacted with nitrous acid. Homolytic cleavage of the carbon-nitrogen bond readily occurs in the case of sp^3 hybridized carbon atoms but less readily in the case of sp^2 hybridized carbon atoms. Thus aryldiazonium ions are sufficiently stable to be used as intermediates for substitution reactions, giving compounds with groups such as halogen, cyanide, or hydroxide replacing the original nitrogen atom. These reactions were presented in Section 14.8 of the text. A specialized reaction of secondary amines with nitrous acid give a nitroso compound. Recall from Chapter 25 that the NO^+ ion is the initial intermediate that is responsible for the reaction of amines with nitrous acid. It is this nitrosonium ion that electrophilically attacks the aromatic ring of activated compounds.

27.9 Azo Compounds

The diazonium ion itself is an electrophile and can attack activated aromatic rings to give a nitrogen-nitrogen double bond of **azo compounds**. Formation of these compounds follows the pattern predicted by the concepts developed for electrophilic aromatic substitution. These compounds are colored and are used as dyes.

Summary of Reactions

1. Formation of Organometallic Reagents

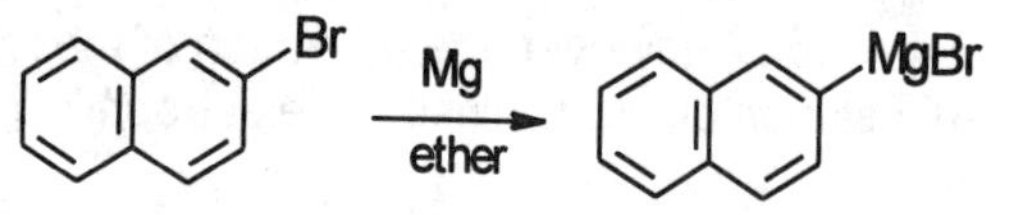

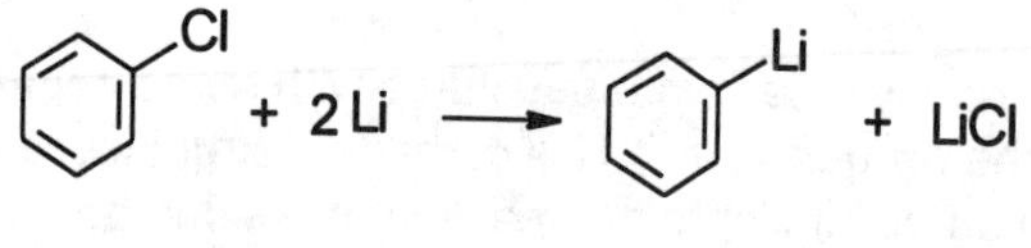

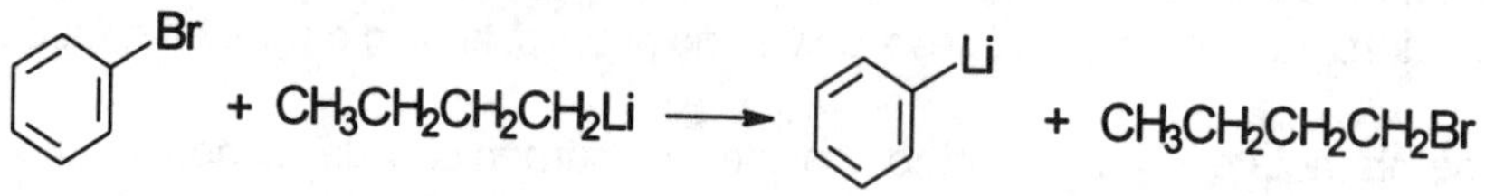

2. Nucleophilic Aromatic Substitution (addition-elimination)

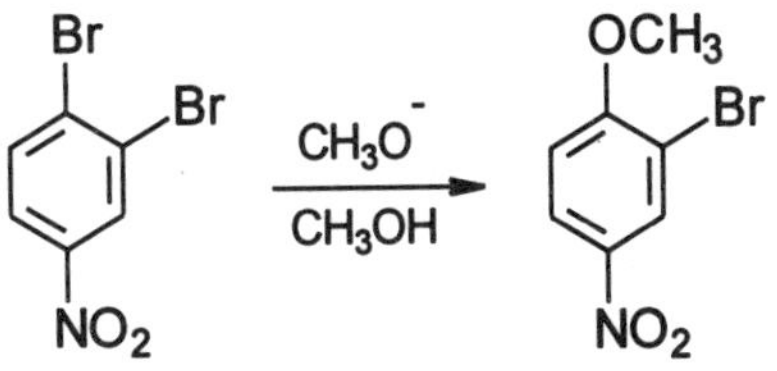

3. Nucleophilic Aromatic Substitution (elimination-addition)

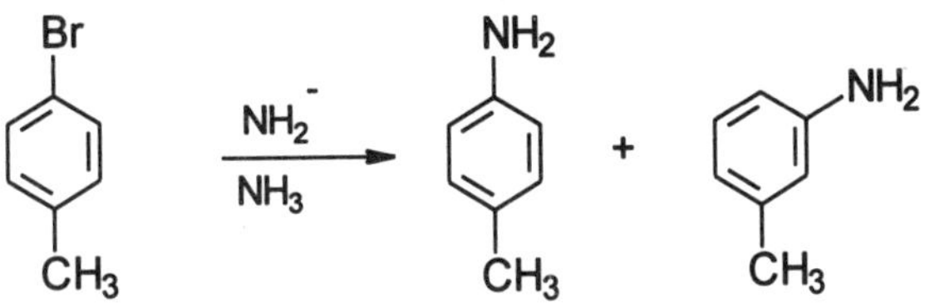

4. Synthesis of Aryl Ethers

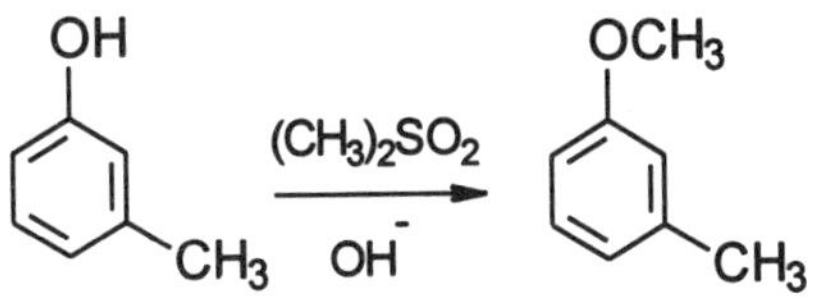

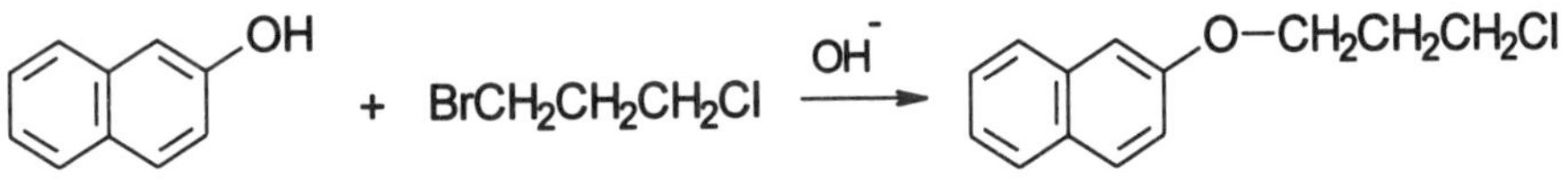

5. Synthesis of Esters of Phenols

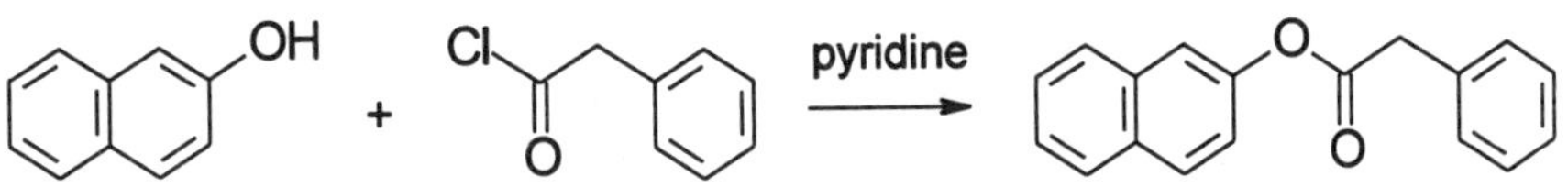

6. Kolbe Synthesis

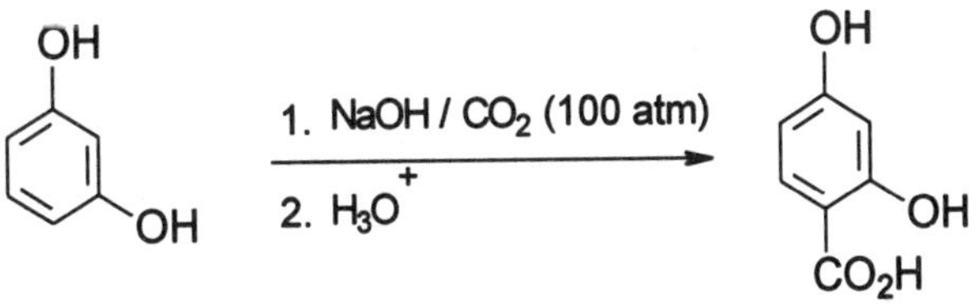

7. Synthesis of Quinones

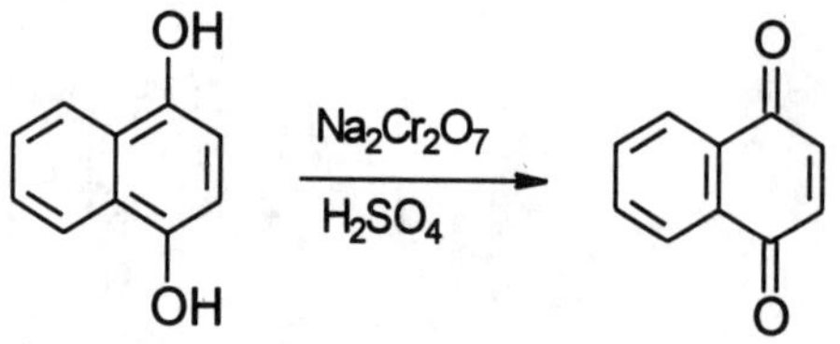

8. Synthesis using Aryldiazonium Ions

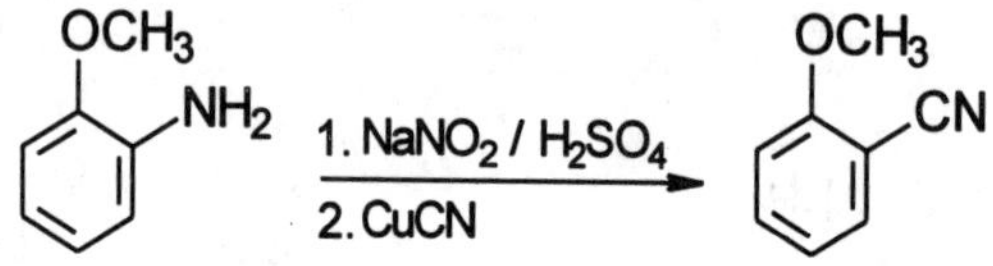

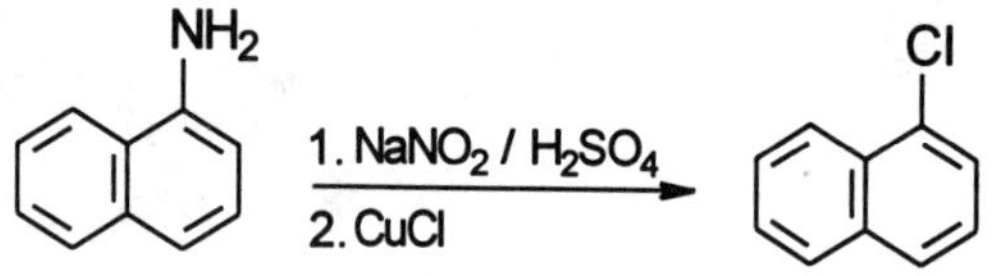

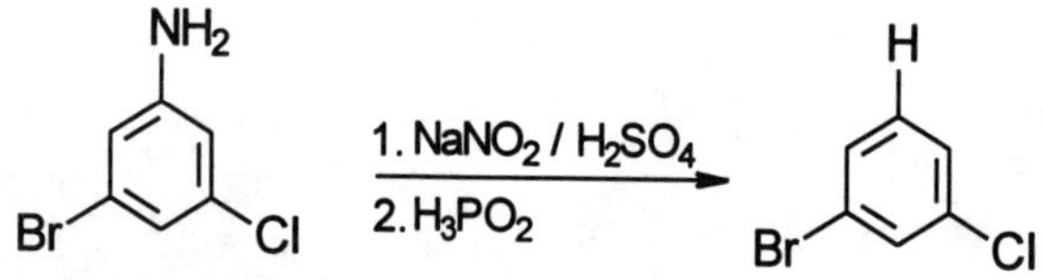

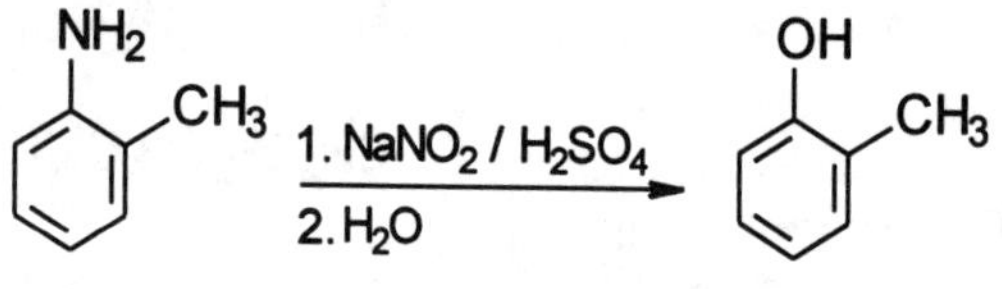

Solutions to Exercises

27.1   The ortho isomer has the hydroxyl groups in close proximity, and an intramolecular hydrogen bond forms as shown below. Intermolecular hydrogen bonding in the meta and para isomers results in an increased boiling point, because there is extensive aggregation of these compounds.

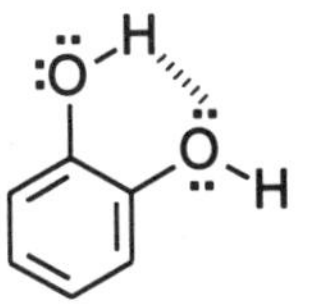

27.2   The compound with the lowest boiling point is the ortho isomer. It forms an intramolecular hydrogen bond between the hydrogen atom of the hydroxyl group and the oxygen atom of the ether. The meta and para isomers form intermolecular hydrogen bonds, resulting in a higher boiling point.

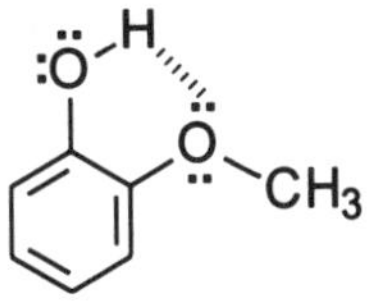

27.3   The negative end of the bond moment for the methyl group bonded to the aromatic ring is toward the aromatic ring, because the methyl group is an inductively electron donating group. The bonding electron pair of the carbon-carbon bond is polarized toward the sp^2 hybridized carbon atom of the aromatic ring. Recall that sp^2 hybridized atoms have a greater % s character, and as a result bonding electrons are held more tightly than for sp^3 hybridized atoms. The negative end of the bond moment for the chlorine atom bonded to the aromatic ring is toward the chlorine atom.   Thus, the two bond moments reinforce one another to give a net dipole moment of 2.1D.

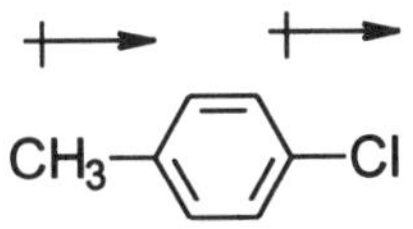

27.4   The negative end of the bond moment for the methyl group bonded to the aromatic ring is toward the aromatic ring, because the methyl group is an inductively electron donating group. The negative end of the bond moment for the hydroxyl group bonded to the aromatic ring is toward the aromatic ring, mainly because oxygen donates electrons to the ring by resonance.   Thus, the two bond moments are opposed, giving a net dipole moment of 1.1D.

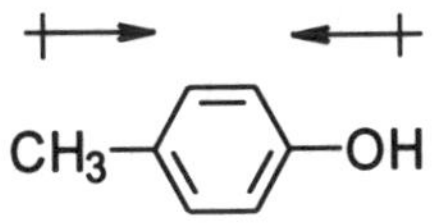

27.5   The para isomer has no dipole moment, because the two bond moments are directly opposed at a 180° angle to give a net resultant of 0 D. The isomer with the larger dipole moment is the ortho compound, because the two bond moments are at a smaller angle and reinforce each other more effectively than for the meta compound. Thus, the net dipole moment resulting from the two moments of the ortho isomer is larger than for the meta isomer.

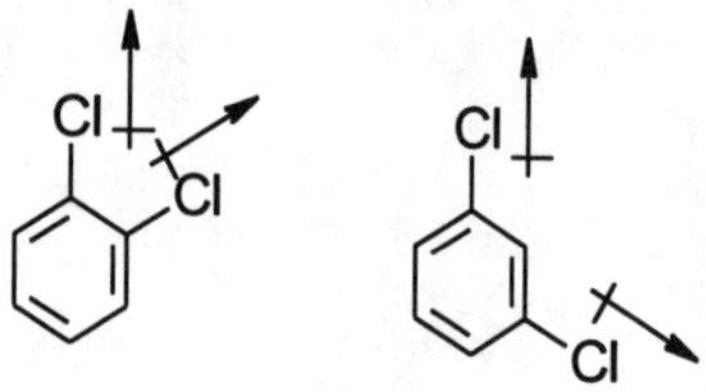

27.6   The negative end of the bond moment for the chlorine atom bonded to the aromatic ring is toward the electronegative chlorine atom. The negative end of the bond moment for the hydroxyl group bonded to the aromatic ring is toward the aromatic ring.   Thus, the two bond moments reinforce one another to give a estimated dipole moment of 3.2 D.

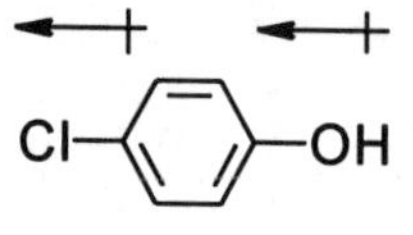

27.7   The cyano group withdraw electrons from the aromatic ring in resonance structures. The amino group can cooperatively donate electrons by resonance, so the double bond character of the C—N bond to NH_2 increases. As a result the C—N bond of p-cyanoaniline will be shorter than C—N bond of p-methoxyaniline, which does not have such double bond character.

27.8   The bonds about the nitrogen atom form a shallower pyramid in aniline compared to cyclohexylamine. As a consequence, the structure of aniline is closer to that of the planar structure at the transition state for inversion, so the activation energy is less.

27.9   The nitrogen atoms of both compounds contribute three electrons to σ bonds. In the case of pyrrolidine, the remaining two valence electrons are present as a nonbonded pair directed away from the ring. In the case of pyrrole, the two valence electrons are incorporated into the aromatic π system. In resonance forms used to depict bonding in pyrrole, there is a decrease in electron density at the nitrogen atom. In short, electrons are drawn away from nitrogen toward the carbon atoms in the ring as a result of formation of the π system.   The opposing dipole moments of pyrrolidine and pyrrole reflect the distribution of electron density in the two compounds. The resonance structures of pyrrole are shown on the following page.

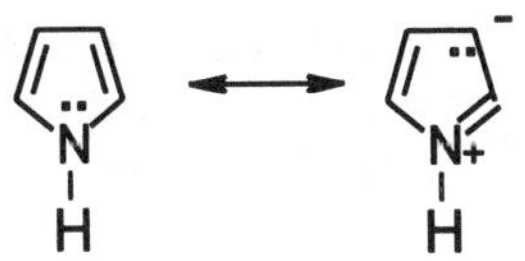

27.10 The negative end of the bond moment for the trifluoromethyl group bonded to the aromatic ring is away from the ring toward the carbon atom of the trifluoromethyl group. The observed dipole moment of p-(trifluoromethyl)aniline is larger than that of trifluoromethylbenzene. Thus, the bond moment of aniline must be in the direction that reinforces the bond moment of the trifluoromethyl group. The negative end of the bond moment for the amino group bonded to the aromatic ring is toward the aromatic ring.

27.11 The aldehyde group can contribute to the delocalization of charge, stabilizing the conjugate base as a result. Stabilization of the conjugate base favors ionization and increases the acid dissociation constant of p-hydroxybenzaldehyde compared to phenol.

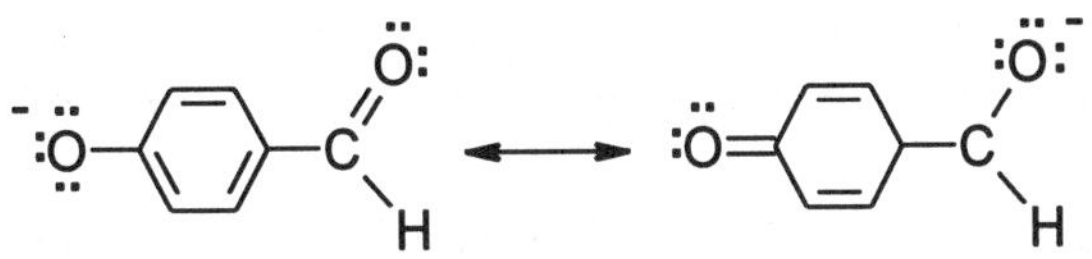

27.12 The nitro groups that are ortho and para to the site bearing the negatively charged ion can stabilize that charge by delocalization as indicated by contributing resonance forms. Such stabilization cannot occur when the nitro group is in a meta position.

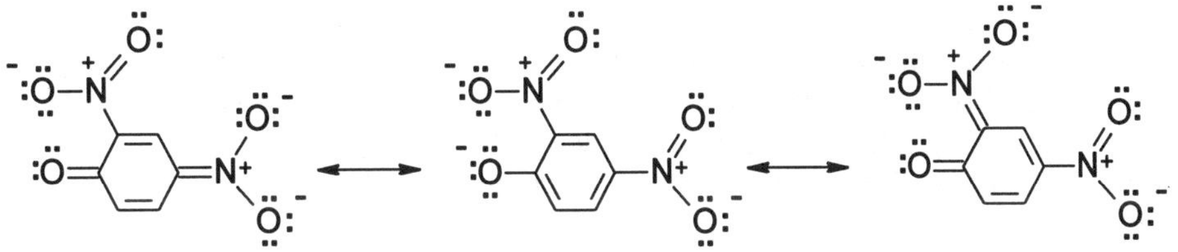

27.13 Five resonance forms with the negative charge on carbon atoms can be written for each conjugate base. However, in the case of 1-naphthol two of them have a formal benzene ring, whereas in 2-naphthol only one of them does. Thus, the conjugate base of 1-naphthol is more stable, so the K_a of this phenol is larger.

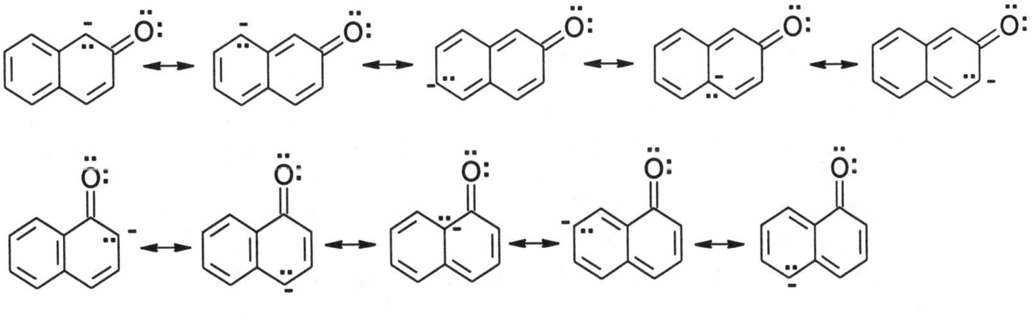

27.14 Compound I is more acidic because the charged nitrogen atom of the ammonium ion is closer to the carbon atom bearing the hydroxyl group. The withdrawal of electron density by an inductive effect is greater for I because the groups are closer. Withdrawal of electron density stabilizes the conjugate base and as a consequence increases the acid dissociation constant of compound I.

27.15 **(a)** The compound is a primary amine with an alkyl group bonded to the nitrogen atom. The K_b should be approximately 5×10^{-4}.
  (b) The compound is a secondary amine with only alkyl groups bonded to the nitrogen atom. The K_b should be approximately 5×10^{-4}.
  (c) The compound is a tertiary amine with only alkyl groups bonded to the nitrogen atom. The K_b should be approximately 5×10^{-4}.
  (d) The compound is a primary amine with an aromatic ring bonded to the nitrogen atom. The K_b should be approximately 4×10^{-10}, like that of aniline.

27.16 **(a)** The compound is a primary amine with an aromatic ring bonded to the nitrogen atom. The K_b should be approximately 4×10^{-10}, like that of aniline.
  (b) The compound is a substituted aniline. The electron donating methyl group should increase the basicity above that of aniline, so the K_b should be larger than 4×10^{-10}.
  (c) The compound is a tertiary amine with only alkyl groups bonded to the nitrogen atom. The K_b should be approximately 5×10^{-4}.
  (d) The compound is a tertiary amine with only alkyl groups bonded to the nitrogen atom. The K_b should be approximately 5×10^{-4}.

27.17 The amide-like nitrogen atom is the weakest base. Of the other two nitrogen atoms, the tertiary amine contained in the five-membered ring on the right is the most basic. The nitrogen atom bonded directly to the aromatic ring is intermediate in basicity compared to the other two.

27.18 For quinine, the nitrogen contained within the aromatic ring should have $pK_b = 8.7$ like that of pyridine. The tertiary amine should have $pK_b = 3.4$. For reserpine, the nitrogen atom within a pyrrole-like ring is a very weak base and $pK_b = 15$. The tertiary amine contained in the six-membered ring should have $pK_b = 3.4$.

27.19 The cyano group is an electron withdrawing group, both by resonance and by an inductive effect. In the para position, both resonance and inductive effects operate to decrease the electron density on the nitrogen atom of the aniline. As a consequence, the K_b is decreased and the K_a of the anilinium ion is larger. This effect is seen in the small pK_a. In the meta position, only an inductive effect operates, so the decrease in the electron density on the nitrogen atom of the aniline is less than for the para isomer. As a consequence the K_b is larger and the K_a of the anilinium ion is smaller than for the para isomer. The effect is seen in the larger pK_a of the meta isomer.

27.20 The methoxy group is an electron donating group by resonance, but electron withdrawing by an inductive effect. In the meta position, the inductive effect operates to decrease the electron density on the nitrogen atom of the aniline. As a consequence, the K_b is decreased and the K_a of the anilinium ion is larger. This effect is seen in the small pK_a. In the para position, the resonance effect counters the inductive effect and increases the electron density on the nitrogen atom of the aniline relative to the meta isomer. As a consequence the K_b is larger and the K_a of the anilinium ion is smaller than for the meta isomer. The effect is seen in the larger pK_a of the para isomer.

27.21 The oxygen atom is inductively electron withdrawing and decreases the electron density on the nitrogen atom, which is reflected in its decreased basicity.

27.22 The resonance forms of the carboxylate ion have negative charge on the two oxygen atoms, but there is no delocalization of charge into the ring. Hence the acidity of carboxylic acids is affected only by the electron density on the carbon atom bearing the carboxyl group, which in turn is affected by the substituents on the ring.

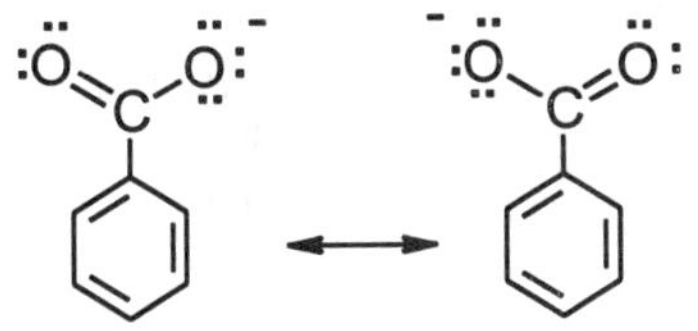

In anilines, the electron pair of the nitrogen atom is delocalized into the aromatic ring, and as a result there is a direct resonance interaction with ring substituents. Thus aniline is a very weak base, and the anilinium ion is a fairly strong acid compared to the methylammonium ion.

27.23 Only the nitro groups in ortho or para positions relative to the site of attack by a nucleophile affect the reactivity. Such groups can stabilize the negative charge of the intermediate by resonance. Groups in the meta position can stabilize the negative charge only by an inductive effect. Compound III is the most reactive. Compound IV is the next most reactive and is slightly more reactive than compound II as a consequence of the additional nitro group in the meta position. Compound I is very much less reactive.

27.24 (a) The carboethoxy groups can stabilize the negative charge of the intermediate formed when the ethoxy ion displaces the bromide ion.

    (b) The two nitro groups that are ortho and para to one of the bromine atoms can stabilize the negative charge of the intermediate formed when the thiol displaces that bromide ion. The other site containing bromine is far less reactive toward nucleophiles.

    (c) The nitro and cyano groups that are ortho and para to the chlorine atom can stabilize the negative charge of the intermediate formed when hydrazine displaces the chloride ion.

    (d) The nitro group that is para to one of the chlorine atoms can stabilize the negative charge of the intermediate formed when the amine displaces that chloride ion. The other site containing chlorine is far less reactive toward nucleophiles.

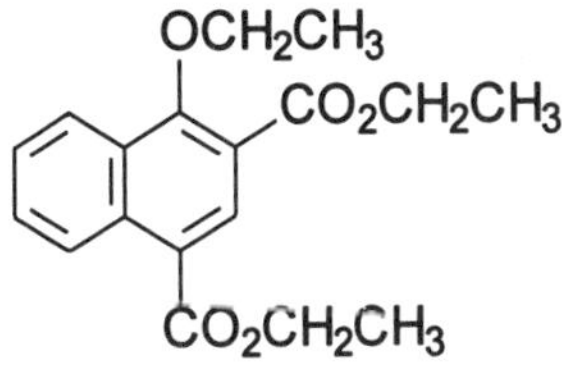

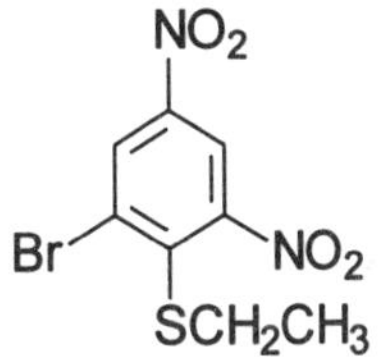

(c)

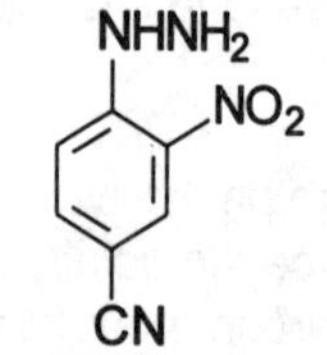

NHNH₂
NO₂

CN

(d)

NO₂

Cl

N

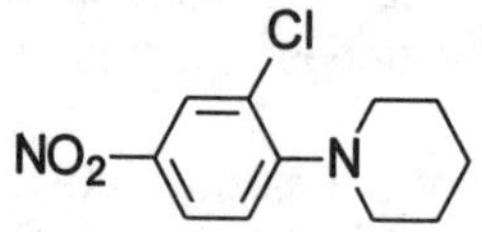

27.25 The negative charge of the intermediate formed in the displacement of chloride from the C-4 atom of pyridine can be resonance stabilized by transfer of the charge to the electronegative nitrogen atom. In the case of 3-chloropyridine, there is no resonance stabilization possible and as a result the transition state for formation of the intermediate has a much higher energy. As a consequence the reaction rate is slow.

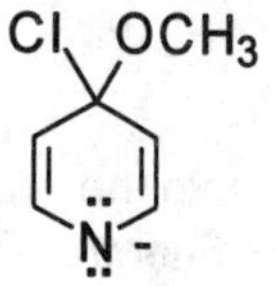

Cl   OCH₃

N⁻

27.26 The two nitro groups can stabilize the intermediate formed by nucleophilic attack by the amine of the amino acid to displace fluoride. The fluorine atom is the most effective of the halogens in the stabilization of the intermediate.

NO₂

NO₂

NH–CH–C–NH–CH–C
R   R'

O   O

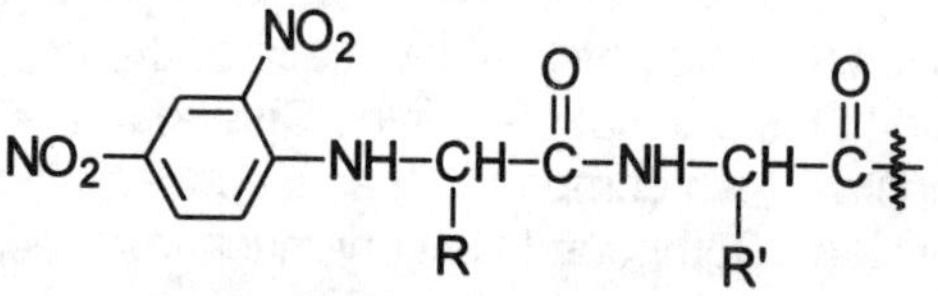

27.27 Three of the fluorine atoms that are ortho and para to the point of attack by methoxide ion are strongly electron withdrawing and can stabilize the negative charge.

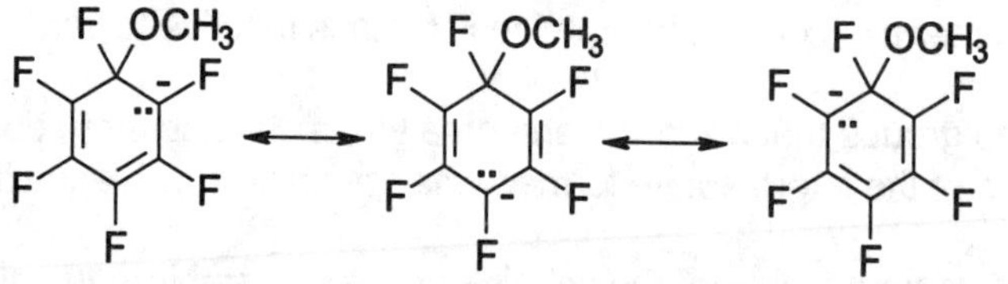

F OCH₃
F   F
F   F
F

⟷

F OCH₃
F   F
F   F
F

⟷

F OCH₃
F   F
F   F
F

27.28 The nitro group can stabilize the intermediate formed by nucleophilic attack by the methoxide ion at positions ortho and para to the nitro group. The products are derived from displacement of a fluoride ion by methoxide ion at either of these two positions.

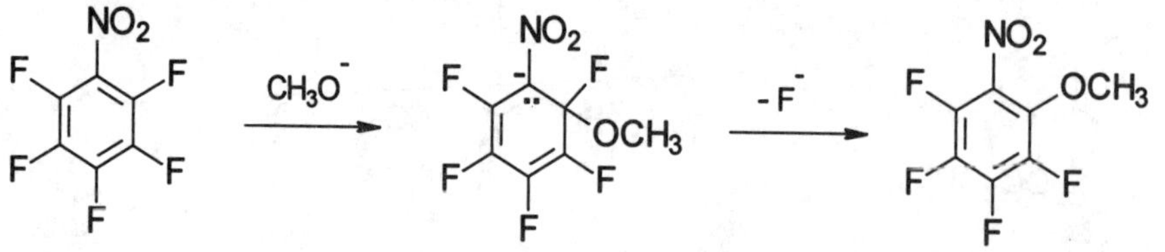

NO₂
F   F
F   F
F

CH_3O^- →

NO₂
F   F
F   OCH₃
F

$-F^-$ →

NO₂
F   OCH₃
F   F
F

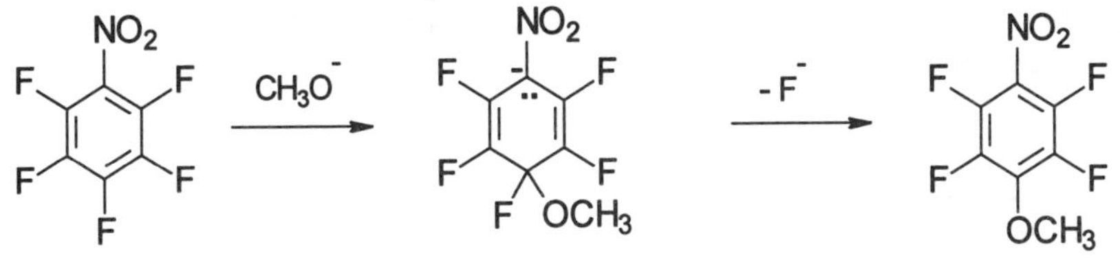

27.29 An alkoxide ion intramolecularly displaces a fluoride ion from the ortho position relative to the side chain. The intermediate is stabilized by fluorine atoms at positions ortho and para to the site of the displacement reaction.

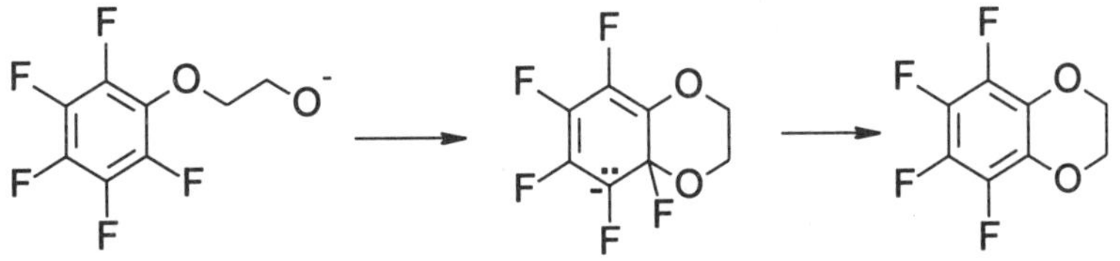

27.30 Reaction with piperidine only occurs by a nucleophilic substitution mechanism in which the piperidine replaces a bromide ion at the C-1 atom. In the presence of sodium amide, the reactions proceeds via a benzyne intermediate. In this case the piperidine can react at either the C-1 or C-2 atoms.

27.31 The benzyne intermediate with a triple bond at the C-2 and C-3 atoms can react to give equal amounts of 2-methyl- and 3-methylaniline. The benzyne intermediate with a triple bond at the C-3 and C-4 atoms can react to give equal amounts of 3-methyl- and 4-methylaniline. As a consequence the 2-methyl, 3-methyl, and 4-methyl compounds are formed in a 1:2:1 ratio.

27.32 The benzyne intermediate with a triple bond at the C-2 and C-3 atoms forms because the proton at the C-2 atom is more acidic than the proton at the C-4 atom. The trifluoromethyl group stabilizes the negative charge of the conjugate base by an inductive effect.

27.33 A benzyne intermediate is formed by elimination of DCl. The amide ion can then add to either of two carbon atoms, and two products result that differ in the location of the deuterium atom that was still present in the benzyne intermediate.

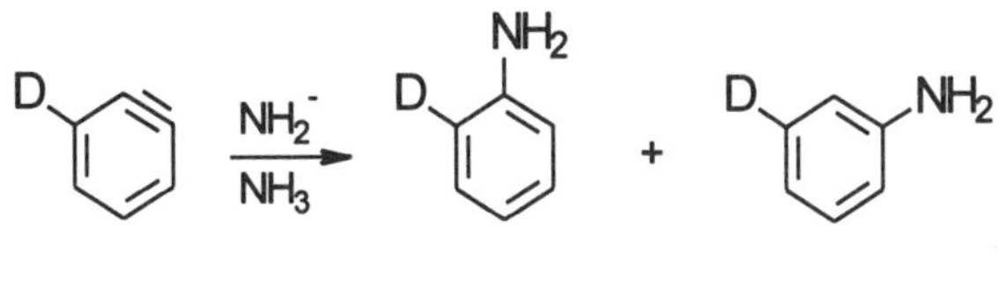

27.34 The benzyne intermediate has a triple bond at the C-2 and C-3 atoms. Reaction of amide ion at the C-3 atom occurs because the resulting charge at the C-2 atom is stabilized by the inductive effect of the methoxy group.

27.35   The carbon atom bonded to the magnesium atom of a Grignard has some carbanionic character. Transfer of those electrons to form a carbon-carbon bond results in the loss of fluoride ion from the adjacent carbon atom.

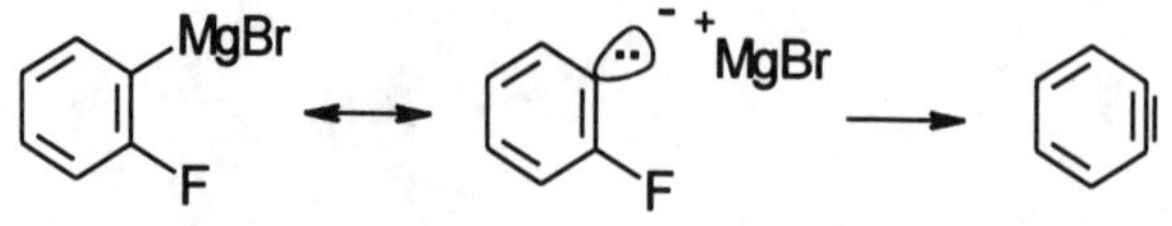

27.36   A diazonium ion forms which can lose nitrogen. The carboxyl group can exist as the carboxylate ion. Loss of carbon dioxide and nitrogen can occur in a concerted reaction.

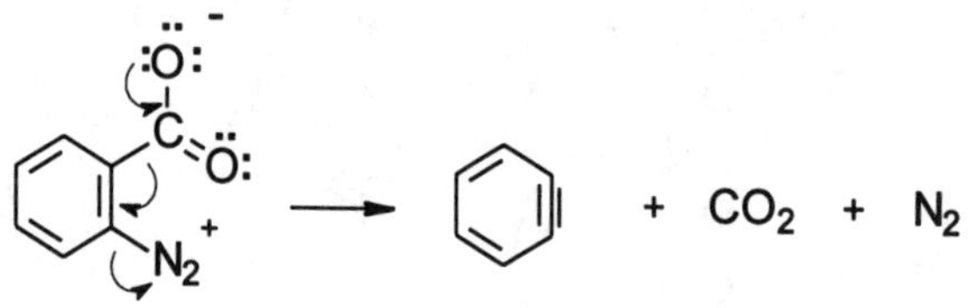

27.37   **(a)** The reaction of this unactivated chlorobenzene occurs via a single benzyne intermediate, and as a result two isomeric phenols are formed.
   (b) The reaction of this unactivated bromobenzene occurs via either of two benzyne intermediates, and as a result three isomeric substituted amines are formed.
   (c) The reaction of this unactivated bromonaphthalene occurs via a single benzyne intermediate, and as a result two isomeric products are formed.

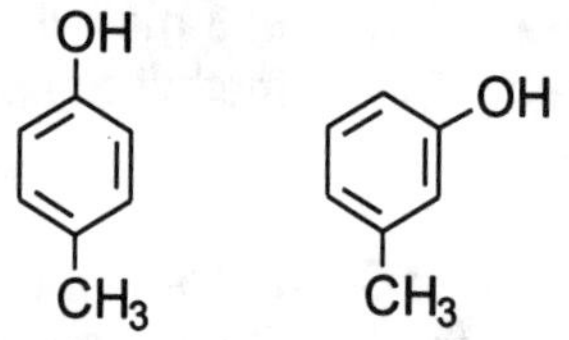

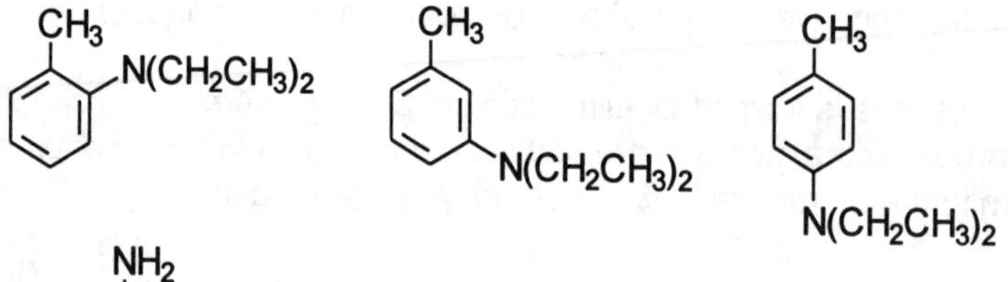

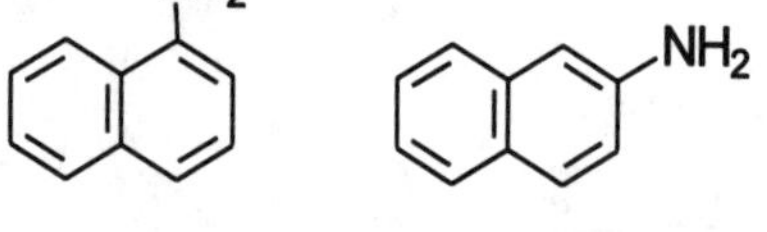

27.38   Two benzyne intermediates are possible if the elimination occurs with amide ion. However, the conjugate base formed from the secondary amine by proton exchange with amide ion can also react. This intramolecular process is favored. The benzyne formed can react with the secondary amine to give a cyclic product shown of the following page.

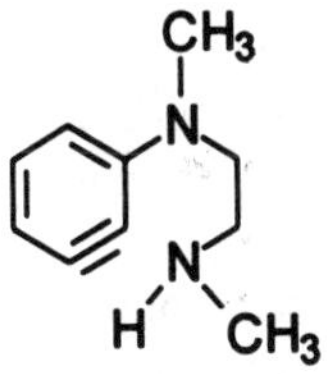

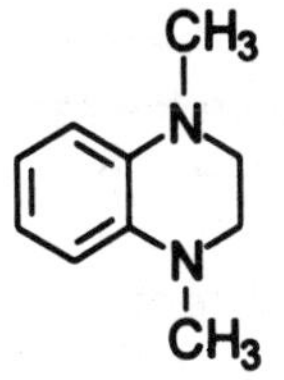

27.39 (a) 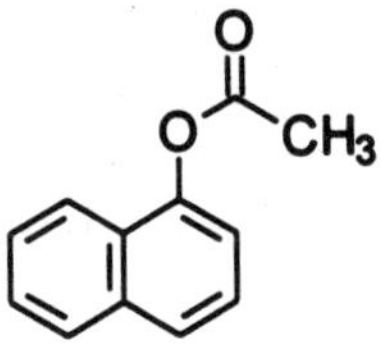 (b) 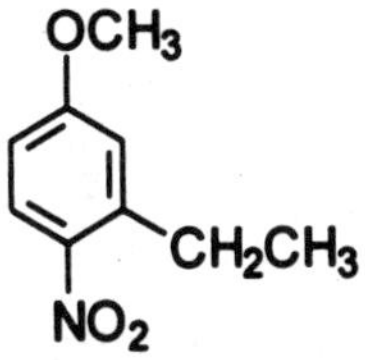 (c) 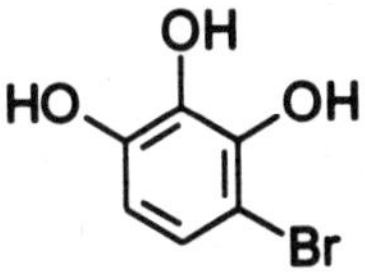 (d)

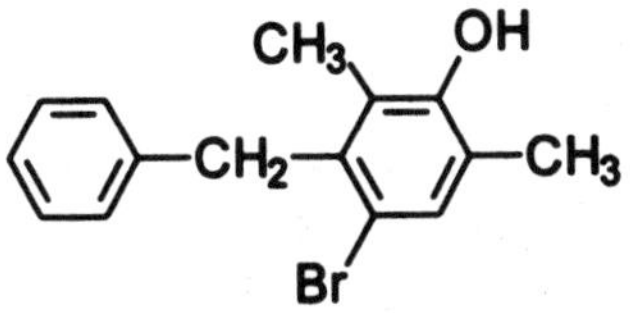

27.40 (a) 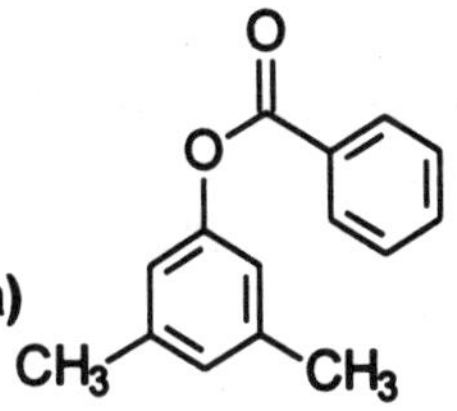 (b) 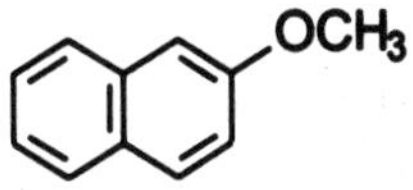 (c) 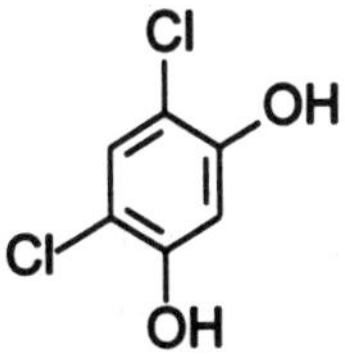 (d)

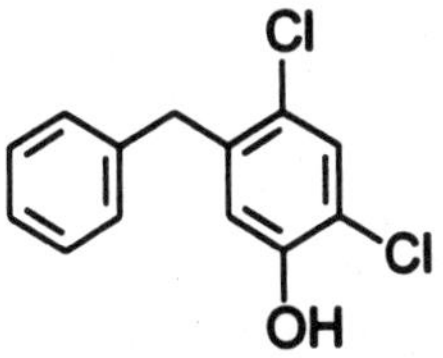

27.41 No, because the carbonyl group of ketones is more stable than the carbonyl group of aldehydes and is less reactive toward nucleophiles.

27.42 (a) (b) 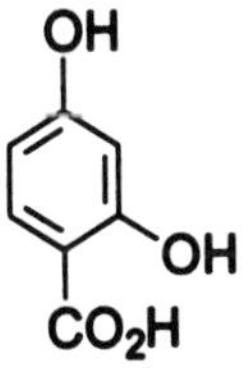 (c)

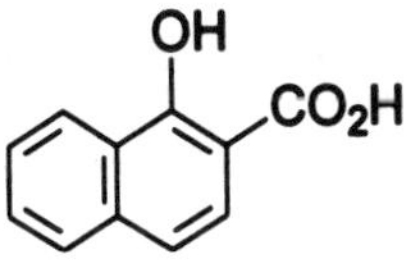

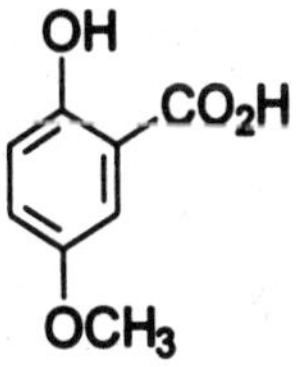

27.43   Electron attracting groups increase the reduction potential. The methoxy groups of ubiquinone are electron donating and should decrease the reduction potential. The methyl groups of plastoquinone should reduce the reduction potential as well, but less than for the methoxy groups, so plastoquinone should have a larger reduction potential.

27.44   Four of the five resonance forms of phenanthrene have a double bond at the C-9 to C-10 atoms, as shown below, and therefore this bond has substantial double bond character. As a result the bond should be susceptible to oxidizing agents.

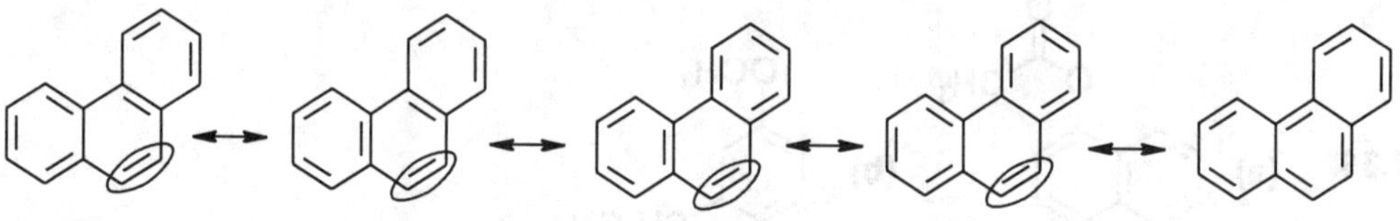

27.45   The reduction potential for 2-methoxy-1,4-benzoquinone is lower than for 2-methyl-1,4-benzoquinone because the methoxy group is a stronger electron donating group than a methyl group. Thus, the reaction should occur in the direction written.

27.46   The reduction potential for 1,2-benzoquinone must be 0.08 V more positive than the reduction potential for 1,4-benzoquinone (0.699 V). The value for 1,2-benzoquinone is 0.78 V.

27.47   **(a)** Cyclohexylamine should be the better nucleophile because the nitrogen atom is more basic. The electron pair on nitrogen for aniline is delocalized by interaction with the π electrons of the benzene ring.
     (b) p-Methoxyaniline should be the better nucleophile because the methoxy group is electron donating by resonance and increases the electron density at the nitrogen atom. The nitro group is strongly electron withdrawing and decreases the electron density at the nitrogen atom.
     (c) The methyl groups of N,N-dimethylaniline increase the electron density at the nitrogen atom and should make the site more nucleophilic. However, these groups also increase the size of the nucleophile and therefore decrease its nucleophilicity.

27.48   Based on the hybridization of the nitrogen atoms, the site of protonation would be the tertiary amine group because its hybridization is sp^3. The lone-pair electrons of the imino group are sp^2 hybridized and are more tightly held to the nitrogen atom, because the % s character of the orbital is greater. However, the effect of resonance stabilization of the base and its conjugate acid is very important. The basic form is resonance stabilized.

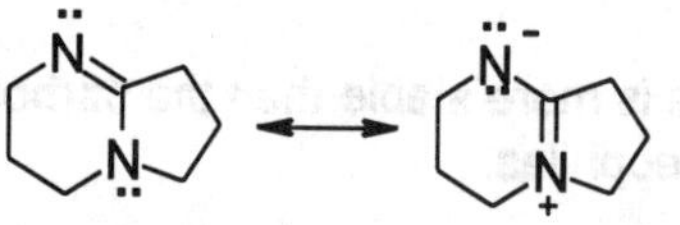

Protonation at the amine group eliminates the possibility of delocalization of electrons.

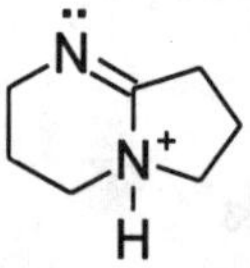

However, protonation of the nitrogen atom of the imino group gives a resonance stabilized conjugate acid, which is the reason why DBN is a stronger base than an amine.

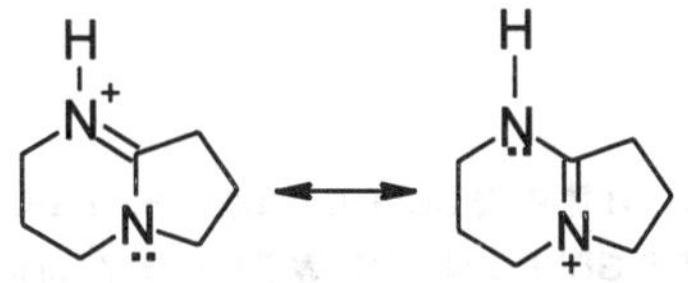

Resonance stabilization of this conjugate acid favors its formation compared to the alternate conjugate acid resulting from protonation of the amine group.

27.49

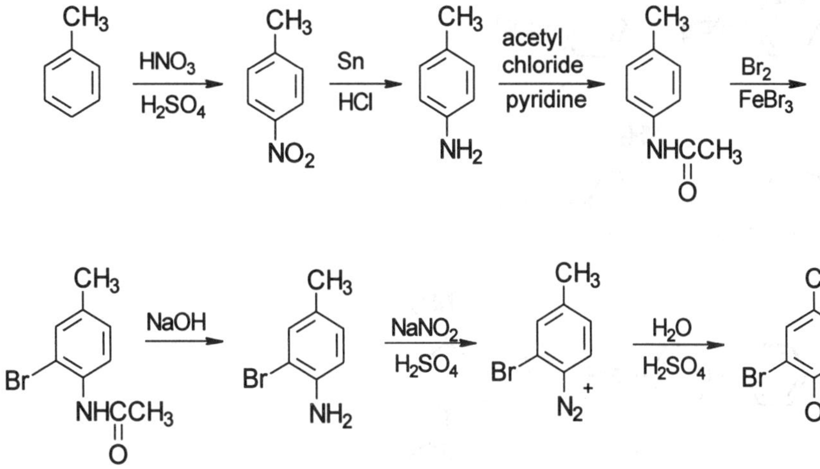

27.50

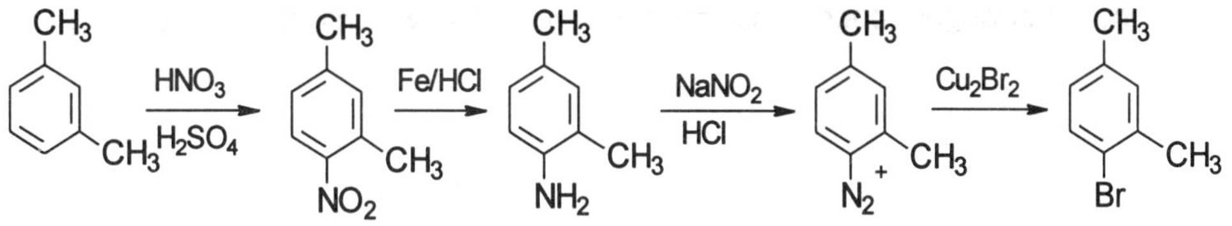

27.51 The nitro group withdraws electrons from the aromatic ring and from the diazonium group, thus increasing its electrophilicity.

27.52 (a) Aniline is more reactive, because the bromine atom of o-bromoaniline deactivates the ring to a small degree toward electrophilic aromatic substitution.
(b) p-Methylphenoxide is more reactive, because the electron pairs of oxygen are more available to the aromatic ring when it is attacked by an electrophile.
(c) N,N-Dimethylaniline is more reactive because nitrogen is a better donor of electrons by resonance than is oxygen.

27.53

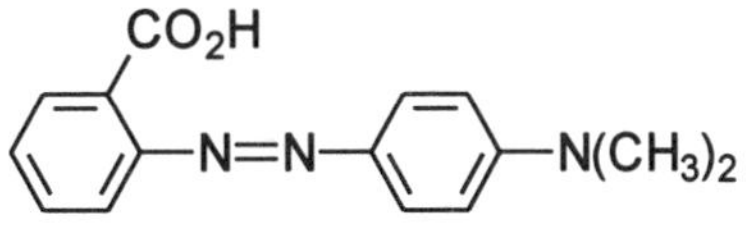

27.54

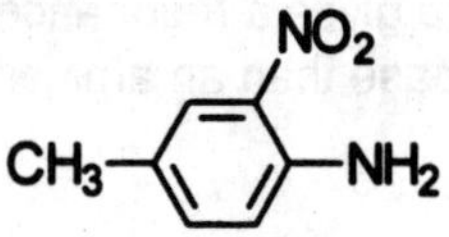

Nitrate toluene and separate the para isomer from the reaction mixture. Reduce the nitro group to give an aniline derivative. Acetylate the amino group with acetyl chloride. Nitrate the amide derivative to place a nitro group at the position ortho to the nitrogen atom. Hydrolyze the amide to obtain the amine.

27.55 (a)

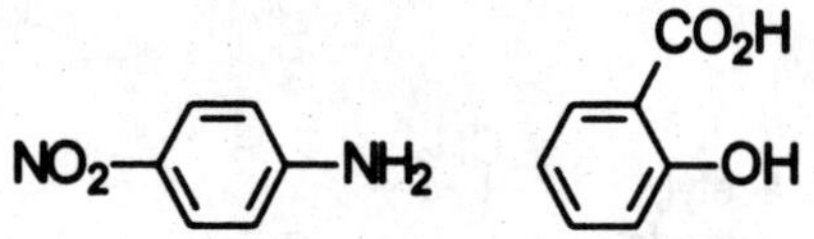

(b)

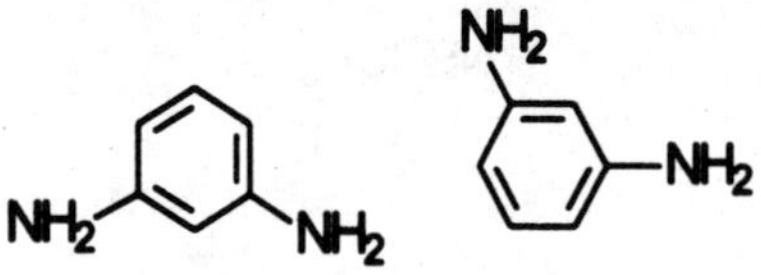

(c)

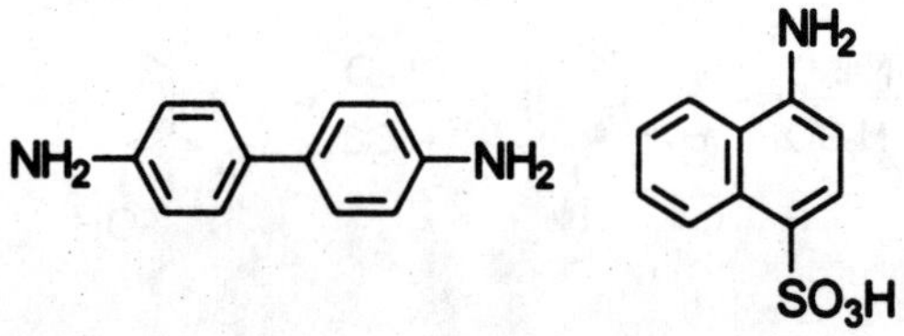

27.56 At pH 9 the phenol exists as a phenoxide, and this group is a much more effective donor of electrons by resonance. At pH 5 the phenol form exists, and the reaction is controlled by the stronger donor of electrons by resonance, which is the amino group in this case.

28

Pericyclic Reactions

This chapter is the most unique in this text, because it deals entirely with molecular orbitals to explain chemical reactions and the only functional groups are π systems. A substantial amount of new terminology is required to describe pericyclic reactions. The "rules" described to account for the observed reactions and to predict new reactions may be memorized, but an understanding of why the rules work based on the identity of the molecular orbitals involved is the preferred method that will allow you to solve problems even when you forget the rules. You should review the molecular orbital concepts presented in Chapters 1 and 12.

Keys to the Chapter

28.1 Concerted Reactions

Concerted reactions occur by the simultaneous formation and cleavage of bonds to give a product in a single step. **Pericyclic reactions** are concerted processes in which only changes in π and σ bonds occur via cyclic transition states. No intermediates form in these reactions.

The energy required for a pericyclic reaction may be supplied by thermal energy or photochemical energy. The differences in the products formed by these two sources of energy reflect the molecular orbitals involved in the reaction. **Thermal reactions** occur using molecular orbitals of the ground state; **photochemical reactions** occur via excited states in which electrons have been promoted to higher energy molecular orbitals.

28.2 Classification of Pericyclic Reactions

Pericyclic reactions are separated into two classes based on the number of electrons that change molecular orbitals in the transition state. Thus, you must account for either $4n$ or $4n+2$ electrons where n is an integer. You must know which molecular orbital is involved in the reaction and recognize its symmetry properties. Mechanistically, pericyclic reactions are divided into three classes: electrocyclic reactions, cycloaddition reactions, and sigmatropic rearrangements.

Electrocyclic reactions interconvert polyenes and isomeric cyclic products having one less double bond. The ends of the polyene join to form the single bond of the ring. These reactions favor the cyclic product because it has two additional single bonds at the expense of losing only one double bond. The $\Delta H°_{rxn}$ for the reaction is thus negative. The reverse of the cyclization reaction is also considered an electrocyclic reaction. In general the reverse reactions are not spontaneous unless there is ring strain that destabilizes the cyclic isomer relative to the polyene containing one more double bond.

Cycloaddition reactions join two components such as a conjugated diene and an alkene (or alkyne). In the Diels-Alder reaction, the ends of the diene join one each to the two carbon atoms of the alkene. The net result of this thermal reaction is the formation of four single bonds and the loss of two double bonds—a process with $\Delta H°_{rxn} < 0$. Although $\Delta S°_{rxn}$ for the cycloaddition is negative, the $\Delta H°_{rxn}$ is sufficiently negative to provide a negative $\Delta G°_{rxn}$. There are 6 π electrons involved in the transition state for the Diels-Alder reaction. Cycloaddition reactions occur between two alkenes in which 4 π electrons are involved in the transition state, but the process occurs only photochemically. Cycloaddition reactions are designated by the number of π electrons in each of the two reactants contained in brackets and separated by a plus sign as in

[4 +2] for the Diels-Alder reaction.

Sigmatropic rearrangements transfer a σ bonded group such as hydrogen from one end of a π system to the other. In the process the positions of single and double bonds are interchanged. These rearrangements are conceptually the most difficult of the pericyclic reactions to recognize and understand.

In addition to recognizing the number of π electrons involved in a pericyclic reaction and the reaction conditions (thermal or photochemical), the stereochemistry of the process, which is a consequence of the first two features, must be understood. The stereochemistry of the reaction based on the symmetry of the reacting molecular orbitals is established by the stereochemistry of attached groups such as the methyl group. As the component atomic orbitals that contribute to the molecular orbital "rotate", the methyl groups move in the same direction.

28.3 Stereospecificity and Molecular Orbitals

Pericyclic reactions occur stereospecifically because the symmetry of the molecular orbitals changes in specific ways described as **conservation of orbital symmetry**. The same results can be described using a **frontier orbital method** that considers the symmetry of only the highest occupied molecular orbital (HOMO) and lowest unoccupied molecular orbital (LUMO).

Pericyclic reactions occur because they are **symmetry-allowed**, which means that the symmetries of the reactants and products match. These reactions occur under relatively mild reaction conditions. Symmetry-disallowed reactions may occur to give the same types of products, but they require substantially higher temperatures.

A molecular orbital is described as symmetric or antisymmetric based on the relationship of the signs of the contributing atomic orbitals with respect to a vertical plane. If the signs on each side of the vertical plane are the same, then the molecular orbital is **symmetric**. If there is a reversal of sign, the molecular orbital is **antisymmetric**. You can determine the symmetry of the molecular orbital by recognizing that π_1 is symmetric for all polyenes and that the symmetry of the molecular orbitals alternate between symmetric and antisymmetric for π_1, π_2, π_3, π_4, and so on. It then remains only to determine which molecular orbital is involved in the reaction.

28.4 Electrocyclic reactions

In an electrocyclic reaction we picture the terminal atomic orbitals rotating to form a σ bond by means of overlap of the p orbital lobes having the same sign. Rotation of the two orbitals in the same direction is **conrotatory**; rotation of the orbitals in the opposite directions is **disrotatory**.

If the π_2 of butadiene is involved in a thermal electrocyclic reaction, it is necessary that rotation occur in a conrotatory sense because the orbital is antisymmetric. However, in a photochemical electrocyclic reaction of butadiene, π_3, which is symmetric, is involved and as a result rotation occurs in a disrotatory sense.

In 1,3,5-hexatriene the results are reversed because the corresponding HOMO and LUMO molecular orbitals that react have different symmetries. The π_3 molecular orbital is required in the thermal electrocyclic reaction, and it is necessary that rotation occur in a disrotatory sense because the orbital is symmetric. However, in a photochemical electrocyclic reaction of the triene, π_4, which is antisymmetric, is involved and as a result rotation occurs in a conrotatory sense.

The quick way to remember this information is to learn one process such as the thermal electrocyclic reaction of butadiene. This compound is a 4n system and it involves a conrotatory process. If either of the features change compared to this reference reaction, then the rotational process changes. Thus if the system has 4n+2 π electrons, the "change" is to a disrotatory process. If you then change the reaction conditions from thermal to photochemical for this new conjugated system, then the direction changes again. In this case a 4n+2 π electron compound reacts photochemically by a conrotatory process. The same features are listed in Table 28.1. Again, make sure that you memorize one process to have a point of reference!

Once you know the direction of rotation of the orbitals, you also know the direction of movement of the attached alkyl groups. You need only have the skill of visualizing where groups are originally and where they will be located after the rotation occurs.

28.5 Cycloaddition Reactions

Cycloaddition reactions occur by the joining of the terminal atoms of one π system to the terminal atoms of another π system. The process is best viewed as the approach of one plane containing one set of carbon atoms to another plane containing another set of carbon atoms. This approach of reacting atoms occurs in **suprafacial** cycloadditions. For bond formation to occur, the symmetry of the appropriate molecular orbitals must the same. Only then will the "signs" of the contributing atomic orbitals match.

Antarafacial cycloadditions are a bit more difficult to visualize, because the ends of one π system must simultaneously bond to different sides of the plane of the second π system. Such processes are not as common because of geometric constraints due to insufficient numbers of carbon atoms to bridge from one side of the plane to the other.

The selection of the appropriate molecular orbitals to "explain" cycloaddition is a little arbitrary. However, the combination of orbitals must include the HOMO of one π system and the LUMO of the other π system. In the text the HOMO of the diene is selected and the LUMO of the alkene. Therefore the diene is viewed as the donor of electrons for the reaction. The HOMO of the diene is antisymmetric, as is the LUMO of the alkene. The Diels-Alder reaction is thus a [4+2] cycloaddition that occurs suprafacially.

The photochemical cycloaddition of two moles of an alkene is a [2+2] reaction. It is visualized by the interaction of one excited state molecule to a ground state molecule. The electrons in the "donor" excited state molecule are in π_2. The available orbital in the ground state molecule is π_2. The symmetries of identical molecular orbitals are of course the same, so the reaction can occur suprafacially.

Table 28.2 in the text summarizes the various possible symmetry-allowed cycloaddition reactions. However, as in the case of electrocyclic reactions, make sure that you know the results of one reaction. Then if any of the characteristics of that type of reaction are different, then reverse the stereochemical consequences for each difference. Memorize that the Diels-Alder reaction is a thermal [4+2] process that occurs suprafacially. Changing from 4n+2 to 4n or changing the conditions from thermal to photochemical changes the stereochemistry to antarafacial. Changing two of the conditions gets you back to a suprafacial reaction.

28.6 Sigmatropic Rearrangements

There are two common types of sigmatropic shifts, [1,5] and [3,3]. In each case a group from one site of a π system migrates to another site at the other "end" of the π system. The thermal process occurs suprafacially in both cases, which are 4n+2 systems.

There are two difficulties in understanding the stereochemical consequences of sigmatropic shifts and why they occur. First it is necessary to consider the molecular orbital over which the migrating group moves. For both the [1,5] and [3,3] systems, there are odd numbers of electrons. Thus, the molecular orbitals are arranged differently than for polyenes, but the lowest energy molecular orbital is still symmetric and the symmetries of additional molecular orbitals in the energy diagram still alternate.

It is convenient to picture the reactant as two radicals, one of which separates at one point and is rejoined with the other radical at the other end of conjugated radical. (This approach does not mean that the process pictured constitutes the actual mechanism. Indeed, it does not, because the actual process is concerted and no radicals are formed) For the pentadienyl radical, the HOMO has a single electron in π_3 and that molecular orbital is symmetric. Thus the migrating group moves along a single face in going from one end of the π system to the other.

For [3,3] sigmatropic shifts, the reaction is visualized as the cleavage and rebonding of two allyl radicals. The HOMO of the allyl radical is π_2 and it is antisymmetric. Because the two radicals are identical, the cleavage and rebonding can occur suprafacially.

Solutions to Exercises

28.1 (a) electrocyclic   (b) electrocyclic   (c) electrocyclic
         (d) sigmatropic   (e) cycloaddition

28.2 (a) cycloaddition   (b) sigmatropic   (c) electrocyclic
         (d) cycloaddition   (e) sigmatropic

28.3 The $\Delta H°_{rxn}$ is estimated using the average energy for the formation of two C—C single bonds and the energy required to "remove" one C—C double bond. Thus, the value should be approximately -90 kJ mole^{-1}, as calculated for *cis*-1,3,5-hexatriene on page 1067.

28.4 The $\Delta H°_{rxn}$ for the reaction is equal to the energy difference between the differently substituted double bonds. Both compounds have a terminal monosubstituted double bond. The first compound has a trisubstituted double bond, as compared to a terminal disubstituted double bond in the second compound. The difference in the average heats of hydrogenation of trisubstituted alkenes and terminal disubstituted alkenes is about 4 kJ mole^{-1} (see Table 6.3). The reaction as written is unfavorable and $\Delta H°_{rxn} = 4$ kJ mole^{-1}.

28.5 The initial product has an enol site that tautomerizes to the more stable ketone.

28.6 The initial product has two enol sites that tautomerize to give the more stable diketone.

28.7 (a) The equilibrium favors the side of the reactant, because the double bonds in that compound are in conjugation with the carbonyl group.
         (b) The equilibrium favors the side of the reactant, because the double bonds are more highly substituted.
         (c) The equilibrium favors the side of the product, because that isomer has more sigma bonds and fewer double bonds.

28.8 (a) The equilibrium favors the side of the product because the double bonds are more highly substituted.
         (b) The equilibrium favors the side of the reactant, because the double bonds of that compound are stabilized by resonance interaction with the lone pair electrons of nitrogen.
         (c) The equilibrium favors the side of the product even though it has more double bonds and fewer sigma bonds. The strain energy of the cyclobutane ring is larger than the difference in energy due to the changes in the types of bonds.

28.9 The $\Delta H°_{rxn}$ is estimated using the bond energies of isolated double bonds. The double bonds of 1,3-butadiene are conjugated, and the compound is stabilized by resonance.

Thus, the $\Delta H°_{rxn}$ should be somewhat smaller that the estimate given in Section 28.2.

28.10   The estimated $\Delta H°_{rxn}$ for cycloaddition reactions given in Section 28.2 is -180 kJ mole^{-1}, so the $\Delta H°_{rxn}$ for the reverse reaction is 180 kJ mole^{-1}. The strain energy of a cyclobutane ring is 111 kJ mole^{-1} (Table 4.4). Thus, the $\Delta H°_{rxn}$ for the conversion of cyclobutane into ethylene is estimated as 69 kJ mole^{-1}.

28.11   Convert $\Delta H°_{rxn}$ into -168,000 J mole^{-1}. When K = 1 the $\Delta G°_{rxn}$ = 0 and therefore we can write $\Delta H°_{rxn} = T\Delta S°_{rxn}$. Substitution of $\Delta H°_{rxn}$ and $\Delta S°_{rxn}$ and solving for T gives 894 K, which is 621 °C.

28.12   The spontaneity of the reaction which converts a single mole of reactant into two moles of product is due in large part to the favorable $\Delta S°_{rxn}$. Note that in Exercise 28.10, the $\Delta H°_{rxn}$ for the conversion of cyclobutane into ethylene is estimated as 69 kJ mole^{-1}. In the indicated reaction of the polycyclic hydrocarbon, the number moles of reactant equals that of the product. Thus, the reaction is not expected to be as favorable. However, the added strain of the small rings in the tetracyclic hydrocarbon should contribute to making the reaction more favorable.

28.13   As shown in the Essay on page 1088 in the text, previtamin D$_3$ is a heptatrienyl system. The HOMO for the seven-electron heptatrienyl radical is π_4, which is antisymmetric. This symmetry requires that the hydrogen atom migrate antarafacially, which is geometrically possible for a system of this size and flexibility.

28.14   The HOMO for the eight-electron octatetraene is π_4, so the molecular orbital into which an electron is promoted for a photochemical reaction is π_5, which is symmetric.

28.15   Refer to Tables 28.2 and 28.3 in the text to relate the number of electrons and type of reaction to the stereochemistry.
   (a) 4 + 6 = 10 = 4n + 2, so the stereochemistry is suprafacial for a thermal cycloaddition.
   (b) 2 + 6 = 8 = 4n, so the stereochemistry is suprafacial for a photochemical cycloaddition.
   (c) 1 + 7 = 8 = 4n, so the stereochemistry is antarafacial for a thermal sigmatropic rearrangement.
   (d) 1 + 3 = 4 = 4n, so the stereochemistry is suprafacial for a photochemical sigmatropic rearrangement.

28.16   Refer to Table 28.1 in the text to relate the number of electrons and type of reaction to the motion of the orbitals as they close or open a ring.
   (a) A triene has 6 electrons (4n + 2), so the motion is disrotatory for a thermal ring closure.
   (b) A diene has 4 electrons (4n), so the motion is disrotatory for a photochemical ring closure.
   (c) The tetraene product has 8 electrons (4n), so the motion is conrotatory for a thermal ring opening.
   (d) The triene product has 6 electrons (4n + 2), so the motion is conrotatory for a photochemical ring opening.

28.17   The reaction is a [4 + 6] cycloaddition. This is a 4n+2 system, which is symmetry allowed. The LUMO of the diene is π_3, which is symmetric. The HOMO of the triene is π_3, which is also symmetric.

28.18 One of the two π bonds of the acetylene group adds to the methylene carbon atom and the C-2 atom of the ring. This analysis represents a tetraene system. The HOMO of this system is π_4 which is antisymmetric. The LUMO of the reacting double bond of the acetylene is π_2, which is antisymmetric. Thus the reaction is symmetry allowed and can occur suprafacially.

28.19 The double bonds are in an s-trans arrangement, so the two "ends" of the π system cannot be bridged by the atoms of a dienophile.

28.20 For a concerted process, the thermal ring opening of a cis-dialkyl substituted cyclobutene occurs by a conrotatory process to give an E,Z diene. In the first compound, this configuration can be accommodated within the cyclic product. 1,3-Cycloheptadiene cannot exist with an E,Z configuration. The product has the Z,Z configuration, so it cannot be the product of a concerted electrocyclic process. The reaction must occur via a multistep process which requires more energy.

28.21 There are two nonequivalent allylic sites from which the hydrogen atom could move. The predominant product should be the compound with the more stable double bond, which is the first of the following two products because its double bond is more highly substituted.

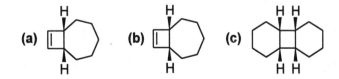

28.22 The thermal [1,7] sigmatropic rearrangement occurs antarafacially in previtamin D, because the open conformation allows the hydrogen atom to transfer from the "top" of one end of the system to the "bottom" of the other end. In the phenyl substituted cycloheptatriene, the "ends" of the system are bonded and a hydrogen atom cannot move from one side of the ring to the other. A photochemical rearrangement occurs suprafacially, which is geometrically possible. The resulting product is a resonance stabilized triene.

28.23 Refer to Table 28.1 in the text to relate the number of electrons and type of reaction to the motion of the orbitals as they close or open a ring. The structures resulting from each reaction are given below.

(a) For a thermal ring closure with 4n electrons, the motion is conrotatory.
(b) For a photochemical ring closure with 4n electrons, the motion is disrotatory.
(c) For a photochemical ring closure with 4n electrons, the motion is disrotatory.

28.24 Refer to Table 28.1 in the text to relate the number of electrons and type of reaction to the motion of the orbitals as they close or open a ring. The structures resulting from each reaction are shown below.

(a) For a thermal ring closure with 4n +2 electrons, the motion is disrotatory.
(b) For a photochemical ring closure with 4n electrons, the motion is disrotatory.
(c) For a thermal ring opening with 4n electrons, the motion is conrotatory.

(a) 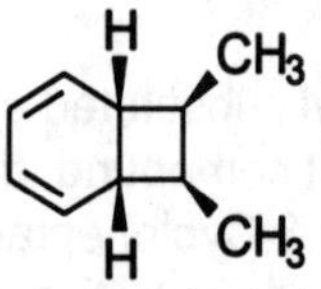 (b) 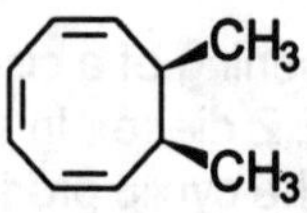 (c)

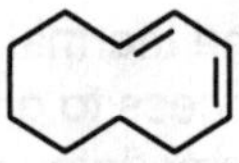

28.25 The conrotatory process could occur in a clockwise or counterclockwise manner to give isomeric dienes. However, the process that moves two hydrogen atoms toward one another is less sterically hindered that the process that moves two methyl groups toward one another. The first of the following two products is the preferred product.

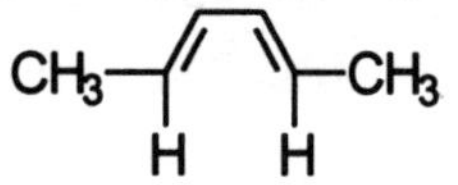

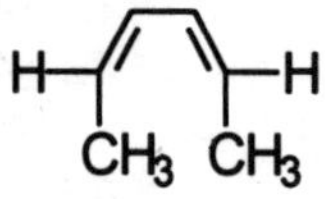

28.26 Compound (b) has an E configuration at the C-4 to C-5 atoms which geometrically prevents the "ends" of the triene system from bonding in a cyclization reaction. Compounds (a) and (c) can react.

(a)    (b) 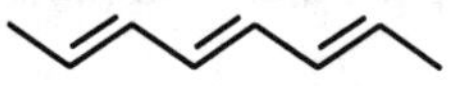   (c)

28.27 (a) 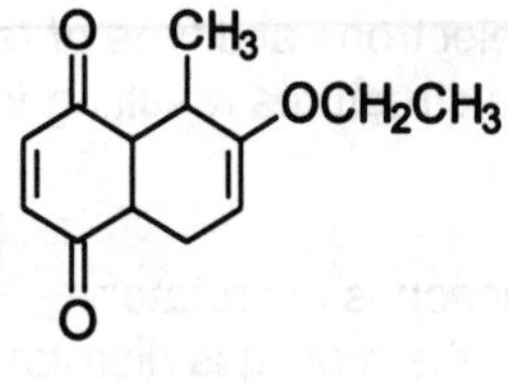   (b) 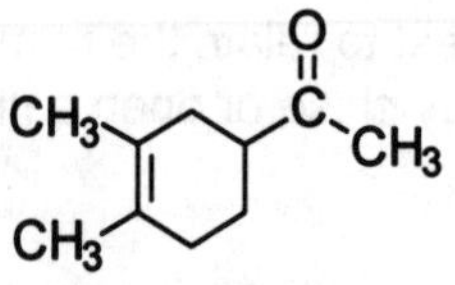   (c)

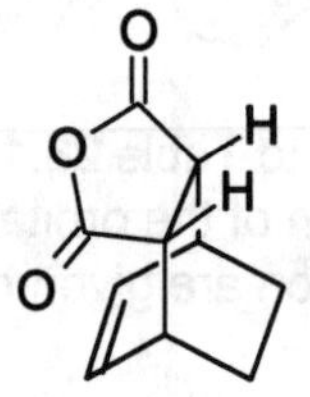

28.28 (a) 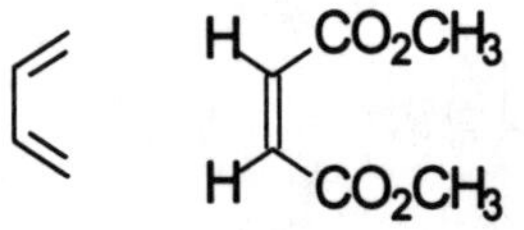   (b) 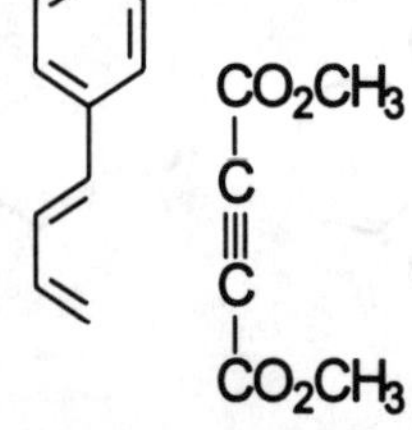   (c)

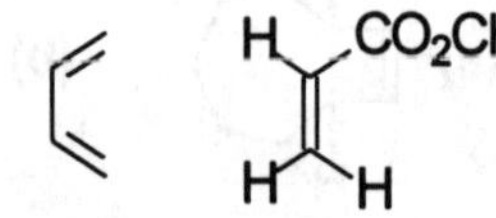

28.29   The two "ends" of the diene system in 1,3-cyclopentadiene are held in a cis arrangement by the ring. The reaction of 1,3-butadiene requires conversion of the more stable s-trans conformation into an s-cis conformation via rotation about the C-2 to C-3 bond before reaction with maleic anhydride can occur.

28.30   The s-cis conformation is sterically hindered by the terminal methyl groups.

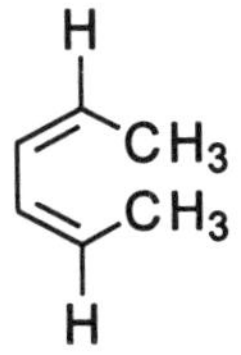

28.31   The dienophile has two nonequivalent double bonds. The more reactive double bond is the one with electron withdrawing groups. In this case, the methyl group is more electron donating than hydrogen is, so the double bond lacking the methyl group is the more reactive.

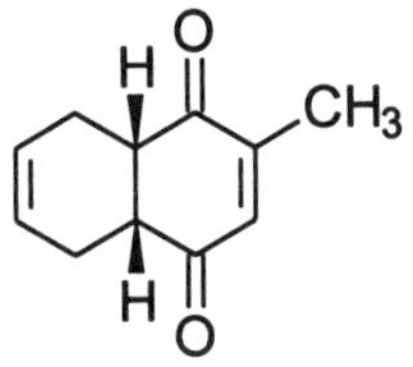

28.32   The dienophile has two nonequivalent double bonds. Both the methoxy group and the methyl group can stabilize their respective double bonds. The methoxy group is more effective in donation of electrons by resonance. Thus, the double bond with the methyl group is the more reactive. The diene can add in two ways to give isomeric adducts.

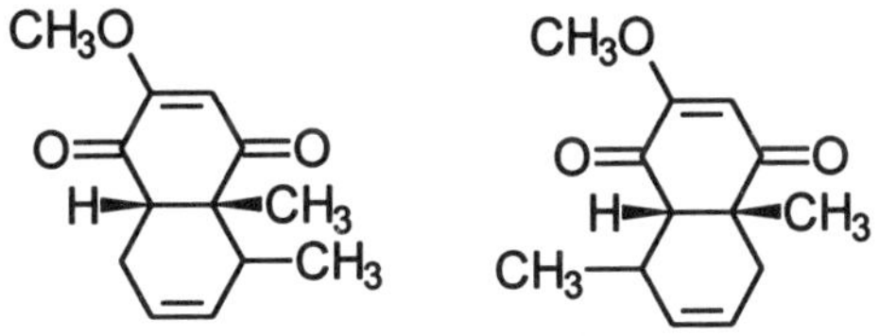

28.33

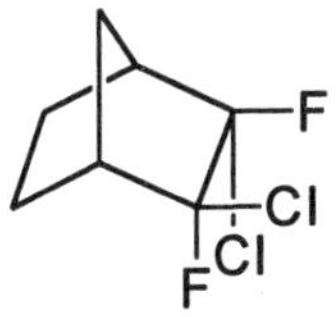

28.34

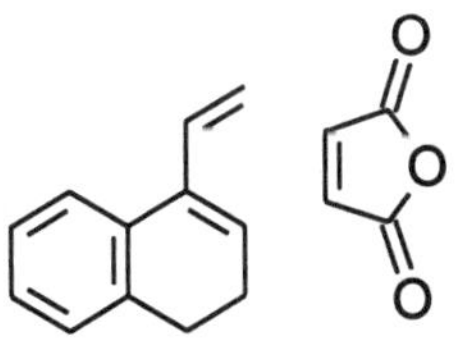

28.35   The two "ends" of the diene that must bond are held in a cis arrangement by the ring in
cis-1,2-divinylcyclobutane. 1,5-Hexadiene is conformational flexible about three sigma
bonds. The reaction of this compound requires a conformation that has an eclipsed
arrangement about the C-3 to C-4 bond.

28.36   The reaction is a [3,3] sigmatropic shift. The driving force for the reaction is the formation
of a stable carbon-oxygen double bond compared to one less stable carbon-carbon
double bond in the reactant.

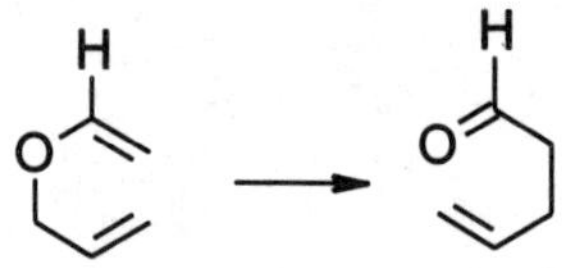

28.37   **(a)**

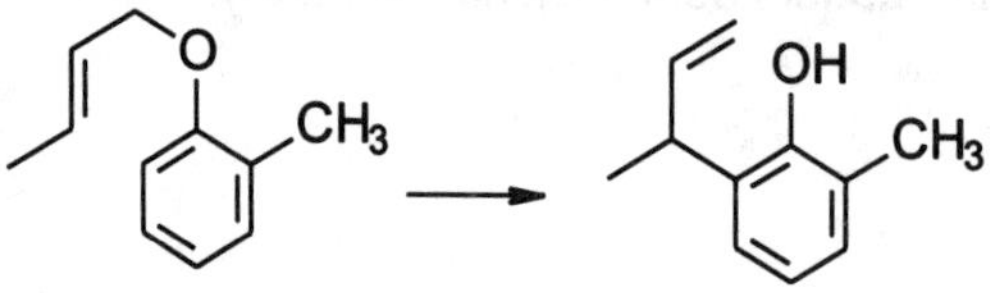

(b)

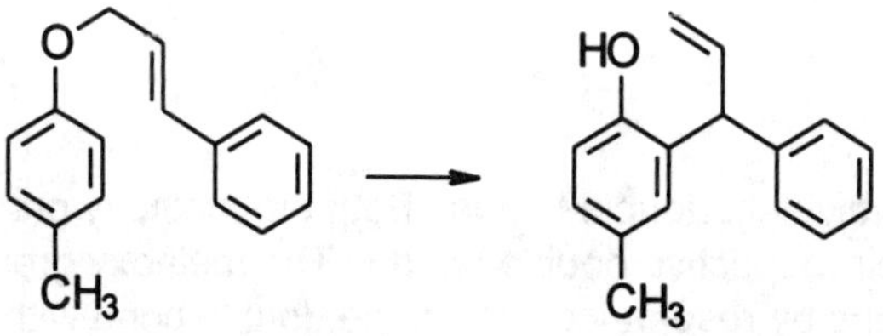

(c)

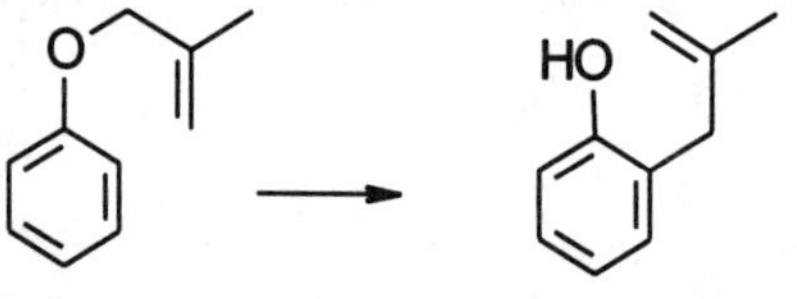

28.38   The reaction is a Claisen-type rearrangement as shown in Problem 28.9. A [3,3]
sigmatropic rearrangement occurs to give an unstable intermediate that undergoes a
tautomerization reaction.

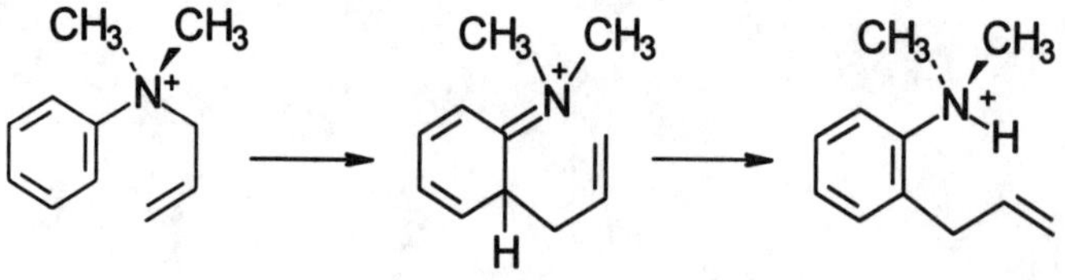

28.39   The reaction involves a [3,3] sigmatropic rearrangement in which a carbonyl group forms
rather than a carbon-carbon double bond as shown on the following page. The driving
force of the reaction is the stability of the carbonyl group.

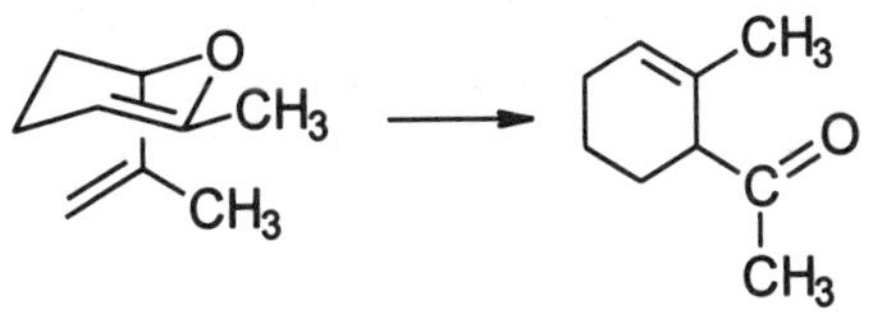

28.40 The reaction involves a [3,3] sigmatropic rearrangement. The driving force of the reaction is the stability of the tetrasubstituted carbon-carbon double bond and the carbonyl group.

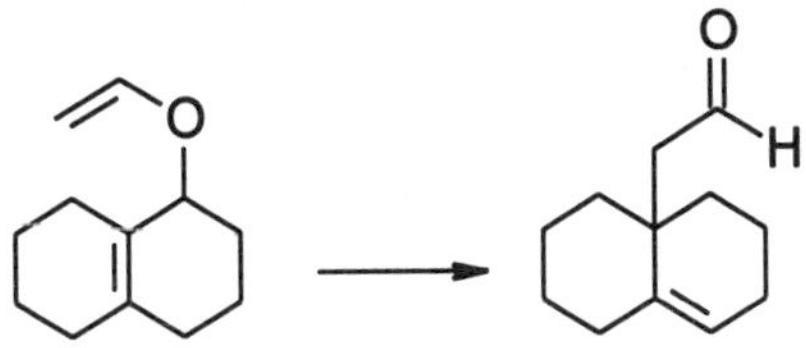

28.41 A [5,5] sigmatropic rearrangement involves 4n + 2 electrons and will occur thermally by a suprafacial process.

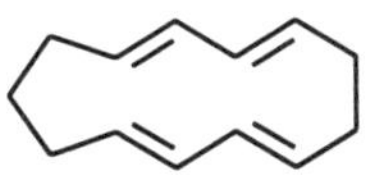

28.42 This [5,5] sigmatropic rearrangement involves two of the π bonds of the aromatic ring as the side chain diene bonds to the para position. To visualize how this occurs, imagine the aromatic ring lying flat and the diene side chain curved above it. The rearrangement occurs in a concerted process in which a π bond forms between oxygen and C-1 of the ring, a σ bond breaks between oxygen and C-5 of the side chain, and a σ bond forms between C-1 of the side chain and C-4 of the ring as shown on the following page. The unstable intermediate that forms then undergoes tautomerization to give a phenol.

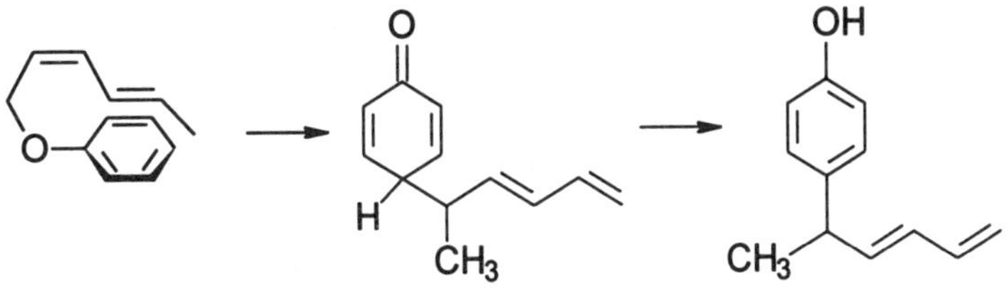

28.43 The reactions are two electrocyclic reactions. The first reaction forms a cyclobutene ring fused to the cyclobutane ring. The double bond is between the two carbon atoms at the points of fusion. Thus, the double bond "belongs" to both rings, and either one can be considered the cyclobutene ring. The second reaction could be the opening of the newly formed ring to give the original diene or an opening of the original ring to give the isomeric diene.

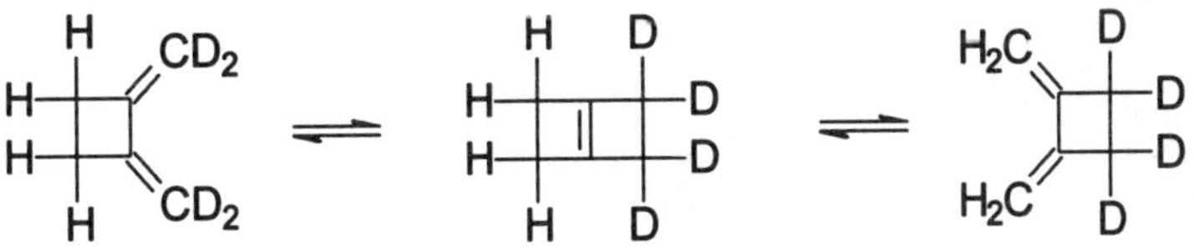

28.44 The reactions are two electrocyclic reactions. The first forms a cyclobutene ring. The second opens the ring. The ring closure and ring opening processes are conrotatory but can occur in either a clockwise or counterclockwise direction to give the products.

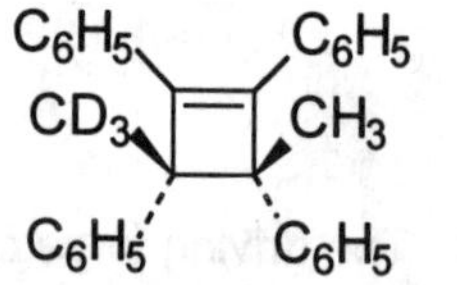

28.45 A [1,5] sigmatropic shift involving the hydrogen atom at the chiral center gives an achiral isomer. Reversal of the reaction can occur by transferring either of two hydrogen atoms from a methylene group. The transfer can occur from the "bottom" in the case of one hydrogen atom or from the "top" in the case of the other hydrogen atom, so the product is a racemic mixture..

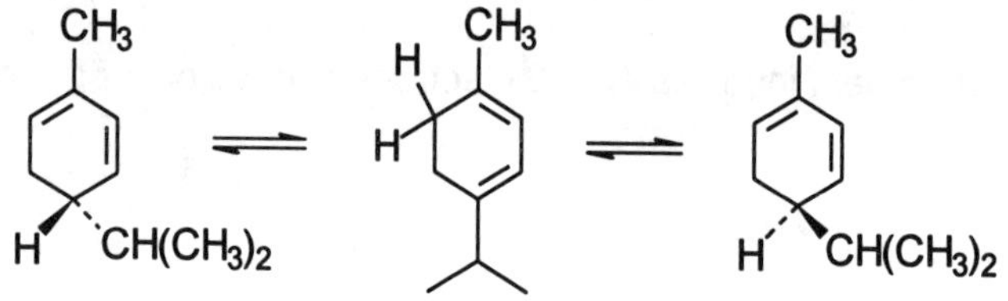

28.46 The intermediates involved are formed by [1,5] sigmatropic shifts. Migration of either a hydrogen atom or a deuterium atom gives similar structures that do not have the resonance stabilization of the benzene ring. The first of the following two structures is derived from transfer of a deuterium atom; the second is from transfer of a hydrogen atom.

Transfer of a hydrogen atom from the first intermediate to either of the equivalent sites in a [1,5] sigmatropic reaction gives one of the isomers.

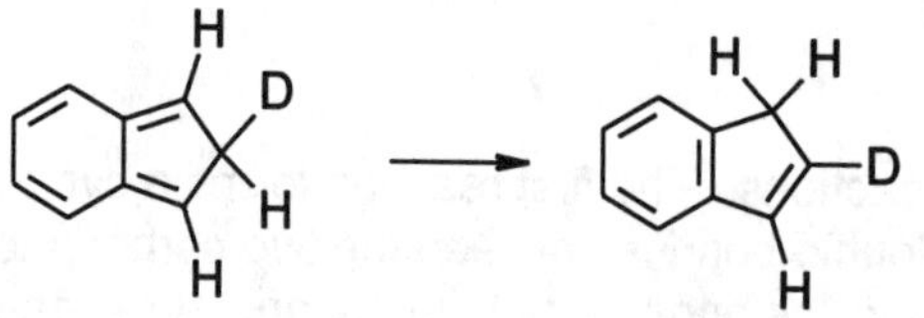

Transfer of a hydrogen atom from the second intermediate to the site not containing the deuterium atom gives the other isomer.

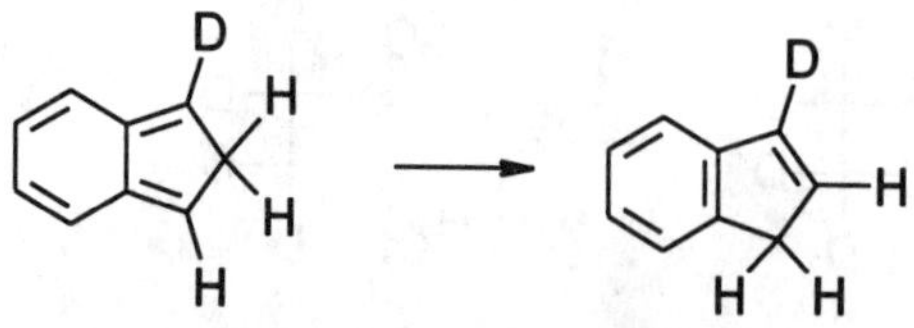

29

Synthetic Polymers

In this last chapter we focus on chemical reactions that we understand in principle but which have been seen only in the context of the synthesis of lower molecular weight compounds. Extension of those principles concerning the reactions of low molecular weight compounds to slightly different compounds called monomers provides an understanding of why these reactants can react over and over again with one another to give polymers.

Keys to the Chapter

29.1 Natural and Synthetic Macromolecules

Large molecules, know as **macromolecules**, occur naturally in carbohydrates and proteins. Chemical industry has learned how to produce similarly structured macromolecules with properties designed for specific uses.

29.2 Physical Properties of Polymers

The physical properties of polymers result from the types of intermolecular attractive forces between chains of the polymer, which in turn have functional groups that are formed from those in the monomer. Hydrogen bonds, dipole-dipole forces, and London forces all play a role in the association of polymer chains.

The types of bonds formed in the polymerization process constitute the primary structure of the polymer. The polymerization process gives mixtures of polymers with different molecular weights, and the properties of the polymer depend on the length of the chains. Reactions that form bonds between chains, known as **cross-linking**, are used to change the properties of polymers.

London forces, although the weakest of the intermolecular forces, are exceedingly important in polymers because of the length of the chain, which allows many "points of contact" between neighboring chains. Branching of chains changes the density of the polymers of ethylene. Polymers containing aromatic rings allow fewer conformations within the chain, and these rings are more polarizable as well. As a result, polymers of aromatic compounds have strong intermolecular forces.

Hydrogen bonding between functional groups such as amides In polyamides gives significant strength to these synthetic polymers. The cumulative effect of a large number of hydrogen bonds is also responsible for the stability of proteins.

Crystalline regions that are the result of regular alignment of polyethylene chains are called **crystallites**. These regions have strong intermolecular attractive forces. Crystallites also occur in polymers that form hydrogen bonds. In each case it is the cumulative effect of numerous individual interactions that stabilizes the crystallite and hence affects the property of the polymer.

29.3 Classification of Polymers

There are three types of polymers: elastomers, plastics, and fibers. **Elastomers** regain their original shape after being distorted in a physical process. The individual chains in elastomers are coiled and can be "straightened" out by stretching the material. However, after the stress is released, the molecule resumes its original coiled conformation. Rubber is an elastomer. It consists of unsaturated units separated by sp^3 hybridized carbon atoms which give the polymer its flexibility.

Plastics are polymers that harden upon cooling. Plastics that soften when heated are **thermoplastics** and can be molded while warm. **Thermosetting plastics** cannot be softened by heating—they are "set". Polyethylene is a thermoplastic. Thermosetting plastics usually have extensively cross-linked chains.

Fibers are thermoplastics that can be spun into materials that resemble natural fibers. One method of generating fibers is to pass molten thermoplastics through tiny pores in a die, after which the material hardens. A second method passes thermoplastics dissolved in a volatile solvent through tiny pores in a die, during which time the solvent evaporates.

29.4 Methods of Polymerization

Addition polymerization is the successive addition of alkene monomers to one another. The addition reaction may occur by radical, cationic, or anionic intermediates. **Condensation polymerization** is a reaction that joins two functional groups such as an alcohol and a carboxylic acid and form a second small molecule such as water.

Addition polymers are **chain growth polymers**, because each intermediate adds another monomer unit one at a time. Condensation polymers are **step-growth polymers**, because condensation may occur between two smaller molecular weight chains. Thus, the joining of oligomers results in a substantial increase in molecular weight in a single step.

29.5 Addition Polymerization

The length of the hydrocarbon chain of polyethylene and the degree of its branching are controlled by reaction conditions. **Termination steps** stop the growth of the polymer chain by destruction of radicals. Two such steps are dimerization, which join two radicals, and disproportionation, which transfers a hydrogen atom between radicals, resulting in an alkane and an alkene.

The regulation of chain length is accomplished using **chain transfer agents**. These substances react by transferring a hydrogen atom to the radical end of a developing polymer chain. But in order to continue polymerization of the monomer, the resulting radical of the chain transfer agent must be sufficiently reactive to initiate another polymerization process by reacting with the monomer.

Inhibitors react with growing polymer chains to transfer a hydrogen atom and give a stabilized radical derived from the inhibitor. This less reactive radical does not allow chain propagation steps to continue.

There are two **chain branching** processes. **Short chain branching** produces butyl side chains to the polymer chain as a result of intramolecular hydrogen transfer by a transition state that contains five carbon atoms and a hydrogen atom. In this process, a primary radical is converted into a secondary radical. **Long chain branching** occurs by random transfer of a hydrogen atom between the reacting primary radical site and some other site within the hydrocarbon skeleton.

29.6 Copolymerization of Alkenes

Mixtures of monomeric alkenes can form **copolymers** containing varying amounts of each monomer and in different sequences. The degree to which the polymer has a random arrangement of monomers or a regular alternation of monomers depends on the reactivity of the radical derived from one monomer toward either itself or the other monomer.

29.7 Cross-linked Polymers

Bonds between polymer chains known as cross-links are formed in two different ways. Some monomers used in copolymerization processes have more than one site for addition reactions to occur. In such cases one site is used in forming the polymer chain and the other site is used to link to another chain by a similar polymerization step. In this process the cross-links are produced while the polymer forms. The second method involves adding a substance after the polymer chain has formed. This material finds reactive sites along the individual chain and chemically reacts to link one chain to another.

29.8 Stereochemistry of Addition Polymerization

Polymerization of substituted alkenes gives polymers with stereogenic centers. The relationship of these stereogenic centers to one another affects the physical properties of the polymer. For example, in polypropylene the methyl groups are on the same side of the chain in an **isotatic** polymer. The regular alternation of methyl groups on each side of the chain is a **syndiotatic** polymer. A random distribution of methyl groups occurs in **atactic** polymers. Ziegler-Natta catalysts are used to control the type of polymer formed.

Diene monomers react to give polymers with one double bond per monomer unit. The stereochemistry of the double bond may be E or Z. The properties of these two polymers differ markedly. Ziegler-Natta catalysts are used to control the type of polymer formed.

29.9 Condensation Polymers

Any of the many reactions used to join two functional groups together and form a second smaller molecule such as water are candidates for condensation polymerization reactions. Two functional groups per monomer unit are required. Two monomers, each containing two units of one of the two possible functional groups, such as dicarboxylic acids and diols, are more commonly used in condensation reactions. Monomers containing one of each type of functional group are more difficult to synthesize and yet prevent from polymerizing

29.10 Polyesters

Polyesters are produced from dicarboxylic acids or their derivative and a diol. Poly(ethylene terephthalate) or PET is made by a transesterification reaction of dimethyl terephthalate and ethylene glycol. Cyclic anhydrides such as maleic anhydride or phthalic anhydride are used to form condensation polymers. Glycerol provides polymer that are cross-linked by reactions of the third hydroxyl group with the acid derivative. Glyptal is a cross-linked polymer of glycerol and phthalic anhydride.

29.11 Polycarbonates

Carbonic acid is an unstable diprotic acid. Its structurally related "diacid chloride",known as phosgene, or the "diester" diethyl carbonate can react with alcohols to form polycarbonates. The reaction with phosgene resembles the reaction of an alcohol with an acid chloride. The reaction of an alcohol with diethyl carbonate is a transesterification process.

29.12 Polyamides

Polyamides can be made by direct reaction of a dicarboxylic acid and a diamine. The resulting salt is heated, and the polyamide forms by loss of water. Nylon is a polyamide of a six-carbon diacid and a six-carbon diamine. Certain lactams can polymerize by a ring opening followed by condensation of the resulting amino acid.

29.13 Phenol-Formaldehyde Polymers

Phenol-formaldehyde polymers result from addition of a carbanion pictured as one of the resonance forms of the phenoxide ion. Addition to the carbon-oxygen double bond of formaldehyde gives a hydroxymethyl derivative, which can dehydrate to give a conjugate ketone, which can undergo conjugate addition reactions that resemble the aldol reaction. As a consequence, a polymeric product known as Bakelite results.

29.14 Polyurethanes

A urethane is an ester of a carbamic acid. Because carbamic acids are unstable, the direct esterification is not possible. Urethanes are made by reaction of alcohols with isocyanates. Polyurethanes are then made by using diisocyanates and diols. The diols may be simple compounds such as ethylene glycol or oligomers.

Solutions to Exercises

29.1 The elastomer has alternating quaternary carbon atoms that prevent the polymer chain from packing closely together in a regular array to give a crystalline-like area.

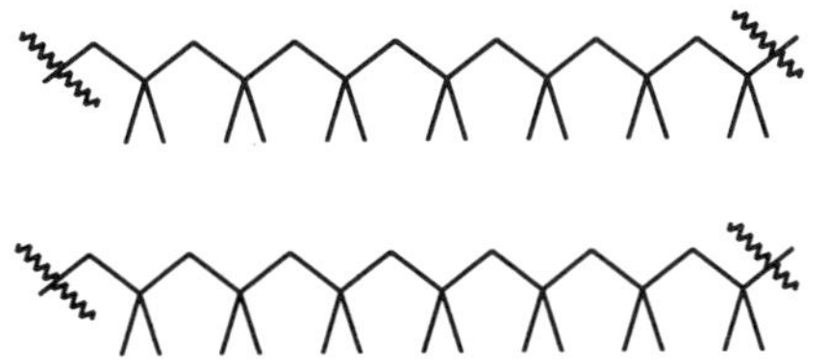

29.2 The secondary amine would give a polymer containing tertiary amide functional groups. This polymer could not form the hydrogen bonds that are responsible for the strong intermolecular forces in polyamides such as nylon 6,6.

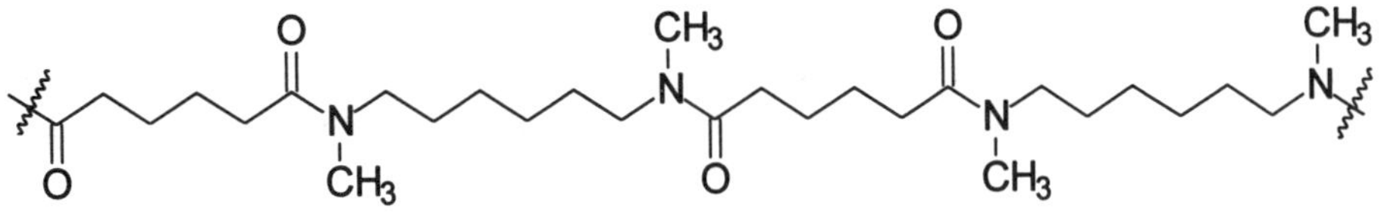

29.3 Reaction of the dianhydride at one of the anhydride sites with a diol gives a polyester that retains an anhydride unit which could react with added monomers to give bridged links between chains. Extensive cross-linking is what gives the thermosetting polyesters their properties.

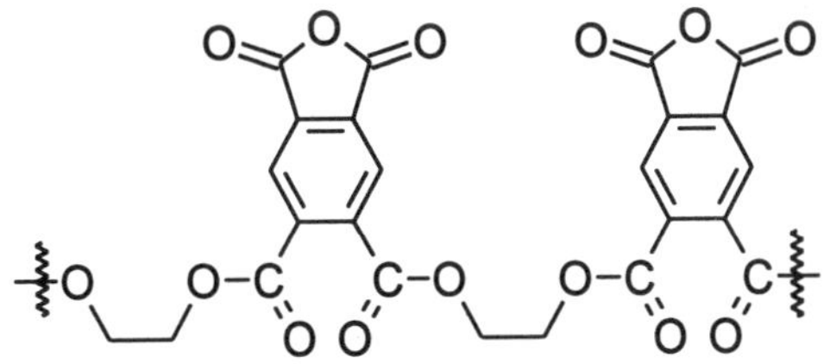

29.4 The additional two methylene units of the diol result in a decrease in London forces and give a polymer that is more flexible. It should have a smaller tensile strength.

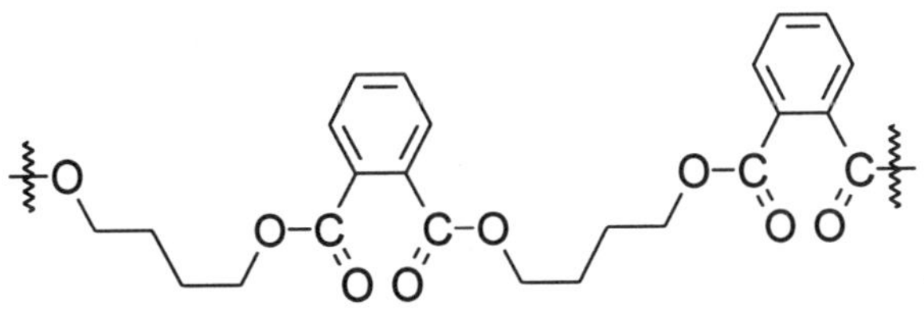

29.5   Neoprene is a polymer of 2-chloro-1,3-butadiene, whereas polyisoprene is a polymer of 2-methyl-1,3-butadiene. The electronegative chlorine atom of neoprene reduces the availability of electrons to oxidizing agents and the large chlorine atom reduces the accessibility of the C—H bonds to oxidizing agents.

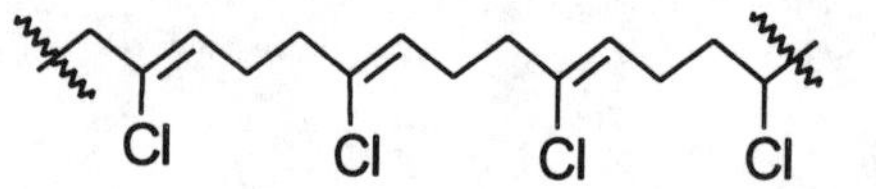

29.6   The electronegative fluorine atom reduces the availability of electrons to oxidizing agents. There are no C—H bonds to be attacked by oxidizing agents.y

29.7   In the following structure, the symbol OAc represents the acetate group.

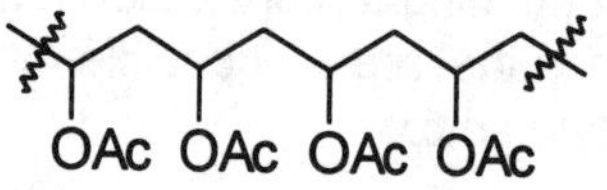

29.8   The required monomer would have to be vinyl alcohol, which does not exist because vinyl alcohol tautomerizes to ethanal (acetaldehyde). Hydrolysis of polyvinyl acetate however, does produce polyvinyl alcohol.

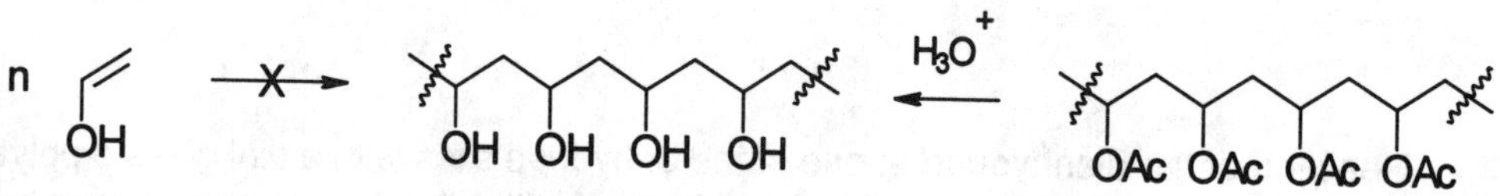

29.9   There is a repeating unit every two carbon atoms. Thus, the polymer is formed from a substituted ethylene compound. One carbon atom has a chlorine and fluorine atom and the other carbon atom has two fluorine atoms.

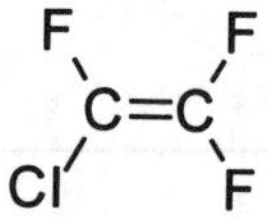

29.10

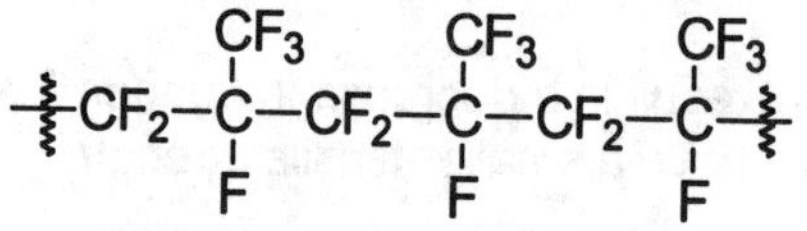

29.11

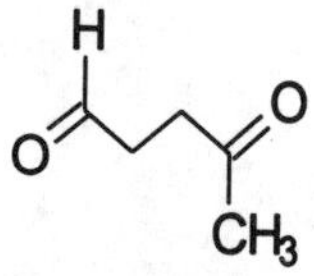

29.12 The structure of the polymer shown below gives 2,5-hexanedione. The diene required to form the polymer is 2,3-butadiene.

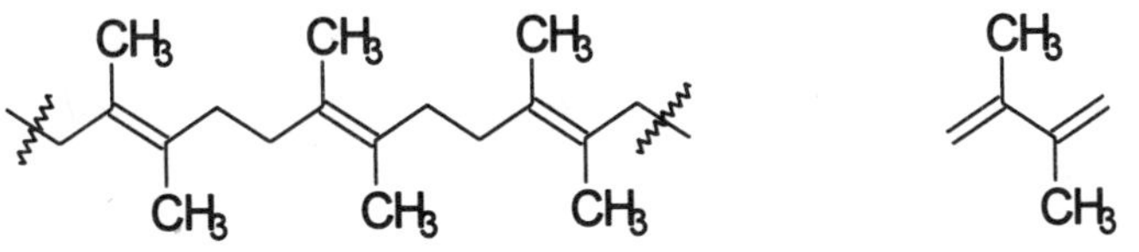

29.13 The short-chain branching reaction involves intramolecular abstraction of a hydrogen atom, resulting in a new radical site where polymerization continues. The branch formed is 2,4-diphenylbutyl.

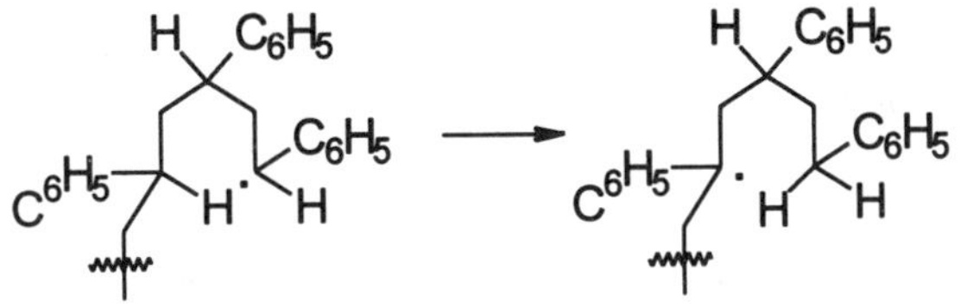

29.14 The side chain on the polymer is a butyl group, and the reaction center is a secondary radical. Abstraction of a hydrogen atom from the methyl group could occur via a six-membered transition state, but a primary radical would result. Abstraction of a hydrogen atom of a methylene group can occur through a five-membered transition state to give a secondary radical. Continued reaction of this radical gives a product with a methyl group at that point.

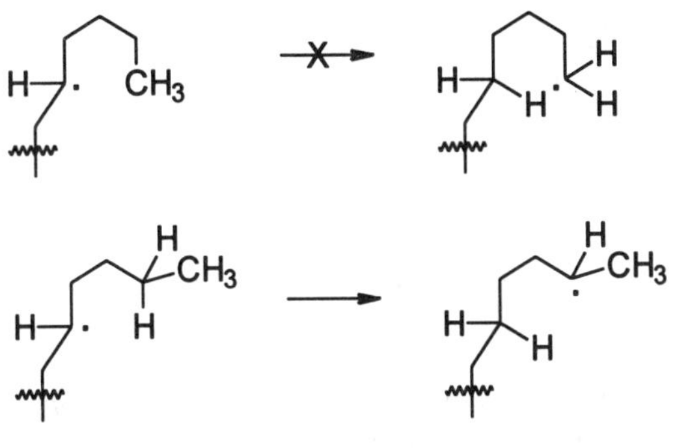

29.15

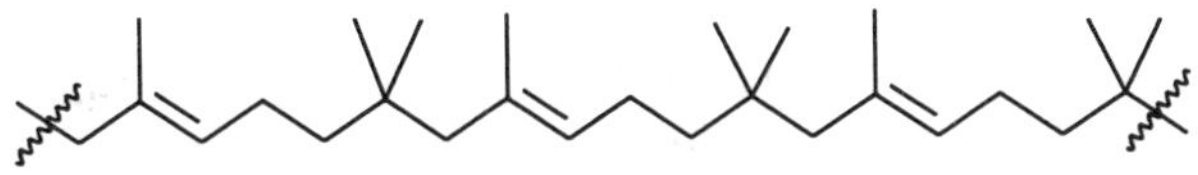

29.16 If the polymer is completely random, then any reactive end has a 50:50 chance of reacting with either of the two monomers.

476 Chapter 29

29.17 The polymer has polar functional groups that resemble those in proteins, and can form hydrogen bonds to the hair molecules.

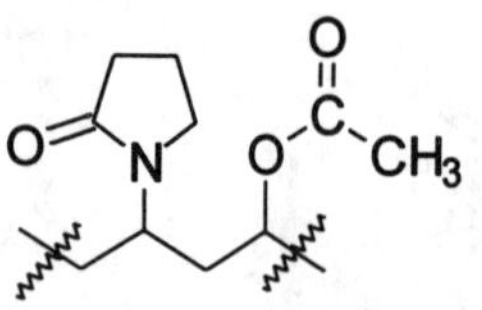

29.18

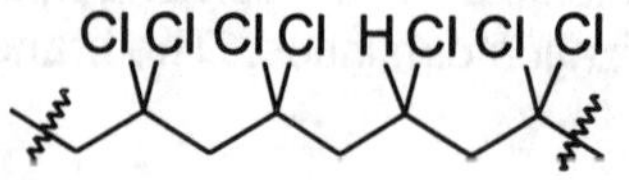

29.19 The monomers are 1,3-butadiene and acrylonitrile, $CH_2=CHCN$.

29.20

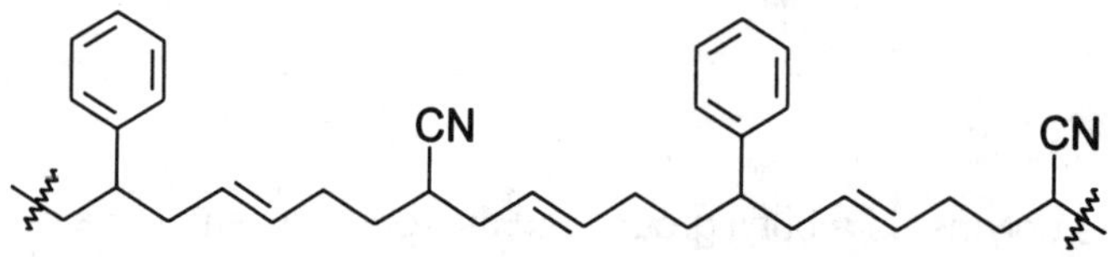

29.21 The number of cross-links in rubber gloves is less, because much greater flexibility is required in gloves than in the rubber of tires.

29.22

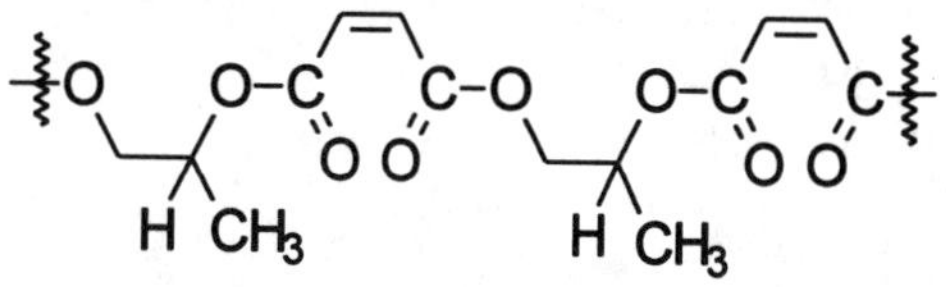

There are double bonds in the polymer that can react with styrene and form cross links containing aromatic rings,

29.23 Only (a) and (d) can give isotactic and syndiotactic polymers. Both 1,1-dichloroethene and 2-methylpropene give polymers with two equivalent atoms or groups of atoms on a carbon atom.

29.24 Neither one is optically active. Both have a mirror plane of symmetry perpendicular to the zigzag chain and passing through the center of the chain of atoms.

29.25 Each tertiary carbon atom of the chain contains a *sec*-butyl group. Each alkyl group is on the same side of the backbone of the zigzag chain.

29.26

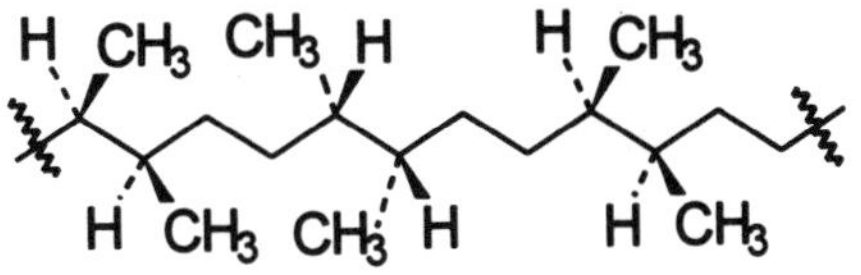

29.27 **(a)**

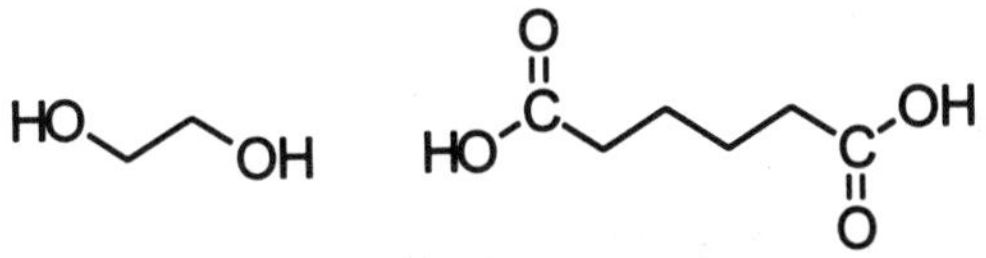

(b)

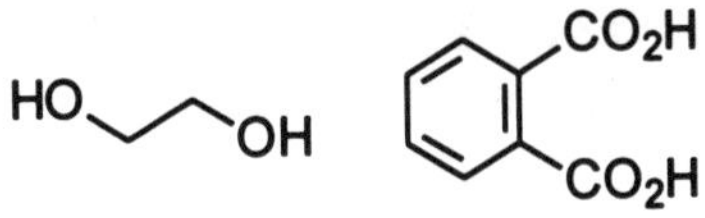

(c)

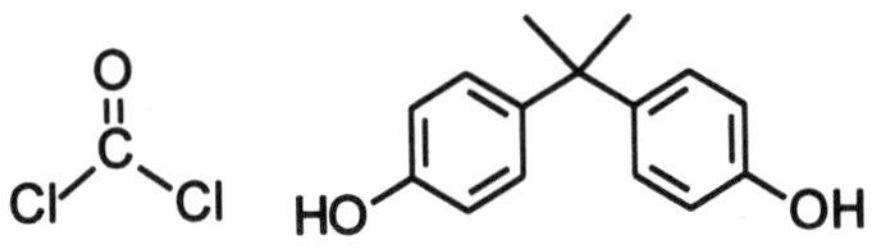

(d)

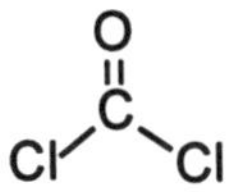

$NH_2(CH_2)_9NH_2$

29.28 **(a)**

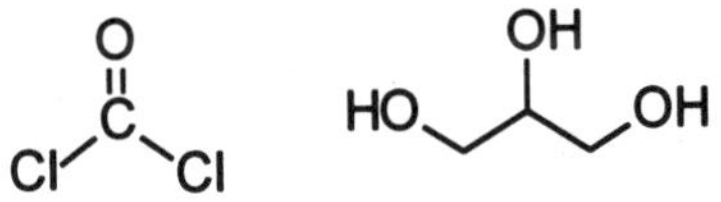

(b)

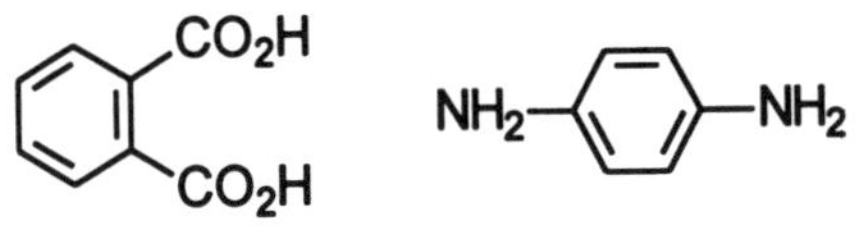

(c) 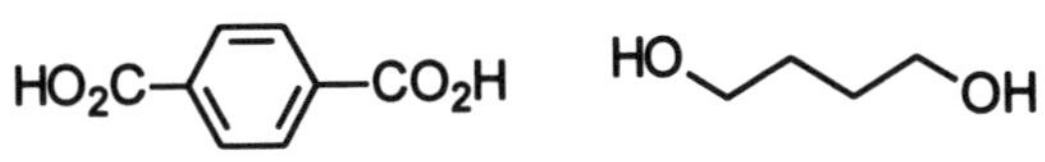 HO_2C-

$-CO_2H$

(d)

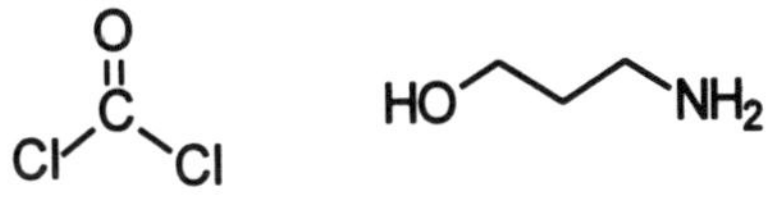

29.29 Lactic acid is CH₃CH(OH)CO₂H.

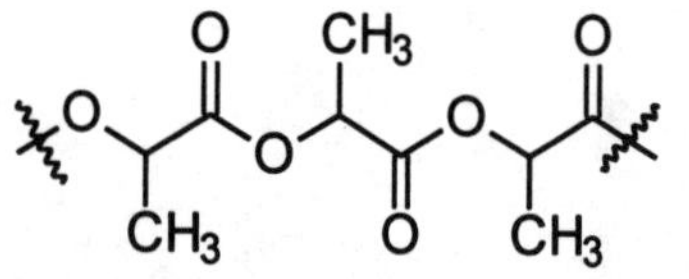

29.30

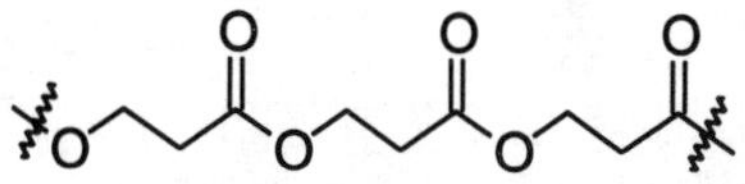

A nucleophile attacks the carbonyl carbon atom giving a tetrahedral intermediate, and an alkoxide ion "leaves" but remains attached to the reacting unit. The alkoxide ion attacks another molecule of the lactone and forms an ester while releasing another alkoxide ion. The reaction continues because the ring strain of the lactone is released.

29.31

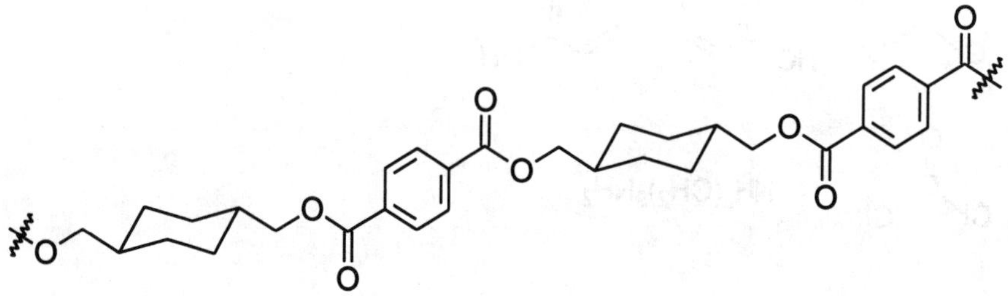

29.32 2-Methyl-1,3-propanediol, phthalic acid and trans-butenedioic acid. The unusual feature is the fact that the copolymer is produced using a mixture of dicarboxylic acids.

29.33 **(a)**

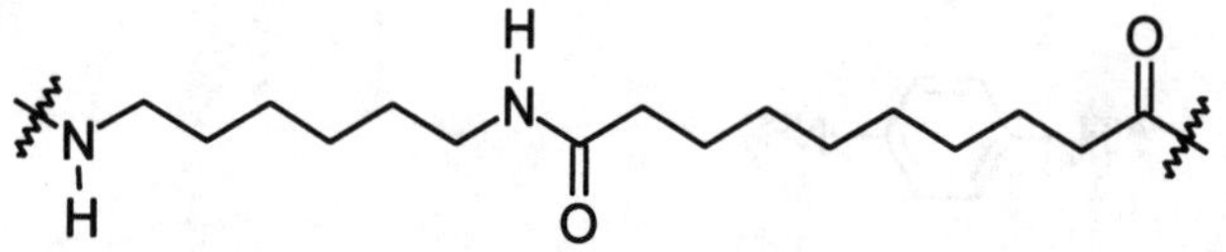

(b)

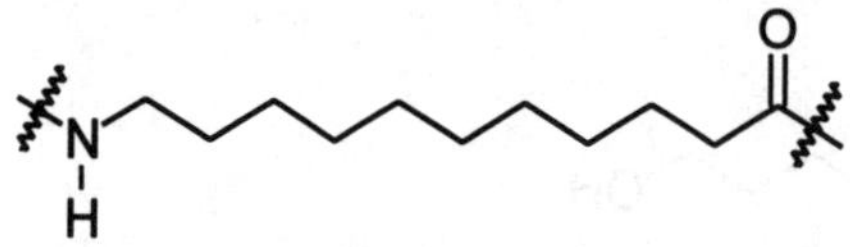

(c)

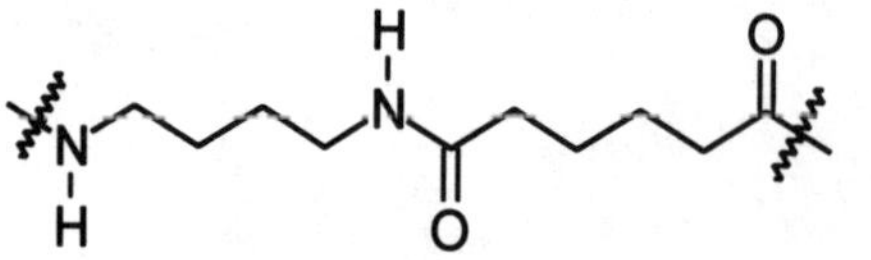

29.34 The diamine portion of the polyamide is relatively rigid. The flexibility of the carbon chain of the dicarboxylic acid affects the properties of the polymer. The structure of the diamine monomer and the dicarboxylic acid with x = 10 are shown.

29.35 A polyamide could be made by ring opening of the lactam. However, the lactam cannot be easily made because of the size of the ring. In fact, if an attempt were made using the following amino acid or some related derivative, it would polymerize.

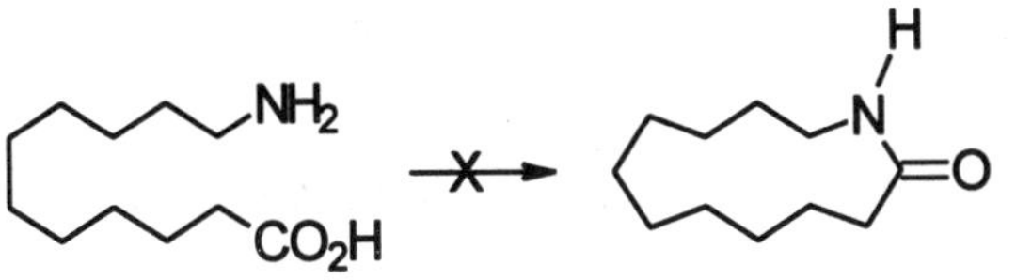

29.36

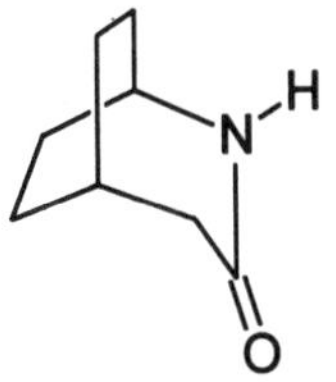

29.37 The ring strain of the epoxide allows the polymerization to occur readily. A nucleophile opens the epoxide ring to give an alkoxide, which in turn opens another epoxide ring.

29.38 Hydroxide ion attacks the C-3 atom, which is primary, to form a chiral alkoxide. This alkoxide in turn opens another epoxide ring.

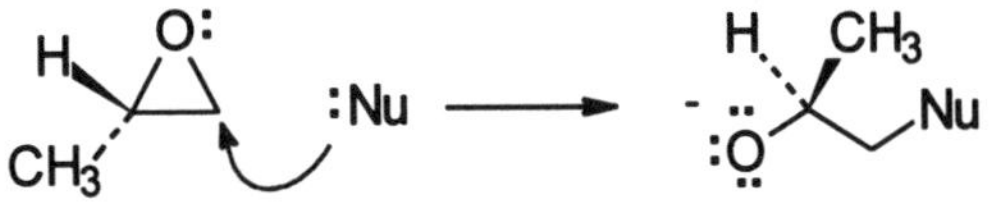

Acid catalysis involves addition of the Lewis acid (shown below with a proton) to the oxygen atom and subsequent ring opening, giving a carbocation that is achiral. Attack of the carbocation on the oxygen atom of another epoxide continues the polymerization process.

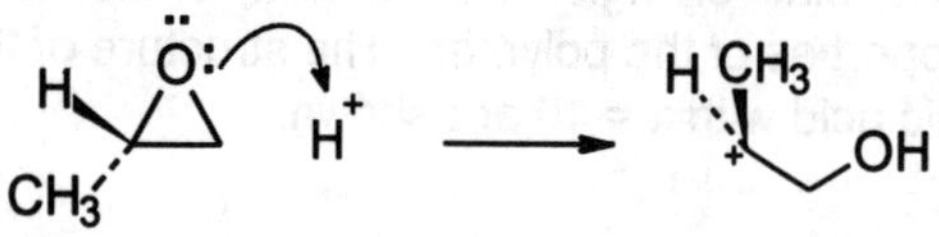

29.39   Protonation of the oxygen atom followed by formation of a carbocation gives an intermediate that can attack another tetrahydrofuran, continuing the polymerization.

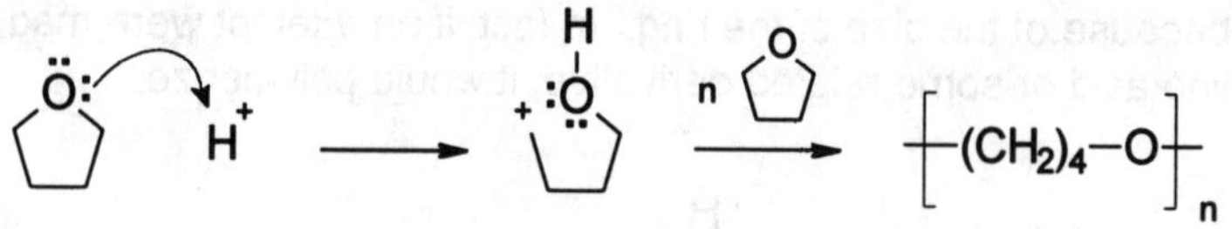

29.40   Hydrolyze the polyvinyl acetate to give a poly alcohol. Form the cyclic acetal from diol units and butanal under acidic conditions.

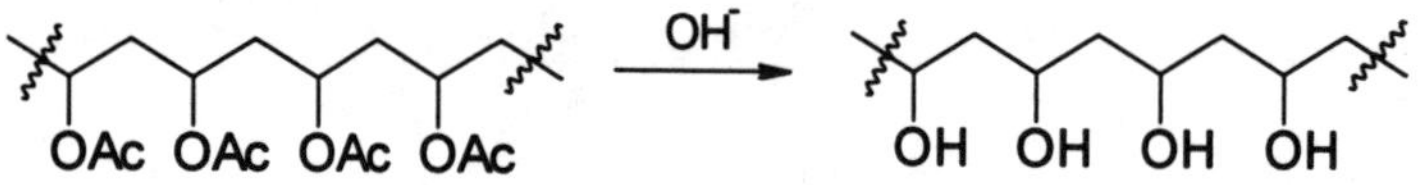

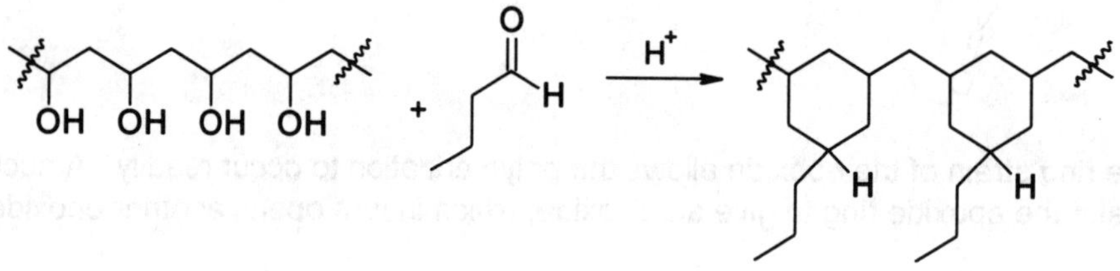

29.41   The glycerol provides a third hydroxyl group that can react with additional toluene diisocyanate to give a cross linked polymer.

29.42

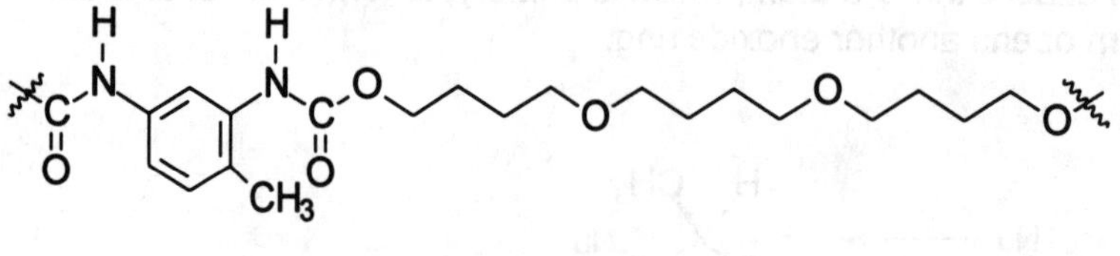

Examination 1

1   Which of the displayed bonds is the longest?

(a) CH_3-CH_3 (b) CH_3-OH (c) CH_3-SH (d) CH_3-Cl (e) CH_3-NH_2

2   Consider the lengths of carbon single bonds to oxygen, fluorine, sulfur, and chlorine. The order of decreasing bond lengths is ____. (Longest first!)

(a) Cl > S > F > O   (b) S > Cl > O > F   (c) O > F > S > Cl ,
(d) F > O > Cl > S   (e) S > Cl > F > O

3   Which of the following bonds is the least polar based on electronegativity values?

(a) N—Cl   (b) C—Cl   (c) C—Br   (d) S—Cl   (e) C—O

4   How many pairs of nonbonded electrons are present in the following carbamate?

(a) 10
(b) 8
(c) 6 $$CH_3-O-\overset{\overset{\displaystyle O}{\|}}{C}-NH-CH_3$$
(d) 5
(e) 3

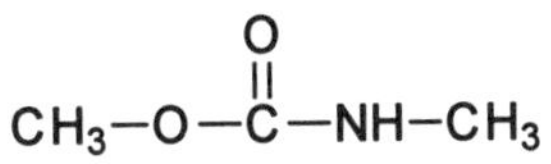

5   How many pairs of nonbonded electrons are present in methionine, an amino acid?

(a) 9
(b) 7
(c) 6 $$CH_3-S-CH_2CH_2\underset{\underset{\displaystyle NH_2}{|}}{CH}-\overset{\overset{\displaystyle O}{\|}}{C}-OH$$
(d) 5
(e) 3

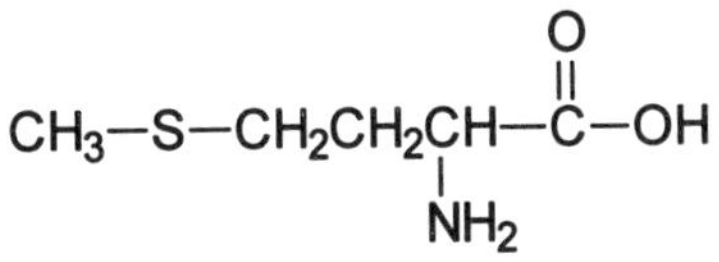

6   The formal charges on carbon and nitrogen atoms in an isonitrile are ____ and ____, respectively.

(a) 0, 0
(b) +1, 0
(c) -1, +1 $$CH_3-N\equiv C:$$
(d) +1, +1
(e) +1, -1

7   The formal charge at nitrogen atoms labeled a, b, and c in methylazide are ____, respectively.

(a) 0, 1, and 0
(b) 1, 1, and 1   a b c
(c) 0, 1, and -1
(d) 0, -1 and 1 $$CH_3-\ddot{N}=N=\ddot{N}:$$
(e) 1, -1 and 1

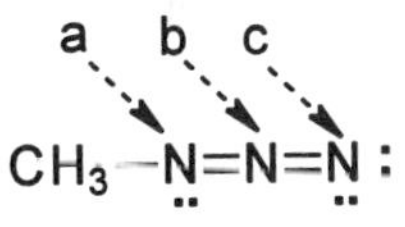

8   Acetylcholine, a compound involved in transmission of nerve impulses, has the following structure. The formal charge on nitrogen is ___ and the net charge of the compound is ___.

(a) 0, 0
(b) +1, 0
(c) 0, +1
(d) +1, +1
(e) +1, -1

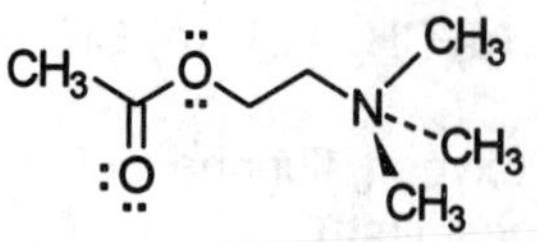

9   Diimide (HNNH) is a reactive reducing agent. In its most stable Lewis structure, there is a ____ bond between the nitrogen atoms and each nitrogen atom has ____ lone pair(s) of electrons.

(a) single; one   (b) double; one (c) triple; one   (d) double; no   (e) triple; no

10   Arginine is a basic amino acid. There are ___ lone pair(s) of electron on the nitrogen atoms.

(a) 8
(b) 7
(c) 6
(d) 5
(e) 4

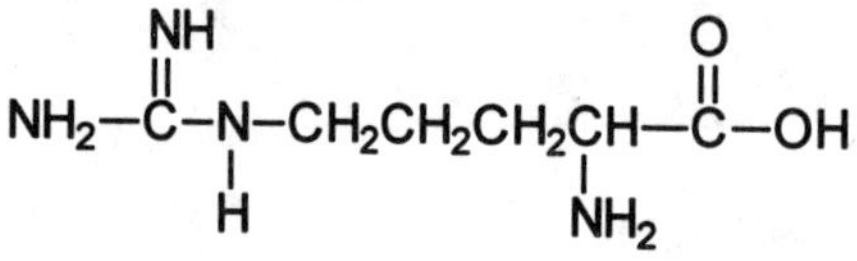

11   Tagamet (generic name cimetidine) is an antiulcer drug. There are ___ electrons in π bonds in the molecule.

(a) 4
(b) 6
(c) 8
(d) 10
(e) 12

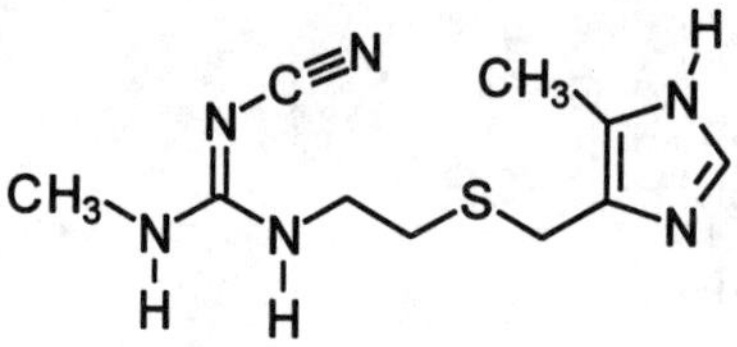

12   The number of π bonds in the following compound, which is part of the defense mechanism of safflowers against nematode infestation, is _____.

(a) 5
(b) 6
(c) 8
(d) 9
(e) 13

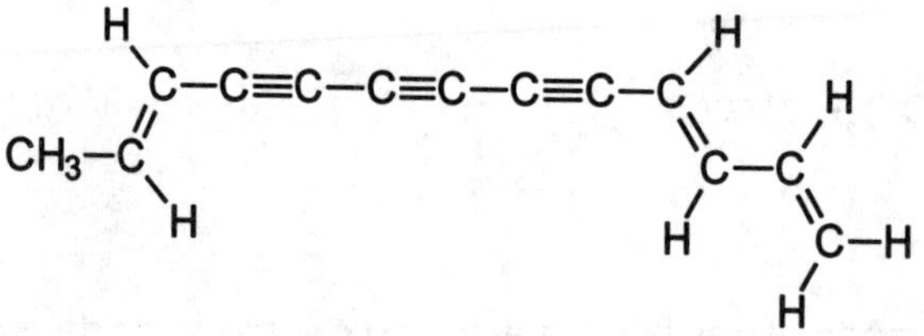

13   In an acceptable Lewis structure for the nitrite ion (NO_2^-) there are ____ N—O bonds and the nitrogen atom has ___ pair(s) of nonbonded electrons.

(a) two single; one
(b) one single and one double; one
(c) two single; two
(d) one single and one double; two
(e) two double; no

14 In an acceptable Lewis structure for the nitrite ion (NO_2^+) there are _____ N—O bonds and the nitrogen atom has ___ pair(s) of nonbonded electrons.

(a) two single; one
(b) one single and one double; one
(c) two single; two
(d) one single and one double; two
(e) two double; no

15 Based on VSEPR theory the expected nitrogen-carbon-oxygen and the sulfur-carbon-nitrogen bond angles of the following compound are _____, respectively.

(a) 90° and 90°
(b) 90° and 180°
(c) 120° and 120° $CH_3—S—\overset{\overset{O}{\|}}{C}—NH—CH_3$
(d) 109° and 109°
(e) none of the above

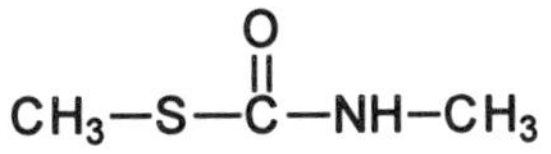

16 Abstraction of a proton from chloroform ($CHCl_3$) yield a conjugate base whose molecular structure is described as ___. The hybridization of the carbon atom is ____.

(a) tetrahedral; sp^2 (b) tetrahedral; sp^3 (c) pyramidal; sp^2
(d) trigonal planar; sp^2 (e) pyramidal; sp^3

17 The formula of formaldehyde is $CH_2=O$. The hybridization of the oxygen atom in the conjugate acid of formaldehyde is _____ and the carbon-oxygen-hydrogen bond angle is ___.

(a) sp^2; 180°   (b) sp^3; 109°   (c) sp; 180°   (d) sp; 90°   (e) sp^2; 120°

18 The formula of the oxime of formaldehyde is $CH_2=NOH$. The hybridization of the nitrogen and oxygen atoms are _____ and ___ , respectively.

(a) sp^2; sp^2 (b) sp^3; sp^2 (c) sp^2; sp^3 (d) sp^3; sp^3 (e) sp; sp^3

19 Arginine is a basic amino acid. There are ___ sp^2 and ___ sp^3 hybridized atoms in the molecule.

(a) 4; 6
(b) 4; 7 $\overset{\overset{NH}{\|}}{NH_2—C—\underset{\underset{H}{|}}{N}—CH_2CH_2CH_2\underset{\underset{NH_2}{|}}{CH}—\overset{\overset{O}{\|}}{C}—OH}$
(c) 3; 6
(d) 3; 7
(e) 2; 6

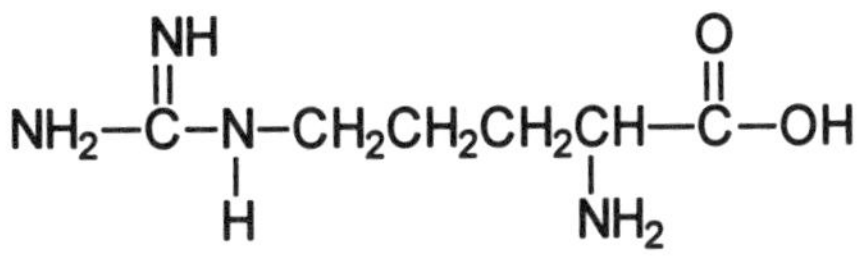

20 The C—Cl bond length of chloroethane (CH_3CH_2Cl) and chloroethene ($CH_2=CHCl$) are approximately 178 pm and 172 pm, respectively. Assuming that bond lengths are linearly related to the % s character of the carbon atom, the best estimate of the C—Cl bond length of chloroethyne ($HC\equiv CCl$) in picometers is

(a) 175   (b) 160   (c) 166   (d) 170   (e) 155

21 Which of the following statements are true?

A A carbon-carbon double bond is stronger than a single bond.
B A pi bond is stronger than a sigma bond.
C A triple bond is shorter than a double bond.

(a) A   (b) B   (c) C   (d) A and B   (e) A and C

22 The molecular formula of isoimpinellin, a carcinogen found in diseased celery, is ___.

(a) $C_{13}H_{10}O_5$
(b) $C_{12}H_{10}O_5$
(c) $C_{13}H_8O_5$
(d) $C_{12}H_8O_5$
(e) $C_{14}H_{10}O_5$

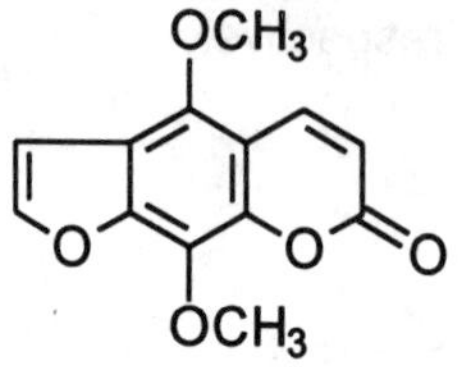

23 The molecular formula of aflatoxin B_1, a carcinogen found in moldy food, is ___.

(a) $C_{17}H_{12}O_6$
(b) $C_{17}H_{14}O_6$
(c) $C_{16}H_{10}O_6$
(d) $C_{15}H_{12}O_6$
(e) $C_{15}H_{14}O_6$

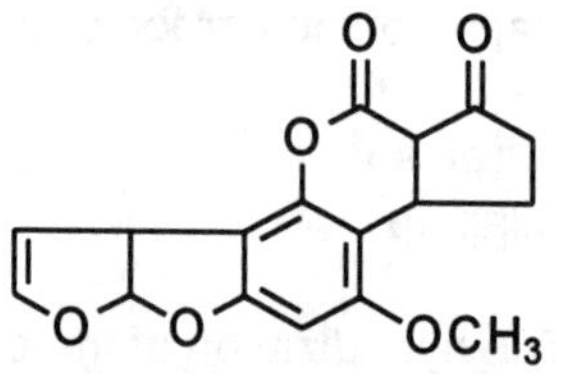

24 A common component of cephalosporins, which are antibiotics, is the following cyclic structure. The functional groups which it contains are ____.

(a) ketone, amine, and thiol
(b) ketone, amine, and thioether
(c) amide and thiol
(d) amide , ketone, and thioether
(e) amide and thioether

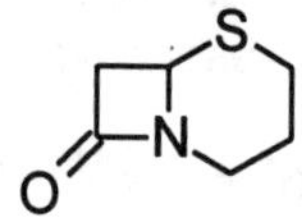

25 The oxygen-containing functional group of the sex attractant of the grape berry moth is ___.

(a) an ether
(b) a ketone
(c) a carboxylic acid
(d) an ester
(e) an aldehyde

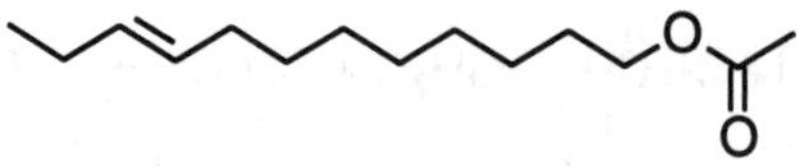

26 The oxygen-containing functional group(s) of the following male sex hormone are ___.

(a) an ether and a ketone
(b) an alcohol and a ketone
(c) an alcohol and a carboxylic acid
(d) an ether and an ester
(e) an alcohol and an aldehyde

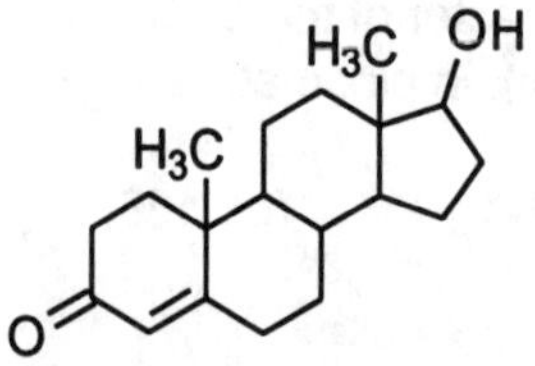

27 Which of the following compounds has the lowest boiling point?

(a) $CH_3-\underset{\underset{CH_3}{|}}{\overset{\overset{CH_3}{|}}{C}}-CH_3$

(b) $CH_3-\underset{\overset{|}{CH_3}}{CH}-O-CH_2-CH_3$

(c) $CH_3-\underset{\overset{|}{CH_3}}{CH}-CH_2-O-CH_3$

(d) $CH_3-CH_2-O-CH_2-CH_3$

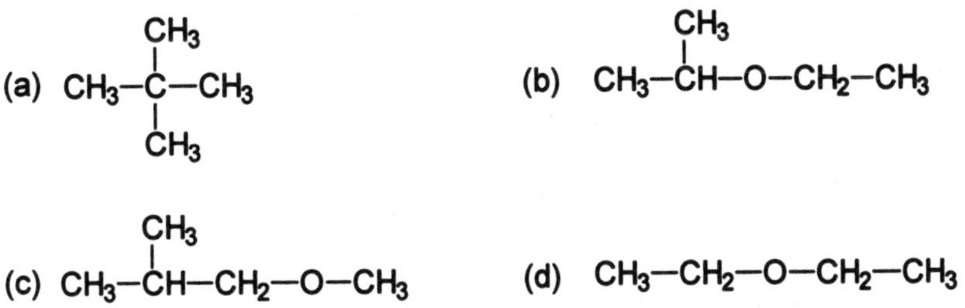

28 Which of the following compounds has the highest boiling point?

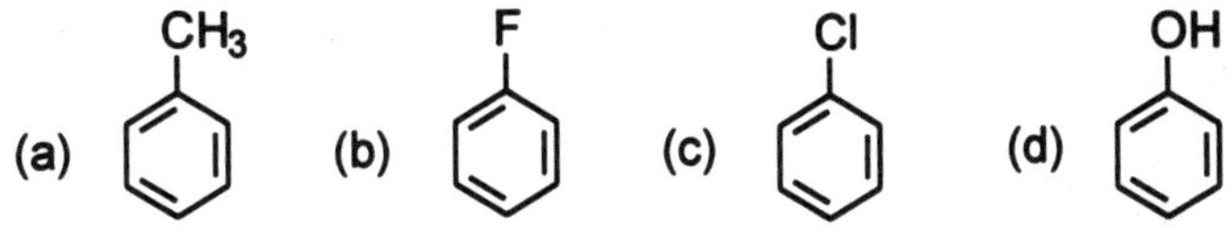

(a) [benzene with CH_3]   (b) [benzene with F]   (c) [benzene with Cl]   (d) [benzene with OH]

29 Which of the following compounds has the lowest boiling point?

(a) [benzene with OH and CHO ortho]   (b) [benzene with OH and CHO meta]   (c) [benzene with OH and CHO para]

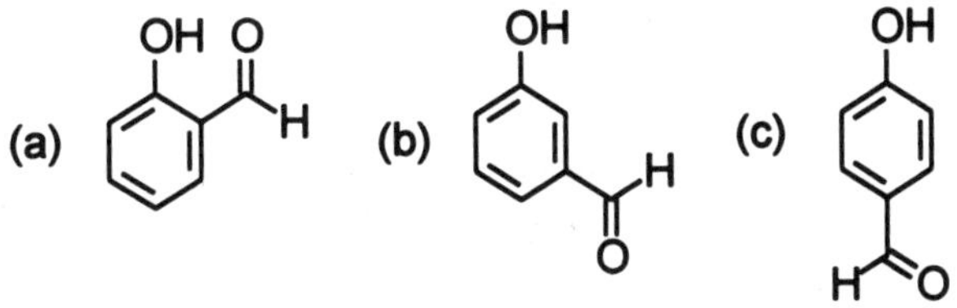

30 Which of the following should be the least soluble in water?

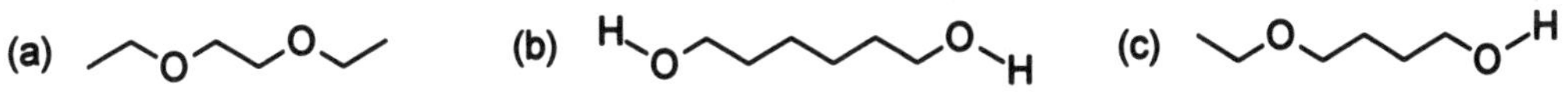

(a) [structure]   (b) [structure]   (c) [structure]

31 The antibacterial prontosil, a prodrug, is metabolized according to the following unbalanced equation, which is classified as a/an _____ reaction.

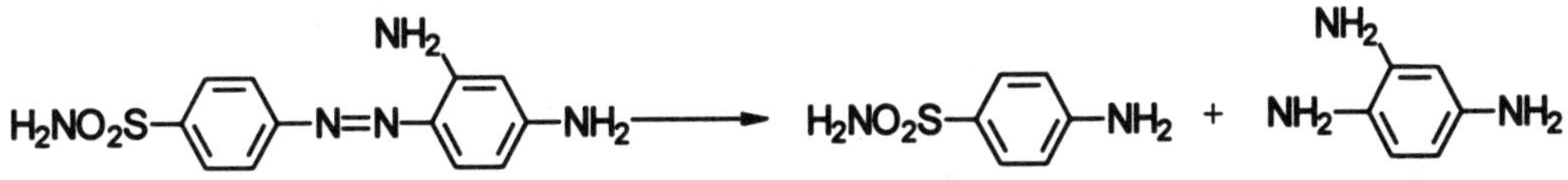

H_2NO_2S-[benzene]$-N=N-$[benzene with NH_2]$-NH_2 \longrightarrow H_2NO_2S-$[benzene]$-NH_2 + NH_2-$[benzene with NH_2]$-NH_2$

(a) hydrolysis   (b) oxidation   (c) reduction   (d) elimination   (e) condensation

32 Two of the steps in the metabolism of the sedative-hypnotic chloral hydrate are as follows. The first and second steps are classified as ____ and ____ reactions, respectively.

(a) elimination; oxidation
(b) oxidation; reduction
(c) hydrolysis; reduction
(d) elimination; reduction
(e) hydrolysis; oxidation

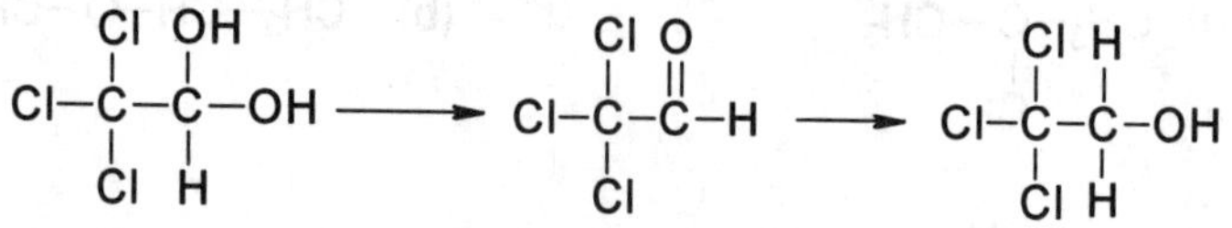

$$Cl-\underset{\underset{Cl}{|}}{\overset{\overset{Cl}{|}}{C}}-\underset{\underset{H}{|}}{\overset{\overset{OH}{|}}{C}}-OH \longrightarrow Cl-\underset{\underset{Cl}{|}}{\overset{\overset{Cl}{|}}{C}}-\overset{\overset{O}{\parallel}}{C}-H \longrightarrow Cl-\underset{\underset{H}{|}}{\overset{\overset{Cl}{|}}{C}}-\underset{\underset{H}{|}}{\overset{\overset{H}{|}}{C}}-OH$$

33 Two of the steps in the series of reactions used in the industrial synthesis of vinyl chloride, a compound used in the production of PVC, are given below. The first and second steps are classified as ____ and ____ reactions, respectively.

$$CH_2=CH_2 \xrightarrow{Cl_2} Cl-CH_2-CH_2-Cl \xrightarrow{heat} CH_2=CH-Cl$$

(a) oxidation; rearrangement   (b) addition; elimination   (c) elimination; reduction
(d) addition; rearrangement   (e) elimination; oxidation

34 The following sequence of reactions occurs in the metabolism of fatty acids. In this sequence the reactions are classified in order as ____ reactions.

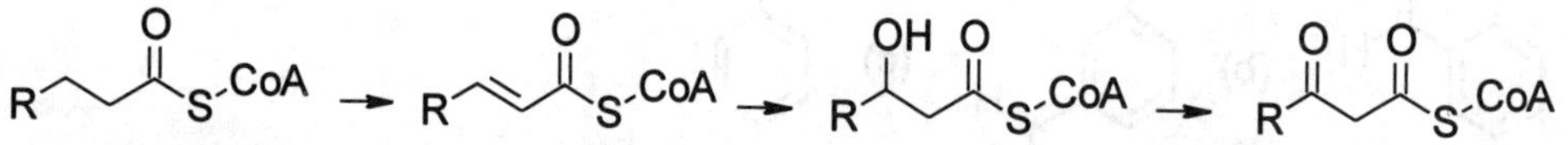

(a) oxidation, oxidation, and reduction   (b) oxidation, addition, and hydrolysis
(c) oxidation, addition, and oxidation   (d) reduction, hydrolysis, and reduction
(e) reduction, addition, and oxidation

35 The pK_a of the conjugate acid of an amine is 11. The K_b of the related amine is

(a) 10^{11} (b) 10^{-11} (c) 10^3 (d) 10^{-3}

36 Which of the following is the weakest base?

(a) $(CH_3)_3CO^-$ (b) CH_3O^- (c) $CF_3CH_2O^-$ (d) $(CF_3)_3CO^-$

37 The pK_a values of $(CH_3)_3COH$ and CCl_3H are 18 and 25, respectively. The K_{eq} for the following reaction is ____.

$$(CH_3)_3COH + CCl_3^- \rightleftharpoons (CH_3)_3CO^- + CCl_3H$$

(a) 10^{43} (b) 10^{-43} (c) 10^7 (d) 10^{-7}

38 Which of the following reactions have $K_{eq} > 1$?

(A) $CH_3O^- + CH_3CH_2SH \rightleftarrows CH_3OH + CH_3CH_2S^-$

(B) $CH_3CO_2^- + CH_3OH \rightleftarrows CH_3O^- + CH_3CO_2H$

(C) $CH_3O^- + CF_3CH_2OH \rightleftarrows CH_3OH + CF_3CH_2O^-$

(a) only A   (b) only B   (c) only C   (d) A and B   (e) A and C

39 Which of the following carboxylic acids has the smallest K_a?

(a) 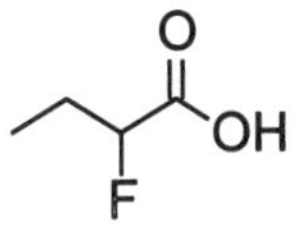 (b) 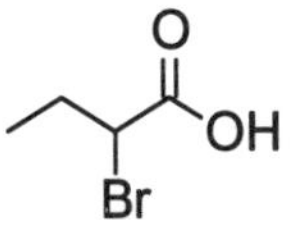 (c) 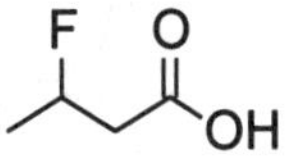 (d)

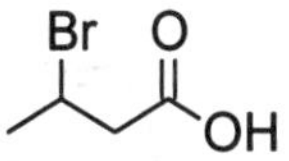

40 Which of the following statements are true?

A An exergonic reaction has $\Delta H° < 0$.
B All solution processes have $\Delta S° > 0$.
C Reactions with $\Delta S° < 0$ become less favorable at higher temperature.

(a) only A   (b) only B   (c) only C   (d) A and B   (e) B and C

41 The $\Delta H°_{rxn}$ and $\Delta S°_{rxn}$ for a reaction are -10.0 kJ mole^{-1} and -30 J mole^{-1} deg^{-1}, respectively. At 27°C, the $\Delta G°_{rxn}$ is _____ kJ mole^{-1}.

(a) -1.9   (b) +1.9   (c) -0.9   (d) -1.0   (e) +1.0

42 The $\Delta H°_{rxn}$ and $\Delta S°_{rxn}$ for a reaction are +7.1 kJ mole^{-1} and +30 J mole^{-1} deg^{-1}, respectively. At 127°C, the $\Delta G°_{rxn}$ is _____ kJ mole^{-1}.

(a) +4.9   (b) -4.9   (c) +2.9   (d) -2.9   (e) -1.0

43 The approximate K_{eq} for a reaction with $\Delta G° = + 11$ kJ mole^{-1} is _____.

(a) 10^2 (b) 10^{-2} (c) 10^{-11} (d) 10^{11} (e) 10^1

44 The $\Delta H°_{rxn}$ and $\Delta S°_{rxn}$ for a reaction are -84 kJ mole^{-1} and -140 J mole^{-1} deg^{-1}, respectively. The temperature in °C at which $K_{eq} = 1$ is _____.

(a) 327   (b) 273   (c) 873   (d) 600   (e) 500

45 The bond dissociation energies for C—Br, Br—Br, C—C, and C=C are 293, 192, 370, and 610 kJ mole^{-1}, respectively. What is $\Delta H°$ (in kJ mole^{-1}) for the following reaction?

$$CH_2{=}CH_2 + Br_2 \rightarrow BrCH_2CH_2Br$$

(a) +154   (b) -154   (c) -48   (d) +48   (e) -92

46 The bond dissociation energies for C—H, C—Br, Br—Br, and H—Br are 438, 293, 192, and 366 kJ mole^{-1}, respectively. What is $\Delta H°$ (in kJ mole^{-1}) for the following reaction?

$$CH_4 + Br_2 \rightarrow CH_3Br + HBr$$

(a) +29   (b) -29   (c) -48   (d) +48   (e) -100

47 The bond dissociation energies for C—H, O—H, C—C, C—O, C=C, and C=O are approximately 400, 490, 370, 380, 610, and 745 kJ mole^{-1}, respectively. What is $\Delta H°$ (in kJ mole^{-1}) for the following isomerization reaction?

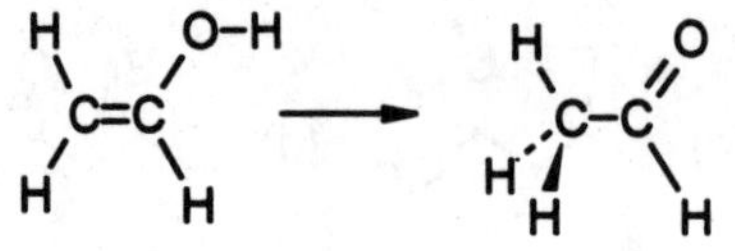

(a) +35   (b) -35   (c) -125   (d) +125   (e) -90

48 The $\Delta S°_{rxn}$ for the formation of cyclooctatetraene (C_8H_8) by a tetramerization reaction of acetylene is approximately _____ J mole^{-1} deg^{-1}.

(a) -500   (b) +500   (c) -375   (d) +375   (e) +250

49 Which of the following reactions has the largest positive $\Delta S°_{rxn}$?

(a) $2\ CH_3CH_2Br \rightarrow CH_3CH_2CH_2CH_3 + Br_2$

(b) $CH_3CH_2Br \rightarrow CH_2=CH_2 + HBr$

(c) $CH_3CH_3 \rightarrow CH\equiv CH + 2\ H_2$

(d) $CH_3CH_2CH_3 \rightarrow CH_2=CH_2 + CH_4$

(e) $2\ Br_2 + HC\equiv CH \rightarrow CHBr_2CHBr_2$

50 As a first approximation, the $\Delta S°_{rxn}$ for an addition reaction should be ___; the $\Delta S°_{rxn}$ for a condensation reaction should be___.

(a) negative; zero   (b) negative; negative   (c) positive; zero
(d) positive; negative   (e) negative; positive

51 As a first approximation, the $\Delta S°_{rxn}$ for an elimination reaction should be ___; the $\Delta S°_{rxn}$ for a rearrangement reaction should be___.

(a) negative; zero   (b) negative; negative   (c) positive; zero
(d) positive; negative   (e) negative; positive

52 The hybridization of the charged carbon atom in a carbocation is ___; the hybridization of the charged carbon atom in a carbanion is ___;

(a) sp^2 ; sp^2 (b) sp^3 ; sp^3 (c) sp^2 ; sp^3 (d) sp^3 ; sp^2

53 Which of the following carbocations is the most stable?

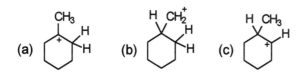

54 Which of the following carbanions is the most stable?

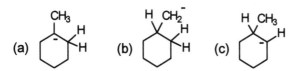

55 Which of the following statements are true?

A The hydroxide ion can serve as an electrophile.
B The reaction of chlorine with methane is a concerted process.
C The first step in the reaction of ethylene with HBr involves electrophilic attack by H^+.

(a) A   (b) B   (c) C   (d) A and B   (e) A and C

56 Which of the following statements is false?

A   reactions rates depend on temperature and reactant structure
B   catalysts shift equilibria toward the side of products
C   molecules must have energy at least equal to the activation energy for endothermic reactions.

(a) A   (b) B   (c) C   (d) A and B   (e) A and C

57 Which of the following reactions occurs at the fastest rate?

(a) one which is exothermic by 130 kJ mole^{-1} and has an activation energy of 75 kJ mole^{-1}
(b) one which is exothermic by 115 kJ mole^{-1} and has an activation energy of 65 kJ mole^{-1}
(c) one which is endothermic by 90 kJ mole^{-1} and has an activation energy of 85 kJ mole^{-1}

58 Assuming that $\Delta S° = 0$ for each of the following reactions, which one has the largest K_{eq}?

(a) one which is exothermic by 130 kJ mole^{-1} and has an activation energy of 75 kJ mole^{-1}
(b) one which is exothermic by 115 kJ mole^{-1} and has an activation energy of 65 kJ mole^{-1}
(c) one which is endothermic by 90 kJ mole^{-1} and has an activation energy of 85 kJ mole^{-1}
(d) one which is endothermic by 190 kJ mole^{-1} and has an activation energy of 125 kJ mole^{-1}

59 The reaction coordinate diagram for a reaction that occurs by a three step mechanism shows ___ transition states and ___ intermediates.

(a) 3, 3   (b) 3, 2   (c) 2, 3   (d) 2, 2   (e) 3, 4

60 The following reaction coordinate diagram for a single step process corresponds to a ___ process with K ___ than one.

(a) exergonic; greater
(b) exergonic; less
(c) endergonic; greater
(d) endergonic; less

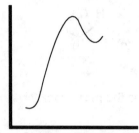

Reaction coordinate

61 The structure for the transition state for the rate determining step for the following reaction coordinate diagram most closely resembles the ___

(a) reactant
(b) product
(c) intermediate

Reaction coordinate

Answers to Examination 1

cbadb   ccdbe   ddbec   eecbb   eabed   badaa   cdbcd   dcede   dbbab

bbcca   ccabc   bbabd   c

Examination 2

1   Which of the following compounds has the largest number of tertiary carbon atoms?

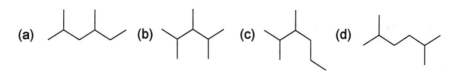

(a)   (b)   (c)   (d)

2   The molecular formula for 2,4,6,8,10,12-hexamethyltetradecane is ____.

(a) $C_{20}H_{42}$ (b) $C_{16}H_{34}$ (c) $C_{14}H_{30}$ (d) $C_{20}H_{40}$ (e) (b) $C_{16}H_{36}$

3   The IUPAC name of the following compound is ____.

(a) 2,6-dimethyl-3-ethylheptane
(b) 3-isopropyl-6-methylhexane
(c) 5-ethyl-2,6-dimethylheptane
(d) 2-methyl-5-isopropylhexane
(e) 3-ethyl-2,6-dimethylheptane

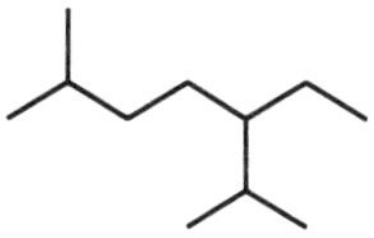

4   The common name of the alkyl group attached to the cyclohexane ring is ____.

(a) isopropyl
(b) butyl
(c) *sec*-butyl
(d) isobutyl
(e) *tert*-butyl

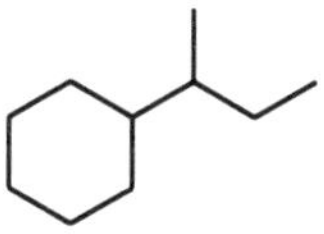

5   The IUPAC name of the alkyl group attached to the left of the benzene ring in the spermicide octoxynol-9 (common name) used in contraceptive products is ____.

(a) 2,2,4,4-tetramethylbutyl
(b) 1,1,3,3-tetramethylbutyl
(c) 2,4,4-trimethyl-2-pentyl
(d) 1,1,3,3-tetramethylpentyl
(e) 2,2,4,4-tetramethylpentyl

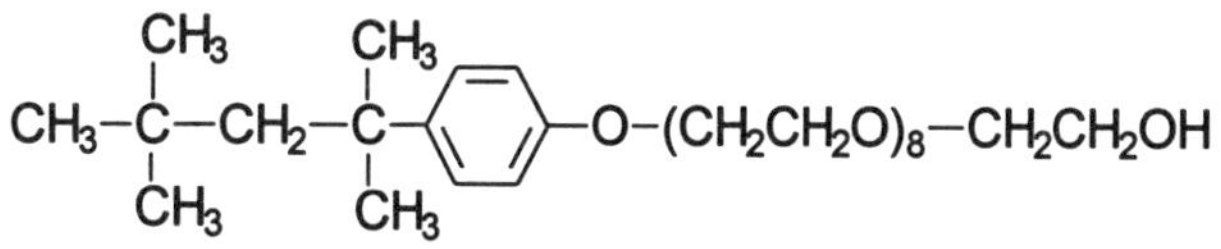

$CH_3-\overset{CH_3}{\underset{CH_3}{C}}-CH_2-\overset{CH_3}{\underset{CH_3}{C}}$ —O–$(CH_2CH_2O)_8$–CH_2CH_2OH

6   The heats of formation of 2-methyloctane and 3-methyloctane are -235.8 and -233.7 kJ mole^{-1}, respectively. The $\Delta H°$ for the following isomerization reaction is ____ kJ mole^{-1}.

(a) 2.1
(b) -2.1
(c) 469.5
(d) -469.5

7   The heat of formation of cyclohexane is -123 kJ mole^{-1}. The heat of formation of cycloheptane is -118.1 kJ mole^{-1}. The strain energy of cycloheptane is approximately ____ kJ mole^{-1}.

(a) 0   (b) 5   (c) 25   (d) 33   (e) 38

8   The heat of formation of cyclohexane is -123 kJ mole^{-1}. The heat of formation of cyclooctane is -124.4 kJ mole^{-1}. The strain energy of cyclooctane is approximately ____ kJ mole^{-1}.

(a) 0   (b) 20   (c) 30   (d) 40   (e) 60

9   The strain energy of cyclopropane is 115 kJ mole^{-1}. The $\Delta H°$ and $\Delta S°$ for the following reaction are ____ kJ mole^{-1} and ____ J mole^{-1} deg^{-1}, respectively.

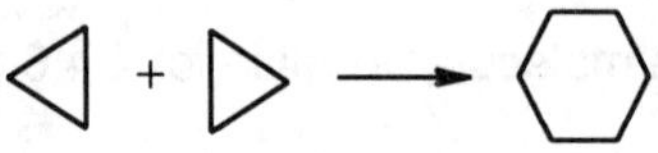

(a) +115, -125
(b) -115, +125
(c) +230, -125
(d) -230, +125
(e) -230, -125

10 The order of increasing strain energy of the following compounds is _____.

(a) A < B < C < D   (b) B < A < C < D   (c) A < D < B < C   (d) A < C < B < D

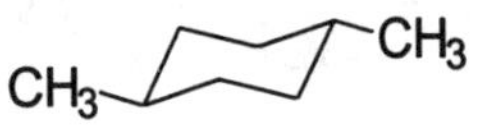

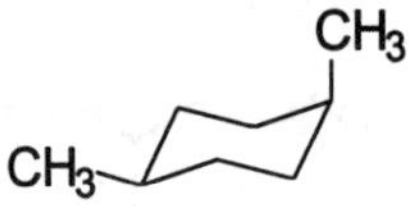

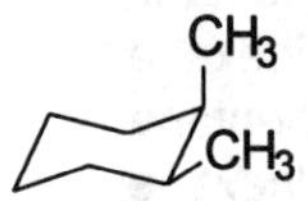

11 Which of the following has an axial methyl group in its most stable conformation?

A *cis*-1,2-dimethylcyclohexane
B *cis*-1,3-dimethylcyclohexane
C *cis*-1,4-dimethylcyclohexane

(a) A   (b) B   (c) C   (d) A and B   (e) A and C

12 The IUPAC name for the following compound is ____.

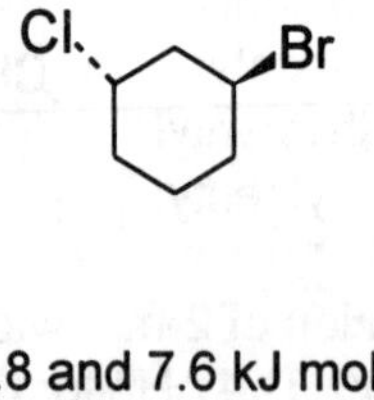

(a) trans-1-bromo-3-chlorocyclohexane
(b) cis-1-bromo-3-chlorocyclohexane
(c) trans-1-chloro-3-bromocyclohexane
(d) cis-1-chloro-3-bromocyclohexane
(e) trans-1-bromo-5-chlorocyclohexane

13 The steric strain of the chloro and methyl groups are 2.8 and 7.6 kJ mole^{-1}, respectively. The conformation on the right of the following equation constitutes approximately ____% of the conformational equilibrium mixture.

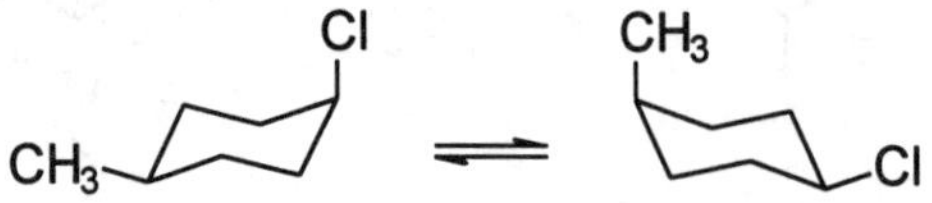

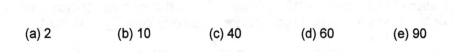

(a) 2   (b) 10   (c) 40   (d) 60   (e) 90

14 Which of the following compounds has the largest heat of combustion?

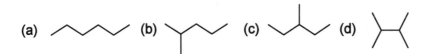

(a)   (b)   (c)   (d)

15 Which of the following compounds is the thermodynamically most stable?

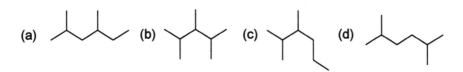

(a)   (b)   (c)   (d)

16 The relative rates of chlorination for 1°, 2°, and 3° hydrogen atoms under a certain set of reaction conditions are 1, 4.0, and 5.5, respectively. The amount of 3° chlorinated product formed from 2,3-dimethylbutane is approximately ___ %.

(a) 48   (b) 52   (c) 36   (d) 58   (e) 32

17 Chlorination of pentane gives approximately 20% 1-chloropentane. The yield of 2-chloro-pentane is expected to be ___ %.

(a) 27   (b) 53   (c) 15   (d) 65   (e) 80

18 The fluorination of 2-methylpropane gives a mixture of 1-fluoro-2-methylpropane and 2-fluoro-2-methylbutane in a 6:1 ratio. The relative reactivity of a tertiary to a primary C-H bond in fluorination is ____.

(a) 1.5   (b) 0.67   (c) 3.0   (d) 0.33   (e) 5.3

19 The rate of reaction of fluorine with hydrocarbons is ___ than the rate of reaction of chlorine with hydrocarbons. Fluorine is expected to be ___ selective than chlorine in its reactions.

(a) faster; more   (b) faster; less   (c) slower; more   (d) slower, less

20 The reaction of the bromine atom with methane is ___ endothermic than the reaction of the chlorine atom with methane. As a consequence the transition state for bromination is more ____ and the bromine atom is ____ selective than the chlorine atom.

(a) less; product-like; more   (b) less; reactant-like; less   (c) less; reactant-like; more
(d) more; reactant-like; less   (e) more; product-like; more

21 The unsaturation number for a substance represented by $C_{10}H_{14}O_3Br_4$ is ___.

(a) 6   (b) 5   (c) 4   (d) 3   (e) 2

22 The unsaturation number for a substance represented by $C_9H_{15}SN_3$ is ___.

    (a) 8   (b) 6   (c) 4   (d) 3   (e) 2

23 Which of the following has the (Z) configuration?

    (a) A, C   (b) B, D   (c) A, B, C   (d) A, B, D   (e) A, C, D

24 The IUPAC name of the following compound is ___.

    (a) (E)-4-ethyl-3-methyl-3-heptene
    (b) (Z)-4-ethyl-3-methyl-3-heptene
    (c) (E)-3-methyl-4-ethyl-3-heptene
    (d) (Z)-3-methyl-4-ethyl-3-heptene
    (e) (Z)-4-ethyl-5-methyl-4-heptene

25 The IUPAC name of the following compound is ___.

    (a) cis-1,5-dimethylcyclohexene
    (b) trans-1,5-dimethylcyclohexene
    (c) 1,3-dimethylcyclohexene
    (d) 1,5-dimethylcyclohexene
    (e) none of the above is correct

26 Which of the following has the largest heat of combustion?

27 Which of the compounds in question 26 has the smallest heat of hydrogenation?

28 Selective partial hydrogenation of 3-allylcyclohexene using the Wilkinson catalyst yields ___.

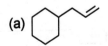 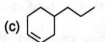 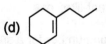 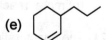

29 Reaction of 4-tert-butylcyclopentene with deuterium gas using a platinum catalyst yields ___.

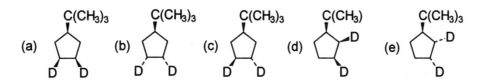

30 The $\Delta H°_{rxn}$ and $\Delta S°_{rxn}$ for the hydration of a specific alkene are -32.0 kJ mole^{-1} and -120 J mole^{-1} deg^{-1}, respectively. At 27°C, the $\Delta G°_{rxn}$ is ____ kJ mole^{-1}.

(a) +8.0   (b) -8.0   (c) -2.0   (d) -4.0   (e) +4.0

31 Reaction of methylenecyclohexane with HBr (in the absence of free radicals) yields _____.

(a) [structure with CH₃ and Br]   (b) [structure with CH₃ and Br]   (c) [structure with CH₂Br]   (d) [structure with Br, CH₃]

32 The structure of the major product obtained by treatment of the 3,3-dimethyl-1-butene with sulfuric acid and water is _____.

(a) [structure with OH]   (b) [structure with OH]   (c) [structure with OH]   (d) [structure with OH]

33 Which of the following products results from th acid catalyzed hydration of 1-methyl-cyclohexene?

(a) [cyclohexane with CH₃ and OH]   (b) [cyclohexane with CH₃ and OH]   (c) [cyclohexane with CH₃ and OH]   (d) [cyclohexane with CH₂OH]   (e) [cyclohexane with CH₃ and O]

34 Reaction of 1-methylcyclohexene with aqueous bromine gives ___.

(a) [structure with OH, CH₃, H, Br]   (b) [structure with OH, CH₃, Br, H]   (c) [structure with Br, CH₃, OH, H]   (d) [structure with Br, CH₃, H, OH]

35 Reaction of a compound with molecular formula C_7H_{12} with ozone followed by reductive workup gives the indicated dicarbonyl compound. The structure of the C_7H_{12} compound is____.

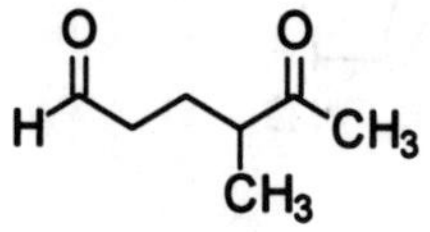

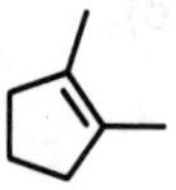

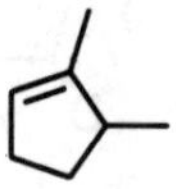

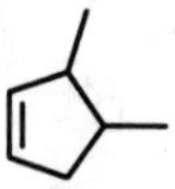

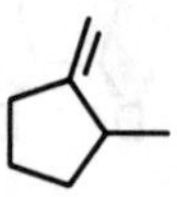

36 Reaction of a compound with molecular formula C_7H_{12} with ozone followed by reductive workup gives the indicated dicarbonyl compound. The structure of the C_7H_{12} compound is____.

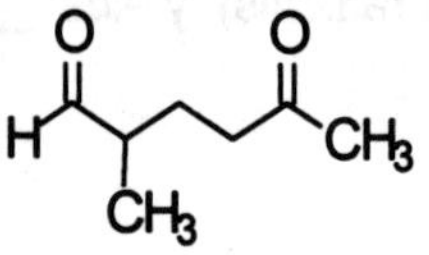

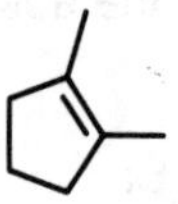

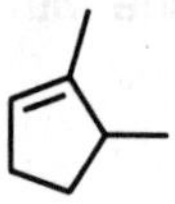

37 Addition of HBr to methylenecyclohexane in the presence of peroxides yields _____.

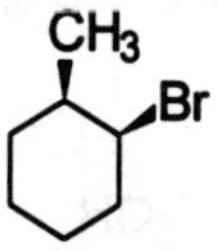

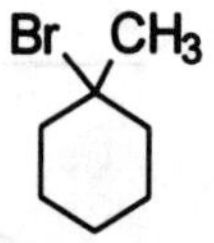

Answers to Examination 2

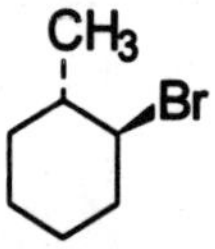

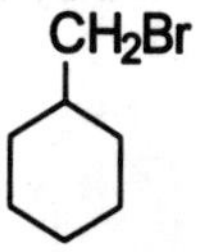

baecb   acdec   eabab   ababe   ecebd   baebe   dcabb   cc

Examination 3

1 Which of the following compounds has one or more stereogenic centers?

(a) A and B
(b) C and D
(c) B and C
(d) only C
(e) only D

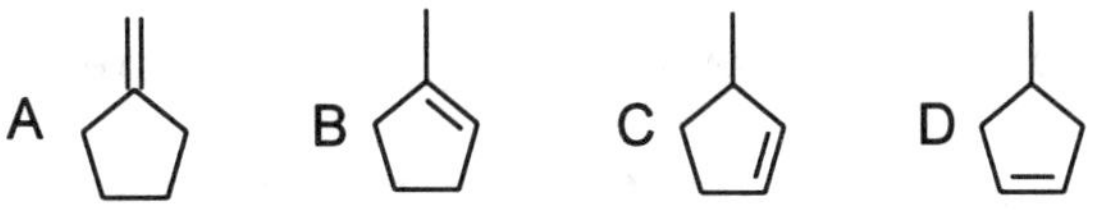

A   B   C   D

2 Chloramphenicol, an antibiotic, has the following structure. How many stereoisomers are possible for this structure?

(a) 16
(b) 8
(c) 6
(d) 4
(e) 2

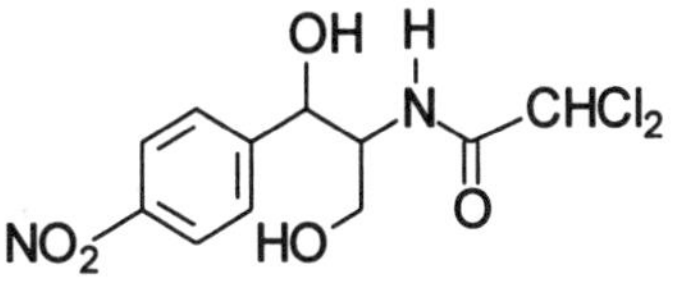

3 There are ___ stereogenic centers in cholesterol.

(a) 9
(b) 8
(c) 7
(d) 6
(e) 5

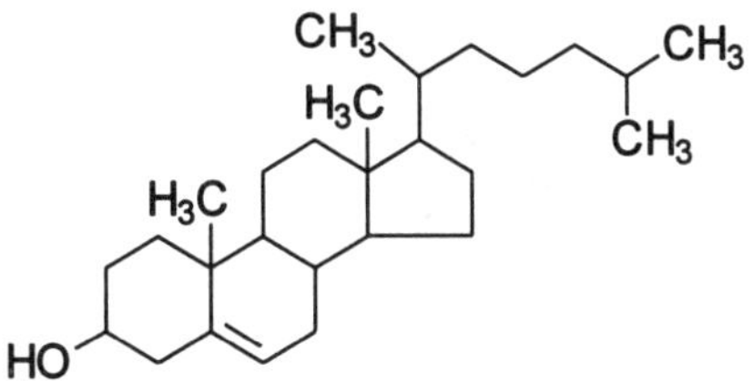

4 Which of the following groups has the highest priority according to Cahn-Ingold-Prelog rules?

(a) —OCH_2CH_3 (b) —SCH_2CH_3 (c) —CF_3 (d) —CH_2CBr_3 (e) —OCF_2CF_3

5 The highest and lowest priority groups at the stereogenic center of ethchlorvynol, a sedative-hypnotic, are _____, respectively.

(a) -OH and -C≡CH
(b) -C≡CH and CH_3CH_2-
(c) -OH and CH_3CH_2-
(d) ClHC=CH- and -OH
(e) ClHC=CH- and -C≡CH

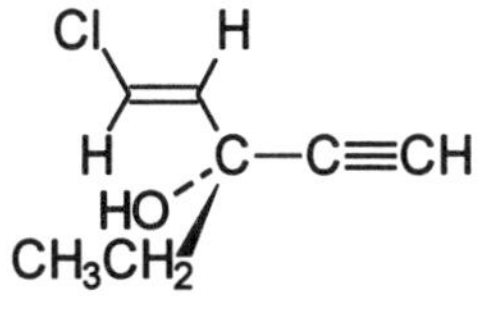

6 Naturally occurring threonine has a 2S,3R configuration. It is enantiomeric with a ____ compound and diastereomeric with a ____ compound.

(a) 2S,3S; 2R,3R   (b) 2R,3S; 2S,3S   (c) 2R,3R; 2R,3S

7   Which of the following substances can have a meso form?

A 2,9-dibromodecane   B 3,8-dibromodecane   C 4,7-dibromodecane

(a) A and B   (b) B and C   (c) A and C   (d) A, B and C   (e) none of them

8   Which of the following substances can exists as four diastereomers?

A 2,3-dibromobutane   B 2,4-dibromopentane   C 2,3-dibromopentane

(a) only A   (b) only B   (c) only C   (d) A and C   (e) A and B

9   Which of the following statements are true?

A Diastereomers have different boiling points.
B A structure with three stereogenic centers has a maximum of six stereoisomers.
C Any mixture of enantiomers has no net optical activity,

(a) A   (b) B   (c) C   (d) A and B   (e) A and C

10 Which of the following has the R configuration?

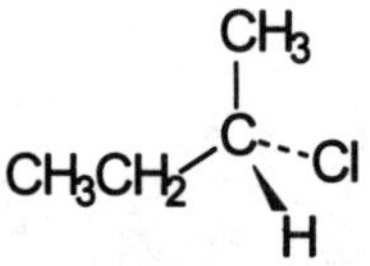

A 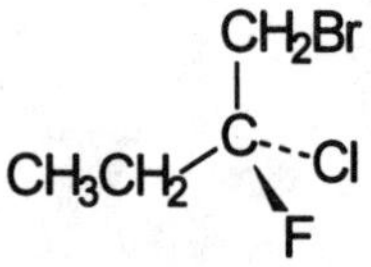   B 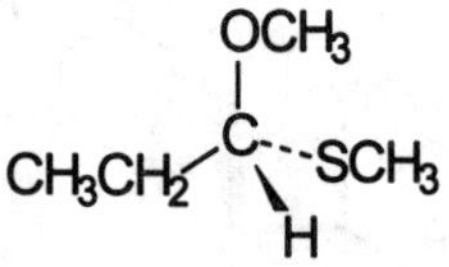   C

(a) A   (b) B   (c) C   (d) A and B   (e) B and C

11 The configuration of the following compound is ____.

(a) 2R,3S
(b) 2S,3S
(c) 2R,3R
(d) 2S,3R

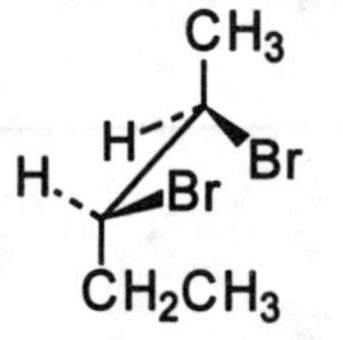

12 The Fischer projection for the structure on the left is ___.

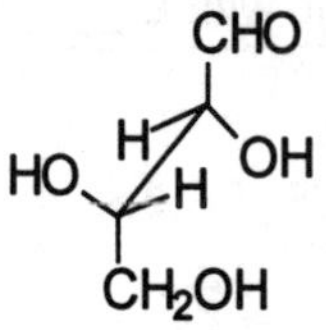

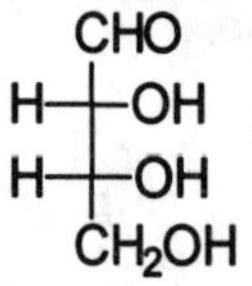

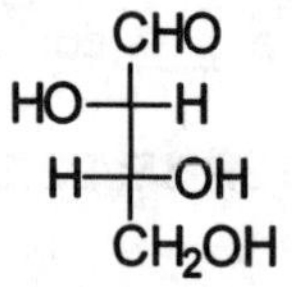

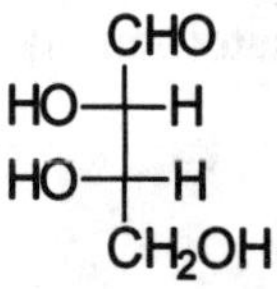

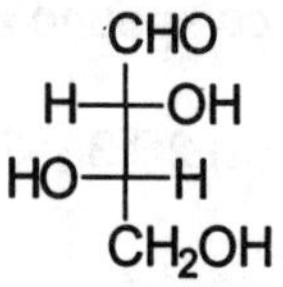

13 The configuration of the following Fischer projection formula is ____.

(a) 2R,3R
(b) 2S, 3S
(c) 2R, 3S
(d) 2S,3R

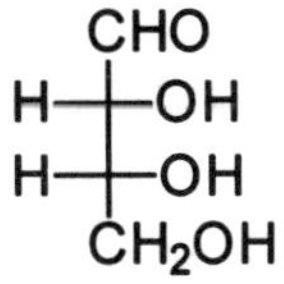

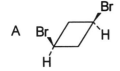

14 Which of the following has a plane of symmetry?

A   B   C   D

(a) only A   (b) only C   (c) A and C   (d) A, B and C   (e) A, B, C, and D

15 Free radical chlorination at the C(2) position of (R)-2-bromobutane gives 2-bromo-2-chlorobutane that consists of

(a) a mixture of diastereomers
(b) a mixture of enantiomers
(c) only the R enantiomer
(d) only the S enantiomer
(e) one specific diastereomer

16 Free radical fluorination at the C(1) position of (R)-2-bromobutane gives 2-bromo-1-fluorobutane that consists of

(a) a mixture of diastereomers
(b) a mixture of enantiomers
(c) one specific diastereomer
(d) only the R enantiomer
(e) only the S enantiomer

17 Free radical chlorination at the C(3) position of (R)-2-bromobutane gives 2-bromo-3-chlorobutane that consists of

(a) a mixture of diastereomers
(b) a mixture of enantiomers
(c) only the R enantiomer
(d) only the S enantiomer
(e) one specific diastereomer

18 Reaction of (Z)-2-butene with bromine yields 2,3-dibromobutane that is ____.

(a) a mixture of diastereomers
(b) a mixture of enantiomers
(c) a meso compound

19 Which of the following compound(s) results from reaction of (R)-4-methyl-1-hexene with HBr?

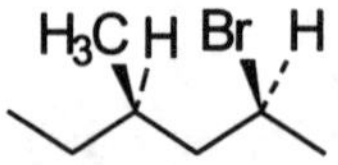

B 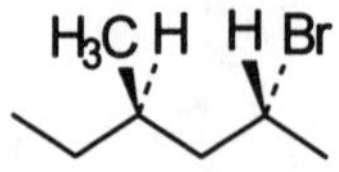   C 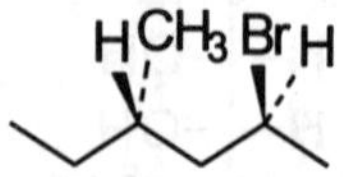   D

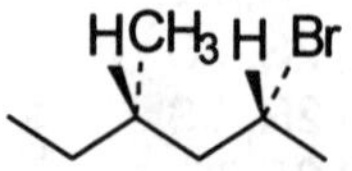

(a) A and B   (b) C and D   (c) A and C   (d) B and D

20 The IUPAC name of the following compound is ____.

(a) *trans*-4-chloro-5-bromocyclohexene
(b) *trans*-4-bromo-5-chlorocyclohexene
(c) *trans*-5-chloro-4-bromocyclohexene
(d) *trans*-5-bromo-4-chlorocyclohexene

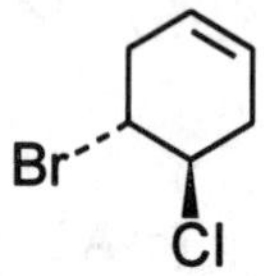

21 The IUPAC name of the following compound is

(a) (E)-2,3-dimethyl-3-decen-6-ol
(b) (Z)-2,3-dimethyl-3-decen-6-ol
(c) (E)-8,9-dimethyl-7-decen-5-ol
(d) (Z)-8,9-dimethyl-7-decen-5-ol

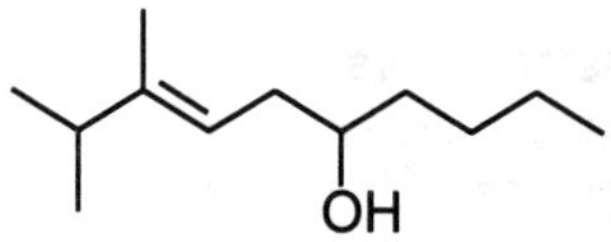

22 Which of the following compounds should have the lowest boiling point?

(a) $CH_3(CH_2)_2CH_2F$ (b) $(CH_3)_3CF$ (c) $CH_3(CH_2)_2CH_2Cl$ (d) $(CH_3)_3CCl$

23 In aqueous solution, the CH_3O^- ion is a ___ base than the CH_3S^- ion and the CH_3O^- ion is a ___ nucleophile than the CH_3S^- ion.

(a) stronger; better   (b) stronger; poorer   (c) weaker; better   (d) weaker; poorer

24 In polar protic solvent, the I^- ion is a ___ nucleophile than the F^- ion. In a polar aprotic solvent, the I^- ion is a ___ nucleophile than the F^- ion.

(a) better; better   (b) better; poorer   (c) poorer; better   (d) poorer; poorer

25 Which of the following statements are true?

A A S_N2 reaction is a two step mechanism.
B Increasing the concentration of the nucleophile increase the rate of an S_N1 reaction.
C The rate of an S_N1 process is faster in more polar solvents.

(a) A   (b) B   (c) C   (d) A and B   (e) A and C

26 Which of the following compounds should give a ratio of enantiomeric alcohols closest to 50:50 in a reaction with water?

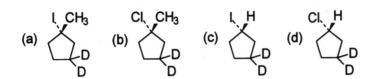

27 Which of the following species reacts fastest with iodomethane?

(a) CH_3OH (b) CH_3O^- (c) CH_3SH (d) CH_3S^- (e) F^-

28 Which of the following species should be the poorest nucleophile in an S_N2 reaction?

(a) $(CH_3)_3CO^-$ (b) CH_3O^- (c) $(CH_3)_3CS^-$ (d) CH_3S^-

29 Which of the following compounds should react fastest with cyanide ion?

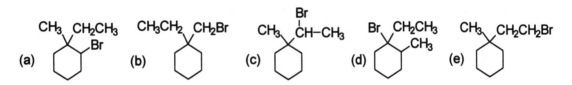

30 Which of the following compound, when reacted with water, will give substantial amount of an alcohol with a rearranged carbon skeleton?

CH₃	CH₃	CH₃	CH₃
$CH_3CHCHCH_2CH_2CH_3$	$CH_3CH_2CCH_2CH_2CH_3$	$CH_3CH_2CHCHCH_2CH_3$	$CH_3CH_2CHCH_2CHCH_3$
Br	Br	Br	Br
A	B	C	D

(a) only A   (b) only B   (c) A and B   (d) C and D   (e) A and C

31 The dehydrogenation of propane to give propene will occur at a high temperature because the $\Delta H°_{rxn}$ is ___ and the $\Delta S°_{rxn}$ is ___.

(a) negative; negative   (b) negative; positive   (c) positive; negative   (d) positive; positive

32 Reaction of 3-bromo-3-methylhexane with $(CH_3)_3CO^-$ can yield a mixture of ___ isomeric alkenes.

(a) 7   (b) 6   (c) 5   (d) 4   (e) 3

33 Reaction of 3-bromo-2,3-dimethylhexane with CH_3O^- can yield a mixture of ____ isomeric alkenes.

(a) 3   (b) 4   (c) 5   (d) 6   (e) 7

34 Which of the following combinations of reactants should give the largest percentage of elimination product(s) compared to the substitution product?

(a) 1-bromobutane and CH_3S^- (b) 1-bromobutane and CH_3O^-
(c) 1-bromobutane and $(CH_3)_3CO^-$ (d) 2-bromobutane and CH_3O^-
(e) 2-bromobutane and $(CH_3)_3CO^-$

35 The number of products formed by dehydrohalogenation of A, B, and C are ____, respectively.

(a) 2, 3, 2
(b) 3, 5, 4
(c) 2, 4, 3
(d) 3, 4, 3
(e) 3, 5, 5

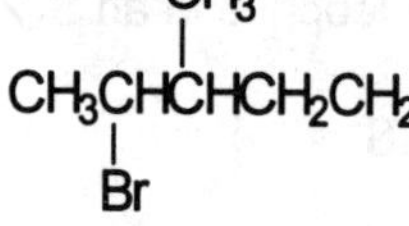

$$CH_3CHCHCH_2CH_2CH_3$$
$$\underset{Br}{\quad}$$

$$CH_3CH_2CCH_2CH_2CH_3$$
$$\underset{Br}{\quad}$$

$$CH_3CH_2CHCHCH_2CH_3$$
$$\underset{Br}{\quad}$$

A 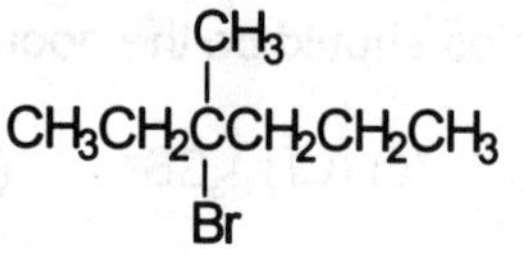   B 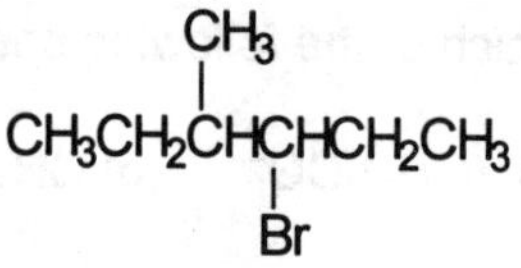   C

36 Dehydration of the following alcohol will give a mixture consisting of ____ isomeric alkenes.

(a) six
(b) five
(c) four
(d) three
(e) two

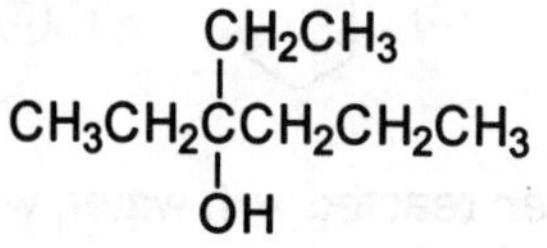

$$CH_3CH_2CCH_2CH_2CH_3$$ with CH_2CH_3 above and OH below the central carbon

37 Dehydration of the following alcohol will give a mixture consisting of ____ rearranged isomeric tetrasubstituted alkenes.

(a) 7
(b) 6
(c) 5
(d) 4
(e) 3

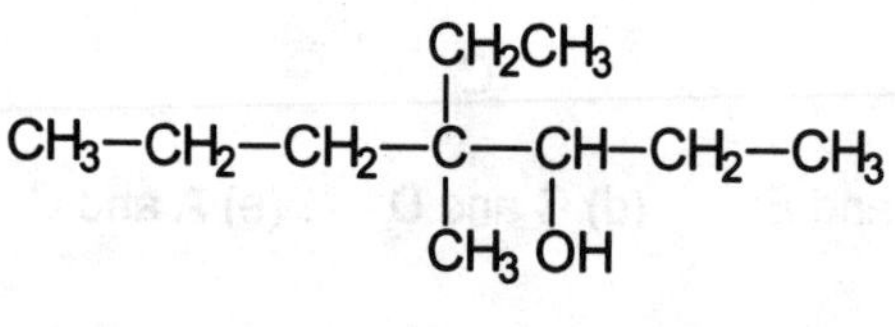

$$CH_3-CH_2-CH_2-C-CH-CH_2-CH_3$$ with CH_2CH_3 above and CH_3, OH below

38 Which of the indicated unsaturated compounds is the major elimination product obtained in the reaction of sodium methoxide with the bromo compound listed to the left of the possible answers?

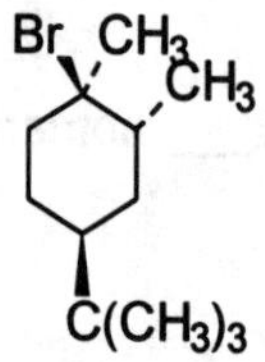

(a) 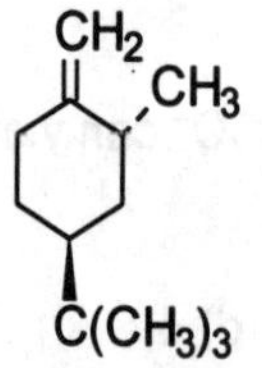   (b) 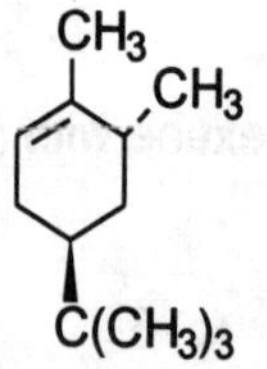   (c)

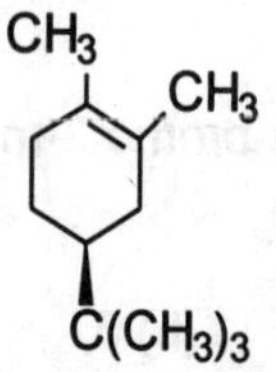

39 Which of the indicated unsaturated compounds is the major elimination product obtained in the reaction of sodium methoxide with the compound listed to the left of the possible answers?

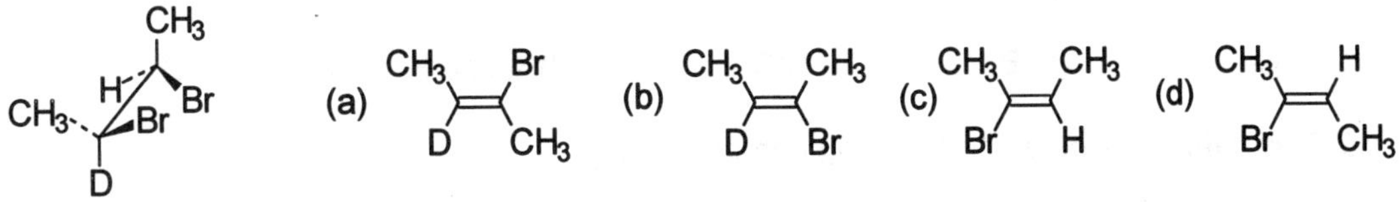

40 In which solvent is the rate of reaction of azide ion with 1-bromobutane the slowest?

(a) $H-\overset{O}{\overset{\|}{C}}-\overset{|}{\underset{CH_3}{N}}-CH_3$ (b) $CH_3-\overset{O}{\overset{\|}{C}}-CH_3$ (c) $CH_3-C\equiv N$ (d) CH_3-CH_2-OH

41 Which of the following combinations of reactants will give the highest yield of substitution product compared to elimination product?

(a) 1-bromodecane in ethanol
(b) 2-bromodecane in ethanol
(c) 1-bromodecane in ethanol with sodium ethoxide
(d) 2-bromodecane in ethanol with sodium ethoxide

42 Assuming that the pK_a values of the following two species are directly proportional to the % s character of the N—H bond, one can predict that the pK_a of the third species should be ___.

R_3NH^+ $R_2C=NH_2^+$ $RC\equiv NH^+$

$pK_a = 9$ $pK_a = 3$ $pK_a = ?$

(a) -6   (b) 15   (c) 12   (d) -3   (e) -9

43 Consider the following reactions and indicate which of them have K > 1.

(A) $NH_2^- + H_2O \rightleftharpoons OH^- + NH_3$

(B) $NH_3 + HC\equiv C^- \rightleftharpoons HC\equiv CH + NH_2^-$

(C) $OH^- + HC\equiv CH \rightleftharpoons HC\equiv C^- + H_2O$

(D) $CH_3CH_3 + HC\equiv C^- \rightleftharpoons HC\equiv CH + CH_3CH_2^-$

(E) $CH_2=CH^- + HC\equiv CH \rightleftharpoons HC\equiv C^- + CH_2=CH_2$

(a) A and E   (b) B and E   (c) C and D   (d) B and D   (e) A, C, and E

44 Which of the following statements are true?

A 1-butyne is more acidic than 1-butene
B Propyne is more reactive toward HBr than is propene.
C An internal alkyne is more stable than a terminal alkyne.

(a) A   (b) B   (c) C   (d) A and B   (e) A and C

45 The IUPAC name of the following compound is ___.

(a) 1-penten-4-yne
(b) 4-penten-1-yne
(c) 1-pentyn-4-ene $H-C{\equiv}C-CH_2-CH{=}CH_2$
(d) 4-pentyn-1-ene

46 The heats of formation of 1-pentyne and 1-pentene are +140 and -22 kJ mole^{-1}, respectively. The $\Delta H°_{rxn}$ for the hydrogenation of 1-pentene is -126 kJ mole^{-1}. The $\Delta H°_{rxn}$ for the hydrogenation of 1-pentyne to give pentane is _____ kJ mole^{-1}.

(a) -186   (b) -215   (c) -238   (d) -270   (e) -288

47 The number of moles of amide ion that are required to produce an alkyne from compounds A, B, C, and D, respectively are _____.

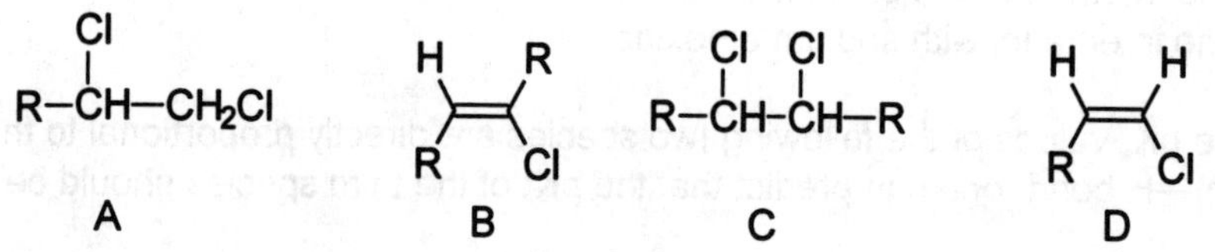

(a) 2, 1, 2, 1   (b) 3, 1, 2, 1   (c) 2, 1, 2, 2   (d) 2, 1, 3, 2   (e) 3, 1, 2, 2

48 Which of the following compounds can be obtained by reaction of propyne with aqueous $H_2SO_4/HgSO_4$?

(a) $CH_3\overset{O}{\overset{\|}{C}}CH_3$ (b) $CH_3CH_2\overset{O}{\overset{\|}{C}}H$ (c) $CH_3\overset{OH}{\overset{|}{C}H}CH_3$ (d) $CH_3CH_2CH_2OH$

49 Which of the following combinations of reactants is most suitable for the synthesis of 2-methyl-3-heptyne?

(a) the conjugate base of 1-pentyne and 2-bromopropane
(b) the conjugate base of 1-pentyne and 1-bromopropane
(c) the conjugate base of 3-methyl-1-butyne and 1-bromopropane
(d) the conjugate base of 3-methyl-1-butyne and 2-brompropane
(e) the conjugate base of 1-butyne and 1-bromo-2-methylpropane

50 The order of the intermediates formed in the lithium/ammonia reduction of an alkyne is ____.

(a) vinyl radical; radical anion; vinyl anion
(b) radical anion; vinyl radical; vinyl anion
(c) vinyl radical; vinyl anion; radical anion
(b) vinyl anion: radical anion; vinyl radical
(c) radical anion; vinyl anion; vinyl radical

51 The final product obtained in the following sequence of reactions is ____.

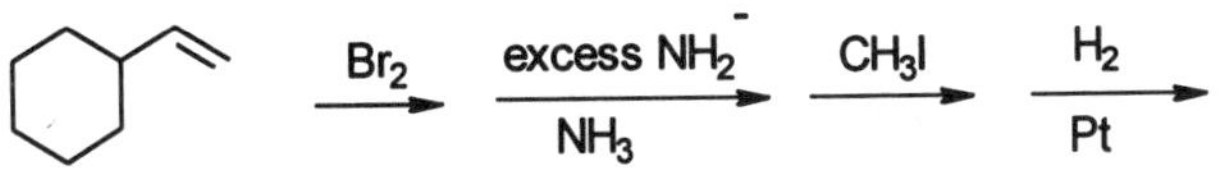

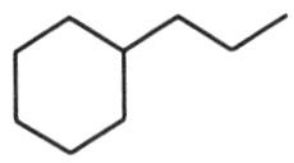

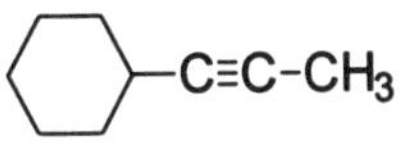

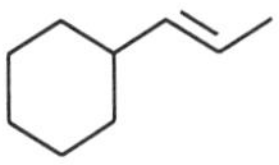

52 The highest occupied molecular orbitals of 1,3-butadiene and 1,3,5-hexatriene are ____, and ____, respectively.

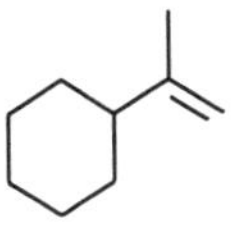

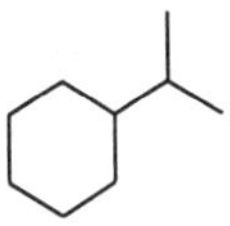

(a) symmetric; antisymmetric   (b) antisymmetric; symmetric
(c) symmetric; symmetric   (d) antisymmetric; antisymmetric

53 The highest occupied molecular orbitals of the allyl carbocation and the allyl carbanion are ____, and ____, respectively.

(a) symmetric; antisymmetric   (b) antisymmetric; symmetric
(c) symmetric; symmetric   (d) antisymmetric; antisymmetric

54 Which of the following has the smallest heat of hydrogenation to give methylcyclohexane?

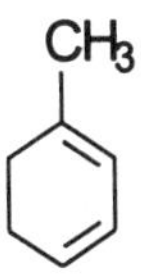

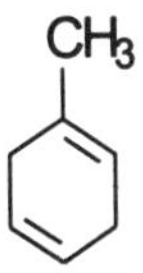

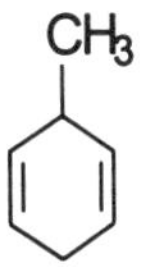

55 The transition state energy for 1,2-addition of HBr to 1,3-butadiene is ____ than for 1,4-addition. The 1,4-addition product is of ____ energy than the 1,2-addition product.

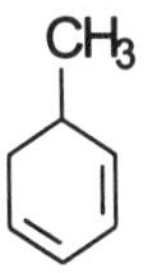

(a) lower; lower   (b) higher; higher   (c) lower; higher   (d) higher; lower

56 Which of the indicated C—H bonds has the lower bond dissociation energy?

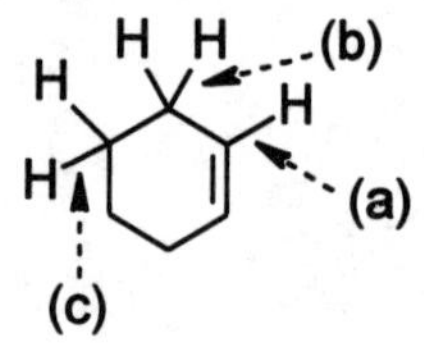

58 Which of the following structures represent optically active compounds?

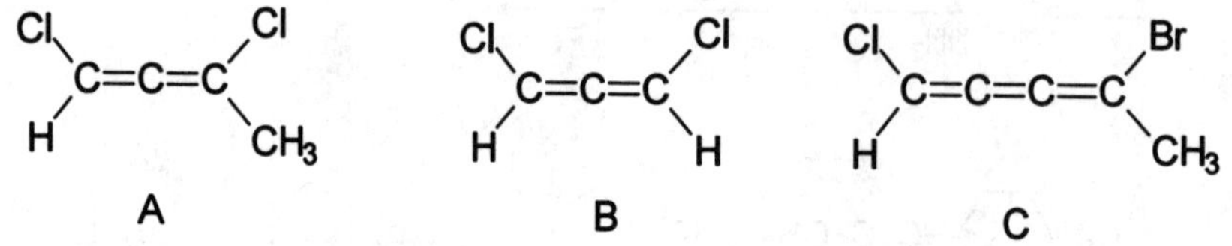

(a) A   (b) B   (c) C   (d) A and B   (e) A and C

Answers to Examination 3

ddbbc   bdcae   ccadb   eaaab   cbbbc   adaee   debeb   dcbbd   aeaea

ecacb   abaaa   bd

Examination 4

1 Consider the following structure for an isomer of benzene. Substitution of a hydrogen atom by bromine can occur to give ___ monobromo compounds and ____ dibromo compounds.

(a) two, three
(b) two, four
(c) three, four
(d) three, five
(e) three, six

2 Electrophilic substitution of C-H bonds of 1,3-diazine and anthracene could give ___ and ___ monosubstituted isomers, respectively.

(a) 3,2
(b) 2,2
(c) 3,4
(d) 4,3
(e) 3,3

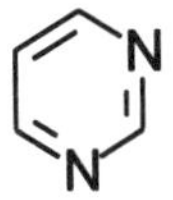

1,3-diazine 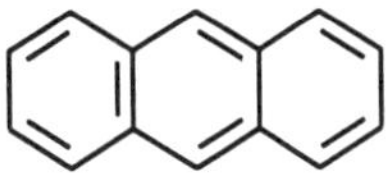   anthracene

3 There are ___ isomeric tribromobenzenes and ___ have(has) a dipole moment.

(a) two, one   (b) two, two   (c) three, one   (d) three, two   (e) three, three

4 Considering the effects of intramolecular and intermolecular hydrogen bonding, which of the isomeric hydroxybenzaldehydes has the lowest boiling point?

(a) ortho   (b) meta   (c) para

5 The $\Delta H°_{rxn}$ for the hydrogenation of cyclohexene is -112.5 kJ mole^{-1}. The $\Delta H°_{rxn}$ for the complete hydrogenation of naphthalene is -332.0 kJ mole^{-1}, respectively. The calculated resonance energy of napthalene is _____ kJ mole^{-1} .

(a) 337.5
(b) 219.5
(c) 230.5
(d) 225.0

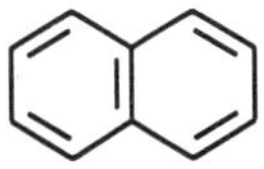

6 The heats of hydrogenation of 1,3-cyclohexadiene and benzene to form cyclohexane are 231 and 208 kJ mole^{-1}, respectively. The ΔH_{rxn} for the following reaction is ____ kJ mole^{-1} and the ΔS_{rxn} is ____.

(a) -23; positive
(b) +23; positive
(c) -23; negative
(d) +23; negative

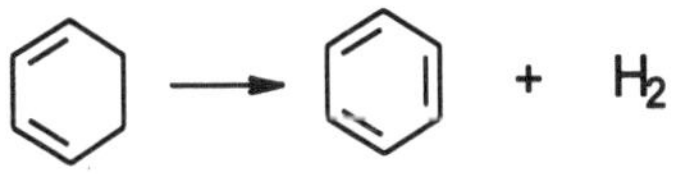

7   The heats of formation of furan and tetrahydrofuran are -35.1 and -184 kJ mole^{-1}, respectively. The ΔH_{rxn} for the hydrogenation of cyclopentene is -110 kJ mole^{-1}. The resonance energy of furan is _____ kJ mole^{-1}.

(a) 71   (b) 39   (c) 109   (d) 220   (e) 150

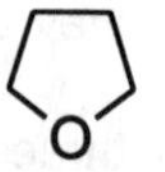

8   The molecular formula for [3,4]benzpyrene, a carcinogen found in smoke from cigarettes, is _____.

(a) $C_{20}H_{14}$ (b) $C_{22}H_{14}$ (c) $C_{18}H_{14}$

(d) $C_{18}H_{12}$ (e) $C_{20}H_{12}$

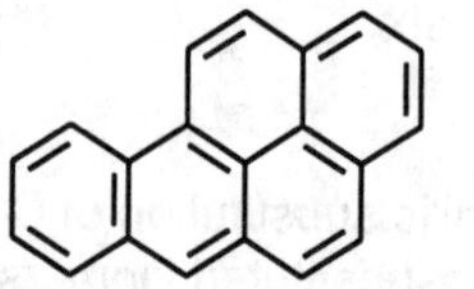

9   The order of increasing rate of reactivity (slowest first) of the following compounds in an S_N1 reaction with methanol is _____.

(a) A < B < C
(b) C < B < A
(c) B < A < C
(d) B < C < A
(e) A < C < B

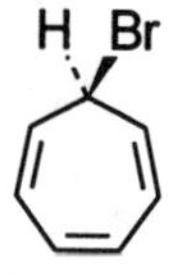

10 Select the heterocyclic compounds that are aromatic.

(a) A, B, C, D, E
(b) A, D, E
(c) C, E
(d) B, C, D, E
(e) B, C, E

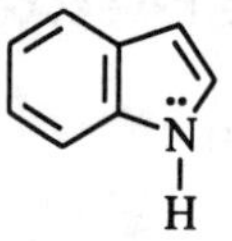

11 Tagamet (generic name cimetidine) is an antiulcer drug. The two nitrogen atoms in the heterocyclic ring contribute a total of ___ electrons to the aromatic ring and those two nitrogen atoms have a total of ___ pair(s) of nonbonded electrons.

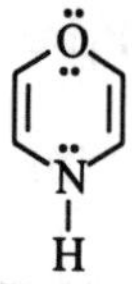

(a) two; one
(b) three; one
(c) two; two
(d) three; two
(e) none of the above are correct

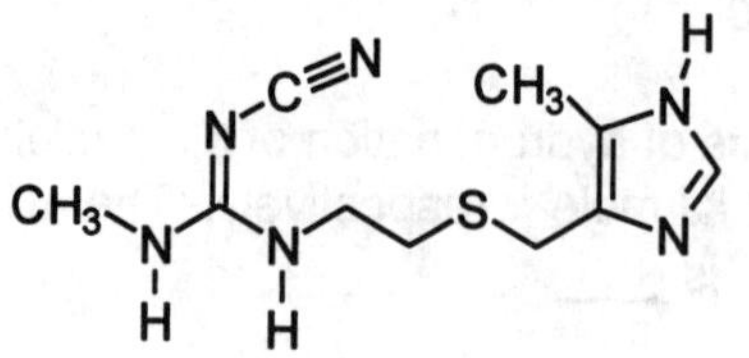

12 Based on molecular orbital theory which of the following orbital energy levels and electron assignments is correct for the cyclopentadienyl cation?

(a)   (b)   (c)   (d)   (e)

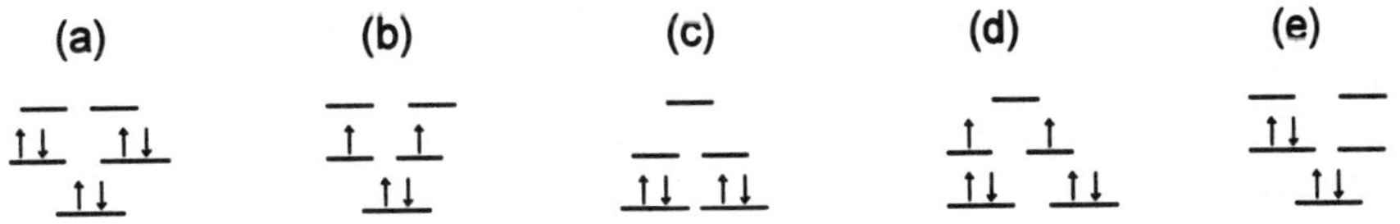

13 The following heterocyclic compound is expected _____ aromatic properties because it _____.

(a) to have; has 4n+2 π electrons
(b) not to have; does not have 4n+2 π electrons
(c) not to have; has severe bond angle strain
(d) not to have; has severe steric hindrance

14 Select among the following species and list those that are aromatic based on the Hückel rule.

A cycloheptatrienyl cation
B cyclononatetraenyl anion
C cyclopentadienyl cation

(a) A and B   (b) B and C   (c) A and C   (d) A, B, and C

15 The nitronium ion, the reactive species in nitration reactions, has its positive charge on _____ and has ___ pairs of nonbonded electrons in its most stable resonance form.

(a) nitrogen, 2   (b) nitrogen, 4   (c) oxygen, 2   (d) oxygen, 4

16 Chlorination of the following compound gives a herbicide. The group bonded to the ring is _____ director and the ring is _____ by the group.

(a) an ortho,para; activated
(b) an ortho,para; deactivated
(c) a meta; activated
(d) a meta; deactivated

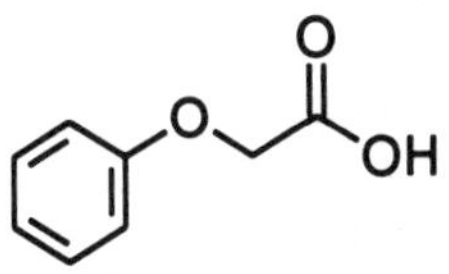

17 Consider the reactivity of the following compound. (Phosphorus is a member of Group V and is a third row element). The group bonded to the ring is expected to be ____ director and the ring is ____ by the group.

(a) a meta; activated
(b) a meta; deactivated
(c) an ortho,para; activated
(d) an ortho,para; deactivated

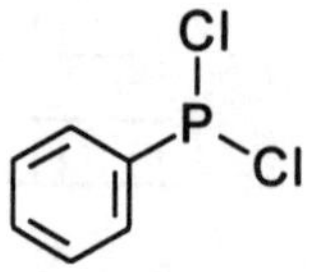

18 Consider the reactivity of the following compound. The nitroso group bonded to the ring is expected to be ____ director and the ring is ____ by the group.

(a) a meta; activated
(b) a meta; deactivated
(c) an ortho,para; activated
(d) an ortho,para; deactivated

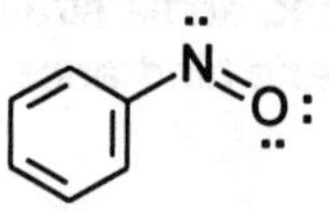

19 Oxygen (period 2) and sulfur (period 3) are both in Group VIA. The CH_3S- group is _____ effective at electron withdrawal inductively and is a ____ effective donor of electrons by resonance than the CH_3O- group.

(a) more, more   (b) more, less   (c) less, less   (d) less, more

20 The order of increasing rate of reactivity (slowest first) of the following compounds in an electrophilic aromatic substitution reaction with bromine is ____.

(a) A < B < C
(b) C < B < A
(c) B < A < C
(d) B < C < A
(e) A < C < B

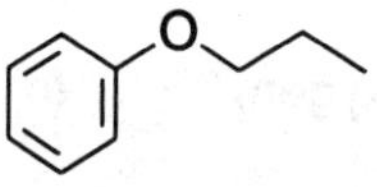

A 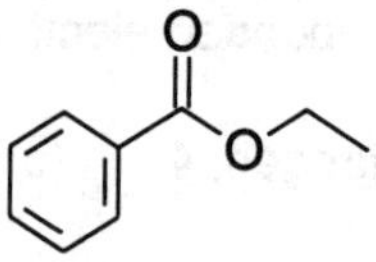   B 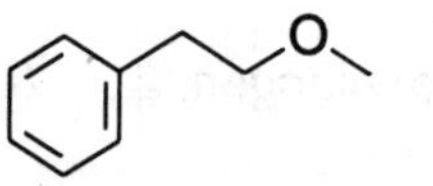   C

21 Arrange the following in order of decreasing reactivity (fastest first) for reaction with typical electrophilic reagents.

(a) B > A > C > D
(b) D > B > C > A
(c) A > B > C > D
(d) C > A > B > D
(e) A > C > B > D

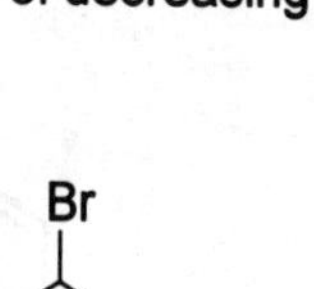

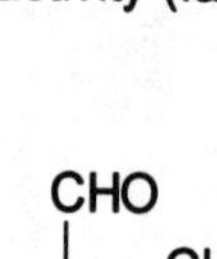

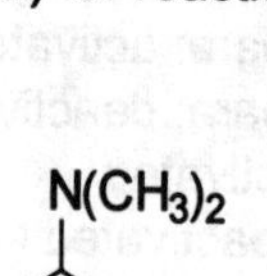

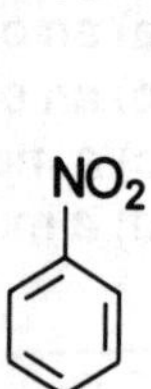

22 Reaction of the following compound with typical electrophilic reagents occurs predominately at position(s).

(a) B and D
(b) A and D
(c) C and D
(d) A and C
(e) B and C

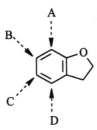

23 Which of the following compounds will give the highest percentage of a single product in an aromatic substitution reaction?

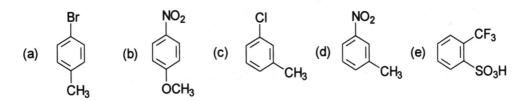

24 Which of the compounds listed in question 23 will react at the slowest rate in an aromatic substitution reaction?

25 Reaction of the following compound with typical electrophilic reagents occurs predominately at the position labeled ___.

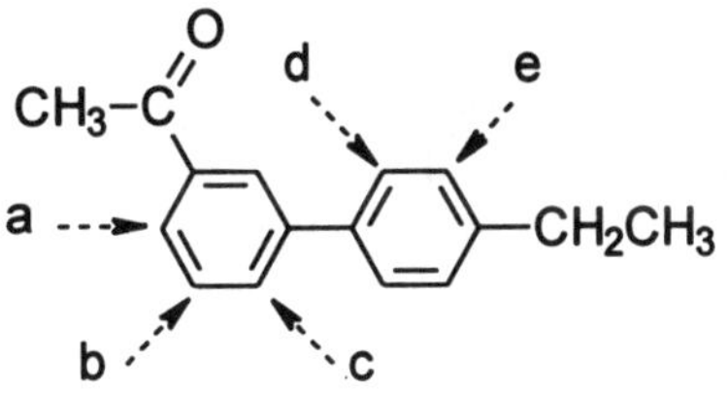

26 Which of the following reaction sequences could be used to synthesize m-nitrobenzoic acid starting from benzene?

(a) HNO_3 /H_2SO_4 ; $CH_3Cl/AlCl_3$; $KMnO_4$
(b) $CH_3Cl/AlCl_3$; HNO_3 /H_2SO_4 ; $KMnO_4$
(c) $CH_3Cl/AlCl_3$; $KMnO_4$; HNO_3 /H_2SO_4
(d) both a and b
(e) both b and c

27 Which of the indicated sites in protosil, an antibiotic, should be most reactive toward an electrophile in an electrophilic aromatic substitution reaction?

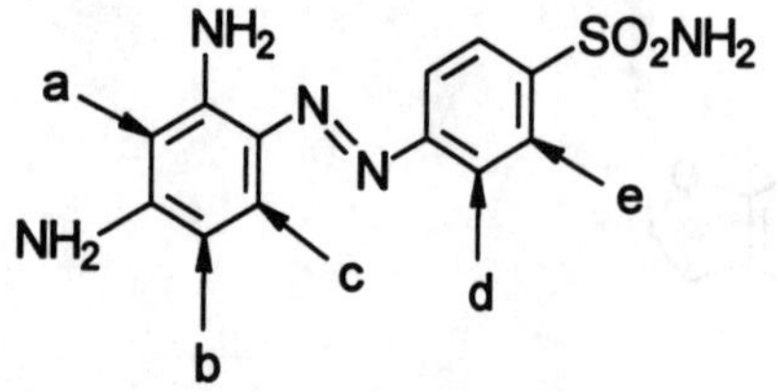

28 Which of the listed reagents could be used for the indicated reaction?

(a) HBr, peroxide
(b) HBr
(c) Br_2 and HBr
(d) Br_2 and $FeBr_3$
(e) NBS

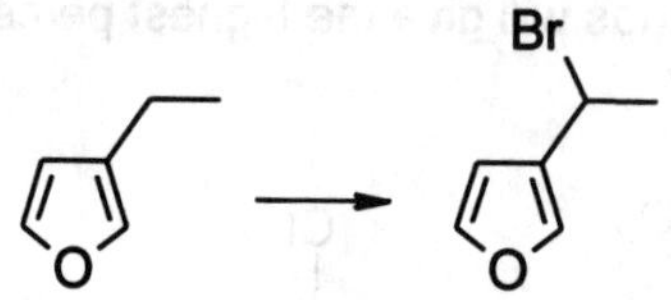

29 Which of the listed reagents could be used for the indicated reaction?

(a) HBr, peroxide
(b) HBr
(c) NBS
(d) Br_2 and light
(e) Br_2 and $FeBr_3$

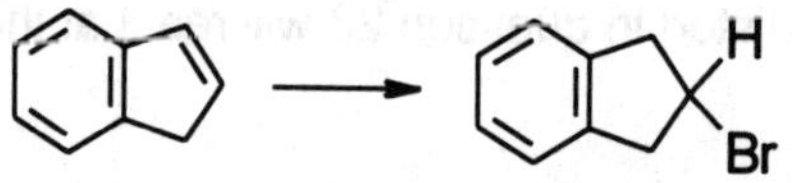

30 Which of the listed reagents could be used for the indicated reaction?

(a) HBr, peroxide
(b) HBr
(c) NBS
(d) Br_2 and light
(e) Br_2 and $FeBr_3$

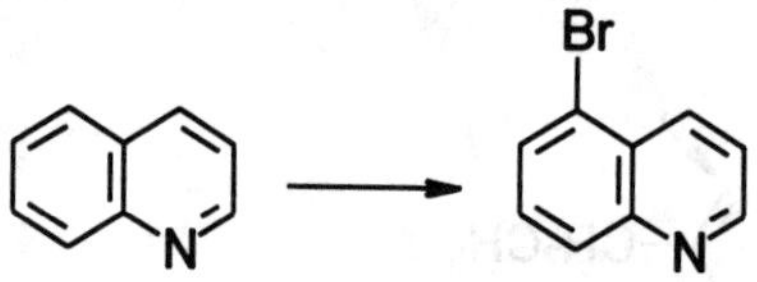

31 Which of the listed reagents could be used for the indicated reaction?

(a) $KMnO_4$
(b) NH_3 and $AlCl_3$
(c) Sn and HCl
(d) HNO_3
(e) $NaNO_2$ and HCl

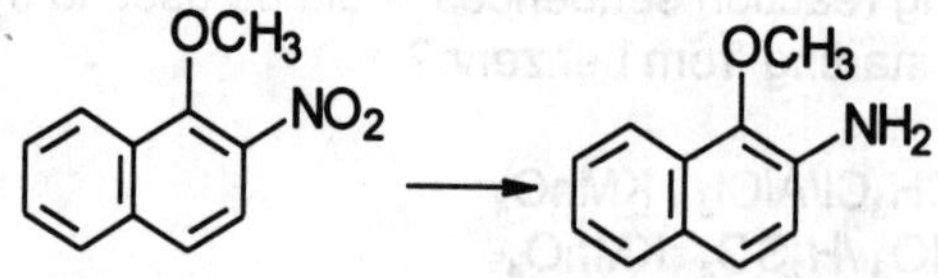

32 Which of the listed reagents could be used for the indicated reaction?

(a) $KMnO_4$
(b) CO_2 and $AlCl_3$
(c) Sn and HCl
(d) H_2 and Pd
(e) $NaNO_2$ and HCl

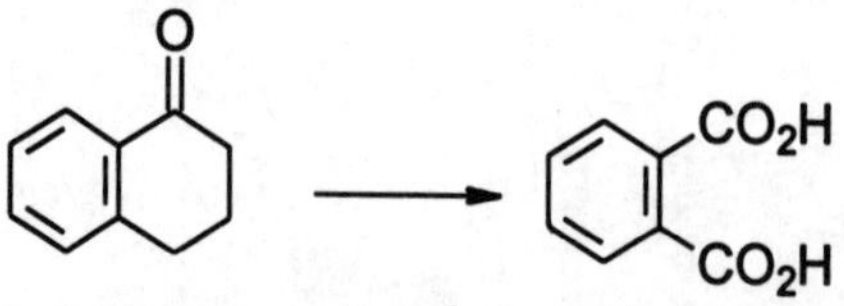

33 Which of the listed reagents could be used for the indicated reaction?

(a) KMnO$_4$
(b) CO$_2$ and AlCl$_3$
(c) Sn and HCl
(d) H$_2$ and Pd
(e) NaNO$_2$ and HCl

34 Which of the following aromatic substitution reactions is readily reversible?

(a) nitration   (b) alkylation   (c) bromination   (d) acylation   (e) sulfonation

35 Consider the following substituents. How many of them seriously limit the utility of the Friedel Crafts alkylation reaction?

—NO$_2$ —NH$_2$ —CO$_2$H —OCH$_3$ —CH$_3$

(a) one   (b) two   (c) three   (d) four   (e) five

36 Which of the following procedures will yield 4-chloro-2-ethylnitrobenzene?

(a) chlorination of o-ethylnitrobenzene
(b) nitration of m-chloroethylbenzene
(c) Friedel-Crafts alkylation of p-chloronitrobenzene

37 Which of the following procedures will yield m-nitrobenzoic acid?

(a) nitration of toluene followed by oxidation with KMnO$_4$
(b) reaction of ethylbenzene with KMnO$_4$ followed by reaction with HNO$_3$/H$_2$SO$_4$
(c) reaction of benzoic acid with HNO$_2$
(d) reaction of nitrobenzene with CO$_2$ and H$^+$
(e) none of the above

38 Which of the following procedures will yield m-chloroaniline?

(a) nitration of chlorobenzene followed by reaction with HNO$_2$ and NH$_3$
(b) nitration of chlorobenzene followed by reaction with SnCl$_2$
(c) chlorination of nitrobenzene followed by reaction with NH$_3$
(d) chlorination of nitrobenzene followed by reaction with Fe and HCl
(e) chlorination of aniline

39 Reaction of the compound on the left with sulfuric acid at 170°C will yield __.

   (a)   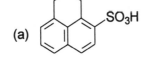   (b)   (c)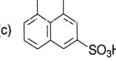

40 Which of the indicated positions is most reactive in an electrophilic aromatic substitution reaction?

41 Reaction of m-nitroaniline with H_2SO_4 / $NaNO_2$ followed by heating with aqueous sulfuric acid gives _____.

(a)   (b)   (c)   (d)   (e)

42 Reaction of the compound on the left with reagents in the sequence (1) HNO_3 and H_2SO_4 ; (2) H_2 / Pd ; (3) HNO_2 ; (4) CuCl gives _____.

(a)   (b)   (c)   (d)

43 Which of the following statements is true?

   A Visible light has a longer wavelength than ultraviolet light.
   B Energy is directly proportional to the frequency of light.
   C Wavenumber values are inversely proportional to frequency.

   (a) A   (b) B   (c) D   (d) A and B   (e) A and C

44 Which of the following has the largest λ_{max}?

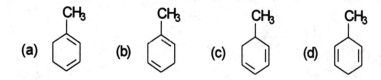

(a)   (b)   (c)   (d)

45 Which of the following has the largest λ_{max}?

(a) 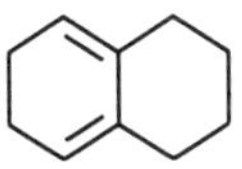   (b) 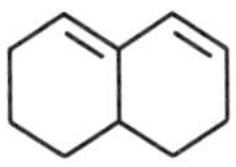   (c) 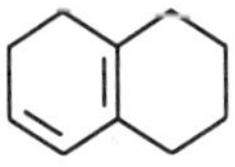   (d)

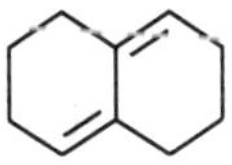

46 Which of the following has an absorption at highest wavenumber?

(a) C–O   (b) H–O   (c) C–H   (d) C=O

47 Which of the isomeric cresols (methylphenols) has the largest number of out-of-plane bending C–H absorptions?

(a) ortho   (b) meta   (c) para

48 Which of the following statements are true?

A The peaks to the left of an NMR spectrum are at highest field.
B The peaks to the right of an NMR spectrum are shielded compare to those on the left.
C Coupling constants depend on the strength of the magnetic field.

(a) A   (b) B   (c) C   (d) A and B   (e) B and C

49 Which of the following compounds has the lowest field absorption in its hydrogen NMR spectrum.

(a) $CH_3\overset{\overset{\displaystyle O}{\|}}{C}CH_3$ (b) $CH_3\overset{\overset{\displaystyle OH}{|}}{CH}CH_3$ (c) $CH_3\overset{\overset{\displaystyle O}{\|}}{C}OCH_3$ (d) $CH_3\overset{\overset{\displaystyle O}{\|}}{C}NHCH_3$

50 Which of the following has the largest number of absorptions in its hydrogen NMR spectrum?

(a) 1,4-dichlorobutane   (b) 2,3-dichlorobutane
(c) 2,2-dichlorobutane   (d) 1,2-dichlorobutane
(e) 1,2-dichloro-2-methylpropane

51 The relative areas in the components peaks of a quintet are ___.

(a) 1:2:3:2:1   (b) 1:2:4:2:1   (c) 1:3:5:3:1   (d) 1:3:6:3:1   (e) 1:4:6:4:1

52 In the hydrogen NMR spectrum of $(CH_3)_2CHOCH(CH_3)_2$ the methyl groups appear as a ___ and the CH groups appear as a ___.

(a) triplet; singlet   (b) quartet; doublet
(c) doublet; heptet   (d) heptet; doublet

53 Which of the following have the smallest number of absorptions in its C-13 NMR spectrum.

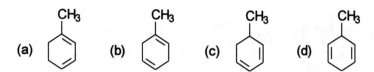

54 Which of the following have the smallest number of absorptions in its C-13 NMR spectrum

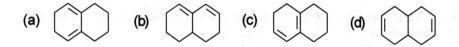

55 In the C-13 NMR spectrum, the lowest field resonance of a compound of molecular formula C_4H_9Cl is a triplet. Which of the following compounds have this property?

A 1-chlorobutane
B 2-chlorobutane
C 1-chloro-2-methylpropane
D 2-chloro-2-methylpropane

(a) A and D   (b) B and C   (c) A and C   (d) C and D   (e) B and D

56 Which of the following compounds has the lowest field absorption in its C-13 NMR spectrum.

(a) CH₃CCH₃   (b) CH₃CHCH₃   (c) CH₃COCH₃   (d) CH₃CNHCH₃

Answers to Examination 4

eedab   aaece   bbdab   addcd   ddbee   cbeae   cacec   bbdcc   aadaa

bbbcd   ecddc   a

Examination 5

1   Which of the following statements are true?

A Thiols have higher boiling points than alcohols having the same number of carbon atoms.
B Thiols are more acidic than alcohols.
C Tertiary alcohols are not as easily dehydrated as secondary alcohols.

(a) A   (b) B   (c) C   (d) A and B   (e) A and C

2   Zinc chloride is used in the reaction of a primary alcohol with HCl in order to

(a) remove the water formed in the reaction and drive the equilibrium to the right.
(b) increase the nucleophilicity of the chloride ion and speed up the reaction.
(c) serve as a Lewis acid to make an oxygen species a better leaving group.
(d) neutralize the acid formed in the reaction.
(e) make the primary carbon atom more nucleophilic.

3   The product of reaction of a primary alcohol with PBr_3 is easily isolated from the reaction mixture because

(a) H_3PO_3 has a high boiling point and is also water soluble.
(b) one of the byproducts is a gas.
(c) phosphorus has a very high affinity for oxygen and forms strong bonds.
(d) PBr_3 is a strong Lewis acid
(e) the bromide ion is a good leaving group.

4   The reaction of a primary alcohol with $SOCl_2$ is driven strongly to the right because

(a) sulfur has a high affinity for oxygen and forms strong sulfur-oxygen bonds
(b) one of the products of the reaction is a gas and makes $\Delta S°_{rxn}$ more positive
(c) chloride ion is a poor nucleophile.
(d) the $\Delta S°$ for the reaction is negative.
(e) pyridine neutralizes the acid formed.

5   Which of the following is the strongest base?

(a) CH_3O^- (b) $(CH_3)_3CO^-$ (c) $(CF_3)_3CO^-$ (d) $CH_3CO_2^-$ (e) Cl^-

6   Which of the following conditions will give the indicated product?

(a) $NaBH_4$ in ethanol
(b) H_2/Pt in ethanol at 1 atm.
(c) $LiAlH_4$ in THF
(d) both a or b
(e) both a or c

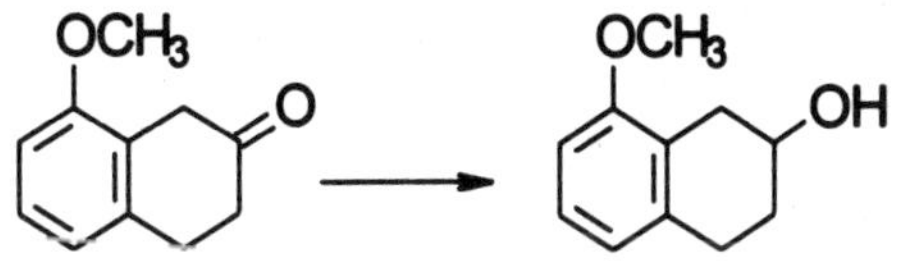

7 What is the major product of the reaction of 3,3-dimethylcyclohexene with BD_3 in THF followed by reaction with basic hydrogen peroxide?

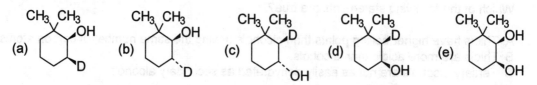

8 Which of the following products results from the reaction of 1-methylcyclobutene with mercuric acetate in water followed by reaction with $NaBH_4$?

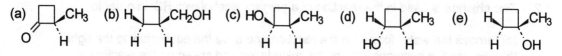

9 What is the major product of the reaction of the unsaturated compound on the left with BD_3 in THF followed by reaction with basic hydrogen peroxide?

10 The number of combinations of an aldehyde or ketone and the required Grignard reagent that can be used to synthesize the alcohols 1-methylcyclobutanol, 2-methyl-2-butanol, and 3-methyl-3-hexanol are _____, respectively.

(a) 1, 2, 3   (b) 1, 1, 2   (c) 2, 1, 2   (d) 1, 2, 2   (e) 2, 2, 3

11 Which of the following reagents would be best for the indicated reaction?

(a) TsCl/pyridine, then NaBr/DMF
(b) PBr_3
(c) $SOCl_2$/pyridine, then NaBr/DMF
(d) both a and b
(e) both b and c

12 Reaction of the phenyl Grignard reagent with ethylene oxide followed by conventional workup and subsequent reaction with Jones reagent will yield ___.

13 Treatment of the following compound with a peroxyacid following by reaction with hydroxide will yield _____ .

(a) 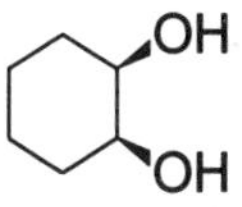 (b) 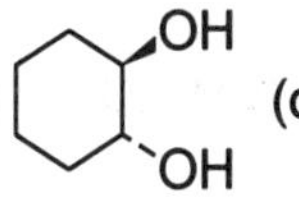 (c) 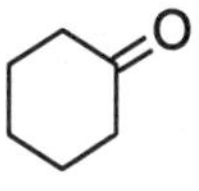 (d) 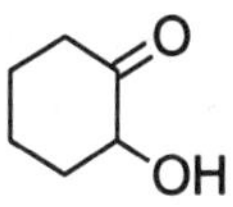 (e)

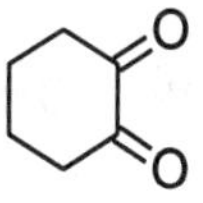

14 The best yield of cyclohexyl isopropyl ether results from

(a) cyclohexanol and NaH followed by reaction with 2-bromopropane
(b) 2-propanol and NaH followed by reaction with bromocyclohexane
(c) cyclohexanone and isopropyl magnesium bromide followed by hydrolysis
(d) cyclohexyl magnesium bromide and acetone followed by hydrolysis
(e) cyclohexene, $HgOAc_2$, and 2-propanol followed by reaction with $NaBH_4$

15 Dehydration of the following alcohol can give three isomeric unsaturated compounds. The order of decreasing yields (highest first) is

(a) A > B > C
(b) B > C > A
(c) A > C > B
(d) C > A > B
(e) B > A > C

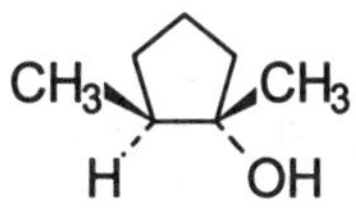

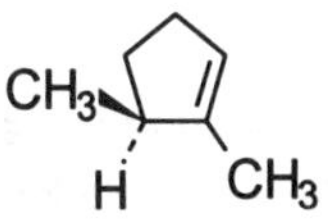

A 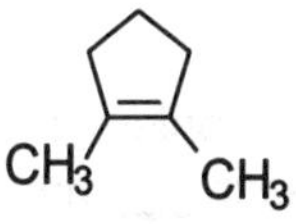   B 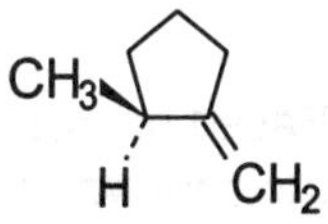   C

16 Reaction of (R)-2-octanol with PBr_3 followed by reaction with sodium methoxide gives 2-methoxyoctane with the ____ configuration. Reaction of (R)-2-octanol with toluenesulfonyl chloride and pyridine followed by reaction with sodium methoxide gives 2-methoxyoctane with the ____ configuration.

(a) R; R   (b) S; S   (c) R; S   (d) S; R

17 Which reagent is best suited for the indicated reaction?

(a) MMPP
(b) PCC in CH_2Cl_2
(c) MCPBA
(d) BH_3 in THF
(e) Jones reagent

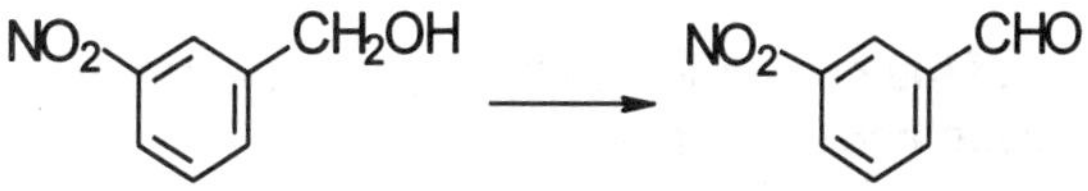

18 Which reagent is best suited for the indicated reaction?

(a) $NaBH_4$ in ethanol
(b) $LiAlH_4$ in ether
(c) MCPBA
(d) BH_3 in THF
(e) Zn/Hg and HCl

19 Reaction of *trans*-4-*tert*-butylcyclohexanol with chlorotrimethylsilane in the presence of an amine followed by reaction of the product with sodium fluoride in DMF gives ____

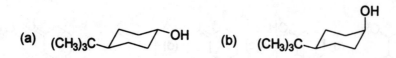

(a) $(CH_3)_3C$—⬡—OH

(b) $(CH_3)_3C$—⬡—OH

(c) 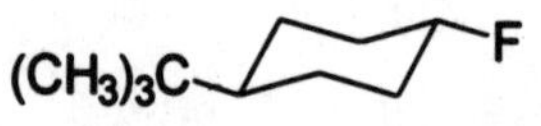 $(CH_3)_3C$—⬡—F

(d) 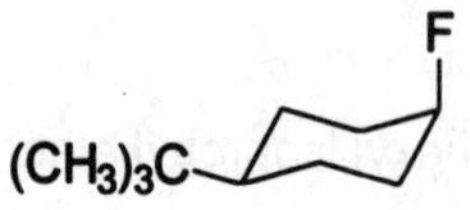 $(CH_3)_3C$—⬡—F

20 Oxidation of alcohols with CrO_3 occurs by a two step mechanism in which a chromium derivative is formed by a nucleophilic substitution on chromium followed by an E2 elimination.

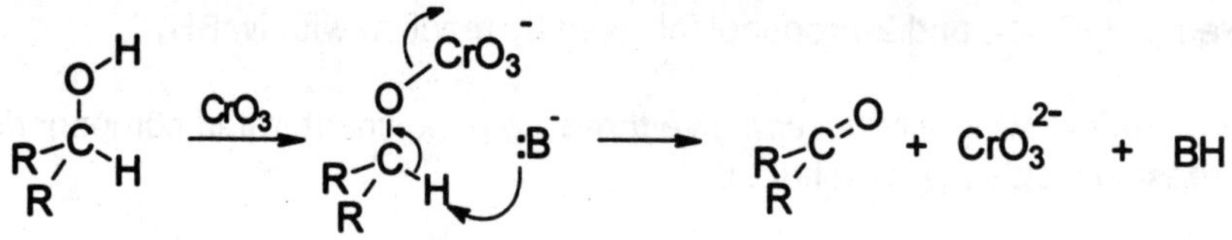

Based on this mechanism, if the first step were rate determining then *cis*-4-t-butylcyclohexanol would react ____ than *trans*-4-t-butylcyclohexanol. If the second step were rate determining then *cis*-4-t-butylcyclohexanol would react ____ than *trans*-4-t-butylcyclohexanol.

(a) faster; faster
(c) slower; faster

(b) faster; slower
(d) slower; slower

21 The initial cleavage reaction of the ether on the left with HBr under S_N2 conditions will yield ____.

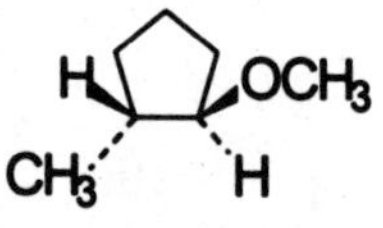

(a) 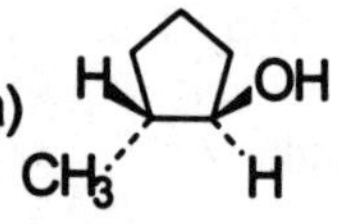 (b) 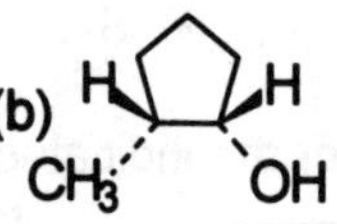 (c) 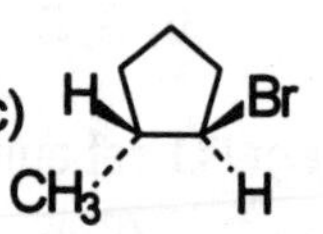 (d)

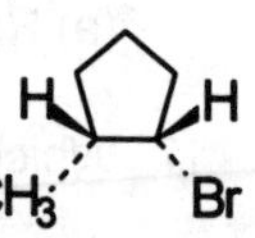

22 The synthesis of the compound on the right by the Williamson method requires sodium hydride as a base and a combination of ____.

(a) cis-4-methylcyclohexanol and methyl iodide
(b) trans-4-methylcyclohexanol and methyl iodide
(c) cis-1-bromo-4-methylcyclohexane and methanol
(d) trans-1-bromo-4-methylcyclohexane and methanol

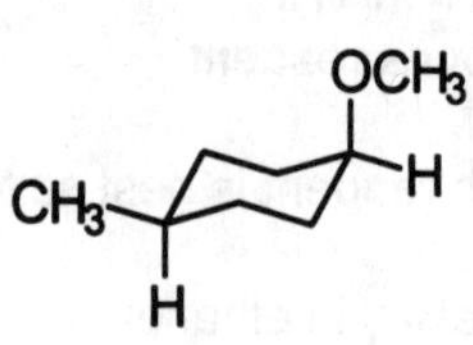

23 The product of the opening of the following epoxide by methanol under acid catalyzed conditions is ____ .

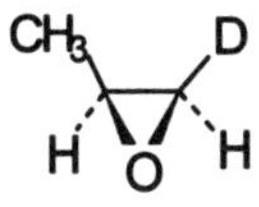

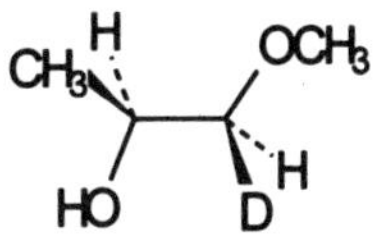

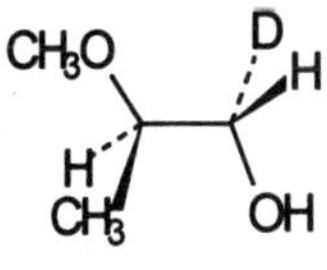

24 The product of the opening of the epoxide of 1-methylcyclohexene using lithium aluminum deuteride is ____ .

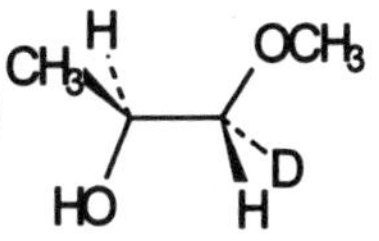

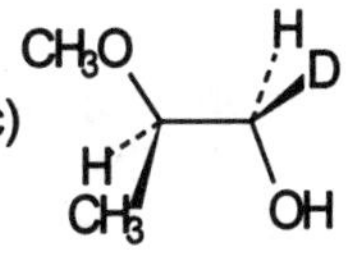

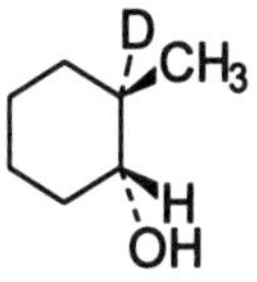

25 Reaction of the diol on the left with the Jones reagent will yield ____ .

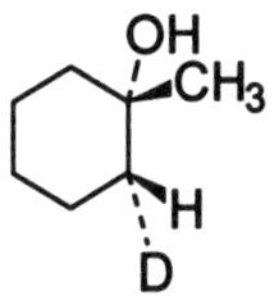

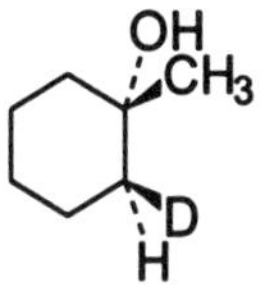

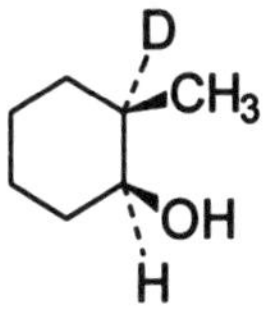

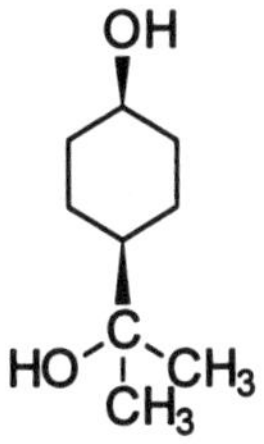

26 The following compound can be produced under acidic conditions using ____ .

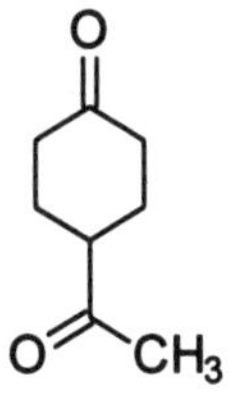

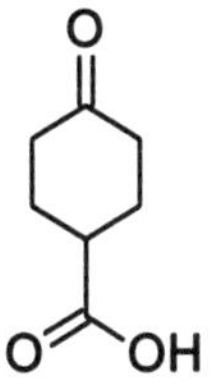

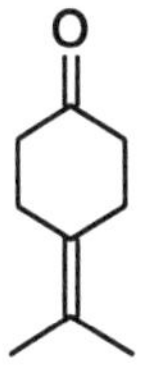

(a) cyclohexanol and 1,2-ethanediol
(b) cyclohexanone and 1,2-ethanediol
(c) cyclopentanone and 1,2-propanediol
(d) 1,3-cyclopentanediol and acetone
(e) cyclopentanone and 1,3-propanediol

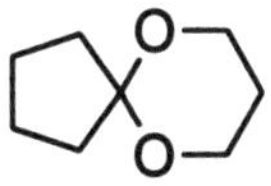

27 Which of the following reagents can be used for the indicated reaction.

(a) Zn(Hg), HCl
(b) NH$_2^-$ /NH$_3$
(c) Ag$_2$O, NH$_3$, H$_2$O
(d) NH$_2$NH$_2$, KOH, diethylene glycol
(e) NaBH$_4$

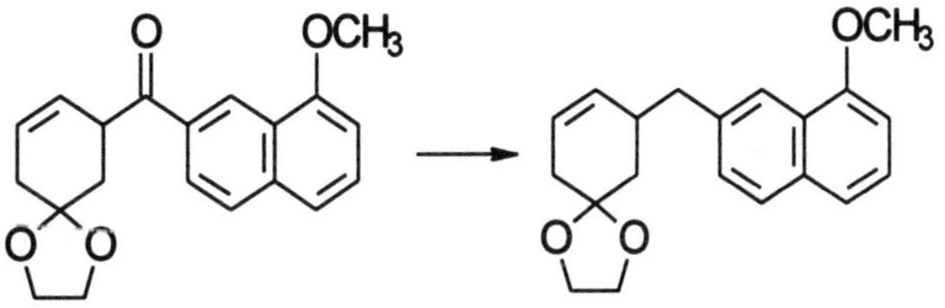

28 Which of the following combination of reagents should give the highest yield of product at equilibrium under comparable acidic reaction conditions.

(a) propanal and ethanol
(b) propanone and ethanol
(c) benzaldehyde and ethanol
(d) propanal and 1,2-ethanediol
(e) propanone and 1,2-ethanediol

29 The following compounds arranged in order of decreasing equilibrium constant (largest first) for formation of a cyanohydrin is given by selection ____

(a) A > C > B > D
(b) B > D > C > A
(c) A > C > B > D
(d) C > B > A > D
(e) C > A > D > B

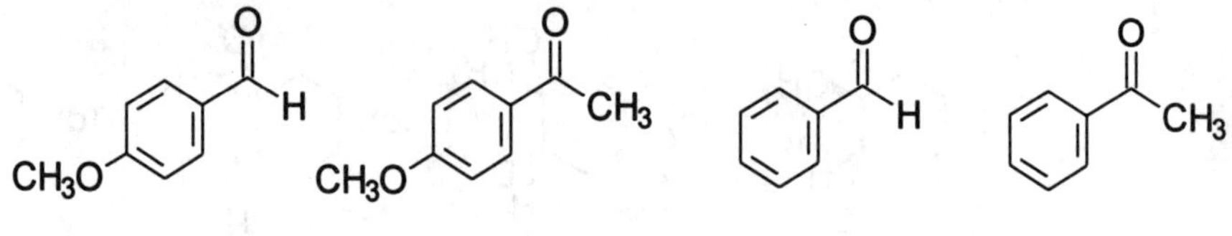

A   B   C   D

30 Which of the following reagents can be used for the indicated transformation?

(a) BH_3 in THF
(b) $KMnO_4$, OH^-
(c) PCC in CH_2Cl_2
(d) Ag_2O, NH_3, H_2O
(e) Zn(Hg)/HCl

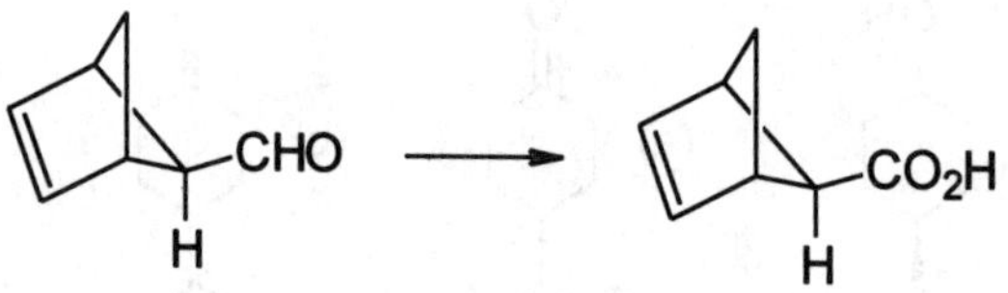

31 Reduction of the following compound by the Clemmensen method gives a compound with the molecular formula ____.

(a) $C_{10}H_{16}O_2$
(b) $C_{10}H_{14}O_2$
(c) $C_8H_{12}O$
(d) C_8H_{12}
(e) C_8H_{14}

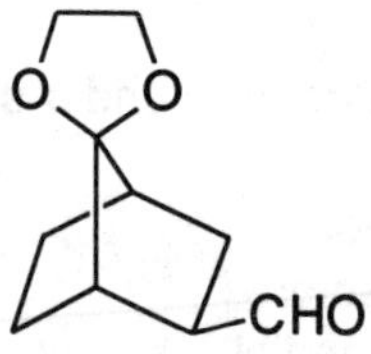

32 The following compounds arranged in order of decreasing equilibrium constant (largest first) for formation of a cyanohydrin is given by selection ____.

(a) C > B > D > A
(b) A > B > D > C
(c) B > A > D > C
(d) B > A > C > D
(e) A > B > C > D

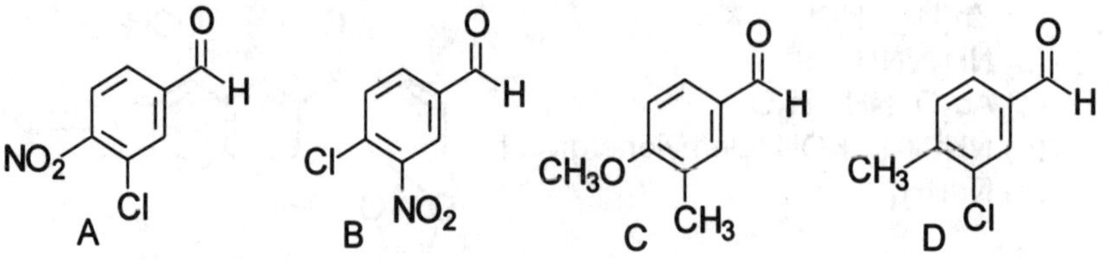

A   B   C   D

33 Which of the following will have the most favorable equilibrium constant for formation of a cyclic ketal with ethylene glycol.

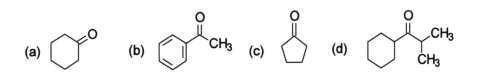

(a) (b) (c) (d)

34 The compound on the left is treated with the following reagents in order. The final product will be ____.

1) chlorotrimethylsilane and triethylamine
2) NaBH₄ in ethanol
3) catalytic H₃O⁺

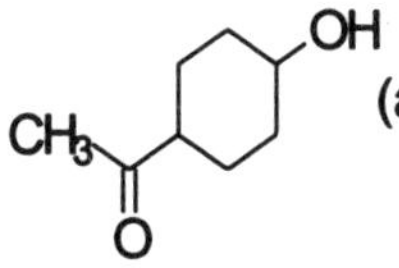

(a) 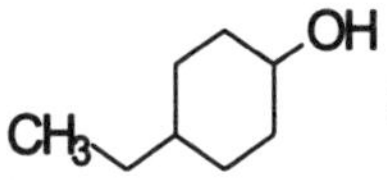 (b) 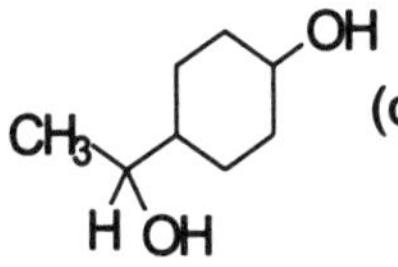 (c) 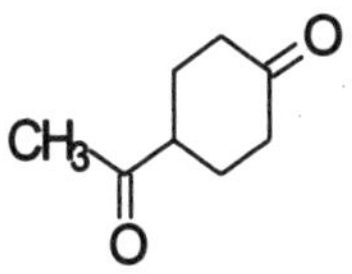 (d)

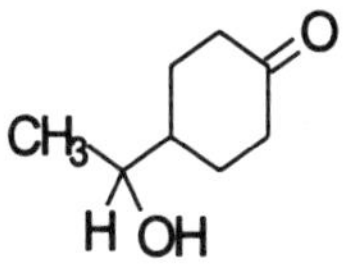

35 The compound on the left is treated with the following reagents in order. The final product will be ____.

1) ethylene glycol and H⁺
2) Mg, ether
3) acetaldehyde
4) H₃O⁺

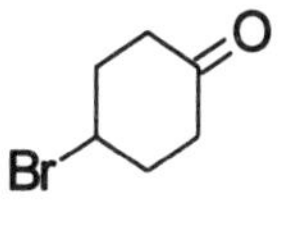

(a) 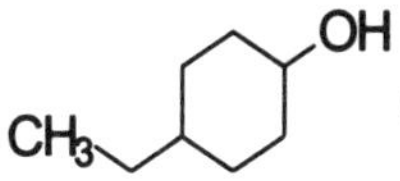 (b) 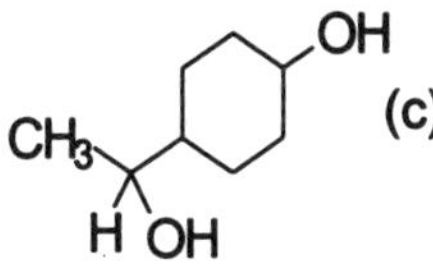 (c) 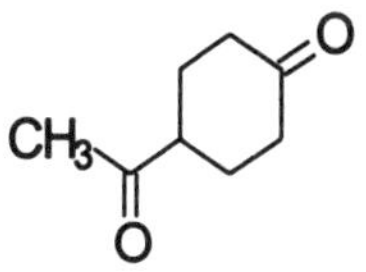 (d)

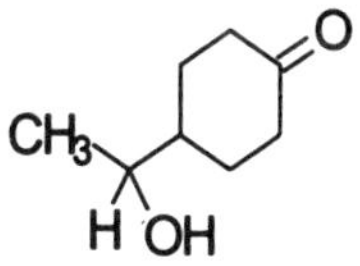

36 Which of the following is a ketopentose?

(a) 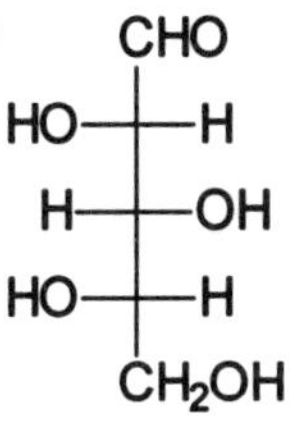
```
      CHO
HO —— H
 H —— OH
HO —— H
     CH₂OH
```

(b) 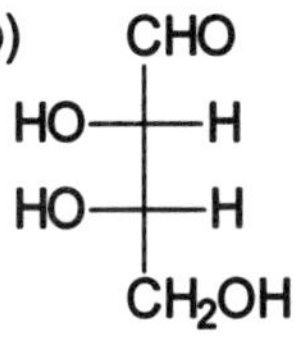
```
      CHO
HO —— H
HO —— H
     CH₂OH
```

(c) 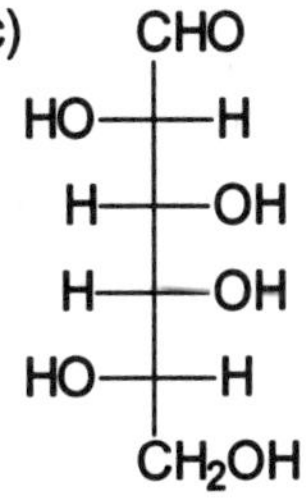
```
      CHO
HO —— H
 H —— OH
 H —— OH
HO —— H
     CH₂OH
```

(d) 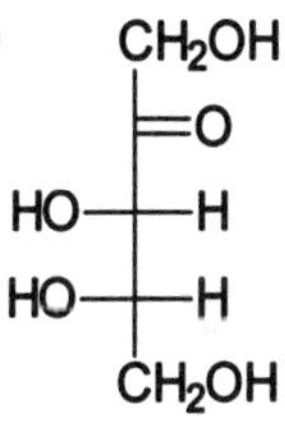
```
     CH₂OH
      ==O
HO —— H
HO —— H
     CH₂OH
```

(e) 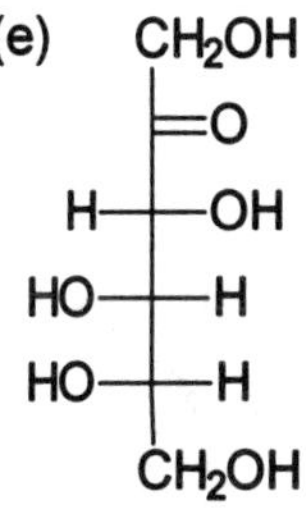
```
     CH₂OH
      ==O
 H —— OH
HO —— H
HO —— H
     CH₂OH
```

37 Which of the following are optically inactive?

(a) 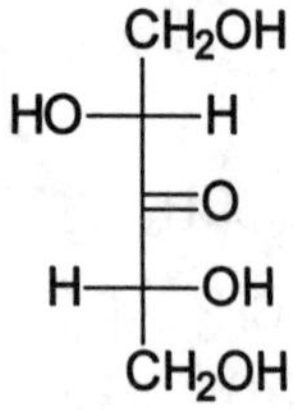
```
      CH₂OH
  HO——H
     ══O
   H——OH
      CH₂OH
```

(b) 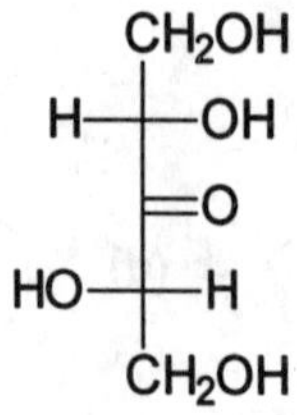
```
      CH₂OH
   H——OH
     ══O
  HO——H
      CH₂OH
```

(c) 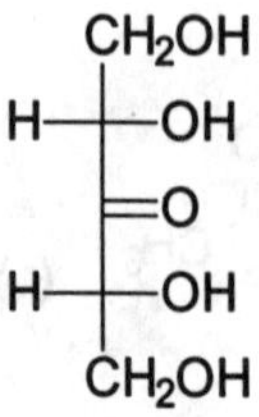
```
      CH₂OH
   H——OH
     ══O
   H——OH
      CH₂OH
```

38 Which one of the following gives an optically inactive alditol upon reduction with sodium borohydride?

(a) 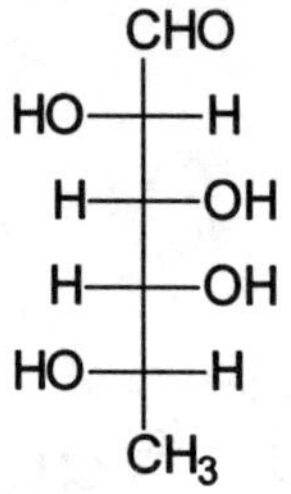
```
       CHO
  HO——H
   H——OH
   H——OH
  HO——H
       CH₃
```

(b) 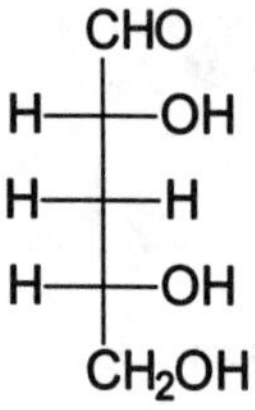
```
       CHO
   H——OH
   H——H
   H——OH
       CH₂OH
```

(c) 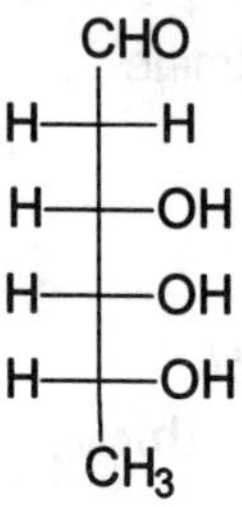
```
       CHO
   H——H
   H——OH
   H——OH
   H——OH
       CH₃
```

(d) 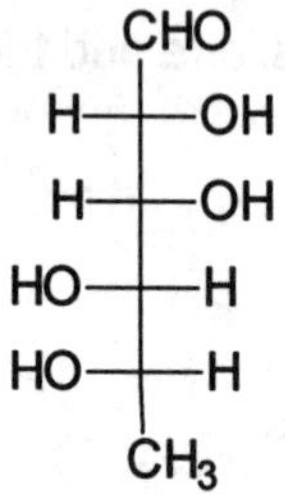
```
       CHO
   H——OH
   H——OH
  HO——H
  HO——H
       CH₃
```

39 Which of the following represents D-glucose in its pyranose form?

(a) 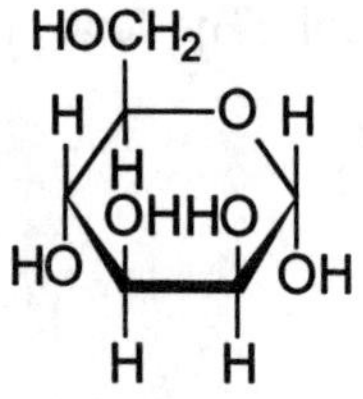
```
        HOCH₂
      H /——O\ H
       H|
      \ OHHO /
   HO \       / OH
       |      |
       H      H
```

(b) 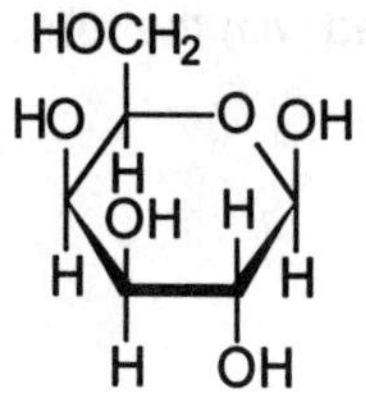
```
        HOCH₂
     HO /——O\ OH
        H|
      \ OH  H /
    H \       / H
       |      |
       H      OH
```

(c) 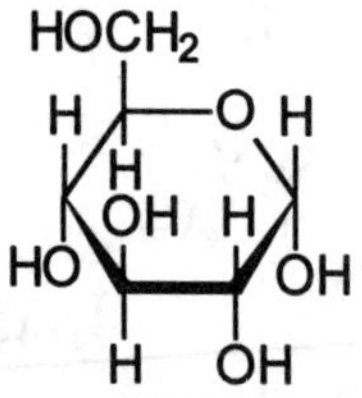
```
        HOCH₂
      H /——O\ H
        H|
      \ OH  H /
   HO \       / OH
       |      |
       H      OH
```

(d) 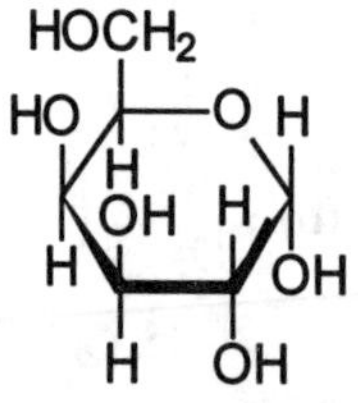
```
        HOCH₂
     HO /——O\ H
        H|
      \ OH  H /
    H \       / OH
       |      |
       H      OH
```

40 Which of the following represents fructose?

(a) 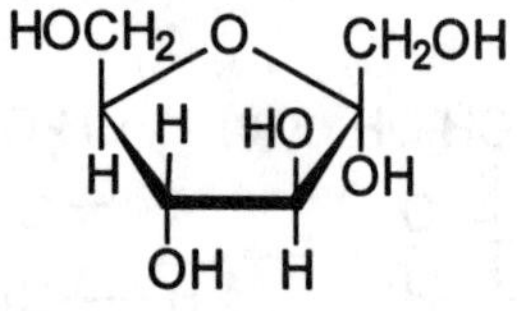
```
   HOCH₂  O   CH₂OH
     \   / \  /
      H   HO
     H|       |OH
      \       /
       OH    H
```

(b) 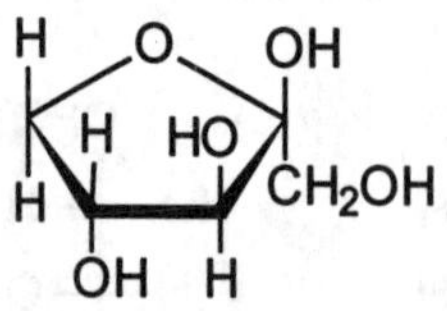
```
    H    O    OH
     \  / \  /
      H   HO
     H|       |CH₂OH
      \       /
       OH    H
```

(c) 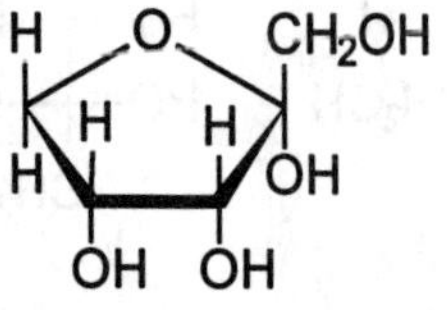
```
    H    O    CH₂OH
     \  / \  /
      H    H
     H|       |OH
      \       /
       OH   OH
```

(d) 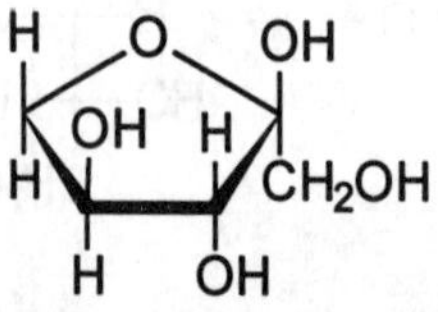
```
    H    O    OH
     \  / \  /
      OH   H
     H|       |CH₂OH
      \       /
       H    OH
```

41 The following structures A and B represent ___ and ___ respectively.

(a) glucose and mannose
(b) mannose and galactose
(c) galactose and mannose
(d) mannose and glucose
(e) galactose and glucose

A

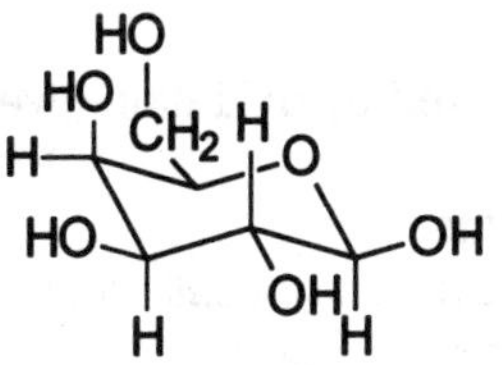

B

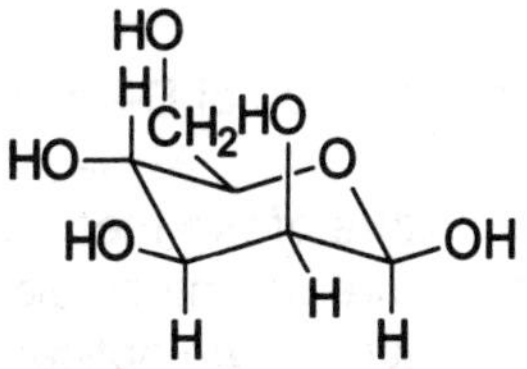

42 Which of the following compounds is a meso compound?

(a)
CO_2H
HO——H
HO——H
H——OH
H——OH
CO_2H

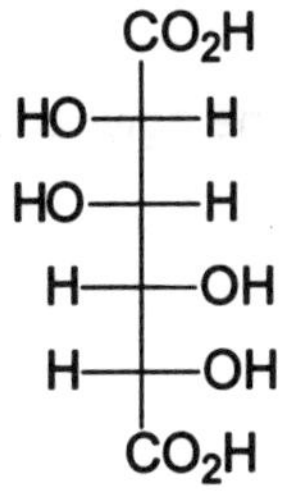

(b)
CO_2H
H——OH
HO——H
HO——H
H——OH
CO_2H

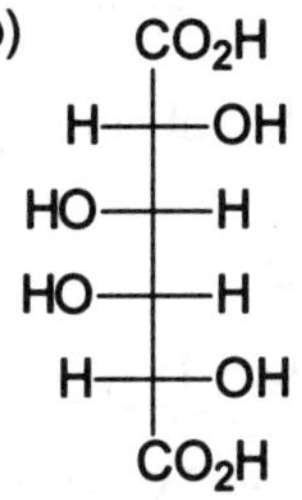

(c)
CO_2H
H——H
H——OH
H——OH
CO_2H

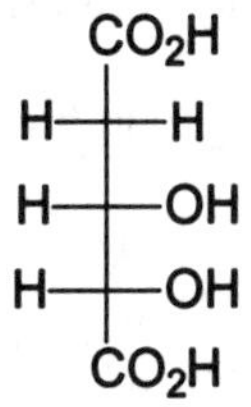

(d)
CO_2H
HO——H
H——OH
H——OH
CO_2H

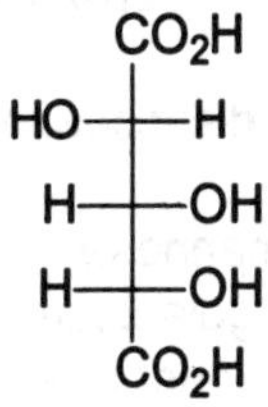

43 Which of the following is a reducing sugar?

(a)

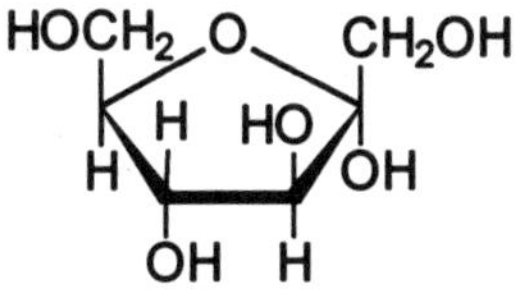

(b)

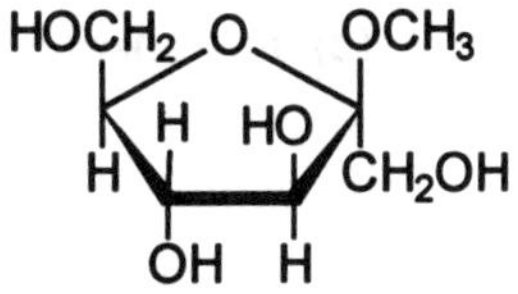

(c)

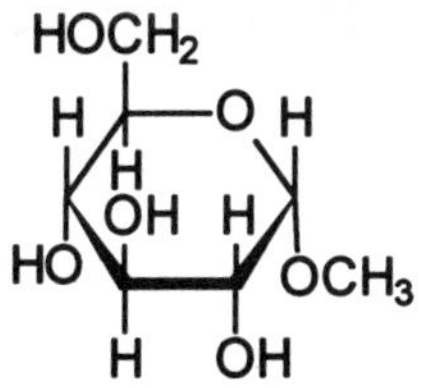

(d)

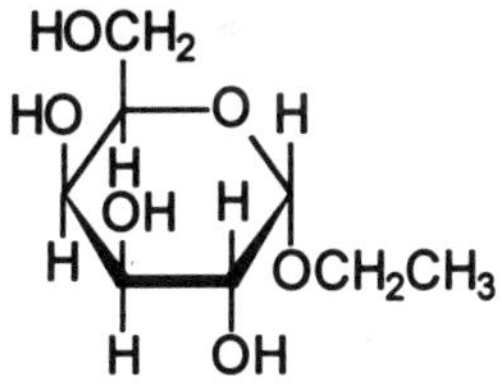

44 The following compound is an ___ methylglycoside of ____.

(a) α ; arabinose
(b) β ; ribose
(c) α ; xylose
(d) α ; ribose
(e) β ; arabinose

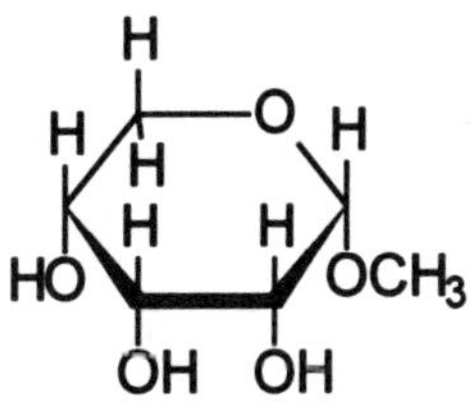

45 Which aldohexose gives the same osazone as galactose?

(a) mannose (b) allose (c) gulose (d) talose (e) glucose

46 Chain extension of D-arabinose followed by oxidation gives ___.

(a) a mixture of two meso aldaric acids
(b) a mixture of a meso aldaric acid and an optically active aldaric acid.
(c) a mixture of two optically active aldaric acids.

47 Lactose is a disaccharide in which ____ is present as an acetal and the configuration at that center is ___.

(a) glucose ; α (b) glucose ; β (c) galactose ; α (d) galactose ; β

48 The following disaccharide has ____ in the form of an acetal and ___ as the aglycone.

(a) glucose; mannose
(b) mannose; galactose
(c) galactose; mannose
(d) mannose; glucose
(e) galactose; glucose

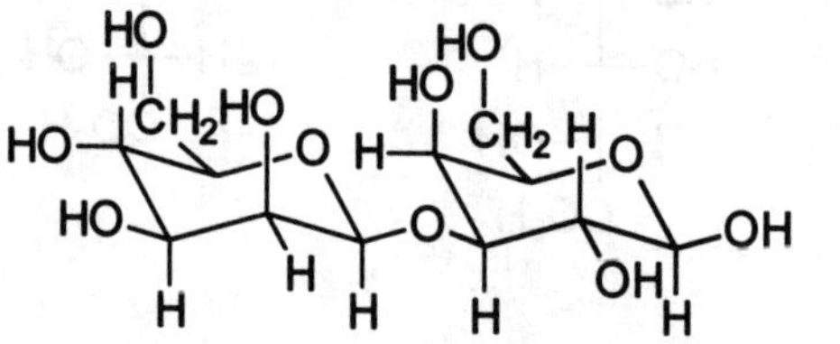

49 Which of the following has the largest degree of branching and shortest chains?

(a) cellulose (b) amylose (c) amylopectin (d) glycogen

Answers to Examination 5

bcabb ededa dcbee cbdac aacbc edded ebabd dcbca cbadd

cdbd

Examination 6

1   Which of the following has the smallest pK_a?

(a) 2-chlorobutanoic acid   (b) 3-bromobutanoic acid   (c) 2-bromobutanoic acid

2   The pK_a of 3-butenoic acid is smaller than the pK_a of butanoic acid because

(a) the double bond provides resonance stabilization
(b) the sp^2 hybridized carbon atom of the double bond is electron withdrawing
(c) the unsaturated acid is more strongly hydrated
(d) the unsaturated acid forms an intramolecular hydrogen bond
(e) the unsaturated acid has a higher dipole moment

3   Arranged in order of decreasing K_a (largest first), the correct order for the following compounds is ___.

(a) A > C > B
(b) C > B > A
(c) B > C > A
(d) A > B > C
(e) C > A > B

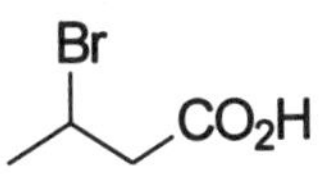

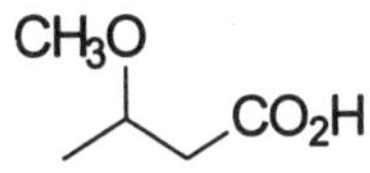

4   Which sequence of reactions is the best way to accomplish the following transformation?

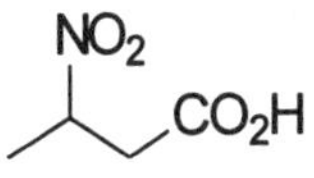

(a) Mg, ether ; CO_2 ; H_3O^+
(b) NaCN, DMF ; OH^- , heat
(c) $NaOCH_3$, CH_3OH ; CO_2 ; H_3O^+
(d) Mg, ether ; HCHO ; H_3O^+
(e) NaH ; HCO_2H

5   Which of the listed conditions can be used for the following transformation?

(a) $LiAlH_4$; Jones reagent
(b) $NaBH_4$; PCC in CH_2Cl_2
(c) $SOCl_2$; $LiAlH[OC(CH_3)_3]_3$
(d) $SOCl_2$; $LiAlH_4$
(e) BH_3 ; Jones reagent

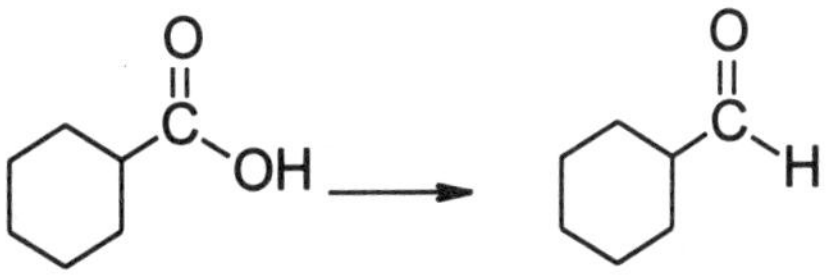

6   Which of the listed conditions can be used to prepare the following compound?

(a) acetophenone + phenyl Grignard ; H_3O^+
(b) methyl acetate + phenyl Grignard (2 equiv) ; H_3O^+
(c) ethyl formate + phenyl Grignard (2 equiv) ; H_3O^+
(d) both a and b
(e) both b and c

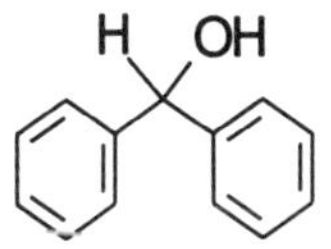

7   The order of decreasing reactivity (most reactive first) with a nucleophile in a nucleophilic acyl substitution reaction is ___.

(a) A > C > B
(b) A > B > C
(c) B > C > A
(d) C > A > B
(e) C > B > A

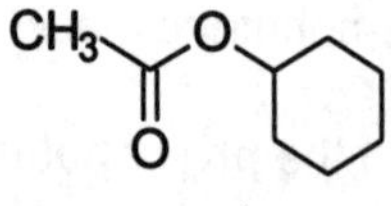

A 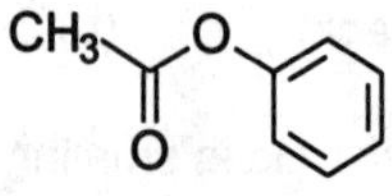   B 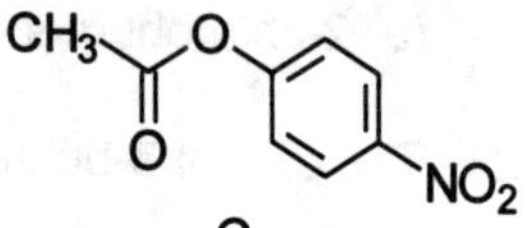   C

8   The order of decreasing reactivity (most reactive first) with a nucleophile in a nucleophilic acyl substitution reaction is ___.

(a) A > D > B > C
(b) D > B > A > C
(c) B > C > D > A
(d) D > A > C > B
(e) B > D > C > A

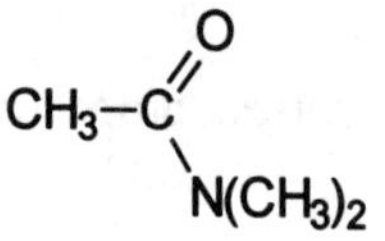

A 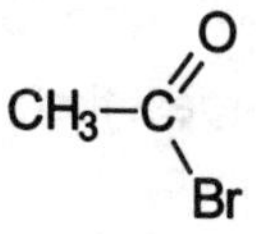   B 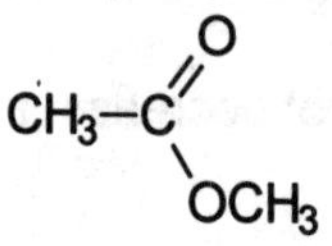   C 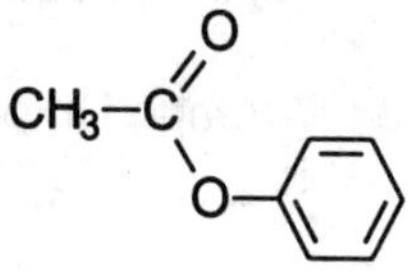   D

9   Based on inductive effects on the stability of a carbonyl group, acetyl fluoride should be ____ reactive than methyl acetate in nucleophilic substitution reactions. Based on resonance effects on the stability of a carbonyl group, acetyl fluoride should be ____ reactive than methyl acetate in nucleophilic substitution reactions.

(a) more; more (b) more; less   (c) less; more   (d) less; less

10 The configuration of the following compound is ___. Reaction of its conjugate base with bromoethane will give an ester with the ___ configuration.

(a) R ; S
(b) R ; R
(c) S ; R
(d) S ; S

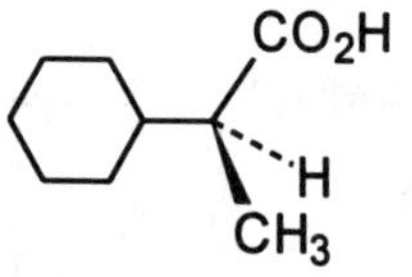

11 Reaction of the following compound containing O^{18} at the indicated atom with acetic acid will give an ester with the ___ configuration that ___ contain O^{18}.

(a) R ; does
(b) R ; does not
(c) S ; does
(d) S ; does not

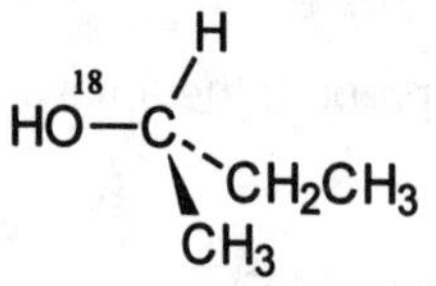

12 Which of the following combinations of reagents gives a reaction with the most negative $\Delta G°$?

(a) acetic acid and ethanol with H^+
(b) acetamide and ethanol with H^+
(c) acetyl chloride and ethanol with pyridine
(d) acetic anhydride and ethanol with sodium hydroxide
(e) all of the above have the same $\Delta G°$

13 Reaction of methyl chloroformate with one equivalent of methylamine is expected to yield ___.

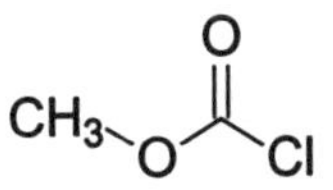

methyl chloroformate

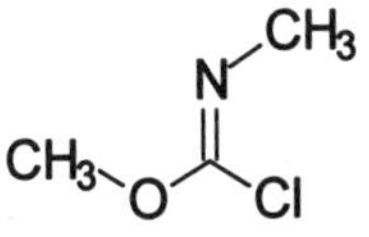

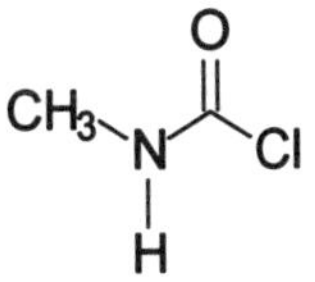

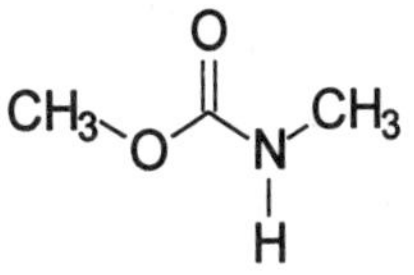

14 Which one of the following compounds is the strongest acid.

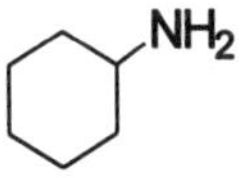

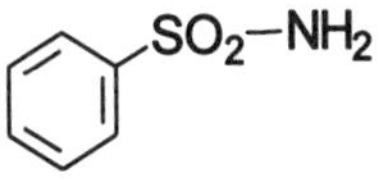

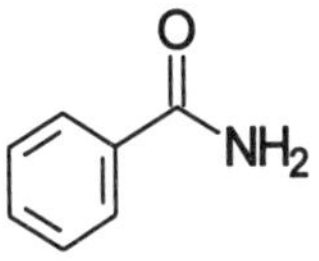

15 Which of the following compounds cannot be used to obtain a monobromo compound by the standard Hunsdieker reaction.

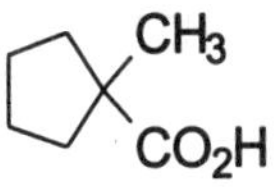

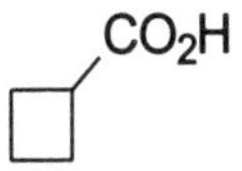

16 Which of the following reagents is used for the indicated reaction?

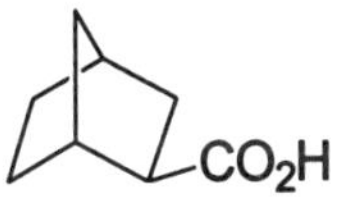

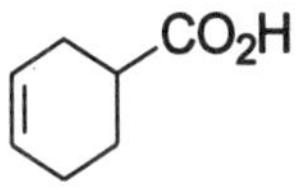

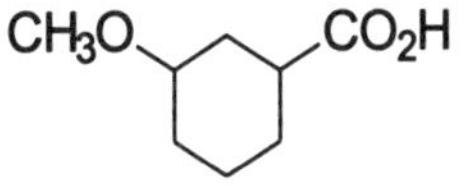

(a) LiAlH[OC(CH₃)₃]₃
(b) DIBAL
(c) NaBH₃CN
(d) LiAlH₄
(e) NaBH₄

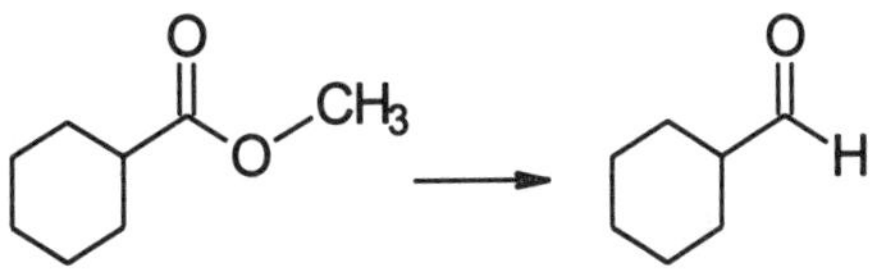

17 Which of the following reagents is used for the indicated reaction?

(a) LiAlH[OC(CH₃)₃]₃
(b) DIBAL
(c) NaBH₃CN
(d) LiAlH₄
(e) H₂ Pt

18 Which of the following reagents is used for the indicated reaction?

(a) LiAlH[OC(CH₃)₃]₃
(b) BH₃
(c) NaBH₃CN
(d) LiAlH₄
(e) H₂ Pt

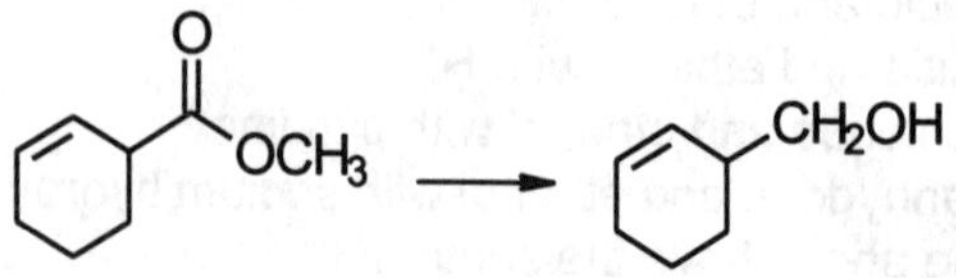

19 Which of the following reagents is used for the indicated reaction?

(a) LiAlH[OC(CH₃)₃]₃
(b) BH₃
(c) NaBH₃CN
(d) LiAlH₄
(e) NaBH₄

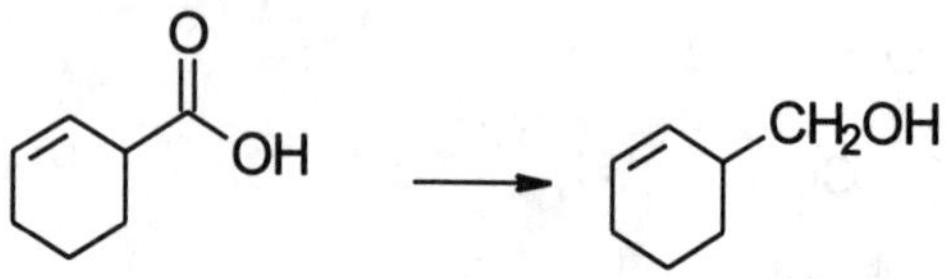

20 Which of the following reagents is used for the indicated reaction?

(a) H₂SO₄
(b) BH₃
(c) H₂ Pt
(d) LiAlH₄
(e) SOCl₂

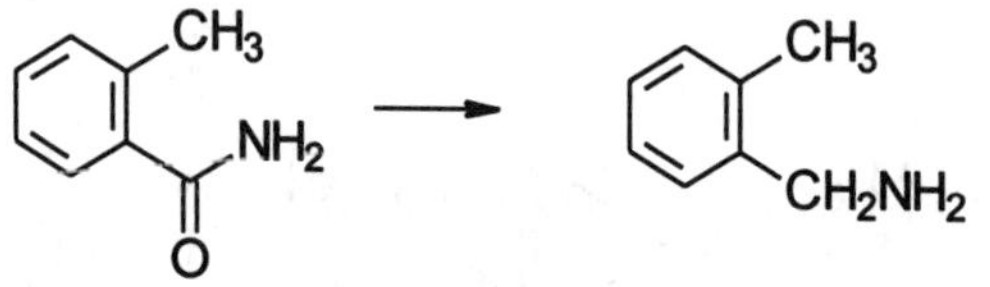

21 Which of the following reagents is used for the indicated reaction?

(a) H₂SO₄
(b) BH₃
(c) SOCl₂
(d) LiAlH₄
(e) H₂ Pt

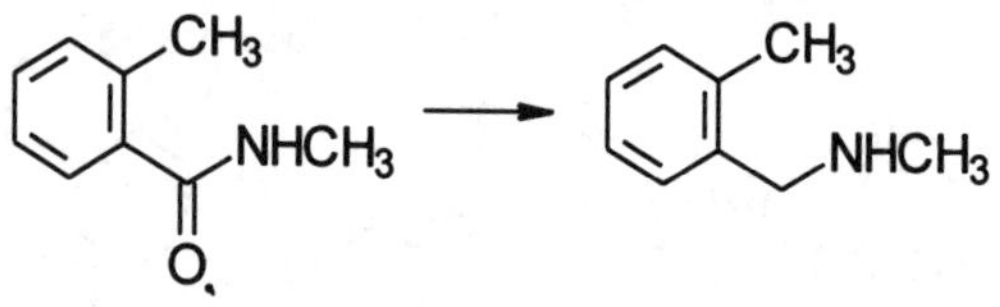

22 Which compound is obtained by reaction of ethyl cyclohexanecarboxylate with one equivalent of DIBAL at -78°C.

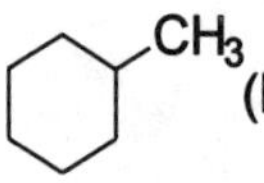

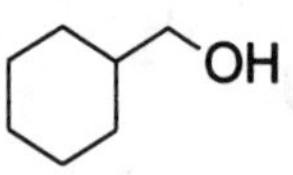

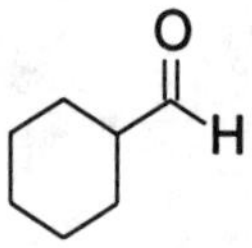

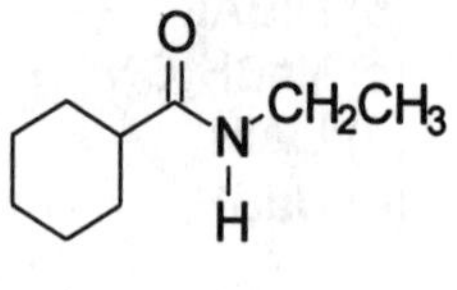

23 Which of the listed conditions can be used for the following transformation?

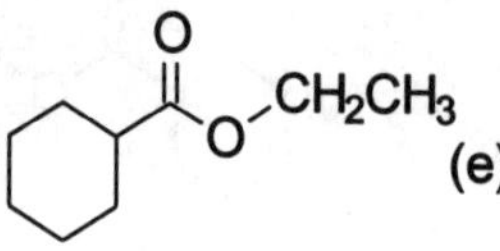

(a) SOCl₂ ; CH₃ CH₂MgBr (1 equiv) ; H₃O⁺
(b) BH₃ ; PCC in CH₂Cl₂ ; CH₃CH₂MgBr ; H₃O⁺
(c) CH₃ CH₂MgBr (1 equiv) ; H₃O⁺
(d) LiAlH₄ ; SOCl₂ ; (CH₃CH₂)₂CuLi
(e) SOCl₂ ; (CH₃ CH₂)₂CuLi

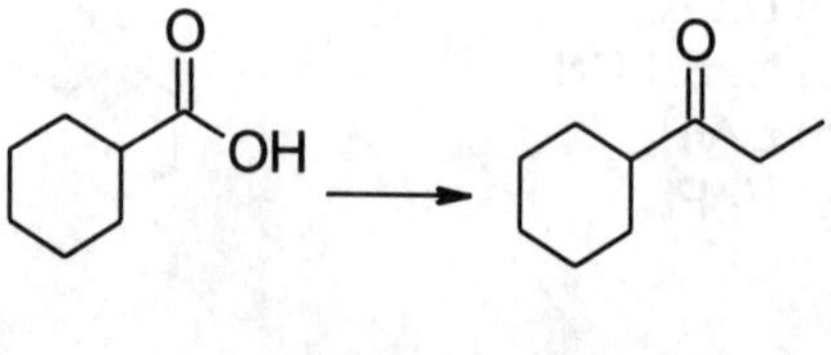

24 Which sequence of reactions is the best way to accomplish the following transformation?

(a) (CH₃)₂CuLi, ether ; HCHO ; H₃O⁺
(b) NaCN, DMF ; H₃O⁺, heat
(c) NaOCH₃, CH₃OH ; CO₂ ; H₃O⁺
(d) Mg, ether ; HCHO ; H₃O⁺
(e) Na₂CO₃

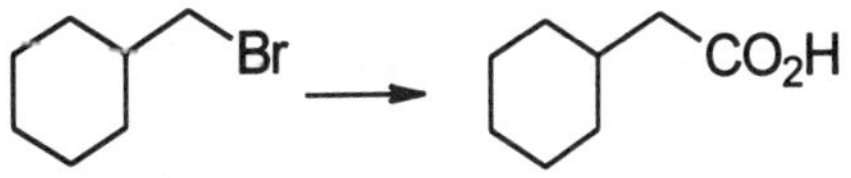

25 The number of combinations of aldehyde, ketone, or ester and a required Grignard reagent that can be used to synthesize the tertiary alcohols 1-methylcyclopentanol, 2-methyl-2-pentanol, and 3-ethyl-3-pentanol are _____, respectively.

(a) 3, 3, and 3   (b) 2, 2, and 3   (c) 1, 2, and 3   (d) 1, 2, and 2   (e) 1, 3, and 2

26 Which of the listed conditions can be used to prepare the following compound?

(a) acetophenone + methyl Grignard ; H₃O⁺
(b) methyl benzoate + methyl Grignard (2 equiv) ; H₃O⁺
(c) acetone + phenyl Grignard (1 equiv) ; H₃O⁺
(d) both a and b will give the product
(e) a, b, and c will all give the product

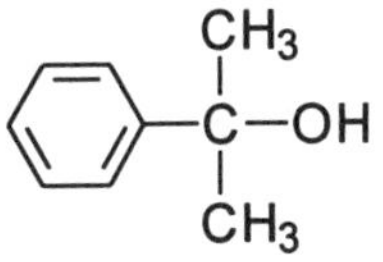

27 Which of the listed conditions can be used to prepare the following compound?

(a) acetophenone + phenyl Grignard ; H₃O⁺
(b) methyl acetate + phenyl Grignard (2 equiv) ; H₃O⁺
(c) ethyl formate + phenyl Grignard (2 equiv) ; H₃O⁺
(d) both a and b
(e) both b and c

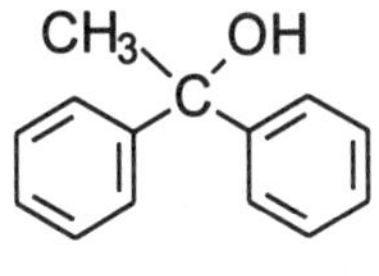

28 Which of the following has the highest equilibrium concentration of its enol tautomer?

(a) 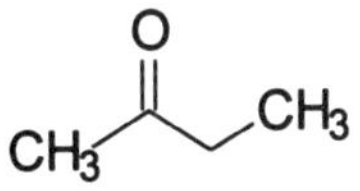 CH₃   CH₃   (b) 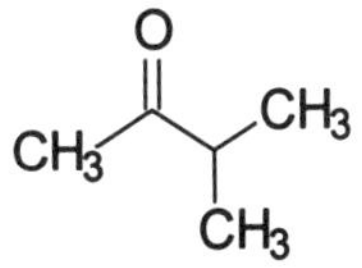 CH₃   CH₃   (c) 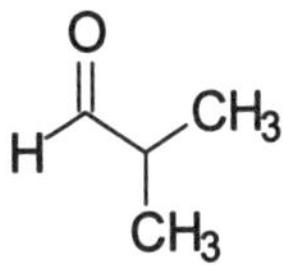 H   CH₃   (d) 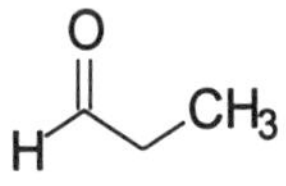 H   CH₃

29 The number of enolizable hydrogen atoms that can be replace by deuterium in a base-catalyzed reaction for compounds A, B, and C are ____, respectively.

(a) 3, 2, and 1
(b) 3, 1, and 1
(c) 4, 2, and 3
(d) 4, 1, and 3
(e) 5, 2, and 3

A CH₃   B H   C CH₃

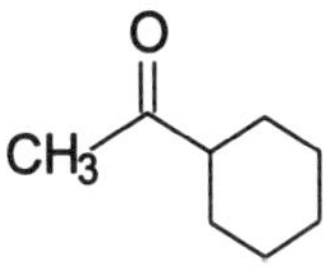

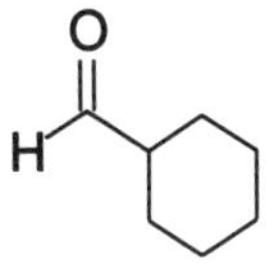

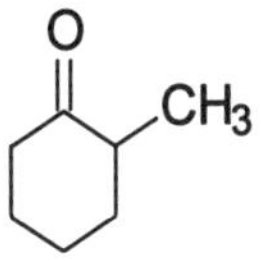

30 Based on the mechanism for the acid-catalyzed halogenation of acetone, bromination should occur at _____ chlorination.

(a) a faster rate   (b) a slower rate   (c) the same rate as

31 Which of the following compound will give iodoform in a reaction with iodine in basic solution?

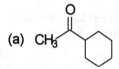

(a)   (b) 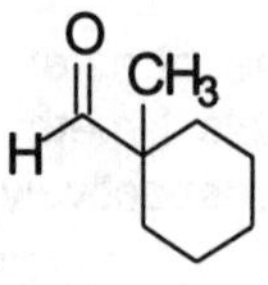   (c) 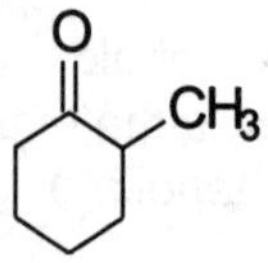   (d)

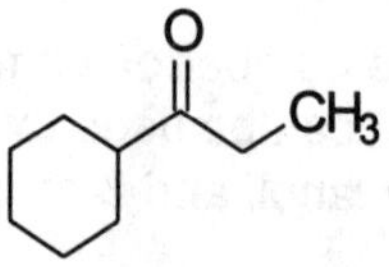

32 The following compound can be synthesized by an intramolecular aldol condensation of ___.

(a) 5-oxoheptanal
(b) 2,6-heptanedione
(c) 2,5-hetpanedione
(d) 4-oxo-5-methylhexanal

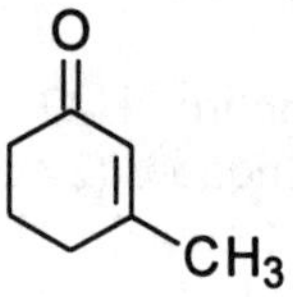

33 Which of the following β dicarbonyl compounds has the largest pK_a?

(a) acetylacetone   (b) ethyl acetoacetate   (c) diethyl malonate

34 The following compound can be synthesized by a _____.

(a) Dieckman condensation
(b) Hell-Volhard-Zelinsky reaction
(c) Knoevenagel condensation
(d) Reformatskii reaction

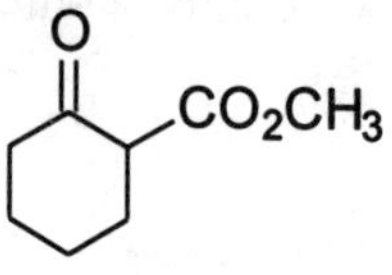

35 The following triglyceride contains _____ acids.

(a) stearic, oleic, and linoleic
(b) palmitic, oleic, and linoleic
(c) stearic, oleic, and linolenic
(d) palmitic, oleic, and linolenic
(e) palmitic, linoleic, and linolenic

$$CH_2-O-\overset{O}{\overset{\|}{C}}-(CH_2)_7CH=CH(CH_2)_7CH_3$$
$$CH-O-\overset{O}{\overset{\|}{C}}-(CH_2)_{14}CH_3$$
$$CH_2-O-\overset{O}{\overset{\|}{C}}-(CH_2)_6(CH_2CH=CH)_2(CH_2)_4CH_3$$

Answers to Examination 6

abbbc   ceead   cccbc   baddd   dcebe   edcdc   abcab

Examination 7

1   Which of the following compounds is the strongest base?

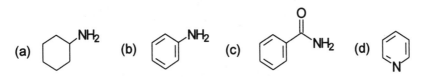

(a)   (b)   (c)   (d)

2   The correct order for the following compounds arranged by increasing pK_b is ____.

(a) A < B < C
(b) A < C < B
(c) B < C < A
(d) C < B < A

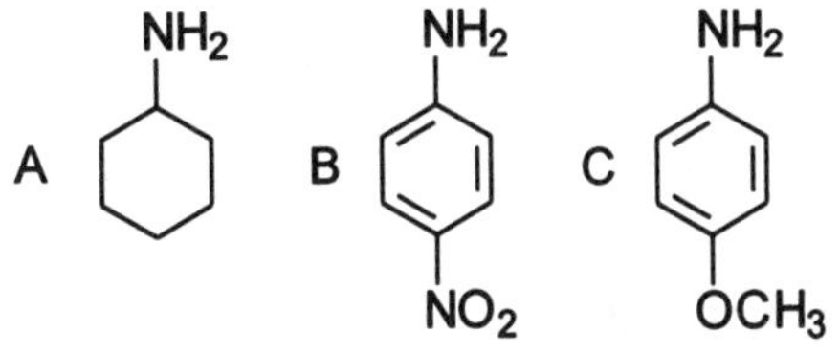

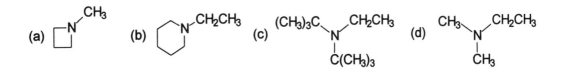

3   Which of the following compounds is the least effective nucleophile?

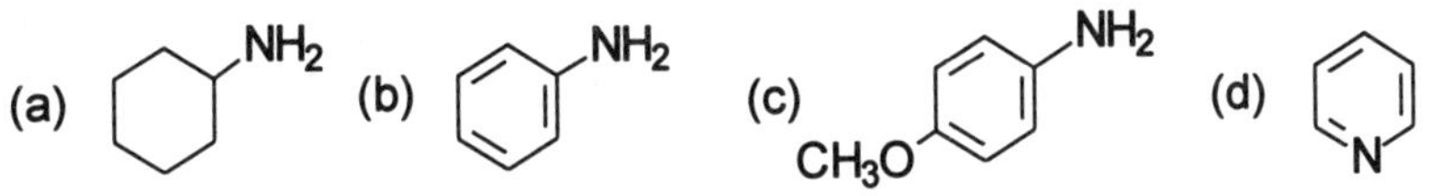

(a)   (b)   (c)   (d)

4   Which of the following compounds has the largest pK_b ?

(a) R_3N (b) $R_2C=NH$ (c) $RC\equiv N$

5   Which of the following amines inverts at the slowest rate?

(a)   (b)   (c)   (d)

6   Which of the following combinations of reagents after appropriate workup conditions could be used to prepare the following compound?

(a) benzonitrile and CH_3CH_2MgBr
(b) benzoyl chloride and $(CH_3CH_2)_2CuLi$
(c) benzene, propanoyl chloride and $AlCl_3$
(d) both a and b
(e) a, b, and c

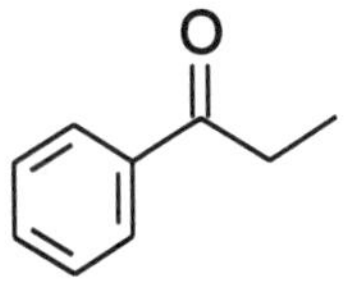

7   Reaction of benzylamine with formaldehyde and $NaBH_3CN$ yields ___.

(a) 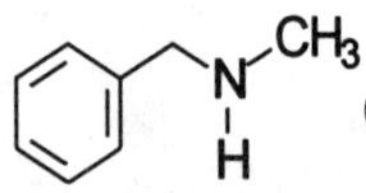 (b) 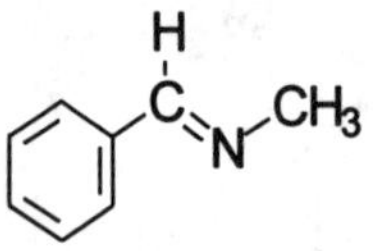 (c) 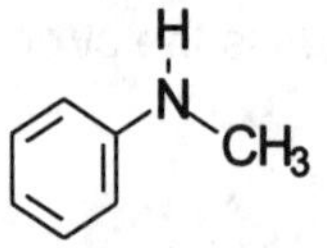 (d) 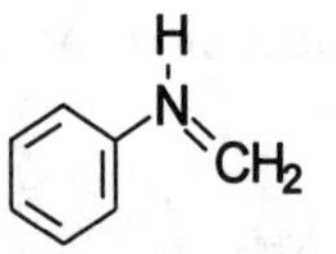 (e)

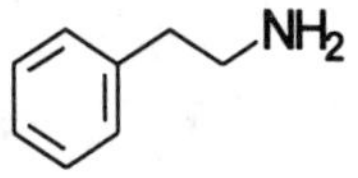

8   Which of the following amines can be made by a Gabriel synthesis?

(a) 2-octanamine
(b) 2-methyl-1-octanamine
(c) N-methyl-2-octanamine
(d) both a and b
(e) a, b, and c

9   A tertiary amine can be made by reduction of an appropriate amide. Amides in turn can be made by reaction of an acyl chloride and an amine. How many combinations of these two reactants can be used to prepare A and B.

(a) three; two
(b) three; three
(c) two; two
(d) one; one
(e) one; two

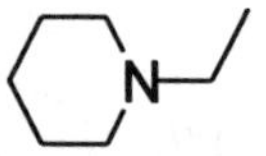

A 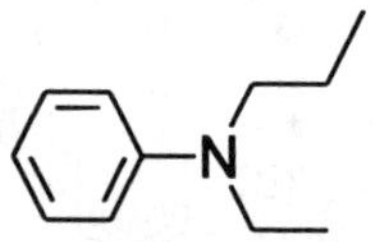   B

10   Reaction of the epoxide of 1-methylcyclohexene with azide followed by reaction with $LiAlH_4$ gives ___.

(a) 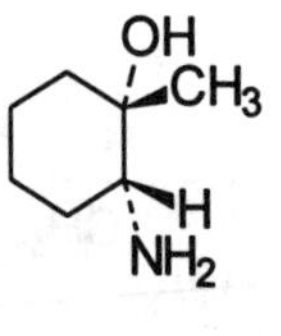 (b) 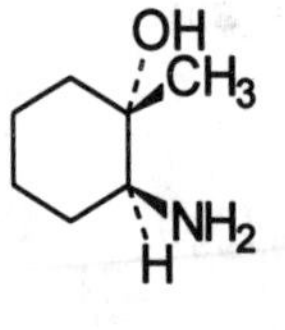 (c) 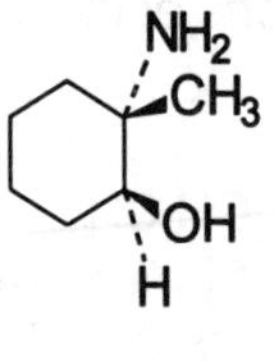 (d)

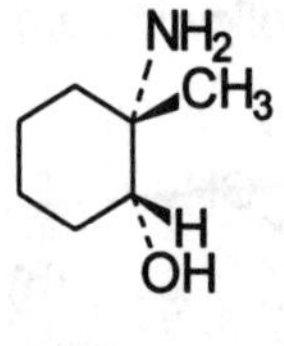

11   Which of the following reagents is used for the indicated reaction?

(a) HgO, Br_2
(b) NaN_3
(c) $SOCl_2$
(d) $LiAlH_4$
(e) Cl_2, OH^-

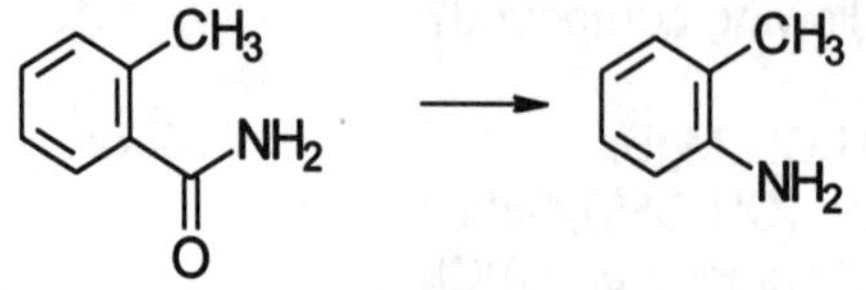

12 The configuration of the following compound is ___. Reaction of this compound with basic aqueous solution containing chlorine will give an amine with the ___ configuration.

(a) R ; R
(b) R ; S
(c) S ; S
(d) S ; R

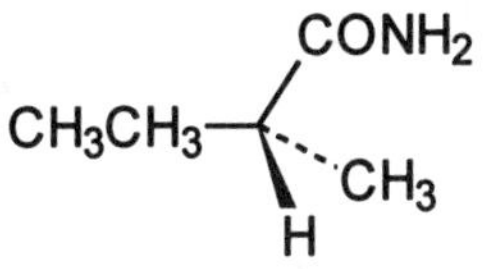

13 Heating the quaternary ammonium hydroxide derivative of the following amine yields ___.

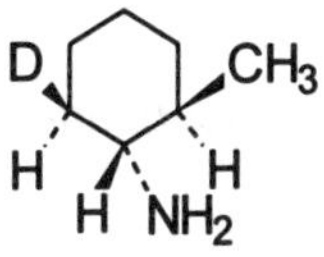

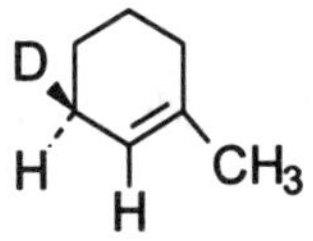

(a) 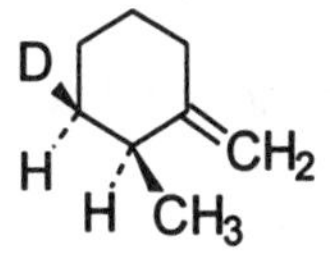   (b) 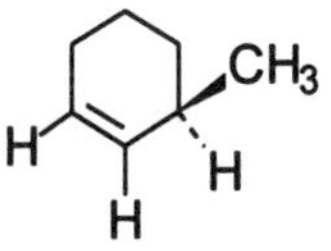   (c) 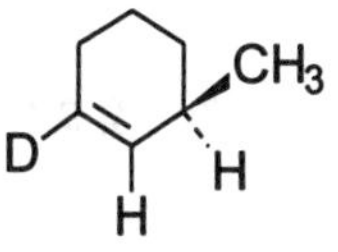   (d) 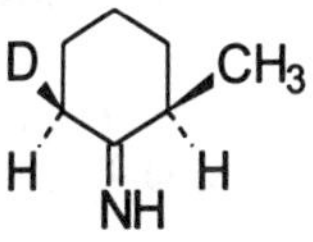   (e)

14 Which of the following compounds reacts with nitrous acid to yield a nitrosoamine?

(a) 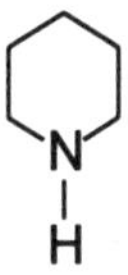   (b) 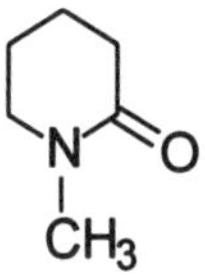   (c) 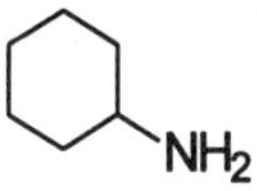   (d)

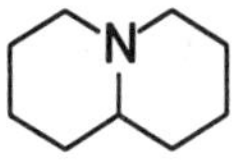

15 p-Methylaniline is treated with the following sequence of reagents listed on the left. The final product is ___.

1) acetyl chloride
2) HNO_3, H_2SO_4
3) OH^-, heat
4) $NaNO_2$, H_2SO_4
5) H_3PO_2

(a) 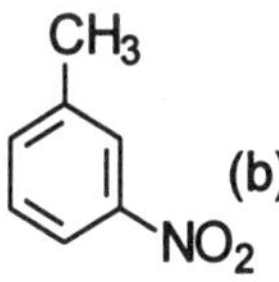   (b) 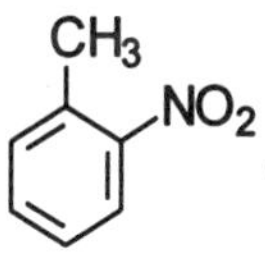   (c) 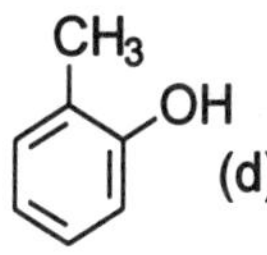   (d)

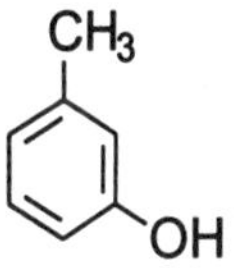

16 p-Nitrotoluene is treated with the following sequence of reagents listed on the left. The final product is ___.

1) Br_2, Fe
2) Fe/HCl
3) $NaNO_2$, H_2SO_4
4) CuCN
5) $LiAlH_4$

(a) 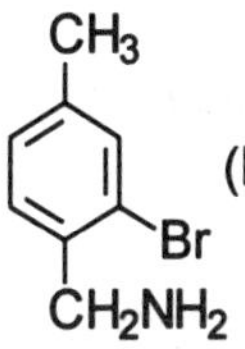   (b) 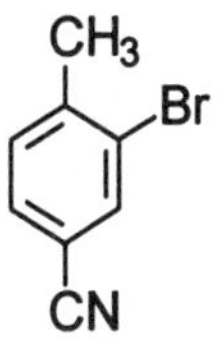   (c) 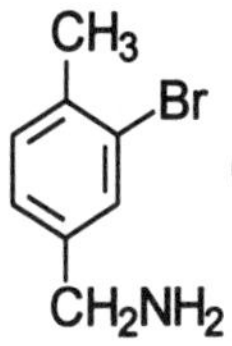   (d) 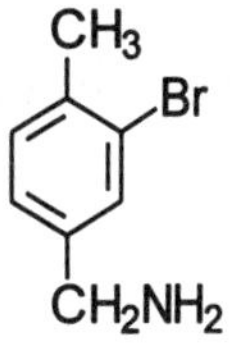   (e)

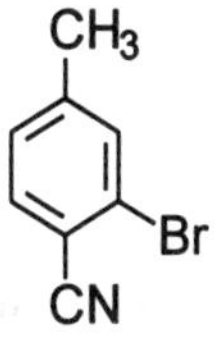

17 Which of the following is the most effective nucleophile?

(a) CH_3-O^- (b) CH_3-S^- (c) 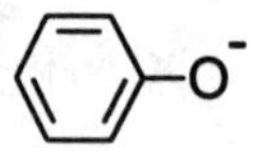 ⟨benzene⟩$-O^-$ (d) 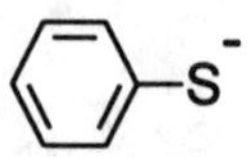 ⟨benzene⟩$-S^-$

18 Reaction of 4-deuteriochlorobenzene with $NaNH_2/NH_3$ gives ___ isomeric deuterium containing aniline(s). Reaction of 3-deuteriochlorobenzene with $NaNH_2/NH_3$ gives ___ isomeric deuterium containing anilines.

(a) two, two   (b) two, three   (c) three, three   (d) three, two   (e) four, two

19 Which of the following products results from the reaction of 2,6-dideuteriochlorobenzene with $NaNH_2/NH_3$?

(a) A
(b) B
(c) C
(d) A and B
(e) A, B, and C

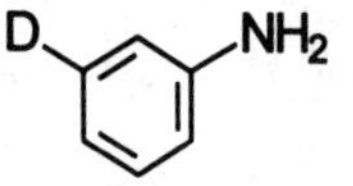

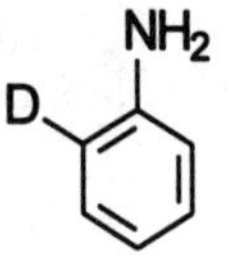

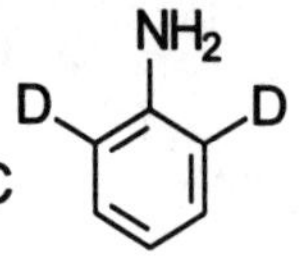

20 Which of the following is the most reactive with methoxide ion?

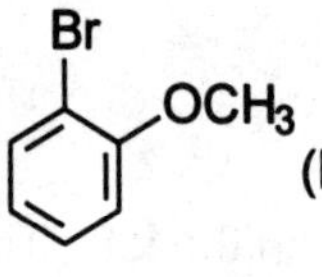

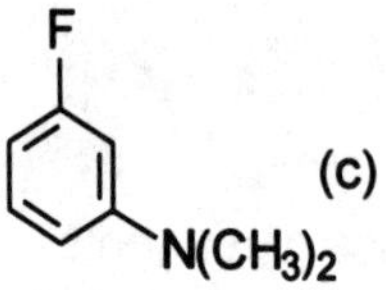

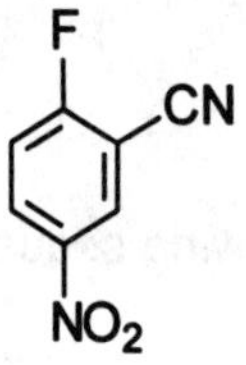

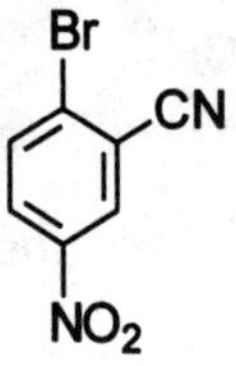

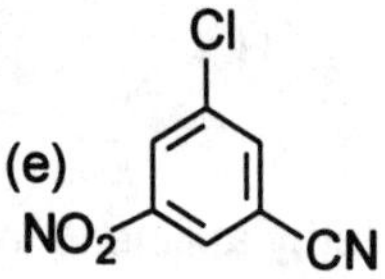

21 The correct order for the following compounds arranged by increasing pK_a is ___.

(a) A < B < C
(b) A < C < B
(c) B < C < A
(d) C < B < A

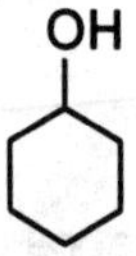

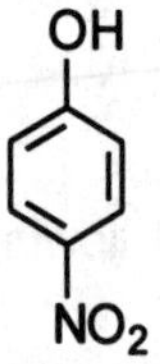

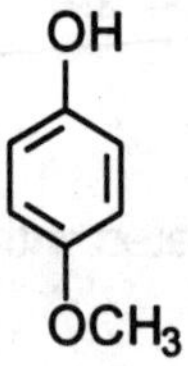

22 Reaction of phenol with $(CH_3O)_2SO_2$ and sodium hydroxide yields ___.

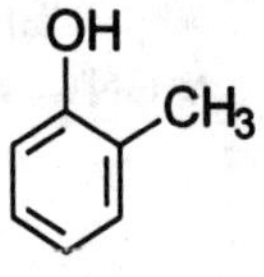

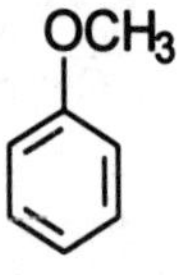

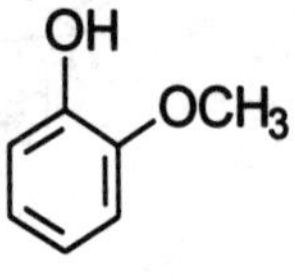

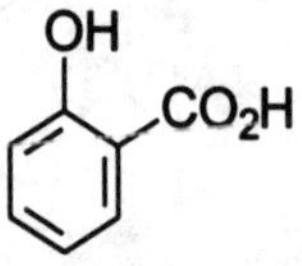

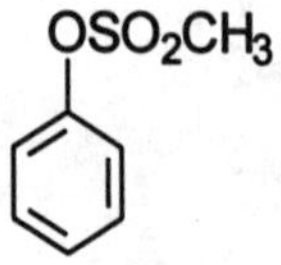

23 Reaction of phenol with carbon dioxide at high pressure yields ___.

(a) 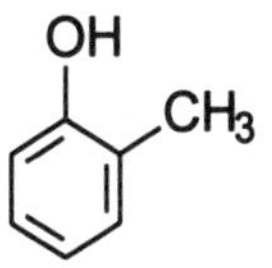 [structure: 2-methylphenol, OH with CH₃]

(b) 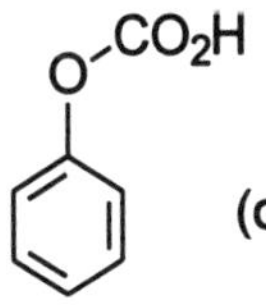 [structure: phenyl with O-CO₂H]

(c) 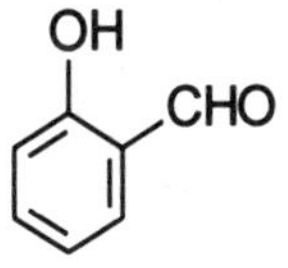 [structure: 2-hydroxybenzaldehyde, OH with CHO]

(d) 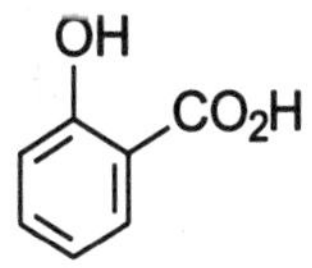 [structure: salicylic acid, OH with CO₂H]

(e) 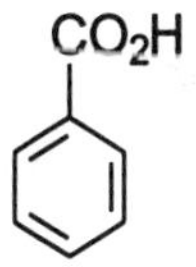 [structure: benzoic acid, CO₂H]

24 Which of the following two reactions is spontaneous?

(a) only A
(b) only B
(c) both A and B
(d) neither A nor B

A   [quinone with Cl + phenol with CH₃, OH ⇌ phenol with Cl, OH + quinone with CH₃]

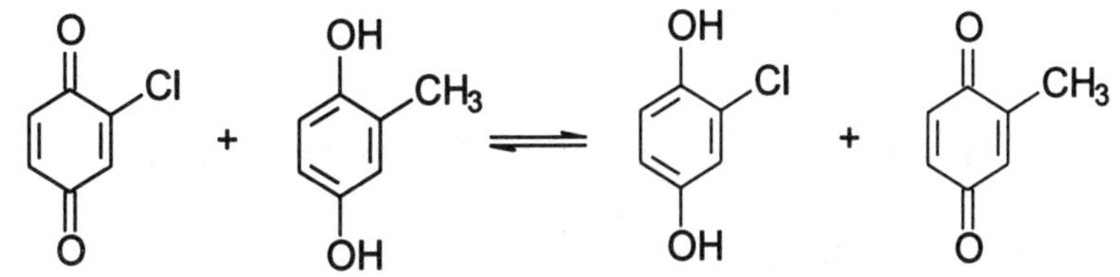

B   [quinone with OCH₃ + phenol with CH₃, OH ⇌ phenol with OCH₃, OH + quinone with CH₃]

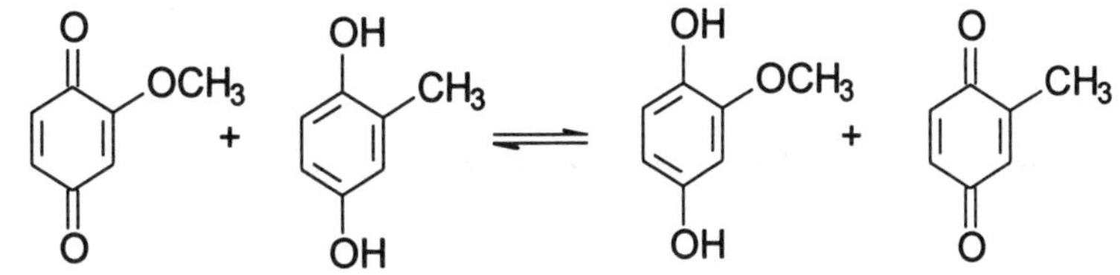

25 The name of the following compound is ___.

(a) L-aspartic acid
(b) D-aspartic acid
(c) L-glutamic acid
(d) D-glutamic acid

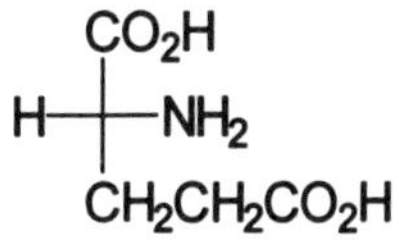

$$CO_2H$$
$$H{-}{-}NH_2$$
$$CH_2CH_2CO_2H$$

26 The name of the following compound is ___.

(a) D-tyrosine
(b) L-tyrosine
(c) D-tryptophan
(d) L-tryptophan

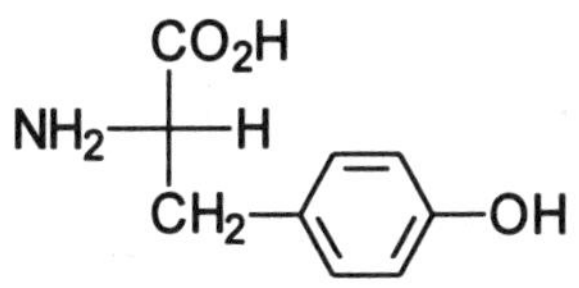

$$CO_2H$$
$$NH_2{-}{-}H$$
$$CH_2{-}\langle\text{ring}\rangle{-}OH$$

27 The name of the following compound is ___.

(a) D-cysteine
(b) L-cysteine
(c) D-methionine
(d) L-methionine

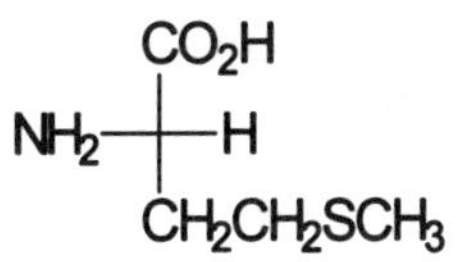

$$CO_2H$$
$$NH_2{-}{-}H$$
$$CH_2CH_2SCH_3$$

28 Which of the following is the zwitterion of serine?

(a) 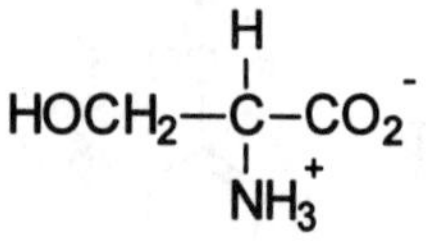 $HOCH_2-\overset{\overset{\displaystyle H}{|}}{\underset{\underset{\displaystyle NH_3^+}{|}}{C}}-CO_2^-$ (b) $(CH_3)_2CH-\overset{\overset{\displaystyle H}{|}}{\underset{\underset{\displaystyle NH_3^+}{|}}{C}}-CO_2^-$

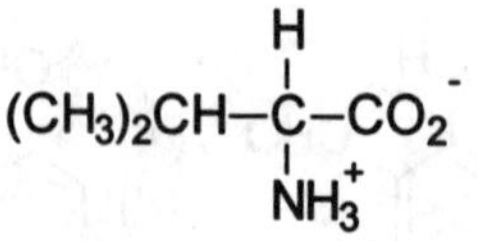

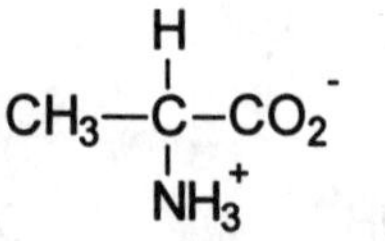

(c) $CH_3-\overset{\overset{\displaystyle H}{|}}{\underset{\underset{\displaystyle NH_3^+}{|}}{C}}-CO_2^-$ (d) 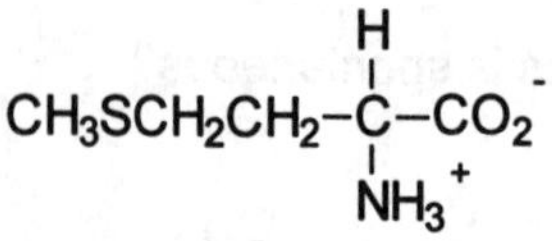 $CH_3SCH_2CH_2-\overset{\overset{\displaystyle H}{|}}{\underset{\underset{\displaystyle NH_3^+}{|}}{C}}-CO_2^-$

29 The following compound can be used to produce an amino acid using ___ .

(a) the Strecker synthesis
(b) reductive amination
(c) acetamidomalonate synthesis

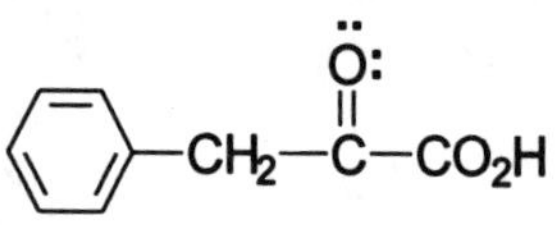

$\underset{}{\bigcirc}-CH_2-\overset{\overset{\displaystyle \ddot{O}:}{||}}{C}-CO_2H$

30 Which of the following compound is alanylserine?

(a) 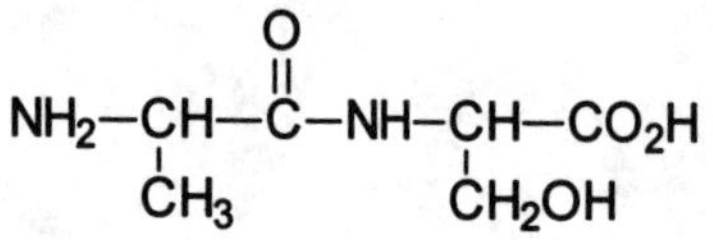 $NH_2-\underset{\underset{\displaystyle CH_3}{|}}{CH}-\overset{\overset{\displaystyle O}{||}}{C}-NH-\underset{\underset{\displaystyle CH_2OH}{|}}{CH}-CO_2H$ (b) $NH_2-\underset{\underset{\displaystyle CH_2OH}{|}}{CH}-\overset{\overset{\displaystyle O}{||}}{C}-NH-\underset{\underset{\displaystyle CH_2SH}{|}}{CH}-CO_2H$

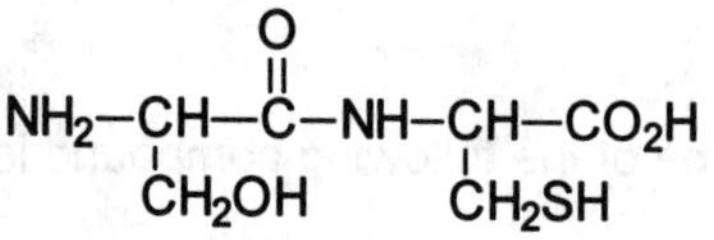

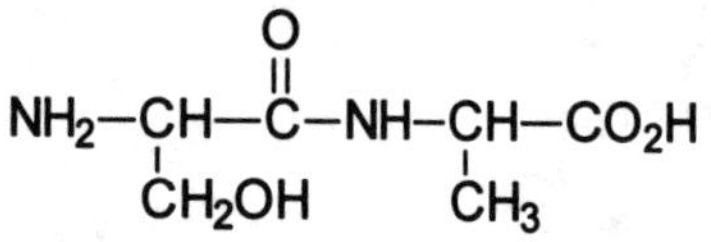

(c) $NH_2-\underset{\underset{\displaystyle CH_2OH}{|}}{CH}-\overset{\overset{\displaystyle O}{||}}{C}-NH-\underset{\underset{\displaystyle CH_3}{|}}{CH}-CO_2H$ (d) 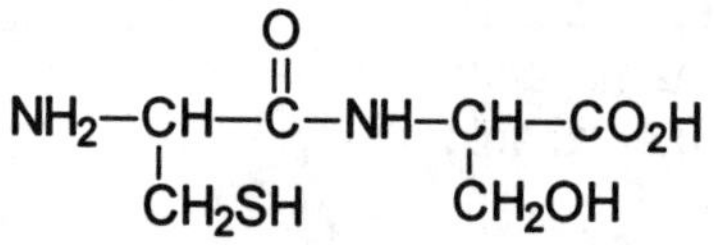 $NH_2-\underset{\underset{\displaystyle CH_2SH}{|}}{CH}-\overset{\overset{\displaystyle O}{||}}{C}-NH-\underset{\underset{\displaystyle CH_2OH}{|}}{CH}-CO_2H$

31 There are ___ isomeric tetrapeptides containing two units of alanine and two units of glycine.

(a) 24   (b) 18   (c) 12   (d) 8   (e) 6

32 The following compound is the phenylthiohydantoin of ____ .

(a) arginine
(b) methionine
(c) threonine
(d) lysine
(e) histidine

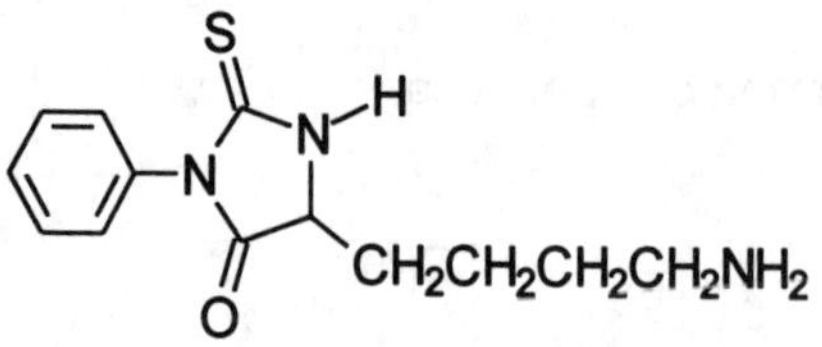

$CH_2CH_2CH_2CH_2NH_2$

33 The following is an example of _____.

(a) a [1,5] sigmatropic rearrangement
(b) a [3,3] sigmatropic rearrangement
(c) a [3+3] cycloaddition
(d) an electrocyclic reaction

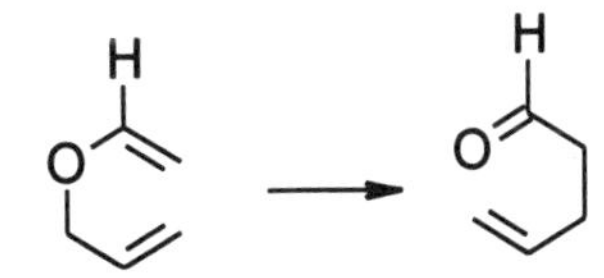

34 The following is an example of a _____.

(a) a [3,3] sigmatropic rearrangement
(b) a [5,5] sigmatropic rearrangement
(c) a [4+4] cycloaddition
(d) an electrocyclic reaction

35 The following is an example of a _____.

(a) a [2,2] sigmatropic rearrangement
(b) a [4,4] sigmatropic rearrangement
(c) a [2+2] cycloaddition
(d) an electrocyclic reaction

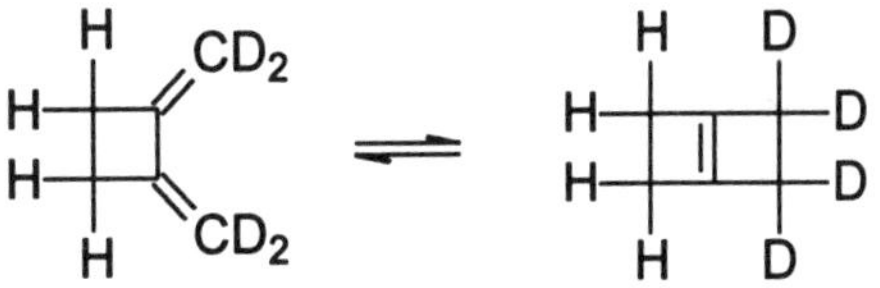

36 The following is an example of a _____.

(a) a [1,5] sigmatropic rearrangement
(b) a [1,3] sigmatropic rearrangement
(c) a [2+2] cycloaddition
(d) an electrocyclic reaction

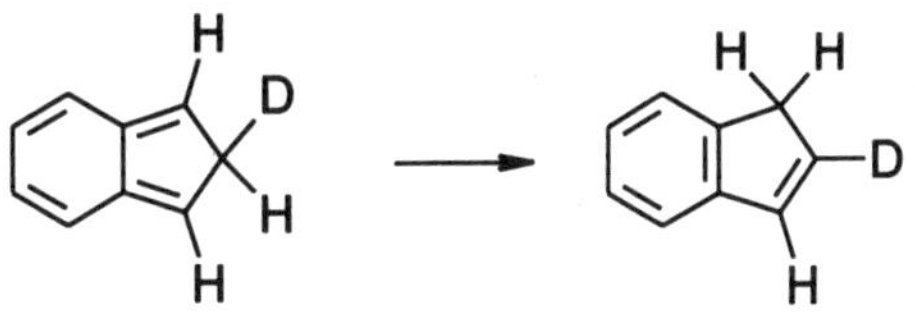

37 The frontier molecular orbital considered for a thermal electrocyclic reaction of a conjugated triene is ___; the reaction occurs via a ___ motion.

(a) π_3 ; conrotatory   (b) π_3 ; disrotatory   (c) π_4 ; conrotatory   (d) π_4 ; disrotatory

38 The frontier molecular orbital considered for a photochemical electrocyclic reaction of a conjugated diene is ___; the reaction occurs via a ___ motion.

(a) π_3 ; conrotatory   (b) π_3 ; disrotatory   (c) π_2 ; conrotatory   (d) π_2 ; disrotatory

39 The frontier molecular orbital considered for a photochemical electrocyclic reaction of a conjugated triene is ___; the molecular orbital is ___.

(a) π_3 ; symmetric   (b) π_3 ; antisymmetric   (c) π_4 ; symmetric   (d) π_4 ; antisymmetric

40 The frontier molecular orbital considered for a thermal electrocyclic reaction of a cyclobutene is ___; the molecular orbital is ___ .

(a) π_3 ; symmetric   (b) π_3 ; antisymmetric   (c) π_2 ; symmetric   (d) π_2 ; antisymmetric

41 Which of the following are allowed cycloaddition reactions?

   A 4+2 thermal
   B 4+4 photochemical
   C 2+2 thermal

   (a) A   (b) B   (c) C   (d) A and B   (e) B and C

42 A [1,7] sigmatropic shift can occur ___ and proceeds by a ___ process.

   (a) thermally; antarafacial   (b) thermally; suprafacial   (c) photochemically; suprafacial

43 The material added to control the chain length of a polymer by interrupting the growth of one chain and initiating the formation of another chain is called ____.

   (a) an initiator   (b) inhibitor   (c) chain transfer agent   (d) cross-linking agent

44 A polymer of propylene with methyl groups in a regular alternating sequence on opposite side of the backbone is ___.

   (a) isotactic   (b) syndiotactic   (c) atatic

Answers to Examination 7

abdca   eabeb   eacaa   cbbdc   bbdad   bdaba   edbbd   abddd   dacb